D0821027

# ORGANIC BUILDING BLOCKS OF THE CHEMICAL INDUSTRY

# ORGANIC BUILDING BLOCKS OF THE CHEMICAL INDUSTRY

**H. HARRY SZMANT**
*Chemical Consultant*

WILEY

**A WILEY-INTERSCIENCE PUBLICATION**

**JOHN WILEY & SONS**

New York / Chichester / Brisbane / Toronto / Singapore

Copyright © 1989 by John Wiley & Sons, Inc.

All rights reserved. Published simultaneously in Canada.

Reproduction or translation of any part of this work
beyond that permitted by Section 107 or 108 of the
1976 United States Copyright Act without the permission
of the copyright owner is unlawful. Requests for
permission or further information should be addressed to
the Permissions Department, John Wiley & Sons, Inc.

**Library of Congress Cataloging in Publication Data:**

Szmant, H. Harry (Herman Harry), 1918–
    Organic building blocks of the chemical industry / H. Harry Szmant.
    p.   cm.
    Bibliography: p.
    Includes index.
    1. Chemistry, Organic—Industrial applications. I. Title.
TP247.S98   1989                                              89–30001
661'.8—dc 19                                                     CIP
    ISBN 0-471-85545-6

Printed in the United States of America

10 9 8 7 6 5 4 3 2 1

# PREFACE

## Objectives

The first objective of this book is to answer questions such as:

- What are industrial organic chemicals, and where do they come from?
- Why should anyone make them on a commercial scale?
- What are the economic and other factors that affect their production levels and pricing?
- What are the likely future trends in the demand and production pathways of a given material?

The second objective is to provide a useful tool for teaching or self-learning of the organic portion of industrial chemistry in a systematic and rather comprehensive manner.

This book should be useful to both chemists and chemical engineers and, in the case of the latter, the book is intended to fulfill a recent recommendation of Chemical Engineering Professor Vlamidir Haensel, who believes that "industrial chemistry, taught with an emphasis on why and how different processes originated, should help to develop a more complete chemical engineer" (see *Chemical and Engineering News*, January 19, 1987—hereafter denoted as *CEN* 1/19/87; see the explanation of abbreviations given below). Also, the book may be useful to chemical engineers who wish to adjust to their "profession in flux" as chemical industry shifts away "from commodity chemicals and petrochemicals and toward high-value-added specialty chemicals" (J. H. Krieger, *CEN* 11/30/87, 7–12).

## Organization

This book consists of three parts. The first, "Background Material," presented in Chapters 1 and 2, deals with some of the economic and historical aspects of industrial chemistry. The systematic review of organic building blocks is divided into two parts: Chapters 3–7 deal with aliphatic structures, discussed in order of increasing number of carbon atoms, whereas Chapters 8–10 deal with aliphatic, aromatic carbocyclic, and heterocyclic structures, respectively. Also, where convenient and appropriate, the systematic coverage of building-block chemistry is interrupted by somewhat expanded treatment of a given topic that goes beyond a cursory discussion and provides an opportunity to introduce such an "interlude." For example, the chemistry of the $C_1$ building block phosgene is the

234730

occasion to present "A glimpse of the world of polyurethanes," and the chemistry of the alkylation of aromatic hydrocarbons provides the occasion to insert a short survey of surfactants (surface-active agents). These "interludes" are highlighted by means of italic type in the detailed outlines found at the beginning of each chapter.

By necessity, the first chapter in Part 2 is rather lengthy because it reflects the predominant industrial practice of assembling large molecules from small building blocks.

The contents of this book are thoroughly indexed and cross-referenced, and where convenient, a code is used that specifies simply the number of the chapter, followed by the numbers of sections and subsections; for instance, 7.5.3 signifies Chapter 7, Section 5, Subsection 3.

## Contents

The material presented in this book is a blend of information concerning the utilization of different building blocks for the assembly of numerous industrial chemicals and their useful properties and the economic rationale for the dominant synthetic pathways. The status of interprocess and interproduct competition and the dynamics of real-world chemical and materials technology are emphasized and illustrated by means of references to readily accessible publications that monitor the pulse of events and trends. Most often, these references are to the *Chemical Marketing Reporter* (*CMR*), *Chemical Week* (*CW*), *CEN*, *Chemical Business* (*CB*), *Chemical Engineering* (*CE*), *CHEMTECH* (*CT*), and *Modern Plastics* (*MP*).

## The Building-Block Principle

The title and practice throughout the book emphasize the recognition that a given organic structure is usually assembled from modules that may differ with respect to the number of carbon atoms and the presence of functional groups. To illustrate: a $C_{12}$ molecule could conceivably be assembled from six $C_2$ units, four $C_3$ units, three $C_4$ units, and so on. The structure of the final product and economic and other pragmatic factors normally determine the preferred synthetic route. A $C_{13}$ molecule, on the other hand, could be derived from the preceding $C_{12}$ structure and an additional $C_1$ building block. I believe that this intellectual approach is useful in order to recognize potential choices among different chemical assembly operations. I hope that repeated applications of this building-block principle develop skill in recognizing synthetic options and strategies.

## Useful Properties, Structures, and Unit Cost of Industrial Chemicals

The chemical industry would not exist if not for the demands of the marketplace, and these demands are a function of properties that render a given material useful for one or more economic activities of modern society. However, properties relevant to commercial applications are usually quite different from those of interest in academia. For example, the physical aspects of a solid, such as its particle size, flow characteristics, and compatibility, are of great importance in mass transfer and manufacturing operations carried out on an industrial scale. In the case of polymers, processability or ease of fabrication of the final product may depend on the morphology of the material. Thus, one may find several grades of supposedly chemically identical materials with certain grades more suitable than others for extrusion, pultrusion, blow molding, or other processing methods. As an illustration of an extreme situation, we cite the fact that in 1983 Du Pont of Canada offered as many as 142 grades of polyethylene (PE)! (*CB* 5/87, 3). In the case of medicinal materials distributed in the form of pills, the mechanical stability of the compressed product is significant because it determines the need for additives (excipients) and the cost of formulation and fabrication. For example, acetaminophen—the active ingredient of several over-the-counter (OTC) pain killers (9.3.2)—is sold in "powdered or granular form" or as a "direct compression grade" with a significantly higher (by $\sim \$1.75/lb$) price tag for the latter. The compatability of physical and chemical properties in a given mixture of materials is very important for many commercial applications because of the widespread use of formulated products. Hence, in the pages that follow the reader will find indications of some major and/or interesting uses of specific chemicals although, admittedly, such aspects cannot be treated exhaustively in a book of this length.

Thomas Edison, the great inventor, insisted on an unprejudiced trial-and-error approach to the search for materials that would perform satisfactorily and achieve the goal pursued by his research organization. Except for cases of serendipity, the Edisonian method has been replaced by molecular engineering, which evolved from a partial but constantly improving understanding of useful property–structure relationships. The term "structure" here means not only the three-dimensional arrangement of atoms in a given molecule, but also three-dimensional arrangements of molecular aggregates that are a function of intermolecular forces and the size and shape of all structural components. Some such structure–property relationships are illustrated in this book, but readers are encouraged to suggest additional correlations for future editions.

To these three parameters—structure, properties, and usefulness—we must add a fourth: unit cost of a given material (expressed, unless stated otherwise, in units of U.S. cents per pound). While certain aspects of pricing of chemicals are discussed elsewhere (Section 2.4), it is clear that rampant interproduct competition, especially when one deals with large-volume purchases, frequently causes such purchases to be decided on small differences between unit prices of competitive materials. Often the choice of materials becomes a compromise between better performance of a more costly material, and a material that carries a lower price tag but still performs satisfactorily within an acceptable range of applications. The reader will find examples of the above-mentioned relationships in this book in order to compensate for the customary omission of such considerations in academic curricula. For a more detailed discussion of the parameters that relate structure–useful properties–applications–unit cost, and a graphic representation of their interdependence, the reader is invited to examine "An Industrial Chemistry Course to Bridge the Academia–Industry Gap," *J. Chem. Ed.* **1985**, 736–741, hereafter abbreviated as *JCE* **1985**, 736–741, and the "Chemistry Plus" columns published in *CHEMTECH* **1984**, pp. 113–114, 184–185, 427–429, 597–599; **1985**, pp. 2–3, 64 (hereafter abbreviated as *CT* **1984**, 113, 184, 427, 597; **1985**, 2, 64).

Approximate unit prices of the more common chemicals are culled from recent (1986–1988) weekly price lists published by the *Chemical Marketing Reporter* (*CMR*). Many other prices not found in *CMR* were generously provided by industrial colleagues and this cooperation is gratefully acknowledged. Some prices of fine and ultrafine chemicals (see below) are culled from special price lists such as Aldrich's "Flavors and Fragrances," April 1988 and identified as *Ald. F&F*. These prices refer to the largest quantity purchases, usually 25 lb, or otherwise specified.

Obviously, the unit prices mentioned here are not binding to the producer(s) or to the author and are not meant to replace price quotations from appropriate suppliers. The purpose of mentioning unit prices is twofold:

- To compensate for the prevalent omission of these considerations in educational curricula.
- To illustrate the economic benefits that result from physical, chemical, and biological transformations of relatively low cost starting materials or intermediates, to products of higher unit value, that is, to higher-value-added products.

Generally speaking, the unit prices cited here are applicable to bulk-quantity purchases; that is, they represent the lower end of the pricing scale. Other limitations of unit prices cited here are discussed elsewhere (2.4). The listing of unit prices should be expanded in future editions of this book through spontaneous collaboration of chemical manufacturers who will have realized that the profit opportunities are not threatened by revelation of unit pricing of their products and who, on the other hand, are expected to benefit from an increased awareness by younger chemists and chemical engineers of relative material costs.

Competitive processes are compared here exclusively on the basis of material costs, and it is left to the reader to consider stoichiometry, selectivity, yield(s) of principal product(s), credits for by-product(s), and other factors (2.4.2).

In this book there are attempts to keep up to date with the flood of changes of company names that accompany the rampant restructuring of chemical businesses characteristic of our times. During 1987 mergers and acquisitions in the world of U.S. corporations gave rise to over 1700 name changes, and this trend continued during the first half of 1988 when 931 name changes were recorded. Any omissions are unintentional and the same is true of unintentional omissions of registered tradenames. The reader is advised to consult the most recent issue of *CMR*'s "OPD Chemical Buyers Directory" or *CW*'s "Buyers Guide Issue" and other such current directories of chemical suppliers. The latter may, by the way, include a convenient source of chemical tradenames.

### Unique Features

Besides the inclusion of unit prices of most materials under discussion, this book does not discriminate against macromolecular organic chemicals. With current sales in the United States of about 50B lb out of an estimated total of about 230B lb for all synthetic organic chemicals, and projected sales to reach 76B lb by the end of the century, it would be unwise to ignore polymers in the discussion of industrial organic chemicals. The worldwide production of polymers in 1986, by the way, reached 160B lb.

Also, while the teaching of industrial chemistry often tends to focus only on the "top 50" or "top 100" large-volume chemicals, this book attempts to present a broader picture of the field. This includes not only the large-volume commodity chemicals, but also other categories of materials characterized, on one hand, by intermediate- and low-volume consumption, but, on the other hand, by increasing unit values. Thus, pseudocommodities, fine, ultrafine, and specialty chemicals, are characterized in 1.1.2 and illustrated as the book follows the downstream transformation of feedstocks through primary, secondary, and higher generations of building blocks to a series of intermediates and arrives, finally, at end products of importance to a modern, "developed" society.

Without apologies to nomenclature purists, I have introduced and employed throughout names and acronyms of chemicals that constitute the language in the world of chemical industry and commerce. The acronyms and abbreviations that refer to company names or government agencies are listed in the Index.

## Clarifications

Most of the annual chemical production and consumption figures refer, generally speaking, to the current (1979–1987) United States marketplace. Over this period of time there occurred small increases, if not some actual decreases, in the automotive, construction, and other manufacturing activities of the United States that consumed large volumes of industrial chemicals. Only lately, in 1988, have the manufacturing activities increased appreciably accompanied by an increasing demand for certain chemicals.

The reader should distinguish between the production capacity of a given chemical and the actual domestic production and consumption. The production may be seriously affected by significant imports and exports in view of the international character of the chemicals industry estimated currently at $725B. Unless stated otherwise, all quantities of chemicals refer to pounds or short tons (t) and the symbols M, MM, and B represent thousands, millions, and billions ($10^9$), respectively.

Most industrial processes are covered by numerous patents, each of which may include a dozen or more "examples" of a given transformation. For this reason, and the existence of proprietary information, process conditions mentioned in this book are meant to indicate representative rather than unique reaction conditions. For specific details and an exhaustive coverage, the reader is advised to consult the original literature (that includes patents) as reported in *Chemical Abstracts*. A convenient and, in the author's opinion, indispensable source of information is the *Kirk–Othmer Encyclopedia of Chemical Technology*, 3rd edition, published over the period 1979–1984 in a set of 26 volumes and a supplement by John Wiley & Sons. Frequent references to specific chemical processes are made to *Faith, Keyes, and Clark's Industrial Chemicals*, edited in 1975 by F. A. Lowenheim and M. K. Moran, also published by John Wiley & Sons, and referred to as *FK&C*. A short but handy tabulation of process conditions published by S. M. Walas in *CE* 10/14/85, pp. 79–83, is referred to as hereafter as *SMW*. Also, the reader must keep in mind the countless monographs that deal with specific topics, and among them are the titles issued by Noyes Publications and Noyes Data Corporation. Among the fine and ultrafine chemicals that are mentioned in this book the reader will find some medicinals, complex materials of natural origin, and so on, and for their clarification the reader is advised to consult *The Merck Index*, 10th edition, published by Merck & Co. in 1984. As a matter of fact, the frequent and routine references to organic name reactions, denoted here by *ONR*-1 to *ONR*-100, pertain to the 100-page-long section of *The Merck Index* where the reader may find additional references concerning a given name reaction.

Finally, I must try to justify the relative neglect in this book of organic building blocks derived from renewable (biomass) resources. For reasons too complicated to explain, *Industrial Utilization of Renewable Resources* was published separately in 1986 by Technomic Publishing Company, and references to it are indicated by *SZM-II*.

## Wishful Thoughts

I hope that all errors that have slipped into the text, as well as suggestions for future improvements, will be brought to my attention. This is particularly true in the case of industrial colleagues, a number of whom have already cooperated by revealing chemical details of production methods, for which I and the readers are extremely grateful.

## Acknowledgments and Dedication

Personal thanks are extended to students of "Industrial Aspects of Chemistry" who suffered through mimeographed handouts when such a course was taught for a dozen of years at the University of Detroit. I am indebted to Ms. Jane Schley for her skillful and intelligent translation of often illegible manuscript pages into a clean, typed document from which this book finally evolved. Also, I wish to thank Senior Editor Jim Smith, Production Editor Margaret Comaskey, and Copyeditor Cathy Hertz for their skillful contributions to the improvement of the manuscript, and Senior Production Editor Rosalyn Farkas for expediting the production of the book.

This book is dedicated to two persons: the late Professor Henry B. Hass (1902–1987) who, a long time ago, awakened my interest in the industrial aspects of organic chemistry by teaching in his unique manner the famous "Whitmore course" at Purdue University; and my wife Nita for her patience and tolerance of the countless hours invested in this project.

H. Harry Szmant

*Sanibel, Florida*
*1989*

# CONTENTS

# LIST OF IMPORTANT REFERENCES AND THEIR ABBREVIATIONS

## Encyclopedias, Handbooks, and Dictionaries

**AIA**  *Acronyms, Initialism & Abbreviations*, 8th ed., E. T. Crowley, Ed., Gale Research Co., Detroit, 1983–1984

**AIAD**  *Acronyms, Initialisms, and Abbreviations Dictionary*, 11th ed., 3 vols., J. E. Towell and H. E. Sheppard, Gale Research Co., Detroit, 1986

**CST**  *Chemical Synonyms and Trade Names*, 8th ed., W. Gardner, Ed., CRC Press, Boca Raton, FL, 1978

**EPST**  *The Encyclopedia of Physical Science and Technology*, R. A. Myers, Ed., Academic Press, Orlando, FL, 1987

**EPST-SZM**  H. H. Szmant, *Organic Chemicals: Industrial Production*, Vol. 10, pp. 26–42

**ETS**  *Encyclopedia of Chemical Tradenames and Synonyms*, H. Bennet, Ed., Chemical Publishing Co., New York, 1984

**HW**  *Houben-Weyl's Methoden Der Organischen Chemie*, 4th ed., 65 vols., George Thieme, New York

**KO**  *Kirk–Othmer's Encyclopedia of Chemical Technology*, 3rd ed., Wiley–Interscience, New York, 1978–1984

**MI**  *The Merck Index*, 10th ed., M. Windholz, Ed., Merck & Co., Rahway, NJ

**ON**  *Organic Name Reactions*, *ONR* 1–100, in *MI*

**OS**  *Organic Synthesis*, ongoing collection of volumes, currently up to Vol. 66, 1987; also *Collective Volumes* I, II, and so on, Wiley, New York

**UL**  *Ullmann's Encyclopedia of Industrial Chemistry*, 5th ed.; 1st ed. in English, W. Gebhartz and Y. S. Yamamoto, Eds., VCH Publishers, up to Ceramic Colorants in Vol. A5 in 1986

## Books

**AL**  H. R. Allcock and F. W. Lampe, *Contemporary Polymer Chemistry*, Prentice-Hall, Englewood Cliffs, NJ, 1981

**AUS**  *Shreve's Chemical Process Industries*, 5th ed., G. T. Austin, Ed., McGraw-Hill, New York, 1984

**BIL**  F. W. Billmeyer, *Textbook of Polymer Science*, 3rd ed., Wiley, 1984

**CHE**  P. J. Chenier, *Survey of Industrial Chemistry*, Wiley–Interscience, New York, 1986

**CM**   C. A. Clausen III and G. Mattson, *Principles of Industrial Chemistry*, Wiley–Interscience, New York, 1978

**CRO**   H. D. Crone, *Chemicals & Society*, Cambridge University Press, Cambridge, UK, 1986

**EM**   W. S. Emerson, *Guide to the Chemical Industry*, Wiley–Interscience, New York, 1983

**FKC**   *Faith, Keyes and Clark's Industrial Chemicals*, 4th ed., F. A. Lowenheim and M. K. Moran, Eds., Wiley–Interscience, New York, 1975

**HT**   M. Harris and M. Tishler, Eds., *Chemistry in the Economy*, American Chemical Society, Washington, DC, 1973

**KCB-I**   K. C. Barrons, *The Food in Your Future*, Van Nostrand Reinhold, New York, 1975

**KCB-II**   K. C. Barrons, *Are Pesticides Really Necessary?*, Regneray Gateway, Chicago, 1981

**KEN**   *Riegel's Handbook of Industrial Chemistry*, 8th ed., J. A. Kent, Ed., Van Nostrand Reinhold, New York, 1983

**KS**   G. B. Kauffman and H. H. Szmant, Eds., *The Central Science: Essays on the Uses of Chemistry*, Texas Christian University Press, Fort Worth, 1984

**ND**   D. C. Neckers and M. P. Doyle, *Organic Chemistry*, Wiley, New York, 1977

**RB**   B. G. Reuben and H. L. Burstall, *The Chemical Economy*, Longmans, London, 1973

**ROD**   F. Rodriguez, *Principles of Polymer Systems*, 2nd ed., McGraw-Hill, New York, 1982

**SC**   R. B. Seymour and C. E. Carraher, Jr., *Polymer Chemistry*, Marcel Dekker, New York, 1981

**SZM-I**   H. H. Szmant, *Organic Chemistry*, Prentice-Hall, Englewood Cliffs, NJ, 1957

**SZM-II**   H. H. Szmant, *Industrial Utilization of Renewable Resources*, Technomic, Lancaster, PA, 1986

**TNJ**   T. M. Tedder, A. Nechvatal, and A. H. Jubb, *Basic Organic Chemistry, Part 5: Industrial Products*, Wiley, New York, 1975

**WA**   K. Weissermel and H. J. Arpe, *Industrial Organic Chemistry*, Verlag Chemie, Weinheim, 1978

**WH**   H. L. White, *Introduction to Industrial Chemistry*, Wiley–Interscience, New York, 1986

**WIS**   *Industrial Organic Chemistry*, P. Wiseman (Ed.), Halsted, Chichester, 1972

**WR-I**   H. A. Wittcoff and B. G. Reuben, *Industrial Organic Chemistry in Perspective, Part One: Raw Materials and Manufacture*, Wiley–Interscience, New York, 1980

**WR-II**   H. A. Wittcoff and B. G. Reuben, *Industrial Organic Chemistry in Perspective, Part Two: Technology, Formulation, and Use*, Wiley–Interscience, New York, 1980

**WRS**   J. Wei, T. W. F. Russell, and M. W. Swartzlander, *The Structure of the Chemical Processing Industries*, McGraw-Hill, New York, 1979

### Frequent Journal References

**CA**   *Chemical Abstracts*, Chemical Abstracts Service, American Chemical Society, Columbus, OH

**CB**    *Chemical Business*, Schnell Publishing Co., New York

**CE**    *Chemical Engineering*, McGraw-Hill, New York

**CEN**    *Chemical & Engineering News*, American Chemical Society, Washington, DC

**CMR**    *Chemical Marketing Reporter*, Schnell Publishing Co., New York

**CT**    *CHEMTECH*, American Chemical Society, Washington, DC

**CW**    *Chemical Week*, McGraw-Hill, New York

**HP**    *Hydrocarbon Processing*, Gulf Publishing Co., Houston, TX

**JACS**    *Journal of the American Chemical Society*, American Chemical Society, Washington, DC

**JAMA**    *Journal of the American Medical Association*, American Medical Association, Chicago

**JCE**    *Journal of Chemical Education*, American Chemical Society, Washington, DC

**JOC**    *Journal of Organic Chemistry*, American Chemical Society, Washington, DC

**MP**    *Modern Plastics*, McGraw-Hill, New York

**SC**    *Science*, American Association for the Advancement of Science, Washington, DC

# ORGANIC BUILDING BLOCKS OF THE CHEMICAL INDUSTRY

# PART ONE

## BACKGROUND MATERIAL

# 1

# INTRODUCTION TO INDUSTRIAL ORGANIC CHEMICALS

## 1.1  ECONOMIC ASPECTS

### 1.1.1  How Many Industrial Organic Chemicals?

The number of truly primary, large-volume organic building blocks obtained more or less directly from petroleum-refining operations or from natural gas, coal, ammonia, carbon dioxide, and renewable resources and utilized in bona fide chemical transformations (rather than as fuels and fertilizers) is limited to less than one dozen. From these materials, referred to as *primary building blocks*, are derived secondary, tertiary, and further *downstream building blocks* and a host of other intermediate and final materials of industrial interest.

It is customary to focus on the top 50 large-volume chemicals when comparing production figures of industrial chemicals from one year to another (for the latest report, see *CEN* 4/13/87). However, examination of such a list reveals that it does not discriminate between primary, secondary, and even tertiary building blocks. A distinction between the different generations of chemicals, listed in descending order of production levels, recorded here as approximate values of billions of pounds (B lb), results in Table 1.1. At a glance, we can now recognize the mother–daughter–granddaughter relationships among the 28 organic compounds that are included in the top 50 listing. A clarification is in order with regard to two chemicals among the 28: methyl-*t*-butyl ether (MTBE) and

3

**TABLE 1.1  Primary, Secondary, and Tertiary Organic Building Blocks among the Top 50 Chemicals in the Economy of the United States**

| Primary Building Block | B lb | Secondary Building Block | B lb | Tertiary Building Block | B lb |
|---|---|---|---|---|---|
| Ethylene | 33 | Ethylene dichloride | 14.5 | Vinyl chloride[a] | 8.5 |
| | | Ethylene oxide | 6 | Ethylene glycol | 5 |
| | | Ethylbenzene[b] | | Vinyl acetate[a,b] | 2.3 |
| Propylene | 18 | Propylene oxide | 2.5 | | |
| | | Acrylonitrile | 2.3 | | |
| | | Isopropyl alcohol | 1.3 | Acetone[b] | 2 |
| | | Cumene[b] | | | |
| Benzene | 10 | Ethylbenzene[c] | 9 | Styrene | 8 |
| | | Cumene[c] | 4 | Phenol | 3 |
| | | | | (Acetone)[c] | |
| | | Cyclohexane | 2 | Adipic acid | 1.5 |
| Methanol | 7.5 | Acetic acid | 3 | (Vinyl acetate)[c] | |
| | | Formaldehyde | 6 | | |
| | | Methyl t-butyl ether | 2.3[e] | | |
| Toluene | 6 | | | | |
| Xylenes | 5.5 | | | | |
| p-isomer | 5 | Terephthalic acid | 8 | | |
| Butadiene | 2.5 | | | | |
| Urea[d] | | | | | |

[a]This chemical can also be a secondary building block.
[b]See duplicate listing below.
[c]See duplicate listing above.
[d]Most of the 12B-lb production is used as fertilizer.
[e]Fuel additive.

urea are not bona fide chemical intermediates. Practically all of MTBE is used as a gasoline enhancer, and only 10% of urea is consumed as a feedstock in the production of other chemicals, while the rest is used as fertilizer. No materials derived from renewable resources are shown among the top 50, although, as we shall see elsewhere, ethanol, glycerine, fatty acids, glucose, and other chemicals are serious contenders for such listing.

The apparent mass balance discrepancies that surface on examination of production levels shown in Table 1.1 are explained by (1) the mass contributions (or losses) attributed to inorganic reactants (or by-products) discussed elsewhere (2.2) and (2) the hundreds, indeed, many thousands of chemicals obtained by additional transformations of the 28 building blocks mentioned so far. Another large gap in the material balance picture is filled by the production levels of different polymers. These, according to recent information (CEN 4/13/87), represent the production levels in Table 1.2. Obviously, almost all of the polymers enumerated in Table 1.2 are derived from one, two, or even three primary, secondary, and tertiary building blocks shown in Table 1.1.

It is estimated (R. M. Busche, *ICN* 5/83) that about 170 chemicals are produced in the United States in excess of 10MM lb/yr, 98% of which are derived from petroleum and natural gas.

TABLE 1.2    Production Levels of Polymers

| Polymer | B lb |
|---|---|
| Polyethylenes: HDPE, LDPE, LLDPE[a] | 16.1 |
| Poly(vinyl chloride) (PVC) | 7.3 |
| Polypropylene (PP) | 5.8 |
| Polystyrene (PS) | 4.5 |
| Polyester fibers (PET) | 3.3 |
| Phenol–formaldehyde resins (PF) | 2.7 |
| Nylon fibers | 2.5 |
| Styrene–butadiene rubber (SBR) | 1.8 |
| Polyolefin fibers (PO) | 1.4 |
| Urea–formaldehyde resins (UF) | 1.3 |
| Unsaturated polyester resins (UPR) | 1.3 |
| Polybutadiene rubber (PBR) | 0.7 |
| Acrylic fibers (PAN) | 0.6 |
| Ethylene–propylene rubber (EPDM) | 0.5 |
| Epoxy resins | 0.4 |
| Melamine–formaldehyde resins (MF) | 0.2 |
| Nitrile rubber (NR) | 0.15 |
| Other polymers, including cellulosics and polyurethanes (PUR) | 2.0 |

[a] High-, low-, and linear low-density polyethylene.

The number of commercially important materials is overwhelming. Thus, in connection with a compilation of toxicity information, the National Research Council's report covers 53,500 individual industrial chemicals that are inventoried by the EPA and the FDA and are, in part, regulated by these government agencies. These materials fall into the categories in Table 1.3 according to their utilization (*CEN* 3/12/84).

Similarly, the National Academy of Sciences listed 53,500 commercially important substances in 1984 (*Toxicity Testing: Strategies to Determine Needs and Priorities*, National Academy of Sciences, Washington, DC, 1984).

The European Community Commission of 12 member states recently published the *European Inventory of Existing Commercial Chemical Substances* (*EINECS*), which lists 100,116 chemical substances marketed during the 1970s (*CMR* 9/21/87, *CW* 9/30/87).

A recent report that deals with specialty monomers (*CMR* 2/15/88) lists 215 different chemicals with 1986 sales of 1.7B lb valued at $1.2 billion for an average unit price of

TABLE 1.3    Commercially Important Materials Listed by NRC

| Category of Utilization | Number[a] |
|---|---|
| Active and inert ingredients of pesticide formulations | 3,350 |
| Cosmetic ingredients | 3,410 |
| Active and inert vehicles of drug formulations | 1,815 |
| Food additives | 8,627 |
| Commercial chemicals used >1MM lb/yr | 12,860 |
| Commercial chemicals used <1MM lb/yr | 13,911 |
| Commercial chemicals of unknown production levels | 21,752 |

[a] A total of 65,725 results from multiple uses.

0.71¢/lb. However, this list includes 60 specialty monomers that carry price tags of $3.00/lb or higher.

The flavor and fragrance sector of the chemical industry is estimated (S. Randel, *CB* 10/87) to depend on the use of over 4000 aroma chemicals, including 400 materials of natural origin that are employed as fragrances (F. V. Wells and M. Billot, *Perfumery Technology*, 2nd ed., Wiley, New York, 1981).

A database that deals with U.S. regulations applicable to hazardous materials lists 3500 chemicals (*CEN* 11/2/87).

The Eastman Kodak Company advertises (*CMR* 11/2/86) that it possesses a "bank of over 900,000 compounds" that can be drawn on "to meet your extra special needs for custom synthesis."

While these numbers are impressively high, let us keep in mind that *Chemical Abstracts* currently recognizes the existence of over 7 million individual substances.

Another measure of the number of industrially significant materials are the entries in popular commercial catalogs. (For our purpose we exclude here catalogs of research chemicals.) For example, the 1980/81 edition of the *OPD Chemical Buyers Directory* published by *CMR* lists 12,946 entries that include mixtures such as anticaking agents, dry-cleaning solvents, fur-dressing greases and oils, liver concentrate, and perfume fixatives. This number increased to 17,995 entries in the 1988 edition.

Exact chemical production and consumption statistics are difficult to ascertain. In this context, the article "Who Keeps Score for the Chemical Industry" by N. Callanan (*CB* 3/10/80) and the editorial by A. R. Kavaler (*CB* 8/85) are very revealing. The U.S. International Trade Commission (USITC) releases production figures for only 6000 synthetic organic chemicals and is hampered in doing so by restrictive policies designed to protect confidential production figures of individual manufacturers. Thus, for example, it does not mention and publish production figures for a given material unless

there are three or more producers, no one or two of which may be predominant. Moreover, even when there are three or more producers, statistics are not given if there is any possibility that their publication would . . . (be a) . . . disclosure of information accepted in confidence.

Another set of restrictions governs minimum production levels: 5000 lb or sales of $5,000, except 50,000 lb and $50,000 in the case of polymers and 1000 lb and $1000 in the case of pigments, medicinal chemicals, flavor and perfume materials, and rubber-processing materials. A useful survey of "Reports from the International Trade Commission" was published by D. G. Michels in *CT* 1/85.

The USITC annual report "Synthetic Organic Chemicals" for 1986 (issued in September 1987) reveals that the total production of all synthetic organic chemicals and primary products derived from petroleum and natural gas amounted to 348.8B lb with each of these sources contributing 235.3B and 113.5B lb, respectively. The sales value of the "primary products from petroleum and natural gas" amounted to $6.02 billion for an average of about 11.3¢/lb, while the average sales value of synthetic organic chemicals was 45.2¢/lb. However, while the increase in the overall average value from 11.3¢ to 45.2¢/lb represents the economic benefits derived from industrial transformations, such economic benefits are not uniformly distributed among all categories of synthetic organic chemicals (Table 1.4).

The difference between production and sales represents mostly intracompany sales or transfers, or "captive use." The volume of intercompany sales illustrates the cliché "the chemical industry is its own best customer."

TABLE 1.4  Average Values of Synthetic Organic Chemicals within Each Category of Materials

| Category | Average Sales Value ($/lb) |
|---|---|
| Cyclic intermediates | 0.32 |
| Dyes | 2.88 |
| Organic pigments | 6.67 |
| Medicinal chemicals | 9.60 |
| Flavor and perfume materials | 6.50 |
| Plastics and resin materials | 0.45 |
| Rubber-processing materials | 1.24 |
| Elastomers (synthetic rubber) | 0.89 |
| Plasticizers | 0.47 |
| Surface-active agents | 0.45 |
| Pesticides and related products | 4.50 |
| Miscellaneous end-use chemicals and chemical products | 0.58 |
| Miscellaneous cyclic and acyclic chemicals | 0.29 |

Certain trade associations publish production and sales statistics for its members. For example, such statistical services are provided by the Society of the Plastics Industry (SPI), the Fatty Acid Producer's Council, and the Pulp Chemicals Association.

### 1.1.2  Categories of Industrial Chemicals

The industrial importance of a given chemical can be judged by either the volume of its demand or the monetary value generated by its production. The domestic demand for a given material usually equals production *plus* imports *minus* exports. Strange as it may seem at first glance, the United States may simultaneously import and export certain chemicals. This can occur because of great transportation distances across the continent or for business reasons (favorable prices of imported materials, intracompany transfers from overseas subsidiaries or parent companies, etc.).

There exists an approximate but very instructive correlation between unit costs and demand in the marketplace for a series of related products, in this case chemicals. This log–log relationship is shown in Fig. 1.1 (see also *WRS*, Section 5.1.1). We note that chemicals are being classified in categories that range from large-volume but low-cost "commodity chemicals," through lower-volume and somewhat higher-priced "pseudo-commodity chemicals," "fine chemicals," "traditional specialty," and "high-tech (high-technology) specialty" materials. Allowing for some overlap, the characteristic demand and unit cost levels of these different categories of materials (applicable to the current 1987/88 economy of the United States) are summarized in Table 1.5, and additional characteristic differences between the categories of chemicals are discussed in the pages that follow.

Commodity chemicals are low-priced materials produced in large, costly ($10^8$–$10^9$), specifically designed production facilities used in a continuous and energy-efficient manner. These circumstances limit the number of producers of commodity chemicals. The chemicals shown in Table 1.1 belong to this category of materials, and below the production level shown for adipic acid one could also add carbon disulfide (450MM at 22¢) and ethyl chloride (270MM at 24¢).

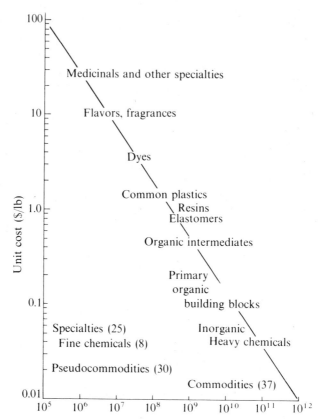

**Figure 1.1** Demand versus unit cost of different categories of chemicals (see p. 21 for meaning of numbers in parentheses).

Table 1.5 suggests that commodity chemicals are characterized by a combination of two parameters: large-volume production and low unit cost, but obviously the quantitative limits are rather arbitrary and are a function of the economy at a given time. Chemicals produced mostly for use as fuels or fertilizers are less costly than chemicals destined to play a role of intermediates because the *principle of economy of scale* (2.4) is strictly applicable in these cases. One could say that the relatively small demands of the chemical industry, as compared to the demands of the fuel and fertilizer industries, cause the chemical industry to "ride piggy-back" on the fuel and fertilizer markets.

Pseudocommodities usually result from a few, relatively simple chemical transformations of commodity chemicals. As a general but approximate rule of thumb, each chemical transformation doubles the unit price of the product relative to that of the precursor(s). As already pointed out in the Preface and discussed in greater detail elsewhere (2.4), however, more accurate estimates of unit cost must take into account stoichiometry, credit for salable by-products, special temperature, pressure, and other process requirements that affect the capital cost of the production facility, and so on. Costs of waste disposal cannot be ignored in these days of environmental concerns (2.4.5). Some chemical transformations are somewhat less costly than predicted by the above-mentioned rule of thumb. Thus, while hydrogen is a relatively costly reactant ($\sim$ \$2/lb), its saving grace is its low

TABLE 1.5   Categories of Industrial Organic Chemicals

| Category | Demand[a] Approximate Range (lb/yr) | Unit Cost[a] Approximate Range ($/lb) |
|---|---|---|
| Commodity chemicals | $10^9$–$10^{10}$ | 0.1–0.75 |
| Pseudocommodity chemicals | $10^7$–$10^8$ | 0.75–2 |
| Fine chemicals | $10^5$–$10^7$ | 2–$10^{1b}$ |
| Ultrafine chemicals | $10^4$–$10^6$ | $10^1$–$10^{4b}$ |
| Traditional specialty products | $10^6$–$10^9$ | 1–$10^{1c}$ |
| High-tech specialty chemicals/devices | $10^5$–$10^8$ | 5–$10^3$ |

[a] Demand and unit cost estimates are applicable to the U.S. marketplace. The unit costs of commoditiy and pseudocommodity monomers and polymers differ little unless special, complex polymerization and fabrication processes are involved.
[b] Unit costs are a function of the number and nature of multistep synthetic processes, purification requirements, the cost of intermediates and reagents, and the patent position of the producer. The production of ultrafine chemicals is more demanding. High-performance specialty polymers can belong to either category.
[c] The range of unit costs usually applies to retail pricing.

molecular weight and the simplicity of most catalytic hydrogenations provided that the catalyst is either inexpensive or not wasted. Similarly, vinyl polymerization can be somewhat less costly than predicted by the rule of thumb. Except for production volume and unit price levels, the other characteristics of pseudocommodities resemble those of the commodity chemicals.

We shall illustrate each category of chemicals by means of specific chemicals that are mentioned elsewhere in this book (please consult the Index).

The demand and unit prices of some pseudocommodities are as follows: glycerol (340MM at about 90¢), methyl bromide (43MM at 57¢), pentaerythritol (115MM at 71¢), TDI (640MM at $1.01), caprolactam (840MM at 86¢), and many others. The polymers in this category of materials are priced somewhat higher than $1.00/lb. Thus, while poly(vinyl acetate) is on the borderline (130MM at $1.05), examples of higher-priced polymers are the epoxies (400MM at $1.40); nylon 6/6 (2.5B at $1.10–$3.00, depending on quality and physical state); polycarbonate pellets (260MM at $1.93), and so on. It is noteworthy that the unit price is a better criterion for membership in the pseudocommodity category of chemicals than the production volume.

Fine chemicals are obtained from the preceding categories of materials by means of additional, often rather delicate chemical and/or biological transformations *or* from natural sources by means of elaborate isolation and purification steps. Generally speaking, fine chemicals are materials of known composition and structure, but they must meet exacting specifications in order to merit optimum unit prices on the marketplace. They are normally produced on a relatively small scale in versatile production facilities, and batch operations may be preferred in order to achieve maximum quality control. Most fine chemicals are employed as either ingredients of a variety of formulations (see discussion of traditional specialties below) or intermediates in the production of still more complex chemicals. The results of a study of the fine chemicals market by Kline & Company were summarized recently (*CMR* 7/20/87) and provide a quantitative idea of the economic significance of this category of industrial chemicals. The more than $23-billion market of fine chemicals in the United States in 1986 is said to be constituted by 40% of intermediates, 14% of synthetic medicinals, 7% of high-purity reagents, and 7%

of precious-metal salts. The remaining 32% include photographic chemicals, food additives, flavor and fragrance enhancers, biochemicals and other high-value reagents, and botanicals. Overall, fine chemicals are said to represent one-sixth of the total industrial chemicals market in the United States that, by this account, would be $138 billion. Also, fine chemicals are said (*CMR* 7/20/87) to be "among the most diffuse and least publicized class of industrial chemicals"—a statement that certainly does not apply to this book.

Many fine chemicals are imported to the United States from throughout the world (see 1.1.3). The unit cost of fine chemicals depends on factors such as the complexity of synthetic procedures, the number of transformations, cost of starting materials and reagents, and purification requirements and thus covers a wide range, say, from $2 to $100 or more per pound. Examples of approximate demand and unit prices of some fine chemicals are ascorbic acid (34MM at $5.80); acetaminophen, *N*-acetyl-*p*-aminophenol, or APAP (24MM at $2.70); aspirin (33MM at $2.45); glycine (4MM at $2.12 USP grade); theophylline (3MM at $5.45 USP grade); theobromine ($63.60); *p*-aminobenzoic acid (PABA) ($5.80); benzocaine (330M at $7.05 USP grade), undecylenic acid ($2.57) and its zinc salt ($4.67); vanillin ($6.00–$6.25, depending on source); ethyl vanillin ($13.50); and isopropyl lanolinate ($2.35).

Proprietary procedures for the production of some fine chemicals may not only raise unit prices into a range of hundreds and even thousands of dollars per pound but also create the impression that one is dealing with members of the high-tech category of materials discussed earlier. Examples of such ultrafine chemicals are vitamin A (synthetic, $15.00), D-α-tocopheryl succinate ($35.70), rhodinol (natural $105, synthetic $15.25), vincristine ($1200/g!), thymol iodide ($40), pilocarpine hydrochloride ($685.00), and substituted pyrazines ($135–$1300/lb depending on structure with 2-acetylpyrazine; e.g., priced at $1480/lb; see *CMR* 7/6/87). Not surprising are the high unit prices of "gourmet chemicals," chemicals that are especially purified and used as standards or in other exacting situations (*CW* 11/19/86). The latter applies to many so-called electronic chemicals.

It has become fashionable to refer to certain industrial materials as "specialties" even though this practice ignores the fact that there are approximately 400 member companies that constitute the Chemical Specialties Manufacturers Association (CSMA), founded in 1914, and that this membership represents about 85% of all chemical specialty firms in the United States. The CSMA definition of a chemical specialty product is as follows:

> A chemical specialty is a chemically formulated product manufactured from basic chemicals or chemical compounds and used, without further processing, by household and industrial consumers for specific purposes.

The CSMA membership includes businesses dedicated exclusively to the formulation, packaging, and distribution of traditional specialty products as well as specialty products divisions of firms that are also engaged in the production of chemical raw materials and/or intermediates. In the case of the latter firms, the CSMA membership may overlap with membership in the Chemical Manufacturers Association (CMA), with its 170-member companies, and the Synthetic Organic Chemical Manufacturers Association (SOCMA). Currently, CSMA consists of six divisions: Aerosol, Antimicrobial Products, Detergents, Industrial and Automotive Specialty Chemicals, Pesticide, and Polishes and Floor Maintenance. As suggested by these names, "traditional" chemical specialty products include countless, carefully formulated products found in most homes, garages, machine shops, and other locations of a nonprimitive society. They function as domestic, institutional, and industrial cleaners; paint removers; polishes; adhesives; sealants; functional liquids such as cutting, brake, and hydraulic fluids; lubricants and lubricating greases; traditional aids in manufacturing operations that deal with textiles, leather, wood, plastics; and so on. They also include nonprescription, over-the-counter (OTC), antiinfectives and antiseptics, personal-care products such as shampoos, shaving creams and after-shaving lotions, suntan and sunscreening products, bath oils, denture adhesives and denture cleansers, hair colorants and conditioners, and the whole armory of cosmetic preparations, and so on and so on. Most libraries contain common reference sources describing the formulation of traditional chemical specialty products. Two such references are

H. Bennett, Ed., *The Chemical Formulary*, Chemical Publishing Co., New York (an ongoing publication effort with 25 or more volumes).

E. W. Flick, *Household, Automotive, and Industrial Chemical Formulations*, 2nd ed., Noyes Publications, Park Ridge, IL, 1984.

The modernization and usually highly advertised improvement of a traditional chemical specialty may be based on the incorporation of one or more ingredient from the category of high-tech specialty materials, but there it is futile to debate regarding which of the two categories of materials a given product may belong. Most likely, products such as lubricants based on the use of Du Pont's Teflon particles, adhesives based on cyanoacrylate esters, silicon- and fluorine-containing materials that function as stain preventatives, and other items in which superior performance is dramatically different from traditional products and is due to high-tech ingredients may belong to the next category of materials. Often, however, the manufacturers of traditional specialties accept the new ingredients on recommendation of the chemical firms responsible for the development.

High-tech chemical specialties are materials and/or devices developed to satisfy the needs of a customer or a group of customers—needs that were unsatisfied, or only poorly satisfied, by traditional products. Often these products are of proprietary origin and composition, and because of their problem-solving capabilities, they command a rather high unit price. The production levels of high-tech chemical specialties depend, of course, on the size of the industrial sector that they serve. High-performance adhesives, plastics, and composites, electronic chemicals, and diagnostic aids employed in clinical laboratories, are among the most important classes of products that belong to this category of materials, but a more complete list is found in Table 1.6. High-tech specialties of highest annual growth rate (*CW* 3/2/88) are (in decreasing order) electronics, diagnostics, adhesives, synthetic lubricants, specialty surfactants, industrial coatings, cosmetic additives, and water-management materials. High-tech specialties may be developed through cooperation between a research and development (R&D) organization and a potential client who can test the performance of the desired product during its evolution. On the other hand, an R&D organization may pursue certain special performance objectives and, once these objectives are attained, offer the proprietary product on a global basis. High-tech specialty materials and devices are the most dramatic examples of the enabling role of chemicals, that is, chemicals that make something possible that was previously impossible or difficult to do. Large corporations that decide to move rapidly into the arena of high-tech materials often do so by acquisition of small, successful organizations that arose around a few inventive and entrepreneurial individuals. The high-tech specialty market is estimated to represent a $45-billion sector of the United States chemicals industry, and "Chemical Specialties Are 'the Place to Be,' but Problems Loom" (*CMR* 2/29/88).

**TABLE 1.6   Some Categories of High-Tech Specialties**

| | |
|---|---|
| Adhesives: sealants, structural | Graphic arts materials |
| Agrichemicals | Hydraulic fluids |
| Alloys: metallic, polymer | Ion-exchange resins |
| Antioxidants | Laboratory reagents |
| Antiozonants | Light absorbers |
| Antistatics | Lubricant additives |
| Audiovisual devices | Lubricants, synthetic |
| Battery materials | Membranes, synthetic |
| Biocides | Metal plating, finishing agents |
| Biomedical materials | Metalworking fluids |
| Biotechnology materials and devices | Mining chemicals |
| Catalysts | Oil-field chemicals |
| Ceramics | Packaging materials |
| Chelating agents | Paint additives, leveling, suspending |
| Cleaners: industrial, institutional, domestic |    agents |
| Coal and fuel additives | Paper additives |
| Coatings, industrial high solids, powder | Photographic materials |
| Composites | Photovoltaic materials and devices |
| Corrosion inhibitors | Pigments |
| Cosmetic additives | Plasticizers |
| Defoamers | Polymer additives |
| Diagnostic materials and devices: | Polymers, high-performance |
|    immunoassay, radiopharmaceuticals, | Printing materials |
|    home medical kits | Refinery–pipeline chemicals |
| Dyes | Reinforcing materials |
| Elastomers | Rubber-processing chemicals |
| Electronic chemicals | Sealants (see "Adhesives") |
| Enzymes | Surface-active agents: amphoteric, ionic, |
| Explosives |    nonionic |
| Flame retardants | Textile-processing agents |
| Flavors and fragrances | Thickening, thixotropic agents |
| Flotation agents | Toners, copying machines |
| Food additives, processing aids | UV screening agents |
| Foundry chemicals | Water-management materials |
| Gases, specialty | |
| Gasoline additives | |

It is not surprising that names of products denoting similar performance appear among both traditional and high-tech specialties because both exist side by side on the marketplace. Thus, for example, an adhesive can be a traditional, starch-based mucilage or a viscous solution of gum arabic, or it can be a high-tech, specially developed polymeric material with a particular affinity for the surfaces that are to be bonded. As already noted in the case of fine and ultrafine chemicals, differences between such materials can occasionally be blurred, and the same problem can exist when one wishes to label other categories of chemicals. However, some of the characteristic differences between extremes in the spectrum of industrial chemicals can be identified as shown in Table 1.7. For additional comments regarding categories of industrial chemicals, see *EPST-SZ* and *CT* **1984** 113, 185, 429, 599; **1985** 3, 64.

**TABLE 1.7   Extreme Differences between Commodity and High-Tech Specialty Products**

| Characteristic | Commodity Chemicals | High-Tech Products |
|---|---|---|
| Production volume and unit costs | Consult Table 1.5 | |
| Number of producers | ~ 250 (members of CMA) | Many ($10^3$) small businesses and divisions of large companies |
| Chemical nature | Simple, standard chemical name | Complex, propietary trade-name |
| Uses | Multiple, well known | Specific, limited market |
| Dependence on state of economy | Very sensitive | Relatively independent |
| Patent situation | Known or propietary or licensed process | Propietary composition and formulation–performance |
| R&D effort | Improved yield and energy utilization, or none if turn-key plant | Lengthy effort to develop synthesis, formulation, and performance |
| Competitive edge | Process economics | Product performance |
| Technical expertise requirements | Engineering ≫ chemistry, maintenance | Chemistry know-how ≫ engineering |
| Production facility | Very large, inflexible | Small, versatile |
| Operation | Continuous, automated economy of scale critical | Batch for better control |
| Capital investment | $10^8$–$10^9$, $1.50–$2 per $ of annual sales | $10^5$–$10^6$, spread over more than one product |
| Labor requirements | Minimal | Costly professional |
| Sales price | ~ 1.5–3 times cost of materials–energy, return on capital investment | What the market will bear, return on R&D costs |
| Profitability | Low, highly competitive market | High |
| Marketing costs | Low, ~ 10% of sales | High promotional efforts |
| Sales practices | Large, long-range contracts; reliable delivery crucial | Technical basis for sales client's problem solving |
| Sales expertise | Business ≫ technical | Technical ≫ business |
| Management style | Hierarchial, conservative, slow, rigid | Entrepreneurial, quick, imaginative |

Based on a table by H. H. Szmant in *CHEMTECH* p. 598 (October 1984). Copyright 1984, American Chemical Society.

Another point that requires clarification is the above-mentioned expression "chemicals and/or devices." We choose to illustrate the meaning of the term "devices" by means of the success story about the superabsorbent, nonbulky, disposable diaper, such as Procter & Gamble's Ultra Pampers described in *CB* 7/87. It illustrates

1. The enabling role of chemicals.
2. Superior performance of a chemical because of a better understanding of the required structure–property relationship.
3. The economic and commercial ramifications (see Section 1.3) that can result from the development of a new chemical of superior performance and available at a reasonable cost.

4. The meaning of a successful device by virtue of the performance of newly developed chemical(s).

5. The contribution of chemicals and chemistry-based devices to the improvement of our "quality of life."

The key chemical in connection with superabsorbent (for aqueous systems) materials is the polymer of an acrylic acid salt, presumably sodium acrylate, cross-linked to a finite, optimum extent with a difunctional acrylate such as ethylene bis(acrylamide), $(-CH_2-NH-CO\cdot CH=CH_2)_2$. The polymer is formed by bulk or suspension polymerization, and the resulting white, granular solid absorbs and retains about 50 times its weight of aqueous fluids. By "retention" one means that the aqueous-fluids absorbent (AFA) does not release the solution on application of reasonable pressures. This critical absorption–retention capability results from an uncoiling of polymeric chains within the stable, three-dimensional gel formed by hydration of metallic ions and by hydrogen bonding between water molecules and carboxylate groups of the polymer. The key polymer, derived from petroleum-based intermediates, is held mechanically within a layer of nonwoven wood fibers—a product of the forestry industry, modified by a petroleum- or natural-gas-derived additive capable of binding cellulose fibers into a fluffy web. This internal layer of the device under discussion is placed between a thin layer formed by a nonwoven web of either PP, or PET fibers and another layer constituted by a continuous, impermeable film of PP. Neither of these outer layers absorb water and thus they feel dry, but they differ in that the first-mentioned layer allows an aqueous solution to pass toward the AFA, while the second-mentioned layer inhibits any escape of the aquous solution. Now, regardless of whether Procter & Gamble is *not* listed (!) among the "Top 100 Chemical Producers" (*CEN*, 6/87, 36–37) or occupies the number one position based on annual sales within the sector "Soaps, synthetic detergents, other cleaning and polishing products" (*CW* 6/3/87; see Section 1.1.4, below here), the fact remains that it introduced this chemistry-based device in February 1986 and became an immediate success in a highly competitive \$3.4B/yr market (17B units/yr), of which about half of all sales are based on superabsorbent polymers, (SAPs). Each diaper described here contains 4–5 oz of AFA, and this translates into a consumption of about 120MM lb of glacial acrylic acid, 50 ¢/lb, and about \$200-million worth of the polyacrylate priced at \$1.40–1.50/lb. The total demand of 210MM–220MM lb for acrylic acid is currently satisfied by five domestic suppliers, and the polymer is also imported from Japan, where, by the way, this type of disposable diaper was introduced 5 years ago. An additional advantage provided by this device is its 50% reduction in bulkiness when compared to that of traditional products, and this favorably affects the cost of packaging, shipping, and storage.

It is debatable whether the device or even the novel cross-linked polyacrylates are high-tech chemical specialties or simply traditional products improved by appropriate physical and chemical modifications of traditional components. The significant point is that AFAs are affecting favorably analogous markets of adult-incontinence garments and outer feminine hygiene pads (that use 12–14 g and 1–1.5 g of the polymer per unit, respectively) and that another large, potential market in our aging population consists of incontinency bedpads for hospitals and nursing homes. Finally, it should be pointed out that the cross-linked polyacrylate AFA is an outgrowth of the "superslurper" (*SZM-II*, 93) developed originally at the USDA Experimental Station in Peoria, IL, then commercialized by General Foods and acquired still later by Henkel. The polyacrylate AFA is apparently competitive with the partially hydrolyzed starch–acrylonitrile graft polymer, but no one should be surprised that it, in turn, will be challenged by a less costly and more effective material.

### 1.1.3   Contribution of Industrial Organic Chemicals to the Economy

It is difficult to determine the exact economic contribution of individual chemicals because of the multitude and variety of their uses. Consider, as an example, the case of ethylene. It is used directly to ripen fruit and for the production of acetylene, but most of it is polymerized to a family of polyethylenes (LDPE, LLDPE, HDPE, UHMWPE) and ethylene copolymers such as EVA, EPDM, and others (4.3.1). Much of it is converted to ethylene oxide (EO) that, in addition of being used as a sterilizing gas, is built into different surface-active agents (detergents, emulsifiers, wetting agents, etc.), antifreeze fluids, solvents, and so forth, while other EO derivatives constitute valuable components of polyurethanes (PUR) (foams, coatings, elastomers that include the rubbery textile fiber Spandex used extensively in the manufacture of athletic support items and stretchable undergarments), and fine chemicals such as phenethyl alcohol, that as the "oil of roses" serves as an ingredient of industrial scents and perfumes. Ethylene is also the feedstock for ethylbenzene—the precursor of styrene (4.3.3), and for vinyl chloride, the building block of PVC and chlorinated (CPVC) (4.3.2). The latter polymers are replacing lead- and copper-based plumbing and facilitate the installation of inexpensive, portable irrigation systems for more productive agriculture. From styrene, on the other hand, there is prepared a family of polymers and copolymers (9.2.2.1) such as PS, high-impact polystyrenes (HIPS), unsaturated polyester resins (UPRs), styrene–acrylonitrile (SAN), SBR, and blends or alloys of PS and poly(phenylene ether) (PPE).

All the above illustrate the major lesson embodied in Fig. 1.1: the magic of chemical transformations (that may include some physical and biological processes) creates more valuable products from less valuable raw materials and intermediates. In other words, real material wealth is created through the transformation of low-cost materials to higher-value-added products that, in turn, are incorporated into countless products for which there is a demand by all sectors of a modern economy (agriculture, mining, construction, manufacture, transportation, government, etc.), and that eventually show up in wholesale and retail commerce. The ubiquitous nature of chemicals in our economy is demonstrated by the recent concern of the EPA (*CEN* 9/22/86) regarding the disposal of *small* amounts of hazardous wastes (< 100 kg per month) that are generated in a variety of economic activities; it is estimated that more than 100,000 businesses will be affected by the contemplated regulations!

An idea of the relative economic values of different classes of synthetic organic chemicals as classified by USITC-a was discussed earlier. These figures, however, are again on the low side because of the restrictive rules by which the USITC operates and because imports are not included in the sales volumes. These limitations—and there are additional ones such as the omission of chemicals obtained from renewable resources (*SZM-II*)—result in a highly underrated picture of the quantitative role that the chemical industry plays in the economy of the United States and, for that matter, in the economy of any nonprimitive society. The very large number of chemicals of industrial importance explains that, at least at this point in history, no country is capable of producing all the chemicals that are needed to enable its economy to function, and it turns out that economic growth is associated with increasing imports of many chemicals (other than fuels and fertilizers). This interesting subject is beyond the scope of the present book.

To illustrate the magnitude of chemical imports, and limiting ourselves to the imports of only benzenoid materials, we learn from USITC-b that in 1983 the United States imported 2.075B lb of 3069 different materials valued at $2.145 billion from 59 different countries. In spite of these and other, similar imports mentioned occasionaly throughout this book, the trade balance in chemicals is still positive in favor of the United States

(> \$11 billion in 1981, then shrinking rapidly to \$7.8 billion in 1985 but regaining some lost ground, reaching \$9.345 billion in 1987 because of the drop in the value of the dollar), while the overall trade balance is negative and, at last, becoming the subject of great national concern.

The ubiquitous need of chemicals in our economy is not appreciated by the general public and by many influential individuals who would be expected to know better. Under duress of World War II exigencies, when the saying "for lack of a nail the kingdom was lost" became applicable, the country suddenly mobilized to become an "Arsenal for Democracy," and the War Production Board orchestrated the production of key chemicals (to make products ranging from SBR and high-octane aviation fuel to nylon 6/6 and penicillin), but this experience soon fell victim to society's short memory.

To focus on the role of industrial chemicals in the U.S. economy, let us examine the statistical methodology employed by the federal government.

The Bureau of the Census breaks down the different sectors of the national economy by means of a Standard Industrial Classification (SIC) code. Thus, while the total value of production and services for 1984—the gross national product (GNP) was about \$3.7 trillion, the contributions of individual sectors to the GNP are summarized in Table 1.8. It is noteworthy that only the first four economic sectors are truly productive in the sense that they give rise to tangible, material products desired in the marketplace. The other economic sectors either facilitate the transfer of material goods from their sources to industrial, business, and individual consumers or provide other services that enrich the collective quality of life of our generally orderly, organized society. It is obvious that a healthy economy must maintain a reasonable balance between the productive and service components.

**TABLE 1.8   Sectors of U.S. Economy, 1984**

| | SIC | Value-Added | Value of Shipments or Services (\$ Billions) |
|---|---|---|---|
| *Productive Sectors of the Economy* | | | |
| Agriculture, forestry, fisheries | 01–09 | 91.1 | |
| Mining | 10–14 | 118.5 | |
| Construction | | | |
| Single \$113 | | | |
| Residential \$184 | 15–17 | 148 | 385 |
| Commercial \$88 | | | |
| Manufacturing | 20–39 | 882 | 2055 |
| Food processing | 20 | 93.4 | 287.0 |
| Tobacco products | 21 | 9.8 | 16.3 |
| Textile mill products | 22 | 21.3 | 53.4 |
| Apparel and other textile products | 23 | 27.3 | 55.4 |
| Lumber and wood products | 24 | 19.5 | 51.3 |
| Furniture and fixtures | 25 | 14.3 | 26.5 |
| Paper and allied products | 26 | 35.6 | 84.6 |
| Printing and publishing | 27 | 60.1 | 92.8 |
| Chemical and allied products | 28 | 86.5 | 183.2 |
| Inorganic industrial chemicals | 281 | 9.2 | 17.76 |

**TABLE 1.8**  (*Continued*)

| | SIC | Value-Added | Value of Shipments or Services ($ Billions) |
|---|---|---|---|
| Plastics, synthetic materials | 282 | 12.4 | 33.0 |
| Drugs | 283 | 19.4 | 27.4 |
| Soaps, cleaners, toilet goods | 284 | 15.5 | 27.0 |
| Paints and allied products | 285 | 4.7 | 10.2 |
| Organic industrial chemicals | 286 | 14.5 | 41.3 |
| Agricultural chemicals | 287 | 5.1 | 13.7 |
| Miscellaneous chemicals | 289 | 5.7 | 12.8 |
| Petroleum and coal products | 29 | 21.0 | 192.6 |
| Rubber and miscellaneous plastic products | 30 | 29.8 | 60.3 |
| Leather and leather products | 31 | 4.9 | 9.8 |
| Stone, clay, glass products | 32 | 25.3 | 48.7 |
| Primary metal industries | 33 | 36.0 | 108.5 |
| Fabricated metal products | 34 | 61.3 | 122.1 |
| Machinery, except electric | 35 | 94.8 | 179.0 |
| Electric, electronic equipment | 36 | 92.5 | 159.2 |
| Transportation equipment | 37 | 99.6 | 244.1 |
| Instruments and related products | 38 | 35.3 | 53.6 |
| Miscellaneous manufacturing | 39 | 13.7 | 26.4 |
| *Economic Infrastructure* | | | |
| Transportation, communications, utilities (electricity, gas, sanitary facilities) | 40–46 | | 342.2 |
| *Service Sectors of the Economy* | | | |
| Trade | | | |
| Wholesale, retail | 50–59 | | 601.9 |
| Finance, insurance, real estate | 60–67 | | 598.1 |
| Health care, education, social services, entertainment | 70–97 | | 529.4 |
| Government and government enterprises | 98 | | 421.9 |
| Other | 99 | | 36.2 |
| | | GNP = | 3662.8 |

*Source: Statistical Abstracts of the United States*, U.S. Department of Commerce, Bureau of Census.

What is defined as the chemical industry, namely, the production and sale of individual chemicals and allied products, is, on the basis of an arbitrary classification, only one of 20 segments of the manufacturing sector, and its SIC code is 28. This unfortunate definition of the chemical industry creates a deficient picture of the essential role that chemicals, chemical processes, chemists, and chemical engineers play in several of the other manufacturing sectors. This becomes obvious if we examine the nature of the remaining 19 manufacturing sectors in SIC 20–39, the value of products manufactured by each segment, and the way that chemicals, especially organic chemicals as they are the subject of this book, contribute to the operations.

It is difficult to imagine any one of the manufacturing segments to function with total indifference to industrial organic chemicals. Some of the contributions of the latter are

SIC 20: food additives (preservatives, flavors, fragrances, FDA-approved coloring matter), decaffeinated coffee and tea, and so on

SIC 21: humefactants, nicotine reduction, cigarette filters

SIC 22: textile fibers, lubricants, detergents, dyes, conditioners

SIC 23: textile finishing materials, plastic buttons and zippers, natural and synthetic fiber thread

SIC 24: wood preservatives, resins for particle boards and plywood, plastic-coated laminates

SIC 25: adhesives, stains, plastic components, cushions

SIC 26: pulping chemicals, paper additives, coloring agents

SIC 27: printing inks, pigments, glues

SIC 29: extraction solvents, catalysts, removal of hydrogen sulfide, hydroalkylation and alkylation reactants, coal-treatment agents

SIC 30: rubber-processing chemicals, additives, molding lubricants

SIC 31: dehairing, tanning, and finishing agents

SIC 32: organic pigments, waterproofing agents

SIC 33: flotation and extraction agents for the separation of desired metallic ingredient of ores, purification of metals

SIC 34: cutting and degreasing fluids, electroplating additives, antirust agents, original equipment protective and decorative coatings, OEC

SIC 35: machining aids, original equipment coatings, OEC

SIC 36: insulators, conductors, semiconductors, electronic materials such as ultrapure reagents, high-performance polymers for audiovisual tapes, compact disks, and body of equipment

SIC 37: rigid, flexible, and semiflexible structural components; cushion and leatherlike materials; coatings; brake, transmission, hydraulic, radiator, and other functional fluids

SIC 38: devices based on specialty materials, plastic structural components.

The chemicals and allied product segment, SIC 28, is subdivided as shown in Table 1.8. This table demonstrates, once more, the arbitrary classification of industrial chemicals since the majority of components except SIC 281 and, in part, components 284, 285, and 287, include organic materials mentioned throughout this book above and beyond those classified as "industrial organic chemicals" of SIC 286. In any case, the value of shipments during 1985 of materials enumerated in SIC 281–289 amounted to $214 billion. This represents 6.7% of the GNP and accounts for an employment of slightly over one million people.

A broader and more rational way to evaluate the role of chemicals, and especially organic chemicals, in the economy is based on the inclusion of the following manufacturing activities:

• Processing of wood to all sorts of wood-derived products (paper, carton, paperboard, tall oil, turpentine, chemical cellulose, etc.)

• Processing of petroleum and natural gas

• Manufacture of polymers (plastics, resins, rubber, or elastomers)

- Manufacture of glass- and clay-derived products and processing of limestone and similar raw materials
- Processing of sugar (sucrose) and refining operations
- Separation of corn components (wet milling)
- Processing of foods and manufacture of beverages
- Dyeing and finishing of textiles
- Tanning and finishing of leather
- Manufacture of dry cells, storage batteries, and semiconductors
- Manufacture of carbon and graphite products
- Manufacture of linoleum and floor tiles

All of these manufacturing operations have three things in common:

1. They utilize large quantities of different organic chemicals.
2. They depend heavily on chemical transformation processes.
3. They employ significant numbers of chemical professionals.

These industries are commonly referred to as the *chemical process industries* (CPIs), and their contribution to the national economy is reflected by the $649-billion value of shipments in 1985, or 20.5% of the GNP, and they employ about 4 million people. While chemical engineers focus their attention on the CPIs, for traditional and self-defeating reasons, chemists tend to limit their vision to the narrowly defined "chemical and allied products" industry. Such split and/or myopic vision is detrimental to the technologically and economically competitive future of the country.

So far, we have examined the importance of industrial organic chemicals in segments 20–39 of the manufacturing sector as listed in Table 1.8. However, chemicals also play a significant role in the functioning of some other, major economic sectors shown in Table 1.8. Some examples are as follows:

SIC 01–09: fertilizers, agrochemicals (herbicides and other pesticides such as insecticides, fungicides, rodenticides), plant growth control and ripening agents, crop preservatives, packaging materials, refrigerants

SIC 15–17: concrete additives, plastic plumbing and roofing, sidings, flooring materials, thermal and electric insulation and installation materials, paints, sealants

SIC 40–49: asphalt and other road-maintenance materials, traffic paint

SIC 50–59: packaging materials, refrigerants, sanitizing formulations

SIC 70–97: health-care products, educational materials, sports equipment

This again demonstrates the ubiquitous nature of industrial chemicals in all facets of a technologically developed society and explains why many chemicals are referred to in a laudatory fashion as enabling, performance, and advanced materials.

### 1.1.4 Shifting Interests of Organic Chemical Producers in the United States

Before we address the subject of this section, it is appropriate to examine how the nature of the producers of chemicals is viewed in the professional literature.

A listing of the "Top 100 Chemical Producers" (*CEN* 6/8/87) shows "Chemical sales as % of total sales" that range from 100% (28 companies classified as "basic chemicals" and concentrated more so among the bottom half of the list), down to a low of 13.7% and still classified as "basic chemicals." But then we find a 100% chemical sales company classified as a "farm cooperative," and there are separate entries among the top 100 chemical producers with classifications "specialty chemicals," "specialty metals," "agrochemicals," "drugs," "photo equipment," "wood products," "petroleum," "natural gas," "dairy products," "machinery," "alcoholic beverages," "auto equipment," "glass products," "petroleum services," and "agricultural supplies." This seems to demonstrate an incongruent attempt to define "who does what" and attach an appropriate industrial classification label to each business. Also, it demonstrates the difficulty and perhaps futility of pinpointing an exact industry classification that causes the statistitions to throw in the towel and settle on a simple label "diversified" in seven cases and "diversified chemicals" in one case.

Another approach to the characterization of producers of chemicals and allied products lists (*CW* 5/28/86, 3/4/87) the financial performance of the 300 top CPIs under the headings of the following "segments":

- Industrial chemicals and synthetic materials
- Glass, cement, lime, abrasives, refractories
- Diversified companies with sales 25–50% of total sales
- Fertilizers
- Food and dairy companies with chemical process operations
- Pulp, paper, packaging
- Specialty chemicals
- Pharmaceuticals
- Soaps, synthetic detergents, other cleaning and polishing products
- Steel, coke, and coal-tar chemicals
- Paints and coatings
- Petroleum and natural-gas processing
- Nonferrous metals and ferroalloys
- Tires, other rubber and plastic products
- Toiletries and cosmetics
- Biotechnology

There is a separate listing of the financial performance in the area of chemicals of

- Diversified companies' chemical segments.
- Diversified companies—chemical sales 25–50% of total sales.
- Multiindustry companies with chemical process operations.

The relatively large number of categories used by *CW* to report the financial picture of the top 300 companies engaged in the production of the broad spectrum of materials is, on one hand, proof of the difficulties encountered in an attempt to pinpoint or "label" the relation of a business organization with the world of industrial chemicals and, on the other hand, proves the ubiquitous presence of chemicals in the economy. As a specific example of this problem, we can cite that Du Pont is listed as one of the "diversified companies—chemical sales 25–50% of total sales" with chemical sales of about $6.7 billion and operating the following businesses: biomedical products, industrial and consumer products, fibers, polymer products, and agricultural and industrial chemicals.

In recent years many of the large chemical corporations in the United States have become increasingly interested in the production of high-tech chemical specialties (that include bioengineered products) with a corresponding loss of interest in the production of commodity chemicals. This caused them to divest certain businesses [by outright sales or through leveraged buyouts (LBOs), i.e., purchases by divisional management or labor groups with borrowed moneys], while acquiring other businesses that fit new goals determined by corporate management. This process of "restructuring," together with a rash of friendly or less friendly ("hostile"), attempted or successful takeovers by outside companies or groups of investors, has produced and continues to produce mergers of past competitors on national and international scale. This all adds up to a state of constant flux and changes of names of chemical industries mentioned in the Preface. A discussion of the dynamics and motivations behind these corporate activities is beyond the scope of this book. However, one factor that contributes to the perturbed panorama of corporate United States is the appearance of first-, second-, and even higher-generation building blocks produced by recently industrialized countries that are blessed with ample petroleum and natural-gas resources. Saudi Arabia leads the way among such hydrocarbon-rich countries and, in the span of a dozen or so years, it has become a large-scale producer of methanol, polyethylene, styrene, ethylene glycol, ethylene dichloride, vinyl chloride monomer (VCM) and PVC, ethyl alcohol, MTBE, and other chemicals. While the bulk of these materials is destined for the European and Far Eastern markets, it still affects the United States production by reducing its own export possibilities. Certain countries that are not endowed by nature with petroleum or natural gas have also made great strides as producers of chemicals. Foremost examples in this group are the Pacific rim countries of Taiwan, South Korea, and Singapore (see H. H. Szmant, "The Role of Applied Chemistry in Economic Growth of the Developing Countries," *International Laboratory*, May/June 1987, pp. 6–9), Brazil, and Finland, and to a large extent their economic progress is based on domestic transformations of imported commodity chemicals or of crude petroleum. Since the chemical marketplace is a truly international business, all such developments produce significant repercussions on the chemical activities of the United States, and the shift in favor of high-tech chemical specialties is one such effect.

A few years ago it was estimated that different categories of chemicals (1.1.2) make the following approximate contributions to the sales by United States companies: commodities, 40%; pseudocommodities, 30%; fine chemicals, 8%; and high-tech specialties, 22%. Traditional chemical specialties were excluded from these estimates because their sales are often in the hands of "specialty products divisions" of larger corporations, or they are handled by autonomous chemical specialty products companies, most of which belong to CSMA.

## 1.2  HISTORICAL PERSPECTIVE

### 1.2.1  Industrial Organic Chemicals and Societal Wellbeing

The frequently quoted saying "man does not live by bread alone" must be balanced by an ancient Byzantine proverb "he who has bread may have many problems, but he who lacks it has only one" (*KCB-II*). Although philosophical discussion of the balance between the nonmaterial and material requirements for the attainment of peace and liberty in the pursuit of happiness is beyond the scope of this book, it is our task here to focus on the evolving contribution of industrial organic chemicals to what is now commonly known as (an implied good) "quality of life."

Naturally, the story of industrial organic chemicals cannot be divorced from the history of humanity's utilization of different kinds of materials, and the birth and growth of industrial chemistry must be perceived in conjunction with the historical phenomenon known as the "Industrial Revolution." In this context the reader is referred to J. Bronowski's magnificent *The Ascent of Man*, Little, Brown and Company, Boston and Toronto, 1973, and *RB*. Another valuable source of historical information is the *Concise Chronology of Science* compiled by Professor of Physics Marshall Walker of the University of Connecticut and published privately in 1978. Table 1.9 offers a glimpse of the material milestones of our civilization.

**TABLE 1.9    Material Milestones of Civilization**

| Previous Eras (approximate years ago) | Development |
|---|---|
| 400,000[a] | Fire → splitting of stones → tools |
| | *Stone Age* |
| 10,000 | Extraction of Cu at ∼1000°C from malachite ore to give soft copper metal |
| 7,000 | Extraction of tin, combination with copper (Persia) to give hard bronze |
| | *Bronze Age* |
| 5,800 | Working of iron (meteorites) at about 3800 B.C., before Egyptian pyramids (2500 B.C.). |
| | *Iron Age* |
| | *Steel Age* |
| 3,000 | India (1000 B.C.) |
| | *Industrial Revolution* |
| 200 | |
| 100 | Macromolecular materials |
| | Naturally occurring→modified→synthetic |
| | Thermoset polymers (resins) |
| | Thermoplastics |
| | Elastomers |
| | Fiber-reinforced polymers |
| | Composites |
| | Polymer blends and alloys |
| | High-performance or advanced materials |
| | *Polymer Age* |

[a] Recent evidence (1988) suggests that the discovery of fire-making dates back 1.5 million years.

TABLE 1.10    Breakdown in Consumption of Different Kinds of Materials
During 1960s[a]

| Material | Percent by Weight |
|---|---|
| Sand and gravel | 21.2 |
| Stone | 20.0 |
| Petroleum | 18.3 |
| Coal | 11.8 |
| Natural gas | 11.8 |
| Other nonmetallic materials | 7.2 |
| Metals | 3.1 |
| Renewable materials | 6.6 |
| Forest products | 96.0 |
| Cotton and other fibers | 1.7 |
| Seed oils | 1.0 |
| Animal fats | 0.9 |
| Natural rubber | 0.4 |

[a] Source: National Commission of Materials Policy Report (June 1973).

A modern, industrially developed society is an avid consumer of materials as well as energy (see Table 1.10). A conversion of the figures from "by weight" to "by value" would probably show some significant changes from the figures cited above because of the shift from the use of coal as a fuel to the greater dependence on petroleum and natural gas and the rapid replacement of metals by polymers.

## 1.2.2  Evolution of Industrial Organic Chemistry

Industrial organic chemistry was born out of ancient technologies such as

1.  Dry distillation of wood to give charcoal and a pyroligneous liquor that contains methyl (wood) alcohol, acetic acid, and other organic constituents, plus water.

2.  Fermentation of all sorts of carbohydrates to ethyl (grain) alcohol. This prehistoric technology, mentioned in the *Old Testament*, was modernized by the use of large-scale, metallic fermentation equipment (first made of copper and later stainless steel), and fractionating columns capable of separating the residue after removal of ethanol. This residue, known as *fusel oil*, was the first source of isoamyl and 2-methyl-1-butyl alcohols as well as smaller quantities of isobutyl and *n*-propyl alcohols and some aldehydes and esters.

3.  Saponification of animal and vegetable lipids (fats and oils). This was a powerful stimulus once the use of soap became fashionable. The need of alkali (satisfied by potash) caused a decimation of the forests of western Europe and led to the development of the Solvay process for the manufacture of caustic soda.

4.  The separation of the volatile by-products of coke required by the emerging steel industry. Interest in industrial organic chemistry increased with development of this technology. The liquid and solid volatile components of coal tar became the building blocks of the first synthetic dyes (such as the dye mauve prepared by W. H. Perkins in 1856) and medicinals.

5.  Development of the internal-combustion engine to power the automobile and other vehicles. This was the next powerful stimulus in the evolution of industrial organic

chemistry. The search for a convenient fuel and lubricants to operate the increasingly popular vehicles caused the maturing of the petrochemical industry (P. Spitz, *Petrochemicals: The Rise of an Industry*, Wiley, New York, 1988) and the evolution of aliphatic chemistry to complement the aromatic chemistry developed for study of coal-tar by-products.

6. Macromolecular technology. The "father of polymer chemistry," Hermann Staudinger, did not lay down the foundation for the understanding of the nature of synthetic macromolecules until the era of World War I (1914–1918). This understanding was required for their phenomenal growth that we are still witnessing. For an account of the development of polymer chemistry "From revolution to evolution" by Herman Mark, the "father of American polymer chemistry," see *CT* 6/87, pp. 328–331.

All these developments occurred practically yesterday in the context of the at-least-400,000-year-old discovery of fire and the approximately 10,000-year-old cultural history of humanity, about which we possess only fragmentary information. While the United States was celebrating the bicentennial of its 1787 Constitution, we must keep in mind that the imagination of chemists was not freed from the mental restraint of a belief in "vital force" until 1804, when Friedrich Woehler succeeded in the synthesis of urea, that a structure of benzene consistent with our current understanding was not proposed by Friedrich Kekulé von Stradonitz until 1865 and that the first synthetically produced dye was not serendipitously synthesized by Perkins until 1856 [see "How the Aniline Dyes Were Discovered," G. B. Kauffman, *Ind. Chem.* **3**, 26–27 (1988)] (see also Table 1.11).

A better understanding and greater appreciation of the countless benefits to humanity contributed by industrial organic chemistry, in spite of its youthful character, is highly desirable in order to understand and excuse some errors of judgment committed by chemical technology and not impede its maturation through excessively punitive legislative actions.

Until the 1920s and 1930s coal continued to be the dominant source of organic chemicals in spite of the rapid growth of the petrochemical industry and some dependence on agricultural and forest resources. Most important among the latter were the fats and oils derived from animal and vegetable sources, fermentation of carbohydrates (mostly molasses—the residue from crystallization of sucrose) to produce acetone, *n*-butyl alcohol, fusel oil in addition to ethanol, cellulose derived from cotton and some "chemical cellulose" (i.e., cellulose destined for chemical rather than textile fiber use) obtained by an appropriate pulping of wood, and the volatile and extractable constituents of pinetrees. The last-mentioned source of turpentine and rosin constituted the naval stores industry, which dates back to the use of wood-constructed ships throughout the world. The U.S. production of oleoresin or gum turpentine actually peaked in 1908 with a production of 2 million barrels (a naval stores barrel contains 55 gal of product) or over 1B lb. The embryonic chemurgic initiatives of the 1920s were stifled by the above-mentioned growth of the petrochemical industry. Dictionaries define "chemurgy" as a branch of chemistry that converts agricultural resources into products other than foods and textile fibers. The current definition of chemurgy should be broadened to include an industrial utilization of products and by-products from *any* renewable resource regardless of whether it is associated with traditional agriculture, silviculture, or aquaculture (*SZM*-II; *CEN* 8/4/75, 9/29/75). An excellent survey of chemurgic possibilities published in 1946 by Wheeler McMillen (*New Riches from the Soil—Progress of Chemurgy*, Van Nostrand, New York) should be reprinted and become required reading for legislators and political leaders who are trying to "save" our agriculture by spending tax moneys for *not* cultivating our fertile

**TABLE 1.11    Some Milestones in the History of Industrial Organic Chemistry**

| | |
|---|---|
| 1823 | Hydrolysis of fats and oils (Chevreal) |
| 1839 | Vulcanization of natural rubber (Charles Goodyear) |
| ~1845 | Wood-pulp cellulose replaces rags in the production of paper |
| 1856 | First synthetic dye "mauve" (W. Henry Perkin, Br. Pat. 1984) |
| 1862 | Cellulose nitrate (A. Parker) |
| 1865 | Kekulé's structure of benzene consistent with current idea |
| 1866 | Alfred Nobel introduces dynamite (nitroglycerine and kieselguhr) |
| 1868 | Cellulose nitrate is plasticized with camphor to give celluloid or "synthetic ivory" (J. W. Hyatt) |
| 1880 | Synthesis of indigo by Adolf von Bayer |
| | George Eastman produces gelatin-covered photographic plate |
| 1882 | Cellulose nitrate-based photographic film |
| 1892 | Production of calcium carbide (and acetylene) and calcium cyanamide |
| 1898 | Production of aspirin by the Bayer Company |
| 1899–1907 | Emil Fischer elucidates structure of carbohydrates and proteins |
| 1905 | Production of viscose rayon |
| 1906 | Aniline-based rubber vulcanization accelerators |
| 1909 | Leo H. Baekland develops phenol–formaldehyde resins (Bakelite) |
| 1910 | Paul Ehrich's Salvarsan (the "magic bullet") produced by Hoechst |
| 1912 | Cracking of petroleum developed by Burton |
| ~1914 | Chaim Weizmann produces acetone, *n*-butanol, and ethanol by fermentation |
| 1918 | Hermann Staudinger clarifies the molecular structure of polymers |
| 1921 | Acetate rayon fibers, film, and molded objects produced by C. and H. Dreyfus |
| 1923 | Midgley introduces gasoline antiknock agent tetraethyl lead, TEL |
| 1925 | Industrial manufacture of PS by IGF[a] |
| 1926 | Introduction of synthetic antimalarial pamaquin |
| | Du Pont introduces synthetic methanol or wood alcohol |
| | Glyptal or alkyd resins introduced by General Electric |
| 1927 | Production of PVC |
| 1929 | Production of UF resins |
| | Fleming discovers penicillin (production begins in 1941) |
| 1931 | Du Pont introduces neoprene rubber |
| 1932 | Domagk introduces first sulfa drug: sulfamidochrysoidine or Prontosil |
| 1933 | LDPE developed by ICI |
| 1935 | Production of ethyl cellulose (EC) |
| 1936 | Production of poly(vinylacetate) (PVA) |
| | Successful nylon 6/6 fiber developed at Du Pont (W. H. Carother's synthesis and J. Hill's cold-drawing technique |
| 1937 | NR (butadiene–acrylonitrile) introduced by IGF[a] |
| | Butyl rubber (isobutylene and some butadiene) introduced by Standard Oil |
| 1938 | Production by Du Pont of nylon 6/6 and poly(tetrafluoroethylene) (PTFE) or Teflon |
| | Introduction of poly(vinylacetal) plastics for safety glass, optically clear poly(methyl methacrylate) (PMM), and cellulose acetate–butyrate for coatings |
| 1939 | Introduction of poly(vinylidene chloride) (PVdC) plastics and melamine–formaldehyde (MF) resins |
| 1939–1944 | War Production Board coordinates production of "Government Rubber Styrene" (GRS) and other materials essential to war effort |
| | Production of LDPE derived from ethanol initiated by ICI (shift to petroleum-derived ethylene does not occur until 1952!) |
| 1942 | Introduction of LDPE and UPRs in the USA |

**TABLE 1.11**   *(Continued)*

| | |
|---|---|
| 1943 | Introduction of silicone resins |
| 1945 | Introduction of cellulose propionate |
| 1948 | Du Pont introduces epoxy resins and poly(acrylonitrile) (PAN) |
| | Introduction of acrylonitrile–butadiene–styrene (ABS) and styrene–acrylonitrile (SAN) |
| 1949 | Introduction of diallyl phthalate (DAP) and diallyl isophthalate (DAIP) |
| | Poly(ethylene–terephthalate) (PET), ICI's Terylene polyester fibers introduced by ICI |
| 1950 | Du Pont introduces its Dacron PET fiber, Mylar PET film |
| 1953 | Production of synthetic detergents, syndets, exceeds the production of traditional soap for the first time in the USA |
| 1957 | PURs developed by Otto Bayer |
| | Introduction of PP and polycarbonate (PC) plastics |
| 1959 | Production of petroleum-derived benzene exceeds the production of coal tar-derived benzene for the first time in the USA |
| 1960 | Worldwide production of synthetic rubber (elastomers) exceeds for the first time the production of natural rubber |
| 1962 | Introduction of poly(aryl ethers), GE's poly(phenylene oxide), or (PPO) and poly(phenylene ether) (PPE) |
| 1963 | Introduction of polyimides (PIs) and poly(ethylene–vinyl acetate) (PEVAc) |
| 1966 | Silane coupling agents, nitrile, and other barrier resins |
| 1970–[b] | Engineering plastics such as poly(butylene terephthalate) (PBT), poly(oxymethylene) (POM), poly(phenylene sulfide) (PPS), poly(ether sulfones) (PESO), poly(benzimidazoles) (PIB), aramids (Du Pont's Kevlar and Nomex); polyarylates (PARs), liquid-crystal polymers (LCPs), and other high-performance specialty thermoplastics (HPSTs) |
| | Thermoplastic elastomers (TPEs); ionomers, including Du Pont's Nafion; metathesis polymerization of dicyclopentadiene |
| | Interpenetrating polymer networks (IPNs), fiber-reinforced polymers (FRPs), polymer blends or alloys, composites |
| | Reaction-injection and reinforced-reaction-injection molding (RIM, RRIM) technology, PAN-based graphite fibers |
| | Use of bound enzymes for industrial production of chemicals |
| | Synthetic membrane technology for large-scale selective separations, including physical resolution of chiral isomers |
| | Large-scale applications of zeolite and homogeneous transition-metal catalysts |
| | Monsanto's methanol to acetic acid conversion process and other applications of syngas (synthesis gas) chemistry |
| | Hock technology that converts cumene to phenol–acetone; many other important developments mentioned in this book |

[a] I. G. Farbenwerke, the large German chemical cartel that was broken up after World War II because of its involvement with Nazi Germany [see W. B. Smith, "Chemistry and the Holocaust," *J. Chem. Ed.* **58**, 836–838 (1982).
[b] 1970 to present: introduction of new technologies mentioned in no particular order.

lands. In the meantime, western Europe (*CW* 1/7–14/87) is mobilizing its excess farm production for industrial utilization, and Japan is initiating similar activities within the United States in view of scarcity of cultivable land on its island homeland.

In his role of President and Director of Research of the Sugar Research Foundation Henry B. Hass's efforts of converting sucrose—the largest-volume renewable source of a pure and cheap organic chemical with an established production base—to a variety of industrial chemicals, collapsed when the sugar industry decided instead to promote an increased consumption through advertisements designed to convince the public that "sugar" is a low-calorie food and does not cause dental caries. Some 30 years later, on an international scale, "worldwide glut of sugar is spurring many producers to expand into (sucrose) derivatives" (see P. L. Layman, *CEN* 12/7/87). No such efforts are apparent on the domestic scene, and in the meantime U.S. taxpayers are being forced to subsidize domestic sugar producers, and while contributing to unemployment compensation funds, the domestic sugar production functions with imported foreign labor under regulations that "hurt Florida farm workers" (P. Chang, *Fort Myers News-Press*, 8/23/87). Having learned little about the development of artificial sweeteners, and apparently oblivious to technical developments looming on the horizon, the Sugar Cane Growers Cooperative in Belle Glade, Florida, a major sugar-production center of the United States, and the "Sugar Association," a Washington DC-based trade association, are undertaking another defensive campaign at a cost of $4 million summarized by the *Fort Myers News-Press* (2/22/88) by the headline "Cane industry plans sweet ads." Currently, there are few domestic applications of sucrose as an industrial feedstock (*SZM-II*, 103–104).

With the advent of World War II the then-established War Production Board coordinated a multifaceted initiative to produce high-octane aviation fuel, synthetic rubber (GRS, or today's SBR), and numerous other products essential for the supply of the Allied war effort. During the postwar era many of these materials, especially plastics, flooded the consumer market. The dramatic change in the origin of industrial organic chemicals is illustrated by the following comparison. While the principal raw materials for the production of organic materials during the 1930s were coal 60%, carbohydrates 30%, and petroleum 10%, by 1960 petroleum and natural gas supplied more than 80% of all industrial organic chemicals, and coal and carbohydrates accounted for about 20 and 1%, respectively. The World War II era was also the beginning of the "plastic age" in our civilization, that is, the gradual replacement of traditional, "natural" materials used by the manufacture and construction sectors of the economy by artificially produced (synthetic) organic macromolecules. At first, the popular mind was suspicious of the quality of "synthetic" materials vis-à-vis "natural" materials, but nylon 6/6 stockings—the first victim of World War II scarcities—helped to convince the female population of the superiority of this synthetic product relative to its traditional cotton counterpart. By now, the stigma of being "synthetic" has practically disappeared from popular thinking. Characteristic of shifting sources for the production of organic chemicals is the history of nylon 6/6. Originally, Du Pont obtained the building blocks of adipic acid and hexamethylenediamine from coal- and petroleum-derived phenol, as well as from furfuraldehyde obtained by hydrolysis of the pentosan constituents of cereal hulls and bagasse—the solid residue from the processing of sugarcane. Currently, however, both building blocks are derived from petrochemical precursors.

The high rate of petrochemical innovation persisted throughout the 1950s and then decreased during the 1960s, when chemical manufacturers seemed to concentrate most of their R&D efforts on an improvement of existing materials and processes. Still, the level of production of industrial organic chemicals continued to increase, fueled by the growth in population and an improvement of its living standards and by exports. For example, the amount of synthetic organic chemicals produced by the end of the 1960s, namely, 105B lb (National Academy of Sciences, *Science and Technology: A Five Year Outlook*, Freeman, San Francisco) increased to more than 186B lb in 1978. Of this amount, approximately 103B lb appeared in the marketplace as intercompany sales and 83B lb were used internally (captive use). The production figures for 1982 and 1983, specifically, 202 and

215B lb, respectively, represented a slowdown in the growth, which was accounted for in part by the end, at the beginning of the 1970s, of the tranquil dependence on inexpensive crude oil ($2–$3 per 42-gal barrel). In 1973 the boycott by Arab members of the Organization of Petroleum Exporting Countries (OPEC) caused serious petroleum shortages in the United States and a consequent rise in the price of crude oil (to > $30/bbl). This scarcity affected the production of industrial organic chemicals and was not remedied by timely government initiatives. On the contrary, while the politicians, media, and the general public were preocupied with political scandals and exaggerated environmental issues, history records a series of delays in the construction of the Alaskan pipeline; issuance of offshore oil-drilling permits; explorations for oil and natural gas on federal lands; construction of additional conventional, fission-based nuclear power plants; development of new coal-conversion processes; and an escalated utilization of renewable (biomass) and other solar-energy-derived sources of materials and energy. Ten years after the boycott, development of the North Atlantic oil fields and excess production by competing OPEC members caused a significant (50%) reduction in the price of crude oil (down to about $13/bbl in late 1986), but these prices began to recover during the early part of 1987. In the meantime, domestic oil exploration and production in the United States was slowed down as a result of the drop in the international oil prices. Currently, the United States again depends on imports for 60% of its fuel needs. The 1987–1988 political tensions in the Persian Gulf created uncertainties with regard to oil supplies and prices for the near future; however, except for the Alaskan pipeline, no remedial measures have been undertaken in order to prevent a new energy and materials crisis in the United States (except for a maintenance of a 530MM-bbl "strategic petroleum reserve", which is equivalent to a 3-month supply of imports).

By contrast, let us briefly examine the Brazilian ethanol program (see 4.4.2) that was initiated in November 1975 by its president, General Ernesto Geisel, on advice of local experts and against the advice of some international experts. Faced with bankrupting petroleum imports, Brazil began to mobilize its agricultural resources to convert sugar cane juice directly to ethanol (rather than resort to the energy- and capital-intensive sugar processing and the use of residual molasses for fermentation). By now, Brazil is the world's largest producer of ethanol and, by law, all vehicles are fueled either with 95% (not "absolute" as practiced in the United States' gasohol!) ethanol ($\sim 2$ million automobiles), or by a mixture of gasoline and 20% ethanol ($\sim 8$ million vehicles). The continued expansion of ethanol production (from a current $10MM\ m^3$—equivalent to 2.65B gal—to a projected $21MM\ m^3$ by 1995) assures Brazil's 144-plus million people of a domestic supply of fuel; permits ethanol exports (to the United States and elsewhere) that earn "hard" currency; creates a feedstock for the production of ethylene (currently from about 5% of ethanol); provides massive employment for a poor rural population that is paid in local "soft" currency; and, equally important, creates local, technological know-how that stimulates further industrial development. By comparison, the production of the highly subsidized (currently by payments of $54 million) corn-based ethanol in the United States reached the 1B-gal mark at the start of 1987, it supplies the 10% "absolute ethanol" content of our gasohol that contributes about 10% to the total fuel demand. Recently Brazil has discovered its own source of petroleum, but apparently this, per se, is not affecting its ethanol production program, which, in turn, is subject to constant scrutiny and improvements.

The most recent and prospective changes in sources and production of specific industrial organic chemicals are discussed in 2.2.1 and elsewhere in this book.

# 2

# SOURCES, PRODUCTION PATHWAYS, AND PRICING OF INDUSTRIAL ORGANIC CHEMICALS

## 2.1 GENERAL INTRODUCTION

In Chapter 1 we described the demands for materials by an industrially developed economy and some historical events that affected the production of industrial organic chemicals. In this chapter we discuss in greater detail three topics enumerated in its title.

## 2.2  PRINCIPAL SOURCES OF INDUSTRIAL ORGANIC TRANSFORMATION PRODUCTS

A graphic overview of the principal sources of industrial organic chemicals is presented in Fig. 2.1. Figure 2.1 diagram indicates the major conversion paths of the three principal fossil feedstocks—natural gas, petroleum, and coal—to the synthetic organic chemicals reported by USITC-a, and it also includes an estimated contribution made by renewable feedstocks. The latter are discussed in greater detail elsewhere (*SZM-II*), but specific examples are mentioned throughout this book since it is important to realize that many compounds of industrial interest are derived from *both* renewable and nonrenewable feedstocks. (Please consult the Index for the items mentioned in Fig. 2.1.)

The contributions of inorganic materials to the final mass of organic transformation products are vaguely shown in Fig. 2.1 but are partially summarized in Table 2.1.

Inorganic materials not only provide essential catalysts (acids, bases, metallic complexes that function as homogeneous catalysts, surface catalysts, etc.) but also in

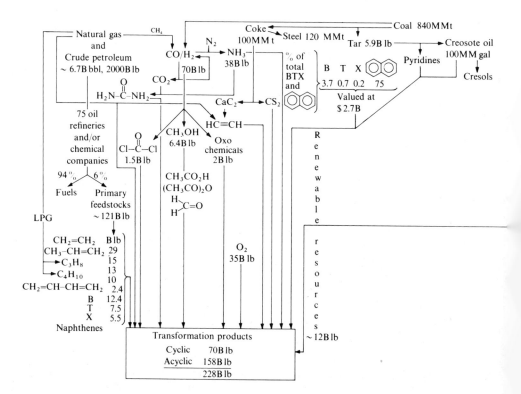

**Figure 2.1**  Principal sources of synthetic organic chemicals (USA).

many instances become structural components of the ultimate organic products. Consider, for example:

1. The oxygen content is derived from atmospheric oxygen or water in most alcohols, ethers, and other oxygen-containing organic compounds (see, e.g., the inconspicuous contributions of oxygen to the transformations of building blocks shown in Table 1.1).

2. The nitrogen content of amines, amides, nitro compounds, isocyanates, and so on is traceable to atmospheric nitrogen, ammonia derived from the latter, or nitric acid obtained from ammonia.

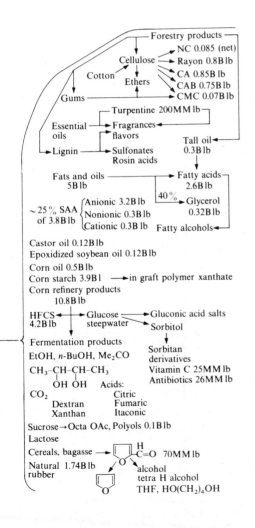

TABLE 2.1   Inorganic Chemicals Used in the Production of Industrial Organic Materials in the United States[a,b]

| Elemental Composition | Chemical | U.S. Production (B lb) |
|---|---|---|
| C, H, O | Acetic acid | 3.3 |
| | Acetic anhydride | 1.7 |
| | Acetone | 2 |
| | Alkyl acrylates | 1.20 |
| | Adipic acid | 1.5 |
| | Ethanol (synthetic) | 1.4 |
| | Ethylene glycol | 5.3 |
| | Ethylene oxide | 6 |
| | Formaldehyde | 6.2 |
| | Isopropyl alcohol | 1.3 |
| | Methanol | 6.8 |
| | Methyl $t$-butyl ether | 2.3 |
| | Phenol | 3.5 |
| | Phthalic anhydride | 1.55 |
| | Propylene oxide | 2.5 |
| | Terephthalic acid | 8 |
| | Vinyl acetate | 2.3 |
| | Oxo chemicals | ∼2.0 |
| C, H, N | Acrylonitrile | 2.5 |
| | Urea | ∼12 |
| | Caprolactam | 1.25 |
| | Polyamides | ∼3.0 |
| | Isocyanates | ∼1.5 |
| | Anilines | ∼1.0 |
| C, H, Cl | Ethylene dichloride | 14.5 |
| | Vinyl chloride | 8.5 |
| | Chlorinated methanes | ∼2.1 |
| | Chlorobenzenes, etc. | ∼1.0 |

[a] Chemicals listed whose annual U.S. production equals or exceeds 1B lb. Pure hydrocarbons are excepted.
[b] Expanded and up-dated table originally published by J. E. Lyons, *HC*, November 1980.

3. The sulfur content of sulfonic acids or organic sulfates, and other sulfur compounds is derived from elementary sulfur, hydrogen sulfide, sulfur dioxide, and sulfuric acid.

4. The iodine, bromine, chlorine, or fluorine content of halogenated organic compounds is derived from alkali or hydrogen halides, elementary halogens or some of their derivatives.

5. Organic compounds that contain silicon, phosphorous, tin, zinc, lead and other inorganic elements are commonly obtained by means of the corresponding halides.

While it would be tedious, and rather pointless, to estimate the quantitative contribution (by weight or by value) of the inorganic reactants responsible for the formation of the desired heteroatom-containing organic products, we must not overlook the fact that inorganic materials play a critical role in the production of industrial organic materials. In this context, contrary to traditional thinking, carbon monoxide and carbon dioxide are considered in this book to be important *organic* building blocks of industrial organic materials.

### 2.2.1  Natural Gas

Natural gas consists primarily of methane ($\sim 85\%$), with minor concentrations of ethane ($\sim 9\%$), propane ($\sim 3\%$), butanes (1%), and nitrogen (1%). In some regions of the world natural gas may contain sufficient helium to make separation of the latter commercially attractive. As discussed in Chapter 3, natural gas has become the principal feedstock for the production of one-carbon chlorinated chemicals, synthesis gas—a mixture of CO and $H_2$ used in the synthesis of methanol and oxo chemicals, hydrogen (for the synthesis of ammonia), hydrogen cyanide, carbon black, and acetylene (by way of cracking). Significant differences in the boiling points of the constituents of natural gas provide a facile separation of ethane and of the readily condensable, liquefiable petroleum gases (LPGs)— a mixture of the $C_3$ and $C_4$ constituents [also referred to as natural-gas liquids (NGLs)].

|  | b.p. (°C, atmospheric pressure) |
|---|---|
| Methane | −161.4 |
| Ethane | −88 |
| Propane | −42 |
| n-Butane | −0.5 |

Until recently the exploration of new natural-gas sources was stymied by regulatory policies of the U.S. government. Deregulation caused the production of natural gas to increase in 1979 for the first time since 1972. The number of new gas and oil wells drilled increased from 49,101 in 1979 to 62,704 in 1980 as compared with 57,111 wells drilled in 1956. Also, the proven natural-gas reserves increased by 35% between 1978 and 1979, and they are now reported (*CEN* 9/8/82) at 201.7T (trillion) scf (standard cubic feet, i.e., volume at 1 atm and 60°F), while more optimistic estimates mention 1000T scf obtained from conventional, relatively shallow wells. The natural-gas reserves must be compared with the current annual consumption of about 23T scf for fuel and chemical uses (3.2). Long-range prospects for the supply of natural gas in the United States are even more optimistic: unconventional reserves obtained from coal seams and Devonian shale, and gas basins of abiogenic origin located at great depth (below 15,000 ft) at the Eastern and Western Overthrust Belts along the western spine of the Appalachian Mountains and along the Rocky Mountains. Speculations about the existence of a relatively unlimited abiogenic methane beneath the crust of the earth create great enthusiasm among some geologists (*CW* 12/15/82, 9/16/87), and experiments on horizontal drilling are under way (*CEN* 2/23/87). In 1986 Sweden announced an exploratory drilling project to test the veracity of the abiogenic methane theory. Another potential source of methane is the existence of methane hydrate deposits in the cold regions of the world and in offshore locations near the American continent (*CMR* 5/9/83, *CW* 5/30/83). The abundance of natural gas, and prospects for an even greater bonanza in the future, are creating interests in the direct use of methane as an automotive fuel (*CW* 9/14/83) and in direct methane-to-liquid-fuel conversion processes (*CEN* 4/27/87).

In recent years new sources of natural gas were discovered in various parts of the world, notably New Zealand, the Patagonia region of South America, Malaysia, and

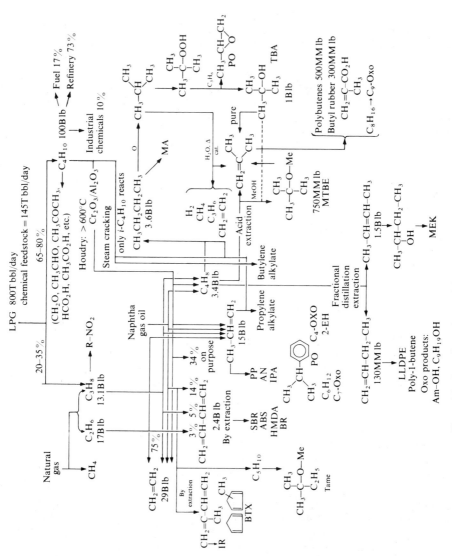

**Figure 2.2** Interdependence of the lower aliphatic hydrocarbons.

34

elsewhere, and the installation of local methanol plants is having a profound effect on the methanol market in the United States (2.4.4.4, 3.3.2). Also of importance are the rising imports of LPGs (*CE* 11/23/87).

Figure 2.2 depicts the interdependence of chemical transformations of the lower hydrocarbons. We note that the largest-volume building blocks of industrial organic chemicals, namely, ethylene and propylene, can be obtained by cracking of the constituents of natural gas or by cracking of the higher-molecular-weight petroleum components (naphtha and gas oil). The choice of feedstocks depends on availability and relative costs, and to a minor extent, on interest in by-products such as butadiene and isoprene. The cracking of higher-molecular-weight feedstocks is more energy-intensive and hence more costly. Dehydrogenation processes for the production of ethylene and propylene from the corresponding alkanes are beginning to be used.

In 1980 the production of LPG was about 1.26MM bbl/day, but only 21% of this huge amount of hydrocarbons was used for chemical transformations. The latter does not include the use of propane and butanes in the manufacture of gasoline (by alkylation of the corresponding olefins; see 5.4.1 and 6.3.1) since this is considered to be a typical petroleum refinery rather than a "chemical" operation. The advent of linear low-density polyethylene (LLDPE) production (4.3.1) and the interest in methyl *t*-butyl ether (MTBE) (6.3.1.3) require the separation of 1-butene and isobutylene, respectively. These and other examples of interdependence of the lower aliphatic hydrocarbons are mentioned in the pages that follow, and the reader will have ample opportunity to refer to Fig. 2.2.

## 2.2.2  Petroleum

Petroleum is a complex mixture of saturated hydrocarbons, sulfur compounds, and other constituents present in minor quantities. Its composition varies from one region to another and depends on its geologic history and the nature of biomass from which it was presumably derived. The primary objective of the variety of refining processes is to convert crude petroleum into acceptable fuels; gasoline for high-compression automobile engines and diesel and jet fuels used mostly for trucks and airplanes, respectively. Also, the production of lubricating oils is of major concern to the petroleum refining industry. Details of petroleum refining processes are left for specialized sources of information. (See, for example, G. D. Hobson, Ed., *Modern Petroleum Technology*, 5th ed., Wiley, New York, 1984.)

Basically, crude petroleum is stripped of its volatiles (LPG) and then subjected to atmospheric and vacuum distillations that separate the following fractions:

|  | ~ b.p. (°C, atmospheric pressure) |
| --- | --- |
| Light naphtha, casinghead or straight-run gasoline | 50–100 |
| Heavy naphtha ($\leqslant C_{10}$) | 150–200 |
| Kerosene, jet fuel ($\leqslant C_{16}$) | 175–275 |
| Gas oils, diesel fuel ($\leqslant C_{25}$) | 200–400 |
| Lubricating oils | 350+ |
| Light fuel oil | |
| Heavy fuel oil | |
| Bunker oil | |
| Industrial fuel oil | |
| Residue: asphalt | |

Gas oil and the higher-boiling fractions are subjected to vacuum distillation, and appropriate fractions are first dewaxed by means of solvents such as toluene and methyl ethyl ketone (MEK) to remove paraffin wax and then are extracted with solvents such as furfuraldehyde to remove aromatic components to produce, finally, reasonably good lubricating oils. During the production of high-quality, that is, high-octane gasoline, the lower-boiling fractions of petroleum are subjected to several fundamental refining operations enumerated as follows.

Cracking serves to reduce molecular weight and hence to increase the volatility of the feed. It can be carried out in the presence of steam or hydrogen (hydrocracking) or at high ($600-800°C$) temperatures (thermal cracking) and in the presence of acidic catalysts (usually aluminum silicates that contain other metallic oxides).

Alkylation processes combine the olefinic fragments produced by cracking with low-molecular-weight alkanes to yield branched, mostly $C_7-C_8$ alkanes that possess high octane numbers. An example of such a product is isooctane, or 2,2,4-trimethylpentane with the assigned octane number of 100.

Isomerization processes serve to rearrange less branched to more highly branched alkanes that have higher octane numbers.

Reforming converts aliphatic and alicyclic (naphthenic) components to aromatic hydrocarbons [mostly benzene, toluene, and the xylenes (BTX)] by a combination of cyclization, dehydrogenation, and aromatization (dehydrocycloaromatization).

Hydrodesulfurization is the removal of sulfur as hydrogen sulfide by means of hydrogen since sulfur-containing ("sour") petroleum fractions tend to deactivate ("poison") catalysts required in the preceding processes and catalytic converters mandated by the government to reduce air pollution.

Only about 50% of each barrel of crude oil is converted to marketable gasoline. Proven reserves of crude petroleum in the United States in 1982 were reported (*CEN* 9/8/82) to be about 29.4B bbl.

The annual pool of commercial gasoline required by the United States amounts to approximately 108B gal, 2.6B bbl, or 365MM t. It is a blend of about 20% straight-run gasoline, 40% catalytically and hydrocracked gasoline, and 20% of isomerized alkylate and reformed gasoline. The consumer price of gasoline includes heavy state, federal, and even local taxes. The demand for gasoline is a function of the public's preference for high-powered, large vehicles ("gas guzzlers") and only high retail prices, or gasoline scarcities, seem to damp this preference. Apparently the fact that the United States currently imports about 60% of its petroleum demand has no effect on the continued use of high-powered, large automobiles. The latter also contribute heavily to U.S. imports and to the consequent trade deficits: in 1986, out of the 387.1B total of imports, the category machinery and transport accounted for 42.9% or $166 billion (U.S. Department of Commerce). Also, apparently, it has not occurred to either Congress or the White House or the fiscal experts who advise both branches of government that a graduated luxury tax (a "frivolous spending disincentive") on high-powered, large automobiles would bring about four desirable effects:

1. Reduce our petroleum consumption and hence imports.

2. Reduce our disastrous trade deficit.

3. Produce revenues from a sector of the population that can easily afford it without interference with transportation needs of the general population.

4. Save lives.

Only as an aftermath of the 1973 OPEC embargo (2.4.2), and during the 1979–1980 period when the EPA-mandated shift from leaded to unleaded gasoline was intensified with a consequent rise in the price of gasoline (because of the higher cost of BTX used to replace TEL as an octane enhancer), there occurred a temporary 10% decrease in the consumption of gasoline. Otherwise, the United States continues to be " . . . Unprepared for Next Energy Crisis" (H. Krieger, *CEN* 6/23/86), and it is heading relentlessly "Back to the Energy Crisis" (H. Crawford, *SC* **235**, 2/6/87, pp. 626–627).

In addition to jet and diesel fuels, heating oils, and lubricating oils (70MM bbl), petroleum refineries produce considerable quantities of by-products: ethylene, propylene, butylenes, and BTX. However, the demand for ethylene and propylene as building blocks of industrial organic chemicals is greater than the supply of refinery by-products. Generally, chemical producers like to be assured of a steady supply of these building blocks, and this explains why some are engaged in their own petroleum processing operations and why others acquire petroleum refining companies (note Du Pont's acquisition of CONOCO).

As indicated in Fig. 2.2, the yield of ethylene from ethane is about 78% while the cracking of propane produces about 42% ethylene in addition to 17% propylene, 3% aromatics, 1% butylenes, and 3% butadiene. As one switches to higher-molecular-weight petroleum feedstocks, the yields of BTX and butylenes tend to increase somewhat two- to threefold), but the yield of ethylene is decreased to 20–34%, while that of propylene is relatively unchanged. Thus, it is clear that the price of ethylene, in particular, depends not only on the cost of crude petroleum (and natural gas) but also on the choice of the petroleum fraction subjected to cracking (2.4.2).

### 2.2.3 Coal

The United States is blessed with abundant reserves of coal, estimated at about 1.7 T t. Nearly 10% of the current coal production of over 800MM t (a level exceeded for the first time in 1980) is exported overseas, and special loading and shipping facilities used for that purpose have been expanded in recent years. The current abundance of natural gas and the recent drop in the price of petroleum-derived fuel oil, as well as the preoccupation with acid-rain-producing pollutants (of special concern to our Canadian neighbors) have damped somewhat the use of coal as a fuel by utilities, but technology to control the emission of sulfur dioxide is being improved. Still, as a result of a high cost of petroleum during the 1970s, electric power plants nearly doubled the use of coal and in 1980 used 677MM t or 81% of the total coal production. The price of coal remained rather constant between 1950 and 1965 (~ $4.75/t fob mine), but then began to rise and reached the $28/t level in 1981–1984. The cost to utilities has always remained higher ($35/t CIF during 1982–1984). The 1984 production of coal of 890MM t was valued at $25 billion, for an average price of $28/t. Coke production in 1984 was only about 31MM t because of increased imports of finished steel and was priced at about $105/t.

Figure 2.3 summarizes some of the conventional and newer uses of coal as a fuel, precursor of coke, and a source of synthesis gas (syngas) and other chemicals.

A discussion of current novel developments of coal as a fuel are beyond the scope of this book and are mentioned here only briefly to the extent that they have a bearing on the role of coal as a source of chemicals. For an extensive coverage of this field, see H. H. Lowry and M. A. Elliott, *The Chemistry of Coal Utilization*, Wiley, New York, 1981, 5600 pp.

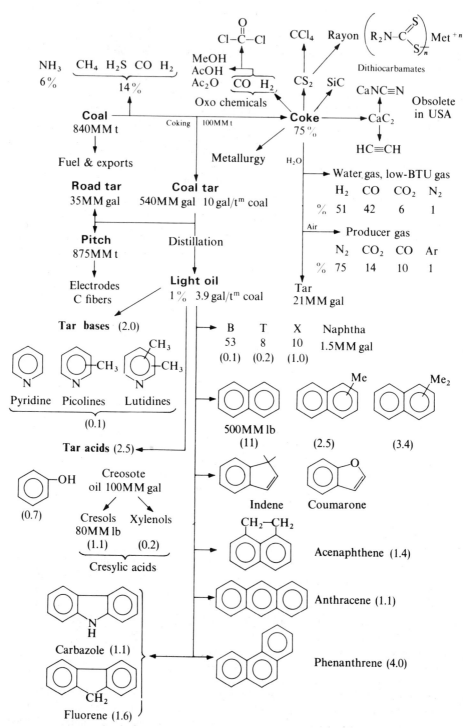

**Figure 2.3** Traditional role of coal as a source of chemicals. (Numbers in parentheses indicate approximate percent concentration in coal tar.)

*Gasification* of coal (also see 3.5.1.3) in the presence of steam produces a mixture of carbon monoxide and hydrogen that can serve not only as a fuel of relatively low energy content but also as a source of crude syngas (3.5.1). For the current status of coal gasification, see a short summary by P. Savage in *CW* 7/15/87. The Texaco process is being employed successfully by Tennessee–Eastman in Kingsport, Tennessee, where 900 t/day of coal is gasified to syngas and converted by way of methanol and methyl acetate to acetic anhydride (3.2.4). On the other hand, the gasification of lignite at Beulah, North Dakota cannot be described as a successful venture. It is designed to recombine carbon monoxide and hydrogen to give a gas of high energy content, namely, synthetic natural gas (SNG). Initiated in the mid-1970s by five industrial partners who invested over $500 million and received $1.5 billion in federal loan guarantees, this Great Plains Coal Gasification Project has produced fuel at about double the current cost of natural gas. In July 1985 the government refused to extend additional loan guarantees, the industrial partners defaulted on $1.56 billion loans, and presently the Department of Energy (DOE) is trying to sell the facility or is faced with operating it by itself (M. Crawford, *Science* 8/16/85, *CMR* 6/23/86).

The coal gasification projects in the United States are reminiscent of the Fischer–Tropsch process used by Germany during World War II and by the similar current projects of South Africa located at Sasol.

Various coal *liquefaction* processes in which coal is subjected to direct hydrogenation or treated with hydrogen-donor solvents (such as tetrahydronaphthalene) to give petroleum-like oil are being developed (see *CEN* 4/28/86). They are an outgrowth of the Bergius process used by Germany during World War II to obtain about 5.5MM t of oil with a conversion of 1 t of coal to 3.3 bbl of oil.

Figure 2.3 displays the impressive spectrum of chemicals that are produced during the traditional coking of coal. Unfortunately, the supply of the organic constituents of coal tar is a function of the production of coke, and the demand for coke depends on the level of steel production, and as stated above, the United States is being inundated with processed iron products. Consequently, the 1984 production of coke represents less than half of the 75MM t produced in the late 1970s. Among the high-volume industrial chemicals obtained from coal tar (see Fig. 2.1) there are minor amounts of BTX, naphthalene and methylnaphthalenes, and rather crude tar "acids" and tar "bases." Most of the relatively pure, low-volume coal-tar chemicals are imported from West Germany, and in the future they are likely to be imported from the People's Republic of China, another coal-rich country.

Coke is a crude form of elementary carbon and, as such, is the precursor of carbon disulfide (3.7) and, indirectly by way of calcium carbide, also of acetylene (4.7). However, over-half of the demand for the acetylene building block is satisfied by isolating it as a by-product of cracking processes designed to produce ethylene.

## 2.3   SOME CRITERIA IN THE CHOICE OF POTENTIAL PRODUCTS AND TRANSFORMATION PATHWAYS

Although some specific factors that enter into the decision-making process with regard to this topic are discussed in the pages that follow, a brief summary of the principal criteria may be helpful at this point. Some of the considerations depend on the category of chemical (1.1.3) considered for potential production.

### 2.3.1  Favorable Demand

A favorable *demand for the product* on the marketplace is essential for a successful business initiative. This must be determined by a careful market study that differs if one projects the production of a new product, or whether one proposes to share the market with already existing products. In case of a new material, one must consider the means to "educate" potential customers concerning the special or superior virtues of the proposed product. If one proposes to enter into competition with existing products, the growth possibilities of the given product must be determined: is the growth rate to follow the not too exciting rise in the GNP and population (as is true for a "mature" market item), or is the demand expected to rise above these parameters; are there export possibilities, or, on the contrary, is the domestic market likely to be undermined by imports?

### 2.3.2  Reliable Supply

A *reliable supply and predictable price range* of materials required for the production of the proposed product must be ascertained. Since "the chemical industry is its best customer," the performance of potential suppliers must be scrutinized. The importance of this consideration is illustrated by advertisements of certain manufacturers who assure potential clients of their unique dedication to the production and timely delivery of a given building block. Also, it may explain, at least in part, why chemical companies engage in either upstream integration by outright acquisition of a business that is a source of the required feedstock, or downstream integration that utilizes existing internal production operations for the proposed venture. Under some circumstances, especially when one is dealing with a special chemical building-block requirement, these may lead to a merger or the creation of a new business organization that joins the established source of the special chemical with a company that promotes the new product venture. In this context, an illustration of potential complications that can arise is the recent legal entanglement that resulted when the producer of aspartame entered into an arrangement with a supplier of L-phenylalanine (who had to mobilize his resources for this particular operation), but then the aspartame producer decided to manufacture phenylalanine independently. It took 2 years to settle the matter (*CW* 8/5/87).

### 2.3.3  Technological Know-how

An adequate reservoir of technological know-how is essential for the financial success of the proposed venture. Delays in production are costly in view of the fact that interests are due on capital investments and the clock on overhead expenditures continues to run regardless of whether the product is being sold. This consideration is sufficient reason for a chemical company to acquire an already functioning business or to contract a reliable development–construction organization to provide the required services that prepare the production facility and essentially deliver to the client a turnkey plant. The alternative, of course, is to invest in one's own R&D efforts or license the technology from others. Two examples of technological fiascos are presented here to illustrate the consequences of underestimating the importance of technological know-how.

A common pitfall that must be guarded against is a hasty transfer of laboratory experience to a full-sized production facility without intermediate scaleup experience.

A plant built to convert ethylene directly to ethylene glycol (EG) on a 800MM-lb/yr scale (by means of a mixture of bromine, acetic acid, and water in the presence of a tellurium catalyst; see 4.3.6) was described in a somewhat exaggerated manner by *CEN* (12/3/79) as the "technological disaster of the century." Corrosion problems condemned the full-sized plant to oblivion.

Another example was the West German project to produce tetrahydroanthraquinone (9.3.5.3) by means of a Diels–Alder reaction of naphthoquinone and butadiene. While these operations functioned satisfactorily on a pilot plant-scale, the scaled-up unit designed to produce 12,000–15,000 $t^m$ clogged up after 2 days of production (*CW* 9/26/84), and the $90-million project (a joint project of two very reputable companies) had to be abandoned. Prior to this decision, the plant suffered an explosion on February 15, 1983 killing one person and injuring five others (*CMR* 3/7/83, 9/24/84).

To recognize accurately one's capabilities and limitations is a great virtue, and in this sense one must admire the approach of Saudi Arabia to the phenomenally successful expansion of its chemical production since the end of 1979 (H. Tandy, *CB* 9/84, A. N. Ghazaii, *CB* 11/86) by means of joint ventures with several successful international producers who became responsible for the construction of production facilities and for the training of local operators. Recently, most plants reported production levels exceeding nameplate capacities. Also admirable is the resolve and execution of Brazil's initiatives to develop its chemical production facilities not only in the case of ethanol (1.2.2) but also covering a broad range of other industrially important materials. Contrasting situations are numerous and too embarassing to mention here.

## 2.3.4  Profitability

Profitability of the proposed business initiative will depend not only on the accuracy of calculated manufacturing costs and capital investments (see 2.4.2) but also on the projected pricing vis-à-vis the prices and manufacturing costs of competitive products. The cost of capital required to install the production facility, the rate of permissible depreciation, the predicted life cycle of the product, and the expected return on investment (ROI) must be taken into account. Processes that require very high temperatures and pressures and the use of costly catalysts and solvents, produce corrosive conditions, and create hazardous wastes are not very attractive. The formation and disposal of harmful wastes and the liability and insurance costs cannot be ignored as they may have been in the past during the juvenile phase of chemical industry and, regardless of whether one likes to admit this, these considerations may lead to the establishment of the production facility outside the jurisdiction of the EPA, OSHA, and other "watchdog" organizations—in other words, in places where concerns for economic growth and employment overshadow exaggerated or imaginary protectionary concerns of an affluent society (2.4.4.6.B).

## 2.3.5  Diversification Potential

The potential for internal utilization (captive use) for downstream integration of the proposed product in order to create one or more additional higher-value-added products may be an enticing reason for a company to embark on a new business venture. A beautiful illustration of this factor is the recent decision by Celanese, now Hoechst Celanese, to enter the highly competitive over-the-counter (OTC) pain-killer market (9.3.2) with the manufacture of acetaminophen (APAP), by means of a novel synthetic pathway based on the intermediacy of p-hydroxyacetophenone. Several additional industrial chemicals with promising industrial uses (9.3.5c) have thus become available because of this development.

### 2.3.6  Merchandising Capabilities

These must be considered in the choice of a new chemical business venture, and these will vary drastically from one category of chemical to another (see 1.1.3). Standard quality commodity chemicals are sold primarily on the basis of pricing and guaranteed delivery, and purchases may be subject to long-range contracts. Fine chemicals are sold on the basis of quality as much as pricing, and pseudocommodities fall somewhere in between. The availability of fine chemicals is widely advertised in trade and professional publications and, in the case of novel (developmental) chemicals, distribution of free samples and technical literature stimulates potential customers to experiment with applications of their choice. The sale of traditional specialty materials is based on extremely high advertising budgets expended on influencing popular thinking, and often, in the case of health and personal-care items, sensational and unwarranted claims are inhibited by the FDA.

## 2.4  PRICING OF INDUSTRIAL ORGANIC CHEMICALS

### 2.4.1  Introduction

The interdependence between the cost of a material and the demand for it on the marketplace, and hence production, employment, and other benefits to the economy is indicated in Fig. 1.1. However, the pricing of an industrial chemical is not a simple matter, and it depends to a great extent on the category of chemical one has in mind. Some parameters that influence pricing are

1. Cost of principal feedstock and other materials such as reactants, solvents, and catalysts but including credit for coproducts and by-products.
2. Cost of energy and utilities (see 2.4.3).
3. Amortization of original capital investment for the site and production facility.
4. Operating costs such as labor, maintenance, quality control, property and income taxes, administration, insurance, and waste disposal.
5. Cost of sales.
6. Potential risks due to perturbances of the demand–supply equilibrium (2.4.4).
7. Potential risks due to extraneous factors.
8. Profits that generally speaking should be commensurate with risks.

What follows is an attempt to dissect these parameters, not necessarily separately or in this order, as they relate to the different categories (1.1.3) of industrial organic chemicals.

As indicated in the Preface, most of the prices of chemicals cited here are culled from the weekly price lists published by *CMR*, and in this connection it is advisable to know "the terminology of the marketplace" reprinted (with permission of *CMR*) as Table 2.2. A consultation of one of the weekly *CMR* price lists reveals that it consists of "spot quotations and/or list prices obtained from the suppliers." Generally, contract prices are slightly higher than list prices. However, the list prices still do not reveal all the facts of the business transactions because of the possibility that the actual prices are reduced because of contractual arrangements between producer and client and occasional promotional discounts referred to on the marketplace as "TVAs" (temporary, voluntary allowances). Table 2.2, or better yet a copy of *CMR*, reveals that prices of chemicals are often qualified

TABLE 2.2  The Terminology of the Chemical Marketplace

# ABBREVIATIONS

## THE TERMINOLOGY OF THE CHEMICAL MARKETPLACE

a/alpha
alld./allowed
amorph./amorphous
AMP/American melting point
anhyd./anhydrous
AOAC/Association of Official Agricultural Chemists
a.p.a./available phosphoric acid
approx./approximately
artif./artificial
ASTM/American Society for Testing & Materials

b/beta
Be/Baume
bbls./barrels
b.g./beta-gamma
bgs./bags
bls./bales
bots./bottles
b.p./boiling point
b.p.l./bone phosphate of lime
b.r./boiling range
bxs./boxes

C./Centigrade
cbys./carboys
c.c./cubic centimeters
CD/completely denatured
c.i.f./cost insurance freight
c.l./carload
cns./cans
coml./commercial
conc./concentrated
cp/chemically pure
cps./centipoises
cryst./crystalline
cs./cases
ctns./cartons
cyls./cylinders

d./dextro
dbl./double
denat./denatured
dest. dist./destructively distilled
dl/dextro-laevo
dist./distilled
distr./distributor
dlvd./delivered
dms./drums
dom./domestic

E/East
e.p./end point
equald./equalized
exp./expressed
extr./extracted

F./Fahrenheit
f.a.s./free alongside
ferment./fermentation
f.f.a./free fatty acid
f.f.c./free from chlorine
f.f.p.a./free from prussic acid
fib./fiber
f.o.b./free on board
f.p./freezing point
frt./freight

g./gamma
gal./gallon
g.p./general purpose
gran./granular
grd./ground

i.b.p/initial boiling point
imp./imported

incl./included
indust./industrial

kgs./kegs

l./laevo
lb./pound
l.c.l./less carload
l.t.l./less truckload
liq./liquid

m-/meta
m.a.p./mixed aniline point
mcg./microgram
mfrs./manufacturers
min./minimum
molt./molten
m.p./melting point

N/nitrogen
n-/normal
nat./natural
neut./neutral
NF/National Formulary
No./number
Nom./nominal

o-/ortho
ord./ordinary
oz./ounce

P/phosphorus
p-/para
Pac./Pacific
pt./proof
phos./phosphate
photo./photographic
pkgs./packages
powd./powdered
precip./precipitated
prod./producer
pt./point
pulv./pulverized
purif./purified

redist./redistilled
refd./refined
refy./refinery
resub./resublimed
ret./returnable

SD/specially denatured
s.d./single distilled
SE/Southeast
sec./secondary

secs./seconds
sp.g./specific gravity
ship't/shipment
soln./solution
std./standard
syn./synthetic

tanks/railroad tank cars
tech./technical
tert./tertiary
t.l./truckload
ton/refers to short ton of 2,000 pounds
TVA/temporary voluntary allowance
t.w./tankwagons

USP/United States Pharmacopeia

vis./viscosity
VM&P/varnish makers & painters

W./West
whse./warehouse
w.w./water-white

NOTE: A unit-ton is 1 percent of 2,000 pounds of the basic constituent or other standard of the material. The percentage figure of the basic constituent multiplied by the unit-ton price shown in Chemical Marketing Reporter gives the price of 2,000 pounds of the material.

[a] Reprinted with permission from weekly "Chemical Prices" section, Chemical Marketing Reporter.

according to grade, physical state, origin, and so on. Also, a price list shows differences in prices for some chemicals according to the geographic location of the supplier and customer. In the case of materials that must be refrigerated in transit (e.g., liquid oxygen), transportation costs beyond a certain maximum distance become prohibitive.

### 2.4.2  Material Costs and Distribution of Production Costs

In the distribution of production costs, the first factor we must consider is the cost of feedstocks and other materials. Precentagewise, the low-priced, large-volume commodity chemicals are most sensitive to fluctuations in the cost of crude petroleum and natural gas—currently the dominant feedstocks for this category of materials. The price of natural gas in the United States has dropped in recent years (2.2.1) but now remains at a rather constant level. The price of crude petroleum has varied significantly up and down. Effects of these changes during 1983–1988 on the unit prices of first-generation building blocks ethylene, propylene, and benzene; the second-generation building blocks vinyl chloride and styrene; and some polymers derived from these materials are shown in Fig. 2.4.

We note that the price of crude oil remained at a rather constant level during 1983 and 1984 and then began to decrease, reaching a low during the third quarter of 1986 that represented a 42% drop, and then began to recover to a second-quarter 1987 level that is still 63% below that of 1983/84. The unit prices of ethylene and propylene generally followed the crude-oil price trend, with only minor deviations due to temporary displacements of the demand–supply equilibrium. On the other hand, the deviations in the price of benzene were more pronounced and registered a steep increase in price during the last quarter of 1986 apparently fueled by a rather strong demand for styrene, a demand that had been building up since the third quarter of 1986. The demand for styrene had a somewhat delayed but relatively strong effect on the demand for ethylene since the latter coincided with a strong demand for the polyethylenes, while, except for a momentary large impact, the unit price of benzene was affected only weakly.

Since 1987, the prices of ethylene, propylene, and benzene have been rising, not so much because of a higher cost of petroleum but because of increasing demand (2.4.4) that results from favorable business conditions and the lower value of the U.S. dollar that, in turn, has a favorable effect on exports. As we shall see below, the profit margins in the production of commodity chemicals are very meager and the producers try to catch up on earnings during periods of near scarcities. The extent to which this is possible depends on the tolerance of the market place mentioned in Section 2.4.6. The rather mature PVC market implies a generally constant demand subject mostly to fluctuations in the overall economy and the construction industry, in particular. This explains why the prices of vinyl chloride monomer (VCM) and PVC showed remarkably small fluctuations compared to those of the other chemicals. With regard to the other commodity polymers, only PS exhibited price changes that paralleled those of styrene indicative of the fact that it is PS (rather than other large-volume polymers derived from styrene, such as SBR) that were responsible for its strong demand. A comparison of the price changes of the polyethylenes and polypropylene suggests that the demand for the latter polymer is more supportive of a higher price structure because of its relatively higher rate of growth, and this, indeed, is the news from the marketplace.

Two additional observations based on the data shown in Fig. 2.4 are noteworthy: (1) the transformation of simple vinyl monomers to simple polymers causes a 1.5–3-fold increase in value; and (2) the increase in unit value that accompanies the conversion of 0.86 lb of benzene and 0.32 lb of ethylene to 1.00 lb of styrene does not make sense unless

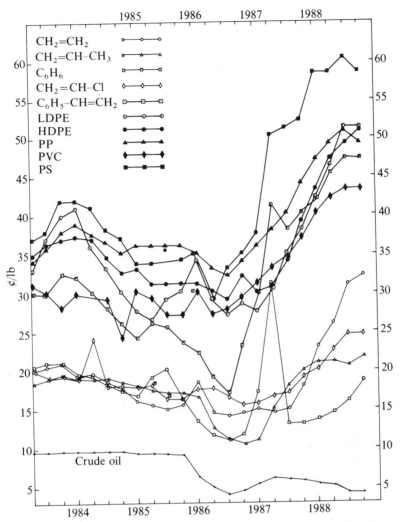

**Figure 2.4**  Fluctuations in unit prices of some polymers and the relevant building blocks as a function of crude oil prices. *Source*: Recent issues of *Modern Plastics*.

we consider that this two-step process generates a valuable by-product, namely, hydrogen gas, which costs about $2/lb.

In the context of the preceding discussion it is desirable to recall the results of a study by A. D. Little (*CEN* 3/29/76) that analyzed the value-added benefits derived from the petrochemical production activities of 17 companies in Texas. It was concluded that the separation of crude petroleum into desired fuel fractions caused a 1.4-fold increase in value. Additional transformations of the appropriate fractions to give the building blocks of industrial organic chemicals gave, on the average, a 7-fold increase in value of the original crude, but the greatest economic benefit, namely, a 105-fold increase in value, was accrued when the crude oil was converted to final consumer products. These results illustrate dramatically the powerful, positive effect on the economy of physical and chemical transformations that convert raw materials into consumer end products.

With regard to the distribution of production costs, a recently published breakdown (R. Leaversuch, *MP* 10/86) of this aspect of ethylene and styrene production is very revealing. For ethylene obtained from ethane, the market price of 15.0¢/lb during the third quarter of 1985 was a composite of a raw material cost of 5.9¢, cost of utilities of 4.5¢, other costs of 4.0¢, and a profit of 0.6¢, while the market price of 14.0¢/lb during the third quarter of 1986 was attributed to a material cost of 4.7¢, a utility cost of 3.2¢, other costs of 3.9¢, and a profit of 2.2¢/lb. We note that the lower prices of ethane feedstock and energy were mainly responsible for the higher profit margin in 1986. In the case of styrene, the market price of 25.5¢/lb during the third quarter of 1985 was composed by raw material costs of 20.4¢, utility costs of 2.0¢, other costs of 2.0¢, and a profit of 1.1¢, while the market price of 17.0¢/lb during the third quarter of 1986 was composed of a much lower materials and utilities costs of 12.3¢ and 1.4¢, respectively, identical other costs of 2.0¢, and resulted in a somewhat larger profit margin of 1.3¢/lb. Here again, the costs of raw materials and utilities (energy) were mostly responsible for a larger profit margin in spite of a lower market price. The depreciation of the production facilities was excluded from these calculations.

The high initial capital investments in production facilities and the relatively small profit margins imply that commodity chemicals be produced in large plants operating in a continuous manner, near full capacity ($\sim$ 85–90% of designed nameplate capacity), and in a highly efficient fashion. This also implies that the economy-of-scale principle is applicable: the cost of the production facility is proportional, on the average, to the (production capacity)$^{0.67}$. The economy of scale principle is built in automatically when the production of a given chemical "rides piggyback" on the shoulders of one of the "four giant F industries": food, feed, fuel, and fertilizer.

As we proceed from large-volume commodities to the intermediate-volume pseudo-commodities and then to the small-volume fine chemicals, we note, generally speaking, the following differences:

1. Unit prices of the products are affected less by the cost of the basic feedstocks.
2. Profit margins become more generous to the producer.
3. The principle of economy of scale is less applicable because the high-quality requirements of the products may cause a preference for batch operations in rather standard, versatile production facilities.
4. Rigorous high-quality requirements and increasing chemical complexity of the products imply greater input of *chemical* know-how vis-à-vis *engineering* know-how that is so essential in the smooth, nearly automated operations of large-volume production facilities of commodity chemicals.

To illustrate the first point, the unit price of neopentyl glycol (3.4.3.3) increased by only 1¢ between November 1982 and July 1987 to the current level of 59¢; the price of diisobutyl ketone (5.5.4) remained constant at 60¢ between March 1983 and January 1989, and the price of USP-grade salicylic acid powder (3.7) rose from $1.53 to only $1.68 between December 1981 and January 1989.

### 2.4.3  Cost of Energy

Generally, the production of chemicals is energy-intensive because it usually involves circulation of heat-transfer media for either heating, cooling, or refrigeration, as well as stirring, grinding, pumping, distillation, drying, and other typical chemical engineering unit operations, not to mention transportation to and from storage facilities within the plant perimeter ("battery limits"). As a consequence of this fact of life, large plants are preferably located near sources of cheap energy although extensive natural-gas pipelines can now deliver this energy source to some distant locations. Of particular importance to the chemical industry is the cost of electricity (*CE* 8/20/84). This is especially true in the case of electrochemical processes and very-high-temperature operations that require electric furnaces and explains why in the United States the chloralkali industry, the production of calcium carbide, calcium cyanamide (now abandoned), carborundum, and some other materials were originally clustered around the Niagara Falls region. The Arab embargo of 1973 had a salutory effect on the use of energy by the chemical industry since this manufacturing sector became the leader in energy-saving measures and reduced its energy requirements by 34% between 1972 and 1985 (*CEN* 4/4/88). The practice of cogeneration—a simultaneous generation of steam and electric power—is spreading rapidly across the United States. The utilization of excess thermal energy (beyond the generation of electricity) from nuclear power plants has been stymied by the extraordinary rise in their construction costs, the popular fear of nuclear installations, and peripheral problems such as the disposal of nuclear wastes. In this respect, United States citizens could learn a lesson from relatively small, densely populated France, where about 80% of all domestic energy needs are derived from nuclear power plants without accompanying hysteria and/or paranoia.

While hydroelectric energy in the United States is rather fully utilized, this is not so in other parts of the world. Two examples are the Guru dam in Venezuela and the Itaipu hydroelectric installation at the border junction of Brazil, Argentina, and Paraguay.

### 2.4.4    Dislocations of the Demand–Supply Equilibrium

The demand–supply parameter is, undoubtedly, the most powerful factor that affects pricing of any material in a free-market economy, and thus it is instructive to examine the nature of dislocations of that critical equilibrium that, for example, produced a rise in the price of styrene from 18¢/lb in early 1986 to 47¢/lb in March 1987, where it has remained for over a year (as of January 1989).

#### 2.4.4.1    Cyclic and/or Seasonal Fluctuations of the Economy

It is an accepted fact of life in the United States that the demand for gasoline rises during the summer travel and vacation season and that there is an accompanying increase in gasoline prices. As long as benzene and toluene are employed as octane enhancers, this may also cause a seasonal increase in their prices.

The unit cost of phenol is sensitive to the strength of the construction industry in the United States. The latter is measured by the annual rate of housing starts, which may range from a low of about 1.5 million to a high of about 2.5 million units and is a function of "consumer confidence" in the economy and, something more tangible, mortgage rates. Large quantities of phenol are used in phenolic resins employed for the manufacture of plywood, particle boards, laminates, and other construction materials. The interesting aspect of changes in the demand of phenol is that it also affects the supply of acetone—the coproduct of the Hock process (5.5.1).

Weakness in the agricultural economy causes havoc in the demand for fertilizers (that include urea), herbicides, insecticides, and other agrichemicals.

### 2.4.4.2  Competition from Imports

The steady expansion of worldwide chemical production facilities, the relative strength of the U.S. dollar relative to the Third World currencies, and the attraction of the world's largest consumer market, stimulate foreign producers to sell their chemicals in the United States. This, of course, softens the prices of domestically produced chemicals with detrimental effects on the national economy. The USITC responds to complaints lodged by U.S. producers against specific cases of "dumping" by investigating each case and, if the foreign producers are found guilty of unfair competition (usually because of state-subsidized production), an import tax is levied on such imported product. An example of state-subsidized or state-controlled prices that resulted in dumping of urea (3.9.1) by Eastern Block countries was the increase of imports from 2% of the domestic demand in 1982 to 11% in 1985 and to about 18% in 1986. In September 1986 the concerned domestic producers filed a complaint with the USITC against unfair trade practices by East Germany, Romania, and the Soviet Union, and the USITC is reviewing the case. A somewhat different situation arises in the case of developing countries that plan to export a part of their chemical production to the United States in order to finance their infant chemical industry. As long as prices are not distorted by unfair subsidies, these imports do not represent a dilemma to the long-range national interests of the United States since the strengthening of the economies of friendly countries (especially those located in this hemisphere) is in line with U.S. foreign policy, and as pointed out earlier, the United States is, to start with, a large importer of chemicals. On the contrary, the United States should promote the internal consumption of these chemicals by means of wise measures that stimulate economic growth of other industrially less developed countries so that they may become, in the long run, economically stronger, internally more stable, better trading partners, and more reliable allies.

### 2.4.4.3  Technological Breakthroughs

Occasionally there appears a technological development that has a profound effect on an existing demand–supply equilibrium of a chemical, usually by creating a more efficient and more economical supply. The following are some recent examples that are discussed more fully elsewhere in this book.

1. Ammoxidation of propylene inaugurated by SOHIO in 1963 replaced a tedious route to acrylonitrile (AN) by a one-step process, boosted the commercial importance of AN to the rank of one billion pounds per year building block, revolutionized the supply of hydrogen cyanide, and stimulated the application of ammoxidation to the production of other chemicals (5.4.3.1).

2. Phenol to aniline conversion (9.6) inaugurated in 1982 by U.S. Chemicals (since 1986 first renamed USX and then becoming Aristech) at Haverhill, Ohio, circumvented the use of nitrobenznene or chlorobenzene intermediates, boosted the importance of phenol, caused a sensational drop in the price of aniline that persisted for several years, and stimulated the application of the same technology to other phenolic compounds (9.3.4). In order to conquer immediately a portion of the aniline market, U.S. Chemicals not only distributed "No one does it the way we do it" buttons, but, more importantly, lowered the prevailing price of aniline (45¢) by about 25% to the consternation of its competitors. Eventually, of course, the unit price of aniline reached a higher equilibrium value of 58¢ (in July 1986) driven by a growing demand for the polyurethane (PUR) building-block MDI.

3. Cumene to phenol/acetone conversion (Hock process; see 5.4.6), known since 1944, has gradually eclipsed tedious, energy-consuming, traditional routes to phenol involving benzenesulfonic acid or chlorobenzene intermediates, raised the importance of cumene to a 1B-lb/yr chemical, provided a new source of acetone, stimulated the application of the same technology to the production of hydroquinone and other phenolic compounds, and stimulated the use of cumene and analogous isopropyl aromatics as precursors to related peroxides and isopropenyl compounds.

4. The Unipol process—a continuous, low-pressure vinyl polymerization technology, developed by Union Carbide in conjunction with a highly efficient Shell catalyst is displacing on a worldwide scale (under licensing by the inventor) the previously employed high-pressure, batch polymerization used in the production of polyethylenes (LLDPE in particular; see 4.3.1), and polypropylene (5.4.1).

5. Ethylene to acetaldehyde conversion (Wacker–Hoechst process; 4.5.1) developed in the late 1950s, eclipsed the use of more costly acetylene for the production of acetaldehyde and probably stimulated the improvements of the analogous ethylene to vinyl acetate conversion that led to successful gas-phase processes in the late 1970s by Hoechst, Bayer, and USI (4.3.5).

6. Catalyst development never ceases by the chemical process industries in the expectation of finding more selective, efficient, and economical catalysts either for existing or for potentially new transformations. The following recent announcements illustrate some activities in this broad area of industrial chemistry.

(i) A new catalyst developed by UOP and used in conjunction with conventional dehydrogenation catalysts reduces the production cost of styrene by 1¢/lb (*CE* 11/86; see 9.2.2, below). It is obvious that such savings are very significant when one deals with multimillion-pound-per-year plants.

(ii) A liquid-phase aluminium chloride catalyst developed by Monsanto and Lummus Crest, and already employed by Shell and Rhône–Poulenc, improves the yield of cumene by promoting transalkylation of di- and triisopropylbenzenes (*CEN* 11/14/85; 9.2.1).

(iii) Asahi reports the development of a highly selective catalyst for the methylation of phenols that allows the process to be operated in fluidized-bed reactors. The new catalyst is derived from iron–vanadium oxides by addition of promoters and has been used since 1984 for the production of *o*-cresol and 2,6-xylenol (*CW* 11/12/86; 9.3.4).

(iv) The importance of zeolite catalysts can be appreciated from its multiple entries in the Index. A consequence of outstanding importance is the development of the methanol to gasoline process by Mobil that has already become an industrial reality in New Zealand. Another interesting development is the zeolite-catalyzed production of pure *p*-methylstyrene that can replace the previously known mixture of isomeric vinyltoluenes. Also of interest is the isomerization and separation of isomeric xylenes to isolate *p*-xylene, the precursor of the terephthalic acid system.

(v) Homogenous catalysis has made a tremendous impact on industrial organic chemistry and several examples are found elsewhere in this book. Additional illustrations are provided by G. W. Parshall and W. A. Nugent in their three articles, "Making Pharmaceuticals via Homogeneous Catalysis" (*CT* 3/88, 184–190), "Functional Chemicals via Homogeneous Catalysis" (*CT* 5/88, 314–320), and "Homogeneous Catalysis for Agrochemicals, Flavors, and Fragrances" (*CT* 6/88, 376–383).

7. Butane to maleic anhydride conversion began to be implemented during the early part of the 1980s, and this released a significant amount of benzene—the traditional raw material for the production of maleic anhydride (MA) (6.2.4).

8. Hydrogen peroxide from the elements, a process announced by Du Pont (*CW* 12/9/87; *CE* 3/14/88), will affect the anthraquinone-based process (9.3.5.4) used for several decades and facilitate the expanded use of hydrogen peroxide in industrial organic chemistry (6.3.1.7) that has caused a 50% increase in one year from the 1986 level to 619MM lb. The Pd-catalyzed process can be carried out in small installations located near potential markets. Fortunately for the recently expanded hydrogen peroxide production facilities, a delay of several years before Du Pont's technology is likely to be scaled up will allow the demand to catch up with current supply.

### 2.4.4.4   The Bandwagon Syndrome

Occasionally several producers react more or less simultaneously to an increased demand for a chemical by engaging in its production, thus creating an escalation of supply above and beyond what could be considered a normal growth requirement. The result is a glut of the market and a sharp drop in price that affects both the new and old producers.

Several years ago such miscalculation of future ammonia requirements caused a depression of the price of ammonia that lasted for a long time until the demand could catch up with excess production capacity. During this time of adjustment, low rates of utilization of production facilities, or outright plant closures, costly storage of unsold product, and payments due on capital investments become a financial burden to the industry.

More recently, numerous natural gas to methanol conversion (3.3.1) production facilities were started worldwide at about the same time, and additional ones are in various stages of construction. This situation caused Du Pont, the pioneer of synthetic methanol production, to dismantle its methanol marketing operation, and it is limiting its reduced methanol production to captive use. In January 1987 Texaco announced the mothballing of its methanol plant in Delaware and a withdrawal from the methanol market, and by April 1987 the United States shut down about 5.4B t of its methanol production capacity and, from an exporter of methanol as recently as 1983, it became an importer of this chemical from Trinidad, Malaysia, Bahrain, Canada, Saudi Arabia, and other countries that had "jumped on the methanol bandwagon." The list price of methanol dropped in July 1987 to 30¢/gal or 4.53¢/lb fob Gulf of Mexico ports but recovered to 53¢/gal by February 1988. Only a few years ago Celanese raised the price of its methanol to 79.5–94.5¢/gal (*CW* 6/9/82) depending on the point of delivery.

A situation similar to that of methanol is apparently developing with regard to the worldwide production of ethylene (*CW* 4/6/88) and as mentioned above (2.4.4.3, paragraph 8) with respect to hydrogen peroxide.

During recent decades the complaints from less developed countries (LDCs) focused on their economic dependence on exports of cheap agricultural and mineral commodities (coffee, sugar, copper, bananas, tin, etc.), the prices of which are controlled by distant, international speculators, while they were forced to import costly manufactured goods from industrialized countries. The case of methanol—a product of billion-dollar production facilities erected with borrowed money—demonstrates that commodity chemicals have joined the ranks of traditional export commodities from the LDCs. Economic benefits that result from such large-scale, risky ventures are limited unless there exists an internal demand for the chemical, or if the large-scale venture is accompanied by downstream integration that produces a variety of other useful materials from the newly accessible feedstock.

### 2.4.4.5  Unintentional Interruptions of Supplies

Accidental fires, explosions, and other mishaps (attributed to *force majeure*) of large production facilities can seriously affect the supply side of the demand–supply equilibrium and cause an immediate rise in the unit price of a given chemical. Thus, for example, a fire in April 1986 at the 1B-lb/yr plant of American Hoechst caused a 3.5¢/lb rise in the price of styrene. An explosion of one of Taiwan's two 66,000-t$^m$/yr acrylonitrile (AN) production units, on top of an already tight demand–supply balance, caused an immediate rise in the price of AN from 35.7¢ to 37.5¢/lb (*CW* 5/6/87).

Similarly, unfavorable weather conditions or pest infestation can raise havoc with the supply of essential oils that feed the flavors and fragrance industry and the supply of alkaloids, gums, and other natural products.

### 2.4.4.6  Extraneous Factors

In between the controllable and uncontrollable factors that affect the demand–supply equilibrium of industrial chemicals are the factors listed in this section.

1. External political factors can be illustrated, to start with, by OPEC decisions. Thus, when 5 of the 13 members of OPEC decided (on June 28, 1987) to reduce the production of crude petroleum during the fourth quarter of 1987 from the previously proposed quota of 18.3MM bbl/day to 16.6MM lb/day, the immediate reaction on the New York Mercantile Exchange was an increase of 60¢/bbl that raised the contract price for crude oil from $20.24 to $20.74. It is generally assumed that a rise of $1/bbl of petroleum causes an increase of 3.8¢/gal or $3.8/7.36 = 0.51$¢/lb in the price of benzene. As an importer of 60% of its current needs of petroleum, the United States is at the mercy of prices manipulated by either OPEC or non-OPEC countries such as Great Britain, Norway, and Mexico. Another interesting example of political effects on the supply of chemicals is the case of naturally produced vanillin. The installation of a Marxist regime on the island of Madagascar resulted in gross neglect of harvesting of the vanilla bean and a sharp increase in the price of this raw material.

2. Internal political factors include the consequences of legislation produced by government agencies and/or politicians who react to either factual, proven causes or unproven but public opinion-supported causes. The United States offers numerous examples of legislated decreases in the utilization of certain materials, and the list of endangered chemicals seems to be growing daily. Three cases, and the accompanying ripple effects on the chemical industry, are mentioned at this point.

(i) The phaseout by EPA of leaded gasoline caused the major producers of tetramethyl (TML) and tetraethyl lead (TEL) (Ethyl and Du Pont) to abandon production with the corresponding shrinkage in demand of the corresponding alkyl chlorides and ethylene dibromide. The remaining TML and TEL production serves the rapidly shrinking domestic requirements and exports. On the other hand, this EPA decision caused an increased role of toluene and benzene as octane enhancers in unleaded gasoline, stimulated the use of highly subsidized fermentation ethanol (used to formulate 10% ethanol-containing gasohol actually mandated, for obvious political reasons, by some grain-producing states), and caused a rapidly growing demand for MTBE. Most recently some political figures began to promote (see "Capitol Hill

Stumps for Alcohols," *CMR* 8/10/87) the use of methanol or blends of methanol, ethanol, and gasoline as an automotive fuel. The fact that the energy content of methanol is only one-half of traditional gasoline, that it is hygroscopic, volatile, and rather toxic to humans, seems to be ignored in this case. The Merck Index describes the biological effect of methanol as follows:

> Poisoning may occur by ingestion, inhalation or percutaneous absorption. *Acute Effects*: Headache, fatigue, nausea, visual impairment or complete blindness (may be permanent), acidosis, convulsions, mydriasis, circulatory collapse, respiratory failure, death.

The most redeeming feature of both methanol and ethanol as constituents of gasoline is that they reduce air pollution caused by exhaust gases (*CEN* 8/10/87).

(ii)  The phaseout of some chlorinated hydrocarbons. The first victim of the infamous "Delaney Ammendment" (1958) was trichloroethylene, until recently the favorite dry-cleaning fluid employed by numerous commercial dry-cleaning establishments. The Delaney Ammendment ignores the indisputable fact that everything in this world is subject to concentration effects and, furthermore, that physiological effects in test animals subjected to absurd, massive doses of a given chemical cannot be equated in a routine fashion with physiological effects in humans. The unsavory situation in the United States caused by the Delaney Ammendment used to regulate chemicals that are blamed for the incidence of human cancer has stimulated an editorial in *Science*, the official organ of the American Association for the Advancement of Science—written by its deputy editor who until recently, and for several decades, was the editor of this prestigious journal, Dr. Philip H. Abelson. This editorial is reprinted (with the permission of *Science*) in its entirety for the sake of readers who may have missed it.

> **Cancer Phobia.**    For more than 10 years, the public has been subjected to a media barrage leading to widespread, misinformed fear of chemicals. Through the use of questionable evidence, many major substances have been labeled carcinogens. If data are adjusted to eliminate effects of cigarette smoking, there has been no overall increase in cancer due to other factors. The highly publicized cancer epidemic that was predicted earlier has not materialized.
>
> With increasing use of synthetic chemicals, it was desirable to screen them for possible carcinogenicity by tests on shorter lived animals such as mice and rats. The general procedure is to subject the mice or rats to massive doses of the chemical to be tested, usually by feeding, gavage, or inhalation. Preliminary experiments determine a maximum tolerated dose. This amount, repeated over a 2-week period, usually leads to a noticeable but tolerated reduction in weight. A chronic dose regimen is then employed in experiments that last for the animal's lifetime. The levels used vastly exceed those to which humans are likely to be exposed.
>
> Many of the experiments that have been cited as proving a potential carcinogenicity of a chemical for humans have been performed on inbred strains of mice that have a natural incidence of liver tumors. In humans, there are taboos against inbreeding, which often leads to genetic impairments. Thus the use of inbred mice, though convenient experimentally, is suspect. More important is the fact of high natural incidence of liver tumors in the test mice. The usual response of these animals to massive doses of a chemical is to develop an even higher incidence of liver tumors. When this happens, the chemical is labeled a potential carcinogen in humans. It so happens that in humans primary liver cancer is rare with the exception of alcoholics and those who have suffered from hepatitis. With these exceptions, incidence of liver cancer in the United States decreased substantially during the times when use of industrial chemicals expanded greatly. Thus extra liver tumors in a naturally tumorigenic mouse is of dubious relevance to humans.

Results from the massive doses are extrapolated linearly to low doses, and the assumption is made that humans are as sensitive as the most cancer prone of the species of animals tested. Use of a linear extrapolation to low doses implies that humans do not have repair mechanisms against injury.

Countless millions of animals have been sacrificed in testing chemicals. Comparatively little effort has been devoted to studying the mechanisms of chemical carcinogenesis. An important exception is work conducted at the Chemical Industry Institute of Toxicology (CIIT). The CIIT, which has achieved an excellent reputation among toxicologists for its careful work, has examined the detailed mechanisms of the interaction of formaldehyde with nasal passages of rats and has shown that the linear model is not correct in predicting nasal cancer in the rats. Rather, a more relevant measure is the extent of interaction of formaldehyde with the nasal DNA. The amount of binding of formaldehyde to DNA decreases much faster than the concentration of formaldehyde in the nasal passages and is a better predictor of carcinogenesis.

Another important study of CIIT has been an investigation of the mechanism of the carcinogenesis of branched-chain hydrocarbons in male rats. These hydrocarbons are key components of unleaded gasoline. The studies showed that the hydrocarbons interfere with the mechanism for excreting a low-molecular-weight protein by the male rat. Research at CIIT indicates that this may be the cause of kidney cancer in male rats exposed to gasoline. A similar mechanism with related cancer does not exist in female rats, in male or female mice, or in humans. This type of result calls into question the practice of the regulatory agencies in selecting for extrapolation of humans data from the most sensitive animal.

Animals differ by more than a factor of 1000 in the levels required for lethal response to dioxin. Rodents and humans are known not to be identical in their biochemistry. To attain a realistic estimate of the hazard—if any—presented by a chemical, specific information about its metabolism and physiological effects is needed. The two examples involving formaldehyde and hydrocarbons illustrate the power of an approach in which detailed mechanisms of chemical carcinogenicity are examined.

Reprinted with permission from Abelson, P. H., "Cancer Phobia" (editorial), *Science* **237**, 473 (1987). Copyright 1987 by the American Association for the Advancement of Science.

By association with trichloroethylene, the EPA is currently planning to reclassify tetrachloroethylene as perchloroethylene or perc from a "possible human carcinogen" (class C material) to a "probable human carcinogen" (class B material), again on the basis of laboratory tests with rats and mice forced to inhale perc. The Halogenated Solvents Industry Alliance (HSIA) is presently arguing against such reclassification on the grounds of "serious and unwarrented adverse impacts" (*CMR* 8/3/87). At the same time, however, a labor union is promoting the prohibition of the use of methylene chloride, methylene dichloride (MDC) by hair-care professionals and claims that the failure of the FDA to ban MDC in cosmetic products as "unconscionable" (*CMR* 7/20/87). Riding on the high tide of public opinion, some manufacturers of decaffeinated, vacuum-dried coffee solids have abandoned the use of methylene chloride as an extractant and make "chemophobia" (*SZM CEN* 11/16/81; see also E. Efron, *The Apocalyptics*, Simon & Schuster, Inc., 1984; and H. D. Crone, *Chemicals and Society: A Guide to the New Chemical Age*, Cambridge University Press, 1986, reviewed by L. G. Nickell in *CEN* 6/8/87) the basis of television advertisements to promote their particular product. One is left wondering how much methylene chloride, b.p. 40°C, could possibly be present in a cup of hot coffee and how much harm could be caused by insignificant traces when MI reports an $LD_{50}$ of 1.6 mL/kg in young rats forced to accept it orally. The opening shot in the methylene chloride war was fired near the end of 1986, when the Assistant Secretary of Labor and head of OSHA issued an advanced notice concerning possible changes in current OSHA standards and invited public comment. This issue is still undecided.

(iii) Ethylene glycol is a well-recognized poison because it is metabolized to oxalic acid, and the latter, among other physiological effects, causes a fatal decrease in the calcium content of blood. EG caused a number of fatalities half a century ago, before the FDA was able to retrieve a sulfa drug medication formulated as an elixir in EG and distributed by a pharmaceutical company. While EG is the dominant ingredient of radiator fluids and this use has not been restricted so far, OSHA has reduced permissible exposure limits of several glycol ethers and their corresponding acetates to 25–200 ppm (see *CMR* 12/15/86 for specific information) by workers exposed during an 8-hour workday on the basis of adverse health effects observed in test animals. As a precautionary measure, some traditional manufacturers of EG and its derivatives are shying away from the use of the ethylene oxide building block (4.3.6) in favor of analogous propylene oxide derivatives (5.4.2).

Other examples of extinct or endangered chemical species are mentioned throughout this book, together with their problematic circumstances. Some consequences of these legislative actions, in addition to those implied above, are worth noting.

Recently Dow abandoned the production of chlorobenzene and its chlorinated derivatives (including *p*-dichlorobenznene, known to consumers as Paradow and used as a moth repellent and deodorant in public restrooms).

Du Pont announced (*CMR* 12/1/86) the dismantling of the production facility at Corpus Christi, Texas that supplied chlorine-based raw materials for the production of some of their Freon line of fluorocarbons. This decision was based on the conclusion that it is more advantageous to buy carbon tetrachloride, chloroform, and perchloroethylene elsewhere than to produce these chemicals in-house. Du Pont also announced that it was considering what to do with its cyclohexane plant at Corpus Christi; perhaps it suspected then that the EPA was to announce a few months later (*CW* 8/5/87) an inquiry on the impact of cyclohexane emissions on industrial employees and the general public. This, apparently, is part of the EPA attempt to reduce evaporative hydrocarbon emissions and, in particular, to reduce the vapor pressure of gasoline (*CW* 4/8/87).

There is always hope that the infamous and embarrassing Delaney Ammendment will eventually be repealed, or at least properly modified, by the U.S. Congress, which is, after all, responsible for having implemented it. There is even hope that the EPA— the agency designated by Congress to implement its laws, good or bad, will modify its *modus operandi* in view of the "harsh assessment" embodied in a report produced by an "agency task force appointed by (EPA) Administrator Lee M. Thomas" (*CW* 3/4/87). This report concludes that "the priorities of the EPA are more closely attuned by public, rather than expert, consideration of environmental risks."

These changes may be accelerated by recent rulings against the EPA by the judicial branch of the U.S. government. Thus, the U.S. District Court for the District of Columbia ruled the EPA to be at fault (*CEN* 8/31/87, 2/29/88; *CMR* 2/29/88) when it negotiated an agreement with the producer of chlordane to phase out this pesticide gradually rather than stop, at once, its distribution. Judge Louis F. Oberdorfer ruled that the pesticide law (engendered by the irrational and rigid mentality of the Delaney Ammendment) does not permit flexibility in the decision-making power of the EPA. Also, a three-member panel of the U.S. Court of Appeals for the District of Columbia rejected (see J. Long, *CEN* 11/9/87) a so-called *de minimis* exception to the Delaney Ammendment invoked by the EPA in the case of some food and cosmetic dyes that in the words of Judge S. F. Williams conclude that "a person would have to be exposed to more than 40 million chemicals with carcinogenicity equivalent to that of Orange No. 17 to reach one hundreth the health risk involved in cigarette smoking."

### 2.4.5  Formation and Disposal of Wastes

In the short discussion that follows, let us distinguish between hazardous wastes and by-products of chemical transformation processes.

The good news about hazardous wastes generated by the CPIs is that their formation is decreasing (212.1MM t in 1985, representing a 21.8% decrease as compared to 1981) in spite of a 10% increase in overall manufacturing operations (*CE* 6/22/87). The bad news about hazardous wastes is their disposal costs. It is estimated (*CE* 12/8/86) that the disposal of toxic materials on land sites costs about \$250/t while incineration costs even more, between \$250 and \$1500 per ton. Hopefully, innovative methods of toxic waste disposal are in the making and are likely be based on microbiological methods. "Superfund" taxes are currently financing the cleanup by EPA of hazardous waste dumps created during the adolescent stage of the chemical industry (1.2).

It is obviously advantageous to utilize by-products of principal chemical trans-formations and credit the income from their sales against the total manufacturing costs. Du Pont is providing some ingenious examples of this practice.

The oxidation of cyclohexane or cyclohexanol to adipic acid (7.2.1) also yields some degradation products such as the $C_5$ and $C_4$ dicarboxylic acids, namely, glutaric and succinic acids, respectively. The mixture of the three dicarboxylic acids is purified by fractional distillation of the corresponding dimethyl esters, and an intermediate fraction that consists of 89% dimethyl adipate and 10% dimethyl glutarate is marketed as Du Pont's Dibasic Esters, DBE solvent blends, at an attractive price of 42¢/lb fob Bayport, Texas. By 1987 Du Pont also became a source of pure glutaric acid, and the corporation created a "new business development group" to "accelerate new product and new application development." Its bulletin *New Horizons*, Vol. 1, No. 4, winter 1988, features DBE, Dytek A (see below), glutaric acid, and linear $C_{12}$ diamine (see 7.4).

Another Du Pont example is the utilization of the $C_6$ dinitrile that contains a 2-methyl substituent. It is an isomer of the linear dinitrile that is the intermediate in the production of hexamethylenediamine (HMDA), the traditional building block of nylon 6/6 and other nylons identified by the first numeral six. While HMDA sells for \$1.07/lb, the branched diamine marketed by Du Pont as Dytek A sells for only 89¢/lb. While this diamine by-product may not be useful in the assembly of tough textile fibers (because branching reduces intermolecular attractive forces desirable in such a product), the branched diamine functions well in the production of amorphous polyamides, epoxy resins, polyurethanes, corrosion inhibitors, wet-strength paper additives, and so on. Also, the branching decreases crystallinity of the diamine, and it is a liquid at room temperature while HMDA melts at about 41°C, and, furthermore, steric inhibition caused by the presence of the 2-methyl substituent differentiates reactivity at the two amino terminals. Thus Dytek A is more readily handled than HMDA during large-scale mass-transfer operations and also offers the possibility of better control of a process based on two-stage reactions [see, e.g., comments concerning reaction-injection molding (RIM) technology] than its more costly isomer in which both amino terminals are identical.

It is noteworthy that, under certain circumstances of industrial applications, mixtures of chemicals can function better than a similar single component. Examples of such circumstances are formulations of plasticizers, ingredients of polyester resins, coating formulations, and foundry core binders.

Ingenious utilization of unplanned or unexpected by-products is a sign of maturing of the chemical industry (1.2). Not long ago, dumping and burning was the primitive way of disposing of by-products. Flaring of gaseous by-products in petroleum refineries was a monument to lack of human imagination, knowledge, and business initiative. Similar practices continue to exist in connection with various chemical operations that involve renewable resources (*SZM-II*). Three examples are as follows:

- Lignin by-product of pulping is utilized mostly as a fuel rather than taking advantage of its phenolic nature.
- Practically all the whey by-product of cheese manufacture is dumped rather than utilizing its lactose content.
- In tropical and subtropical societies most of the agricultural residues (e.g., the stalks and leaves of harvested bananas) are allowed to rot.

Only a combination of dire necessities and/or enforced environmental regulations have stimulated initiatives to subject human and animal wastes to anaerobic fermentation that gives methane fuel for domestic use and fertilizers–soil conditioners from the fermentation residues. Similar treatments of municipal and industrial wastes (such as the notorious by-products of rum-manufacturing plants) are being developed in technologically advanced countries.

### 2.4.6 Adjustment of Market Prices

It is appropriate to complete this discussion of pricing of industrial organic chemicals with some thoughts about the mechanism by which producers adjust sales prices of their products. The prevalent rule is to charge customers whatever the marketplace will bear. As mentioned above, even very small changes in the cost of commodity chemicals have serious economic implications because of large-volume transactions, national and international competition, and meager profit margins (see B. Nadel, "Petrochemicals: Patience Pushed to the Limit," *CB* 3/88, 12–14). For reasons enumerated in Section 2.4.1, a supplier may attempt to raise a given price, but such attempt is not always successful. Consider, as an illustration, the headline in *CMR* of December 1, 1986: "Phenol Discord Hurts Two Price Initiatives." The story begins with the sentence "Twice this quarter, phenol producers have failed in efforts to raise pricing despite rising feedstock costs" and continues for 12 inches of column space to tell details of the "tug of war" between producers and buyers. Obviously, at that point in time, the producers failed to get the desired increase in price because the marketplace did not wish to tolerate it. Eventually, as stocks of a given chemical are depleted, the resistance of the buyers breaks down and the forces of demand and supply produce a new price that reflects the temporary equilibrium situation.

Competitive pricing also determines the cost of pseudocommodities and other downstream categories of industrial organic chemicals (1.2); however, as mentioned previously, prices are not subject to relatively great fluctuations unless one deals with "exotic" biomass products that may be affected by occasional changes in growing conditions. Proprietary specialty materials are priced at a level that recoups the cost of

R & D invested on the long road that leads from a gleam in the eye of a researcher to the final product ready to be launched into the world.

Pricing levels during the introductory phase of the life cycle of a new product (J. B. Frey, *CT* 1/85, 40–41) must be fair to the supplier, but they should not discourage potential clients from exploring uses of the new material. Once the virtues of the new product are well established, prices can be allowed to creep upward during its growth phase (characterized by an increase in demand that is severalfold greater than the change in GNP and population). If the profits being made are excessively attractive, and patent protection is about to expire, then competitors are attracted to join in on the feast. During the declining phase of a given product's life cycle, when a superior or less costly competitive product appears on the marketplace and begins to chip away the demand for the older material, the resulting decline in sales can be held back somewhat by a reduction in price by offering TVAs, improving quality and performance, reducing production costs, and similar defensive measures. When the product has conquered a share of a specific market but its demand increases only at a rate equivalent to GNP and population changes, the material is considered to be "mature." For example, the per capita consumption of sulfuric acid—the "barometer of industrial development," a term coined by President Herbert Hoover, is a mature material in highly industrialized societies while it continues to be a valid econometric indicator in developing countries.

The dynamic nature of the chemical industry, and especially the segment that deals with the constantly expanding world of industrial organic chemicals, requires the attention of marketing experts who comprehend the technological basis for the interdependence among the numerous commercially significant materials. Only then can they guide business decisions on the basis of sound predictions and an understanding of scientific and technological developments that impact the production and applications of chemicals. Furthermore, the thinking of such econochemical professionals must encompass a glocal scenario because the chemical industry has definitely become an international endeavor.

# PART TWO

# ALIPHATIC BUILDING BLOCKS

# 3

# C$_1$ BUILDING BLOCKS

## 3.1  INTRODUCTION: PAST, PRESENT, AND FUTURE IMPORTANCE OF C₁ CHEMISTRY

The length of this chapter is witness to the importance of one-carbon building blocks in the assembly of industrial organic structures.

The term "C₁ chemistry" was coined in recent years because of the successful, large-scale conversions of carbon monoxide and methyl alcohol to industrially important materials. Foremost among these are acetic acid and its anhydride (3.3.2.4), and high-octane gasoline from methanol (see 3.3.2.7.B and J. Haggin, *CEN* 6/22/87). However, there exist numerous other possibilities (J. Haggin, *CEN* 5/19/86; D. A. Fahey, Ed., *Industrial Chemicals via C₁ Processes*, ACS Symposium Series 328, 1987) with methane, carbon monoxide, and methanol occupying the center stage in these developments and creating great appreciation of the catalytic capabilities of homogeneous transition-metal complexes and the zeolite family of aluminium-containing silicates (P. B. Weisz, *CT* 6/87, 368–373).

Symptomatic of the worldwide research effort in this area of $C_1$ chemistry was the announcement by Japan (J. Haggin, *CEN* 6/8/81, 23–24 and 11/16/81, 57–58) of a 7-year R&D program coordinated by the Ministry of International Trade and Industry (MITI) with the goal to develop, among other things, the technology for the production of ethylene glycol (EG), ethanol, acetic acid, and lower olefins. Also symptomatic was the appearance in 1983 of $C_1$ *Molecular Chemistry: An International Journal*, Harwood Academic Publishers, New York. The enthusiasm for $C_1$ chemistry went so far as to predict "the demise of the present petrochemical industry . . . within the next 10 years . . . and its replacement . . . (by) . . . $C_1$ chemistry" (*CEN* 6/28/82).

The eager pursuit of developments based on new $C_1$ chemistry should not obscure the fact that there exists a wealth of traditional chemistry based on such building blocks as formaldehyde, phosgene, formic acid, and carbon dioxide. The last-mentioned chemical led to the production of aspirin a century ago (3.7), and the announcement (*CE* 11/24/86) of a development of poly(alkylene carbonates) is evidence of the rising industrial importance of $CO_2$. The hydroformylation (oxo) and the related hydrocarbonylation reactions (3.5.3.4, 3.5.3.5) are responsible for the production of about 2B lb of industrially important oxo chemicals in the United States alone.

As shown in the detailed outline at the beginning of this chapter, the discussion of the different $C_1$ building blocks follows an order of ascending oxidation states, and, where applicable, the analogous sulfur-containing building blocks are dealt with side by side with their oxygen analogs.

## 3.2   NATURAL GAS AND METHANE

For an introduction to natural gas and methane, see Section 2.2.1.

In the United States the quantities of natural gas are reported in terms of either scf units or British thermal units (BTUs). In Europe, on the other hand, volumes of natural gas may be reported in terms of cubic meters measured at 0°C and 1 atm. For conversion of volumes or thermal equivalents to mass units one can employ the following approximate relationships (*WR-I*):

$$10^6 \text{ scf of gas} = 2635 \text{ lb} \times \text{molecular weight (MW) of gas}$$

$$10^3 \text{ scf of gas} = 26.8 \text{ m}^3$$

$$1 \text{ scf} = \sim 10^3 \text{ BTU (average thermal equivalent of natural gas)}$$

The current consumption of natural gas in the United States is about 23T scf. This is equivalent to

$$? \text{ lb CH}_4 = 23 \times 10^{12} \text{ scf CH}_4 \times \frac{2635 \text{ lb} \times 16}{10^6 \text{ scf CH}_4} = \sim 975\text{B lb}$$

Some natural gas is imported from Canada and Mexico. Unlike the United States, Japan and other markets in the Far East such as those of Taiwan and South Korea are importing increasing amounts of liquefied natural gas (LNG) from Indonesia, Alaska,

Malaysia, Brunei, and Abu Dhabi (*CW* 6/22/87). The same is true in the case of France, Belgium, and Spain, with Algeria being an important supplier.

Starting with 1954, the Federal Energy Regulating Commission (FERC) maintained an artificially low price of natural gas destined for the interstate market while allowing the supply–demand equilibrium to determine prices of gas consumed within the gas-producing state. This stymied exploration for new sources of natural gas until the deregulation in 1979, when increased supply stabilized prices at the average level of $3.03/M scf. Actually, prices range between $2.77 and $4.05 per thousand standard cubic feet depending on the region within the United States with lowest prices in the Southeast Central and Corn Belt regions, and the highest in the Southeast, Southwest, and Northeast regions. Canadian natural gas reaches our border at about $3.00/MM BTU. On either basis, the unit price of methane, assuming that the natural gas is 100% methane, turns out to be 7–9¢/lb. As the American Gas Association (AGA) points out (*CMR* 5/27/85), the cost of natural gas to domestic customers ($5.50/MM BTU) is very favorable as compared to the cost of electricity ($23.75/MM BTU).

Currently, the consumption of 23T scf of natural gas in the United States is utilized as follows:

| | |
|---|---|
| Fuel | 72% |
| Energy-intensive inorganic chemicals (ammonia, etc.) | 15% |
| Carbon black | 1% |
| Industrial organic chemicals | 12% |
| | 100% |

The use of natural gas for the production of industrial organic chemicals peaked in 1980 when about 3.1T scf were employed for that purpose. At one time the production of ammonia was consuming as much as 80% of all nonfuel uses of methane, but this is decreasing as a result of imports (2.4.4.2). Nearly 80% of the 820B scf of hydrogen that is produced from methane is consumed by petroleum refineries. The maximum domestic production of methanol was 1.3B gal in 1981, but it has been shrinking since then again because of imports from hydrocarbon-rich countries (2.4.4.4). Carbon black (see below) is obtained now mostly from residual petroleum fractions. An overview of the production of industrial organic chemicals from methane is presented in Fig. 3.1, while Fig. 2.2 demonstrates the interdependence of all components of LPG.

The picture of industrial uses of natural gas may change dramatically in the future if and when a technologically and economically feasible method materializes that converts methane directly to liquid fuels (J. Haggin, *CEN* 6/8/87). This would circumvent the energy- and capital-intensive production of intermediate syngas (3.2.2).

However, the activation of C–H bonds of methane is rather difficult, but progress in this area of chemistry is encouraging (K. Brooks, *CW* 10/21/87; J. Haggin, *CEN* 1/18/88). Another procedure that is being investigated (*CE* 4/13/87) is the high-temperature ($\approx 1150°$C) reaction of methane with chlorine that produces ethylene and vinyl chloride monomer (VCM). Obviously, the R&D incentives will escalate as new sources of natural gas mentioned in 2.2.1 become a reality.

Methane, long known as marsh gas because of its formation in the course of anaerobic fermentation of biomass, is now produced locally in an increasing number of organic waste-disposal plants (J. Chowdhury, *CE* 9/17/84) together with the carbon dioxide coproduct. Novel, selective membrane separation systems allow the isolation of methane in sufficiently high concentration to be suitable as a high-energy fuel.

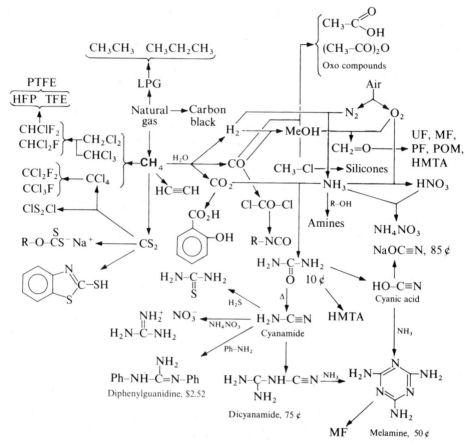

**Figure 3.1**   Some derivatives of natural gas, air, and water.

### 3.2.1   Thermal Decomposition of Natural Gas

The pyrolysis or cracking of methane gives unstable fragments such as $CH_3\cdot$, $CH_2\colon$, $\cdot CH_2\cdot$ (methyl, carbene, and methylene radicals, respectively), the methine radical, $\cdot \overset{\cdot}{C}H\cdot$, and even carbon atoms stripped of all hydrogens. These fragments recombine on cooling to give a broad spectrum of products.

The most complex among them is undoubtedly carbon black, a colloidal-size aggregate of highly aromatic carbon structures probably described best as amorphous graphite or soot. Methane, and other alkanes, for that matter, are subjected to partial combustion that generates the high temperatures (1300–1400°C) required for the fragmentation process. A discussion of the process details that lead to different grades of carbon black are beyond the scope of this book, but the reader may wish to know that there are nine grades mentioned in the *CMR* price list ranging in price between 25¢ and 32¢ per pound. Carbon black acts as a reinforcing agent of rubber and tires may contain as much as 40% of C. Since about 60% of all carbon black is employed in compounding rubber for the manufacture of tires, many automobiles purchased in the United States are imported together with their tires, and the average size of automobiles has decreased and the so-called spare tire has been replaced by a miniature facsimile, there is currently less demand for carbon black ( ~2.5B lb). The use of carbon black in the manufacture of inks, plastic products, and the like consumes about 40% of demand.

At the other end of complexity spectrum among the recombination of methane fragments we find acetylene (4.7). The Wulff process consists of two cycles: methane and other alkanes are first pyrolyzed at about 1300°C, and then the gas is passed through a refractory brick reactor at temperatures below 400°C. Acetylene is, thermodynamically speaking, an unstable compound and tends to trimerize to benzene at about 600°C and to decompose otherwise at about 780°C. Thus, the yields of acetylene are low ( ~20%). In the Sachsse process, the cracking of methane and other alkanes at about 1500°C is attained by partial combustion and by quickly cooling the hot gases by means of cold water. The yields of pure acetylene tend to be lower than those of the Wulff process. In the arc process, methane along with other alkanes is passed through a low-voltage arc and the resulting gas is cooled rapidly to give about 20% acetylene. The electric power requirement is in excess of 10 kWh/kg of acetylene, but this process promises to be a future economical source of acetylene wherever hydroelectric power and a supply of natural gas happen to coexist. Other sources of acetylene and its uses as a chemical feedstock are discussed elsewhere (4.7).

The thermal decomposition of a mixture of methane and sulfur at about 600°C gives carbon disulfide (3.7.3). This transformation is catalyzed by silica, alumina, and magnesia impregnated with other metallic oxides.

### 3.2.2   Steam Reforming of Natural Gas: The Production of Syngas

Thermal decomposition of natural gas in the presence of steam and appropriate catalysts, known as *steam reforming*, is the major source of synthesis gas, now commonly referred to as syngas.

$$CH_4 + H_2O \xrightarrow[\text{Ni, 800°C, 13 atm}]{} CO + 3H_2$$

This endothermic process is balanced thermally by the exothermic, partial combustion of methane:

$$CH_4 + 1.5O_2 \rightarrow CO + 2H_2O + energy$$

Additional details about steam reforming of methane are presented in Section 3.5.1.1.

### 3.2.3   Chlorination Products of Methane and Their Principal Uses

The thermal chlorination of methane is difficult to control because this free-radical process is exothermic and, at the same time, temperatures of 250°C or higher are required to dissociate chlorine molecules into atoms. Hence, many reactors have been designed to optimize the formation of the desired products.

Photochemical dissociation of chlorine improves the control of the process. Thus, a mixture of excess methane and chlorine at 350–370°C activated by means of a mercury arc lamp produces about 60% methyl chloride and about 30% methylene dichloride (MDC). A higher yield of MDC can be obtained by recycling the methyl chloride through the reactor. The desired distribution of chlorination products is complicated by the fact that chlorination is facilitated as chlorine substituents are introduced because of the increasing resonance stabilization of the intermediate radicals:

$$CH_3 \cdot \; < ClCH_2 \cdot \; < Cl_2CH \cdot \; < Cl_3C \cdot$$

Thus, in order to produce reasonably good yields of methyl chloride one must use a high (e.g., 10:1) ratio of methane to chlorine and operate at low conversions. Also, in order to produce chloroform it previously was and may still be practical to allow the chlorination to proceed all the way to $CCl_4$ and then reduce the latter by means of iron filings and steam.

With the advent of cheap methanol, methyl chloride is prepared preferentially from the alcohol and hydrogen chloride—an undesirable by-product of chlorination processes. For this reason methyl chloride is considered to be a derivative of methanol, and its uses are discussed elsewhere (3.3.2.3).

As we shall see on several occasions (4.3.3.1, 9.2.2, 9.2.3), industry avoids the accumulation of hydrogen chloride and recycles it by means of *oxychlorination*. Aqueous hydrochloric acid is sold at 2.7–5¢/lb depending on the region of the United States, while anhydrous hydrogen chloride is sold at 13.5¢/lb when delivered in the buyer's tube trailers. The oxychlorination of methane is carried out by passing methane, hydrogen chloride, and oxygen through molten $CuCl_2$–KCl, where the oxidation of HCl to chlorine is brought about by cupric ions and the resulting cuprous ions are reoxidized by oxygen.

The future of chlorinated methanes and, for that matter, the companion $C_2$ compounds, in the United States is very bleak, indeed. A steady, almost linear decline in production has been observed between 1974 and 1986. Thus, the production of carbon tetrachloride dropped from nearly 1.2B lb to about 600MM lb, that of chloroform from 1.5B to 1.0B lb, that of MDC from 2.1B to about 1.6B lb, and that of methyl chloride from 2.6B to about 2.0B lb. This decline is due to imports and existing as well as prospective EPA restrictions (2.4.4.6.B).

The major, traditional consumption of chlorinated hydrocarbons is summarized (numbers are percentages) as follows (E. Goldbaum and J. F. Dunphy, *CW* 9/23/87):

Methylene dichloride, 550MM lb, used for

| | |
|---|---|
| Aerosols | 29 |
| Paint stripping | 27 |
| Polyurethane foams | 10 |
| Metal cleaning (degreasing) | 9 |
| Electronic industry | 8 |

1,1,1-Trichloroethane, 565MM lb, used for

| | |
|---|---|
| Metal cleaning | 44 |
| Dry cleaning | 20 |
| Aerosols | 11 |
| Adhesives | 9 |
| Electronic industry | 6 |
| Fluorocarbons (see below) | 4 |

Trichloroethylene (TCE) 180MM lb, used for

| | |
|---|---|
| Metal cleaning | 85 |
| PVC production (4.3.3.1) | 7 |

Tetrachloroethylene, 610MM lb, used for

| | |
|---|---|
| Dry cleaning | 56 |
| Chlorinated fluorocarbons | 29 |
| Metal cleaning | 11 |

The Consumer Product Safety Commission (CPSC) is currently concerned about the hazardous nature of these materials on the basis of some incidences of cancer in laboratory animals exposed to chlorinated solvents. Apparently warning labels and intensified educational programs directed at consumers and workers exposed to MDC are acceptable measures to CPSC while the EPA is determining risks. The Halogenated Solvents Industry Alliance (HSIA) and the National Paint and Coatings Association oppose the warning labels on the grounds of insufficient evidence of cancer hazards to humans (and hamsters). At this point in time, 1,1,1-trichloethane is exempt from legislative scrutiny and its use is growing as a replacement of other endangered chlorinated solvents.

Also, as the belief that chlorofluorocarbons (CFCs) are responsible for the depletion of the atmospheric ozone layer is strengthening, we can expect an additional drop in the demand for chlorinated methanes since the latter are the precursors to many CFCs. It is expected (CEN 9/21/87, CW 9/23/87) that by 1991 the use of CFCs will be reduced by 50% as the result of the global accord promoted by the United Nations and signed in Montreal, Canada to freeze their production at current levels and then to begin to reduce the consumption. The Montreal accord was signed by all major producing countries except the Soviet Union and will go into effect in 1989 provided that two-thirds of signatories responsible for CFC production ratify it. Under these circumstances, Du Pont and Allied-Signal have already announced a voluntary phaseout of their CFC production, while the search for substitute materials is being intensified (CEN 1/11/88, 2/8/88; CE 1/18/88; CMR 9/26/88).

The perturbance of the a priori complex atmospheric chemistry (L. R. Ember et al., CEN 11/24/86, 14–64) caused by CFCs with the resulting formation of the "Antarctic ozone hole" is subject of intensive studies (P. S. Zurer, CEN 8/17/87, 7–13), and at this time the results suggest a "complex picture" (P. S. Zurer, CEN 11/2/87, 21–26) that is slowly being unraveled (P. S. Zurer, CEN 11/30/87).

The use of CFCs as propellants in aerosols and as blowing agents in the manufacture of foams has been limited in the United States to certain industrial uses for which substitute gases have not yet been found. The worldwide production of CFCs amounts to about 1.5B lb. Currently, about equal amounts (35%) of the CFCs are employed in compressors that operate refrigeration–air-conditioning units and in industrial foam-blowing operations. About 18% are utilized as industrial solvents, especially by the electronics industry. The negative pressure on the demand side of the demand–supply equilibrium concerning chlorinated methanes is balanced somewhat by the expanding market of fluorine-containing polymers. Thus, while chloroform is the precursor to the endangered refrigerants (produced at a worldwide level of about 1.5B lb), namely, dichlorofluoromethane, (FC 21) and chlorodifluoromethane (FC 22), it is also the starting point for the production of tetrafluoroethylene (TFE), the monomer of the first common fluorocarbon polymer, namely, Du Pont's Teflon, poly(tetrafluoroethylene) (PTFE), and about 90% of the demand of chloroform is consumed for that purpose. The relationship between the above-mentioned chlorinated hydrocarbons and the fluorocarbons under discussion is summarized in Scheme 3.1.

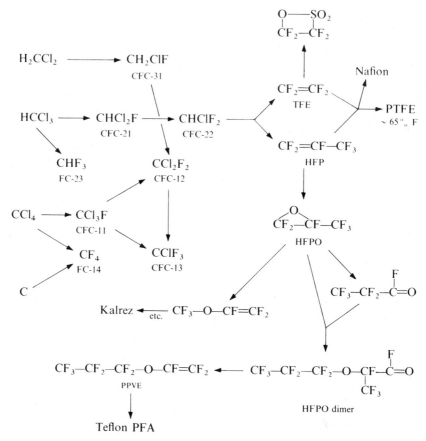

**Scheme 3.1**

The conversion of FC 22 to TFE deserves a special comment. The exposure of FC 22 to 700–900°C temperatures causes the cleavage of the weakest bonds and this leaves behind a transient difluoromethylene (or difluorocarbene), most of which dimerizes to give TFE. However, an important by-product of this reaction is derived from the insertion of the carbene into the molecule of TFE with the resulting formation of hexafluoropropylene (HFP), which, as we shall see below, is an important monomer for the formation of more complex polymers than PTFE.

The current demand for PTFE in the United States is about 10MM lb, and in the worldwide endeavor to supply PTFE, Du Pont is now joined by competitors such as ICI (Fluon), Sumitomo, Daikin, and Ausimont. The 3M company produces a relatively low-molecular-weight PTFE (Dynamar) employed as a processing aid (internal lubricant) of polyolefins.

The inclusion of certain amounts of HFP comonomer during the polymerization of TFE lowers the processing temperature of the resulting modified PTFE without fundamentally affecting the chemical nature of the polymer that still retains only C–F bonds. Thus, one obtains a series of truly thermoplastic PTFEs. A minor use of TFE is the production of about 1MM lb of the copolymer with ethylene: poly(ethylene–tetrafluoroethylene) (ETFE).

Du Pont's Teflon has become a household term because of the nonsticking cooking utensils that were introduced first in France and then in the United States in 1960. More recently still, the adjective "Teflon" has even entered the popular vocabulary to describe the personality of certain politicians who are not "wetted" by events that would normally be considered prejudicial to political careers.

Characteristic of changes in pricing of new materials, Teflon was offered for $19/lb when first introduced in 1957. However, 3 years later the price dropped to $11.60, and during 1961 there were two additional reductions in price, first to $9.60 and then to $6.60 when Du Pont began to operate its 15MM-lb/yr production facility. Presently, the $6.60 price applied to the truly thermoplastic form of Teflon of the relatively low *glass-transition temperature* $T_g$ (see 4.3.3.2) of 126°C, so that it can be fabricated by injection molding, extrusion, and so on. A granulated Teflon resin was priced at $3.25–$5.05/lb in 1961 for use as a coating material processed by means of a powder metallurgy technique; particles of the resin are applied under some pressure to the given surface and then subjected to heat at about 400°C. Once the particles are sintered into a continuous layer, the resulting surface is insoluble, thermally stable (below the temperature at which pyrolytic breakdown takes place). Teflon coatings are not wetted by either aqueous (hydrophilic) or oily (lipophilic) materials. We note that the price of PTFE has not changed markedly since 1961.

The chemistry of aliphatic fluorine compounds stands apart from that of compounds containing the other halogens. It is characterized by

- Some unique synthetic methods.
- Physical properties that are distinctly different from those of the analogous chlorine compounds or those of the parent hydrocarbons.
- A peculiar nomenclature that has evolved in industrial circles (as we have already seen).

Because the field of aliphatic fluorocarbons is of great industrial importance, it calls for a special interlude.

### A Broader View of Nonaromatic Fluorocarbons

At the outset, it is noteworthy that the uniqueness of the fluorocarbons can be attributed to three dominant features:

1. The exceptional strength of the C–F bonds (about 104 kcal/mol versus about 87, 69, and 63 kcal/mol for C–H, C–Cl, and C–C bonds, respectively.
2. The exceptionally high electronegativity of fluorine (the top value of about 4.0 on the Pauling scale) versus about 3.5 and 3.0 for the runner-up oxygen followed by chlorine and nitrogen, respectively.
3. The exceptionally low polarizability of fluorine in spite of the existence of the three outermost nonbonding electron pairs and the reinforcement of this property when multiple fluorine substituents are present in clusters.

The speculation about the origin of the preceding characteristic features of fluorine compounds is beyond the scope of this book, but the consequences are not.

Let us start with what is referred to above as the peculiar *nomenclature* of industrially significant fluorocarbons. The prefix FC or CFC is employed to signify fluorocarbon and chlorofluorocarbon, respectively, and "perfluoro" refers to molecules in which all the hydrogen substituents are replaced by fluorine. Following the prefix FC, CFC, or Freon, we find a numerical code. The latter has the following meaning:

1. The first digit (always examining the set of numbers from right to left) represents the number of fluorine substituents.
2. The second digit indicates the number of hydrogens that remain in the molecule plus one (in order to avoid a zero).
3. The third digit indicates the number of carbon atoms minus one, but the zero is omitted in the case of one-carbon molecules.
4. The presence of bromine atoms in replacement of chlorines is indicated by a "B" and the number of bromines, and this information is denoted to the right of the preceding numerical symbols.
5. A prefix "c" denotes a cyclic structure.

For additional information of this kind, consult A. J. Gordon and R. A. Ford, *The Chemist's Companion*, Wiley-Interscience, New York, 1972 and J. H. Fletcher et al., *Nomenclature of Organic Compounds* (Advances in Chemistry Series), American Chemical Society, Washington, DC, 1974.

The use of the above-mentioned nomenclature, together with other information of interest, is offered in Table 3.1.

Next, let us examine some of the more common *synthetic methods* that are unique to nonaromatic fluorocarbons.

**1. Halogen Exchange by Means of SbF₃ and Related Reagents.** The replacement of *multiple* chlorine substituents by means of $SbF_3$ is usually carried out in the presence of HF that serves to regenerate the fluorinating agent from $SbCl_3$. Thus, for example, carbon tetrachloride is converted to a mixture of FC 11 and 12 (as a function of stoichiometry), but chloroform is converted only to FC 22 because $SbF_3$ is unable to replace a *single* chlorine by fluorine. In order to achieve the latter transformation, one also employs $SbCl_5$ that is converted by HF to the more powerful fluorinating agent $SbCl_2F_3$, and now, for example, chloroform gives FC 23.

**2. Fluorination by Means of Chlorination in the Presence of HF (the 1969 Montedison Process).** The replacement of hydrogen by fluorine substituents occurs because thermal chlorination at about 370–470°C generates chlorine atoms that react with HF to liberate fluorine atoms, and the latter combine with carbon radicals generated in the course of the free-radical chlorination chain reaction. Thus, for example, the exposure of methane to the preceding conditions gives a mixture of FC 11 and 12.

**3. Direct and Indirect Fluorination.** Direct fluorination of hydrocarbons by means of elementary fluorine is more exothermic than combustion or chlorination because of the implied changes in bond energies:

$$-\overset{|}{\underset{|}{C}}-H + F-F \rightarrow -\overset{|}{\underset{|}{C}}-F + H-F + 104 \text{ kcal/mol}$$

**TABLE 3.1    Nomenclature of Nonaromatic Fluorocarbons**

| Structure | FC Number | Comments b.p. (°C) | $/lb |
|---|---|---|---|
| $Cl_3CF$ | 11 | 24 | 0.57 |
| $Cl_2CF_2$ | 12 | -12 | 0.68 |
| $ClCF_3$ | 13 | | |
| $HCCl_2F$ | 21 | 9 | |
| $HCClF_2$ | 22 | -41 | 1.05 |
| $H_2CClF$ | 31 | | |
| $HCF_3$ | 23 | -82 | |
| $CF_4^d$ | 14 | -128 | |
| $CF_3Br$ | 13B1 | | |
| $CF_2Br_2$ | 12B2 | | |
| $Cl_3C-CClF_2$ | 112[a] | 91 | |
| $Cl_2FC-CClF_2$ | 113[a] | 46 | 0.89 |
| $ClF_2C-CClF_2$ | 114[a] | 4 (m.p. -94°C) | 1.02 |
| $F_3C-CHClBr$ | 123B1[a] (halothane[b]) | 50 | |
| $F_2BrC-CF_2Br$ | 114B2[a] (fire ext.) | | |
| $F_3C-CF_3$ | 116 | -72 | |
| $FCH_2-CF_3$ | 134a (Showa Denko)[c] | | |
| $H_3C-CClF_2$ | 142[a] (Pennwalt's Isotron 142B) | | 1.50 |
| $H_3C-CHF_2$ | 152[c] | | |
| $F_2CH-CF_2-CH_2Br$ | (Halopropane[b]) | 74 | |
| $(CF_2)_4$ | c318 | -6 | |
| $(CF_2)_6$ | | 52 | |

[a] This name does not distinguish between isomeric structures.
[b] Anesthetic.
[c] Aerosol without ozone complications.
[d] A specialty gas used by the electronics industry for plasma etching during the fabrication of very-large-scale-integrated (VLSI) circuits; radio frequencies excite $CF_4$ into a reactive plasma.

Attempts to control such explosive transformations have utilized high dilution (vapor) techniques, very low temperatures, and an aerosol technique (*CEN* 9/1/80).

Among the indirect fluorination methods, the more common ones employ the use of perfluorides, or the Simon electrochemical method. The former method converts a metallic fluoride to a higher oxidation state perfluoride and allows the perfluoride to descend to the "normal" oxidation state by way of fluorination. For example $CoF_2$ is first converted to cobaltic fluoride

$$CoF_2 + F_2 \rightarrow CoF_3$$

and the latter is then heated (at, say, 150–300°C) with the compound subjected to fluorination:

$$CoF_3 + R-H \rightarrow CoF_2 + R-F + HF + 46 \text{ kcal/mol}$$

It is noteworthy that less than half of the bond energies are liberated as compared to the above-mentioned direct fluorination.

The Simon electrochemical method is particularly useful for the preparation of perfluorinated carboxylic and sulfonic acids. The substrate is placed in a steel electrolytic cell equipped with nickel anodes. The cell is filled with anhydrous HF containing some LiF electrolyte in case of nonconducting substrates. The application of 4–5 V at 0°C generates fluorine in a gradual fashion. The carboxylic and sulfonic acids are introduced as the corresponding acid fluorides, but the perfluorinated products are subsequently hydrolyzed to the corresponding acids. Examples of products obtained in this manner are trifluoroacetic acid ($F_3C-CO_2H$), trifluoromethanesulfonic acid (commonly known as triflic acid, $F_3C-SO_3H$; (3M Co., $57.95 in 200-lb quantities) and their higher-molecular-weight homologs, perfluorinated tertiary amines $[(R_F)_3N]$, where the $R_F$ group represents a perfluorinated alkyl moiety, and so on.

Among the products available from the 3M Company are the "Fluorad" intermediates such as

Perfluoro-*n*-octanesulfonyl fluoride (FX-8, $21.00)

Perfluoro-*n*-octanesulfonamide (FX-12, $25.00)

Perfluoro-*n*-octyl alcohol (FC-10, $29.15)

Perfluoro-*n*-octyl acrylate (FX-13, $30.75)

**4. The Replacement of Oxygen by Fluorine in Carbonyl and Carboxylic Functions by means of Sulfur Tetrafluoride.** Sulfur tetrafluoride is generated from sulfur dichloride and sodium fluoride in the mildy dipolar aprotic solvent acetonitrile:

$$SCl_2 + NaF \rightarrow SF_4 \ (+OSF_2, S_2Cl_2, NaCl)$$

Gaseous $SF_4$ (b.p. −38°C) selectively replaces the oxygen moieties by fluorine substituents:

$$R-CH{=}O \rightarrow R-CHF_2$$

$$R-CO-R' \rightarrow R-CF_2-R'$$

$$R-CO-OH \rightarrow R-CF_3$$

Carbon–carbon double bonds are not affected:

$$CH_2{=}CH-CO\cdot OH \rightarrow CH_2{=}CH-CF_3$$

**5. Formation of the Perfluorinated Epoxide of HFP and Some of Its Reactions.** The ease with which HFP is oxidized by means of molecular oxygen to the corresponding epoxide, HFPO:

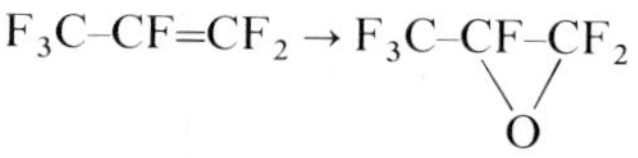

$$F_3C-CF{=}CF_2 \rightarrow F_3C-\overset{\displaystyle \diagdown \diagup}{\underset{\displaystyle O}{CF}}-CF_2$$

and the highly unusual, CsF catalyzed rearrangement, is the source of perfluoro(methyl vinyl ether)—an important monomer for the synthesis of Du Pont's melt-processible, high-performance specialty polymer developed in the early 1970s and called Kalrez:

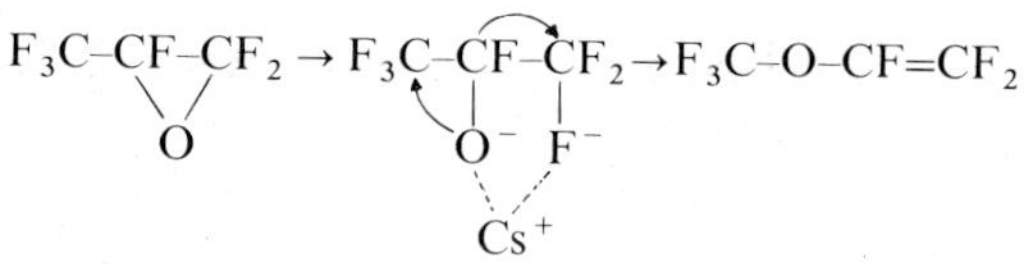

$$F_3C-\overset{\diagdown \diagup}{\underset{O}{CF}}-CF_2 \rightarrow F_3C-\underset{\underset{Cs^+}{O^-}}{CF}-\underset{F^-}{CF_2} \rightarrow F_3C-O-CF{=}CF_2$$

The assembly of Kalrez also involves TFE and another monomer of the perflurovinyl ether family that contains a reactive site suitable for curing the resulting elastomer. Thus, the structure of Kalrez can be represented as follows:

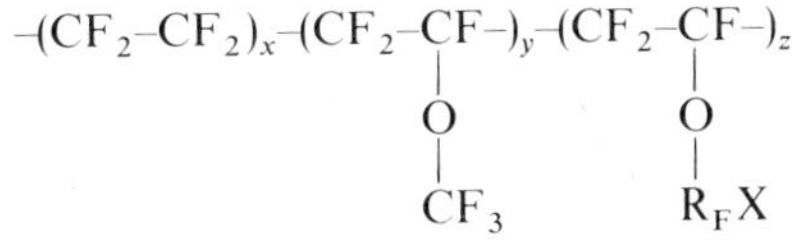

$$-(CF_2-CF_2)_x-(CF_2-CF-)_y-(CF_2-CF-)_z$$

$$\begin{array}{ccc} & O & O \\ & | & | \\ & CF_3 & R_FX \end{array}$$

where the proportions $x$, $y$, and $z$ are chosen in order to obtain the desired thermo-mechanical properties and the X represents reactive groups such as $-CO_2Me$, $-OC_6F_5$, and $-C\equiv N$. The excellent thermal stability of Kalrez and its stability under highly corrosive conditions are accompanied by rather difficult and special processing requirements. Consequently, to conquer a market for this extraordinary and costly ($>$ \$1000/lb) material, Du Pont became a custom fabricator of objects desired by its customers (gaskets, O rings, valves, pump housings and components, etc.). A production unit of about 28MM lb was established after research of a potential market for applications in which the initial high cost is offset by large maintenance and replacement savings, not to mention downtime of the given operation.

Another rearrangement of HFPO involves the migration of a fluoride and converts the epoxide to perfluoropropionyl fluoride (PFPF):

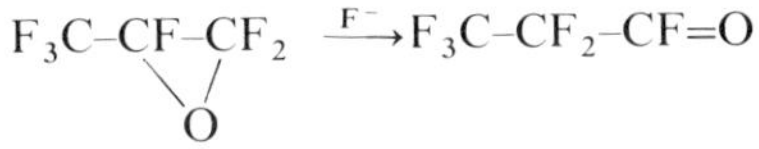

$$F_3C-CF-CF_2 \xrightarrow{F^-} F_3C-CF_2-CF=O$$
$$\begin{array}{c} \backslash / \\ O \end{array}$$

This highly reactive aldehyde (because of the cumulative effect of electron-withdrawing fluorine substituents) can now be subjected to a highly unusual addition reaction with HFPO:

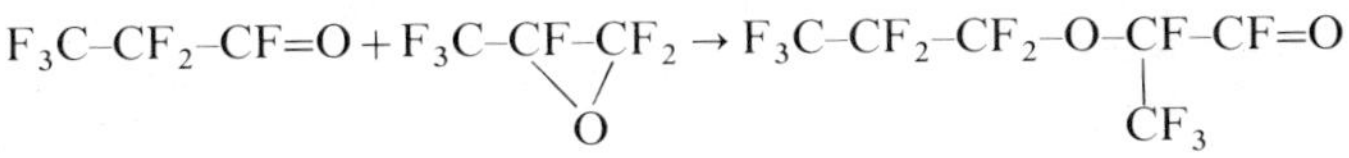

$$F_3C-CF_2-CF=O + F_3C-CF-CF_2 \rightarrow F_3C-CF_2-CF_2-O-CF-CF=O$$
$$\begin{array}{cc} \backslash / & | \\ O & CF_3 \end{array}$$

and the resulting acyl fluoride subjected to decarbonylation and a fluoride elimination reaction (a net loss of carbonyl fluoride) by application of heat in the presence of sodium carbonate gives perfluorinated propyl vinyl ether (PPVE), another useful monomer:

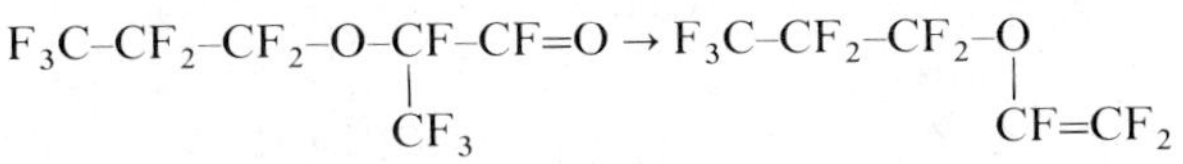

$$F_3C-CF_2-CF_2-O-CF-CF=O \rightarrow F_3C-CF_2-CF_2-O$$
$$\begin{array}{cc} | & | \\ CF_3 & CF=CF_2 \end{array}$$

The same PPVE is formed when a "dimer" of HFPO is decomposed as shown:

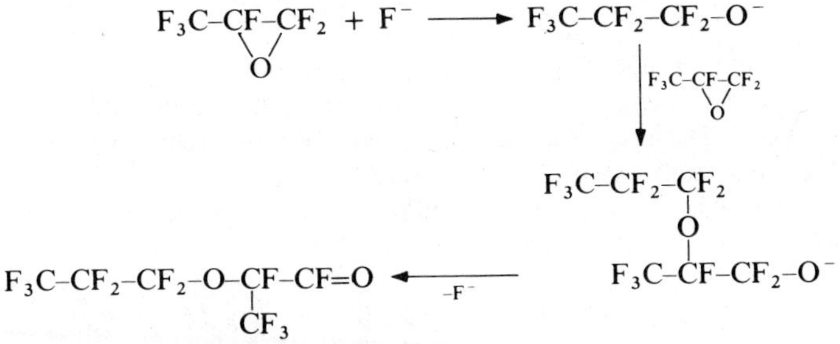

$$F_3C-CF-CF_2 + F^- \longrightarrow F_3C-CF_2-CF_2-O^-$$
$$\begin{array}{c} \backslash / \\ O \end{array}$$

$$\downarrow F_3C-CF-CF_2 \atop \backslash / \atop O$$

$$F_3C-CF_2-CF_2$$
$$|$$
$$O$$
$$|$$
$$F_3C-CF_2-CF_2-O-CF-CF=O \xleftarrow{-F^-} F_3C-CF-CF_2-O^-$$
$$CF_3$$

Alkaline hydrolysis of the perfluoroalkoxy acid fluoride and thermal decomposition of the resulting sodium salt produce a decarboxylation and loss of a fluoride ion to give PPVE:

$$F_3C-CF_2-CF_2-O-CF(CF_3)-CO_2^-\,Na^+ \rightarrow F_3C-CF_2-CF_2-O-CF=CF_2$$

PPVE is the comonomer that, accompanied by TFE, became the building block of Du Pont's Teflon PFA, where the acronym PFA stands for perfluoroalkoxy. The use of perfluoroalkoxy ether side chains turned out to be more advantageous than the use of plain $CF_3$ side chains (introduced by means of HFP) to reduce the crystallinity of PTFE and to create conveniently thermoplastic, modified Teflons. Another important product derived from HFPO is its reaction with the $F-SO_2-CF_2-CF=O$. The latter material is obtained by a fluoride-catalyzed thermal rearrangement of the sultone of TFE:

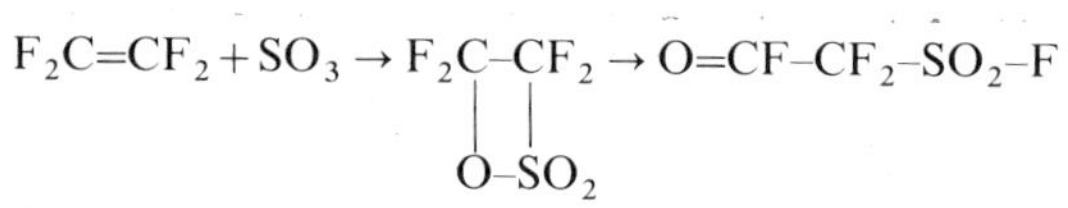

$$F_2C=CF_2 + SO_3 \rightarrow \underset{\overset{|}{O}-\overset{|}{SO_2}}{F_2C-CF_2} \rightarrow O=CF-CF_2-SO_2-F$$

$$\underset{O}{F_3C-\overset{\diagdown}{C}F-\overset{\diagup}{C}F_2} + O=CF-CF_2-SO_2F \xrightarrow{F^-} \underset{F_3C-CF-CF=O}{O-CF_2-CF_2-SO_2F}$$

The fluoride-catalyzed reaction of the acyl fluoride in the presence of additional HFPO gives the polymer

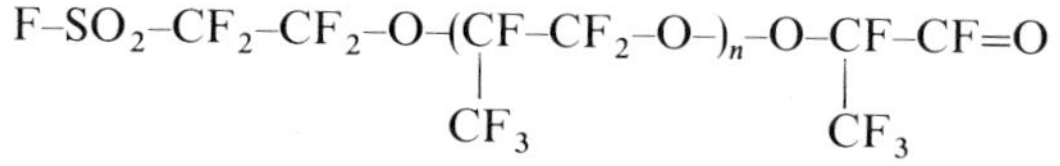

$$F-SO_2-CF_2-CF_2-O-(\underset{CF_3}{CF-CF_2-O-})_n-O-\underset{CF_3}{CF-CF=O}$$

In a manner similar to that shown above, the acid fluoride is converted to a trifluorovinyl terminal by way of alkaline hydrolysis and thermal decomposition of the sodium carboxylate. Now, the trifluorovinyl terminal is capable of polymerization with TFE to give a block polymer (Du Pont's XR resin), and the hydrolysis of the sulfonyl fluoride terminal gives rise to Du Pont's Nafion:

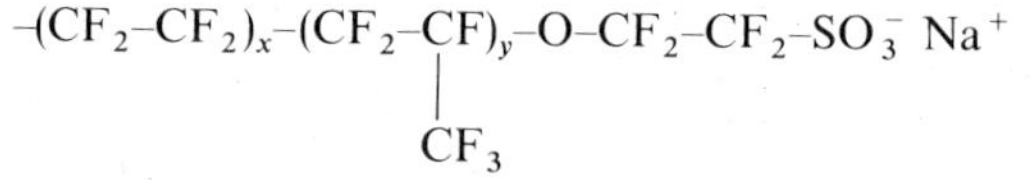

$$-(CF_2-CF_2)_x-(\underset{CF_3}{CF_2-CF})_y-O-CF_2-CF_2-SO_3^-\,Na^+$$

(where $x > y$ and $x = 6-13$). Nafion can be processed into membranes that replace the asbestos diaphragms used traditionally to separate the cathode and anode compartments of chloralkali cells. The ionomeric material is inert to the harsh chemical environment and allows water and electrolyte to migrate during electrolysis.

Asahi Glass of Japan developed a material similar to Nafion except that the ionic terminal consists of a carboxylate salt:

$$-(CF_2-CF_2)_x-(\underset{CF_3}{CF_2-CF})_y-O-CF_2-CF_2-CF_2-CO_2^-\,Na^+$$

The terminal moiety of this ionomer is obtained in the following fashion:

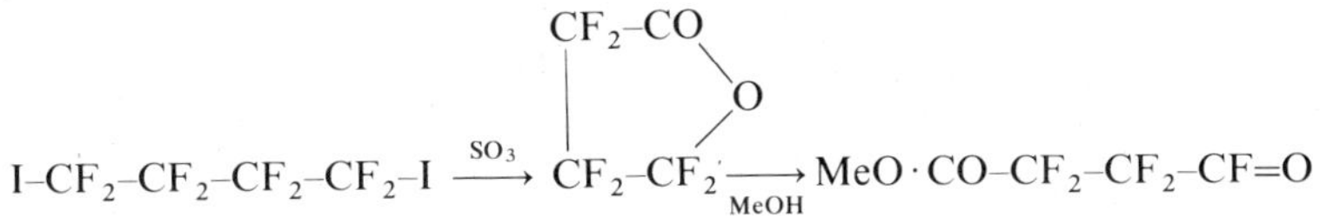

$$I-CF_2-CF_2-CF_2-CF_2-I \xrightarrow{SO_3} CF_2-CF_2 \quad\xrightarrow[MeOH]{} MeO \cdot CO-CF_2-CF_2-CF=O$$

with the epoxide group $CF_2-CO\!\!\!-\!\!\!O$ at the terminus

(see following section), and now the acyl fluoride is allowed to react with HFPO and TFE as shown above.

The important role of fluoride ion in the catalysis of epoxide and acyl fluoride reactions is noteworthy. The behavior of fluoride in perfluoro systems is reminiscent of the catalytic role of hydroxide ions in carbon–hydrogen systems.

For more information about perfluorinated ionomer membranes, see ACS Symposium Series 180, 1982. The inert and strongly acidic sulfonic acid of Nafion can also be used as a catalyst (for details, see F. J. Waller and R. Warren Van Scoyoc, *CT* 7/87, 438–441).

**6. Telomerization of TFE.** A *telomer* is an oligomer of a vinyl system in which the prominent terminal groups perform a desired function. While telomerization is not limited to perfluorinated vinyl compounds, the means of assembling fluorinated telomers is somewhat unique.

The reaction of TFE with iodine alone or in the presence of a iodine perfluoride gives rise to a diiodide such as the $C_4$ system shown above or perfluoroethyl iodide shown below. The iodo substituent attached to a $-CF_2-$ group is exceptionally resistant to nucleophilic substitution reactions because of the bond-reinforcing effect of neighboring fluorines. However, the iodo group is vulnerable to radical-induced reactions, which facilitates the attachment of vinyl monomers. Thus, with additional TFE one obtains a perfluoroalkyl iodide, and in the presence of ethylene, the iodoterminal accepts a hydrocarbon chain:

$$F_2C=CF_2 + I_2 + IF_5 \rightarrow F_3C-CF_2-I$$

and

$$F_3C-CF_2-I + n\,F_2C=CF_2 \rightarrow F_3C-CF_2-(CF_2-CF_2)_n-I$$

or

$$R_F-I + n\,CH_2=CH_2 \rightarrow R_F-(CH_2-CH_2)_n-I$$

Atochem's Foralkyls, $35
(in moderate quantities)

where $R_F$ represents, as before, a perfluoroalkyl group. The nucleophilic susceptibility of the iodine substituent is restored when it is adjacent to the $-CH_2-$ system, and now the iodo group can be replaced by $-SH$, $-O-H$, amino functions, and so on. This chemistry has been utilized to give a series of Atochem's Forafac products. The significance of these materials is discussed below in connection with the unusal physical properties of fluorocarbons.

Another example of telomerization of TFE is a peroxide-catalyzed addition of methanol:

$$2F_2C=CF_2 + CH_3-OH \rightarrow H-CF_2-CF_2-CF_2-CF_2-CH_2-OH$$

2,2,3,3,4,4,5,5-octafluoro-1-pentanol

Some of the "normal" reactions of nonaromatic fluorocompounds, such as simple addition reactions of unsaturated systems and elimination reactions to produce unsaturated systems, are illustrated below in connection with the presentation of additional fluorinated materials of industrial importance.

Finally, we turn to the *unique physical properties* of the nonaromatic fluorine compounds. Probably the most surprising among the physical properties of the fluorocarbons relates to their volatility. All conventional thinking about the relationship between boiling points and molecular weights within similar members of a family of organic compounds (such as the rule of thumb that the boiling points in homologous series increase $\sim 25$–$30°C$ with the addition of each –CH$_2$– group) fall by the wayside when we deal with fluorocarbons. Even though each fluorine contributes 19 units of mass compared to a hydrogen, the substitution of hydrogens by fluorines in a given structure causes the boiling points to decrease. The presence of multiple fluorine substituents exacerbates this phenomenon. Thus, for example, the boiling. points of *n*-heptane and perfluoro-*n*-heptane are 98 and 82°C, respectively, and the boiling points of adiponitrile and perfluoroadiponitrile are 295 and 63°C, respectively. In the latter case, the larger than expected difference can be accounted for by some hydrogen bonding that is eliminated in the perfluorinated analog. Additional examples of usually low boiling points can be encountered among the compounds listed above to illustrate the nomenclature of FCs.

Hand-in-hand with the extremely low intermolecular attractive forces of fluorocarbons is their nearly Ideal Gas Law behavior. Thus, for example, the van der Waals *a* constant of FC 21 is of the same magnitude as that of helium and about 65 times smaller than that of methane.

The other unique physical property of fluorocarbons that contain significant perfluorinated chain moieties, concerns their effect on surface activity. The fluorocarbon chains are not expected to be hydrophilic, but they are neither lipophilic in the conventional sense because of their low affinity for paraffinic materials. As noted above in connection with boiling points, fluorocarbon moieties even exhibit reduced affinity for each other. Consequently, surface-active agents assembled from fluorocarbon structures offer some special properties by functioning in exceptionally low concentrations and by wetting metallic particles, certain polymers, and acting as surfactants in nonaqueous, two-phase systems. The versatility of fluorochemical surfactants is broadened further by combining the perfluoroalkyl chains with conventional hydrophilic, or especially chosen organophilic moieties, to give the usual categories of non ionic and ionic classes of surface-active agents (see Nonionic Surface-Active Agents and HLB in 4.3.6.1). In addition to the versatility of fluorochemical surfactants, we deal here with substances that are chemically and thermally rather stable.

The 3M Company supplies a spectrum of Fluorad surfactants for many specific applications, and more recently, Atochem has also entered this area of applied chemistry with its Forafac products.

### 3.2.4  Prospective Developments in Methane Utilization

The conversion of methane to syngas (3.2.2) is a highly energy intensive, and, hence, costly process. The conversion of syngas to methanol also requires high temperatures (350–400°C) and pressures (250–350 atm), and this implies high capital investments. A more economical route from syngas to methanol is claimed to be a liquid-phase process under development by Air Products (*CE* 10/28/85) and by a research group at Brookhaven National Laboratory (*CEN* 8/4/86).

Naturally, there is also great interest in processes that convert methane *directly* to methanol and other "functionalized methane" products. It appears that iridium, lutetium, and other metallic catalysts activate the otherwise rather inert hydrogen atoms of methane (*CEN* 11/17/83; *CE* 11/28/85; *CW* 4/16/86; *CEN* 6/16/86; G. W. Parshall, *CT* 11/84, 628–638; *CEN* 6/1/87). For a review of different R&D activities in the area of oxidative transformations of methane to methanol and $C_2$ coupling products, see J. Haggin, *CEN* 9/14/87, 19–21.

The chlorine–catalyzed, oxidative pyrolysis (CCOP) of methane is studied by S. Senkan at Illinois Institute of Technology and is reported (*CE* 9/14/87) to produce $C_2$ and higher hydrocarbons.

In the Soviet Union Professor Kh. E. Khcheyan and collaborators continue the study of oxidative methylation represented as follows:

$$CH_4 + R-CH_3 + O_2 \rightarrow R-CH_2-CH_3 + R-CH=CH_2 + R-H$$

[where R represents $C_6H_5-$, $CH_2=CH-$, $CH_2=C(CH_3)-$, or $N\equiv C-$]. For example, the conversion of toluene and methane to styrene occurs at 700–750°C and gives as major by-products benzene, phenol, and cresols. Similarly, the oxidative coupling of methane and acetonitrile gave acrylonitrile and HCN, while the oxidative methylation of propylene gave mostly 1-butene, and isobutylene was converted mostly to isoamylene and isoprene.

In Japan, the oxidative coupling of methane to $C_2$ compounds led to the production of aromatic compounds, and this technology is expected to become an industrial reality in the near future.

In Israel, research in cooperation with Sandia National Laboratories (in Albuquerque, NM) (*CE* 1/18/88; R&D, February 1988), demonstrates a successful capture of solar energy for the conversion of methane and carbon dioxide to a mixture of CO and $H_2$ and the possibility of transporting the resulting mixture to a site where the energy is regenerated:

$$CH_4 + CO_2 + E \rightleftharpoons H_2 + 2CO$$

Time will tell what industrially practical processes will emerge from all these efforts.

## 3.3  THE ROLE OF METHANOL AS A KEY $C_1$ BUILDING BLOCK

### 3.3.1  Production of Methanol and an Overview of Its Uses

The spot price of methanol listed in *CMR* 12/8/86 as 28¢/gal fob producing point Gulf Coast is equivalent to 4.23¢/lb and translates to a worldwide glut (2.4.4.4). A defensive measure by the worldwide methanol producers was the creation of the International Methanol Producers and Consumers Association (IMPCA) in Vienna (*CMR* 10/12/87, *CW* 10/14/87). By the end of 1977 contract prices of methanol were 53¢/gal, but by early 1988 spot prices rose to 75–78¢/gal as the result of several plant turnarounds and closures for repairs and a strong winter buying trend (see "Methanol Makers Look for Price Peak Soon," *CMR* 3/7/88). The price roller coaster is expected to head downward again during the summer of 1988 because of some plant openings and, indeed, by January 1989 the price decreased to 60¢/gal.

World demand for methanol in 1987 was estimated at 16.5MM $t^m$ while the effective capacity, that is, 90% of nameplate capacity, was about 18.5MM $t^m$. In 1987 the United

States produced methanol slightly above the rated capacity of 1.06MM gal but consumed 1.47MM gal by making up the difference through imports of nearly 400MM gal. The fickle nature of commodity pricing is demonstrated by the fact that only 4 years ago (July 1982) the bulk price of methanol at Bayport, Texas was 79.5¢/gal or 12.05¢/lb. It is predicted (*CW* 8/13/86) that the demand–supply equilibrium will balance until past 1991. On the supply side, such a prediction may be upset if natural gas projects in Argentina, Chile, and other locations moved ahead of schedule. It is of interest to note that in 1982 the Soviet Union announced a plan to increase its 2MM t methanol capacity to 30MM t by 1990 using Western (ICI) technology, and Vladimir V. Listove, Soviet Chemistry Industry Minister, was quoted (*CW* 6/9/82) as saying that "these projects help us get capacity more quickly than we might otherwise and they are profitable for our Western partners." By 1985 Soviet methanol production reached 3.6MM t.

In the United States, Senator Rockefeller of West Virginia announced in October 1986 his intention to reintroduce legislation to promote coal to methanol conversion projects (*CMR* 7/27/87). On the demand side, excessive methanol production capacity could be absorbed if and when methanol (octane rating 110) becomes an additive or substitute for hydrocarbon-based gasoline. California has a fleet of 550 Ford Escorts fueled by a mixture of 85% methanol and 15% gasoline that has been operating since 1983 in order to ascertain problems associated with the use of such a fuel (*CE* 6/22/87). A Canadian experiment to test a fleet of Ford LTD Crown Victoria sedans fueled by the same mixture of methanol and gasoline was announced in September 1986. The demand for automotive fuel in the United States is so huge that a 2% methanol content would require a doubling of the current methanol production and a 3% methanol content would require an additional 3B gal. Because of the closing of much of the methanol capacity in the United States (2.4.4.4) the domestic capacity in 1986 of about 1.26B gal roughly matches the current demand for methanol as a chemical feedstock.

The use of methanol as an automotive fuel has several disadvantages mentioned elsewhere (2.4.4.6.B; see also *CEN* 8/17/87) and would require, among other things, an EPA ruling to accept a relatively high volatility of the resulting fuel. The White House announced (*CMR* 7/20/87) the decision to purchase 5000 vehicles powered by methanol, ethanol, and compressed natural gas and requested the EPA to adjust its volatility rulings accordingly. A potential solution of the volatility of methanol-based fuels is to convert syngas to a mixture of higher $C_2$–$C_5$ alcohols according to a process announced by Dow (*CW* 11/7/84, *CEN* 11/12/84, *CMR* 11/19/84).

The potential role of methanol as a fuel has a direct bearing on its expanding potential as a chemical building block. Overlapping this dual role of methanol (J. Haggin, *CEN* 8/25/86, 24–29) are, of course, the previously mentioned Mobil MTG process (2.4.4.6, 3.3.2.7.B) and the escalating demand for MTBE (3.3.2.2 and 6.3.1.C).

Originally, the Du Pont production of methanol from syngas required high temperature ( ~ 300°C) and high pressure ( ~ 350 atm) conditions and the process utilized alumina and oxides of zinc, chromium, manganese, and other catalysts:

$$CO + 2H_2 \rightarrow CH_3OH + (CH_3)_2O$$

The ICI copper-based catalyst represented a technological advance since it allowed the use of milder reaction conditions (50–100 atm and 250–270°C).

Dimethyl ether (b.p. −23.6°C) is a by-product of synthetic methanol and is used as a propellant in some aerosol dispensers and as a refrigerant. Akzo is building a 55MM lb DME plant in Rotterdam to be operational by 1990 (*CEN* 3/21/88, *CW* 3/23/88).

A traditional source of methanol is a process operated until recently by both Celanese and Union Carbide in which the propane and butane components of LPG were subjected to oxidative cracking that leads to a host of products:

$$C_3H_8, C_4H_{10} \xrightarrow[\text{400-600°C}]{O_2, \text{cat.}} CH_3OH, CH_2O, CH_3CHO, CH_3CO_2H,$$

$$CH_3CH_2CH_2OH, CH_3COCH_3, CH_3CH_2CHO,$$

$$MEK, H_2O, CH_2{=}CH_2, CO_2$$

The major products of this process are formaldehyde (33%), acetaldehyde (31%), methanol (20%), and acetone (4%). Only a large-scale, continuous operation could handle the separation of such a complex mixture. The analogous liquid phase oxidation of butane in the presence of cobalt catalyst gives mostly acetic and formic acids (78 and 6%, respectively) in addition to 4–6% ethyl and methyl alcohols, and the use of a manganese catalyst shifts the yields of the carboxylic acids in favor of formic acid ($\leqslant 23\%$).

Methanol is known as "wood alcohol" because, prior to 1926, it was obtained by dry distillation of wood that produces the so-called pyroligneous liquor that contains, among other things, about 6% acetic acid.

Another potential approach to the production of methanol is the *in situ* conversion of biomass gasification products (D. H. Mitchell et al., *Chem. Eng. Progr.* **1980**, 76, 53; F. Haggin, *CEN* 7/12/82).

The two latter sources of methanol are obviously of little interest at this time except in case of a developing country with abundant biomass resources.

The percentage breakdown of the current (*CMR* 9/86) major uses of methanol in the United States is as follows:

| | |
|---|---|
| Formaldehyde | 27 |
| MTBE and analogous Me ethers | 25 |
| Chloromethanes | 13 |
| Acetic acid | 11 |
| Solvent | 8 |
| Methyl methacrylate | 4 |
| Methylamines | 3 |
| Dimethyl terephthalate, DMT, and miscellaneous other uses | 2 |
| Fuel | 1 |

### 3.3.2  Methanol Transformation Products

The presentation of the conversion of methanol to its transformation products follows, more or less, the order of the preceding tabulation.

#### 3.3.2.1  Formaldehyde

A typical process for the oxidation of methanol to formaldehyde uses air and a silver gauze catalyst:

$$CH_3OH + O_2 \xrightarrow[\text{450-600°C}]{\text{Ag cat.}} H_2C{=}O + H_2O$$

The role of formaldehyde as a $C_1$ building block is discussed elsewhere (3.4).

### 3.3.2.2   MTBE and Other Methyl Ethers

The increasing use of MTBE [motor octane number (MON)$=95\text{--}115$] as an octane-enhancing gasoline constituent (Table 1.1) involves the etherification of isobutylene and is extended to higher olefins as well:

$$(CH_3)_2C{=}CH_2 + CH_3OH \xrightarrow{\text{acid cat.}} (CH_3)_3C{-}O{-}CH_3$$

(32¢/lb)                      MTBE ($\sim$85¢/gal)

$$(CH_3)_2C{=}CHCH_3 + CH_3OH \xrightarrow{\text{acid cat.}} C_2H_5(CH_3)_2C{-}O{-}CH_3$$

Isoamylene                      $t$-amyl methyl ether (TAME)

Similarly, naphtha cracking fractions are etherified with methanol to give three-octane-number higher, olefin-free fuels in West Germany (*CEN* 5/26/86). A bifunctional catalyst consists of an acidic ion-exchange resin impregnated with platinum or other hydrogenation catalyst. While the acidic catalyst promotes the etherification of the more reactive branched olefins, the hydrogenation catalyst removes the troublesome, unreactive olefins that gum up as a result of gradual polymerization during storage of the fuel.

A 1% MTBE content in the U.S. gasoline pool would require the production of about 6B lb of MTBE, while the current production capacity is reaching only 1B lb in 1988 (*CMR* 5/27/88) primarily because of a shortage of methanol.

The EPA currently limits the oxygen-containing ingredients of gasoline to 2% by weight but allows the content of MTBE to be as high as 10% by volume.

A curious footnote to the increasing presence of MTBE in our lives is the discovery that it is capable of dissolving gallstones and thus patients can avoid gall-bladder surgery.

Various methyl ethers are obtained by means of the reaction of the alcohol with appropriate epoxides. The Union Carbide family of Cellosolves and Carbitols dates back to the 1920s, when fermentation ethanol was still the common industrial alcohol and ethylene for the preparation of ethylene oxide (EO) was still obtained by dehydration of ethanol by way of ethylene chlorohydrin:

$$CH_3OH + CH_2{-}CH_2 \rightarrow CH_3OCH_2CH_2OH \qquad \text{methyl cellosolve, 49¢}$$
$$\diagdown O \diagup$$

$$CH_3OH + 2CH_2{-}CH_2 \rightarrow CH_3O(CH_2CH_2O)_2H \qquad \text{methyl carbitol, 55¢}$$
$$\diagdown O \diagup$$

Additional examples of ethers prepared from methanol and epoxides are shown elsewhere (4.3.5.2, 5.4.3). Other methyl ethers, including double ethers of difunctional alcohols (the glymes—see below), are obtained by means of methylating agents such as methyl chloride, methyl bromide, and dimethyl sulfate.

### 3.3.2.3   Halomethanes and Some of Their Derivatives

Methyl chloride is obtained by the gas-phase reaction of methanol and hydrogen chloride at about 345°C:

$$CH_3OH + HCl \rightarrow CH_3Cl + H_2O$$

The reaction is carried out over alumina gel, zinc chloride on pumice, $Cu_2Cl_2$, or activated carbon.

Methyl chlorides (25.5¢) has good growth prospects. While the elimination of lead-containing gasoline is eclipsing the production of tetramethyl lead (TML)

$$4CH_3Cl + Pb-Na \rightarrow (CH_3)_4Pb + NaCl$$

alloy

methyl chloride is the main organic building block of the expanding family of silicones (3.3.2.6.B) and has several other important uses. Thus, it is used to prepare methyl cellulose (MC) valued at \$2.25–\$2.75 (depending on purity and viscosity)—an industrially useful gum obtained from alkali cellulose in such a manner that an average 1.8 methyl groups are introduced for each glucose unit (*SZM-II*, 87) that contains three hydroxyl groups. This ratio provides the desired hydrophilicity of the product.

The dimethyl ethers of glycols are known as *glymes* and are obtained by methylation of the monomethyl ethers such as the Cellosolves and Carbitols shown above and higher ethoxylated products discussed elsewhere (4.3.5.1). The glymes are excellent solvents and, furthermore, since two or more ether functions coordinate alkali metal cations, they increase the nucleophilic reactivity of the (negatively charged) counterion and hence function as relatively inexpensive phase-transfer catalysts.

The methylation of phenols can be illustrated by means of the preparation of anisole (9.3.4):

$$C_6H_5OH + CH_3Cl \rightarrow C_6H_5OCH_3$$

In this reaction the phenol is first converted to sodium phenolate.

A number of industrially important quaternary ammonium compounds that function as bacteriocidal surface-active agents are prepared by exhaustive methylation of fatty amines (*SZM-II*, 56, 80). For example, myristylamine is converted to myristyltrimethyl-ammonium chloride as follows:

$$n\text{-}C_{14}H_{29}NH_2 + 3CH_3Cl \rightarrow n\text{-}C_{14}H_{29}N(CH_3)_3^+ \ Cl^-$$

An exchange of chloride and bromide ions results in the corresponding quaternary ammonium bromide used as a disinfectant and in personal care deodorants.

The methylation of hydrazine (85%, \$1.25) produces $H_2N\text{-}NMe_2$-unsymmetrical dimethylhydrazine, 1,1-dimethylhydrazine, which is used as a rocket fuel (Uniroyal, \$11.00).

Some of the methylation reactions that involve anionic, water-soluble substrates can be carried out more conveniently by means of dimethyl sulfate $[CH_3O)_2SO_2$, 46¢], but only one of the two methyl groups participates in the reaction. An exception to this statement is the replacement of both methyl groups by iodide ions by virtue of the high nucleophilicity of the latter that compensates for the poor leaving-group character of the methyl sulfate anion. Dimethyl sulfate is prepared by vacuum distillation of methanol and oleum (i.e., fuming sulfuric acid, ~4¢/lb).

Methyl bromide (57¢, 42MM lb) is used mostly as a soil and grain fumigant. In warm regions of the United States that are prone to termite infestations, methyl bromide is used to fumigate private residences. It is obtained from methanol and hydrobromic acid (48%, 38.5¢). It is a more reactive methylating agent than the chloride but because of the higher cost, its use is reserved for the methylation of more delicate organic structures. This is even more true in the case of methyl iodide that is prepared conveniently by means of the Finkelstein reaction (ONR-30), which takes advantage of the difference in the solubilities of sodium iodide and the corresponding chloride or bromide in acetone:

$$CH_3Cl + NaI \longrightarrow CH_3I + NaCl \text{ ppt}$$

USP $10.15

(7/87)

Methyl iodide is converted readily to the corresponding Grignard reagent, methyl magnesium iodide ($CH_3MgI$; $10.55/mol in 800-mol quantities as a solution in diethyl ether or THF). The reaction is carried out in absolutely dry solvents and with nonoxidized magnesium shavings. The corresponding bromides and chlorides are prepared in a similar fashion but with increasing difficulty. The cost of the Grignard reagents suggests that Grignard reactions (ONR-37) are reserved for the synthesis of costly fine chemicals. The Grignard reagents derived from methyl chloride or bromide carry lower price tags:

MeMgCl $3.84/lb (2*M* in THF); $3.60/lb (3*M* in THF)

MeMgBr $2.50/lb (2*M* in toluene–THF (1 : 1)); $2.70)/lb (in 100% THF)

(all prices quoted by Orsynex, 3/87, for 5000 kg quantities)

Methylene iodide is also rather expensive because of the intrinsic high cost of iodine (USP $17.00/lb; crude $16.00/kg; 7/87). It is one of the most dense organic liquids known ($d_4^{20}$ 3.325) and can be used to separate minerals. It is obtained by the reduction of iodoform with sodium arsenite [*OS I*, 52 (1921)], while iodoform, in turn, is obtained from acetone by means of the haloform reaction (ONR-56). Iodoform (USP $26.61/lb) is used as a topical antiseptic.

The methyl halides are the precursors of methyl-containing organometallic compounds. For example, tetramethyltin can be obtained from stannic chloride and methyl chloride in the presence of finely divided sodium (Wurtz reaction, ONR-98) or by the reaction of stannic chloride with a methyl Grignard reagent. Tetramethyltin is used by the electronic industry and as a component of certain catalysts.

### 3.3.2.4   Acetic Acid and Anhydride

The production of acetic acid from methanol and carbon monoxide (Monsanto in 1970)

$$CH_3OH + CO \rightarrow CH_3CO_2H$$

illustrates beautifully the importance of homogeneous, transition-metal catalysis, and, of course, it revolutionized the production of acetic acid worldwide. On closer examination, the reaction mechanism involves an alternation of the oxidation states of rhodium so that the methyl moiety is first carbonylated, and the resulting acetyl group is released. The reaction is primed by the formation of some methyl iodide from methanol and hydrogen iodide (see the preceding section), and at the same time, hydrogen iodide and CO convert rhodium chloride to the essential diiododicarbonyl rhodium complex:

$$RhCl_3 \cdot 3H_2O + 2HI + 2CO \rightarrow [Rh(CO)_2I_2]^- H^+ + 3HCl$$

Now the cyclic process summarized by the above, simple equation can begin, as shown in Scheme 3.2.

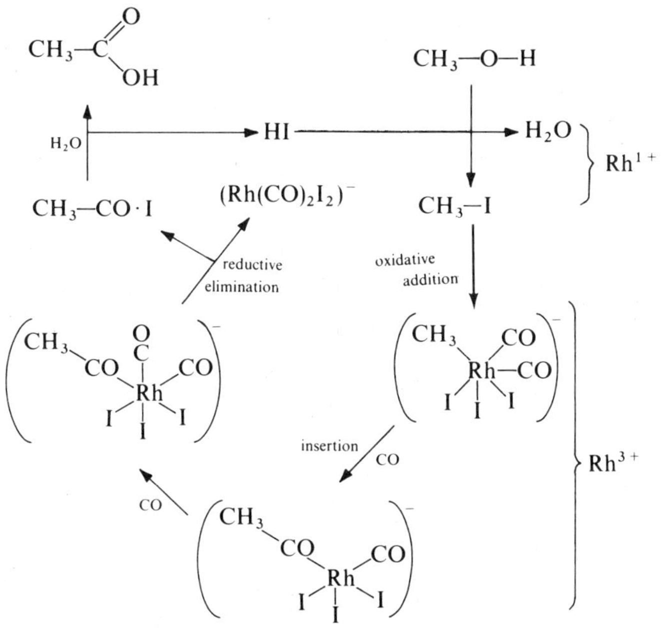

**Scheme 3.2**

Naturally, acetyl iodide is rapidly hydrolyzed to acetic acid and hydrogen iodide, and the latter primes another methanol into action. The evolution of the Monsanto acetic acid process was described by D. Foster in *Advances in Organometallic Chemistry*, Vol. 17, Academic Press, New York, 1979, p. 255. The construction of a 500MM-lb/yr plant by Rhône-Poulenc in Pardies, France at a cost of $100 million (*CEN* 5/18/81) marked the entry of this technology onto the European scene. Near the end of 1985 Monsanto had granted nine licenses to employ its acetic acid process, including two to British Petroleum (*CMR* 9/23/85).

A BASF variation of the Monsanto process employs a cobalt–iodine catalyst and a combination of methanol and dimethyl ether at 250°C and 750 bar. Another variation was announced by Texaco (*CEN* 5/5/86) in which a 1:1 syngas mixture is converted directly to acetic acid by means of a ruthenium–cobalt–iodine catalyst dispersed in molten tetra-*n*-butylphosphonium bromide (m.p. ~ 100°C).

In 1983 Tennessee Eastman inaugurated (in Kingsport, TN) a versatile complex that produces 500MM lb of acetic anhydride and varying amounts of acetic acid coproduct depending on the composition of the $CO$–$H_2$ feed. The key transformations are two carbonylation reactions: methanol to methyl acetate and methyl acetate to acetic anhydride. The pivotal intermediate of the process is the complex of acetic anhydride with a rhodium–methyl iodide–lithium catalyst (*CEN* 3/3/80, *CW* 8/12/81, *CEN* 11/29/82, *CE* 12/23/85). The other unique feature of this plant is the fact that CO is obtained by gasification of 900 t/day of coal. Modifications of the plant are possible to produce vinyl acetate and acetaldehyde also. These possibilities are summarized in Scheme 3.3.

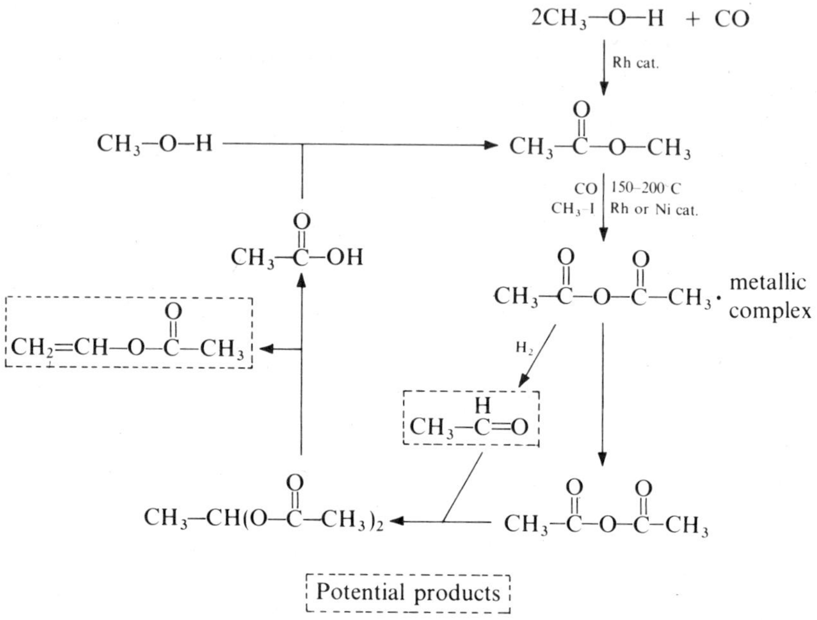

**Scheme 3.3**

### 3.3.2.5   Methyl Methacrylate (MMA), Dimethyl Terephthalate (DMT), and Other Methyl Esters

A time-honored process for the production of MMA that dates back to pre-World War II is the conversion of acetone, hydrogen cyanide, and methanol to the monomer that yielded poly(methyl methacrylate) (PMM), commonly known as Rohm & Hass's Plexiglass and Du Pont's Lucite. The one-pot process actually consists of three successive reactions: the formation of acetone cyanohydrin, its dehydration to give methacrylonitrile, and the methanolysis of the nitrile function:

$$(CH_3)_2C=O + HCN + CH_3OH \rightarrow CH_2=C(CH_3)CO \cdot OCH_3$$

$$\quad\;\; 24¢ \qquad\quad 50¢ \qquad 4.5¢ \qquad\qquad\qquad 62¢$$

The reaction is carried out in the presence of sulfuric acid that is converted to ammonium hydrogen sulfate or ammonium sulfate (5¢).

We note that the C$_5$ structure of MMA is assembled here from a C$_3$ and two C$_1$ building blocks, and since the popularity of PMM has not diminished with time, other processes are described elsewhere (6.3.1) for the production of the C$_4$ portion of MMA. However, regardless of how the C$_4$ skeleton is assembled, methanol must still be employed in the formation of the methyl ester function. The demand for MMA in the United States has remained for a number of years at the 1B-lb level. Besides the production of PMM, MMA is also employed as a comonomer to improve the impact resistance of other vinyl polymers, and by transesterification one also obtains higher esters of methacrylic acid.

Another large-volume consumption of methanol is accounted for by the production of DMT, the building block of poly(ethylene terephthalate) (PET) and poly(butylene terephthalate) (PBT). Both of these materials are thermoplastic polyesters. The former is processed into the commonly known "polyester" textile fibers, Du Pont's Mylar films, and other plastic products. The production of quality terephthalate polyesters requires the absence of isophthalic impurities (that would be responsible for nonlinear macromolecules), and since these dicarboxylic acids are obtained from the corresponding xylenes (9.2.2.2, 9.4.2), crude terephthalic acid is usually purified by distillation of the dimethyl esters to give purified terephthalic acid (PTA). Also, since the condensation–polymerization of the terephthalic moiety with the appropriate glycol occurs most readily by means of a transesterification reaction, DMT and PTA are listed jointly in production–consumption statistics.

The same principle of purification of rather nonvolatile carboxylic acid by fractionation of the corresponding methyl esters is applied in the case of the naturally occurring complex mixtures of fatty acids (*SZM-II*, 61, 77, 78). The mixture of fatty methyl esters is obtained most conveniently by methanolysis of the triglycerides.

Direct esterification of an acid, or the use of the corresponding acid anhydrides or chlorides, gives rise to numerous methyl esters of industrial importance. Some of them are listed here according to the functionality and in alphabetic order together with a recent unit price, and many are mentioned again throughout this book.

Methyl abietate, $9.40

Methyl abietate, hydrogenated $10.00

Methyl acetoacetate, 85¢

Methyl acrylate, 65¢ (lachrymator, b.p. 80°C)

Methyl anthranilate, $1.41

Methyl benzoate, 25¢, 99.9% perfume grade, $1.65

Methyl cinnamate, $4.65

Methyl formate, pure, tanks, 29¢

Methyl *p*-hydroxybenzoate, methyl paraben, NF $4.60

Methyl methacrylate, 62¢
Methyl parathion, $1.65

Methyl phenylacetate, $3.60

Methyl salicylate, oil of wintergreen, NF $1.99

Dimethyl anthranilate, $15.80

Dimethyl carbonate, 90¢

Dimethyl dichlorovinyl phosphate, dichlorophos, DDVP, $1.80

Dimethyl phthalate, 65¢

Dimethyl sebacate, $2.48

Dimethyl sulfate, 46¢

We note that the unit prices depend on several factors such as the unit price of the acid component, purity, scale of production, difficult distribution problems, and so on.

3.3.2.6 Nitrogen, Silicon, Sulfur, and Metallic Derivatives of Methanol

**A. METHYLAMINE BUILDING BLOCKS.** The production of the methylamines is accomplished by means of dehydrogenation/hydrogenation of the alcohol in the presence of ammonia and an appropriate catalyst:

$$CH_3OH + NH_3 \rightarrow MeNH_2 + Me_2NH + Me_3N$$

$$55¢ \qquad 55¢ \qquad 54.5¢$$

This reaction, referred to as *alcohol amination*, is related to the reductive amination of carbonyl compounds since the dehydrogenation–hydrogenation catalyst establishes an equilibrium

$$RR'CH-O-H \leftrightharpoons RR'C=O + H_2$$

and in the presence of ammonia (or primary or secondary amines), the carbonyl compound is converted to the corresponding imine that becomes hydrogenated with the aid of the same catalyst that is responsible for the dehydrogenation–hydrogenation equilibrium. The alcohol amination or reductive amination of aldehydes or ketones is carried out at 350–500°C, 10–20 bar, over $Al_2O_3 \cdot SiO_2 \cdot AlPO_4$ and other catalysts. This approach to the production of simple amines on an industrial scale is preferred over classical alkylation of ammonia by means of alkyl chlorides. It illustrates an unwritten rule of industrial organic chemistry: whenever possible, *do not use halogenated intermediates unless the halogen is part of the final product.*

Some of the industrial uses of the individual methylamines are as follows:

$MeNH_2 + CO_2$ (3.7) $\rightarrow$ MeNHCONHMe (*sym*-dimethylurea)

Butyrolactone (4.7) $\rightarrow$ *N*-methylpyrrolidone (NMP)

Phosgene (3.8) $\rightarrow$ MeNCO (methyl isocyanate, MIC)

$CH_2O/HCN/H_2O$ (3.4, 3.10) $\rightarrow$ $MeNHCH_2CO_2H$ (sarcosine)

$Me_2NH + CO/H_2$ (3.5.3.7) $\rightarrow$ $Me_2NCHO$ (*N,N*-dimethylformamide, DMF)

$Ac_2O$ (4.6) $\rightarrow$ $Me_2NCO \cdot CH_3$ (*N,N*-dimethylacetamide, DMA)

$CS_2$ (3.7) $\rightarrow$ $Me_2NCS \cdot SH$ (*N,N*-dimethyl dithiocarbamate)

$CO_2$ (3.7) $\rightarrow$ $Me_2NCONMe_2$ (tetramethylurea)

$ClCH_2CH_2Cl$ (4.3.2) $\rightarrow$ $Me_2NCH_2CH_2NMe_2$

(tetramethylethylenediamine, TMDEA)

$Me_3N + ClCH_2CH_2OH$ (4.3.5.4) $\rightarrow$ $Me_3\overset{+}{N}CH_2CH_2OH$ $\quad$ $Cl^-$ (choline chloride)

$C_6H_5CH_2Cl$ (9.3) $\rightarrow$ $C_6H_5CH_2\overset{+}{N}Me_3$ $\quad$ $Cl^-$

(benzyltrimethylammonium chloride)

The total demand for the family of methylamines in the United States is about 200MM lb, and the relative consumption decreases in the order secondary > primary > tertiary. Among the above-mentioned compounds, we find several dipolar aprotic solvents (3.5.3.1), namely, NMP, DMF, DMA, and tetramethylurea; the food and feed additive choline chloride; the bacteriocide benzyltrimethylammonium chloride; the pesticide building blocks MIC and $N,N$-dimethyldithiocarbamate; and sarcosine, the building block of certain surfactants.

**B. SILICON COMPOUNDS.** The demand for silicones, specifically poly(dimethyl-siloxanes) (PDMS), $(Me_2SiO)_x$, consumes about 80% of the methyl chloride production and is increasing at an annual rate greater than 10%. This demand is driven by the construction and automotive industries.

The most common silicone materials are assembled from methylated chlorosilanes prepared from methyl chloride and a silicon–copper alloy by means of a reaction that is usually optimized to yield dimethyldichlorosilane:

$$MeCl + Si{-}Cu \rightarrow Me_2SiCl_2 + Me_3SiCl + MeSiCl_3 + MeSiCl_2H$$

As we shall see below, each product of this high-temperature reaction of methyl chloride plays a useful role in the industrial applications of silicon compounds.

Dimethyldichlorosilane, a material produced in excess of 250MM lb, is the chain-forming monomer responsible for the formation of poly(dimethylsiloxane) (PDMS), commonly referred to as *dimethylsilicone*, by means of a controlled hydrolysis:

$$nMe_2SiCl_2 + nH_2O \rightarrow [O{-}Si(Me)_2{-}O{-}Si(Me)_2{-}]_m + 2nHCl$$
$$\text{PDMS}$$

where $m = n/2$. The dimethylsiloxane unit can form chains or rings and, depending on the molecular-weight distribution, the products can be either liquids or solids. The distribution of molecular weights of the linear structures is controlled by addition of the chain-terminating trimethylchlorosilane. On the other hand, the degree of cross-linking of the silicone polymer is controlled by addition of methyltrichlorosilane. Thus, by careful control of the composition of the methylated chlorosilanes that are subjected to hydrolysis (and condensation reactions), one obtains a series of silicones used as heat-transfer media and dielectric fluids (in replacement of the outlawed PCB transformer fluids; see 9.2.3 and *CEN* 6/29/87), a variety of functional liquids such as brake fluids, lubricants, and ingredients of cleaning and polishing formulations. A small amount of cross-linking produces elastomeric PDMS materials that include the well-known "silly putty," caulking, sealing, and adhesive compositions. Silly putty is valued at $2.85/lb, a price representative of most simple PDMS materials. Finally, an appropriately high degree of cross-linking produces rather flexible solids that, among other things, are used in the manufacture of prosthetic devices, under-the-hood automotive tubing, roofing materials, slow-release medication patches based on diffusion-rate-controlling silicone membranes, protective and water-repellent coatings on textiles, concrete, and glass or plastic objects. Emulsions of liquid silicones are defoaming agents, and some liquid silicones are internal plasticizers and mold-release additives employed during the fabrication of plastic parts.

The great variety of applications of PDMS materials and their structural modifications can be attributed to the following factors:

• The structural possibilities alluded to above that provide room for much imaginative "molecular engineering"

- The thermal stability of PDMS materials as compared to analogous carbon structures (as a result of the relatively high Si–O–Si bond energies)
- Chemical stability (except under harshly alkaline conditions), that explains the inocuous behavior of PDMS materials with respect to living tissues
- The hydrophobic surface created by exposed methyl groups
- The flexibility of –O–Si–O–Si– bonds, that is, small rotational and bending energies as compared to analogous carbon structures, due to the relatively large silicon bond radius
- The ability of silicon to expand its outermost electron shell beyond the octet

Chlorine-substituted methylsilanes retain the chemical reactivity of the halogen substituents that account for their hydrolysis and condensation reactions to give Si–O–Si bonds. Thus, the chlorines can be replaced by methoxy and other alkoxy groups, by acetate groups, by amino groups, and so on. In fact, in some of the above-mentioned practical applications of the simple silanes, the liberation of corrosive HCl is avoided by using, instead, the methoxy, acetoxy, or amino derivatives. These silicon compounds are also employed as surface modifiers of a variety of materials that possess reactive hydroxy groups. A common practice, for example, is to subject glass objects, and other mineral materials, to a methoxy-group-containing methylsilanes, as shown in Scheme 3.4.

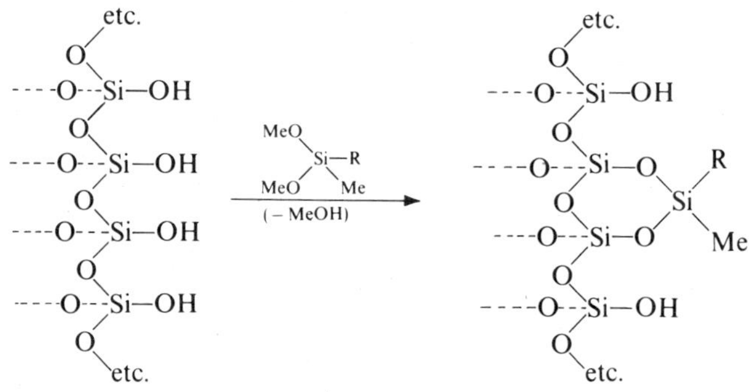

Scheme 3.4

It is clear that such a treatment converts a hydrophilic surface of a silicate into a hydrophobic one, and this concept is utilized in the preparation of certain packings for chromatographic columns. On a larger scale, it also increases the compatibility of mineral fillers with polymeric materials and thus the fillers become reinforcing agents. Even in the absence of reactive hydroxyl groups, the surface of solid object can be modified by depositing a silicone film. An excellent example of this practice is the use of an antimicrobial, quaternary ammonium compound for the treatment of textiles. The quaternary ammonium structure is part of a trimethoxysilane moiety (assembled from simple building blocks as shown below) with a typical structure represented by

$$\{(MeO)_3Si\text{--}CH_2\text{--}CH_2\text{--}CH_2\text{--}N(Me_2)\text{--}n\text{-}C_{18}H_{37}\}^+\ Cl^-$$

*n*-octadecyldimethyl[3-(trimethoxysilyl)-propyl]ammonium chloride,
Petrarch Systems, 50% in MeOH, 100 mL/$86 ($d_y^{20} = 0.89$)

The appropriate application of a methanolic solution of such an antimicrobial agent to a given surface and the removal of methanol by drying create a polymeric silicone film in which the quaternary ammonium terminals ward off microbial proliferation and guard against discoloration and/or development of undesirable odors due to microbial decomposition products.

Finally, methyldichlorosilane, a by-product of the production of the fundamental methyl-containing chlorosilanes, is subjected to addition reactions of terminal double-bond-containing compounds. This hydrosilylation reaction is catalyzed most conveniently by chloroplatinic acid ($H_2PtCl_6$) (Speier's catalyst when used for this purpose) and is not limited to the silicon–hydrogen compound under discussion:

$$Q-CH=CH_2 + H-SiXYZ \rightarrow Q-CH_2-CH_2-SiXYZ$$

As we shall see in 5.4.3.2, the hydrosilylation reaction plays an important role in the development of coupling agents.

The hydrosilylation reaction of trichlorosilane and acrylonitrile, followed by methanolysis of the adduct, gives cyanoethyltrimethoxysilane [$(MeO)_3Si-CH_2-CH_2-C\equiv N$] that, on hydrogenation to the corresponding aminopropyl compound, is converted to the desired quaternary ammonium compound by means of stearyl and methyl chlorides to give the above-mentioned antimicrobial agent.

Another example of hydrosilylation is the reaction of methyldichlorosilane ($MeSiCl_2H$, Petrarch Systems, $75/2 kg) with allyl alcohol, which introduces a 3-hydroxypropyl substituent onto silicon followed by a reaction with dimethyldichlorosilane (Petrarch Systems, $58/3 kg), and some trimethylchlorosilane (Petrarch Systems, $51/2.5 kg), and finally by subjecting the alcohol function to ethoxylation and/or propoxylation. This synthetic route explains the origin of the hybrid silicone–polyol ether hybrid, dimethicone polyol:

$$Me_3Si-O-(SiMe_2-O)_x-SiMe-CH_2-CH_2-CH_2-O-(CH_2-CH_2-O)_m-(CHMe-CH_2-O)_nH$$
$$| $$
$$O-SiMe_3$$

Such a material is used as an ingredient of cosmetics and personal-care products (lotions, hairsprays, hair-styling gels and fixatives), plasticizers, and so on.

Trimethylchlorosilane can be treated with triflic acid (3.2.3) to give trimethylsilyl triflate, trimethylsilyl trifluoromethyl sulfonate (Petrarch Systems, $72/100 g, $CF_3-SO_2-O-SiMe_3$. The last-mentioned compound is employed as a trimethylsilylating agent (because of the excellent leaving-group character of the triflate anion) that converts nonvolatile and hydrophilic polyhydroxylic systems to more volatile and hydrophobic trimethylsilylated derivatives. In the case of compounds of relatively low molecular weight, these derivatives can be handled better in gas-phase chromatographic analyses than can the hydrogen-bonded parent structures, and in the case of high-molecular-weight materials, the trimethylsilyl groups tend to orient themselves at the molecular surfaces where they affect compatibility properties.

The discussion of organosilane chemistry that involves other alkyl or aryl derivatives is beyond the scope of this book (except for the mention of some coupling agents in 5.4.3.2).

In closing, it is instructive and, at the same time ironic, to mention the fact that Frederick S. Kipping, the English chemist who dedicated most of his adult life to establishing the fundamentals of organosilicon chemistry, was duly honored in 1937 at the age of 74 for his pioneering contributions. Unfortunately, he included in his award address a dubious remark to the effect that "the prospects of any immediate and important advance in this section of organic chemistry does not seem to be very hopeful." There must be a lesson in the fact that the industrial production of silicones began seven years later and has led to countless applications of organosilicon compounds to our "quality of life."

**C. SULFUR COMPOUNDS.** The reaction of methanol with hydrogen sulfide (13¢), an abundant by-product of hydrodesulfurization (HDS) of petroleum (2.2.2), leads to a series of C$_1$ sulfur compounds. The initial product, methyl mercaptan, is readily oxidized by air in the presence of iron and other metallic catalysts to dimethyl disulfide (DMDS; Pennwalt, 59¢). DMDS is used for the "presulfiding" of cobalt and molybdenum catalysts employed during hydrotreating operations designed to remove sulfur from "sour" petroleum and natural-gas fractions. Extensive chlorination of DMDS produces trichloromethylsulfenyl chloride, perchloromethyl mercaptan (Cl$_3$C–S–Cl), the building block of the fungicide captan. In addition, the sulfenyl chloride can also be the precursor to thiophosgene or thiocarbonyl chloride (S=CCl$_2$), which is converted to isothiocyanates (R–N=C=S) or to thiocarbamyl chlorides (R$_2$N–C(=S)Cl) by means of primary and secondary amines, respectively. Both of the latter compounds can be converted to substituted thioureas or to thiocarbamates by reaction with either amines or alcohols:

$$R-N=C=S + R'-NH_2 \rightarrow R-NH-CS-NH-R'$$

$$R_2N-C(=S)Cl + R'-NH_2 \rightarrow R_2N-CS-NH-R'$$

$$R-N=C=S + R'-OH \rightarrow R-NH-CS-OR'$$

$$R_2N-C(=S)Cl + R'-OH \rightarrow R_2N-CS-OR'$$

The oxidation of methyl mercaptan with hypochlorite produces methanesulfonic acid (MSA, b.p. 165°C; Pennwalt, 100%, $2.23) used frequently as a gentle acid catalyst in place of sulfuric acid because, unlike the latter, it is not an oxidizing agent. MSA is converted to the corresponding sulfonyl chloride, methanesulfonyl or mesyl chloride (MSC; Pennwalt, $1.08) by means of phosphorus pentachloride or, on a smaller scale, by means of thionyl chloride (55¢). MSC is a useful reagent in the synthesis of fine chemicals for the purpose of blocking hydroxylic functions while other transformations are being carried out. In the case of chiral compounds, the formation of mesylate esters occurs without racemization and either the configuration at the mesylated carbon can be retained, or, under conditions of nucleophilic substitution of the mesylate group, the substitution takes place with inversion (Walden inversion; ONR-94). An example of the use of a mesylate derivative to preserve the configuration of a chiral alcohol center while other transformations are being carried out, followed by a replacement of the mesylate group with retention of the original stereochemistry, is the stereospecific synthesis of the antiinflammatory *d*-2-(6-methoxy-2-naphthyl)propionic acid, or naproxen, derived from naturally occurring dextrarotatory *S*-lactic acid (also referred to as L-lactic acid). The other example also starts with *S*-lactic acid, but this time it gives rise to the optically active herbicide ethyl 2-(2'-methyl-4'-chlorophenoxy)propionate of *R* configuration because of inversion during the nucleophilic displacement by the phenolate ion of the mesylate group, as shown in Scheme 3.5.

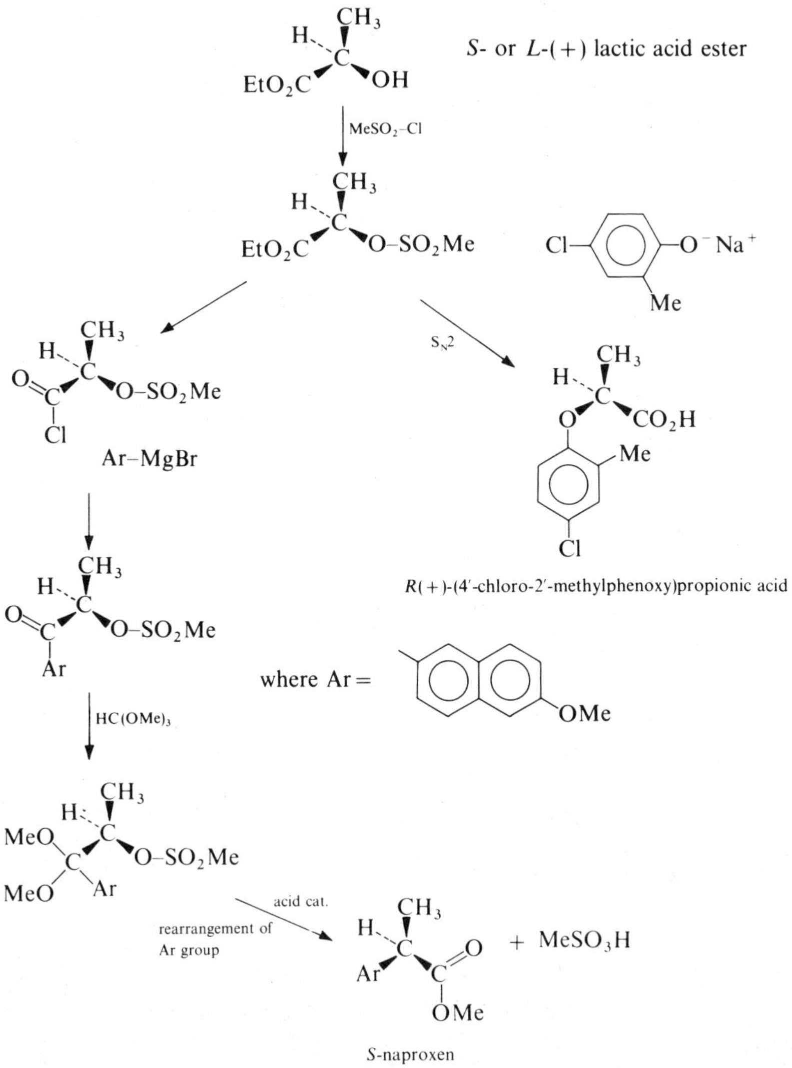
S- or L-(+) lactic acid ester
MeSO$_2$-Cl
S$_N$2
Ar-MgBr
HC(OMe)$_3$
where Ar =
acid cat.
rearrangement of Ar group
+ MeSO$_3$H
R(+)-(4'-chloro-2'-methylphenoxy)propionic acid
S-naproxen

**Scheme 3.5**

Simon fluorination of MSC gives trifluoromethanesulfonyl fluoride (3.2.3) that is hydrolyzed to triflic acid (b.p. 162°C). The latter can be converted to methyl triflate by means of dimethyl sulfate, and the lithium salt is used as a battery electrolyte (the 3M Company's Fluorad FC-124, \$45.10). The use of trimethylsilyl triflate for trimethylsilylation of hydroxylic compounds is mentioned in the preceding section.

A vigorous treatment of methanol with a source of sulfur (e.g., carbon disulfide over alumina at 375–535°C) produces dimethyl sulfide (DMS) (Pennwalt, 59¢). DMS is also a by-product of the pulping industry since the heating of lignin with sodium sulfide tends to cleave some of the phenolic methoxy groups (*SZM-II*, 154–155). DMS is oxidized by means of oxygen in the presence of a small amount of nitrogen oxides ($NO_x$) to dimethyl sulfoxide (78¢), an excellent dipolar aprotic solvent (b.p. 189°C) (see 3.5.3.1), as well as a useful reactant. Consult, for example, the following references:

D. Martin and H. G. Hanthal, *Dimethyl Sulfoxide*, Van Nostrand Reinhold, New York, 1975, 498 pp.

H. H. Szmant, *Chemistry of DMSO*, Chapter 1, in S. W. Jacob, E. E. Rosenbaum, and D. C. Wood, Eds., *Dimethyl Sulfoxide*, Marcel Dekker, New York, 1971, pp. 1–97.

*Dimethyl Sulfoxide Technical Bulletin*, Crown-Zellerbach (now Gaylord Chemical), P.O. Box 4266, Vancouver, WA 98662.

Dimethyl Sulphoxide is one of few organic liquids capable of diffusing through living tissues when applied topically, and this property may be hazardous when solutions of harmful materials are handled carelessly by uninformed individuals. Otherwise, DMSO is known to be an effective remedy for pulled muscles and minor tendon injuries, swollen ankles, and so on when applied topically as a 70% solution in pure water, and injured athletes and race horses have been treated in this fashion for some time. In spite of extensive literature concerning the biological effects of DMSO (see some references listed below), the FDA has not approved therapeutic uses of DMSO except for interstitial cystitis, and the problem seems to be the difficulty associated with carrying out blind control tests. One physiological effect of DMSO is the rapid formation of small amounts of its reduction product, DMS—a malodorous degradation product of DMSO. An ordinary placebo would be recognized during a double-blind test program by not releasing the telltale DMS. A party interested in FDA approval of DMSO as a useful therapeutic agent would have to go to the trouble of incorporating a slow-release system for DMS in the placebo and, so far, this has not been done. In the mean time, DMSO is accessible to the general public on a "black market" (as a relatively costly "industrial solvent") with the accompanying risk of having been mishandled during dilution and packaging operations.

For some general references that deal with the biological effects of DMSO, see

S. W. Jacob and R. Herschler, "Biological Actions of Dimethyl Sulfoxide," *Ann. N Y Acad. Sci.* **243**, 5–508 (1975).

S. W. Jacob, E. E. Rosenbaum, and D. C. Wood, Eds., *Dimethyl Sulfoxide: Basic Concepts*, Marcel Dekker, New York, 1971, pp. 99–479.

C. D. Leake, "Biological Actions of Dimethyl Sulfoxide," *Ann. N Y Acad. Sci.* **141**, 1–671 (1967).

B. Tarkis, *DMSO—the True Story of a Remarkable Pain-Killing Drug*, Morrow, New York, 1981, 225 pp.

The major transformation routes of C₁ sulfur compounds are summarized in Scheme 3.6.

**D. SODIUM METHOXIDE OR METHYLATE.** The syntheses of fine chemicals that involve strongly basic catalysts may utilize the sodium salt of methanol for that purpose. With the cost of metallic sodium (tanks, works) at 70¢, the unit price of sodium methylate is about 97¢.

### 3.3.2.7    More Recent Developments Based on Methanol

**A. ORTHOALKYLATION OF PHENOLS AND ANILINES.** Since the 1960s Ethyl has been engaged in the orthoalkylation of phenols and anilines, and more recently the reaction

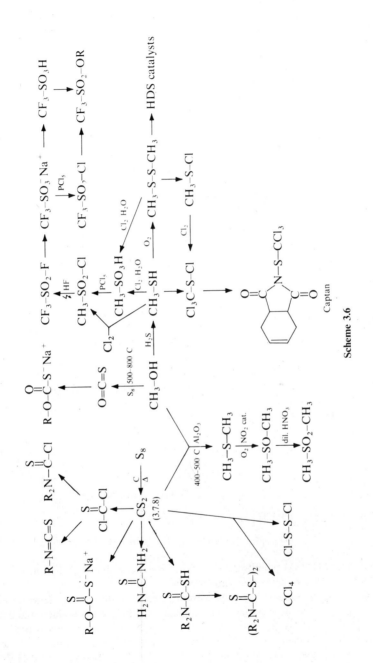

**Scheme 3.6**

95

was extended to thiophenóls. The process involves methanol (also other alcohols and olefins) and the aromatic substrate, and either aluminum methoxide, or a gas-phase reaction over alumina at 300–400°C and 40–70 bar. The main product of the ortho-methylation of phenol is 2,6-dimethylphenol, and some of the by-products are *o*- and *p*-cresol and 2,4-dimethylphenol. The mechanism that explains the selectivity of the reaction, applications of this reactions to other alcohols and olefins, and the industrial importance of some of the products are described elsewhere (9.3).

**B. THE METHANOL TO GASOLINE (MTG), PROCESS AND VARIATIONS ON THAT THEME.** This serendipitous (W. H. Land, *CEN* 9/28/87) development was considered newsworthy (*The New York Times*, 2/19/76) even though the story was lacking the health- or environmental-threatening component favored by the media. It climaxed a 5-year research effort led by W. M. Kaeding at Mobil that involved the treatment of methanol with an acidic, selective-pore-size zeolite catalyst known by its code name H-ZSM-5 to directly yield a high-octane gasoline of research octane number (RON) 90–95. For a review of several aspects of the MTG process, see C. D. Chang and A. J. Silvestry, *CT* 10/87, pp. 624–631.

The first large-scale, $1.5-billion, production unit went onstream in New Zealand during March 1986. It is designed to produce 5MM bbl/yr to satisfy one-third of New Zealand's gasoline needs. Natural gas is first passed over a zinc oxide catalyst at 350–400°C to remove traces of sulfur compounds and then is subjected to steam reforming (2.2.1) at 800–900°C over a nickel-based catalyst. The resulting syngas is converted to methanol over a copper-based catalyst, and water and dimethyl ether are the main by-products. This mixture is finally converted to gasoline over Mobil's propietary H-ZSM-5 catalyst at an estimated cost of $0.67–0.87¢/gal. An unexpected, but highly welcome, by-product isolated from the gasoline is durene, or 1,2,4,5-tetramethylbenzene, which is converted (9.4.2) to a valuable building block of some HPSTs.

The sequence of transformations of methanol to the gasoline components can be visualized as follows:

$$2MeOH \rightarrow MeOMe$$

$$MeOMe \rightarrow 2CH_2: + H_2O$$

$$2CH_2: \rightarrow CH_2 = CH_2$$

$$CH_2: + CH_2 = CH_2 \rightarrow CH_2 = CHCH_3 \quad \text{(insertion)}$$

$$nCH_2: + CH_2 = CHCH_3 \rightarrow C_{n+3}H_{2(n+3)}$$

$$C_{n+3}H_{2(n+3)} \rightarrow \text{alkanes} + \text{cycloalkanes} + \text{methylbenzenes}$$

The pore size of H-ZSM-5 limits the size of the products to about six carbon alkanes and cycloalkanes and tetramethyl derivatives of benzene.

Naturally, improvements of this process are already in the making. Thus, a fluidized-bed reactor operating at 700–800°F and a pressure below 60 psi replaces the fixed bed reactor of the Mobil pilot plant (*CEN* 12/17/84) in a project cosponsored by the Rheinische Braunkohlen Kraftstoffe, DOT, and the West German Ministry for Research and Technology. A demonstration unit operated in Houston, Texas since 1984 by Haldor Topsoe A/S combines the separate steps of the Mobil process into a continuous one that avoids the condensation and evaporation of methanol (*CE* 8/18/86).

**C. METHYL GLUCOSIDE.** In 1982 the A. E. Staley Company, a large producer of cornstarch, announced (*CW* 8/25/82) the purchase of a plant in Van Buren, Arkansas in order to adapt it to the first "biopetrochemical" production facility for conversion of glucose to methyl glucoside (*SZM-II*, 7, 105). The rationale for this venture is the need of polyfunctional alcohols by the chemical industry and the likelihood that the tetrafunctional methyl glucoside can compete with petrochemical-derived pentaerythritol, trifunctional 1,1,1-trimethylolpropane, and similar materials. In 1986 A. E. Staley formed a subsidiary Horizon Chemical to develop the market for methyl glucoside (Sta-meg 104) and its derivatives, such as the transesterification product with soybean oil and alkyl polyglycosides. The unit price of methyl glucoside is estimated (*CEN* 8/10/83) to be 15–50¢/lb depending on the purity. In 1988 A. E. Staley was acquired by Tate and Lyle of England.

Amerchol utilizes methyl glucoside for the production of a distearate that, on reaction with EO and/or PO serves, among other things, as an emollient in the formulation of personal-care products and as a viscosity enhancer.

Corn Products has been coining the term "carbochemicals" in recent advertisements to denote corn-derived chemical feedstocks. Thus, Horizon's "biopetrochemicals" join H. B. Hass's "sucrochemicals" (J. L. Hickson, *Sucrochemistry*, ACS Symposium Series 41, 1977) to denote the growing, alas very slowly, family of biomass-derived organic chemicals of industrial importance.

**D. DIMETHYL CARBONATE.** An oxidative carbonylation of methanol in the presence of a copper catalyst produces dimethyl carbonate (DMC):

$$MeOH + CO/O_2 \rightarrow (MeO)_2CO$$

The reaction occurs at 110–130°C and 20–30 atm (*CW* 4/13/83, *CE* 5/2/83), and the process was developed in Italy by EniChem. The company increased its production capacity to nearly 20MM lb by 1988 (*CEN* 11/21/87).

Dow claims (*CEN* 9/27/87) to have improved the oxidative methanol carbonylation technology by the use of a vapor-phase process and a cupric chloride–activated-carbon catalyst. The mechanism is believed to involve the following steps:

$$CuCl_2 + 2MeOH \rightarrow Cu(OMe)Cl + MeCl + H_2O$$

$$Cu(OMe)Cl + CO \rightarrow Cu(CO–OMe)Cl \quad \text{(a CO insertion)}$$

$$Cu(CO–OMe)Cl + Cu(OMe)Cl \rightarrow O{=}C(OMe)_2 + Cu_2Cl_2$$

$$Cu_2Cl_2 + \tfrac{1}{2}O_2 + 2MeOH \rightarrow 2Cu(OMe)Cl + H_2O$$

Virginia Chemical markets DMC at 90¢/lb. The production of DMC by means of CO signals the current general trend to avoid the use of phosgene (3.8). DMC can be used for the preparation of carbonates derived from higher alcohols, for the preparation of methyl carbamates and then the corresponding isocyanates:

$$RNH_2 + DMC \rightarrow RNHCO{\cdot}OMe \rightarrow RN{=}C{=}O + MeOH$$

It can also function as a methylating agent (in place of dimethyl sulfate) of nucleophilic substrates.

### 3.3.2.8   Prospective Developments Based on Methanol

**A. METHANOL TO OLEFIN (MTO) CONVERSION.** A modification of the MTG process that occurs at about 400°C and also uses zeolite catalysts (Rao & Gormby, *HP* 11/80) yields ethylene and propylene. BASF has constructed a 15-t/month pilot plant in Ludwigshafen, West Germany that produces the two lower olefins starting with a coal feedstock.

An ingenious method for the conversion of methanol to ethylene takes advantage of the classical, pyrolitic elimination reaction of acetates. First, methyl acetate is treated with syngas to give ethyl acetate, the latter is thermally decomposed to ethylene, and the regenerated acetic acid is esterified with methanol to methyl acetate to repeat the cycle.

**B. MISCELLANEOUS METHANOL TO HYDROCARBON CONVERSION PROCESSES.** Among various projects designed to convert methanol to hydrocarbons, one can enumerate the following:

1. Shell Development Company is investigating (C. Kim, M. M. Wald, and S. S. Brandenberger, ACS Meeting, Houston, 1980) the transformation of methanol to a mixture of hydrocarbons by means of molten zinc halides ($ZnBr_2$ at 220°C; $ZnI_2$ at 190°C). The product is a 1:1 mixture of gasoline and gas oil, and the former is a high-octane gasoline because of the presence of 2,2,3-trimethylbutane, or triptane.
2. Metal cluster catalysts (J. Haggin, *CEN* 10/5/87, 31–44) such as $Ru_3(CO)_{12}$, $Os_3(CO)_{12}$, and $Ir_4(CO)_{12}$ convert methanol to linear hydrocarbons containing 3–30 carbon atoms. Since the monomeric metal carbonyls do not function as catalysts, it is believed that two or more metallic sites are involved in the generation of transient methylene intermediates that join to give the alkanes.

**C. CONVERSION OF METHANOL TO HIGHER ALCOHOLS AND ACETIC ACID HOMOLOGS.** This homologation reaction has been known for some time (M. Orchin, *Advances in Catalysis*, Vol. 5, Academic Press, New York, 1950) and occurs when methanol is allowed to react with hydrogen-poor syngas at 185°C and about 400 atm in the presence of $Co_2(CO)_8$:

$$MeOH + 2CO + H_2 \rightarrow MeCH_2OH$$

Other developments in the area of homologation include the use of $Fe(CO)_5$ (*CEN* 9/20/82), rhodium–cobalt carbonyls (*CEN* 4/7/80), and other experimental conditions discussed, for example, by H. J. Chen and J. W. Rathke [*JACS* **104**, 7346 (1982)], W. Keim (*Catalysis in C$_1$ Chemistry*, D. Reidel, 1983; *Chemicals from Syngas and Methanol*, Symposium, Petroleum Chemistry Division, ACS Meeting, New York, April 13–18, 1986), and others.

The homologation process can be also applied to acetic acid, and ruthenium catalysts seem to be particularly effective (J. F. Knifton, *CT* 11/81).

There is no evidence that any homologation processes have become an industrial reality, so far, although Gulf announced a few years ago (*CEN* 4/7/80) the conversion of methanol and syngas to ethyl fuel—a mixture of 80% ethanol, 10% methyl acetate, 4% dimethyl ether, 1% diethyl ether, and 3% miscellaneous constituents.

A different approach to the conversion of methanol to ethanol is being explored by BASF (*CE* 8/19/85). Crude acetic acid obtained by the preceding carbonylation of methanol is esterified by methanol and ethanol (as the latter builds up in the reaction mixture), and the esters are then hydrogenated catalytically under conventional, rather low pressure conditions to give additional ethanol that is recycled as indicated. The cost of the resulting fuel-grade ethanol is calculated to be \$0.64–\$0.90/gal provided the price of natural gas is in the \$1–\$2-million BTU range.

**D. METHANOL AS A CONVENIENT SOURCE OF HYDROGEN.** The decomposition of methanol to CO and hydrogen, combined with the shift reaction of CO (3.5.1.6), was proposed (*CEN* 1/28/85) by Haldor Topsoe of Denmark as a mobil source of hydrogen for the production, for example, of fertilizer ammonia on locations remote from sources of natural gas. The same propietary catalyst functions for the decomposition of methanol (a reversal of its synthesis from CO and H$_2$) and the shift reaction and practically pure hydrogen is delivered at 525°F and under 210–300 psig (pounds per square inch gage).

**E. CONVERSION OF METHANOL TO ETHYLENE GLYCOL.** A free-radical-catalyzed conversion of methanol to ethylene glycol (J. Kollar, *CT* 8/84, 504) is initiated by *t*-butoxide radicals generated by the thermal decomposition of di-*t*-butyl peroxide (6.2) and the formation of hydroxymethyl radicals. The reaction cycle is represented as follows:

$$(t\text{-BuO})_2 \rightarrow 2t\text{-BuO·}$$

$$t\text{-BuO·} + CH_3OH \rightarrow t\text{-BuOH} + \cdot CH_2OH$$

$$t\text{-BuO·} + \cdot CH_2OH \rightarrow t\text{-BuOH} + H_2C{=}O$$

$$\cdot CH_2OH + H_2C{=}O \rightarrow HOCH_2CHO·$$

$$HOCH_2CH_2O· + CH_3OH \rightarrow HOCH_2CH_2OH + \cdot CH_2OH$$

The termination step of this chain reaction is the dimerization of the hydroxymethyl radicals to ethylene glycol:

$$2 \cdot CH_2{-}OH \rightarrow HOCH_2CH_2OH$$

Another route to ethylene glycol involves oxidative carbonylation similar to that shown for the production of dimethyl carbonate (3.3.2.7.D) except that it employs a Pd catalyst:

$$2CH_3OH + 2CO + O_2 \rightarrow (CO_2CH_3)_2 \qquad \text{dimethyl oxalate}$$

The dimethyl oxalate is hydrogenated to EG and MeOH, and the latter is recycled.

**F. METHANOL/TOLUENE TO STYRENE CONVERSION.** A one-step production of styrene according to the equation

$$C_6H_5CH_3 + CH_3OH \rightarrow C_6H_5CH{=}CH_2 + H_2 + H_2O$$

is very enticing since it avoids the use of the more costly benzene and the two-step process (9.2.2.1). Interest in the one-step process has been expressed by Monsanto, but so far there is no evidence that is has become an industrial reality.

**G. AMMOXIDATION OF METHANOL TO HCN.** While hydrogen cyanide (3.10) has become a valuable by-product of the ammoxidation of propylene to acrylonitrile, there exists the following alternative source explored, among others, by Monsanto:

$$CH_3OH + NH_3 + O_2 \rightarrow HCN + H_2O$$

The catalysts for most ammoxidations are ferric molybdate or manganese pyrophosphate.

**H. SINGLE-CELL PROTEIN (SCP).** Not long ago some European producers of methanol, ICI in particular, were quite interested in the conversion of methanol into SCP by means of *Methylophilus methylotropus.* This high-protein material could replace soybean and fishmeal proteins in animal and poultry feed, and it is obtained in about 50% yield from the feedstock.

## 3.4 FORMALDEHYDE AS A BUILDING BLOCK OF INDUSTRIAL ORGANIC CHEMICALS

### 3.4.1 Introduction: Reactivity and Hazards

The exceptional reactivity of formaldehyde, as compared to most other aldehydes and ketones (except multi-halogen-substituted ones such as chloral ($Cl_3CCHO$) or hexa-fluoroacetone [$(F_3C)_2CO$], in which the presence of electron-attracting halogens exacerbates the electrophilicity of the carbon terminal of the carbonyl group), is attributable to the nearly unobstructed access by nucleophilic reactants to the carbonyl–carbon terminal. The intrinsic polarization of the $H_2C=O$ molecule explains the ease with which it forms oligomers and polymers (including the industrially important poly(oxymethylene) (POM; the backbone of polyacetal resins) and the ease of the formation of additon products such as the hydrate. The relationship between these species is summarized in Scheme 3.7.

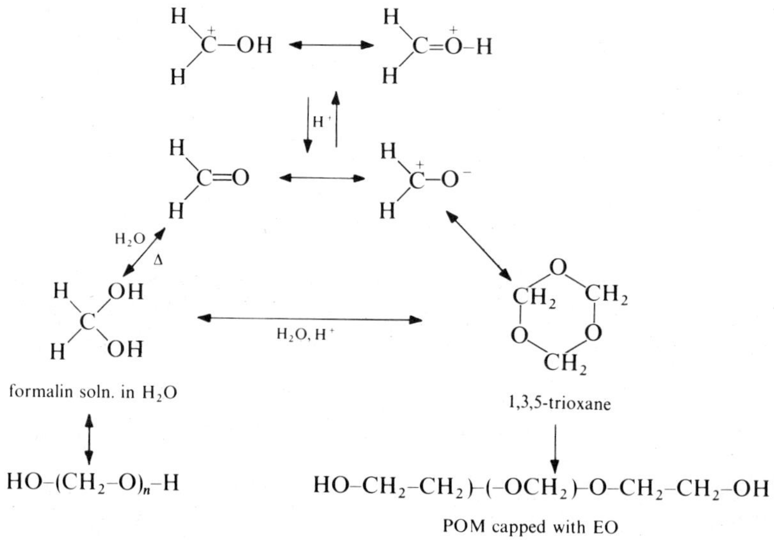

formalin soln. in $H_2O$

1,3,5-trioxane

$$HO-(CH_2-O)_n-H$$

$$HO-CH_2-CH_2-)-(-OCH_2-)-O-CH_2-CH_2-OH$$

POM capped with EO

**Scheme 3.7**

Formaldehyde also condenses readily with amino compounds (and other reactants mentioned below), and these include urea that forms the urea–formaldehyde (UF) resin, in which each formaldehyde binds two (or more) urea moieties by elimination of water (3.4.3.6). However, partial hydrolysis of the UF chains, or still more likely, incomplete polymerization, creates hydroxymethylamino or $N$-methylol terminals that can slowly liberate gaseous formaldehyde:

$$-NH{\cdot}CO{\cdot}NH{\cdot}CH_2OH \rightarrow -NH{\cdot}CO{\cdot}NH_2 + H_2C{=}O$$

It is well known that formaldehyde reacts readily with proteins (it is used, after all, as an embalming fluid) and that high concentrations of gaseous formaldehyde are toxic. The liberation of gaseous formaldehyde from improperly formulated UF foams used during the energy crisis of the 1960s and 1970s to insulate homes raised the specter of health hazards to residents of such insulated residences, and the highly publicized formaldehyde toxicity scare spilled over to phenol–formaldehyde (PF) or UF-treated wood-derived construction products. The latter have been manufactured for several decades by binding particles or sheets of wood with these thermosetting–adhesive resins without producing any complaints from manufacturers or consumers. However, when the UF foam hazards were making newspaper headlines across the country, even the personnel of automotive manufacturers engaged in the construction of wooden models for future automotive designs discovered effects harmful to their health and attributed them to formaldehyde.

On February 22, 1982 the Consumer Product Safety Commission banned the use of UF foam insulation even though a study of Du Pont employees exposed to formaldehyde between 1957 and 1979 did not provide any statistical evidence of increased cancer incidence that could be attributed to such exposure (*CEN* 5/31/82; *CMR* 6/2/82). Also, a study of morticians who provide an excellent human sample of long-standing, occupational use of aqueous formaldehyde, formalin, did not reveal any health hazards. In April 1983 a Court of Appeals overturned the UF foam insulation ban. In December 1986 OSHA reopened the examination of formaldehyde hazards caused by the use of formaldehyde-derived foundry resins (that bind sand particles in molds used to cast iron pieces) and promised the announcement of revised exposure limits by October 1987. Current human exposure limits over an 8-hour period are 3 ppm, and these concentrations may produce irritation of the mucous membrane. Also during December 1986 the EPA announced its intention to classify formaldehyde as a probable carcinogen on the basis of nasal tumors in animals exposed to high concentrations even though recent epidemiological studies of humans did not demonstrate carcinogenic effects. In January 1987 the 10th Federal Circuit Court of Appeals upheld (*CMR* 1/26/87) the acceptability of the 0.4-ppm formaldehyde concentration limit permitted by Wisconsin and Minnesota Department of Housing and Urban Development (HUD) in mobile homes. In April 1987 The EPA reclassified formaldehyde from "possible" to "probable human carcinogen" (Group B-1) because "there is limited evidence to indicate that formaldehyde may be a carcinogen in humans" on the basis of "nine studies reported statistically significant associations between site-specific respiratory neoplasm and exposure to formaldehyde or formaldehyde-containing products" in rats and mice of both sexes. The Formaldehyde Institute opposes the B-1 classification of formaldehyde, and in October 1987 the United Auto Workets filed suit against OSHA demanding the issuance of final exposure standards to formaldehyde. Time will tell how this issue is settled.

### 3.4.2 Production and an Overview of Industrial Uses

The conversion of methanol to formaldehyde (3.3.2.1) has been practiced for a long time, but there is still room for optimization of reaction conditions. Thus, for example, the development of a fluidized bed of $Fe_2O_3$–$MoO_3$ catalyst deposited on microspheroidal silica (*CEN* 11/3/80) is promising: a stream of methanol and air (possibly enriched by oxygen) at 340°C and 6 bar gave a 98% conversion of methanol with 90–98% selectivity. Also, a "ring-shaped" catalyst developed by Swedish Perstop Corporation will be utilized in the world's largest formaldehyde plant under construction by Du Pont (*CB* 8/87) designed to produce 150MM lb of formaldehyde.

Formaldehyde is one of the many products of oxidative cracking of LPG (6.6.2.2).

The current demand for formaldehyde is about 6B lb, and its price is 8.8¢ (37%, methanol-free), 10.15¢ (44–45%, 1% methanol to inhibit polymerization). Formalin usually contains 40 g of formaldehyde in 100 mL of solution.

More than 50% of the formaldehyde consumption is dedicated to the production of polymers: 25% for UF, 20% for PF, 5% for POM, and some for melamine–formaldehyde (MF). All except POM are thermoset resins.

Other major chemicals assembled from formaldehyde and described further on are 1,4-butanediol, 1,1,1-trimethylolpropane, pentaerythritol, hexamethylenetetramine, and 1,4-methylenedianiline, the precursor of the diisocyanate MDI and its oligomeric derivative PMDI. It is noteworthy that all are building blocks of polymeric materials.

Minor but interesting applications of formaldehyde include, besides embalming fluids, tanning aids, hardeners of photographic gelatin layers, fingernail hardening agents, disinfectants, germicides, and textile treatment formulations. In some of these applications formaldehyde is employed as the labile sodium bisulfite adduct, $HOCH_2SO_3Na$.

### 3.4.3 The Role of Formaldehyde as a Building Block of Industrial Organic Compounds

3.4.3.1 Introduction

Formaldehyde contributes the following structural features or modifications:

(a) The methylol group $-CH_2OH$, attached to C, N, O, and S atoms that contain a labile (somewhat acidic) hydrogen.

(b) The chloromethyl group, $-CH_2Cl$, when the former reaction is carried out in the presence of hydrochloric acid.

(c) The methylene bridge, $-CH_2-$, between two moieties that contain a labile hydrogen (as in a).

(d) Reduction of an aldehyde to the corresponding alcohol by virtue of the facile oxidation of formaldehyde to the formate ion in the presence of alkali (a Cannizzaro reaction (ONR-16), known since 1853). The combination of a and d gives rise to polyfunctional alcohols shown in 3.4.3.3.

(e) Addition to unsaturated carbon systems (3.4.3.8) occurs by virtue of the dipolar nature of formaldehyde.

3.4.3.2 Acetals and Hemiacetals

Examples of simple acetals are the acid-catalyzed formation of methylal, or formal, and the cyclic acetal of ethylene glycol:

$$CH_2O + 2MeOH \rightarrow CH_2(OMe)_2 + H_2O$$

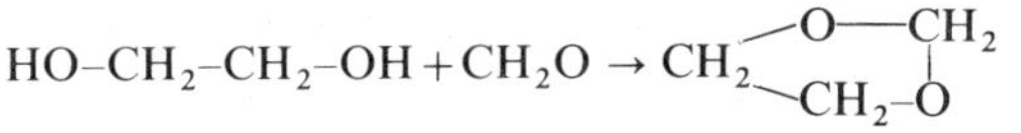

$$HO-CH_2-CH_2-OH + CH_2O \rightarrow CH_2 \underset{CH_2-O}{\overset{O-CH_2}{<}}$$

1,3-dioxacyclopentane or dioxolane

A useful property of these and other acetals is their stability under basic conditions. Six-membered cyclic acetals are also formed from appropriate diols. Thus, for example, poly(vinyl alcohol) (PVOH; $1.05) is converted to poly(vinyl formal) (PVF), employed as a wire-coating material:

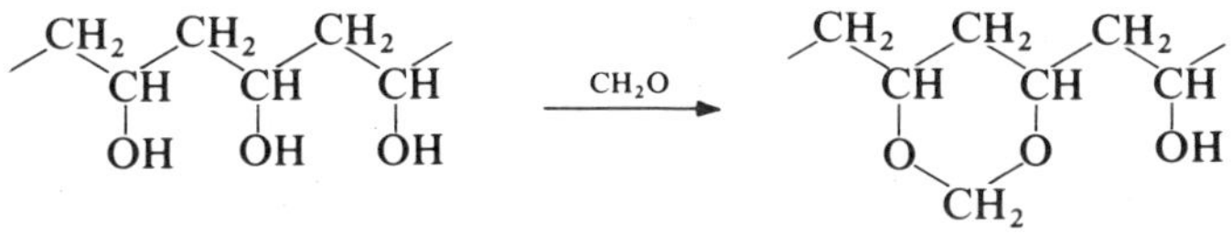

The analogous acetal of PVOH and butyraldehyde, poly(vinyl butyral) (PVB), competes with PVF for the automotive safety-glass market.

A clever use of a formal intermediate is the conversion of methyl cellosolve (3.3.2) to glyme ($2.25; German Patent 2,434,057, 1974):

$$2MeOCH_2CH_2OH + CH_2O \rightarrow (MeOCH_2CH_2O)_2CH_2$$
$$\downarrow H_2, Ni$$
$$MeOCH_2CH_2OMe + MeOCH_2CH_2OH$$
recycle

The polymeric acetal of formaldehyde (POM) must be "capped" in order to prevent the "unzipping" of terminal hydroxymethyl groups:

$$HO(CH_2OCH_2O)_nH \rightarrow HO(CH_2O)_{n-1}H + H_2C=O$$

This is achieved by treatment of the unstable polymer with either acetic anhydride or ethylene oxide. Actually, when formaldehyde is polymerized in the presence of EO, some ethylene glycol units are also introduced in a random fashion along the oxymethylene chain. This helps to decrease the crystallinity (and brittleness) of the material and produces a tough engineering resin. The same role is played by the above-mentioned dioxolane. In the same fashion BASF incorporates THF (4.7.2, 10.2) during the acid-catalyzed polymerization to produce a tough acetal copolymer.

Du Pont introduced the first homopolymeric POM resin as Delrin in 1960, and its current domestic consumption is about 125MM lb at $1.64/lb. By 1988 there will be three domestic manufacturers of POM. In 1983 Du Pont revealed (*CMR* 3/7/83) the availability of two alloys of POM with an elastomeric component that is intimately associated with the poly(oxymethylene) chains and promotes impact resistance and thus gives rise to Delrin T and Delrin ST, where T and ST stand for "tough" and "supertough," respectively. It is claimed that these are "the world's toughest metal-like engineering plastics." In 1987 Du Pont announced (*CMR* 6/1/87, *CW* 6/17/87) the production of Delrin II, an acetal homopolymer with improved thermal stability due to the incorporation of a proprietary stabilization technology and still priced at $1.64. The current worldwide market for acetal plastics is estimated at about 600MM lb with a growth rate of about 9%/yr.

One traditional building block of aliphatic sulfide or thiokol elastomers is the formal of 2-chloroethanol. The polymer is produced by means of a nucleophilic substitution reaction by sodium polysulfide:

$$CH_2(OCH_2CH_2Cl)_2 + Na_2S_x \rightarrow (-CH_2CH_2OCH_2OCH_2CH_2S_x-)_n$$

The resulting organic polysulfide is converted to a caulking material by reaction with $PbO_2$ and is also an ingredient of solid rocket propellants.

Generally, hemiacetals of formaldehyde tend to decompose and exist mainly in solution just as its hydrate, $CH_2(OH)_2$. The addition of some methanol to aqueous formaldehyde tends to prevent its oligomerization to solid but unstable paraformaldehyde [$HO(CH_2O)_nH$, where $n = \sim 30$].

### 3.4.3.3 Methylols, Products of Aldol and Aldol–Cannizzaro Reactions and Methylene-Bridged Derivatives of Aliphatic Systems

Almost any carbon compound that contains a sufficiently acidic hydrogen (one that is ionized in the presence of an appropriately strong base) undergoes an aldol-type reaction (ONR-2) with formaldehyde to give the corresponding methylol.

Methylol derivatives of nitroparaffins (5.2) are as follows:

$$(CH_3)_2CHNO_2 \longrightarrow (CH_3)_2C(CH_2OH)NO_2$$

2-nitropropane, 55¢              2-nitro-2-methyl-1-propanol, $1.25

$$\downarrow H_2, Ni$$

2-amino-2-methyl-1-propanol, 89¢

$$CH_3CH_2CH_2NO_2 \longrightarrow CH_3CH_2C(CH_2OH)_2NO_2$$

1-nitropropane              2-nitro-2-ethyl-1,3-propanediol

$$\downarrow H_2, Ni$$

$$CH_3CH_2C(CH_2OH)_2NH_2$$

2-amino-2-ethyl-1,3-propanediol, $1.82

$$CH_3CH_2NO_2 \longrightarrow CH_3C(CH_2OH)_2NO_2$$

nitroethane, $2.50              2-nitro-2-methyl-1,3-propanediol

$$CH_3NO_2 \longrightarrow (HOCH_2)_3CNO_2$$

nitromethane, $2.37              tris(hydroxymethyl)nitromethane, 60.5¢

$$\downarrow H_2, Ni$$

$$(HOCH_2)_3C-NH_2$$

tris(hydroxymethyl)aminomethane, 50¢

The condensation reactions of formaldehyde with acetylene, and especially the formations of 1,4-butanediol, which consumes about 7% of the formaldehyde demand, are shown elsewhere (4.7).

In the case of aldehydes that contain one or more α-hydrogen atoms the aldol condensation is accompanied by a concurrent Cannizzaro reaction (ONR-16). This double-header reactivity of formaldehyde accounts for about 300MM lb or 5% of its consumption. The by-product of the Cannizzaro reaction is sodium formate (20¢). Some polyols of industrial importance obtained in this fashion are as follows:

$C(CH_2OH)_4$, pentaerythritol (71¢), derived from acetaldehyde

$CH_3CH_2C(CH_2OH)_3$, 1,1,1-trimethylolpropane (76¢), derived from n-butyraldehyde

$(CH_3)_2C(CH_2OH)_2$, neopentyl glycol (solid, 60¢), derived from isobutyraldehyde

These polyfunctional alcohols rose to prominence as modifiers of the traditional glyptal resins that, as indicated by the name, were prepared originally from glycerol and phthalic anhydride to yield amorphous, thermoset polyesters used in baked-on coatings of metallic objects. Actually, historically speaking, pentaerythritol provided the first available (before the advent of petrochemicals) replacement of glycerine, and its molecular rigidity promoted improved thermal stability and hardness of the resulting coatings. The evolution of the glyptal resins into alkyd resins and unsaturated polyester resins (UPR) is discussed elsewhere in this book (see 6.2.2).

The above-mentioned polyols also serve as precursors to multifunctional monomers, such as the acrylates and allyl ethers, which have become important cross-linking agents for radiation-polymerized coatings, high-speed printing inks, and so on (5.4.3.2 and 5.4.5). Among such derivatives we should also mention fire-retarding additives of the phosphorus family. An example of such a fire-retardant is the ester of methylphosphonic acid [Me–P(=O)(OH)$_2$] and 1,1,1-trimethylolpropane:

$$Me-P(=O)\left[O-CH_2-C\begin{smallmatrix} CH_2-O \\ \\ Et \quad CH_2-O \end{smallmatrix}\begin{smallmatrix} \\ P(=O) \\ Me \end{smallmatrix}\right]_2$$

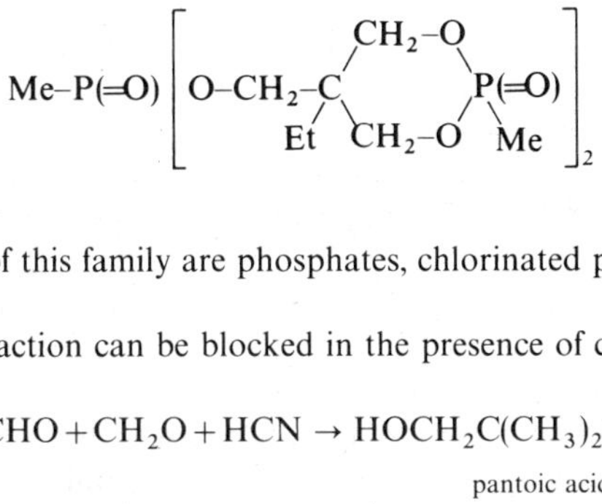

Other fire-retardants of this family are phosphates, chlorinated phosphates, and related compounds.

The Cannizzaro reaction can be blocked in the presence of cyanide:

$$(CH_3)_2CHCHO + CH_2O + HCN \rightarrow HOCH_2C(CH_3)_2CH(OH)CN$$

pantoic acid nitrile

The hydrolysis of the nitrile produces pantolactone that is converted to pantothenic acid upon treatment with β-alanine:

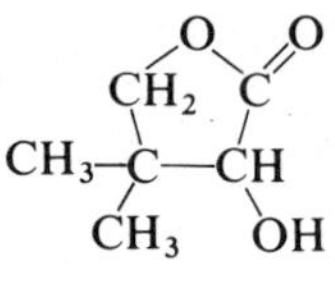

$$CH_3-\underset{\underset{OH}{|}}{\overset{\overset{CH_2}{|}}{C}}-\underset{CH_3}{\overset{}{CH}} \quad + H_2NCH_2CH_2CO_2H \rightarrow HOCH_2C(CH_3)_2CH(OH)NH(CH_2)_2CO_2H$$

as Ca salt, 4.4 MM lb

d,l-Ca pantothenate, feed grade $3.63
d-calcium pantothenate, USP $6.15

Pantothenic acid is a member of the vitamin B complex.

The methylol adduct of HCN can react with one, two, or three hydrogen atoms of ammonia to yield aminoacetonitrile, $H_2N-CH_2-C\equiv N$, iminodiacetonitrile,

$HN(CH_2-C\equiv N)_2$, and nitrilotriacetonitrile, $N(CH_2-C\equiv N)_3$, respectively. Hydrolysis of the nitrile function then gives the corresponding carboxylic acids. The latter can be also assembled from chloroacetic acid and ammonia (4.3.2.3).

*Cyanoacrylate Adhesives*

The aldol condensation of ethyl cyanoacetate, followed by nearly spontaneous dehydration is intended to give ethyl cyanoacrylate, but this highly reactive monomer undergoes an immediate polymerization to poly(ethyl cyanoacrylate):

$$(-CH_2C(CO_2Et)-CH_2C(CO_2Et)_n \underset{H_2O\ cat.}{\overset{\Delta}{\rightleftharpoons}} 2n\ CH_2=C(CN)CO_2Et$$
$$\underset{CN}{|}\qquad\qquad\underset{CN}{|}$$

However, this polymerization is reversible and the polymer is depolymerized on heating to give the ethyl cyanoacrylate monomer, while the latter is again poymerized by traces of moisture. These properties constitute the basis of the exceptionally strong cyanoacrylate adhesives known to the public by such names as "miracle glue," "crazy glue," and similar. It is of interest to note that the first application of this remarkable adhesive occurred under battlefield conditions in Korea when the material was used as a surgical tool. Current efforts in microsurgery sponsored by the FDA suggest the use of cyanoacrylate polymers to mend eyes and brain blood vessels by means of adhesives that "set" in 0.1 of a second and 7 sec, respectively (*CB* 4/87).

The success of cyanoacrylate adhesives in gluing cracked solid objects must be attributed to the application of the material as a low-viscosity monomer capable of penetrating and filling the crevices of matching surfaces and the *in situ* polymerization to form a thin film that fills the spaces between the surfaces and binds the parts. Here is an example where more is *not* better. Other esters of the cyanoacrylate family (methyl, *n*-butyl),and oligomers that set less rapidly are also on the market.

The methyl groups in the alpha and gamma positions of pyridine are somewhat acidic and hence react with formaldehyde at elevated temperatures to give the corresponding vinylpyridines by way of a dehydration of the intermediate methylol. The formation and uses of these monomers and polymers are described elsewhere (10.3).

A high-temperature, forced aldol condensation–dehydration between formaldehyde and acetaldehyde gives acrolein (5.4.4):

$$CH_2O+CH_3CHO \rightarrow CH_2=CHCHO$$

This reaction is carried out over a fluidized bed of manganese oxide-containing silica gel at about 300°C.

### 3.4.3.4   Chloromethylation Products

Chloromethylation, or the Blanc reaction (ONR-12), introduces a $ClCH_2$–substituent onto an aromatic ring through the formation of a transient chloromethyl carbocation, $Cl-CH_2^+$. The latter is usually generated from formaldehyde, hydrochloric acid, and zinc chloride in the presence of the aromatic substrate that does not resist electrophilic substitution reactions. The resulting, rather reactive, benzylic chlorides are intermediates in additional synthetic transformations illustrated elsewhere (9.2.1, 10.2).

*Ion-Exchange Resins*

Partial chloromethylation of polystyrene (PS) can be the first step in the production of a common class of *ion-exchange resins*. Actually, for this purpose the PS is somewhat cross-linked [by incorporating a few percent of divinylbenzene (DVB) during the polymerization of styrene] in order to create an insoluble PS gel before it is subjected to chloromethylation (9.2.1, 9.3.1).

If we represent the partialy chloromethylated PS by PS–CH$_2$Cl, we can illustrate the preparation of an *anion-exchange resin* as follows:

$$PS-CH_2Cl + Me_3N \rightarrow PS-CH_2-NMe_3^+ Cl^-$$

Since only the chloride ions are mobile in the resulting quaternary ammonium resin, a treatment of the latter with alkali will convert it to the corresponding quaternary ammonium hydroxide resin, PS–CH$_2$–NMe$_3^+$ OH$^-$. Such a basic resin can now be used to neutralize protons of an acidic solution that is passed through a column of the ion-exchange resin, while the anions of the acidic solution are trapped by the resin. The basic form of the resin can be regenerated by another washing with alkali. Another member of the anion-exchange resin family is obtained by nitration of the somewhat cross-linked PS followed by reduction of the nitro to amino groups (see 9.2.5 and 9.3.6.2). This gives rise to PS–NH$_2$, which represents a weakly basic anion-exchange resin because of the intrinsically weak basicity of the aniline system and the relatively low concentration of PS–NH$_3^+$ OH$^-$.

It is convenient to describe at this point a partner *cation-exchange resin*. Sulfonation (9.2.4) of the above-mentioned, somewhat cross-linked PS results in the formation of PS–SO$_3^-$ H$^+$. Now only the proton of the strongly acidic sulfonic acid function is mobile, and should one treat such a material with a solution of sodium chloride; the latter is converted to HCl while the resin becomes neutralized as PS–SO$_3^-$ Na$^+$.

It is obvious that successive treatments of a saline solution with both cation- and anion-exchange columns will deionize the solution (if the two columns are initially in the acid and basic states, respectively). Once the ion-exchange capacity of the columns is exhausted, they can be regenerated by treatment with sodium chloride (rock salt, 1¢–3¢).

Typical ion-exchange resins for water-treatment applications are priced as illustrated:

Rohm & Hass's Amberlite strongly acidic cation-exchange resins, \$72–\$97 per truckload of 700 ft$^3$ minimum (50–54 lb/ft$^3$), either H$^+$ or Na$^+$ ionic form

Rohm & Hass's Amberlite strongly basic anion-exchange resins, \$209–\$237 per truckload of 700 ft$^3$ minimum (42–45 lb/ft$^3$), either Cl$^-$ or HO$^-$ ionic form

A variety of other types of ion-exchange resins are available in the marketplace, but their description is beyond the scope of this book. Ion-exchange resins are used as catalysts, for the recovery of valuable metallic materials for the removal of radioactive constituents in water-cooled nuclear reactors and so on.

Chloromethylated PS is also used for the immobilization of costly enzymes used to catalyze industrial processes (*SZM-II*, 34). The same is true of costly transition-metal catalysts.

For an introduction to the preceding topics, see the following references:

N. K. Mathur, C. K. Narang, and R. R. Williams, *Polymers as Aids in Organic Chemistry*, Academic Press, Orlando, FL, 1980, 258 pp.

D. Naden and N. Streat, *Ion Exchange Technology*, Wiley, New York, 1984, 724 pp.

### 3.4.3.5   Methylene-Bridged Derivatives of Aromatic Systems

Aromatic systems such as phenols and anilines are susceptible to electrophilic reactants (see 9.3.4 and 9.3.6.1) and thus are readily attacked by formaldehyde under acidic conditions. For phenol one first obtains ortho- and para-substituted methylols. Then the acidic conditions induce condensation reactions with additional phenol molecules to give a relatively low-molecular weight, soluble, thermoplastic novolac polymer in which the phenol : formaldehyde ratio is about 1 : 1. Such a polymer can be treated with additional formaldehyde [usually in the form of hexamethylenetetramine, hexa, or $(CH_2)_6N_4$ (3.4.3.6)] to change the phenol : formaldehyde ratio to 0.8 : 1, and under these conditions the polymer becomes a cross-linked, thermoset resin that normally also contains fillers, pigments, reinforcing agents, lubricants, and other additives for compression molding of the desired fabricated item. Another approach to the production of the thermoset resin is to condense a 1.5 : 1 mole ratio of formaldehyde and phenol under aqueous alkaline condition to yield a water-soluble resin known as *resol*. Heating the latter in the presence of the preceding additives produces a cross-linked thermoset PF resin. The complex chemistry of PF resins was elucidated in 1907 by Baekeland, who demonstrated that the condensation–polymerizations could be carried out in one or two stages to give the first synthetically produced resin, named Bakelite in his honor.

A simple representation of PF chemistry is illustrated in Scheme 3.8.

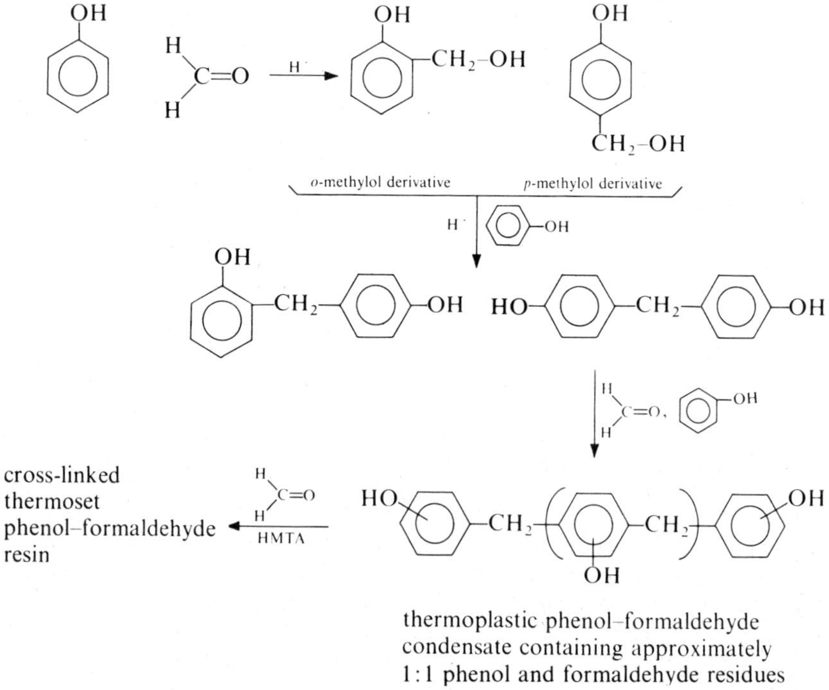

**Scheme 3.8**

Phenol–formaldehyde resin production is about 2.7B lb. This amount includes a great variety of fillers and additives since the resin content per se averages about 55%. They are used mostly ($\sim 47\%$) as binding resins in the manufacture of laminates such as plywood and as adhesives (28%) in the manufacture of glass fiber and other types of insulation materials, particle boards and similar construction materials, and abrasives. PF resins are inexpensive; they average about 45¢ on the basis of 100% resin content, but for obvious reasons their consumption is very sensitive to the needs of the construction industry.

There exist numerous modifications of PF resins in which substituted phenols, and even terpenes (*SZM-II*, 110–148), are copolymerized with phenol and formaldehyde. One common substituted phenol used for this purpose is *p*-nonylphenol (9.3.4.1), which promotes the formation of oil-soluble phenolic resins suitable for the manufacture of paints and other coatings.

The reaction of anilines with formaldehyde produces important precursors to isocyanates required for the manufacture of polyurethanes (PUR). Thus, under relatively acidic conditions that deactivate the aromatic rings to multiple formaldehyde condensations, there are formed 4,4′-methylene dianiline (MDA) or bis(*p*-aminophenyl)methane (crude $1.75, purified $2.25) and 2,4-bis(*p*-aminobenzyl)dianiline (40¢). The first of the two aniline derivatives is converted to the corresponding diisocyanate (MDI) and together with TDI (see 3.8.2, 9.2.5, 9.3.6), are the main PUR building blocks. If MDA is subjected to a more extensive treatment with formaldehyde, one obtains a methylene-group-bridged oligomeric or polymeric MDA, and the latter, when converted to the multifunctional isocyanate, gives polymeric MDI (PMDI). Unlike MDI, this is an ingredient of *rigid* PUR materials.

Until the dioxin scare (9.3.1) related to the production of 2,4,5-trichlorophenol, the condensation product of the latter with formaldehyde, namely, hexachlorophene, was a common germicidal ingredient of deodorants and other personal-care products. The analogous condensation product of *p*-chlorophenol is dichlorophen, and this material is still widely used as a germicide and agricultural fungicide. The structures of these two products are

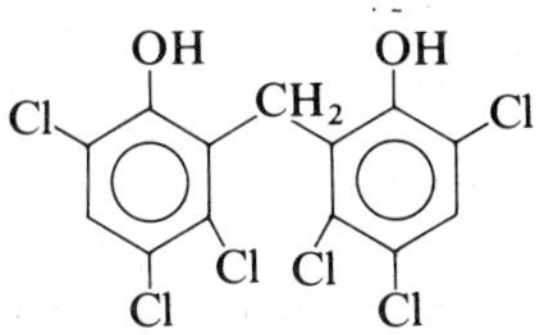

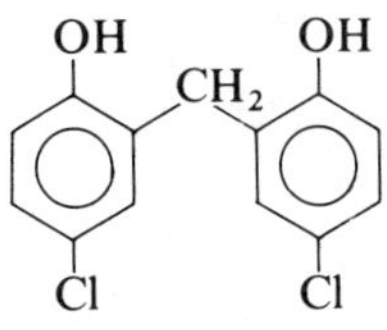

For additional examples of diarylmethane compounds obtained by bridging two aromatic rings by means of formaldehyde see 9.2.4 and 9.3.3.

### 3.4.3.6   Bridging of Amino Groups by Formaldehyde

The two large-volume materials that belong to this category of formaldehyde derivatives are the thermoset urea and melamine resins UF and MF, respectively, referred to collectively as "amino resins." The consumption in 1986 of both classes of resins was 1.4B lb, with the majority represented by UF in accord with the price differential between the monomers:

Urea, 10¢

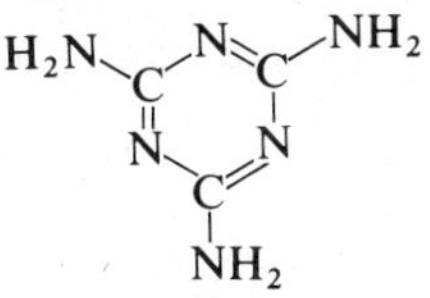

Melamine 50¢

Even though each hydrogen of the N–H moiety can be potentially converted to a *N*-methylol group, it suffices to have a formaldehyde–amino group ratio somewhat larger than 1:1 in order to have a thermoset resin.

About two-thirds of the amino resins are used as adhesives of wood products, textile treatments, and coatings, and under these circumstances the free *N*-methylol groups are allowed to react with hydroxyl groups of cellulose, amino groups of appropriate textile fibers, and so on in order to bind the components of the final product by chemical means. MF is capable to be molded into attractive and shatterproof dinnerware long known to the general public but this usage employs only about 20MM lb.

In Section 3.4.1 it was implied that the likely cause of much of the formaldehyde health hazard scare can be attributed to the improper applications of UF foam during thermal insulation of homes. The foam was produced at the site of the application and include some reactive materials (such as resorcinol; see 9.3.4.1) to avoid the presence of unreacted *N*-methylol groups in the final product. If present, the latter are sure to be hydrolyzed in time with the liberation of formaldehyde. Even though the machines used at the site of UF foam application have monitoring devices designed to control the ratios of all reactants that include, of course, foam-producing $CO_2$, the hundreds of operators of UF-foam-insulating companies that emerged throughout the United States could not have been expected to be perfect. Thus, human errors should have been anticipated among such a large number of nonchemical personnel. This is only one among many situations in which "enabling chemicals" are allowed to be handled by chemically unskilled people with detrimental consequences for the whole chemical industry and its image, about which so much is said and written.

A variety of *N*-methylol compounds are available in the marketplace for use as methylene-bridging reactants. An example of such a compound is *N*-methylol methacrylamide [$CH_2{=}C(CH_3)CO{\cdot}NHCH_2OH$, Rohm \$1.50], which can be co-polymerized with other vinyl monomers to yield a polymer capable of combining with appropriate substrates in which active hydrogen centers can form methylene bridges. Another compound that performs in a similar manner is *N*-methylolacrylamide [$CH_2{=}CHCO{\cdot}NHCH_2OH$, National Starch, NMA 48%, 42¢]. A reversal of strategy calls for the acid-catalyzed attachment of the *N*-methylol compounds to an appropriate polymeric material and then cross-linking, with the aid of peroxide catalysts, such a modified material with other vinyl monomers. This approach is illustrated by the *N*-methylol acrylamide-treated cellulose, a "reactive cellulose," produced by Horizon Chemical as Stalink.

As mentioned elsewhere in this chapter (3.8.2), difunctional monomers in which the two functional groups follow a different "chemical drummer" are referred to here as "heterodifunctional monomers." This nomenclature serves to distinguish between the above-mentioned NMA and the related *N*,*N*-methylene-bisacrylamide, Cyanamid's MBA (\$3.75, in less than truckload quantities) in which both polymerization sites are chemically alike and the material is simply a difunctional acrylic system.

A *N*-methylol group tends to react in an intramolecular fashion with a nearby active hydrogen group to give, preferably, a five- or six-membered cyclic structure. An example of this behavior is the formation of 3-methyl-1,3-oxazolidine from *N*-methylethanolamine and formaldehyde:

$$MeNHCH_2CH_2OH + CH_2O \longrightarrow \underset{\underset{\displaystyle CH_2{-}CH_2}{|\qquad\ |}}{MeN\underset{}{\overset{\displaystyle CH_2}{<}}\ O}$$

The ultimate condensation product of formaldehyde and ammonia is hexamethylene-tetramine, hexamine (hexa, HMTA, 55¢ or 60¢, granular or powder form, respectively), mentioned in 3.4.3.5 as a source of both additional formaldehyde and catalytic ammonia during the two-stage production of PF resins. This use consumes about 50% of the HMTA demand. While the bird-cage structure of HMTA is readily hydrolyzed to its simplest components under conditions of PF and the below-mentioned nitrilotriacetic acid (NTA) formation, under more gentle conditions of substitution reactions one obtains derivatives of the intermediate six-membered heterocyclic ring (see below).

Nitrilotriacetic acid (60¢ as trisodium salt) is obtained from about 25% of the 100MM-lb total production of hexa by means of sodium chloroacetate (cost of acid = 56¢):

$$\text{[HMTA structure]} + 4ClCH_2CO_2Na \longrightarrow N(CH_2CO_2Na)_3$$

A competitive process for NTA is the reaction of ammonia with hydroxyacetic or glycolic acid—the carbonylation product of formaldehyde (3.5.3.7.A).

Nitrilotriacetic acid was the first substitute proposed for replacement of phosphate "builders" in household detergent formulations when the United States suffered the "eutrophication scare" in the early 1970s. In other words, the phosphates (specifically sodium tripolyphosphate) were suspected of promoting the growth of algae in waste-waters, causing oxygen concentration of lakes and rivers to decrease below the survival level of fish and other aquatic creatures. Rather than legislating the removal of phosphates (generated primarily by human wastes) from municipal wastewaters, certain states (e.g., Michigan, Minnesota, Wisconsin, Virginia and North Carolina, $\sim 30\%$ of the total U.S. population) prohibited the use of phosphates in laundry detergents, and the capability of NTA to chelate calcium ions became the basis for its use as a "detergent builder." NTA (trisodium salt) is inexpensive (40% solution, 25¢; dry powder 60¢) and prevents the deposition of heavy-metal compounds on laundry (the famous "ring around the collar") and, generally speaking, economizes on detergents and energy (because of the use of lower temperatures during laundry cycles). The EPA withdrew its initial objections to the use of NTA in 1980, but once a given chemical is placed in the limelight where it becomes recognizable by certain environmental groups and the media, its long-range survival becomes questionable. Thus, in 1982 the anti-NTA campaign was initiated in the State of New York and the New York State Environmental Conservation Department proposed a statewide ban of NTA (*CW* 12/1/82) on the allegation that it causes "an adverse effect on human health." This controversy lingers on (*CEN* 5/18/87, 3).

A derivative of HMTA that originates with the loss of only one methylene group is *N,N*-dinitrosopentamethylenetetramine, obtained from hexa and nitrous acid:

$$(CH_2)_6N_4 + HNO_2 \longrightarrow \text{[dinitroso structure]}$$

This product can also be obtained by condensation of the stoichiometric quantities of formaldehyde and ammonia in the presence of nitrous acid. The dinitroso compound is a foaming agent used in the manufacture of foam rubber objects.

The treatment of HMTA with fuming nitric acid produces the military explosive cyclonite, RDX, which is more powerful than TNT (9.2.5) by about 50%:

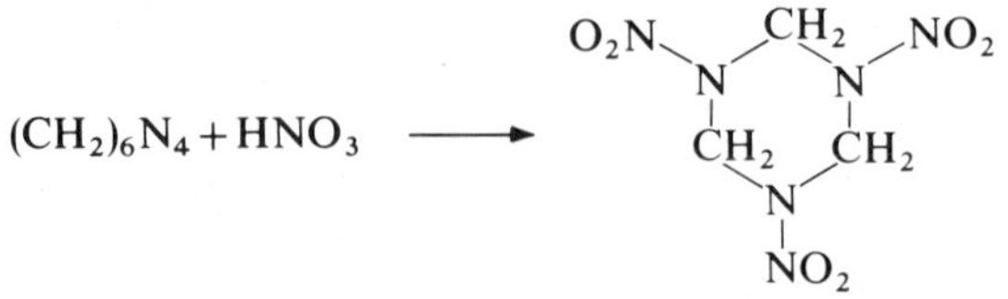

$$(CH_2)_6N_4 + HNO_3 \longrightarrow$$

Cyclonite (Research Department explosive) is used in conjunction with the analogous $N,N'$-dinitro-$N$-acetyltriazine.

The next examples of formaldehyde–amino compound derivatives resemble the preceding cyclic structures, but they may be prepared more conveniently by the assembly of the appropriate building blocks rather than by intercepting by chemical means the partial hydrolysis product of HMTA. Thus, the condensation of formaldehyde with ethylamine gives Uniroyal's Trimene Base, one of the many catalysts to control the course of PUR formation. Similarly, formaldehyde and 3-dimethylaminopropylamine (5.4.2.2) form the expected $N,N',N''$-tris(3-dimethylaminopropyl)-1,3,5-hexahydrotriazine, another PUR catalyst. The structures of the two latter compounds are shown as follows:

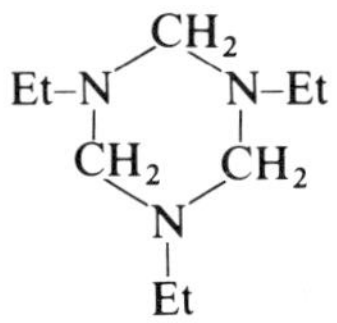

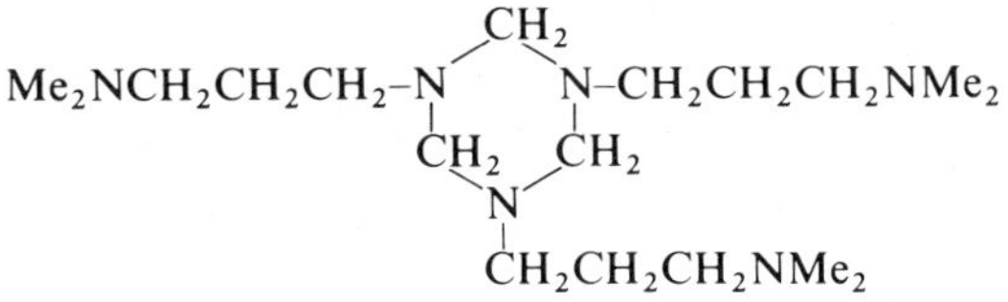

The announcement from Argonne National Laboratory (*CW* 4/8/87) that the ferrous–HMTA complex removes nitrogen oxides ($NO_x$) emitted from coal-burning power plants and may also aid in the removal of $SO_2$ promises to be a bonanza for HMTA if large-scale tests confirm the laboratory experiences, and if a more rigorous control of acid rain-producing emissions is mandated. The same claim (*CEN* 4/18/88) is made for cheaper urea.

The lone electron pair of tricovalent phosphorus compounds is a good nucleophile and hence it reacts with the carbonyl group of formaldehyde. For example, a fire-retardant tetrakis(hydroxymethyl)phosphonium sulfate is prepared from toxic and spontaneously inflammable phosphine and four formaldehyde molecules in the presence of sulfuric acid:

$$:PH_3 + 4CH_2O \rightarrow (HOCH_2)_4P^+ \quad \tfrac{1}{2}SO_4^{2-}$$

Thiols or mercaptans react with formaldehyde and hydrochloric acid to produce chloromethyl sulfides, and the latter can then be subjected to substitution reactions of the chlorine function. Thus the soil insecticide terbufos ($C_9H_{21}O_2PS_3$) is obtained from $O,O$-diethyl phosphorodithioic acid, formaldehyde, and hydrogen chloride, followed by reaction with $t$-butyl mercaptan:

$$(EtO)_2PS \cdot SH + CH_2O - HCl \rightarrow (EtO)_2PS \cdot SCH_2Cl \rightarrow (EtO)_2PS \cdot SCH_2SC(CH_3)_3$$

Terbufos is the active ingredient of American Cyanamide's pesticide Counter.

Analogous behavior of phosphorus trichloride (35¢) gives rise to two interesting agrichemicals obtained by means of the reaction of aminoacetic acid or glycine (see below), technical grade $1.88, USP grade $2.12, with a phosphonic acid $[R–PO(OH)_2]$ that results from the $PCl_3$–$CH_2O$–HCl reaction followed by hydrolysis of the phosphorus–chlorine bonds in the intermediate ion $(Cl_3PCH_2Cl)^+$. They are glyphosate or $N$-phosphonomethylglycine and glyphosine or $N,N$-bis(phosphonomethyl)glycine:

$$(HO)_2PO{\cdot}CH_2NHCH_2CO_2H \quad and \quad [(HO)_2PO{\cdot}CH_2]_2NCH_2CO_2H$$

Monsanto's Roundup, the isopropylamine salt of glyphosate, is used to formulate a broad-spectrum postemergence herbicide, while Monsanto's Polaris is formulated with glyphosine and functions as a sugarcane ripener that is alleged to increase the yield of sucrose by 8–15% (*CEN* 8/28/72). A recent announcement by Monsanto (*CMR* 6/6/88) reveals research on genetically modified tomato plants that are tolerant to the use of Roundup by the insertion of a specific gene from petunia.

The chloromethylphosphorus intermediate of interest in connection with the preceding two agrichemicals can be also prepared from a trialkyl phosphite $[(RO)_3P]$ and $CH_2O$–HCl by means of the Michaelis–Arbuzov reaction (ONR-61), which yields $(RO)_2PO{\cdot}CH_2$–Cl.

The nonbonding electron pair of tricovalent sulfur compounds is nucleophilic with respect to formaldehyde as illustrated by the reaction of sodium dithionate $[^-O_2\ddot{S}\text{–}SO_2^-\ 2Na^+]$ and formaldehyde to give sodium formaldehyde sulfoxylate, sodium hydroxymethanesulfinate, or Rongalite ($1.04), a mild reducing agent employed to generate soluble, reduced vat dyes (9.3.6) during dyeing operations:

$$2Na^+\ ^-O_2\ddot{S}\text{–}SO_2^- + CH_2O \rightarrow HOCH_2SO_2^-\ Na^+$$

Sodium dithionate is obtained from $SO_2$ by reduction with either Zn dust or sodium formate.

### 3.4.3.7  Mannich and Analogous Reactions of Phosphorus and Sulfur Compounds

The Mannich reaction (ONR-57) introduces a methylene bridge (donated by formaldehyde) between a N–H-containing moiety and a carbon atom of a potential carbanion or enol. Regardless of the mechanistic details of this reaction, which can occur under either acid- or base-catalyzed conditions, the net effect is

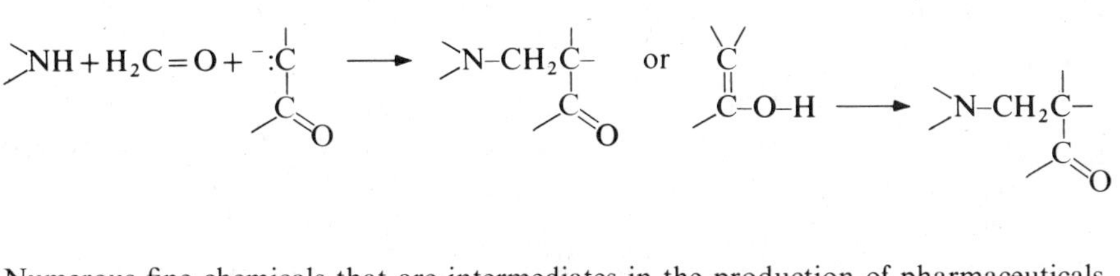

Numerous fine chemicals that are intermediates in the production of pharmaceuticals, pesticides, and other products are assembled from the three building blocks indicated above. Some examples are shown in Scheme 3.9 and elsewhere in this book.

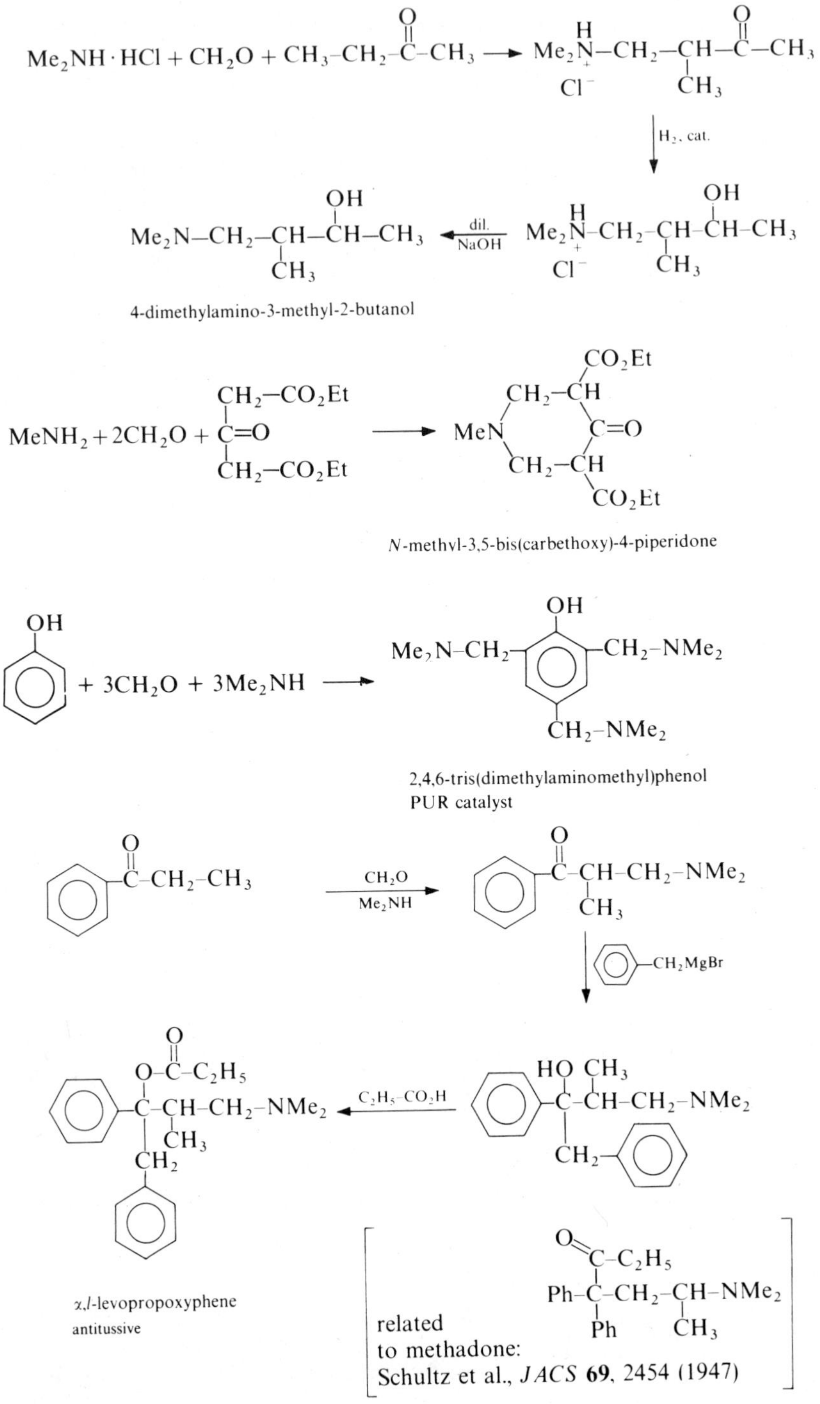

$$Me_2NH \cdot HCl + CH_2O + CH_3-CH_2-\overset{\overset{\displaystyle O}{\|}}{C}-CH_3 \longrightarrow Me_2\overset{\overset{\displaystyle H}{\mid}}{\underset{Cl^-}{N^+}}-CH_2-\underset{CH_3}{\overset{\mid}{CH}}-\overset{\overset{\displaystyle O}{\|}}{C}-CH_3$$

$\Big\downarrow$ H₂, cat.

$$Me_2N-CH_2-\underset{CH_3}{\overset{\mid}{CH}}-\overset{\overset{\displaystyle OH}{\mid}}{CH}-CH_3 \xleftarrow[\text{NaOH}]{\text{dil.}} Me_2\overset{\overset{\displaystyle H}{\mid}}{\underset{Cl^-}{N^+}}-CH_2-\underset{CH_3}{\overset{\mid}{CH}}-\overset{\overset{\displaystyle OH}{\mid}}{CH}-CH_3$$

4-dimethylamino-3-methyl-2-butanol

$$MeNH_2 + 2CH_2O + \underset{CH_2-CO_2Et}{\overset{CH_2-CO_2Et}{\overset{\displaystyle |}{\underset{\displaystyle |}{C=O}}}} \longrightarrow MeN\underset{CH_2-\underset{CO_2Et}{\overset{\mid}{CH}}}{\overset{CH_2-\overset{CO_2Et}{\overset{\mid}{CH}}}{\Big\langle}}C=O$$

N-methyl-3,5-bis(carbethoxy)-4-piperidone

$$+ 3CH_2O + 3Me_2NH \longrightarrow$$

2,4,6-tris(dimethylaminomethyl)phenol
PUR catalyst

$\xrightarrow[Me_2NH]{CH_2O}$

$\Big\downarrow$ ⟨⟩—CH₂MgBr

α,l-levopropoxyphene
antitussive

$\xleftarrow{C_2H_5-CO_2H}$

related to methadone:
Schultz et al., JACS **69**, 2454 (1947)

**Scheme 3.9**

The cyanide ion can also participate in a Mannich reaction, and this behavior is illustrated by the preparation of sarcosine ($CH_3NHCH_2CO_2H$, 50¢):

$$CH_3-NH_2 + H_2C=O + NaC\equiv N \rightarrow CH_3-NH-CH_2-C\equiv N \rightarrow CH_3-NH-CH_2-CO_2H$$

Sarcosine is a building block of liquid soap surfactants of the type $R_fCO_2N(CH_3)CH_2CO_2H$ (where $R_fCO_2H$ represents a fatty acid; see *SZM-II*, 162). Sarcosine is also converted to enzyme inhibitors added to some toothpastes.

The analogous reaction in which ammonia replaces methylamine gives rise to the nitriles of glycine ($1.88–$2.12), iminodiacetic acid, and nitrilotriacetic acid. Similarly, by starting with ethylenediamine one obtains (on hydrolysis of the initial nitrile) the powerful chelating agent ethylenediaminetetracetic acid, editic acid, or simply EDTA (37¢ for 40% solution of ammonium salt), $(HO \cdot CO-CH_2)_2N-N(CH_2-CO \cdot OH)_2$.

Iminodiacetic acid undergoes some interesting reactions. For example, dehydration by heating in DMF gives rise to poly(*N*-carboxymethyl) derivative of Nylon-2:

$$x\,HN(CH_2-CO_2H)_2 \rightarrow -(N-CH_2-C(=O)-)_x$$
$$\underset{\displaystyle CH_2-CO_2H}{\overset{\displaystyle |}{}}$$

The reaction of iminodiacetic acid with epichlorohydrin produces another chelating agent:

$$(HO_2C-CH_2)_2N-CH_2-CH(OH)-CH_2-N(CH_2-CO_2H)_2$$

1,3-diamino-2-propanol tetraacetic acid

Hydrogen cyanide can be employed in the formation of hydroxyacetic or glycolic acid

$$H_2O + H_2C=O + HCN \rightarrow HO-CH_2-CO_2H$$

but the latter is prepared more conveniently and economically by means of carbon monoxide (3.5.3.7).

As in the case of glycolic acid, the above-mentioned glycine can be prepared from ammonia, formaldehyde, and hydrogen or sodium cyanide or by means of the Reppe hydrocarbonylation reaction discussed below (3.5.3.5). In the United States the current demand for glycine is about 7MM lb, of which as much as 1MM lb is imported. It is used as a sweetener to mask the taste of saccharine in soft drinks, pharmaceutical formulations, and food products, as well as in the preparation of the above-mentioned glyphosate and glyphosine.

### 3.4.3.8    Addition Reactions with Unsaturated Systems

The cycloaddition of gaseous formaldehyde to ketene (4.6.4) produces β-propiolactone:

$$\begin{array}{ll} CH_2=C=O & CH_2-C=O \\ CH_2=O & \rightarrow\; | \quad\; | \\ & CH_2-O \end{array}$$

The extremely strained ring system of β-propiolactone induces a high degree of reactivity. Alkylating reactions provoke carcinogenic properties; consequently, the material is made only for captive use in the preparation of special acrylic esters of alcohols that are sensitive to the usual esterification procedures. $\beta$-Propiolactone is probably the only organic compound that reacts with sodium chloride and gives sodium $\beta$-chloropropionate.

The Prins reaction (ONR-72) is initiated by the addition of protonated formaldehyde to an olefinic double bond. In the case of isobutylene the initial product is a 1,3-dioxane that is pyrolyzed at about 250–350°C to isoprene with partial recovery of formaldehyde. The course of this C$_4$ + C$_1$ assembly of isoprene is shown in Scheme 3.10.

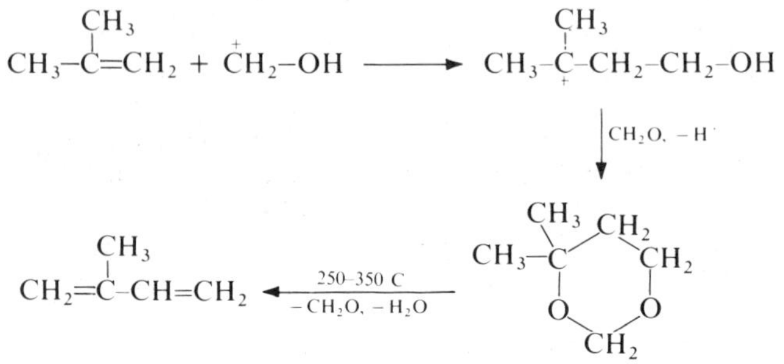

**Scheme 3.10**

A modification of the preceding process gives rise to prenyl alcohol and the corresponding prenyl chloride, $(CH_3)_2C=CHCH_2X$, where X stands for OH and Cl, respectively. The prenyl system is a building block of synthetic pyrethroids.

The Prins reaction applied to $\beta$-pinene produces nopol—a compound with a camphor-like scent (*SZM-II*, 129).

### 3.4.4  Some Prospective Uses of Formaldehyde

1. Electrohydrodimerization (EHD) of formaldehyde is a potential source of EG. This process is being developed by Electrosynthesis of Amherst, New York (*CEN* 2/2/84) and resembles the EHD process that Monsanto uses to convert acrylonitrile to adiponitrile (5.4.2.2).

2. Reductive carbonylation to ethylene glycol represents another approach to the preparation of EG. The addition of CO to formaldehyde under high pressure and in the presence of a rhodium catalyst generates a transient species $^-O–CH_2–C(=O)^+$ that is stabilized by hydrogenation to the desired EG. This process is of interest to Union Carbide.

### 3.5  CARBON MONOXIDE AND SYNTHESIS GAS

In recent years carbon monoxide has become a giant performer on the stage of industrial chemistry as witnessed by the production of about 70B lb of syngas derived from natural gas (2.2.1 and 3.2) or gasification of coal (2.2.3). The production of syngas in the United States is about double that of ethylene and of the same magnitude as that of sulfuric acid. Fortunately for its future prospects, there exist additional sources of carbon monoxide, as indicated below. The role of syngas as the primary source of methanol (3.3.1), acetic acid and its anhydride (3.3.2.4), gasoline by the MTG process (3.3.2.7.B), and its other potential contributions by way of methanol (3.3.2), were already mentioned as indicated.

### 3.5.1  Sources of Syngas

#### 3.5.1.1  Steam Reforming of Natural Gas and Petroleum

This source of syngas was discussed briefly in Section 3.2.2. One may add that oil shale, tar sands, bitumen, and high-molecular-weight, high-sulfur-containing sources of petroleum could also serve as sources of syngas when the appropriate technology is developed.

#### 3.5.1.2  Water and Producer Gas from Coke and/or Coal

The production of these "dirty" mixtures of CO, $H_2$, $CO_2$, and many impurities derived from the constituents of coal (*AUS*, 94–95) is now obsolete because of the availability and the distribution facilities of natural gas. In the "good old days," municipalities operated their own "city gas" production facilities and early-day-movies occasionally pictured suicides by exposure to gas used for cooking purposes.

#### 3.5.1.3  Gasification of Coal

To the preceding discussion of coal gasification (2.2.3) we may add here that the most common process (Lurgi) employs either a fixed or moving bed of crushed coal at about 1000°C. In the Texaco modification a slurry of coal is mixed with a stream of oxygen, and Exxon claims that a temperature of only 150°C suffices when some KOH catalyst is present. The Koppers–Totzek process uses pulverized coal and a stream of oxygen and steam at 1400–1600° at atmospheric pressure. Large, experimental coal gasification units were constructed in connection with the U.S. Synthetic Fuel Program as an aftermath of the petroleum embargo of 1973 but neglected when the immediate crisis subsided.

In Cologne, West Germany the gasification of 240 $t^m$/day of pulverized coal in the presence of oxygen and steam at atmospheric pressure is being tested using a molten-iron bath at 1400–1500°C. It is reported (*CE* 7/8/85) to convert 98% of coal and to produce 65–70% CO, 25–35% $H_2$, and less than 2% of $CO_2$.

Underground gasification of coal ("underground burn") will be used by Energy International of Pittsburgh at a site in Rawlins, Wyoming for the production of 142,000 t/yr of ammonia that will be converted into 153,000 t/yr of urea. The construction of the facility is to be completed in July 1989 (*CE* 10/12/87). The pioneering efforts in the Soviet Union during the 1950s did not turn out to be economical (*CEN* 10/13/80). For progress in underground coal gasification (UCG), see J. Palenik, "Bright Prospects for in situ Gasification," *CW* 12/23/87, pp. 34–35.

#### 3.5.1.4  Gasification of Biomass

Cellulosic waste materials, wood residues, and the like, can be gasified at high temperatures to give a mixture of CO–$H_2$, ethylene, acetylene, propylene, benzene, toluene, and other minor constituents [see, for example, Prakacs, Barclay, and Bhatia, *Pulp Pap. Mag. Can.*, **72**, T-199 (**1977**)]. So far, this source of syngas has not become an industrial reality, although the French Energy Conservation Agency (AFME) reports (*CW* 11/11/85, *CE* 2/17/86) that the gasification of biomass in the presence of oxygen and steam at 700–800°C and 30 bar, followed by a reforming treatment at 1300°C of the initially produced gas, gives a mixture of 39% CO, 31% $H_2$, and 27% $CO_2$. It is estimated that at a $25/$t^m$ cost of 45% water-containing biomass, a $90-million plant could produce 500 $t^m$/day of methanol at a cost of $230–$250/$t^m$ or 10¢–11.4¢/lb.

### 3.5.1.5   By-product of Metallurgical and Ammonia Processes

A hitherto unexploited source of CO may be the production of aluminum metal by means of the electrolysis of $AlCl_3$, which uses 30% less electricity than the traditional Hall–Heroult process. For this purpose bauxite is treated with chlorine and coke:

$$Al_2O_3 + 6C + Cl_2 \rightarrow 2AlCl_3 + 6CO$$

It is estimated that the production of elementary phosphorus by means of a reduction of $P_2O_5$ with coke could yield as much as 1MM t of 80% pure CO.

The production of steel in basic oxygen furnaces generates CO when oxygen is blown through a molten mass of scrap iron and lime, and the same is true in the case of electric furnaces used to produce alloy steel. The recovery of CO is apparently practiced in Japan. In 1987 Norway announced (*CW* 1/7–14/87) a $90-million project to produce 250MM lb/yr of acetic acid by 1989 based on imported methanol and CO recovered from the manufacture of pig iron.

The vent gases from ammonia reformer units contain, in addition to some $CH_4$, $N_2$, and $H_2$, also about 50% CO.

### 3.5.1.6   Separation, Purification, and the Shift Reaction of Syngas

Certain uses of syngas call for a hydrogen-enriched mixture in which case some of the initially formed CO is sacrificed in the water gas-shift reaction:

$$CO + H_2O \rightleftharpoons CO_2 + H_2$$

Studies of this important equilibrium are continuing [*JACS* **104**, 1444 (1982)]. At 298° K, the thermodynamic parameters are $\Delta G° = -6.82$ kcal/mol, $\Delta H° = 9.84$ kcal/mol, and $\Delta S° = -10.1$ eu. The catalyst $Fe_3O_4$–$Cr_2O_3$ promotes a rapid equilibrium at 350°C, and, while a CuO–ZnO catalyst functions well at lower temperatures, it is readily "poisoned" by sulfur impurities. A more recent basic catalyst contains $K_2CO_3$, CoO, and $MoO_3$ on alumina support and is more resistant to poisoning (C. L. Aldridge, U.S. Patent 3,580,840, 1974, assigned to Exxon). For the development of a homogeneous transition metal-based catalyst, see P. C. Ford, *Accts. Chem. Res.*, **14**, 31 (**1981**).

For nonfuel uses, the gaseous mixture obtained from the gasification of coal, or from the reforming of natural gas and petroleum, and even from naturally occurring gas deposits, usually requires the separation and purification of the CO–$H_2$ component. Apart of the recently developed energy-saving membrane separators, other methods may include

(a) Removal of $CO_2$ by absorption in mono- and diethanolamines promoted by $K_2CO_3$ by trapping in methanol (the Lurgi Rectisol process) or by means of solvent separation such as the Selexol process developed by Norton. The latter employs a dimethyl ether of poly(ethylene glycol) that allows the isolation of components of the complex mixture (3.7.1) found in the La Barge Anticline of southwestern Wyoming [J. J. H. Johnson and A. C. Homme, Jr., *Energy Progress* **4**, 241–248 (**1984**)].

(b) Fractional distillation of the $CH_4$–CO condensate since their boiling points ($-161.4$ and $-191.5°C$, respectively) are significantly different. Hydrogen is difficult to liquefy and is separated first.

(c) Selective absorption of CO in either a mixture of $Cu_2Cl_2$ and $AlCl_3$ to give a transient (Cu–Co) complex that is soluble in toluene (Tenneco Cosorb process) or in $Cu_2Cl_2$ and ammonia (Uhde process). The CO is subsequently removed by thermal decomposition of the complexes.

### 3.5.2  Chemical Versatility of Carbon Monoxide

In the presentation that follows, we shall start with the well-established, traditional uses of CO in the synthesis of organic chemicals and then proceed to the more recent applications, some of which are yet to be exploited on an industrial scale.

The chemical versatility if CO can be understood in terms of its electronic structure:

$$:C=\ddot{O}: \leftrightarrow \,\,^-:C\equiv O:^+$$

Both resonance structures represent unsatisfactory electron distributions: one lacks an octet of electrons, and in the other the polarization is contrary to the relative electronegativities of the two elements in question. These problems explain the good ligand, good base, and good electrophilic behavior of CO. Also, the electronic structure explains why CO is an easy victim of an attack by free radicals (symbolized here by Q·). All these behavior patterns of CO are illustrated in Scheme 3.11.

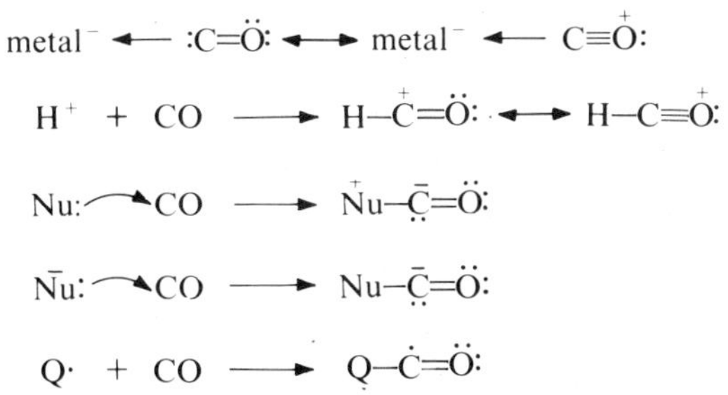

$$\text{metal}^- \leftarrow :C=\ddot{O}: \leftrightarrow \text{metal}^- \leftarrow C\equiv \overset{+}{O}:$$

$$H^+ + CO \longrightarrow H-\overset{+}{C}=\ddot{O}: \leftrightarrow H-C\equiv\overset{+}{O}:$$

$$Nu: \frown CO \longrightarrow \overset{+}{Nu}-\bar{C}=\ddot{O}:$$

$$\bar{Nu}: \frown CO \longrightarrow Nu-\bar{C}=\ddot{O}:$$

$$Q\cdot + CO \longrightarrow Q-\dot{C}=\ddot{O}:$$

Scheme 3.11

### 3.5.3  Established Uses of the CO Building Block

#### 3.5.3.1  Formic Acid, Esters, and Amides

In the presence of their conjugate acids, hydroxide and alkoxide ions add to CO to give the formate salts and esters, respectively. In actual practice, caustic soda or lime are treated with CO at 115–150°C and pressures up to 30 bar to give sodium formate (20¢) and calcium formate, respectively.

Because of the ease of air oxidation, rapid fusion of sodium formate (m.p. 256°C) and heating up to 400°C produces sodium oxalate (45¢). The latter is isolated by quenching the reaction mixture with lime water, isolation of insoluble calcium oxalate, and treating the latter with dilute sulfuric acid.

A Finnish process by Kemira Oy produces formic acid (90%, 36.5¢; 95%, 51.5¢) by way of a transitory formation of methyl formate (29¢) as shown in Scheme 3.12.

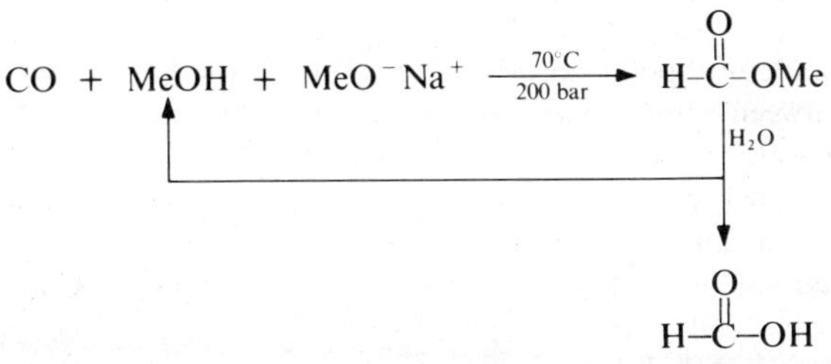

$$CO + MeOH + MeO^- Na^+ \xrightarrow[\text{200 bar}]{\text{70°C}} H-\overset{\displaystyle O}{\overset{\|}{C}}-OMe$$

$$\downarrow H_2O$$

$$H-\overset{\displaystyle O}{\overset{\|}{C}}-OH$$

Scheme 3.12

The industrial potential of formic acid is discussed in Section 3.6. The formation of *N*-methyl formamide and DMF is shown elsewhere in connection with the corresponding amines (3.3.2.6.A).

This is an excellent opportunity to compare the solvent behavior of the two last-mentioned formamides, as well as that of the nonsubstituted formamide, and briefly indicate some of the broader implications of these three compounds as solvents.

The molecular weight and approximate boiling and melting points are as follows:

|  | $H \cdot CO \cdot NH_2$ | $HCO \cdot NHMe$ | $HCO \cdot NMe_2$ |
|---|---|---|---|
| Molecular weight | 45 | 59 | 73 |
| Boiling point (°C) | 210 | 198 | 153 |
| Melting point (°C) | 2.6 | −40 | −61 |

We note that both the boiling and melting points decrease as methyl groups replace the nitrogen-attached hydrogens that participate in hydrogen-bonded intermolecular associations. These trends are not surprising because the effects of hydrogen bonding have been well recognized since the World War I era. On the other hand, it was only since World War II that solvents of the dipolar aprotic type have risen to prominence, and this topic is mentioned here because they play an important role in industrial chemistry.

*Dipolar Aprotic Solvents*

Dipolar aprotic solvents exhibit very strong intermolecular attractive forces due to the built-in dipole. In the case of DMF, the dipole is best understood if we examine one of the resonance structures:

$$Me_2N^+ = C - O^-$$
$$|$$
$$H$$

The strength of dipole–dipole attractions can be appreciated if we compare the boiling point of DMF with the boiling points (in Celsius) of other substances of similar molecular weights (shown in parentheses):

| | |
|---|---|
| *n*-Butyraldehyde (72) | 75 |
| Tetrahydrofuran (72) | 66 |
| Pentane (72) | 36 |
| *n*-Butylamine (73) | 77 |
| Ethyl formate (74) | 54 |
| Benzene (78) | 80 |
| Cyclohexane (84) | 81 |

The built-in dipole of dipolar aprotic solvents also gives rise to strong intermolecular associations between solutes that are attracted to either the positive and/or negative terminals of the solvent molecules. Molecules with double bonds are among those that are attracted to the positive terminals of dipolar aprotic solvents. Thus, for example, a mixture of benzene and cyclohexane (note that 1°C difference in boiling points) can be separated on addition of DMF or another member of the dipolar aprotic solvent family to be mentioned below and by subjecting the resulting solution to *extractive distillation*. The polarizable benzene molecules are retained by the dipolar aprotic solvent while cyclohex-ane is distilled, and eventually the associations between benzene and the dipolar aprotic solvent are broken apart at moderately elevated temperatures. In the same fashion, polarizable, conjugated double-bond-system-containing butadiene or isoprene can be separated from alkanes.

Furfuraldehyde (b.p. 162°C) is mentioned (2.2.2) as a solvent used to refine higher petroleum fractions for use as lubricating oils by the removal of unsaturated and aromatic impurities, and its relatively high boiling point and utility may be attributed to the dipole that is built in across the whole molecule because of the contribution of the following resonance structure:

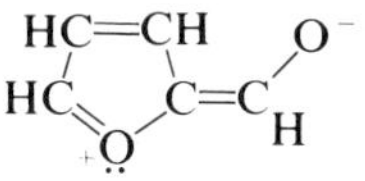

$$HC\!\!=\!\!CH \qquad O^- $$
$$HC\diagdown_{\underset{O}{}}C\!\!=\!\!C\diagup_{H}$$

A cluster of dipolar aprotic solvent molecules (symbolized by $\xrightarrow{+\ -}$ tends to surround metallic cations of ionic reactants and consequently the reactivity of the "naked" anions is greatly increased in nucleophilic substitution reactions:

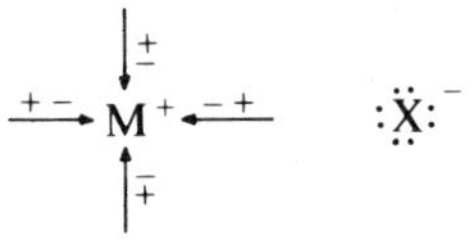

$$\xrightarrow{+\ -} M^+ \xleftarrow{-\ +} \qquad :\ddot{X}:^-$$

Some other members of the dipolar aprotic solvent family are enumerated as follows:

Dimethyl sulfoxide, DMSO, 78¢, b.p. 189°C

*N*-Methylpyrrolidone, NMP, $1.32, b.p. 202°C

Sulfolane, $(CH_2)_4SO_2$, Phillips $1.74 anhydrous, b.p. 285°C

Ethylene carbonate (3.7.2.5), b.p. 238–245°C

Propylene carbonate (3.7.2.5), 52.5¢, b.p. 240°C

Tetramethylurea (3.8.6), b.p. 176°C

Hexamethylphosphoramide, $(Me_2N)_3PO$, HMPA, b.p. 105–107°C (11 mmHg)

### 3.5.3.2 Formylation of Aromatic Systems

Under strongly acidic conditions, CO assumes the role of a formyl cation, $O\!\!=\!\!\overset{+}{C}\!-\!H$, and the latter reacts with an aromatic ring susceptible to electrophilic substitution to give the classical Gattermann–Koch reaction (ONR-35) that converts, for example, benzene to benzaldehyde, NF $1.25, technical grade 73¢. The reaction is catalyzed by chloroaluminic acid:

$$C_6H_6 + CO\ (+HAlCl_4) \rightarrow C_6H_5CH\!\!=\!\!O$$

The nonbonding electrons of the aldehyde form a complex with the acid; thus, a minimum of an equimolar quantity of $AlCl_3$ must be used to catalyze the reaction.

An interesting application of the Gattermann–Koch reaction is the Mitsubishi process for the production of terephthalic acid (9.2.2.2) from toluene by way of *p*-methyl-benzaldehyde that is oxidized subsequently to the desired dicarboxylic acid.

### 3.5.3.3  Production of Methanol

As discussed in Section 3.3 of this chapter, the production of methanol has expanded tremendously since the 1926 pioneering venture of Du Pont, and the prospects for further growth are excellent because of the assured supply of different sources of syngas.

### 3.5.3.4  Hydroformylation or the Oxo Reaction of Olefins

In simplest terms, this revolutionary process, discovered in 1938 by Otto Roelen and developed in Germany during World War II, is represented by

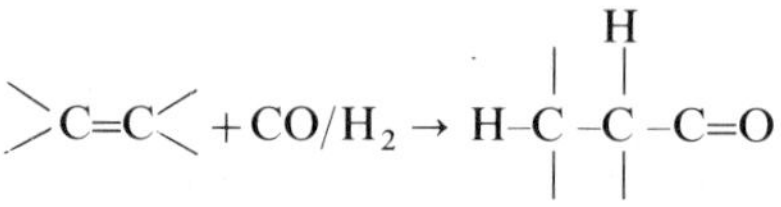

$$>C=C< \;+\; CO/H_2 \;\rightarrow\; H-\underset{|}{\overset{|}{C}}-\underset{|}{\overset{H}{\overset{|}{C}}}-C=O$$

The resulting carbonyl compound (*Oxo Verbindung* in German) can be hydrogenated *in situ* to the corresponding alcohol if one uses an excess of hydrogen or oxidized in a subsequent operation to the corresponding carboxylic acid. The one-carbon-higher alcohol, aldehyde, and carboxylic acid are considered here to be the first-generation oxo chemicals derived from the initial olefin.

The industrial importance of hydroformylation can be judged by the approximately 2B-lb annual production of oxo chemicals in the United States. The *first-generation* large-volume oxo products derived from the C$_2$ and C$_3$ olefins are as follows:

| | | |
|---|---|---|
| CH$_3$CH$_2$CHO | CH$_3$CH$_2$CH$_2$CHO | (CH$_3$)$_2$CHCHO |
| propionaldehyde | *n*-butyraldehyde | isobutyraldehyde |
| (35.5¢) | (750MM lb, 29.5¢) | (150MM lb, 35¢) |
| CH$_3$CH$_2$CH$_2$OH | CH$_3$CH$_2$CH$_2$CH$_2$OH | (CH$_3$)$_2$CHCH$_2$OH |
| *n*-propyl alcohol | *n*-butyl alcohol | isobutyl alcohol |
| (50¢) | (760MM lb, 38¢; also by fermentation) | 150MM lb, 37¢) |
| CH$_3$CH$_2$CO$_2$H | CH$_3$CH$_2$CH$_2$CO$_2$H | (CH$_3$)$_2$CHCO$_2$H |
| propionic acid | butyric acid | isobutyric acid |
| (33¢) | (44.5¢) | (150MM lb, 75¢) |
| (see 3.5.3.5) | | |

Some examples of *second-generation* oxo products derived from isobutyraldehyde are

- Isobutyl acetate (44¢).
- Isobutyl isobutyrate (through a Tishchenko reaction, ONR-90) (42.5¢).
- Isobutyl acrylate and methacrylate (80¢ and 87¢, respectively).
- Isobutyl phenylacetate ($3.10).
- Isobutyl salicylate ($3.45).

Similarly, there are various second- and higher-generation oxo chemicals derived from the other two families of compounds but they are mentioned elsewhere. Higher-molecular-weight oxo chemicals are derived from dimers, trimers, and tetramers of propylene; from oligomers of ethylene; and from other, readily accessible olefins. They serve as precursors to plasticizers, amines, surfactants, and so on.

The approximate distribution of the 2B-lb production of oxo alcohols is as follows:

50%: butyl alcohol and 2-ethylhexanol, 2-EH (6.5)
28%: iso-$C_{10}$ ($3xC_3 + C_1$), iso-$C_{13}$ ($4xC_3 + C_1$)
5%: $C_{11}$–$C_{13}$ ($5xC_2 + C_1$ and $6xC_2 + C_1$)
4%: n-propyl alcohol
13%: miscellaneous ($C_7 = 2xC_3 + C_1$; $C_9 = 2xC_4 + C_1$; $C_8 = C_3 + C_4 + C_1$; etc.)

*Homogeneous Catalysis by Transition-Metal Complexes*

The hydroformylation reaction is an excellent illustration of the role that transition-metal complexes play in homogeneous catalysis. The original hydroformylation catalyst was cobalt carbonyl, but it was later shown that the active species, hydride cobalt carbonyl, was formed by an initial hydrogenation occurring under the reaction conditions of 100–200°C and 200–450 atm:

$$Co_2(CO)_8 + H_2 \rightarrow 2HCo(CO)_4$$

Both preceding species are electronically saturated in the sense that all of the outermost $s$, $p$, and $d$ orbitals of Co (a $d^9$ metal) are filled by electrons of the metal itself and those contributed by the ligands (represented here by L). Thus, the next step in the mechanism is the activation of $HCoL_4$ by *extrusion* of one CO molecule:

$$HCoL_4 \rightarrow CO + HCoL_3$$

Now the electronically unsaturated complex accepts the $\pi$-electron pair of the olefin, and a concerted shift of the hydride and the rotation of the olefinic moiety produce a cobalt–carbon bond, and at the same time a new electronically unsaturated site is generated at the cobalt:

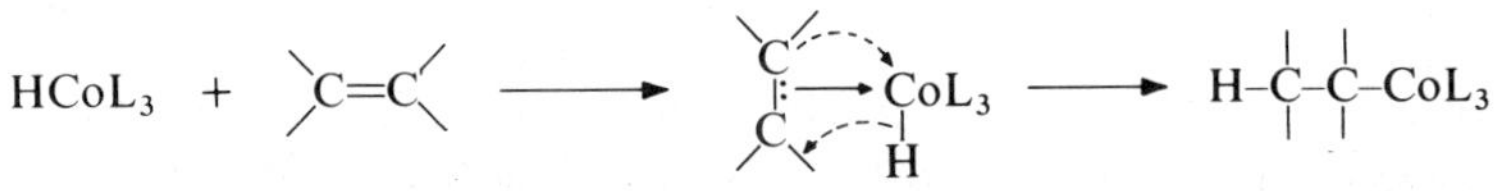

It is at the point of rotation that an unsymmetrical olefin forms the carbon–cobalt bond at either the less or at the more crowded carbon, and as one may well expect, the less crowded terminal of the original double bond is favored. As we shall see below, the reluctance to form a more crowded bond can be reinforced by means of ligands that are more bulky than CO. The newly electronically unsaturated site at cobalt becomes saturated when another CO molecule is coordinated to give a pentacovalent complex H–C–C–Co(CO)$L_3$, and now an *insertion* of a CO into the carbon–cobalt bond gives H–C–C–Co–Co$L_3$. The last step of the reaction, known as *reductive elimination*, consists of hydrogenation that regenerates the active species $HCoL_3$ and liberates a molecule of the hydroformylation product:

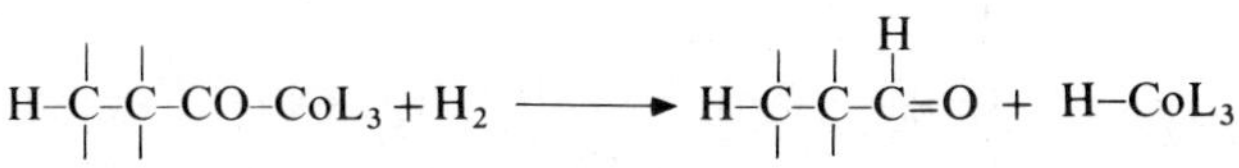

The use of Co$_2$(CO)$_8$ with propylene under the experimental conditions mentioned above produces a 4:1 n-butyraldehyde:isobutyraldehyde ratio. The replacement of one CO ligand by bulky triphenylphosphine (Ph$_3$P) or tri-n-butylphosphine ligands increases the preceding ratio to 8:1, but the reaction is also slowed down considerably. The development of the analogous rhodium catalyst by Union Carbide removes this handicap because of the relatively larger atomic radius of Rh compared to that of Co. Thus, for example, the n-butyraldehyde:isobutyraldehyde ratio is 30:1 when the triphenylphosphine or tribenzylamine ligands are present on the Rh catalyst, and, also, the reaction is 10$^3$–10$^4$ times faster at 80–120°C and 7–25 bar. These results explain why rhodium catalysts have been employed since 1975, and their advantages are more pronounced as the size of the olefinic reactant is increased. The latter point is illustrated in the case of the formylation of 1-octene:

| Catalyst and reaction conditions | n-:iso- Product Ratio |
|---|---|
| Co$_2$(CO)$_8$, 180°C, 350 atm | 2.5:1 |
| Rh$_2$(CO)$_8$, 130°C, 14 atm | 3:7 |
| Rh cat. with Ph$_3$P, 110°C, 7 atm | 16:1 |

Mechanical losses of costly rhodium are avoided by attaching the catalytic complex through an appropriate ligand to a cross-linked, insoluble polymer matrix (3.4.3.4). The milder reaction conditions of the Rh-catalyzed hydroformylations allow this reaction to be applied to olefins that contain other functional groups. Shell has developed cobalt carbonyl catalysts that contain trialkylphosphine ligands (that are more electron-donating than triarylphosphines), and these catalysts promote a direct reduction of the hydroformylation products to the corresponding alcohols.

### 3.5.3.5  Reppe Hydrocarbonylation

$$CH_2=CH_2 + CO/HOR \rightarrow CH_3-CH_2-CO \cdot OR$$

This process is similar to the preceding hydroformylation except that now the prefix "hydro" refers to water or to alcohols, amines, and other reactants that have an active hydrogen atom. The reaction occurs with olefins and acetylenes; it is catalyzed by Ni(CO)$_4$ and Fe(CO)$_5$ (catalysts that, incidentally, are inactive in hydroformylations) or carbonyls of Co, Ru, Pd, and so on. Acetylenes react more readily than olefins, and, in the case of substituted acetylenes, the addition occurs in a trans fashion. Hydrogen required to give some initial metallic hydride is believed to be generated by the water–gas-shift reaction (3.5.1.2), but after the initial formation of the acyl CH$_3$CH$_2$CO–metal complex, hydrogen is replaced when the complex reacts with the nucleophilic portion of the active hydrogen-containing reactant:

$$\begin{bmatrix} CH_3CH_2CO\text{–metal} \\ RO\text{–H} \end{bmatrix} \rightarrow CH_3CH_2CO + H\text{–metal} \\ \qquad\qquad\qquad RO$$

Some examples of the Reppe hydrocarbonylation follow:

$$CH_2=CH_2 + CO/H_2O \xrightarrow{\;Ni(CO)_4\;} CH_3CH_2CO_2H$$

[propionic acid (33¢) is prepared by this process rather than by hydroformylation (3.5.3.4)]

$$RCH{=}CH_2 + CO/HCl \xrightarrow[\substack{PdCl_2(PPh_3)_2 \ SnCl_2 \\ \text{U.S. Patent 3,880,898 (Texaco)}}]{} RCH_2CH_2CO \cdot Cl$$

oleic acid esters $+ CO/H_2O \rightarrow$ 9-, 10-carboxystearic acid (at 110–140°C and about 300 bar (*J. Am. Oil Chem. Soc.* **50**, 210, 455 (1973))

$$HC{\equiv}CH + CO/H_2O \xrightarrow[\substack{Ni(CO)_4 \\ 35\text{–}80\,^\circ C,\ 1\ atm}]{} CH_2{=}CH{-}CO_2H$$
(4.7)    acrylic acid, 60¢ (4.7.4)

$$HC{\equiv}CH + CO/ROH \xrightarrow[Ni(CO)_4]{} CH_2{=}CH{-}CO \cdot OR$$
acrylates

$$Cl{-}CH_2C{\equiv}CH + CO/H_2O \xrightarrow[Ni(CO)_4]{} HO_2CCH_2C({=}CH_2)CO_2H$$
propargyl chloride    itaconic acid, $1.45
(4.7.2)    U.S. Patent 3,025,320 (Montecatini)

In the last-mentioned example, the Reppe hydrocarbonylation at one end of the molecule is accompanied by a caboxylation of an organometallic reactant [such as occurs in a Grignard reagent (3.7)]. As interesting as this example may be, it is noteworthy that itaconic acid and its anhydride are obtained in a more economic fashion by the thermal decomposition of citric acid (89.5¢) or by fermentation (*SZM-II*, 42).

Currently, the major industrial use of the Reppe hydrocarbonylation reaction is the production of propionic and acrylic acids and the esters and substituted amides of the latter. The unsubstituted acrylamide (solution 100% basis, 73¢) is obtained in a more economic manner from acrylonitrile (36¢).

### 3.5.3.6    Production of Phosgene and an Overview of Its Uses

Phosgene is produced by means of a typical free-radical chain reaction between chlorine and carbon monoxide:

Initiation step:      $Cl_2 \rightarrow 2Cl\cdot$

Propagation steps:   $Cl\cdot + CO \rightarrow Cl{-}\dot{C}O$

                     $Cl{-}\dot{C}O + Cl_2 \rightarrow Cl{-}CO{-}Cl + Cl\cdot$

Termination steps:   $2Cl\cdot \rightarrow Cl_2$

                     $Cl\cdot + Cl{-}CO\cdot \rightarrow Cl{-}CO{-}Cl$

Minor by-product:    $2Cl{-}CO\cdot \rightarrow Cl{-}CO{-}CO{-}Cl$
                     oxalyl chloride

While phosgene was discovered and was given its name because of the reaction being catalyzed by sunlight, it is currently manufactured at a minium temperature of 50°C, at a pressure of 5–10 atm, and in the presence of activated carbon. Phosgene was used by Germany during World War I as a war gas and caused numerous casualties. The toxicity of phosgene is an incentive to replace its synthetic applications by less noxious chemicals.

The Bhopal, India disaster in December 1984 resulted from a massive release of methyl isocyanate (MIC) and the latter happened to be prepared from methylamine and phosgene. This tragic event provided an additional stimulus to avoid the use of phosgene and the production and transportation of MIC and to arrive at the desired carbamates methyl carbamates [MeNH–C(=O)–R] by alternate means. The issues of guilt, the specific cause of the release of MIC, indemnities for the families of casualties and surviving victims, the degree of responsibility of Union Carbide (which owns 51% of the Indian subsdiary), and related issues remain undecided before the courts at the time of this writing. In any case, the outcome of these issues may have a profound effect on future capital investments in developing countries by chemical companies of the industrialized world.

The capacity of the United States to manufacture phosgene stands at about 2B lb, and its unit price is 55¢. The gas is packaged in 1-t returnable containers that cost about $3000 each. Most of the demand of about 1.6B lb is consumed in the production of PUR followed by the production of polycarbonates (PC), and for this reason the nature of these two classes of polymers is discussed in some detail in Section 3.8.

The fluorine analog of phosgene, carbonyl fluoride, $O=CF_2$, is formed from CO and elementary fluorine.

### 3.5.3.7 Carbonylation Reactions

The almost inseparable consideration of CO and its immediate hydrogen-containing descendants—formaldehyde and methanol, as building blocks of organic industrial chemicals, is responsible for the mention of some carbonylation reactions in the preceding pages. Here, we amplify those comments and introduce additional building block applications of CO.

**A. HYDROXYACETIC OR GLYCOLIC ACID.** A typical carbonylation reaction of formaldehyde is responsible for the production of glycolic or hydroxyacetic acid (70%, 49.5¢; 100%, 70.7¢):

$$H_2C=O + :C=O: \rightarrow (^-OCH_2-C^+=O) \xrightarrow[H_2O]{} HOCH_2-CO_2H$$

Glycolic acid is used in tanning operations, processing of textiles, brightening of metals preceding and during electroplating, and so on.

There exist obvious alternative routes for the production of glycolic acid. One (mentioned in 3.4.3.7) involves a two-step process and relatively costly hydrogen cyanide (50¢), or even more costly sodium cyanide (77¢). Another alternative is the alkaline hydrolysis of chloroacetic acid (56¢); however, this route is not economical if we calculate the cost of the starting material on a molar basis since much of the mass becomes sodium chloride worth about 3¢. This potential process illustrates the previously postulated rule (3.3.2.6.A) that chlorine-containing intermediates should be avoided unless chlorine is part of the final product. Finally, one could also consider an oxidation of ethylene glycol (31¢) with, say, dilute nitric acid, (∼ 10¢), but such a process would also lead to glyoxylic and oxalic acids, $H·CO·CO_2H$ and $HO_2C–CO_2H$, respectively, as well as glyoxal, $H·CO–CO·H$, and is difficult to control. Clearly, the carbonylation of aqueous formaldehyde is the best industrial route to glycolic acid.

**B. FORMAMIDE AND HOMOLOGS.** Formamide and the methyl-substituted deriva-tives, as well as the general topic of dipolar aprotic solvents, are mentioned in Section 3.5.3.1.

The United States and world production capacity of DMF is 120MM and 400MM lb, respectively, and most of it is used for solution spinning of acrylic fibers and as a solvent for acetylene (stored as a solution in tanks) and for vinyl resins. Minor quantities of DMF are consumed in the synthesis of fine chemicals. One such synthetic application is the introduction of formyl groups onto aromatic systems susceptible to electrophilic attack (3.6.2) by means of the Vilsmeier–Haak reaction (ONR-92) in which phosphorus oxychloride (40¢), transforms DMF into an electrophilic reagent:

$$Me_2NCH=O \rightleftharpoons Me_2\overset{+}{N}=CH-O^- + OPCl_3 \rightarrow Me_2\overset{+}{N}=CHCl + {}^-O_2P(OH)_2$$

**C. CARBONYLATION OF OLEFINS.** Olefins capable of forming relatively stable car-bocations react with CO in the presence of aqueous acids to give the family of *neoacids*. This name is derived from neopentanoic or pivalic acid (93¢) obtained in the case of isobutylene:

$$(CH_3)_2C=CH_2 + CO/H_2O \rightarrow (CH_3)_3C-CO_2H$$

The name Koch acids and the term "carboxylation" are also used because this reaction is reminescent of

(a) The Koch–Haaf carboxylation (ONR-51) in which a carbocation generated from an alcohol is converted to a carboxylic acid in a manner similar to the Reppe hydrocarbonylation of nonbranched olefins (3.5.3.4).

(b) The Kolbe synthesis of phenolic acids by the reaction of phenolates with carbon dioxide (3.7.2.2 and 9.3.4).

A higher tertiary carboxylic acid such as neodecanoic acid (50¢) is derived from the adduct of isobutylene and isoamylene; as shown in Scheme 3.13.

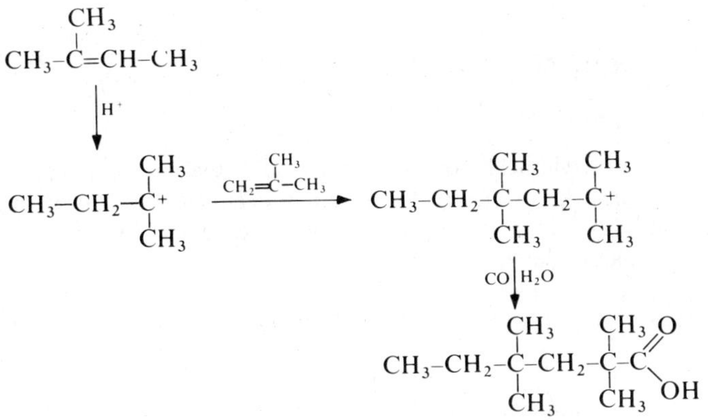

**Scheme 3.13**

The branching at the alpha position of neocarboxylic acids increases their solubility in hydroxylic solvents relative to isomeric nonbranched acids. While the latter tend to exist as hydrogen-bonded dimers

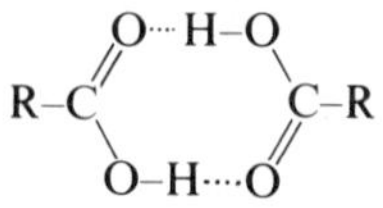

$$
\begin{array}{c}
\quad\quad O\cdots H-O \\
R-C \quad\quad\quad\quad C-R \\
\quad\quad O-H\cdots O
\end{array}
$$

steric inhibition in neoacids causes them to exist preferrably as monomeric species that form hydrogen bonds with solvents capable of this type of association. The same phenomenon also explains the greater volatility of neoacids as compared with that of their linear isomers; for example, the boiling points of pivalic and the nonbranched valeric acids are 164 and 186°C, respectively. The esterification of neoacids is somewhat more difficult than that of the corresponding nonbranched isomers, but once the ester is formed, their resistance to hydrolysis is of practical significance when they are employed as lubricants and other products exposed to heat and atmospheric moisture.

The carbonylation of ethylene at 150–300°C and 200–500 atm or under similar conditions in the presence of some hydrogen and cobalt oleate, yields diethyl ketone (DEK):

$$2CH_2{=}CH_2 + CO/H_2 \rightarrow CH_3{-}CH_2{-}CO{-}CH_2{-}CH_3$$

**D. BISCARBONYLATION OF ALKYL HALIDES TO α-KETO ACIDS.** The biscarbonylation of alkyl halides to α-keto carboxylic acids by means of a propietary catalyst was advertized recently by Ethyl Corporation:

$$R{-}Br + 2CO/H_2O \rightarrow R{-}CO{\cdot}CO_2H$$

Similarly, the following transformations are possible:

$$Cl(CH_2)_nBr \rightarrow (CH_2)_{n-1}CH{-}CO{\cdot}CO_2H \quad\quad \text{alicyclic product}$$

$$Br(CH_2)_7Br \rightarrow HO_2C{-}CO{-}(CH_2)_7{-}CO{-}CO_2H$$

$$C_6H_5{-}CH_2CH_2{-}Br \rightarrow C_6H_5{-}CH_2CH_2CO{\cdot}CO_2H$$
  phenethyl bromide          2-keto-4-phenylbutyric acid

The 2-keto carboxylic acids are readily converted to the corresponding 2-hydroxy and 2-aminocarboxylic acids by means of hydrogenation either without or in the presence of ammonia and with the aid of rhodium or nickel catalysts, respectively, and can be subjected to reductive dimerization to produce

$$
\begin{array}{c}
RCH_2CH(OH)CO_2H \\
| \\
RCH_2CH(OH)CO_2H
\end{array}
$$

in the presence of TiCl₃ and acetic acid. Also, as is the case with other keto carboxylic acids, they are converted to a variety of heterocyclic structures (Chapter 9).

**E. ADIPIC ACID BY CARBONYLATION OF 1,3-BUTADIENE.** In view of a perduring growth in the demand of nylon 6/6, efforts continue to find additional sources for adipic acid (8.3). BASF reported (*CE* 3/4/85) the design of a 130MM-lb/yr plant at Ludwigshafen, West Germany to produce dimethyl adipate from 1,3-butadiene, CO, and methanol [F. J. Weller, *J. Molec. Catal.* **31**, 128 (1985); *CW* 7/4/84]:

$$CH_2=CH-CH=CH_2 + CO-MeOH \rightarrow MeO_2C-CH_2-CH=CH-CH_2-CO_2Me$$

The ester is subjected to hydrogenation to remove the residual unsaturation and is then hydrolyzed in order to recycle methanol and to isolate adipic acid (57¢).

### 3.5.3.8   The Fischer–Tropsch (FT) and Related Processes

This topic is included among the "established uses of CO" because a modification of the original FT process lives on in Sasol, South Africa (F. Haggin, *CEN* 10/26/81; M. E. Dry, *HP* 8/82). The first Sasol unit delivers 250MM $t^m$ of hydrocarbons composed of gasoline, diesel fuel, and higher parafinic compounds. The second (Sasol II) unit is designed as an integrated complex that produces 1.5MM $t^m$ of hydrocarbon fuels, 150MM and 100MM $t^m$ of ethylene and propylene, respectively, and 200MM $t^m$ of coal-tar products (*CW* 12/4/79). The consumption of 30.3MM t of subbituminous coal per year produces 50,000 bbl of gasoline and diesel fuel per day. The coal is gasified in a battery of 30 Lurgi Mark IV converters and consumes 31,000 t of coal, 9000 t of oxygen, and 31,000 t of steam daily. A large amount of electricity is required to power the air and oxygen compressors. The cost of the hydrocarbon products is estimated at $70–$90/bbl, and eventually the project is to be expanded to produce 4MM t of gasoline per year.

The original FT process (German Patent 484,337, 1925) called for a heterogeneous catalyst of Co, Th, and Mg oxides deposited on kieselguhr. The process can be visualized to depend on the generation of methylene radicals or carbenes, $\cdot CH_2 \cdot$ or $CH_2:$, which combine to give mostly linear olefins, and eventually become hydrogenated to the corresponding alkanes. Thus, the primary FT products must be subjected to conventional petroleum refining operations in order to improve the quality of gasoline.

Some attempts to improve the FT technology began during the Arab-imposed petroleum embargo.

The Dow-modified FT process uses an alkali-containing molybdenum catalyst on a carbon support and is claimed to give a 70% yield of $C_2$–$C_5$ hydrocarbons that are either reformed in the presence of a H-ZSM-5 catalyst to give BTX, or they are cracked to olefins.

The "isosynthesis process" uses a $ThO_2$-based catalyst with Ce and Zr oxide additives and temperatures up to 450°C, and produces isobutane and other branched alkanes as well as naphthenes and aromatic hydrocarbons. This process is advantageous because $ThO_2$ is not readily poisoned by sulfur-containing impurities and it leads directly to high-octane gasoline.

### 3.5.4   Prospective Uses of Carbon Monoxide

The same introductory statement presented in Section 3.5.3.7 applies here, except for the additional proviso that it is difficult to predict when an industrial exploration becomes an industrial reality. It is clear that a direct conversion of syngas to materials for which there exists a demand is more expedient than the use of one of the syngas descendants such as formaldehyde, methanol, or the methylamines. However, the success of such transformations depends on the *in situ* formation of the desired products without complex by-products. The following examples were reported during the Chemicals from Syngas and Methanol Symposium conference sponsored by the Division of Petroleum Chemistry, American Chemical Society, held in New York City on April 13–18, 1986:

- J. A. Marsella and G. P. Pez of Air Products described the ruthenium-catalyzed formation of formamide, $N$-methylformamide (NMF), and trimethylamine in sulfolane at 230°C and about 272 atm of $CO-H_2-NH_3$ pressure.

- J. F. Knifton of Texaco described the direct formation of acetic and propionic acids and the corresponding methyl, ethyl, and propyl esters from $CO-H_2$ in the presence of a ruthenium/cobalt halide catalyst dispersed in tetrabutylphosphonium bromide (m.p. 100°C).

- D. J. Elliott and F. Pennella of Phillips described the formation of a mixture of oxygen-containing products in the presence of a $Cu-ZnO-Al_2O_3$ catalyst at 285° and 65 atm of $CO-H_2$ (2:1 ratio). The resulting mixture was composed mostly of methanol (76.6%) but also contained significant amounts of ethanol (7%), isobutyl alcohol (3.9%), n-propyl alcohol (3.7%), and methyl acetate (2.2%). Minor quantities (0.21–1.38%) of methyl formate, methyl propionate, and $C_4-C_8$ ketones were also present. While such a complex mixture may not be a practical source for any individual chemical, it can serve as an oxygen-containing gasoline additive.

- B. Whyman of ICI described the formation of $C_2$ oxygen-containing products such as methyl and ethylene glycol acetates when a 1:1 ratio of $CO-H_2$ was allowed to react in the presence of Ru–Rh catalysts at 230°C and 500–1000 atm.

- H. S. Kesling, Jr. of ARCO described an oxidative carbonylation of 1,3-butadiene that gives adipic acid by way of its dimethyl ester (see 3.5.4 above) and its homolog obtained from the $C_5$ monocarbonylation by-product subjected to reductive dimerization in the presence of $Co_2(CO)_8$ catalyst to eventually yield decanedioic or sebacic acid ($1.94). Also, the $C_5$ by-product reacts with additional butadiene in the presence of $H_2-Pd°$ to give, on hydrogenation, nonanoic or pelargonic acid (70¢). It is estimated that for every 30MM lb of adipic acid, one obtains 15MM and 12MM lb of pelargonic and sebacic acid, respectively. The traditional source of sebacic acid is castor oil (*SZM-II*, 72), and it is employed in the manufacture of nylon 6/10. Esters of pelargonic acid are used as plasticizers, hydraulic and other functional fluids, and the main source of the acid is the ozonolysis of oleic acid (*SZM-II*, 70). The complex series of transformations is summarized in Scheme 3.14.

The direct conversion of aromatic nitro compounds to the corresponding isocyanates is of great potential interest to the PUR industry because it avoids the production of intermediate amines and also avoids the use of phosgene. There is little evidence that this process is being used by industry although the conversion has been described in the literature:

$$Ar-NO_2 + 3CO \rightarrow Ar-NCO + 2CO_2$$

W. W. Pritchard (U.S. Patent 3,576,836, 1971) uses $PdCl_2$ catalyst at 200°C and 700 atm. On the other hand, Alessio, Vinci, and Mestroni [*J. Molec. Catal.* **22**, 327 (1984)] use $Rh_6(CO)_{16}$–phenanthroline catalyst at 165°C in the presence of water and the reaction

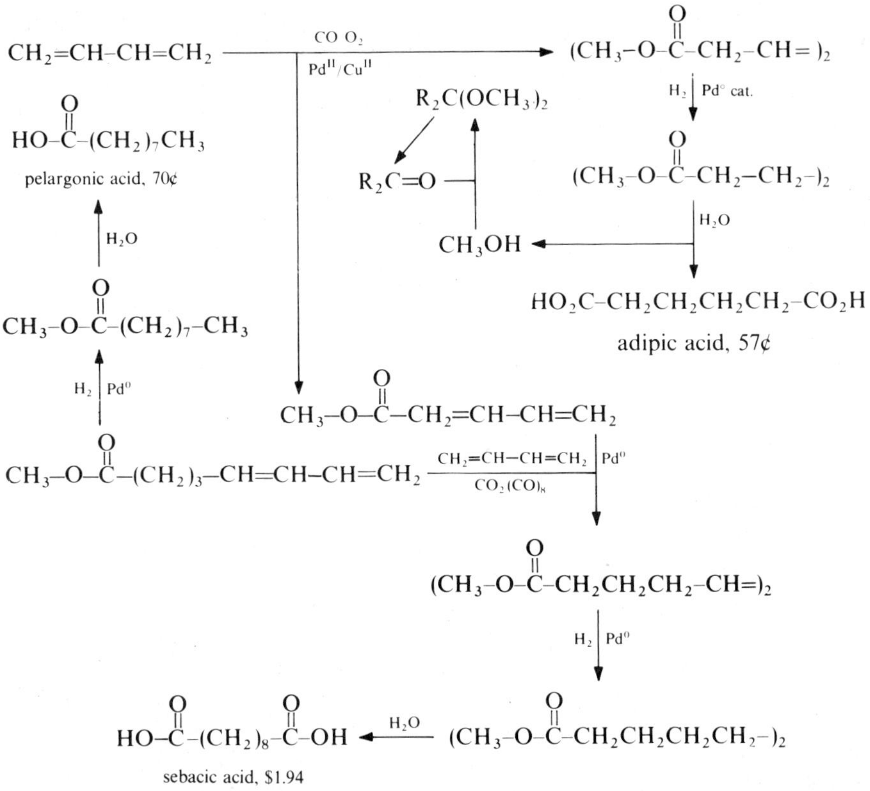

(where $R_2C{=}O$ = cyclohexanone)

**Scheme 3.14**

stops at the intermediate stage:

$$Ar-NO_2 + 3CO + H_2O \rightarrow Ar-NH_2 + CO_2$$

Similarly, carbonylation in the presence of an alcohol, ROH, at 177°C and 1000 bar also provides an intermediate carbamate ester that is thermally decomposed to a isocyanate (CT 6/78, 782; CE 3/4/85, 79):

$$Ar-NO_2 + CO + ROH \rightarrow Ar-NH-CO \cdot OR + CO_2$$

$$Ar \cdot NH-CO-OR \rightarrow Ar-NCO + ROH$$

Additional comments concerned with the production of isocyanates are presented elsewhere (3.8).

"Methanation of syngas" refers to the formation of "synthetic, or substitute, natural gas" (SNG):

$$CO + 3H_2 \rightarrow CH_4 + H_2O + 9 \text{ kcal/mol}$$

This transformation occurs at about 375°C and is catalyzed by Ce–Mo–Al catalysts. Progress in SNG technology that lowers the temperature requirements is reported in *CEN* 4/5/82. The SNG process is favored thermodynamically but currently is not economical since natural gas is abundant and inexpensive. An announcement from the University of Kyoto (*CEN* 10/13/80) that a lanthanide catalyst is capable of co-methanation and CO and $CO_2$ is of interest because it could be applied to the gas mixture without separation of the two constituents. Somewhat along similar lines, the COthane process announced by Union Carbide (*CEN* 8/24/81) is a disproportionation of CO to $CO_2$ and active carbon, and the latter plus steam then give methane and $CO_2$. These reactions occur at 270–300°C in the presence of Ni or Co catalysts. Although 79% of the chemical energy of CO is stored in methane, methanation and comethanation are potential means for reducing pollution caused by the operation of blast furnaces, petroleum refineries, carbon-black production facilities, and so on.

## 3.6 FORMIC ACID

### 3.6.1 Production of Formic Acid and an Overview of Its Uses

The current demand for formic acid in the United States is only about 60MM lb/yr, but the potential uses exceed this amount manifold. Until recently, formic acid was obtained as a coproduct of the oxidation of the butanes together with various other chemicals (6.2.2), and it is also recovered from the aldol–Cannizzaro reactions of formaldehyde (3.4.3.3). Faith in the future of formic acid must be shared by BASF that announced in 1986 the inauguration of a 220MM-lb production facility for the "on-purpose" (intentional) manufacture of formic acid in "the largest, most advanced plant in the world." About the same time, the China National Chemicals Import and Export Corporation of the P.R. of China announced its offer to supply formic acid worldwide. Also, a few years ago, the formic acid process of Kemira Oy of Finland (3.5.3.1) was being installed in the United States by the Leonard Process Company (*CMR* 2/25/80), accompanied by the optimistic prediction (*CMR* 4/5/82, *CE* 7/12/82) that the unit cost of formic acid would drop from the then prevalent level of 30¢/lb to about 5–6¢/lb and thus become competitive with mineral acids such as hydrochloic or sulfuric in the treatment of steel, in the pulping of wood, and so on. The September 1987 price of formic acid was 51.5¢ for 95% grade and 36.5¢ for 90% grade, suggesting that the previously expressed optimism was somewhat premature. At these prices the growth of demand is stymied, and the awakening of the formic acid potential must await massive uses through revolutionary changes of the wood pulping industry. However, the gigantic operations of this tradition-bound industry are not readily modified, and the revolutionary changes implied here are more likely to occur in newly installed pulping operations located in developing countries that are rich in forestry resources and have acquired a technological foothold.

The traditional consumption of formic acid depends on its use in the processing of textiles and leather (~35%) as an acidulant (30%) when circumstances call for a volatile (b.p. 100°C) acid that is 10 times stronger than acetic but weaker than hydrochloric acid and as a building block of fine chemicals (35%). In Europe formic acid is also employed for the preservation of silage.

### 3.6.2 Formic Acid and Derivatives as Building Blocks of Industrial Chemicals

Formic acid is a source of the formyl, –CHO, and the methine, $H\overset{|}{C}=$, moieties, and it also functions as a donor of hydride ions, especially when it exists as a formate ion. The latter property accounts for the reductive role of formic acid in the Leuckart–Wallach (ONR-55) and the Eschweiler–Clarke (ONR-28) reactions represented here by the following general transformations:

$$R-CO-R' + H-NR_2 + HCO_2H \rightarrow RR'CH-NR_2 + CO_2 + H_2O$$

and

$$RNH_2 + 2CH_2O + 2HCO_2H \rightarrow R-N(CH_3)_2 + 2CO_2 + 2H_2O$$

respectively. These mild alkylation reactions of amines are useful in the synthesis of fine chemicals, particularly pharmaceuticals that account for 50% of the synthetic uses of formic acid.

The formylation of an aromatic system uses DMF as the carrier of the formyl moiety in the Vilsmeier–Haack reaction (ONR-92; 3.5.3.7.B) and is illustrated by the conversion of N,N-dimethylaniline to p-dimethylaminobenzaldehyde (9.3.5).

The introduction of the formyl group into an aliphatic system is achieved by the application of the classical Claisen condensation reaction (ONR-18) using formic esters. Thus, for example, methyl formate (29¢) is condensed with methyl acetate in the presence of sodium methylate to give methyl formylacetate:

$$H-CO-OCH_3 + CH_3CO_2CH_3 + CH_3ONa \rightarrow H-CO-CH_2CO_2CH_3 + CH_3OH$$

Two examples of the synthetic utility of the last-mentioned Claisen condensation product are shown as follows.

The first example is the preparation of the pyrimidine nucleus and its incorporation into sulfadiazine, as shown in Scheme 3.15.

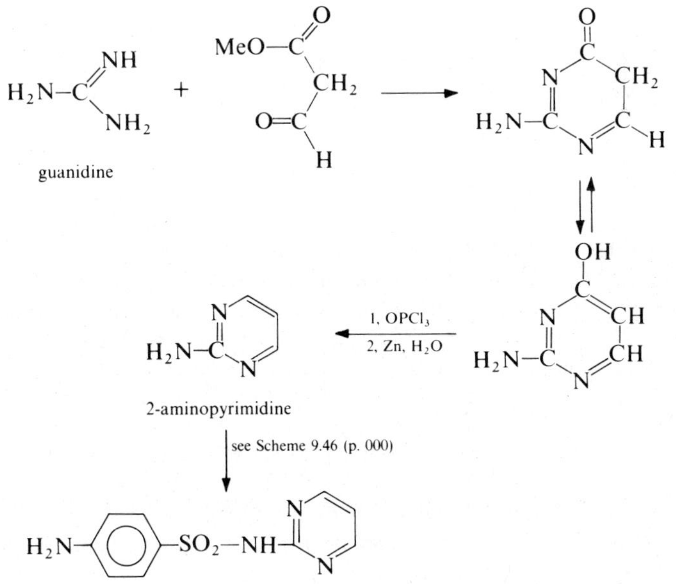

**Scheme 3.15**

The acidity of the sulfonamide group is sufficiently high to permit the formation of sulfadiazine salts: sulfadiazine sodium (USP $18.50) is a water-soluble antibacterial, and the corresponding silver salt is an antibacterial ingredient of creams used to treat burns (Silvadene, retail cost $25.50/lb). The second example is the preparation of 2-thiouracil, a medicinal employed in the treatment of hyperthyroidism, as shown in Scheme 3.16.

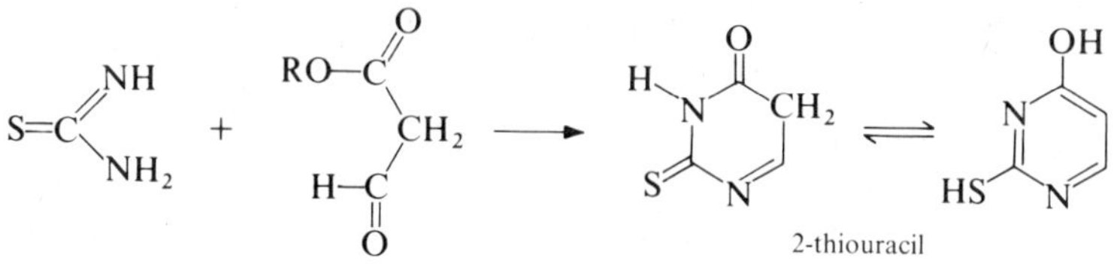

Scheme 3.16

The Claisen condensation of a formic ester with the ester of propionic acid provides the 2-formylpropionic building block for the assembly of thymine, one of the four "bases" of the nucleic acids, as shown in Scheme 3.17.

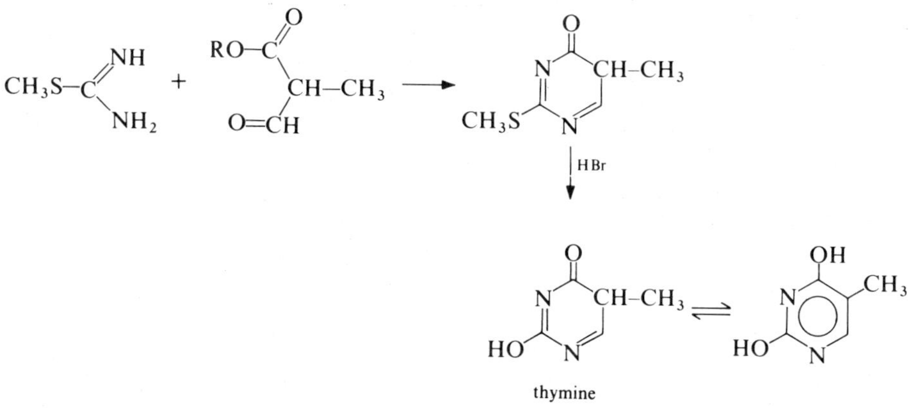

Scheme 3.17

For use as a research chemical thymine costs about $200/lb.

Another means for the introduction of a formyl group into an aliphatic system employs N,N-diphenylformamide, a convenient carrier of the formyl group because of the formation of the relatively inert diphenylamine "leaving group." The reaction is catalyzed by alkoxides that generate the rather reactive carbanion intermediate from the methyl ketone:

$$R-CO-CH_3 + HCO\cdot NPh_2 \rightarrow R-CO-CH_2-CHO + HNPh_2$$

The ring closure to give numerous five- and six-membered heterocyclic systems is achieved when formic acid inserts a methine moiety between two appropriately located functional groups, as shown in Scheme 3.18.

The heating of alkali formates above 400°C gives the corresponding salts of oxalic acid (Goldschmidt process, ONR-36) with a liberation of hydrogen. A low-temperature (30–70°C), palladium-catalyzed process based on this transformation of formates has recently been suggested (CW 4/16/86, CE 7/21/86) as a readily transported source of

o-phenylenediamine + H–C(=O)OH → benzimidazole

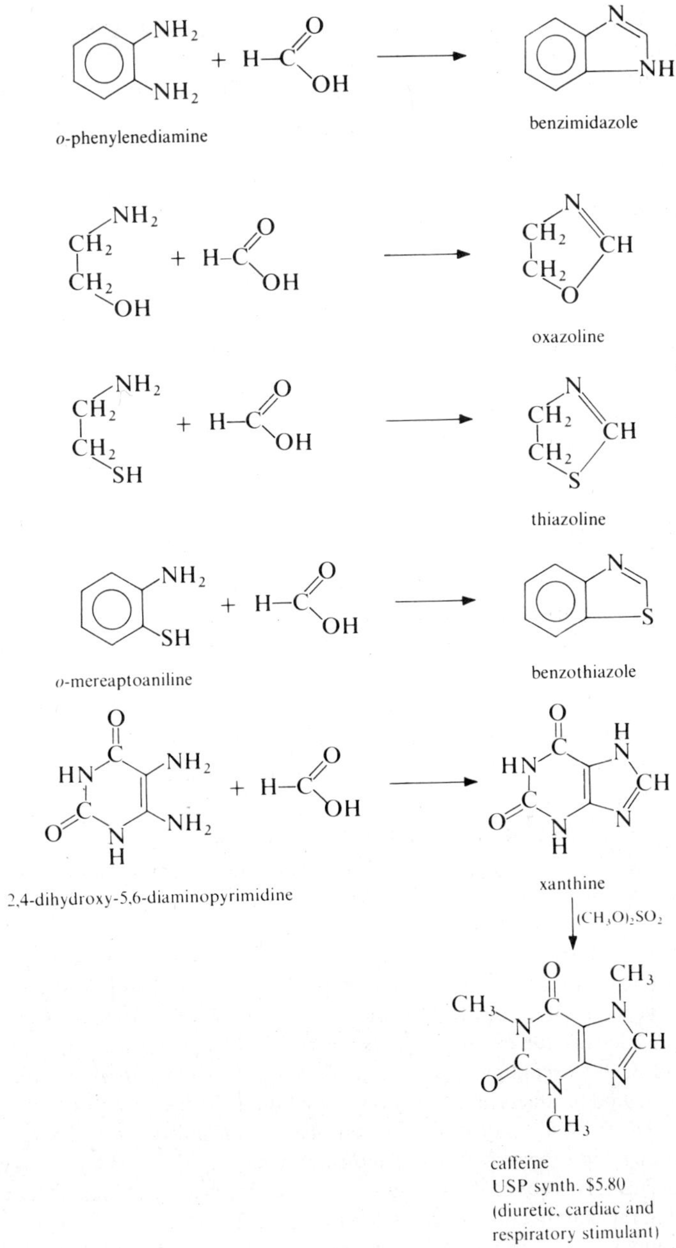

+ H–C(=O)OH → oxazoline

+ H–C(=O)OH → thiazoline

o-mereaptoaniline + H–C(=O)OH → benzothiazole

2,4-dihydroxy-5,6-diaminopyrimidine + H–C(=O)OH → xanthine

$(CH_3O)_2SO_2$

caffeine
USP synth. $5.80
(diuretic, cardiac and
respiratory stimulant)

**Scheme 3.18**

hydrogen:

$$2HCO_2Na \xrightarrow{NaOH} H_2 + NaHCO_3$$

$\longrightarrow CO_2$

The reaction is catalyzed by palladium and carried out in aqueous solution at 30–70°C and 0.5–5 atm.

Formamide could serve as a source of HCN (3.10), but such a process is not favorable economically.

## 3.7   CARBON DIOXIDE AND ITS SULFUR-CONTAINING ANALOGS

### 3.7.1   Sources and an Overview of the Uses of Carbon Dioxide

The wide-scale utilization of $CO_2$ as a building block of industrial organic chemicals continues to be a challenge of the century in view of its readily available sources:

1. The water–gas-shift reaction (3.5.1.6).
2. Naturally occurring $CO_2$ from gas wells.

An example of a thus far, untapped gas deposit is found in southwest Wyoming at the La Barge Anticline, estimated to contain 35T scf of a mixture of 65% $CO_2$, 22% $CH_4$, 8% $N_2$, 4.5% $H_2S$, and 0.5% He. The exploitation of the discovery of such large gas reserve in 1962 at a 15,400-ft depth had to await a satisfactory separation technology (3.5.1.6). When fully exploited, this particular Wyoming gas deposit is expected to yield 25T scf of $CO_2$, 7T scf of methane, 3T scf of nitrogen, 60–200B scf of helium, and 50MM t$^m$ of sulfur.

3. Inudstrial fermentations (*SZM-II*, 33–51).
4. Acidification of soda ash (found as the mineral trona, Kaplan *CE* 11/29/82, 30–31).
5. Stack or flue gases from combustion of carbonaceous fuels (*CMR* 9/8/82; *CEN* 10/11, 12/6/82).
6. Calcining of limestone and production of quicklime.
7. Smelter operations, recovery from the manufacture of certain chemicals such as the oxidation of ethylene to EO (4.3.5), the production of ammonia, and so on.

Currently, sources 1 and 2 account for about 60 and 20%, respectively, of the production of 8.5B and 9B lb of liquid and solid $CO_2$, respectively. Liquid $CO_2$ is priced at about 3.5¢, but the price depends on transportation and storage costs. Carbon dioxide from gas wells is usually used locally for enhanced-oil-recovery (EOR) operations. The approximate utilization of $CO_2$ (excluding EOR) is as follows: 45% for freezing, chilling, and refrigeration of food products (in competition with liquid nitrogen) by means of liquid and solid (dry ice) $CO_2$; 5% for industrial refrigeration; 25% for carbonation of beverages; and 15% for welding and other metal-manufacturing operations that require an oxygen-free atmosphere. This leaves about 15% for production of chemicals, and most of that amount is consumed in the manufacture of about 15B lb of urea (3.9). A relatively small amount of $CO_2$ is consumed in the carboxylation of phenols to salicylic and naphthoic acids (9.3.4).

*Supercritical Fluids*

Two more recent uses of $CO_2$ are extraction operations by means of supercritical fluids (SCFs), among which $CO_2$ plays an important role (E. J. Shimshik, *CT* 6/83, 374–375). Thus, extraction of vegetable oils from oil seeds, such as soybeans, can employ liquid $CO_2$

at 50°C and 5000 psi rather than the conventional hexane (*CMR* 5/25/81). The advantages of this process are the facile recovery of the extracted material and the production of germ-free oil. Other large-scale extraction operations remove caffeine from coffee, cholesterol from eggs, ethanol from "light beer" or wine, and fats from deep-fried foods such as potato chips without adversely affecting food flavors (*CW* 5/27/81 and B. J. Spalding, *CW* 6/24/87, 66–67). The report "$CO_2$ Extraction Featured at Perfumery Conference" (*CMR* 2/15/88) spells out the advantages of this technology in the isolation of ultrafine perfume ingredients.

While $CO_2$ is a highly convenient SCF material, other compounds used for the same purpose include ammonia, ethylene, methane, butane, ethyl acetate, and water. The last-mentioned material behaves quite differently from the accustomed behavior of water under ordinary temperature and pressure conditions. Thus, water at 220 bar and 374°C exhibits hardly any hydrogen bonding, it dissolves non polar organic but not ionic materials, and it becomes highly corrosive.

The other, very interesting, development is the joint effort by ARCO and Air Products (*CE* 11/24/86) to commercialize poly(alkylene carbonate) thermoplastics from $CO_2$ and ethylene and/or propylene epoxides (3.7.2.5). Finally, a potentially large use of $CO_2$ is its role as a plant growth stimulant in commercial greenhouses, and as an aqueous solution by irrigation of field crops (*CMR* 6/16/86); see also B. J. Spalding, "A Silver Lining for the Green House Effect" (*CW* 8/3/88, 41). The methanation process catalyzed by lanthanide oxides is mentioned elsewhere (3.5.4.5).

### 3.7.2   Chemical Uses of Carbon Dioxide

As a building block, $CO_2$ provides the $-CO_2H$, $-O-CO-O-$, and $-CO-$ moieties.

#### 3.7.2.1   Production of Urea

The formation of 1 part-by-weight (pbw) of urea consumes approximately 0.75 pbw of $CO_2$, and the process is usually carried out at 180–200°C and about 200 atm:

$$2NH_3 + CO_2 \rightarrow H_2N-CO-NH_2 + H_2O$$

Temperatures above 120°C tend to induce the formation of biuret:

$$2H_2N-CO-NH_2 \rightarrow H_2N-CO-NH-CO-NH_2 + NH_3$$

and, since the latter is harmful to cattle when urea is employed as a nitrogen supplement, its content is usually maintained below 1%.

The demand for urea hovers about 15B lb, and, of course, it is sensitive to the level of agricultural activities. The unit price of urea is 3.75¢–5¢ for agricultural grade at the Gulf Coast and Midwest, respectively, and 10¢ for the industrial grade. In recent years significant imports from the USSR, East Germany, Romania, Italy, and Venezuela are threatening domestic production and, as a matter of fact, some suppliers are accused of dumping practices (*CMR* 12/22/86). Urea fertilizer is applied either as a solution or in solid form, and a slow-release fertilizer is manufactured by coating prilled urea with sulfur. The whole fertilizer and agrichemical industry is strongly affected by agricultural policies generated in Washington, DC that overlook the possibilities of industrial utilization of farm products (*SZM-II*, 1–6) to supplement and replace materials obtained currently from fossil resources.

### 3.7.2.2  Carboxylation of Phenols

Phenolate anions are sufficiently nucleophilic to compensate for the reluctant electrophilic behavior of $CO_2$ and lead to the Kolbe–Schmitt reaction (ONR-52). Thus, sodium or potassium phenolate and $CO_2$ at about 150°C give salicylic acid, 50MM lb (technical grade $1.23; USP grade crystals or powder ($1.33 and $1.68, respectively), which becomes the starting material for the preparation of aspirin, methyl and phenyl salicylates, salicyl amide, and other materials (9.3.4). A by-product of this reaction is *p*-hydroxybenzoic acid, which is used as a fungicide, for the preparation of dyes, and so on.

The carboxylation of the sodium salt of β-naphthol produces 3-hydroxy-2-naphthoic acid, or oxynaphthoic acid, at 250°C and 80 psi:

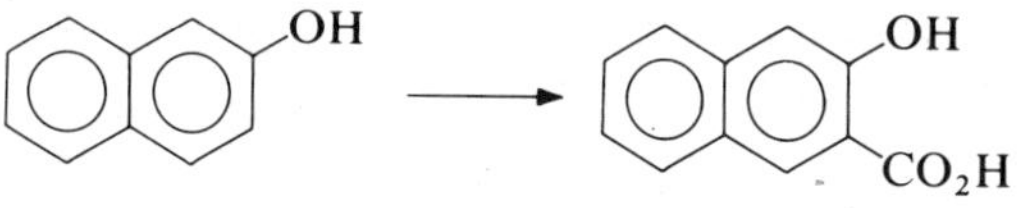

Some industrial uses of this and related naphthols are shown elsewhere (9.3.4).

### 3.7.2.3  Ethylene Urea

Ethylenediamine (4.3.3.3) ($1.30) and $CO_2$ react under pressure [Mulvaney and Evans, *Ind. Eng. Chem.* **40**, 393 (1948)] to yield a cyclic urea, 2-imidazolidinone, commonly referred to as "ethylene urea":

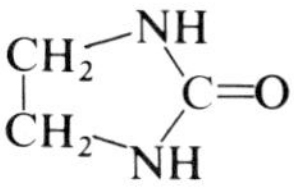

The same product is obtained by elimination of ammonia from urea and ethylenediamine [Schweitzer, *JOC* **15**, 471 (1950)].

### 3.7.2.4  Dimethyl, Dialkyl, and Diphenyl Carbonates

Under pressure and base-catalyzed conditions, $CO_2$ reacts with alcohols to produce the corresponding dialkyl carbonates (compare 3.3.2.7). This has led (*CMR* 10/22/84) to the industrial production of dimethyl carbonate (DMC) (3.3.2.7.D) (90¢, Virginia Chemicals), hailed as a replacement of phosgene (3.8) in the production of polyurethanes and as a replacement of dimethyl sulfate in methylation processes.

Transesterification enables us to produce higher dialkyl carbonates and diphenyl carbonate. The latter was previously synthesized by means of phosgene for use in the assembly of polycarbonate and polyurethane resins.

Dimethyl carbonate is a convenient building block for the introduction of carbomethoxy groups into systems that contain active, that is, acidic hydrogens, and these Claisen-like condensations are catalyzed by sodium methoxide (3.3.2.6.D) or more powerful bases such as sodium hydride ($1.86, 60% dispersion in mineral oil), *n*-butyllithium (15% solution, $14.75), and lithium diisopropylamide. The choice of base depends on the reluctance of the acidic hydrogen-containing reactant to form the required carbanion.

### 3.7.2.5  Alkylene and Poly(alkylene Carbonates)

There exist several patents that describe the formation of alkylene carbonates from epoxides and $CO_2$. For example, U.S. Patent 4,233,221 (issued in 1981 to D. A. Raines and

O. C. Ainsworth and assigned to Dow) claims the formation of ethylene carbonate in 87% yield and with 100% selectivity when the two reactants are passed over the anionic exchange resin Dowex XF 4155 L. Obviously, the production of ethylene carbonate from EO and $CO_2$ is more attractive than the use of the corresponding glycol and phosgene or dimethyl carbonate, as illustrated in Scheme 3.19.

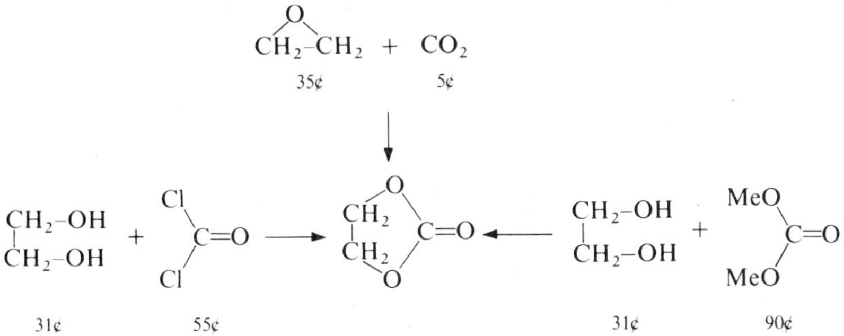

**Scheme 3.19**

The epoxide–$CO_2$ process is the likely reason for Texaco's and ARCO's recent announcements of the availability of ethylene and propylene carbonates, which, incidentally, are two additional members of the dipolar aprotic solvent family (3.5.3.1) (b.p. 245, 240°C, respectively). In addition, ethylene carbonate is a useful hydroxyethylation agent.

A natural outgrowth of the preceding industrial activities is the joint initiative or ARCO and Air Products (*CE* 11/24/86) to develop thermoplastic polymers of the poly(alkylene carbonate) family. These polymers were previously described (*CT* 9/76, 588–594) by S. Inoue of the University of Tokyo, who used diethylzinc and a compound that possesses two active hydrogens as co-catalyst. According to the ARCO–Air Products announcement, the reaction between EO or PO with $CO_2$ is carried out at about 600 psi in the presence of a propietary catalyst, and the price of the resulting polymers is estimated at about $1/lb:

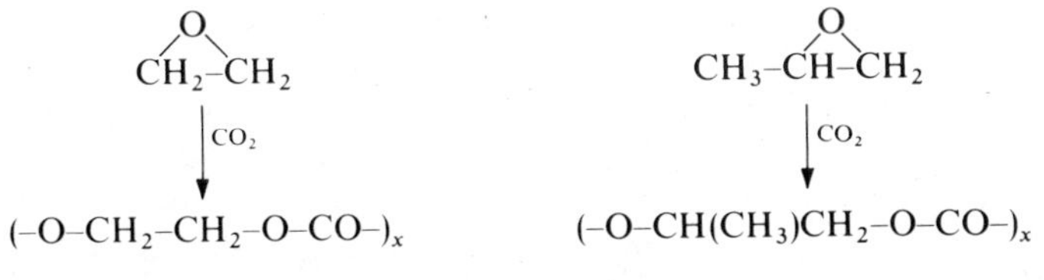

These thermoplastics decompose at relatively low temperatures (200–250°C) but function as an effective oxygen barrier, a desirable property for packaging of food. Also, they burn cleanly to $CO_2$ and $H_2O$, a desirable behavior from an ecological point of view; exhibit effective adhesion to paper and wood; and are stain- and abrasion-resistant. Plans for commercial scaleup are in the making.

### 3.7.2.6   Some Prospective Uses of Carbon Dioxide

The literature suggests other potential uses of $CO_2$ as a building block of polymers. These are attractive possibilities in view of the low price of $CO_2$ and its relatively high mass contribution to the final product. For example, the above-mentioned article by S. Inoue and other authors suggest the following products:

- Poly(arylene ureas) assembled at 40°C from appropriate aromatic diamines and $CO_2$ in the presence of diphenyl phosphite and pyridine [N. Yamazaki et al., *J. Polym. Sci., Chem. Ed.* **13**, 785 (1975)].
- Diallyl carbonate of diethylene glycol, a difunctional monomer:

$$O(-CH_2CH_2-O-CO-O-CH_2CH=CH_2)_2$$

This is obtained according to the Tokuyama Soda Co. [*CA* **93**, 238828w (1980); **94**, 174375g (1981)] when the carbonate dianion, formed by the reaction of $CO_2$ with diethylene glycol at 100°C in the presence of triethylamine, is intercepted with allyl chloride. Other allyl carbonate monomers derived from different glycols can be obtained in a similar fashion:

- Carbamate formation by combining an amine with $CO_2$ in the presence of an alcohol, such as the assembly of ethyl phenylcarbamate:

$$Ph-NH_2 + CO_2 + EtOH \rightarrow Ph-NH-CO \cdot OEt + H_2O$$

This method is similar to the alternative chosen by Du Pont for the synthesis of MIC in the aftermath of the Bhopal disaster (3.8.5).

Judging by the time lag between the publications of S. Inoue and the ARCO–Air Products announcement concerning the poly(alkylene carbonates), we may still expect to hear in due time about additional industrial initiatives that deal with the above and other large scale chemical uses of $CO_2$.

### 3.7.3   Carbon Disulfide

#### 3.7.3.1   Sources of Carbon Disulfide and an Overview of Its Uses

Carbon disulfide (21¢) is obtained by the high-temperature reactions (500–700°C) of elementary sulfur with either coke or methane. The reaction between sulfur and methane is catalyzed by $Al_2O_3-Cr_2O_3$ and is carried out at about 700°C:

$$CH_4 + S_8 \rightarrow CS_2 + H_2S$$

The undesirable $H_2S$ is recycled by means of the Clause process:

$$H_2S + O_2 \rightarrow SO_2 + H_2O$$

is followed by

$$SO_2 + H_2S \rightarrow S + H_2O$$

for a net change

$$H_2S + \tfrac{1}{2}O_2 \rightarrow S + H_2O$$

Between 1979 and 1986 the demand for $CS_2$ decreased from about 450MM lb and leveled off at about 320MM lb mostly because of the weakness of its largest-volume consumption, the cellulosics market. The popularity of rayon textiles is subject to changes in fashions dictated by Paris and New York and is also affected by the popularity of synthetically produced textiles. The demand for cellophane has also decreased because of interproduct competition by PE, PP, and other petroleum-derived materials used in packaging. In the case of cellulosics, $CS_2$ does not enter into the final product and plays a transient role by allowing the xanthates to be processed with a recovery of cellulose in the desired physical form as a filament, film, or sponge (3.7.3.3.A). In 1986 45% of $CS_2$ was consumed in the processing of cellulosics.

The chlorination of $CS_2$ to give carbon tetrachloride and sulfur chlorides has thus far maintained itself at a level of about 90MM lb (or about 30% in 1985), but the prognosis for the domestic carbon tetrachloride market is not encouraging in view of the pressures against the use of CFCs and chlorinated solvents (3.2.3). This leaves us with 25% of the current demand of $CS_2$ that is channeled into the production of thiourea, dithiocarbamates, ammonium and sodium thiocyanates, and some other interesting structures.

### 3.7.3.2 The Role of Carbon Disulfide as a Building Block of Organic Structures

Because of the ease with which a carbon–sulfur double bond undergoes addition reactions, and the facile oxidation of a sulfur terminal to a disulfide structure, carbon disulfide is more versatile than its oxygen analog and contributes the following structural moieties:

$$-CS\cdot S- \qquad -CS\cdot S-S-CS\cdot - \qquad {>}N{-}CS\cdot{-}SH \qquad {>}N{-}CS\cdot S-S-CS\cdot N{<} \qquad {>}N{-}CS{-}N{<}$$

$$-N{=}C{=}S \longleftrightarrow N{=}C{-}S- \qquad -CS\cdot S{-}CS\cdot - \qquad -O{-}CS\cdot{-}S-$$

$$-O{-}CS\cdot{-}S-S-CS\cdot{-}O-$$

The reactivity of carbon disulfide goes hand-in-hand with its instability exemplified best, perhaps, by the spontaneous ignition in air when exposed to sunlight.

### 3.7.3.3 Chemical Uses of Carbon Disulfide

**A. XANTHATES.** In order to convert insoluble, intractable chemical cellulose—that is, cellulose not destined to become paper, carton, and so on, but rather to be used as a feedstock for chemical transformations (*SZM-II*, 83–91), it is first soaked in concentrated sodium hydroxide, caustic soda, until an alkali cellulose gel is formed. The resulting mass is then treated with $CS_2$ to produce a viscous yellow solution of cellulose xanthate:

$$\text{Cell–OH} + \text{NaOH/CS}_2 \rightarrow \text{Cell–O–CS·S}^- \quad \text{Na}^+$$

(where Cell–OH symbolizes cellulose). The cellulose is then regenerated (precipitated) by extruding the xanthate solution into a sulfuric acid bath through a slit to form cellophane film, or through spinnerettes to form rayon filaments:

$$\text{Cell–O–CS·S}^- \ \text{Na}^+ + \text{H}_2\text{SO}_4 \rightarrow \text{Cell–OH} + \text{CS}_2 + \text{Na}_2\text{SO}_4$$

Most of the $CS_2$ is recovered, and the sodium sulfate, rayon grade, is sold as "salt cake" (4.5¢). By incorporating crystals of $Na_2SO_4 \cdot 10H_2O$, or Glauber's salt, into the xanthate solution before addition of dilute sulfuric acid, one obtains cellulose sponges—a $130 million market in the United States (*CB* 11/87). The open-cell nature of cellulose sponges and their hydroxylic nature provide a competitive edge over synthetic products fabricated from PET or PUR materials because the former tend to soak up more water and release it more slowly than do the latter. On the other hand, the wear-and-tear properties favor the totally synthetic products.

The world production of "viscose rayon" is estimated at 2.3MM t. The current price of rayon (89–95¢) is higher by a few cents than polyester (PET) fibers and cotton. Rayon textiles must also compete with acrylic and polyolefin fibres and the latter are very popular in sportswear. The popularity during the middle 1980s of challis cloth—a soft, free-flowing women's wear fabric—favors the rayon industry as long as this product remains fashionable in the continually fluctuating market of women's fashions. In the case of rayon, the most recent advance is the production of fibers from cellulose of about twice the molecular weight of traditional product. This so-called polynosic rayon is character-ized by high wet-modulus strength and gives longer and finer fibers during processing. Obviously, the formation of such product requires a smaller degree of degradation of the naturally occurring cellulose chain during pulping operations of wood or the processing of cotton lint. Practically all of the rayon imported from Japan is polynosic rayon.

The xanthate of starch is insolubilized by means of some cross-linking with epi-chlorohydrin (5.4.8) and then employed to scavenge heavy metal ions during the treatment of wastewater (SZM-II, 92–93). This is particularly true in the case of wastewater originating from electroplating operations, in which case both the metals and insoluble starch are recovered by virtue of the facile oxidation of the xanthate group, as indicated in Scheme 3.20.

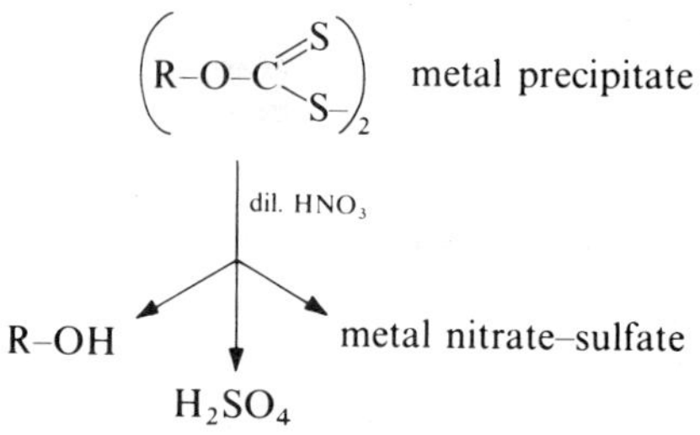

$$\left( R\text{-}O\text{-}C\diagdown_{S}^{S} \right)_2 \quad \text{metal precipitate}$$

$$\downarrow \text{dil. HNO}_3$$

$$\text{R-OH} \longleftarrow \downarrow \longrightarrow \text{metal nitrate–sulfate}$$

$$\text{H}_2\text{SO}_4$$

(where R–O– represents cross-linked starch moiety)

**Scheme 3.20**

Xanthates of some higher alcohols (7.3.1) find hydrometallurgical applications as flotation agents for the separation of valuable metallic constituents of ores from the accompanying silicates.

**B. DITHIOCARBAMATES, THIOUREAS, AND ISOTHIOCYANATES.** Primary and sec-ondary amines react cleanly with carbon disulfide to give the corresponding dithio-carbamates:

$$\text{Et}_2\text{NH} + \text{CS}_2 \rightarrow \text{Et}_2\text{NCS·SH} \rightarrow \text{Et}_2\text{NCS·S}^- \ \text{Na}^+$$

Sodium N,N-diethyldithiocarbamate is a reagent for the separation and quantitative analysis of copper, and analogous thiocarbamates are used as bacteriocides in water-based petroleum processing and storage facilities.

Similarly, ethylenediamine and an excess of $CS_2$ is converted in the presence of ammonia to diammonium ethylene–bis(dithiocarbamate), EDBC:

$$(-CH_2-NHCS \cdot S^- NH_4^+)_2$$

This family of pesticides is currently under review by EPA (*CW* 7/22/87) because of their oncogenic and teratogenic effects on laboratory animals.

The dithiocarbamates are oxidized readily to the corresponding thiurams:

$$Me_2NCS \cdot SH + air \rightarrow Me_2NCS \cdot S-S \cdot CS \cdot NMe_2$$

tetramethylthiuram disulfide<br>thiram, TMTD, $1.35

Thiram is an accelerator in the vulcanization of rubber by means of elementary sulfur, and as many other dithiocarbamate derivatives, it is used a fungicide to disinfect seeds and is present in many antifungal formulations. In particular, the dithiocarbamate fungicides do an excellent job in the control of *Phytophera infestans*, the wind-blown fungus responsible for the historic potato famine in Ireland during the late 1840s. Introduced in the late 1940s, the dithiocarbamates (and other pesticides) increased the yield of potataoes [in hundredweight per acre (cwt/acre)] from 56–82 registered between 1920 and 1949 to 165–208 registered between 1950 and 1969 and to 247 during the 1970s (*KCB-II*, p. 56).

The lability of the disulfide bond explains the thermal extrusion of sulfur that produces thiuram sulfides such as the following rubber vulcanization accelerators and agricultural fungicides:

$(Et_2NCS)_2S$       tetraethylthiuram sulfide, 68¢

$[(Me_2)NCS]_2Zn$     zinc dimethyldithiocarbamate, $1.49

The reaction of $CS_2$ with hydrazine produces 2,5-dimercapto-1,3,4-thiadiazole (9.3.3).

Under more vigorous reaction conditions, $CS_2$ and ammonia or primary or secondary amines give rise to thioureas:

$$2NH_3 + CS_2 \rightarrow H_2N \cdot CS \cdot NH_2 \quad \text{thiourea} \quad (+H_2S)$$

Thiourea, (74¢, 3–4MM) is also obtained in the reaction of hydrogen sulfide (13¢) and sodium thiocyanate [$Na^+$ $\overline{S}CN$, techn. (technical grade) 97¢, or ammonium thiocyanate, techn. cryst. $1.02], which, in turn, are also obtained from $CS_2$ and the appropriate base. Thiourea is used to a limited extent as a photographic chemical and as a rubber vulcanization accelerator. Thiourea is a valuable building block of certain heterocyclic systems (Chapter 10).

Organic isothiocyanates, such as allyl isothiocyanate ($5.40), for example, are prepared from sodium thiocyanate (techn. 97¢):

$$CH_2=CH-CH_2Cl + Na^+ (SCN)^- \rightarrow CH_2=CH-CH_2N=C=S$$

Allyl isothiocyanate is isolated from black mustard and is the prototype of "mustard oils" (RN=C=S). As may well be expected, allyl isothiocyanate is an irritant and is used as a counterirritant when applied to skin in order to relieve a more severe irritation elsewhere on the body. Allyl isothiocyanate is also a potential war gas.

*Hard/Hard–Soft/Soft Theory*

The above-mentioned formation of the isothiocyanate (rather than a thiocyanate, R–SCN) during the nucleophilic substitution by the ambident SCN anion is in line with the "hard/hard–soft/soft theory," which predicts that the "harder," that is, less polarizable, nitrogen terminal reacts preferentially with the relatively "hard" α-carbon of allyl chloride. On the other hand, the reaction of the SCN anion with the relatively "soft" carbon of methyl iodide explains the formation of methyl thiocyanate ($CH_3$–SCN) rather than the isomeric methyl isothiocyanate ($CH_3$–N=C=S). For additional applications of this theory see 3.10.2.2A.

$$2EtNH_2 + CS_2 \rightarrow (EtNH)_2CS, \quad N,N'\text{-diethylthiourea, 58¢}$$

Ethylenediamine, \$1.30, and $CS_2$ produce ethylene thiourea:

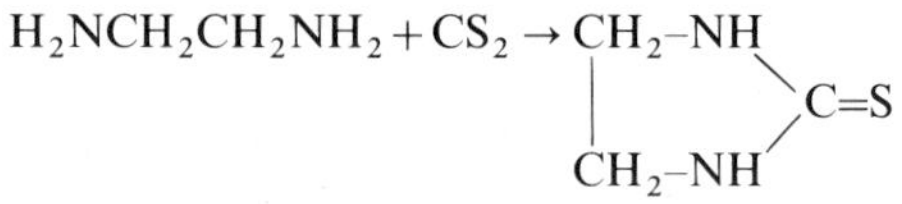

$$H_2NCH_2CH_2NH_2 + CS_2 \rightarrow \begin{array}{c} CH_2-NH \\ | \quad\quad\quad \diagdown \\ \quad\quad\quad\quad C=S \\ | \quad\quad\quad \diagup \\ CH_2-NH \end{array}$$

The reaction of $CS_2$ with aniline to give a series of rubber vulcanization accelerators is discussed elsewhere (9.3.6).

The reaction of $CS_2$ with cyclohexylamine is the first step in the preparation of the most common carbodiimide, namely, dicyclohexylcarbodiimide (DCC):

$$2R-NH_2 + CS_2 \rightarrow R-NH-C(=S)-NH-R \xrightarrow[\text{HgO}]{} R-N=C=N-R + HgS + H_2O$$

(where $R$ = cyclohexyl). DCC is an excellent catalyst for the joining of two amino acids into a peptide by virtue of coordinating the carboxylic and amino functions and causing a loss of water under very mild reaction conditions, as shown in Scheme 3.21.

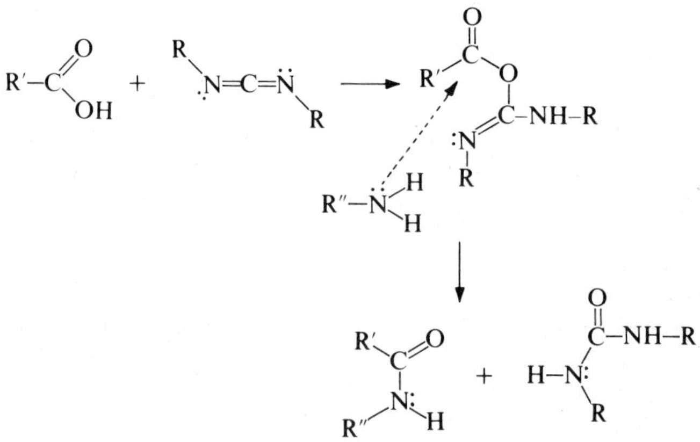

**Scheme 3.21**

This use of DCC is relevant in view of the increasing commercial importance of peptides, but DCC also functions in a similar fashion as an esterification catalyst.

**C. CARBON TETRACHLORIDE.** The reaction of $CS_2$ (21¢) with chlorine (10¢) is the traditional source of carbon tetrachloride (techn. 24¢, CP grade 36¢) because of the market for sulfur monochloride (16.25¢) coproduct and also because, unlike the chlorination–chlorinolysis of low-molecular-weight hydrocarbons, it does not produce other chlorinated carbon compounds:

$$CS_2 + 3Cl_2 \rightarrow CCl_4 + ClS-SCl$$

The reaction is carried out at about 30°C in the presence of an iron catalyst. If the formation of sulfur monochloride is not wanted, it can be utilized to produce additional carbon tetrachloride at about 60°C:

$$2S_2Cl_2 + CS_2 \rightarrow 6S + CCl_4$$

Actually, sulfur monochloride and the related sulfur dichloride,

$$S_2Cl_2 + Cl_2 \rightleftharpoons 2SCl_2$$

$$16.25¢ \quad\quad 10¢ \quad\quad 17.75¢$$

are useful inorganic building blocks of various industrially important organic materials and are also the starting materials for the production of thionyl chloride ($OSCl_2$, 55¢) and sulfuryl chloride ($O_2SCl_2$), reagents encountered throughout this book:

$$S_2Cl_2 + SO_3 \rightarrow OSCl_2 + SO_2 + S$$

$$SCl_2 + SO_3 \rightarrow OSCl_2 + SO_2$$

$$2S_2Cl_2 + Na_2SO_4 \rightarrow O_2SCl_2 + 2NaCl + SO_2 + 3S$$

$$SCl_2 + O_2 \rightarrow O_2SCl_2$$

Thionyl chloride is the reagent of choice for the conversion of carboxylic acids to the corresponding acid chlorides because all the by-products are volatile (if one employs purified thionyl chloride). Sulfuryl chloride, among other things, is useful in the synthesis of diaryl sulfones (9.3.3).

**D. THIOPHOSGENE.** This sulfur analog of phosgene is obtained by a carefully controlled reaction of $CS_2$ with chlorine. It is a versatile reagent for the preparation of fine chemicals (Ald. $45.50 in 100-kg quantities). For example:

• Amines are converted to isothiocyanates, and then, if desired, into unsymmetrical, substituted thioureas:

$$R-NH_2 \rightarrow R-N=C=S \xrightarrow{\quad R'-NH_2 \quad} R-NH-C(=S)-NH-R'$$

• Vicinal diamines give cyclic thioureas, or thionoimidazolidinones:

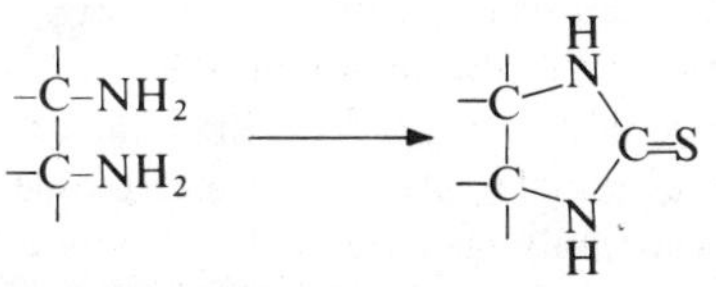

- Vicinal diols give, first, cyclic thiocarbonates, and the latter, if subjected to treatment with trialkyl phosphites, $P(O–R)_3$, are converted to olefins according to the Corey–Winter olefin synthesis (ONR-20):

- Friedel–Crafts (ONR-33) acylation converts aromatic compounds to the corresponding thiobenzophenones:

$$Ar–H + S=CCl_2 \rightarrow Ar–C(=S)–Ar + HCl$$

## 3.8   PHOSGENE, CHLOROFORMATES, AND THEIR DERIVATIVES

Phosgene is a chemical with a very interesting past (3.5.3.6) but with an endangered future. Its use as a building block of isocyanates is being challenged by DMC (3.7.2.4)—an intermediate obtained without intervention of chlorine and by the direct use of carbon dioxide (3.7.2.5, 3.7.2.6). A possible breakthrough in the use of phosgene is the announcement of an allegedly safe facility developed by Rhône-Poulenc and located only some 350 ft away from a residential section of Grenoble, France (*CE* 4/25/88). It is used for the synthesis of hexamethylene diisocyante, one of the PUR building blocks mentioned below.

Additional comments concerning nonphosgene approaches to the production of isocyanates are offered in Section 3.8.4, although the origin of most of the building blocks of PUR, mentioned in Section 3.8.1, "A Glimpse at the World of Polyurethanes" (hereafter referred to as 3.8.1, PUR) is covered in other parts of this book (see the index).

The synthetic uses of chloroformates (Cl–CO·OR) for the preparation of fine chemicals is one area of phosgene chemistry that is likely to survive, although admittedly, it represents a small fraction of the current phosgene demand.

### 3.8.1   Consumption of Phosgene in the United States

As mentioned in Section 3.5.3.6, most ($\sim 90\%$) of the current demand of 1.6 B lb of phosgene is absorbed by the PUR industry, and most of it is produced for captive use because of its hazardous nature and the consequently high shipping and insurance costs.

The PUR polymers represent an extremely versatile family of materials that can be manufactured in the form of rigid but impact-resistant solids, semiflexible or flexible foams, thermoset or thermoplastic elastomers, coatings, adhesives, elastomeric fibers known as spandex, and so on. The products range from bedding and furniture cushions, flooring and roofing materials, automotive components such as bumpers and window gaskets, components of appliances, ingredients of paints, binders for structural or decorative boards and foundry molds, thermal and sound-insulating materials, and so on, and even include bioengineering parts. Flexible foam products constitute about half of PUR consumption. The 1986 production of PURs in the United States was 2.7 B lb, valued at about $2.5 billion, and it is expected to grow in the foreseeable future at a rate that exceeds the growth of the GNP, especially because of the rapidly expanding reaction-injection molding (RIM) and reinforced-reaction-injection molding (RRIM) technologies employed in the highly productive manufacture of large objects such as components of automobiles (*CB* 11/87). The RIM and RRIM technology is not limited to PURs but, worldwide, its application is growing at an average annual rate of 17% (*CMR* 9/2/87).

The discrepancy between the isocyanate and PUR production levels signifies that components other than the isocyanates contribute to the final PUR materials. These other materials, often referred to simply as "polyols," are double- and higher-functionality alcohols, and they are discussed below in some detail because they, too, determine the properties and uses of the PURs.

The formation of a single urethane bond by the addition reaction of hydroxylic and isocyanate functional groups is represented simply by

$$R–N=C=O + R'–OH \rightarrow R–NH–CO·O–R'$$

Before we proceed further, it may be useful to recall that the somewhat confusing traditional nomenclature calls structures such as $H_2N–CO·O–R$ *carbamates* and $Cl–CO–O–R$ *chloroformates*, while substitution at both N and O terminals gives *urethanes*.

The great variety of PUR-based products, and the nearly endless variations of the fundamental PUR chemistry, call for a separate, concise summary (3.8.1) of the parameters that can be manipulated to control the nature of the end products and their processing technology. In order to do this coherently we must involve chemicals whose origin is explained elsewhere in this book.

The other important family of polymers dependent on the phosgene building block is that of the polycarbonates (PCs) (3.8.3). Their production ($\sim$350MM lb) consumes approximately 6–7% of the phosgene demand. The high-performance characteristics of the PCs also promise continued growth albeit at a lower level than that of the PURs because the PCs are not employed to the same extent as the PUR in the manufacture of large objects. On the other hand, special grades of PCs are utilized in the manufacture of compact disks (CDs) and other popular electronic devices.

The remaining few percent of the current phosgene demand are employed in the production of valuable pesticides of the carbamate family (3.8.2) and other fine chemicals. The use of carbamate pesticides is likely to continue in order to protect agricultural and silvicultural productivity, but the impact of the Bhopal disaster (3.5.3.6) is a powerful incentive for utilizing, wherever possible, nonphosgene technology (3.8.4), especially when this takes place in the industrially less-developed countries.

### A Glimpse at the World of Polyurethanes

The principal variables and parameters that can be chosen for the production of the above-mentioned great variety of PUR products are enumerated as follows.

*1. THE CHOICE OF PUR PROCESSING TECHNOLOGY.* This depends on the nature of the end product, convenience, and safety considerations. It can consist of a *one-step* ("*one-shot*") *process* in which the major building blocks of the desired PUR (isocyanate and alcohol or some other ingredient that reacts with isocyanate) are mixed together with catalysts, fillers, reinforcing and coloring agents, blowing agent (if a foam is to be produced), and other minor constituents of the reaction mixture to give the final solid product (most commonly as blocks, pads, and otherwise-shaped objects of thermoset elastomers or rather rigid foams). An alternative is a *two-step process* in which a relatively high-molecular weight *PUR prepolymer* is assembled in such a fashion that it contains reactive terminal groups (most commonly isocyanate-terminated or hydroxyl-terminated structures obtained by the use of an excess of one or the other component). The prepolymer can then be subjected to a reaction with a chain extender, a rather reactive and relatively low-molecular-weight diol, diamine, and so on, that links molecules of the prepolymer to give larger two- or three-dimensional PUR end products. The two-step process is the basis of RIM (or RRIM) technology, but it is also used to avoid the handling of hazardous isocyanates [particularly tolylene or toluene diisocyanate (TDI)] during shipment or during final processing operations. The control during the formation of PUR prepolymer is not as simple as it may sound: only recently (*CW* 1/6–13/88) Air Products announced a new technology that produces a "perfect prepolymer."

A modification of the two-step reaction concept is the use of blocked isocyanates. This entails the conversion of the isocyanates to a derivative that liberates the isocyanate under relatively mild thermal conditions and allows it to react with the partner component while the blocking agent is either volatilized or also incorporated into the end product. Blocked isocyanates are obtained by the addition of a reagent that offers a good leaving group on thermal activation of the adduct. For example:

$$R-N=C=O + Q-O-H \rightarrow R-NH-CO-O-Q \xrightarrow{\text{heat}} R-N=C=O\ (+Q-O-H)$$

where Q–O–H represents a phenol, an oxime, $R'_2C=N-OH$, and so on.

### 2. THE CHOICE OF THE ISOCYANATE COMPONENT

*a. Aromatic Diisocyanates.* Tolylene or toluene diisocyanate is the traditional "work horse" of the PUR industry. Because of its origin, which starts with a nitration of toluene (9.2.5), ordinary TDI is a 80:20 mixture of 2,4- and 2,6-diisocyanatotoluene ($1.01). The minority component of this mixture reacts slower than the 2,4-isomer because both isocyanato groups are sterically inhibited, and hence the 100% pure 2,4-isomer is also available at a premium price of $1.60 (Mobay's Mondur TDS). The pure 2,4-isomer of TDI is likely to give an overall more linear PUR end product. The consumption of TDI is affected strongly by the consumer demand for furniture (cushions, mattresses, etc.) and carpet pads, and thus it is sensitive to the general state of the economy and the construction sector, in particular.

Methylene di-*p*-phenylene diisocyanate, diphenylmethane 4,4'-diisocyanate, or simply MDI, and its polymeric analog, PMDI (91¢) have recently caught up and exceeded the 760MM-lb demand for TDI and are consumed in the United States at a level of about 900MM lb (see A. Agoos and D. Hunter, "The Pace Quickens for Urethane Intermediates," *CW* 3/9/88). Both are derived from aniline and formaldehyde (3.4.3.5, 9.3.6) and are listed together in statistical reports. The growing demand for MDI and PMDI by the PUR industry is responsible for the strong aniline market since they consume about 60% of the latter. Both MDI and PMDI give rise to more rigid PUR end products than does TDI and are used for the manufacture of insulating panels for refrigerators and other appliances, automotive components such as bumpers and fascia, and in the constantly expanding RIM and RRIM operations. The polyfunctionality of PMDI is responsible for cross-linking of the resulting PURs and the formation of rigid end products. On the other hand, because of the nearly linear structure and a relatively restricted freedom of rotation about the central methylene group, MDI alone is used to assemble the elastomeric textile fiber Spandex known to many consumers of support hosiery and many types of elastic garments (see discussion of thermoplastic elastomers below). This elastomeric textile fiber was commercialized by Du Pont in 1962 under the tradename "Lycra."

The structures of these large-volume aromatic isocyanates and the more specialized building blocks of the PURs mentioned in paragraphs b–d below and identified as (a)–(i), are shown in Scheme 3.22.

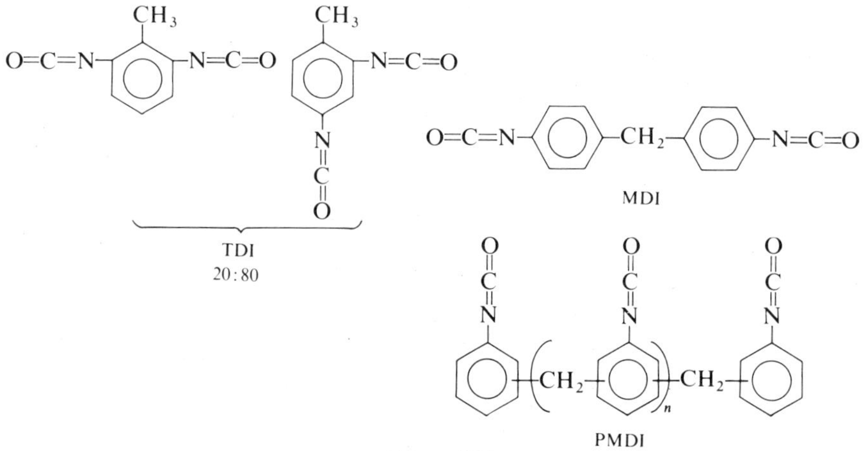

**Scheme 3.22**

2,6-Naphthtalene or naphthylenediisocyanate (NDI) is a PUR building block used mostly in Europe.

All aromatic isocyanate building blocks are susceptible to light-catalyzed discoloration, and hence in coating and similar applications are replaced by aliphatic isocyanates.

*b. Nonaromatic Isocyanates.* These are structures in which the isocyanato functions are not attached directly to an aromatic ring. Most often they are also referred to as "aliphatic isocyanates" even though they may actually be benzylic isocyanates such as

*m*-Xylenediisocyanate, Sherwin–William's MXDI, $3.25 (5.4.3.1, 9.2.1)

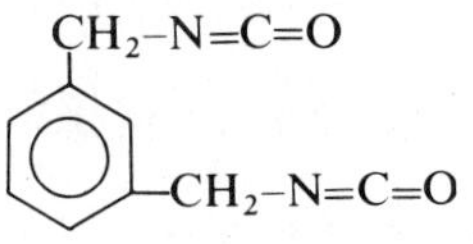

(a)

and the isomeric *p*-isomer, Takeda's XDI

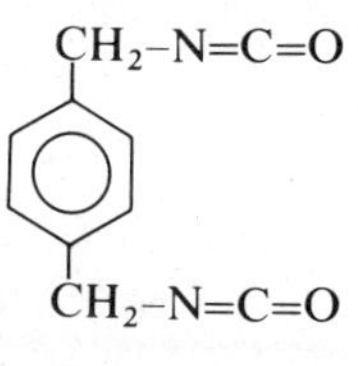

(b)

Tetramethyl-*m*-xylenediisocyanate, Cyanamid's TMXDI

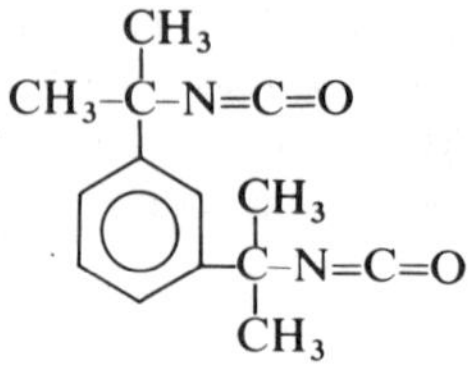

(c)

and the isomeric *p*-isomer (9.2.1)

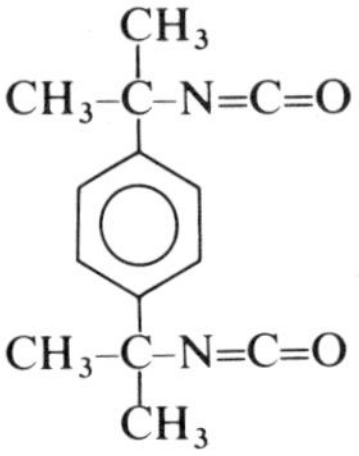

(d)

The analogous isopropenyldimethylbenzylisocyanate, Cyanamid's TMI,

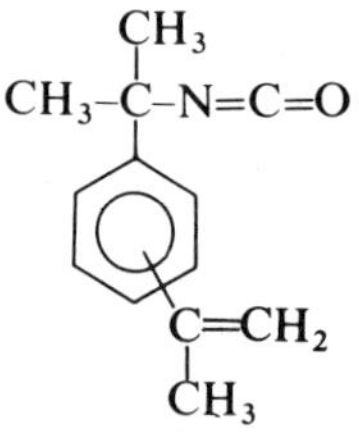

(e)

TMI is useful for "capping" hydroxy-terminated PURs or, for that matter, also capping amino- or carboxy-terminated prepolymers. TMI can also serve as a *heterodifunctional* monomer (see 3.d and 8, below) if the PUR structure is to be combined with a polyvinyl chain.

The problem of discoloration of PURs derived from benzylic isocyanates is not solved completely because of slow oxidation at the benzylic hydrogens.

*c. Alicyclic Isocyanates.* These include the following:

Isophoronediisocyanate (5.5.4), Huels's IPDI

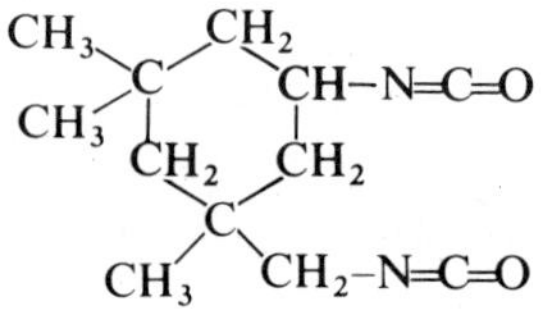

(f)

1,4-bis(isocyanatomethyl)cyclohexane (8.3), Mobay's *p*-Desmodur

$$O{=}C{=}N{-}CH_2{-}\underset{S}{\bigcirc}{-}CH_2{-}N{=}C{=}O \qquad (g)$$

and the corresponding 1,3-isomer, Takeda's $H_6XDI$, \$3.50

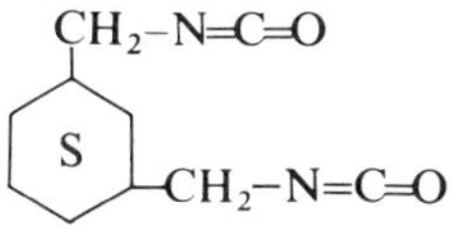

$$\qquad (h)$$

methylene bis(4-cyclohexylisocyanate) (8.3), Mobay's Desmodur W, \$3.00, also known as $H_{12}MDI$, RMDI (reduced MDI), and PACM [bis-*p*-aminocyclohexyl methane]

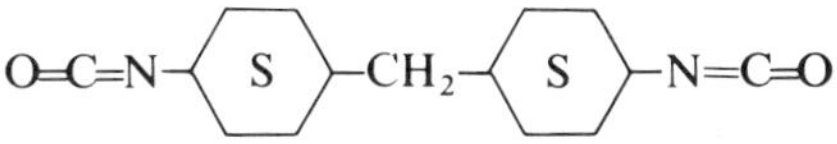

$$O{=}C{=}N{-}\underset{S}{\bigcirc}{-}CH_2{-}\underset{S}{\bigcirc}{-}N{=}C{=}O \qquad (i)$$

*d. Truly Aliphatic Isocyanates.* These include hexamethylenediisocyanate derived from acrylonitrile (5.4.3.2), (HDI):

$$OCN{-}(CH_2)_6{-}NCO \qquad (j)$$

and trimethylhexamethylenediisocyanate, actually a 1:1 mixture of 2,2,4- and 2,4,4-trimethylhexamethylenediisocyanate (7.4), Nuodex's (a Huels subsidiary) TMDI:

$$OCN{-}CH_2(CH_3)_2CH_2CH(CH_3)CH_2CH_2{-}NCO$$
$$OCN{-}CH_2CH(CH_3)CH_2C(CH_3)_2CH_2CH_2{-}NCO$$

Multiple methyl substituents in TMDI inhibit intermolecular associations of the urethane segments derived from these building blocks and hence promote flexibility in the resulting PURs. An interesting heterodifunctional (see 3.d and 8 below) isocyanate is Dow's isocyanatoethyl methacrylate (4.3.6.2) (IEM) [$CH_2{=}C(CH_3)CO{\cdot}O{-}CH_2CH_2{-}NCO$], since it facilitates the grafting of vinyl oligo or polymer side chains when the isocyanato group is attached first to an appropriate prepolymer. Also, vice versa, it facilitates a build-up of an oligo or polymer side chain attached by way of the isocyanato group when the methacrylic moiety is copolymerized first with another vinyl system. Similar possibilities are also available by means of the heterodifunctional isopropenyldimethylbenzylisocyanate (TMI), mentioned above (2.b).

*3. THE CHOICE OF THE POLYOL COMPONENT*

*a. Short-Chain Diols.* Examples are as follows:

Ethylene glycol (EG) (4.3.6.1)
Diethylene glycol (DEG) (4.3.6.1)

Propylene glycol (PG) (5.4.2.2)

Dipropylene glycol (DPG) (5.4.2.2)

Neopentyl glycol (3.4.3.3)

1,4-Butanediol, tetramethylene glycol (6.7)

1,6-Hexamethylene glycol derived from adipic acid (7.4.3)

*N*-substituted diethanolamines (4.3.6.2)

*N*-methyldiethanolamine (MDEA) in particular, and others

*Note*: Short-chain diols are used as chain extenders (see paragraph 1 above), especially when both hydroxyl groups are primary. Secondary hydroxyl groups react more slowly with isocyanates but can first be capped (or "tipped") with EO to give the primary hydroxy-terminated derivatives.

*b. Long-Chain Polyether Diols.* Examples of these diols, of molecular weight $10^2$–$10^3$, are

Poly(ethylene glycols) (PEG) (4.3.5.2).

Poly(propylene glycol) (PPG) (5.4.3.2)—in this case, since only one hydroxyl group is a primary one, PPG is usually capped with some EO in order to convert the secondary hydroxyl terminal also to a primary hydroxyl.

Poly(tetramethylene ether) glycol (PTMEG), also known as polytetrahydrofuran (PTHF) (6.7.1, 10.2.1, and *SZM-II*, 102).

*Note*: The contribution by the above-mentioned polyether diols of similar molecular weight to the flexibility of the PUR end product increases in the order PEG < PPG < PTMEG. The high degree of rotational freedom in PTMEG makes it an ideal polyol for the assembly of elastic PURs such as spandex (see 5 below) and its worldwide demand is estimated to be about 100MM lb.

*c. Long-Chain Polyester and Acid Anhydride Diols.* These include the a priori hydroxy-terminated or EO-capped derivatives of simpler building blocks

Poly(caprolactone), (8.3)

$$HO-(CH_2)_5CO[O-(CH_2)_5-CO]_n-O-CH_2CH_2-OH$$

Poly(ethylene glycol terephthalate) (9.2.2 and 9.3.2.2)

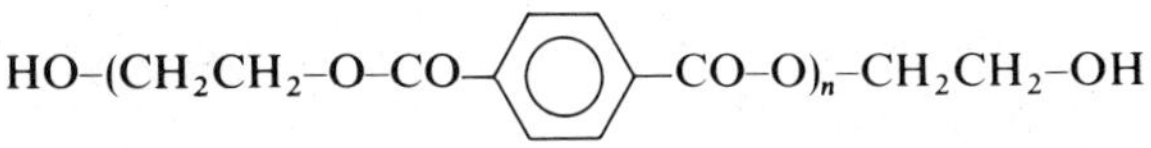

$$HO-(CH_2CH_2-O-CO-\langle\bigcirc\rangle-CO-O)_n-CH_2CH_2-OH$$

Poly(ethylene glycol adipate) (8.3.1)

$$HO-[CH_2-CH_2-O-CO-(CH_2)_4-CO-O]_n-CH_2CH_2-OH$$

Poly(carbonate-linked ethylene glycols) of the general structure

$HO-(R-O-CO\cdot-O)_n-R-OH$ (4.3.5.1), PPG's Duracarb 120, ($2.85), where R represents a difunctional linear, aliphatic moiety such as PPG, MW 850–1500, suitable for the preparation of flexible PURs, while the analogous diol in which R represents an alicyclic moiety, PPG's Duracarb 140 ($2.93), MW 600–1000, is suitable for the preparation of rigid PURs.

Polyazelaic polyanhydride (7.3.1 and *SZM-II*, 70), Emery's PAPA, $HO-CO-(CH_2)_7-CO-[O-CO-(CH_2)_7-CO]_n-OH$.

*Note*: Generally, the contribution of intermolecular attractive forces operating within PUR end products increases in the order polyethers < polyesters < polycarbonates or polyanhydrides.

The polyether diols derived from polyesters seem to represent a compromise between cost and their contribution to superior elongation and tensile properties as compared to plain polyether diols, and the former account for about 10% of the flexible PUR production. Examples of useful products formed from hydroxy-terminated esters of terephthalic acid (formed, in turn, by the reaction of an excess of glycols with DMT or PTA—see 9.3.2) and TDI or MDI–PMDI are flexible ski clothing, gaskets, rollers for printing presses, and so on.

*d. Miscellaneous Trifunctional and Higher Alcohols.* These include the following:

Glycerol (5.4.2.2 and 5.4.8), 90¢

1,1,1-Trimethylolpropane (3.4.3.3), 76¢

Pentaerythritol and dipentaerythritol (3.4.3.3), 71¢ and $1.42, respectively

Triethanolamine (4.3.6.2), 35¢ and the polyols of more complex structure; (i) Sorbitol *SZM-II*, 106–109), powder (68¢), (ii) Methyl glucoside (3.3.2.7 and *SZM-II*, 7, 105), (iii) Sucrose polyether polyols (*SZM-II*, 104), Dow's Voranols (78¢).

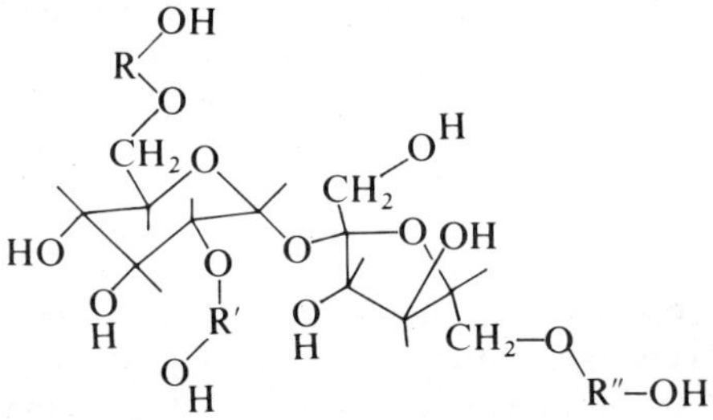

where R, R′, R″ represent polyether polyol chains introduced by means of propoxylation and ethoxylation. The propoxylated derivatives of sorbitol, sucrose, and the relatively most recent addition to this family, namely, methyl glucoside, are responsible for about 30% of polyols consumed in the production of rigid PUR foams.

*4. CHOICE OF ISOCYANATE-REACTION PARTNERS THAT FORM BONDS OTHER THAN URETHANE BONDS.* These include the following families.

*a. Primary Amines.* These react to give ureas

$$R-N=C=O + H_2N-R' \rightarrow R-NH-CO-NH-R'$$

Commonly employed for this purpose are

> Ethylenediamine (4.3.3.3), $1.30, a popular chain extender.
> 4,4'-Methylene dianiline (9.3.6), crude $1.75, purified $2.25, the precursor of MDI.
> Toluenediamines (the precursors to TDI, 9.2.5).

An amine can be generated *in situ* by addition of a controlled amount of water that reacts with the isocyanate to yield a spontaneously decomposed carbamic acid with a release of $CO_2$ and formation of foam:

$$R-N=C=O+H_2O \rightarrow R-NH-CO-OH \rightarrow R-NH_2+CO_2$$

*b. Carboxylic Acids.* These react to give amides and, at the same time, carbon dioxide is released (foam formation):

$$R-N=C=O+HO \cdot CO \cdot R' \rightarrow R-NH-CO-R'+CO_2$$

The above-mentioned (3.c) caprolactone and polyazelaic acid (PAPA) can be employed in this fashion.

*Note*: Intermolecular attractive forces between individual PUR chains increase in the order

$$\text{Aliphatic--NH--CO--R} < \text{aromatic--NH--CO--R} < -NH-CO-NH-$$

$$<-NH-CO-NH-CO-NH-$$

(see 6.c). It is clear that an admixture of significant amounts of amines, amine-generating water, or carboxylic acids gives rise to *hybrid* urethane–urea or even urethane–urea–biuret (see 6.c) products. The extent to which such hybrid marcomolecules are formed will affect the thermomechanical properties of the end products.

*5. CHOICE OF THE DEGREE OF CROSS-LINKING.* This determines whether one produces an elastomer or a thermoset solid.

A small degree of cross-linking provides "elastic memory," meaning that the initially rather unorganized molecular chains tend to recover their original chaotic arrangement on removal of an external force (compression or stretching) that was applied to the material. Elastic memory causes the material to behave like a rubber or an elastomer. Cross-linking of macromolecules that creates elastic memory can be achieved in two different ways:

1. The use of a relatively small amount of polyfunctional reactants—in the case of the PURs these can be either isocyanates or "polyols".
2. The incorporation into the polymeric chains *blocks* of both highly disorganized, flexible, "soft" segments, and "hard" segments that exhibit strong intermolecular attractions for each other (by way of Coulombic forces among ionic centers, hydrogen bonding, dipole–dipole, or even dipole-induced dipole forces). This is achieved by the assembly of *block polymers*, that is, polymeric chains that contain alternating hard and soft segments.

The most significant difference between these two modes of achieving cross-linking is that the first generates a traditional *thermoset elastomer*, while the second gives rise to a *thermoplastic elastomer* (TPE).

The concept of traditional thermoset elastomers was pioneered by Goodyear's discovery in 1839 that heating natural rubber with some sulfur converted the tacky material when warm and brittle when cold into a "vulcanized rubber" that was conveniently useful over a wide temperature range. Cross-linking of the macromolecules of rubber (see 6.8.2 and *SZM-II*, 144–148) with sulfur bonds endowed the naturally occurring material with some elastic memory and caused it to behave as we have come to expect elastomers to behave. Excessive sulfur cross-linking converts the stretchable, compressible, bouncy rubber into hard rubber such as the material found in the heads of mallets used in machine shops to pound sheet metal into desired shapes (after unfortunate "fender-benders"). A small dose of cross-linking prevents the macromolecules of naturally occurring rubber to crystallize at low temperatures and turn into a brittle solid and to become a tacky, sticky semifluid at elevated temperatures.

Thermoplastic elastomers (TPEs) constitute a relatively recent class of elastomers that offers several practical advantages when compared to the traditional, thermoset materials:

- Ease of processing because injection and extrusion molding is more efficient than curing of materials confined to a mold
- Facile recycling of scrap and recovered TPE materials
- Greater opportunity for molecular engineering, that is, the designing of materials for specific applications

Historically speaking, the first TPEs belonged to the PUR family, but since 1950 they were joined by *block polymers* assembled from hard blocks of polymeric styrene, attached to soft blocks of polymeric ethylene and butylene, butadiene, or isoprene monomers, and in the 1960s there appeared hard blocks of polymeric aromatic polyesters joined to soft blocks of oligomeric or polymeric ethers. Still later, hard polyamide blocks derived from aromatic dicarboxylic acids were combined with soft blocks of aliphatic components, and the world of TPEs continues to expand.

In the case of PUR block polymers, the hard segments may consist, for example, of a combination of aromatic diisocyanates and short-chain diols, and the more linear MDI is preferred over a less linear TDI. Such a hard segment may be assembled by allowing a short-chain diol of relatively inhibited freedom of rotation to react with an excess of MDI to give us, first, an isocyanate-terminated prepolymer PUR segment. This can then be treated with a small amount of a chain extender (see above) such as ethylenediamine, for example, to lengthen the hard segment while still retaining the isocyanate-terminated nature of the prepolymer. Finally, the latter is allowed to react with a relatively long-chain (MW = $10^2$–$10^3$) of a highly flexible diol, and thus one would obtain a block polymer of structure:

Soft polyol segment–hard PUR segment–soft polyol segment

The elastic textile fiber Spandex (Du Pont's Lycra) is an excellent example of a successful TPE material (U.S. Patent 2,692,873 issued in 1954 but not commercialized by Du Pont until 1962). Another example is the development of Du Pont's Adiprene PU rubber (see M. E. Schroeder, *CT* 2/87), which is assembled by the reaction of 3 parts of PTMEG (an oligomer of 9–11 tetramethylene ether units) with 4.3 parts of TDI to give an isocyanate-terminated prepolymer that is subsequently treated with 0.3 part of TDI and some water to form an urea-linked end product.

A recently announced ($CW$ 8/19/87) development of a water-retaining hydrogel (W. R. Grace's Hypol) illustrates the many possibilities of molecular engineering within the realm of PURs. In this case, a cross-linked polyol is treated with TDI, for example, to give a hydrophilic semisolid that can be employed as a carrier of medicinals for the treatment of surface wounds, fragrances, in personal-care products, and so on.

6. *CHOICE OF CATALYSTS.* This determines not only the rate of PUR formation by means of the fundamental reaction between isocyanates and their partner reactants but also the contribution of some of the more "exotic" reactions of the isocyanates. The rate of PUR formation is, of course, of critical importance to the RIM and RRIM technology. Among the exotic reactions of the isocyanates we list the following:

*a. Formation of Isocyanurates (10.3.3) by Trimerization.*

$$3R\text{–}N\text{=}C\text{=}O \rightarrow \quad \begin{array}{c} O \\ \| \\ R\text{–}N\text{–}C\text{–}N\text{–}R \\ | \qquad | \\ O\text{=}C\text{–}N\text{–}C\text{=}O \\ | \\ R \end{array}$$

If we keep in mind that when each R group contains at least one more isocyanate function, it is clear that even a small degree of isocyanurate formation creates rigid centers that bind two or more PUR chains and create a focus of mechanical and thermal strength. A high degree of isocyanurate formation leads to polyisocyanurates (PIR), which can be manufactured to give flame-retardant, resilient, rigid materials.

*b. Formation and Some Reactions of Carbodiimides.* See Section 3.7.3.3B:

$$2R\text{–}N\text{=}C\text{=}O \rightarrow \quad \begin{array}{c} R\text{–}N\text{—}C\text{=}O \\ | \qquad | \\ O\text{=}C\text{—}N\text{–}R \end{array} \quad \rightarrow R\text{–}N\text{=}C\text{=}N\text{–}R + CO_2$$

a carbodiimide

Actually, the carbodiimide can react with another molecule of an isocyanate, and the resulting highly strained four-membered ring intermediate undergoes further reactions that need not be detailed here. Suffice it to say that these reactions lead to an oligomerization of isocyanates.

An interesting and practical illustration of this behavior is the case of MDI that starts out by being a solid of melting point 37°C and thus is not easy to handle on a large, industrial scale. However, a small degree of carbodiimide formation depresses the melting point to give the much more convenient "liquid" MDI.

*c. Formation of Substituted Biuret Structures.* By analogy with the well-known formation of biuret by heating urea (3.9):

$$2H_2N\text{–}CO\text{–}NH_2 \rightarrow H_2N\text{–}CO\text{–}NH\text{–}CO\text{–}NH_2 + NH_3$$

the analogous biuret structure can be produced when moisture decomposes some isocyanate to the corresponding amine (see 4.a) and the latter adds to *two* intact isocyanates:

$$R-NH_2 + 2O=C=N-R \rightarrow R-NH-CO-NR-CO-NH-R$$

It is clear that a biuret moiety offers greater hydrogen-bonding opportunities than the urea or carbamate moieties (see 4.b).

*d. Formation of Allophanates.* The reaction of the N–H moiety of one urethane function with another molecule of isocyanate gives rise to an alophanate:

$$R-NH-CO-O-R' + O=C=N-R'' \rightarrow R-N-CO-O-R'$$
$$|$$
$$CO-NH-R''$$

The alophanates represent a hybrid of the urea and urethane structures, and their hydrogen-bonding capability lies between those of the parents.

There are numerous catalysts employed by the PUR industry, and in practically all situations they function in a poorly understood manner. Consequently, we shall list some of the more common catalysts without a pretense to clarify the mechanisms that explain their involvement.

Among the organometallic catalysts the most prominent are di-*n*-butyltin diacetate and dilaurate:

$$n\text{-}Bu_2Sn(O\cdot CO-R)_2$$

where R is $CH_3$ or $n\text{-}C_{11}H_{23}$, respectively. Numerous tertiary amines, substituted amino alcohols, and quaternary ammonium salts (with appropriate anions since these catalysts function as dipolar ion pairs (or "zwitterions") have been shown to have a selective catalytic effect on some of the reactions of isocyanates. For example, *N'*-hydroxyethyl-*N,N,N'*-trimethylenediamine [$Me_2N-(CH_2)_3-N(Me)-CH_2-CH_2-OH$] is said to promote the cyclotrimerization of an isocyanate to the corresponding isocyanurate, while simple amines 2-dimethylaminoethanol, or dimethylaminoethanolamine, and triethylenediamine (4.3.3.2, 4.3.6.2), the original Air Product's DABCO, catalyze the formation of urethanes more so than cyclotrimerization. The latter is believed to require a tertiary ethylenediamine system with a primary hydroxyl group (also see Scheme 4.7, p. 211).

*7. CHOICE OF FOAM-PRODUCING CONDITIONS.* This is important in view of the wide use of PURs as thermal or shock- and sound-absorbing insulators. The liberation of $CO_2$ by means of the reaction of isocyanates with water or carboxylic acids was already noted above (4.b). The density of the foam end product can be controlled by the amount of added gas-producing reactants. This approach to foam formation can be magnified by means of formic acid (*CMR* 2/28/83):

$$R-N=C=O + H-CO\cdot OH \rightarrow R-NH_2 + CO_2 + CO$$

Foam can also be generated by means of blowing agents. During the relatively low-temperature processing of PURs, the blowing agent is normally an appropriately low-boiling solvent that is volatilized because of the heat of the reaction. Liquids that have been commonly used for this purpose are

> Trichlorofluoromethane, Du Pont's Freon 11, fluorocarbon FC 11, b.p. 24°C, 57¢.
>
> Dichlorodifluoromethane, Du Pont's Freon 12, FC 12, b.p. 29°C, 68¢.
>
> Methylene chloride (3.3.2.3), b.p. 40°C, 35¢.

The restrictions on the use of CFC's as industrial blowing agents, and the search for substitutes, are discussed in Section 3.2.3.

The liberation of $CO_2$ during polymerization creates spongelike, *open-cell* foams used in cushions and padding in a variety of shock-absorbing applications. On the other hand, the use of inert fluorocarbons or methylene chloride produces *closed-cell* foams in which the cells have trapped some gaseous solvent. Particularly in the case of the fluorocarbons, characterized by their low thermal conductivity, the resulting foams have superior insulating capacity.

Many consumer products are fabricated as one-piece objects composed of an interior cellular core covered by an integral, high-density "skin" formed in a single molding operation. MDEA (3.a) is an excellent reactant–catalyst to promote the formation of "self-skinning" PURs.

Foam stabilizers are important additives during the foam manufacturing operation, and the polysiloxane–polyglycol ethers, such as

$$CH_3Si[O-SiMe_2O_x(O-CH_2-CH_2-)_y-(O-CH_2-CHMe-)_z-O-n-C_4H_9]_3$$

are used for this purpose.

*8. CHOICE OF COMBINING OTHER POLYMER FAMILIES WITH PURs.* This option creates countless additional molecular engineering possibilities, including polymer blends and alloys and interpenetrating polymer networks (IPNs). An extensive discussion of these possibilities is beyond the scope of this succinct glimpse at the world of the PURs.

However, in order to accomplish the objective stated in the heading of this segment, we can mention the use of a heterogeneous dispersion of vinyl polymers or copolymers, and other kinds of polymers, use them as fillers that provide impact resistance and other useful properties to the end product. The next step may be the use of blends of PURs with other kinds of polymers shown to be compatible. The resulting polymer blends or alloys combine in an advantageous fashion some properties of both polymer families. Then, there exists the option of assembling a PUR network with a very significant presence of urea, isocyanurate, ester, organic carbonate, and other structural moieties distributed in a random fashion, and thus different from the block polymers (5 above) in which the chemical links other than urethane are segregated in separate macromolecular segments.

Finally, one can incorporate into the PUR network special monomers capable of reacting (during the polymerization process) with either terminal isocyanate or terminal hydroxyl functions. Once attached to the growing PUR structure, these special monomers provide opportunities for copolymerization with other monomers that polymerize by a different mechanism from that of the traditional PUR formation, and hence the two polymerization modes do not interfere with each other. The preceding examples of heterodifunctional monomers (2.b, d) that contain polymerizable isocyanato and vinyl groups offer such opportunities.

Another example of a versatile monomer is triglycidyl isocyanurate (TGIC; structure as that shown in 6.a above except that each R = $-CH_2-CH-CH_2$). It can initiate the *in situ* formation of an epoxy component, while the epoxy function can react not only with hydroxyl but also with isocyanate groups. TGIC is employed in the formation of thermoset powder coatings that are applied electrostatically on metallic surfaces of automotive and appliance components and are cured by heat.

An example of a hybrid PUR–UPR resin is Amoco's Xycon (*MP* 3/88) produced at a 20MM-lb level at Pasadena, Texas. The material results from the combination of a styrene solution of a prepolymer (obtained from isophthalic acid, maleic anhydride, ethylene glycol, diethylene glycol, and neopentyl glycol) that contains unreacted hydroxyl groups, with multifunctional isocyanates (PMDI and others) and curing the resulting mixture with such vinyl polymerization catalysts as benzoyl peroxide (BPO), or *t*-butyl perbenzoate while tertiary amines and/or di-*n*-butytin dilaurate catalyze the isocyanate–hydroxyl group reaction (see H. R. Edwards, "Polyester/Urethane Hybrids Are Suitable for Conventional Molding," *MP* 5/87, 66–74).

*Ionomers* are polymers that contain either anionic or cationic centers, and an example of a cationic PUR is the product of HDI (2.d) or MDEA (3.a), which once incorporated in a PUR, is converted by alkylation of the tertiary amine centers to give a quaternary ammonium structure:

$$[-O-CH_2CH_2-NMe-CH_2CH_2-O-CO\cdot NH-(CH_2)_6-NH-CO\cdot-]$$
$$\downarrow \text{R X}$$
$$[-O-CH_2CH_2-\overset{+}{N}Me(R)-CH_2CH_2-O-CO\cdot NH-(CH_2)_6-NH-CO\cdot-] \quad X^-$$

The end product is a germicidal and antistatic polymer when R–X = benzyl chloride, or a bioplymer when R–X represents a halogenated nucleotide.

### 3.8.2  Monofunctional Isocyanates

Other than the monofunctional isocyanates mentioned in connection with the discussion of PURs (3.8.1, paragraphs 2.b and 2.d), there exists a large family of carbamate insecticides that have traditionally been prepared from methyl isocyanate (MIC), which, in turn, is derived from phosgene. The most famous member of this family is carbaryl, 1-naphthyl *N*-methylcarbamate. This widely used (50MM lb/yr) insecticide is manufactured by Union Carbide and BASF under the tradenames "Sevin" and "Dicarbam," respectively, and is formed from MIC and 1-naphthol:

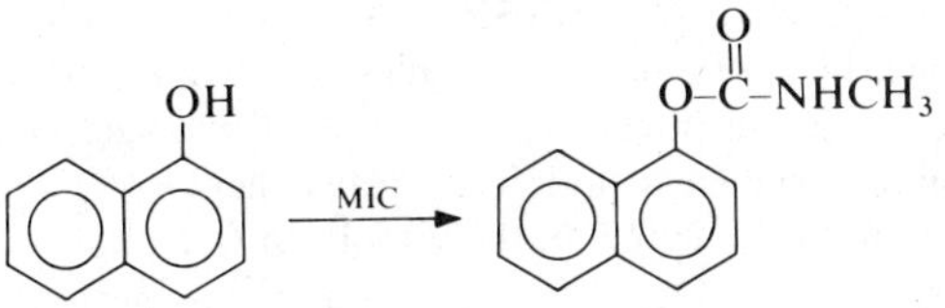

Other members of the carbamate family of pesticides are as follows:

Aldicarb, Union Carbide's Temik         $CH_3-S-\underset{\underset{CH_3}{|}}{\overset{\overset{CH_3}{|}}{C}}-CH=N-O-\overset{\overset{O}{\|}}{C}-NHCH_3$

Carbofuran, FMC's Furadan

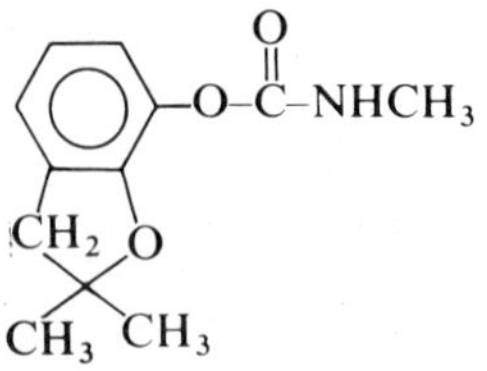

Methomyl, Du Pont's Lannate

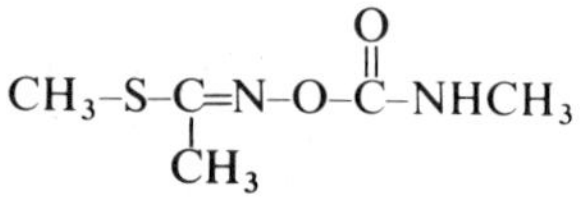

$$CH_3-S-C=N-O-\overset{\overset{\displaystyle O}{\|}}{C}-NHCH_3$$
$$\underset{\displaystyle CH_3}{|}$$

Oxamyl, Du Pont's Vydate

$$(CH_3)_2N-\overset{\overset{\displaystyle O}{\|}}{C}-C=N-O-\overset{\overset{\displaystyle O}{\|}}{C}-NHCH_3$$
$$\underset{\displaystyle S-CH_3}{|}$$

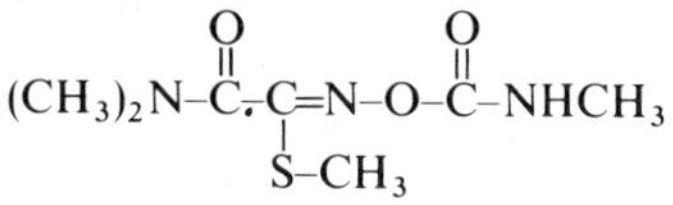

Phenmedipham

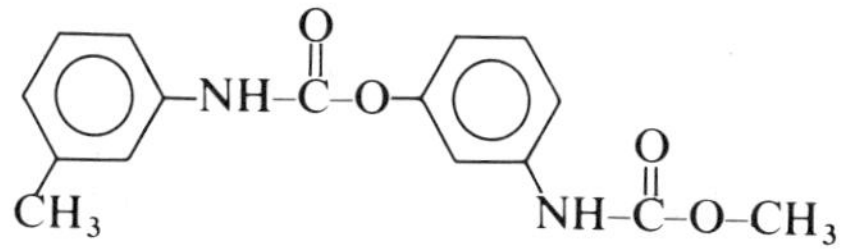

We note that all examples except the last have the *N*-methylcarbamate structure in common, and this is the portion of the molecule believed to possess anticholinesterase activity. However, the last-mentioned material is actually used as a herbicide.

## Use of Pesticides

The use of pesticides by consumers in the United States is estimated (*CW* 6/24/87) to range between some 660MM and 1.2Blb/yr. More than 60% of the pesticides are represented by herbicides (see H. J. Sanders, *CEN* 8/31/81, 20–35; S. Randel, *CB* 5/87, 32–34; and 3.8.5), and the rest are insecticides, fungicides, rodenticides, and so on.

Weed control is essential to mechanized agriculture, and the importance of insecticides can be illustrated by the grasshopper plague in 1985, the greatest in the United States since the late 1930s (*CW* 5/13/87). The damage caused by grasshoppers to food, forage, and forest crops averages about $390 million per year, but in 1985 the USDA was forced to treat nearly 14 million acres infested by an estimated 55 million adult grasshoppers. This time, an integrated pest-control management (IPM) program was used, which combines chemical and biological defenses (*KCB*, 39–47, 74): carbaryl, acephate, and malathion (a member of the thiophosphate family of insecticides; see 6.6.8) were used in combination with a grasshopper parasite *Nosema locustae*. At the time of this writing, the locust plague that is sweeping across Africa (April 1988) is likely to cause greater famine than that experienced in recent history, and the appropriate United Nations agencies are even considering use of the highly maligned chlordane to control the disaster.

Insecticides and fungicides must be employed to prevent postharvest damage to fruit estimated at about 20% of the crop (*CW* 5/7/87). Nematodes cause an estimated annual crop loss valued at $4 billion in the United States (*CW* 8/18/86).

The traditional route to carbaryl and other methylcarbamate pesticides involves the reaction of MIC with the given substrate:

$$CH_3-NH_2 + Cl_2CO \rightarrow CH_3-N=C=O \quad MIC \ (+2HCl)$$

$$CH_3-N=C=O + Q-O-H \rightarrow CH_3-NH-\overset{\overset{\displaystyle O}{\|}}{C}-O-Q$$

The nonphosgene routes to the preparation of isocyanates are discussed in 3.8.4.

### 3.8.3 Organic Carbonates and Polycarbonates (PCs)

The traditional route to simple organic carbonates, such as dimethyl carbonate (DMC), ethylene and propylene glycol carbonates, and other carbonates, involves the reaction of the corresponding hydroxy compounds with phosgene, but efforts are currently being exerted to replace the use phosgene by carbon dioxide (3.7.2.4, 3.7.2.5).

As mentioned elsewhere (3.7.1), the production of the most prominent polycarbonate polymer based on bisphenol A (see below) is the second largest consumer of phosgene, while poly(alkylene carbonates) are obtained from epoxides and $CO_2$ (3.7.2.5).

The production of *the* most common "polycarbonate resin" pellets ($1.93) involves the reaction of bisphenol A, BPA (5.5.3 and 9.3.4, polycarbonate grade 86¢, epoxy grade 82¢) and phosgene (55¢). The reaction is carried out in a convenient solvent such as methylene chloride, and this process offers an excellent illustration of phase-transfer catalysis.

*Phase-Transfer Catalysis*

Since the efficiency of the reaction requires that BPA be used in the form of water-soluble phenolate anions, while phosgene must be dissolved in a chemically inert and hence water-insoluble solvent, the desired reaction would have to depend on the diffusion rate of the two reactants to the interface between the immiscible solvents. The area of the interface can be increased by vigorous stirring of the two-phase reaction mixture, but a more efficient way to accelerate the process is to induce one of the reactants to migrate into the phase that a priori is not a particularly receptive to it. In this example, the sodium phenolate ions are in equilibrium with a phase-transfer catalyst such as tetra-*n*-butylammonium chloride, and while one of the products of the equilibrium (sodium chloride) remains in the aqueous phase, the other product of the equilibrium (the BPA anion–tetra-*n*-butylammonium cation ion pairs) are of sufficient covalent character to migrate into the nonaqueous phase where they encounter phosgene and the reaction takes place. The process is represented in Scheme 3.23.

The disadvantage of the preceding process is the insolubility of the polymer as its molecular weight increases. An alternative process that avoids this limitation involves the reaction of molten BPA (m.p. 155–157°C), with a convenient simple organic carbonate that produces a volatile by-product. Diphenyl carbonate has been employed for that purpose because phenol is sufficiently volatile (b.p. 182°C at atmospheric pressure) and provides a good leaving group. Another advantage of this process is the absence of traces of HCl from the end product. Diphenyl carbonate can be prepared from phenol by means of phosgene or by transesterification with DMC. Similarly, one can use di-*n*-butyl

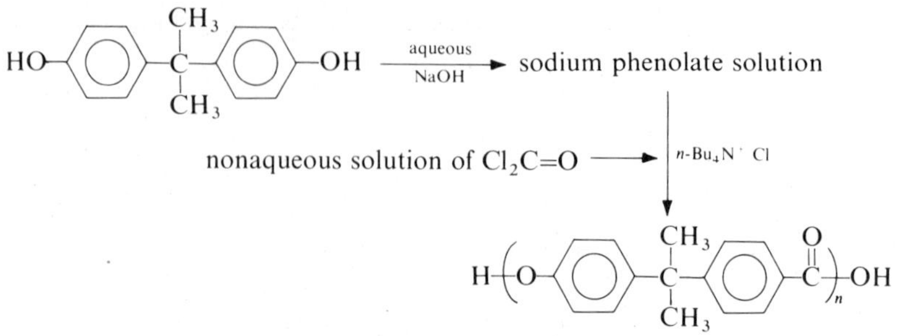

**Scheme 3.23**

carbonate. The use of organic carbonates in place of phosgene allows a better control of the molecular weight distribution of the resulting PC.

The current demand for PC is about 300MM lb and is increasing at about 8% per year. PC is an engineering-quality thermoplastic and forms useful blends with PBT, ABS, PS, and other polymers. Its structural uses include the manufacture of components of typewriters and other office equipment, including computers and scratch-resistant eyeglasses (a $2-billion market; see *CMR* 9/28/87). Until recently, a special high-quality PC required for the manufacture of compact disks and other electronic and tele-communication devices had to be imported from Bayer in West Germany. Fortunately, Mobay is now manufacturing such material in the United States. According to an announcement by Mobay (*CMR* 10/12/87, *CW* 10/14/87), the coextrusion of PC–PVDC films is now possible to give excellent moisture barrier properties in plastic films suitable for packaging moisture-sensitive products.

### 3.8.4  Nonphosgene Routes to Isocyanates, Carbamates, and Urethanes

The Bhopal disaster in December 1984 has certainly intensified the search of replacements of phosgene and the avoidance of isocyanate intermediates, although such efforts predate this event (see, e.g., N. Yamazaki, *CEN* 4/16/78). The above-mentioned use of diphenyl carbonate is an approach useful not only in the case of PCs but also for the assembly of PURs, as shown in Scheme 3.24.

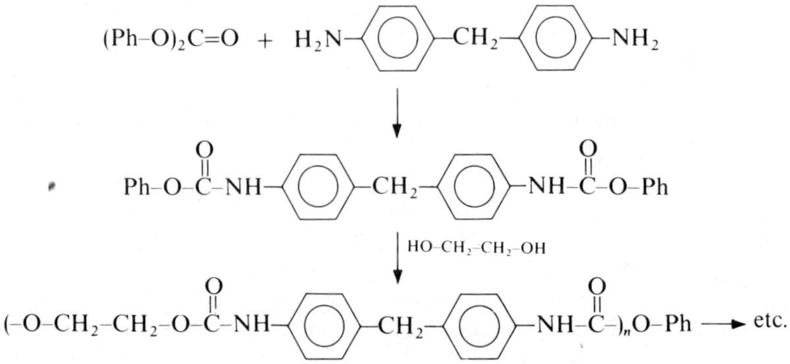

**Scheme 3.24**

The reduction of nitroaromatics to the corresponding anilines is mentioned in Section 3.5.4. An extension of this transformation in the presence of an alcohol is reported (*CT* 6/78, 782; *CEN* 4/4/85, 79) to lead to the formation of urethanes:

$$Ar-NO_2 + CO + R-OH \rightarrow Ar-NH-CO \cdot OR$$

when the reaction is carried out at 177°C, 1000 bar, and in the presence of selenium or selenium oxide catalyst. Once such an aromatic urethane is formed, it can be subjected to a condensation reaction with formaldehyde and the resulting oligomeric or polymeric product can then be decomposed thermally to give PMDI, as shown in Scheme 3.25.

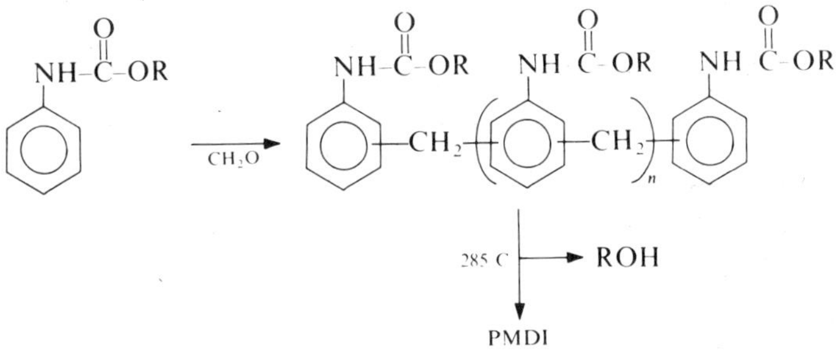

**Scheme 3.25**

The use of toxic selenium catalyst is a drawback to a large-scale application of this process and explains the ARCO decision (*CEN* 2/2/81) to postpone the construction of a 150MM-lb PMDI plant in Channelview, Texas. Along similar lines, we note the announcement (*CE* 1/19/87) by Catalytica of Mountainview, California, of the joint effort with Haldor Topsoe of Denmark and Nippon Kokan K. K. of Tokyo to produce the monourethane from aniline, CO, and an alcohol in the presence of a homogeneous ruthenium catalyst. This is followed by a treatment of the product with formaldehyde to give the diurethane and, finally, by the thermal decomposition of the latter to MDI and some PMDI by-product.

Regardless of how one arrives at a primary amine, the reaction with the currently available DMC provides a phosgene-free entry to the corresponding methyl carbamate, and this material can, as shown above, be converted to the corresponding isocyanate:

$$R-NH_2 + MeO-CO-OMe \rightarrow R-NH-CO-OMe \rightarrow R-N=C=O$$

The Du Pont announcement (*CE* 6/24, 8/19/85) of a continuous-loop oxidative dehydration process that converts *N*-methylformamide to MIC at 550–650°C:

$$H-CO-NHMe + O_2 \rightarrow (HO-CO-NHMe) \rightarrow O=C=N-Me + H_2O$$

and the immediate conversion of the latter to the insecticide methomyl shown in Section 3.8.2 is symptomatic of efforts to avoid the use of phosgene. This process is apparently preferable to the reaction of methylamine and DMC because of the facile formation of *N*,*N*′-dimethylurea.

A totally different approach to the production of isocyanates from olefins that are likely to give carbon cations is being exploited by Cyanamid (*CMR* 9/8/86, *CW* 9/10/86). An isopropenyl aromatic compound, symbolized here by Ar–C(CH$_3$)=CH$_2$, in the presence of an acidic catalyst, reacts with, say, methyl carbamate to give the corresponding urethane:

$$Ar\text{–}C(CH_3)\text{=}CH_2 + H_2N\text{–}CO\cdot OMe \rightarrow Ar\text{–}C(CH_3)_2\text{–}NH\text{–}CO\cdot OMe$$

Thermal decomposition of the urethane produces the expected "aliphatic" isocyanate such as TMXDI and TMI mentioned in Section on PURs, 1.d, if the operation starts with the diisopropenyl derivatives of benzene. The acid-catalyzed reaction of olefinic systems with amino group-containing reactants is not unexpected in view of the well-known Ritter reaction (6.3.1.4).

### 3.8.5  Miscellaneous Synthetic Uses of Phosgene

The preparation of chloroformates (Cl–CO–OR) cannot avoid the use of phosgene, and compounds such as

- Ethyl chloroformate, Cl–CO–OC$_2$H$_5$
- Diethylene glycol bischloroformate, (Cl–CO–O–CH$_2$CH$_2$)$_2$O
- Myristyl chloroformate, $n$-C$_{14}$H$_{29}$–O–CO–Cl
- $o$-Isopropylphenyl chloroformate, $o$-iso-PrC$_6$H$_4$–O–CO–Cl

and many others, serve as organic building blocks of interest. One industrially important application of chloroformates is the synthesis of peroxy-group-containing vinyl polymerization catalysts discussed in some detail elsewhere (6.3.1.7). Examples of such syntheses follow:

$$R\text{–}O\text{–}CO\text{–}Cl + Na^{+\,-}O\text{–}O\text{–}t\text{-}Bu \rightarrow R\text{–}O\text{–}CO\text{–}O\text{–}O\text{–}t\text{-}Bu$$

a *t*-butylperoxy carbonate

$$R\text{–}O\text{–}CO\text{–}Cl + Na_2O_2 \rightarrow R\text{–}O\text{–}CO\text{–}O\text{–}O\text{–}CO\text{–}O\text{–}R$$

a peroxydicarbonate

Thiols react with phosgene to give substituted thiolchloroformates, R–S–CO–Cl, and these, with amines, give the expected substituted thiol carbamates, R–S–CO–NR$'_2$. Some of the thiol carbamates are important herbicides:

*S*-Ethyl di-*n*-porpylthiocarbamate, EPTC, Et–S–CO–N(*n*-Pr)$_2$
*S*-Ethyl di-isobutylthiocarbamate, butylate, Et–S–CO–N(iso-Bu)$_2$

Disubstituted dithiolcarbonates, R–S–CO–S–R, can be obtained from either phosgene or DMC. The same is true of tetrasubstituted ureas such as tetramethylurea, Me$_2$N–CO–NMe$_2$ (b.p. 176°C), one of the dipolar aprotic solvents (3.5.3.1).

Alpha-amino acids react with phosgene to give the corresponding *N*-carboxyanhydrides, and the latter can be polymerized to polypetides, as shown in Scheme 3.26.

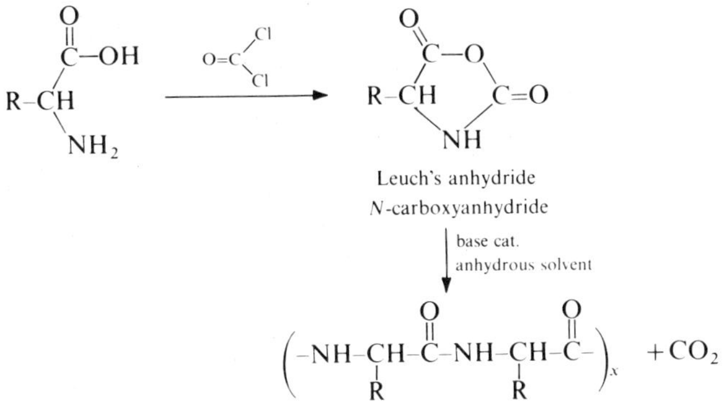

**Scheme 3.26**

A costly derivative of phosgene or DMC is $N,N'$-carbonyl diimidazole (CDI, \$68/lb). It introduces the –CO– building block under very mild conditions because of the superior leaving-group character of the imidazole anion. It can be used to esterify a carboxylic acid in a two-step operation carried out at low temperatures, as demonstrated in Scheme 3.27.

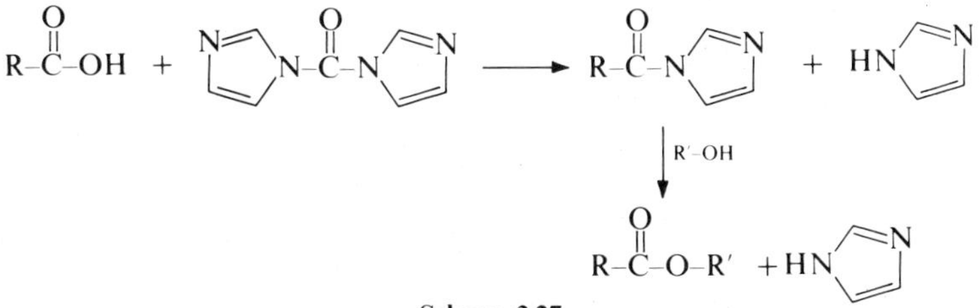

**Scheme 3.27**

Obviously this reagent is employed in synthetic operations involving very delicate and costly end products such as some pharmaceuticals, antibiotics, and polypeptides, (see e. g., *CMR* 3/18/85).

Finally, phosgene can participate in a stepwise manner in Friedel–Crafts acylations of aromatic systems (9.3.2) to produce either an intermediate acid chloride or a diaryl ketone (9.3.2). Aromatic systems extremely susceptible to electrophic substitution reactions produce the ketone directly without catalytic assistance of aluminum chloride or other Lewis acid catalysts. Thus, for example, $N,N$-dimethylaniline (\$1.03) is converted to Michler's ketone, a dyestuff intermediate, as shown in Scheme 3.28.

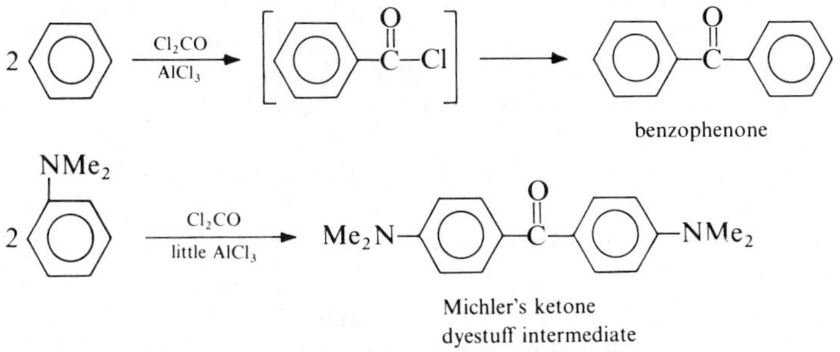

benzophenone

Michler's ketone
dyestuff intermediate

**Scheme 3.28**

## 3.9   UREA, CYANAMIDE, THIOUREA, AND OTHER UREA DERIVATIVES

### 3.9.1   Production of Urea and an Overview of Its Uses

As stated elsewhere (Table 1.1 and Section 3.7.1 above), the production of about 12B lb of urea is governed primarily by fluctuations in the demand of nitrogenous fertilizers. The economy of scale of this large market for urea creates an inexpensive feedstock for the production of a broad spectrum of organic chemicals that range from the relatively low-cost UF resin, the somewhat more costly melamine–urea (MF) resins, to still higher-value-added medicinals such as the barbiturates and certain sulfa drugs. The current relatively low cost of urea (10¢) is another case of price advantages associated with a situation when chemical needs "ride piggy back" on massive demands by other sectors of the economy (2.4).

Fertilizer and animal feeds consume about 90% of the urea demand. In recent years the production of urea by the United States has been threatened by imports. To illustrate, the price of U.S.-manufactured urea during the early part of 1986, namely, $100–$110/t (5¢–5.5¢), dropped to $74–$78/t in December 1986, when urea imported from the Soviet Union, East Germany, and Romania was offered at $61/t, while urea from Italy and Venezuela was priced at $67 and $78 per ton, respectively. It is estimated (*CMR* 12/22/86) that a price range of $84–$88 per ton is necessary to avoid shutdowns of domestic production facilities. Early January 1987 (*CW* 1/7–14/87) the Commerce Department imposed preliminary import duties of 144.1, 84.9, and 53.71% on the value of urea imported from East Germany, the Soviet Union, and Romania, respectively, in order to compensate for the price-cutting policies of those countries. Final judgment on punitive import tariffs will depend on the conclusions derived from a study by the USITC of true production costs. The urea import pressures have caused a shrinkage of domestic production capacity from over 8MM t to about 7.5MM t in 1987.

Obviously, agricultural disincentive policies of the Federal Government do not help to maintain a healthy demand for urea and, for that matter, threaten the whole domestic fertilizer and agrichemical production (*SZM-II*, 1–5).

Of the 10% of nonagricultural urea consumption, the two largest markets for chemical urea are the above-mentioned UF and MF resins produced at the level of about 1.4B lb.

The use of UF foams as domestic insulating materials experienced a formaldehyde scare (3.4.1) a few years ago, but the use of UF as an adhesive in the fabrication of particle and laminated construction boards is firmly established, as is its use in paper and textile treatments. Urea is a source of melamine (see below)—the building block of MF resins—and the latter maintain a small but steady demand (∼45MM lb) for the production of light-shade, decorative consumer products such as unbreakable dishes, and other household items.

At high temperatures urea suffers the loss of either ammonia (at ∼400°C) or water (at ∼450°C) to give either cyanic acid (H–O–CN) or cyanamide ($H_2N$–CN), respectively. The former becomes the parent of various inorganic cyanates, while the latter is the source of melamine, thiourea, guanidine and substituted guanidines, and dicyanamide (see below).

The logical route to urea is the combination of $NH_3$ and $CO_2$ (3.7.2.1), and this preparative approach often becomes a part of an integrated facility for the conversion of methane (natural gas), air, and water to ammonia, methanol, urea, nitric acid, and ammonium nitrate as summarized in Fig. 3.1 (p. 66).

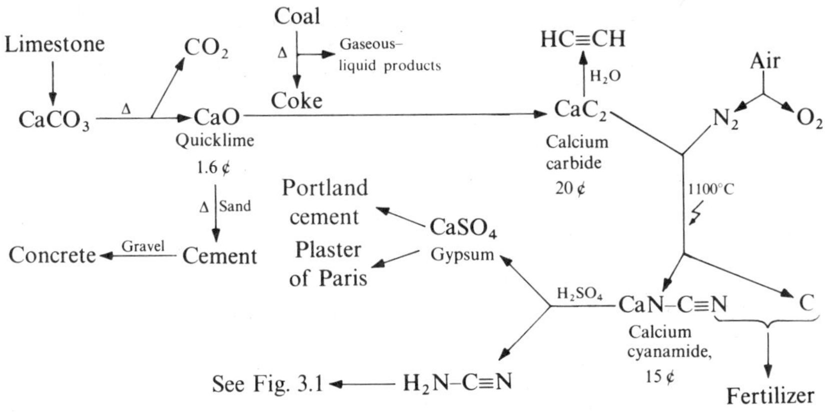

**Figure 3.2**  Some transformation products of limestone, coal, and air.

An older, high-energy route to urea and some of its derivatives is the nitrification of calcium carbide. The latter is also a product of energy-intensive transformations. In the case of the United States, an integrated production scheme based on this route (see Fig. 3.2) is currently of only historical importance because of the limitated availability of inexpensive hydroelectric energy. The major reason for the production of calcium carbide in the United States is its conversion to acetylene (4.7) for use mainly in acetylene torches for metal welding and cutting operations, with only a small amount employed for chemical purposes by Air Products in the coal-rich region of Kentucky. Echoes of the historical role of calcium carbide and calcium cyanamide linger in the names of companies such as Union Carbide and American Cyanamid that were born several decades ago in the Niagara Falls region. However, there exist other regions of the world where hydroelectric energy is either not exploited or wasted, and where operations such as those shown in Fig. 3.2 could be revived.

### 3.9.2  The Amino Resins: UF and MF

A formaldehyde:urea mole ratio somewhat higher than a $1:1$ creates a cross-linked UF resin of a structure symbolized by

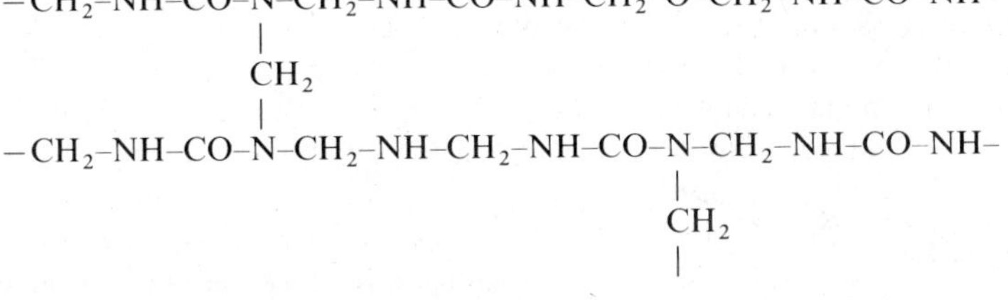

$$-CH_2-NH-CO-N-CH_2-NH-CO-NH-CH_2-O-CH_2-NH-CO-NH-$$
$$|$$
$$CH_2$$
$$|$$
$$-CH_2-NH-CO-N-CH_2-NH-CH_2-NH-CO-N-CH_2-NH-CO-NH-$$
$$|$$
$$CH_2$$
$$|$$

The condensation polymerization occurs by way of methylol- and dimethylolurea, $H-CH_2-NH-CO-NH_2$ and $HO-CH_2-NH-CO-NH-CH_2-OH$, respectively, and this accounts for occasional methylene ether linkages as the one indicated above. The nature of the UF resin end product is very sensitive to the pH of the reaction mixture, the ratio of the two major components, and the presence of additives such as lower alcohols (methyl, butyl, etc.) or fillers and reinforcing agents such as cellulose and glass fibers. Thus, careful control of the polymerization conditions produces UF coating resins, adhesives and binders, thermoset molding compositions, and films that serve as *microencapsulation* materials. The last-mentioned product is one of the basic reasons for the growth of controlled-release technology applied in numerous areas of the economy. A recent example of the latter is the announcement (*CW* 9/23/87) of a project to develop microencapsulated biopesticides produced by Mycogen of San Diego, California and release them, where needed, by means of Monsanto's controlled-release technology. Another example of the use of microencapsulation is the color-imaging system developed by Mead of Dayton, Ohio (see. S. Jones Ainsworth and A. Slakter, "Mead Co. Brings Color to Business," *CW* 12/23/87, 32–33).

The possibility of potential emissions of gaseous formaldehyde from improperly formulated and processed UF foams (see 3.4.1) generated at the site of homes being serviced for thermal insulation is likely when one considers the delicate controls that must be exercised in order to avoid the presence of terminal methylol groups in the UF end product.

The chemistry of MF resins is similar to that of the UF system except that melamine (3.9.3) has more reactive sites than urea for coupling by means of methylene bridges. However, a 1:1 melamine:formaldehyde mole ratio suffices to produce a linear polymer and little additional formaldehyde creates cross-linked, thermoset products. The exceptional clarity and toughness of MF resins accounts for their use in the manufacture of laminated products and a 1:2 melamine:formaldehyde mole ratio is commonly used for that purpose. MF resins produce adhesives and binders that are superior than UF resins, but their higher cost (due to the relative pricing of melamine and urea, 50¢ and 10¢, respectively), accounts for the fact that mixtures of the two building blocks are employed as a compromise between performance and costs (2.4). The current price of MF resin is 55¢, and formulated molding compounds are priced at 46.5¢. An important use of MF resins is the treatment of textiles to induce wrinkle resistance.

The production of UF and MF resins in 1986 is reported (*CW* 6/24/87) to be 1.27B and 173MM lb, respectively.

### 3.9.3  Urea Derivatives Other Than UF–MF Resins

The group of materials discussed in this section includes melamine, barbiturates, sulfamic acid, azodicarbonamide blowing agents and polymerization initiators, guanidine, cyanide, dicyanamide, ethylene urea, and acetylene diureine.

The thermal decomposition reactions of urea to cyanamide and cyanic acid and the subsequent transformations of these two initial products to materials of industrial importance are summarized in Fig. 3.1. Of these materials, the most important in terms of volume is the above-mentioned melamine ($\sim$135MM lb). Melamine Chemicals of Donaldsonville, Louisiana, one of two domestic manufacturers of melamine, has announced (*CW* 8/19/87) a proposed shift from the traditional low-pressure catalytic process (*FKC*, 519–523) used to convert urea to melamine to a more efficient and capital cost-saving continuous, high-pressure, noncatalytic process in which molten urea (m.p. 132°C) is converted directly to "liquid" melamine, ammonia, and CO$_2$. Since melamine is reported to sublime below 250°C, the high-pressure conditions and the absence of catalyst that reduces the need of purification seem to be the novel features of the process. The 30MM lb plant, constructed at a cost of $7.25 million, is expected to be onstream by early 1989, with an output complementing that of the existing 70MM-lb/yr plant.

The second domestic supplier of melamine and also a prominent supplier of cyanamide and dicyanamide (see below) is American Cyanamid.

The condensation of urea with substituted malonic acid esters (5.6.5) leads to *barbiturates*, medicinals that function as hypnotics and sedatives.

The reaction of urea and sulfuric acid produces sulfamic acid (38¢):

$$H_2N-CO-NH_2 + H_2SO_4 \rightarrow H_2N-SO_3H$$

Sulfamic acid is used in electroplating and boiler cleaning operations and as a herbicide in the form of ammonium sulfamate.

The reaction of urea with hydrazine (hydrate 85%, $1.25) leads to the formation of 1,1-azobisformamide (azodicarbonamide, $2.70):

$$H_2N-CO-NH_2 + H_2NNH_2 \rightarrow H_2N-CO-NHNH-CO-NH_2$$

$$\xrightarrow{\text{air}} H_2N-CO-N=N-CO-NH_2$$

The latter can be also prepared as shown in Scheme 3.32.

Traces of alkali catalyze the oxidation step, and the preparation of azodicarbonamide illustrates the ease of oxidation of 1,2-disubstituted hydrazines to the corresponding azo compounds. The nonaromatic azo compounds, R–N=N–R, represent one class of blowing agents, that is, materials used to release gas during foam manufacturing operations.

*Azo Blowing Agents*

Azo blowing agents are gas-producing materials employed by the plastics processing industry in the manufacture of low-density, injection-molded or -extruded objects such as pipes, sheets, wire and textile coatings, and flooring pannels. The decomposition of the blowing agents during the processing temperatures (166–232°C) creates uniformily dispersed, small, gaseous closed cells, but the decomposition temperatures can be lowered in the presence of catalysts (activators). The solubility characteristics (required for the production of a unform cell structure) favor the use of azodicarbonamide in the processing of ABS, acetal, acrylic, PPO, HDPE, LDPE, PP, PS, TPE, vinyl, and some other polymers. Another azo blowing agent is azobisisobutyronitrile, $N\equiv C-C(CH_3)_2-N=N-C(CH_3)_2-C\equiv N$, AIBN, derived from acetone (5.5.2) and HCN (3.10.2). Similar homologous azo nitriles are derived from higher ketones:

$$2R-CO-R' + 2HCN + H_2N-NH_2 \rightarrow N\equiv C-CRR'-NH-NH-CRR'-C\equiv N$$

$$\downarrow \text{air, alkali cat.}$$

$$N\equiv C-CRR'-N=N-CRR'-C\equiv N$$

On heating, these azo dinitriles decompose to give relatively long-lived radicals (that function as free-radical polymerization initiators) and nitrogen gas (which creates a plastic foam):

$$N\equiv C-CRR'-\ddot{N}=\ddot{N}-CRR'-C\equiv N \rightarrow 2(:N\equiv C-\dot{C}RR' \leftrightarrow \cdot\ddot{N}=C=CRR') + :N\equiv N:$$

A wide selection of azo polymerization initiators is available commercially, and among other factors, the choice depends on the temperature at which a given azo compound decomposes to initiate the desire vinyl polymerization, solubility of the azo compound in the polymerization mixture, and the compatibility of the residual fragments in the end product. Thus R and R' can, for example, be

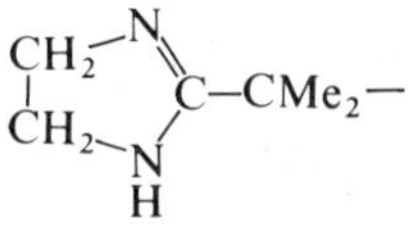

$$\underset{\underset{H}{\overset{\displaystyle CH_2-N}{\underset{CH_2-N}{\big|}}}}{} C-CMe_2-$$

2'-imidazolinyl-2-propyl

$$Me-O-CO-CMe_2-$$

2-carboxymethyl-2-propyl

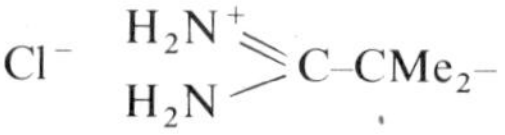

$$Cl^- \quad \underset{H_2N}{\overset{H_2N^+}{\diagup}}C-CMe_2-$$

hydrochloride of 2-amidino-2-propyl

It is noteworthy that whatever the terminal group may be, the azo group is usually attached to a tertiary carbon in order to facilitate the liberation of nitrogen gas.

In connection with the aliphatic azo blowing agents, we may mention two other common families of blowing agents:

1. Arylsulfonylhydrazides, $Ar-SO_2-NH-NH_2$ (9.3.3).
2. N-Nitroso compounds of general structure $RR'N-N=O$ obtained by nitrosation of $RR'NH$ precursors.

All three families of blowing agents generate nitrogen gas (and possibly some other gases such as CO in the case of azodicarbonamide) and, as implied above, a variety of solid decomposition products that become incorporated in the polymeric end product.

The reaction of urea with ammonium nitrate produces *guanidine* or, more properly, guanidinium nitrate:

$$H_2N-CO-NH_2 + NH_4^+\,NO_3^- \;\rightarrow\; (H_2N)_2C=NH_2^+\,NO_3^-$$

This nitrate (see below) decomposes explosively and can be used as a propellant. However, an ion-exchange resin can convert the nitrate to nonexplosive guanidinium chloride or sulfate, and these salts can then be used with impunity in synthetic operations such as those shown below. Guanidine is a relatively strong base because protonation results in the formation of the highly resonance stabilized guanidinium ion $(H_2N)_3C^+$. The stability of the latter accounts for the availability (Chemie Linz) of guanidine carbonate,

$$2(H_2N)_3C^+\,CO_3^{2-}$$

*Cyanamide* is a versatile building block of thiourea, S-substituted thioureas [$H_2N-C(=NH)-S-R$], guanidine, and substituted guanidines such as diphenylguanidine (Fig. 3.1). The latter compound is formed by virtue of the facile addition of the first aniline to the cyano group of cyanamide, followed by a facile substitution of the remaining amino group of the intermediate phenylguanidine by another aniline molecule. As mentioned above, unsubstituted guanidine can be obtained more directly from urea and ammonium nitrate.

*Dicyandiamide*, also known as "dicy" or cyanoguanidine (75¢, Cyanamid Canada), is a dimer of cyanamide obtained from the former in the presence of aqueous ammonia or slaked lime. It is the industrial source of guanidine nitrate (see above):

$$H_2N-C(=NH)NH-C\equiv N + 2NH_4NO_3 \rightarrow 2[H_2N-C(NH_2)=NH_2^+ \ NO_3^-]$$

The presence of amino, imino, and cyano functional groups in dicyandiamide provides numerous opportunities for synthetic uses. Formaldehyde yields monomethyloldicyandiamide:

$$HO-CH_2-NH-C(=NH)NH-C\equiv N$$

that forms resinous products with or without addition of casein (*SZM–II*, 30, 161), urea, and other amino-containing materials. Dicy is also an intermediate in the production of melamine and barbiturates (see above), but perhaps its most unique property is that of *intumescence*, that is, thermal decomposition to give a charred, swollen, foamy surface layer on a coating to which it was added. The char becomes an insulator of the bulk of the material and thus dicy functions as a fire-retardant.

*Guanidine* is the building block of sulfa drugs of the sulfadiazine family. These are assembled from 2-aminopyrimidines (10.3.2) obtained, in turn, by means of a condensation reaction between guanidine and 1,3-dicarbonyl compounds, as shown in Scheme 3.29.

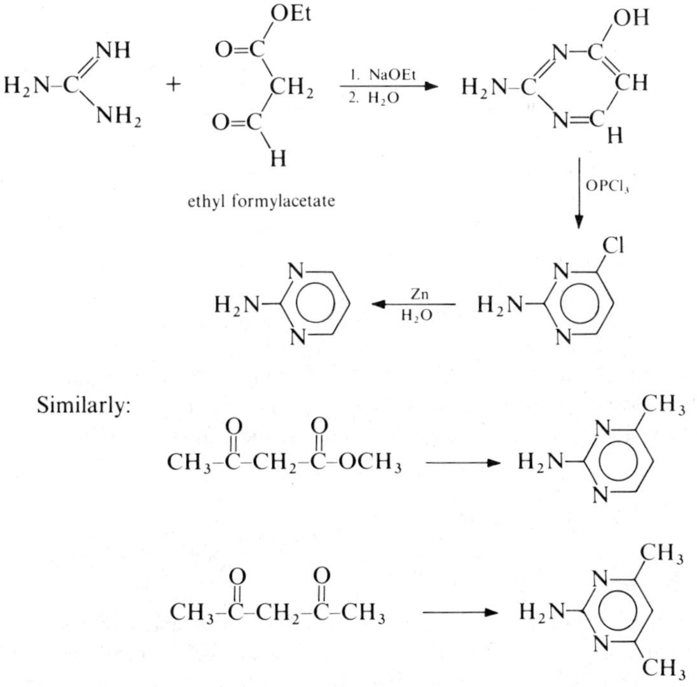

ethyl formylacetate

Similarly:

Scheme 3.29

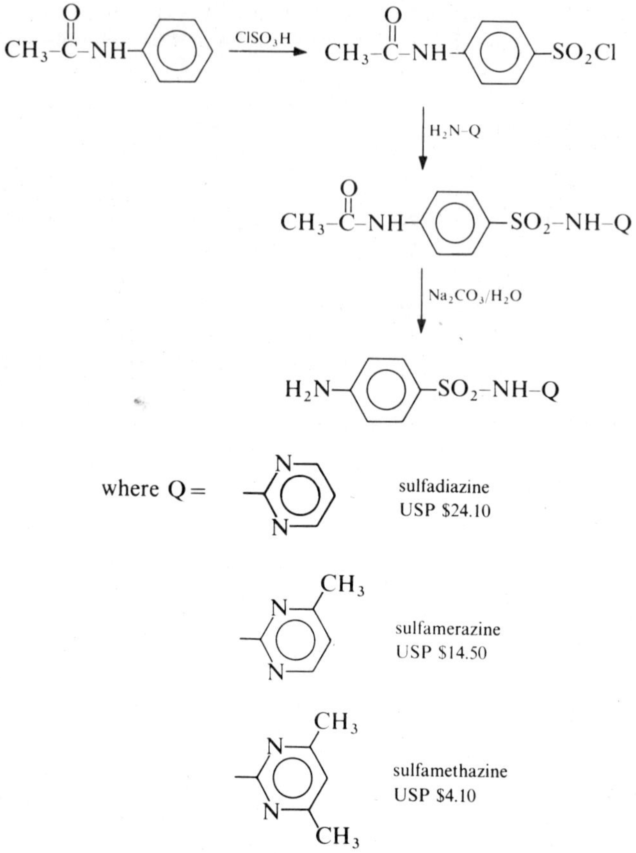

$$\underset{\overset{\displaystyle O}{\parallel}}{CH_3-C-NH-}\underset{}{\bigcirc} \quad \xrightarrow{ClSO_3H} \quad \underset{\overset{\displaystyle O}{\parallel}}{CH_3-C-NH-}\underset{}{\bigcirc}-SO_2Cl$$

$$\downarrow H_2N-Q$$

$$\underset{\overset{\displaystyle O}{\parallel}}{CH_3-C-NH-}\underset{}{\bigcirc}-SO_2-NH-Q$$

$$\downarrow Na_2CO_3/H_2O$$

$$H_2N-\bigcirc-SO_2-NH-Q$$

where Q =

sulfadiazine
USP $24.10

sulfamerazine
USP $14.50

sulfamethazine
USP $4.10

**Scheme 3.29**    (*Continued*)

Obviously, guanidine is also responsible for the formation of another sulfa bacteriocide, namely, sulfaguanidine:

$$4-H_2N-C_6H_4-SO_2-N=C(NH_2)_2$$

The hydrazino analog of guanidine can be obtained by the addition of hydrazine followed by a treatment with formic acid that also produces ring closure to yield 3-amino-1,2,4-triazole, a postemergence, nonselective herbicide known as "amitrole":

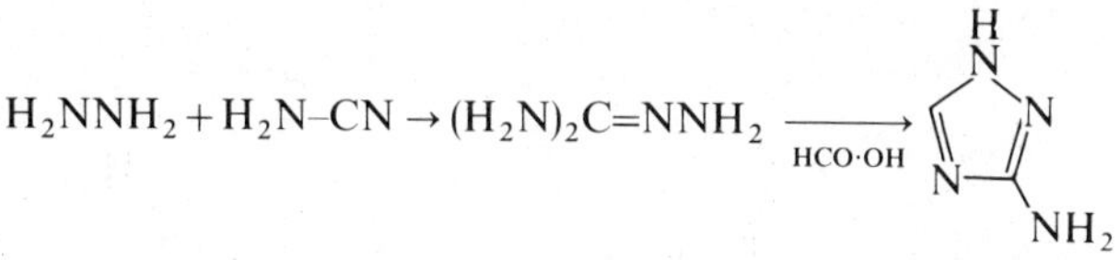

$$H_2NNH_2 + H_2N-CN \rightarrow (H_2N)_2C=NNH_2 \xrightarrow{HCO\cdot OH}$$

The formation of *ethylene urea* from ethylenediamine and either $CO_2$ or urea is mentioned in Section 3.7.2.3. This product is imported by BASF from West Germany and is priced at $4.20/lb. The dimethylol derivatives can be used in the treatment of textiles and leather and as modifiers of formaldehyde-derived resins.

An extension of the preceding reaction is the formation of acetyleneurea or *acetylene diureine*, obtained by the reaction of two moles of urea with glyoxal (4.5.3) 40%, 44.5¢. Acetylene diureine is an excellent scavenger for small amounts of formaldehyde, and its trimethylol derivative, trimethylol acetylene diureine, (TMDAU, National Starch, $60.00), is a specialty cross-linking agent for UF or MF resins. It stabilizes compatible polymers against degradation by UV light, but its most important use is the manufacture of "durable-press" clothing. The formation of both acetylene diureine and its trimethylol derivative is shown in Scheme 3.30.

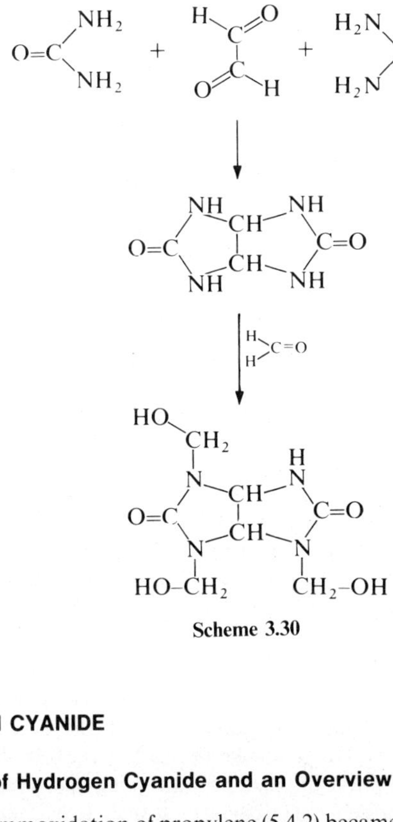

Scheme 3.30

## 3.10   HYDROGEN CYANIDE

### 3.10.1   Sources of Hydrogen Cyanide and an Overview of Its Uses

Until 1963, when ammoxidation of propylene (5.4.2) became Sohio's answer to the search of an efficient source of acrylonitrile, hydrogen cyanide was obtained by either acidification of sodium cyanide or high-temperature reactions of ammonia with methane. Both of these traditional sources of HCN are highly energy intensive:

(a)  Na, 70¢, (molten at 200–300°C) + $NH_3$, anhydrous, 5¢ → $NaNH_2$

$$NaNH_2 \text{ (sodamide), } \$3.75 + C \text{ (coke)} \xrightarrow{800°C} NaCN \xrightarrow{H_2SO_4} HCN$$

(b)  Oxidative ammination of methane or natural gas (Andrussov oxidation; ONR-3), occurring at about 1400°C on the presence of a platinum or platinum–rhodium catalyst:

$$CH_4 + NH_3 + 1.5O_2 \rightarrow HCN + 3H_2O$$

Of these processes, (b) is more economical. However, neither can compare with formation of HCN as a 15% by-product of the ammoxidation of propylene since the main product, acrylonitrile, is in high demand (5.4.2).

Hydrogen cyanide can also be prepared by means of a dehydration of formamide (3.5.3.1), (39¢), but this transformation does not seem favorable economically (HCN is currently priced at 50¢) unless it is carried out for captive use by a producer of formamide from syngas.

A potential source of HCN is the ammoxidation of methanol, mentioned elsewhere (3.3.2.8.G). Whatever source of HCN one has in mind, it is most convenient that it be used near the site of production because of rather stringent shipping regulations (boiling point of HCN 25.6°C, very toxic). Although special railroad cars are designed for more distant shipments, but there is no guarantee against derailment.

The current demand for HCN in the United States slightly exceeds the 1B–lb level, and about 25% of this demand is satisfied by ammoxidation of propylene since HCN represent about 10% of the output of acrylonitrile (AN).

Before we discuss the organic uses of HCN, it is noteworthy that the flow of chemicals represented by the above-mentioned traditional source (a) has now been reversed: about 5% of HCN is converted to sodium (and potassium) cyanide. On a molar basis, sodium cyanide is about 1.5 times more costly than HCN:

$$HCN, 50¢ + caustic\ soda \rightarrow NaCN, 71¢$$

73% liquid, 10¢
76% flake, 25¢

Potassium cyanide ($1.32) is more costly than NaCN, in accord with the higher price of caustic potash (45% liquid, 13¢; 88–92% flake, 43.35¢) and the smaller production volume. In addition to organic synthetic uses in which one may employ NaCN as an *in situ* source of HCN (see below), NaCN is used in the production of gold and silver by means of hydrometallurgical processes and in electroplating operations.

Currently, about 40% of HCN is used in production of the building blocks of nylon 6/6, namely, hexamethylenediamine and adipic acid, by way of adiponitrile obtained from 1,3-butadiene and HCN (3.10.3.6, 6.8.1). About 30% of HCN is consumed for the production of methyl methacrylate (MMA) (3.10.3.1)—a 1B-lb building block of poly(methyl methacrylate) (PMMA) and methacrylate copolymers. About 10% of HCN is converted to cyanogen chloride (Cl–CN), which, in turn, is the building block of the trimeric cyanuric chloride (3.10.2.3, 10.3.3). The remaining 20% of the HCN demand is channeled into the production of inorganic cyanides; chelating agents such as ethylene-diaminotetraacetic acid, editic acid, or EDTA (4.3.3.3); and nitrilotriacetic acid (NTA) (3.4.3.7 and below); glycine (3.4.3.7, 4.6.3); glycolic acid (3.4.3.7); sarcosine (3.4.3.7); and the synthesis of fine chemicals such as isophorone diisocyanate (IPDI) (3,8.1 and below), methionine, and other materials mentioned below.

### 3.10.2   The Role of HCN as a Multifaceted C$_1$ Building Block

An orderly survey of the multitude and variety of HCN reactions requires a systematic approach based on different reaction patterns. In view of the objective of this presentation, mechanisms are invoked only for pragmatic reasons without evidence of their veracity. Also, together with the characteristic reactions of HCN and/or cyanide ion, the initial cyano-containing product may be shown to undergo some of its characteristic transformations such as (a) hydrolysis to the amide and finally to the corresponding carboxylic acid and (b) hydrogenation to the expected aminomethyl compound.

### 3.10.2.1  Addition of HCN–CN to Polar, Multiple-Bond-Containing Systems

1. The addition of HCN to carbonyl groups and the analogous carbon–sulfur and carbon–nitrogen multiple bonds are a common behavior pattern of HCN and even more so of its conjugate base, the cyanide ion commonly employed as a catalyst. The result of this addition to aldehydes and ketones are the corresponding cyanohydrins that can subsequently be hydrolyzed to $\alpha$-hydroxycarboxylic acids or converted to $\alpha$-aminocarboxylic acids (Strecker amino acid synthesis; ONR-87) because of the facile replacement of the hydroxy by an amino function as shown in Scheme 3.31.

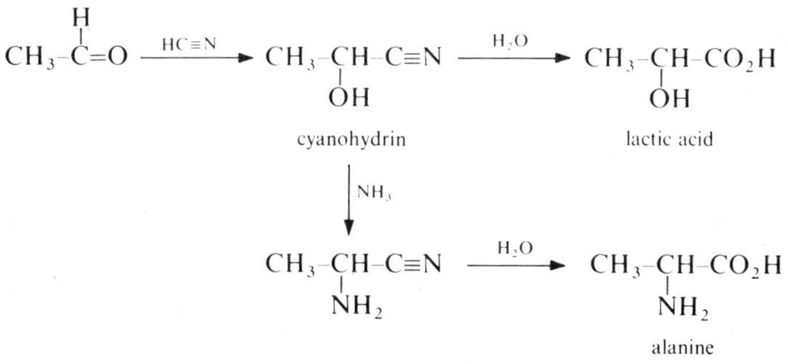

$$CH_3\text{-}\overset{\overset{\displaystyle H}{|}}{C}=O \xrightarrow{\ HC\equiv N\ } CH_3\text{-}\underset{\underset{\displaystyle OH}{|}}{CH}\text{-}C\equiv N \xrightarrow{\ H_2O\ } CH_3\text{-}\underset{\underset{\displaystyle OH}{|}}{CH}\text{-}CO_2H$$

cyanohydrin            lactic acid

$$\downarrow NH_3$$

$$CH_3\text{-}\underset{\underset{\displaystyle NH_2}{|}}{CH}\text{-}C\equiv N \xrightarrow{\ H_2O\ } CH_3\text{-}\underset{\underset{\displaystyle NH_2}{|}}{CH}\text{-}CO_2H$$

alanine

**Scheme 3.31**

The cyanide ion gives rise to Mannich-like products in the presence of ammonia or amines (3.4.3.7), and hence it is responsible for products such as glycine and NTA.

2. A minor modification of the preceding case is the use of hydrazine in place of ammonia. This is illustrated by the formation of azodiisobutyronitrile introduced in 3.9.3 as one of the original azo blowing agents and free-radical polymerization initiators. A competitive blowing agent, namely, azodicarbamide ($2.70), is prepared from methyl carbamate or urea and hydrazine and is commonly used in the manufacture of plastic foams. The preparation of both azo blowing agents is shown in Scheme 3.32. Other examples of HCN addition to multiple carbon–nitrogen systems are the reactions of oximes and nitrile oxides, that result in the formation of $\alpha$-hydroxylaminonitriles and $\alpha$-oximinonitriles, respectively, as shown in Scheme 3.33. A mixture of aqueous sodium cyanide and ammonium carbonate is an effective source of HCN, $CO_2$, and $NH_3$, and an initial carbonyl reactant is converted to the corresponding substituted hydantoin by means of the Bucherer–Bergs reaction (ONR-15). This reaction is illustrated by the preparation of the famous anticonvulsant (antiepileptic) phenytoin, better known as "Dilantin" (see 10.2.2), as well as by the preparation of dimethylhydantoin. The latter is converted to mono- and dimethylol dimethylhydantoins, MDMH and DMDH, respectively, which serve as preservatives of cosmetic formulations and as precursors of

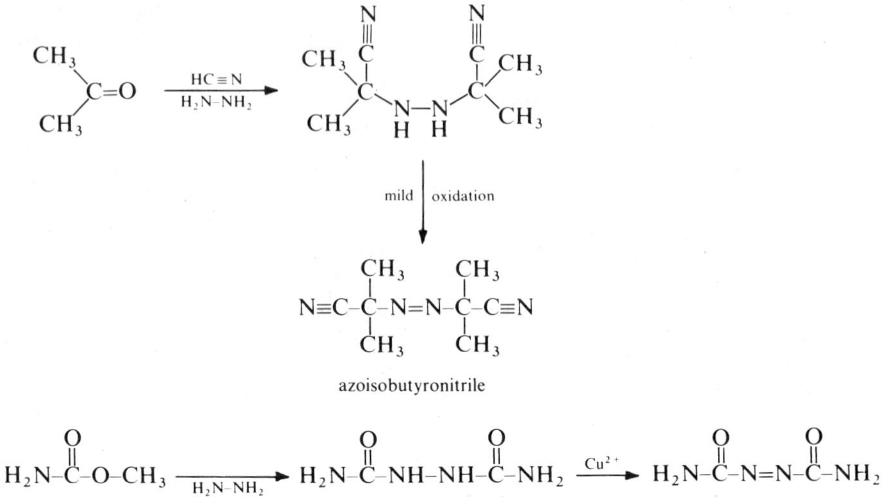

**Scheme 3.32**

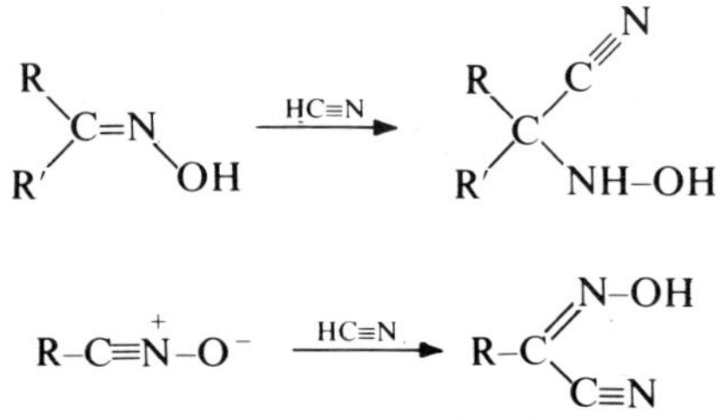

**Scheme 3.33**

condensation products used in hair lacquers. The preparations of these hydantoins are shown elsewhere (10.2.2). The hydantoin parent compound is an important starting point for the synthesis of several $\alpha$-amino acids such as, for example, phenylalanine required for the assembly of aspartame (see 6.3.1.3). In place of the carbonyl system, one obtains $\alpha$-mercaptonitriles from C=S systems, as illustrated by the case of CS$_2$ that readily gives dimercaptomaleonitrile as the result of the intrinsic instability of carbon–sulfur double bonds (3.7.3), as shown in Scheme 3.34.

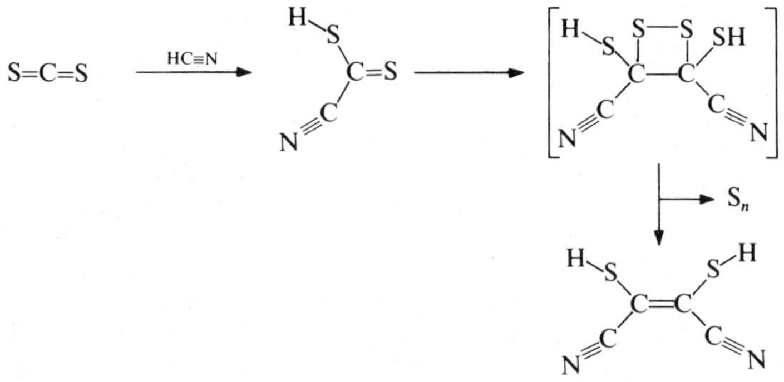

**Scheme 3.34**

3. 1,4-Addition of HCN–CN$^-$ to a conjugated carbon double-bond system activated by carbonyl, cyano, sulfone, and other electron-withdrawing groups. For example:

$$>C=C-C=O \rightarrow N\equiv C-C-CH-C=O$$

$$>C=C-C\equiv N \rightarrow N\equiv C-C-CH-C\equiv N$$

$$>C=C-SO_2-R \rightarrow N\equiv C-C-CH-SO_2-R$$

A relatively recent industrial application of this type of chemical behavior of the HCN–CN$^-$ reactants is the DSM process for the preparation of tetramethylenediamine, the lower homolog of the well-established hexamethylenediamine (3.8.1). It involves the 1,4-addition of HCN to acrylonitrile or, in other words the cyanoethylation (5.4.2.2) of HCN, followed by catalytic hydrogenation:

$$N\equiv C-H+CH_2=CH-C\equiv N \rightarrow N\equiv C-(CH_2)_2-C\equiv N \rightarrow H_2N-(CH_2)_4-NH_2$$

Tetramethylenediamine is expected to give rise to a new series of nylons with a first digit "4".

### 3.10.2.2   Nucleophilic Behavior of Cyanide Ions and HCN

**A. RITTER REACTION (ONR-77).** Interestingly, the cyanide system is an ambident nucleophile (*ND*, 782–783; see also 3.7.3.3.B), but the two partially negative terminals differ in polarizability, in accord with differences in electronegativity between carbon and nitrogen. Thus, the less electronegative carbon terminal is more polarizable and is considered to be "soft," while the more electronegative nitrogen terminal is less polarizable and is consequently "hard." According to the "hard/hard–soft/soft" theory (see 3.7.3.3.B), a hard acid, such as a (solvated) proton or a fullfledged carbon cation, will seek out the hard nitrogen terminal, while a soft acid, such as an electrophilic carbon system, will seek out the soft carbon terminal. The former situation is illustrated by the Ritter reaction, in which HCN and a potential carbon cation, generated in the presence of a mineral acid from an olefin or an alcohol, gives rise to a new carbon–nitrogen bond:

$$Me_3C-OH+H^+(H_2O)_n+:N\equiv CH \rightarrow (Me_3C-\ddot{N}=CH^+) \rightarrow Me_3C-NH-CH=O$$

Hydrolysis of the resulting *N*-substituted formamide converts, in fact, the initial alcohol (or olefin) to a *t*-alkyl primary amine:

$$Me_3C-NH-CH=O \rightarrow Me_3C-NH_2+HCO\cdot OH$$

This is the most practical way to obtain *t*-butylamine ($1.17) and its higher *t*-alkylamine homologs, such as *t*-octylamine ($2.60), $(CH_3)_3C-CH_2-C(CH_3)_2-NH_2$ (6.3.1.2). Readily protonated olefins behave in an analogous manner, with C–N bond formation occurring at the most likely location of a carbon cation.

A more common situation is encountered when cyanide ions and HCN are allowed to react with soft carbon systems such as that described in Section 3.10.2.2.B.

**B. SATURATED CARBON SYSTEMS THAT CONTAIN A REASONABLY GOOD LEAV-ING GROUP.** These systems (*ND*, 413) are represented here by LG:

$$:N\equiv C:^- + -\overset{|}{\underset{|}{C}}-LG \rightarrow :N\equiv C-\overset{|}{\underset{|}{C}}- + LG^-$$

where LG can represent all halogens except fluorine, sulfonate moieties such as mesylate ($MeSO_3^-$), tosylate ($4\text{-}MeC_6H_4SO_3^-$), and other groups that form relatively stable anion $LG^-$ but that are not attached to the carbon structure by a very strong bond (as is true in the case of the C–F system).

**C. RING-OPENING ADDITION REACTIONS.** An example is

$$N\equiv C^- / HCN + \overset{\diagdown}{\underset{\diagup}{C}}\underset{O}{-}\overset{\diagup}{\underset{\diagdown}{C}} \rightarrow N\equiv C-\overset{|}{\underset{|}{C}}-\overset{|}{\underset{|}{C}}-OH$$

epoxide         cyanohydrin

**D. ADDITION TO ACETYLENE SYSTEMS.** These systems (4.7) are generally more electrophilic than analogous olefinic systems:

$$R\text{-}C\equiv C\text{-}H + C\equiv N^- / HCN \rightarrow RCH=CH\text{-}C\equiv N$$

**E. SUBSTITUTION OF HYDROXYL GROUPS IN *N*-METHYLOL SYSTEMS.** This differs from the Ritter reaction pictured above because the potential electrophilic carbon holding a protonated hydroxyl group is soft as a result of electron delocalization from the adjacent nitrogen, represented as follows:

$$R_2N\text{-}CH_2\text{-}OH + H^+(H_2O)_n \rightarrow R_2N\text{-}CH_2\text{-}OH_2^+(H_2O)_{n-1} + H_2O$$

$$\Updownarrow$$

$$R_2N^+=CH_2(H_2O)_n$$

Consequently, *N*-methylol groups are converted to substituted aminoacetonitriles, $R_2N\text{-}CH_2\text{-}C\equiv N$, and these serve as stepping stones to several important families of derivatives depicted in Scheme 3.35.

### 3.10.2.3   Electrophilic Behavior of Protonated HCN

The protonated HCN species ($H\overset{+}{C}=NH$) seeks out electron-rich reactants, as shown by the following examples:

$$(H\overset{+}{C}=NH) + R_2NH \rightarrow HC(=NH_2^+)NR_2 \quad \text{(a formamidinium ion)}$$

$$(H\overset{+}{C}=NH) + H_2S \rightarrow H\text{-}C(=NH_2)^+SH \rightarrow HC(=S)NH_2 \quad \text{(thioformamide)}$$

The classical Gattermann aldehyde synthesis (ONR-35; 9.3.5.1) falls into the same category of behavior and occurs with aromatic systems that are susceptible to electrophilic attack.

### 3.10.2.4   Cyanogen Chloride and Its Derivatives

Chlorination of HCN produces cyanogen chloride:

$$H\text{-}CN + Cl_2 \rightarrow Cl\text{-}CN + H\text{-}Cl$$

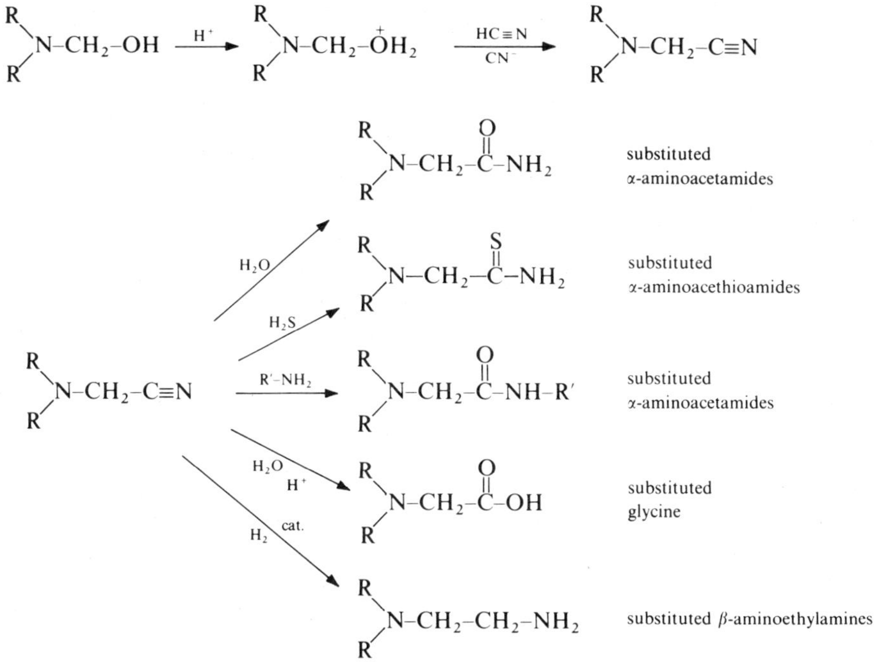

$$\underset{R}{\overset{R}{>}}N-CH_2-OH \xrightarrow{\ H^+\ } \underset{R}{\overset{R}{>}}N-CH_2-\overset{+}{O}H_2 \xrightarrow[CN^-]{\ HC\equiv N\ } \underset{R}{\overset{R}{>}}N-CH_2-C\equiv N$$

$$\underset{R}{\overset{R}{>}}N-CH_2-\overset{\overset{O}{\parallel}}{C}-NH_2 \qquad \text{substituted}\ \alpha\text{-aminoacetamides}$$

$$\underset{R}{\overset{R}{>}}N-CH_2-\overset{\overset{S}{\parallel}}{C}-NH_2 \qquad \text{substituted}\ \alpha\text{-aminoacethioamides}$$

$$\underset{R}{\overset{R}{>}}N-CH_2-C\equiv N \xrightarrow{R'-NH_2} \underset{R}{\overset{R}{>}}N-CH_2-\overset{\overset{O}{\parallel}}{C}-NH-R' \qquad \text{substituted}\ \alpha\text{-aminoacetamides}$$

H₂O, H₂S (reagents on branches)

$$\underset{R}{\overset{R}{>}}N-CH_2-\overset{\overset{O}{\parallel}}{C}-OH \qquad \text{substituted glycine}$$

(H₂O / H⁺)

$$\underset{R}{\overset{R}{>}}N-CH_2-CH_2-NH_2 \qquad \text{substituted}\ \beta\text{-aminoethylamines}$$

(H₂, cat.)

**Scheme 3.35**

Cyanogen chloride (b.p. 14°C) is toxic but, in proper hands, serves as a synthetic reagent for the introduction of cyano groups onto electron-rich substrates. Its most important application, however, is the trimerization to cyanuric chloride, or 2,4,6-trichloro-1,3,5-triazine, which occurs at elevated temperatures:

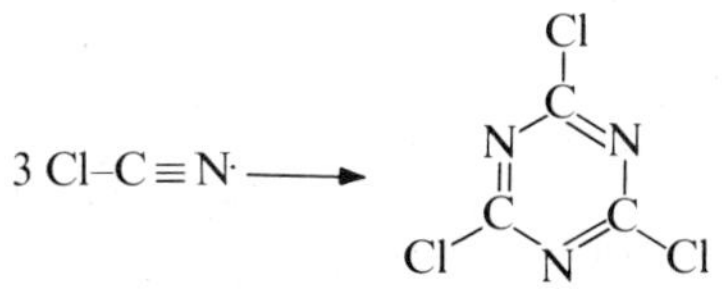

$$3\ Cl-C\equiv N \cdot \longrightarrow$$ (2,4,6-trichloro-1,3,5-triazine ring structure)

Cyanuric chloride (m.p. 146°C, b.p. 190°C) is an important building block of pesticides (3.10.3), dyes that can be attached to the substrate by chemical means, detergent brighteners, and so on (see Fig. 10.11). Cyanuric chloride is useful because of the reactivity of all three chlorine substituents and the stability of the aromatic triazine ring (10.3.3). This provides an opportunity to join as many as three different moieties by chemical means.

Complete hydrolysis of cyanuric chloride yields cyanuric acid ($1.16). The latter can also be obtained by heating urea, biuret, or carbonyldibiuret—the product of phosgene and biuret, $(H_2N-CO-NH-CO-NH)_2CO$. Cyanuric acid is the trimer of cyanic acid (HOCN), shown to be one of the decomposition products of urea. Cyanuric acid is used to stabilize cyanuric chloride when the latter is employed as a swimming-pool chlorinating agent. A recent claim (*CMR* 12/22/86, *CW* 12/24/86) that cyanuric acid is capable of eliminating nitrogen oxide pollutants suggests the potential for future large-scale uses.

Another interesting reaction of cyanogen chloride is its thermal condensation with acetonitrile (4.6.2) to produce the versatile malononitrile (Lonza, $7.85/lb):

$$N\equiv C-Cl + CH_3-C\equiv N \rightarrow N\equiv C-CH_2-C\equiv N + HCl$$

Some synthetic uses of malononitrile are shown elsewhere (5.6.4).

### 3.10.2.5 Oxidation to Cyanogen

A mild oxidation of cyanide ions produces cyanogen:

$$2C\equiv N^- + 2Cu^{2+} \rightarrow N\equiv C-C\equiv N + Cu_2^{2+}$$

Cyanogen can be hydrolyzed to oxalic diamide ($H_2N-CO-CO-NH_2$), and both of these compounds can serve as $C_2$ building blocks (see, for example, Scheme 10.34, p. 601). In the presence of cupric nitrate an oxidative hydrolysis of HCN leads directly to oxamide, and because of its slow hydrolysis, oxamide has been proposed as a slow-release fertilizer.

### 3.10.2.6 The Catalytic Role of Cyanide Ligands

The relative absence of obstruction because of its linear structure and its electron-withdrawing nature (by electrostatic and electron delocalization interactions) explain the importance of the cyanide ion as a ligand when it is clustered around heavy metals. Some of these complex ions are excellent catalysts as illustrated by the double addition of HCN to 1,3-butadiene (6.8.1) to give adiponitrile according to the Du Pont process introduced in 1977, as shown in Scheme 3.36.

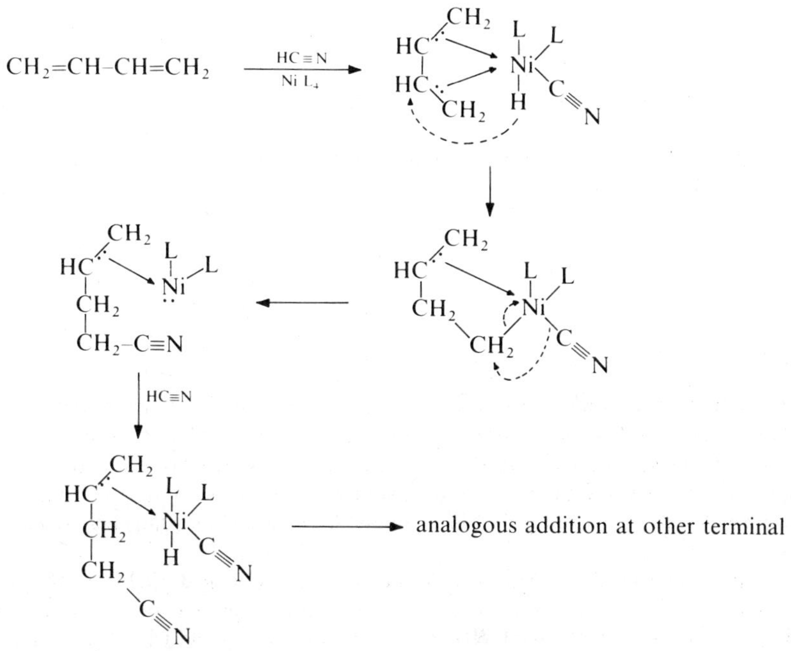

[where $L = R_3P$, $(Ph-O)_3PO$, etc.; reaction promoted by $AlCl_3$, $ZnCl_2$, etc.]

**Scheme 3.36**

The industrial importance of adiponitrile and the competitive processes for its production are summarized elsewhere (3.10.2.6 and 8.3.1).

### 3.10.3  Production of Major Chemicals Derived from HCN

3.10.3.1  Methyl Methacrylate (MMA) and Other Methacrylic Systems

The traditional source of methyl methacrylate (MMA, 62¢) is the one-pot reaction of acetone (24¢), HCN (50¢), methanol (30¢/gal or 30¢/gal × 6.60 lb/gal = 4.5¢/lb), and sulfuric acid ($68.15/t in the Southeast region of the United States, or 3.41¢/lb). This remarkable process combines the reaction pattern described in Section 3.10.2.1, paragraph 1 (see 3.10.2), with dehydration of an intermediate cyanohydrin of acetone, hydrolysis of the initial cyano group, and esterification of the resulting carboxylic acid. However, the continued demand for PPM and copolymers of MMA used extensively as impact modifiers has generated competitive processes that start with a $C_4$ building block (see 6.3.1.4) in place of the $C_3 + C_1$ process (see 5.5.2).

The analogous process in the absence of methanol gives methacrylonitrile. The partial hydrolysis of the latter generates methacrylamide and its complete hydrolysis produces methacrylic acid (78¢). Again, this preparative approach is challenged even more seriously than the above-mentioned MMA process by the use of $C_4$ building blocks (6.3.1.4).

3.10.3.2  Isophorone Diisocyanate (IPDI)

This specialty PUR building block (see 3.8.1) is obtained from the aldol condensation product of three acetone molecules (5.5.4), namely, phorone, which is subsequently subjected to a cyclization reaction to give isophorone (81¢) (3.8.1):

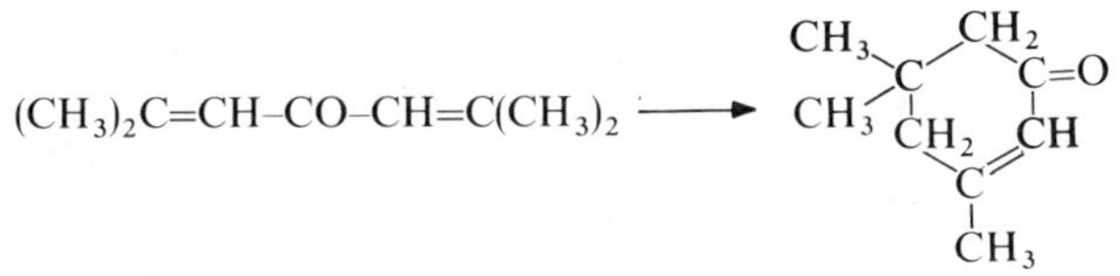

$$(CH_3)_2C=CH-CO-CH=C(CH_3)_2 \longrightarrow$$

*Reductive Amination and Alcohol Amination*

The structure shown immediately above is the starting material for the intervention of HCN, which gives a typical 1,4-addition (3.10.2.1, paragraph 3). The resulting cyano derivative is subjected to a *reductive amination* that converts the remaining carbonyl group to an amine

$$\text{>}C=O + H_2/NH_3 \rightarrow \text{>}CH-NH_2$$

while, at the same time, the cyano group is hydrogenated to the corresponding aminomethyl group—a possibility mentioned at the beginning of Section 3.10.2. The reaction of the resulting isophoronediamine with phosgene produces IPDI, a building block of light-stable PURs (3.8.1).

It is noteworthy that the above-mentioned reductive ammination of carbonyl compounds is related to *alcohol amination* mentioned several times throughout these pages since, in the presence of a hydrogenation–dehydrogenation catalyst, the carbonyl and corresponding alcohol systems are in equilibrium (see 4.3.3.3).

Alcohol amination is utilized in the case of chelating agents of the EDTA type (see 3.10.1) that can be assembled from NaCN, $CH_2O$, and ammonia to give aminoacetonitrile ($H_2NCH_2C \equiv N$), nitrilotriacetonitrile [$N(CH_2C \equiv N)_3$], and imino(diacetonitrile) [$HN(CH_2C \equiv N)_2$]. The reaction of the last-mentioned compound with ethylene dichloride (EDC), followed by hydrolysis of the nitrile to give NTA, is competitive with the alcohol amination route, although the latter most likely is more costly. Thus, for example, the use of glycolic acid (70%, technical, grade 49.5¢) and ammonia or ethylenediamine ($1.30) can be used to produce glycine ($H_2NCH_2CO_2H$) (Hampshire glycine, technical grade $1.88, USP-NF grade $2.12), and iminodiacetic acid [$HN(CH_2CO_2H)_2$] (Hamshire IDA Acid, $3.00), or EDTA (Hamp-ene Acid, $1.35), respectively. The tetrasodium salt tetrahydrate of NTA is also available (Hamp-ene 220, $1.10). Similarly, the use of glycolic acid and diethylenetriamine leads to the formation of diethylenetriaminepenta-acetic acid, $HO_2CCH_2N[CH_2CH_2N(CH_2CO_2H)_2]_2$, (Hamp-ex Acid, $2.50). The N-hydroxyethyl derivatives of glycine ($HO-CH_2CH_2-NH-CH_2CO_2H$) (Hampshire N-hydroxyethylglycine, sodium salt) and ethylenediaminetriacetic acid [$HO-CH_2CH_2-N(CH_2CO_2H)CH_2CH_2N(CH_2CO_2H)_2$] (Hamp-ol Acid, $3.50), are additional members of this family of chelating agents.

The advantage of the alcohol amination process, as compared to the use of chlorine-containing building blocks (such as EDC or sodium chloroacetate) for the assembly of the above-mentioned chelating agents is the total absence of inorganic impurities from the final products. This difference is important in biomedical and other sensitive applications of chelating agents.

### 3.10.3.3 Methionine and Its Hydroxy Analog and Lysine

Methionine, dl- or racemethionine (USP $2.96, feed grade $1.07) is an essential amino acid used widely as a food and feed supplement. The demand in the United States amounts to about 100MM lb/yr, and it is on the rise because of demands in the poultry industry. The worldwide annual demand for methionine is estimated at about 220MM lb.

The synthesis of methionine commences with a 1,4 addition of methyl mercaptan, methanethiol (3.3.2.6) to acrolein (5.4.4) (62¢), and the resulting β-methylmercaptopro-pionaldehyde ($MeS-CH_2CH_2CH=O$) is then subjected to a hydantoin synthesis as pictured in the case of the synthesis of Dilantin (3.10.2.1, paragraph 1).

Even though biological processes are normally very specific to chirality of the substrates, methionine metabolism involves an oxidative deamination of its α-amino group and, obviously, the resulting α-keto derivative of methionine is identical regardless of which enantiomer one starts with. This accounts for the fact that racemic methionine is the material found on the marketplace and also explains why Monsanto has been producing the so-called methionine hydroxy analog (MHA), $MeSCH_2CH_2CH(OH)CO_2H$, priced competitively at 86–88¢ (86–88% activity). The latter material is also metabolized to the above-mentioned α-keto derivative of methionine and is prepared from β-methylmercaptopropionaldehyde by a hydrolysis of its simple HCN adduct.

Another essential amino acid used as a food and feed supplement is the six-carbon, terminal amino-group-containing lysine. The carbon skeleton is assembled by means of catalyzed addition of acetaldehyde to acrylonitrile. In other words, cyanoethylation of acetaldehyde (see 5.4.2.2) results in the $C_5$ structure $N \equiv C-CH_2CH_2CH_2-CH \equiv O$, which is subjected to a reaction with HCN–NH₃ and subsequent hydrolysis of the amino cyano terminal (3.10.2.1, paragraph 2), followed by hydrogenation of the acrylonitrile-derived cyano group, produces lysine:

$$H_2N-CH_2CH_2CH_2CH_2CH(NH_2)CO_2H$$

Unlike the case of methionine, only the *l*-isomer of lysine is biologically active, and this suggests either of two alternatives: (1) the synthetic racemic mixture is subjected to resolution, or (2) the desired enantiomer is prepared by a biological method that produces L-lysine, food-grade monohydrochloride ($5.50; feed grade $1.40).

It happens that the propietary Ajinimoto acetic acid fermentation process produces the desired lysine enantiomer and around 80% of the worldwide demand of about 100MM lb is supplied in this way. However, there also exists a combination of a synthetic process that starts with cyclohexene but involves the use of L-hydrolase for a bio-technological resolution of a racemic intermediate (*SZM-II*, 164).

*Physical Resolution of Racemic Mixtures*

The possibilities of large-scale physical resolution of racemic mixtures of industrial importance depend on progress in separation technology and on economic consider-ations. In the case of $\alpha$-amino acids, Daicel Chemical offers a high-performance liquid chromatographic (HPLC) technology based on silica-gel particles coated with cellulose in which the three free hydroxyl groups of each glucose unit are converted to $-O-CH_2CH_2CH_2NH-CHR-CO_2H$ side chains.

In the presence of cupric ions, the racemic mixture of an amino acid represented here by $R'-CH(NH_2)CO_2H$, a copper complex of the structure shown in Scheme 3.37 forms.

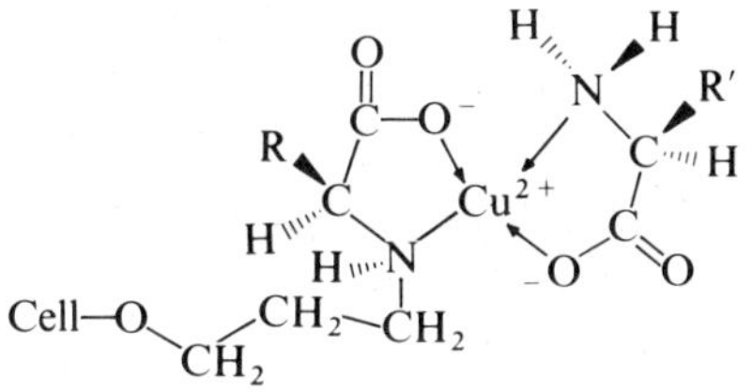

and

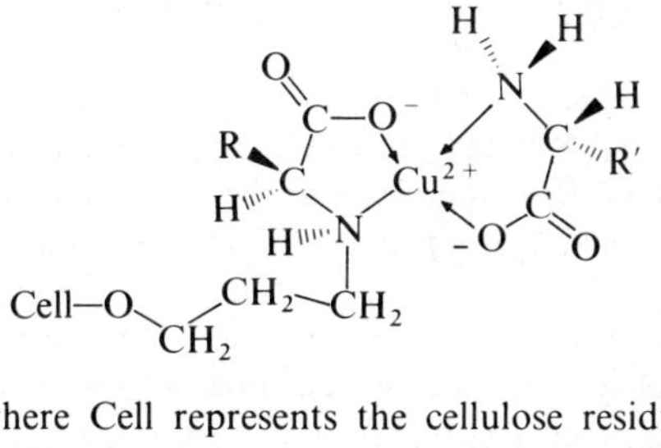

(where Cell represents the cellulose residue and
$R'-CH-CO_2H$ is the racemic mixture subjected to resolution)
$|$
$NH_2$

**Scheme 3.37**

However, the stability of the complexes produced by enantiomers of the amino acid subject to separation depends on the difference in the three-dimensional structure at the chiral centers. Thus, elution of the racemic mixture with 0.25 m$M$ aqueous copper sulfate produces a separation due to preferential ligand exchange.

Different column packings exist for HPLC separations of other racemic mixtures. The largest, packed column available from Daicel measures 10 × 50 cm and costs between $30,000 and $45,000, depending on the nature of the packing. Also, the column packings themselves are available at a cost of $40,000–$60,000 per 5000 g.

Another relatively recent method of resolution depends on the stereospecificity of enzymatic reactions. Thus, in the case of α-amino acids, the racemic mixture is esterified and this racemic mixture is subjected to enzymatic hydrolysis. There results a readily separated mixture of the *l*-amino acid and of the nonhydrolyzed ester of the *d*-isomer. The latter is isolated, racemized by chemical means, and recycled.

### 3.10.3.4 Lactic and Mandelic Acids and Cyanazine

Lactic acid [$CH_3CH(OH)CO_2H$] (88%, food grade $1.06, technical grade %1.03) is produced by fermentation (*HHS-II*, 41), but it can be also obtained from HCN (50¢) and acetaldehyde (37¢), although the synthetic approach does not appear to be competitive.

Mandelic acid [$C_6H_5$–$CH(OH)CO_2H$] ($3.65) is obtained from HCN and benzaldehyde (NF $1.25, technical grade 73¢). The glucoside of the intermediate mandelonitrile was alleged to be an anticancer drug known to the gullible public as Laetrile. Mandelic acid is used as a urinary antiseptic either alone or in conjunction with hexamethylenetetramine (3.4.3.7) and as a building block of some fine chemicals.

The monosubstitution product of cyanuric chloride and ethylamine can be combined with the product of acetone, HCN, and ammonia to give the selective, preemergence herbicide cyanazine (Shell's Bladex) of the structure

$$CH_3-CH_2-NH-\underset{\underset{Cl}{\displaystyle\bigvee_{N}^{N}}}{\overset{N}{\bigwedge}}-NH-\overset{\overset{\displaystyle CH_3}{|}}{\underset{\underset{\displaystyle CH_3}{|}}{C}}-C\equiv N$$

### 3.10.3.5 Nucleophilic Substitution Products of the Cyanide Ion

There are numerous fine chemicals that are prepared in part by substitution reactions that follow the behavior pattern described in 3.10.2.2. Some specific examples follow.

**A. PHENETHYLAMINE, PHENYLACETIC ACID, AND DERIVATIVES.** The reaction of benzyl chloride (9.2.3) (54¢) with sodium cyanide (71¢) gives phenylacetonitrile or benzyl cyanide, which, in turn, can be hydrogenated to phenethylamine ($C_6H_5CH_2CH_2NH_2$) ($1.50). Partial hydrolysis of benzyl cyanide yields phenylacetamide, which is used in the production of penicillin II or G (*HHS-II*, 43). Complete hydrolysis gives phenylacetic acid ($4.50), a building block of synthetic fragrances, antibiotics, and other fine chemicals. An analogous family tree can be constructed from almost any aromatic chloromethylation (9.2.1) product.

**B. CYANOACETIC ACID AND DERIVATIVES.** The reaction of sodium cyanide with chloroacetic acid (4.6.3) (56¢) or better yet, sodium chloroacetate, gives, on acidification, cyanoacetic acid ($N\equiv C-CH_2-CO_2H$). Esters of cyanoacetic acid react with formaldehyde to produce the family of polymeric cyanoacrylate adhesives mentioned previously (3.4.3.3). Also, the doubly activated methylene group of a cyanoacetic ester readily undergoes condensation and alkylation reactions and thus provides the opportunity to assemble a variety of more complex carbon skeletons. The alkylated derivatives of cyanoacetic esters can be converted to the corresponding derivatives of malonic ester by a combination of hydrolysis and esterification. For example, the doubly ethylated methyl or ethyl cyanoacetate produces the substituted malonic esters that become the building blocks of the hypnotic barbital (NF $10.25, sodium salt NF $10.40), as shown in Scheme 3.38.

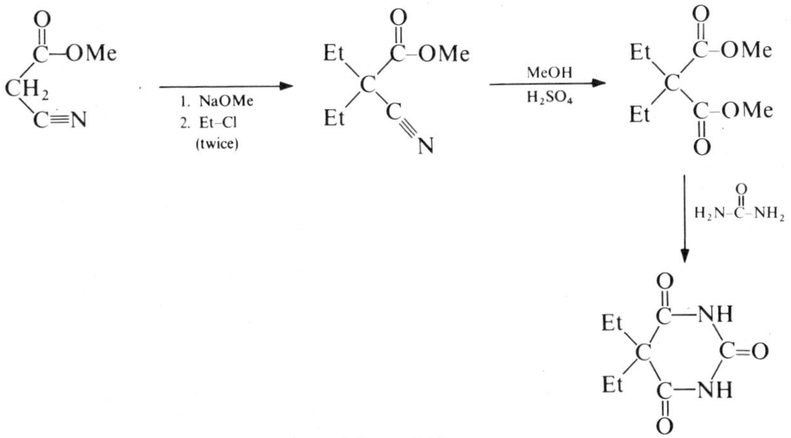

Scheme 3.38

### 3.10.4  Some Prospective Uses of HCN

In addition to the greater exploration and exploitation of the chemistry sketched so far, there are many new vistas on the horizon of HCN chemistry. A few are as follows:

1. Nippon Soda of Tokyo has explored the tetramer of HCN named in a picturesque manner "DAMN" (for diaminomaleonitrile) and obtained under oxidative conditions. Its formation can be rationalized as shown in Scheme 3.39.

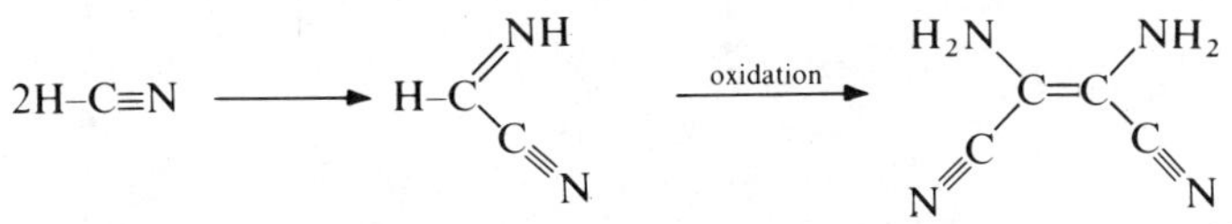

Scheme 3.39

DAMN is the building block of various heterocyclic systems represented by the examples shown in Scheme 3.40.

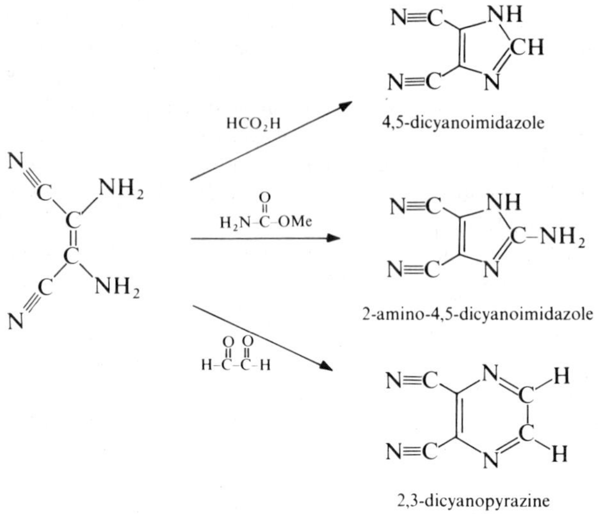

**Scheme 3.40**

It is noteworthy that the preceding structures can be transformed to the corresponding carboxylic acids by hydrolysis of the cyano groups and that the vicinal dicarboxylic acids of stable heterocycles can be decarboxylated to monocarboxylic acids. Also, the carboxamides obtained by partial hydrolysis can be converted to amines by means of the Hofmann reaction (ONR-45), and in the case of the six-membered heterocycle one can obtain substituted pyrazines, which are valuable fragrance-producing materials (9.3.3).

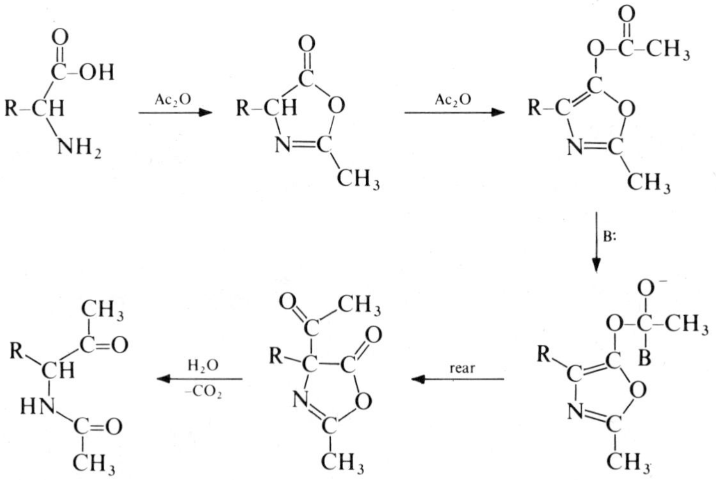

**Scheme 3.41**

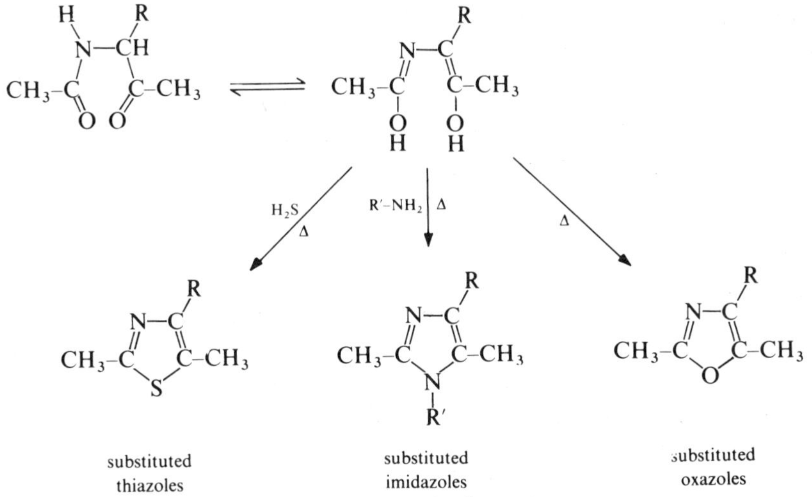

substituted thiazoles

substituted imidazoles

substituted oxazoles

**Scheme 3.42**

2. W. R. Grace & Company advertises that it makes "new products happen with cyano chemistry." Among other things, W. R. Grace announces the preparation of linear nitriles, $R-CH_2CH_2C\equiv N$, from HCN and alphaolefins by means of a high-temperature, metal-catalyzed reaction reminiscent of the addition of HCN to 1,3-butadiene (see 3.10.2.6). These linear primary nitriles are competitive with the traditional source of such compounds, namely, the conversion of naturally occurring fatty acids, by way of their methyl esters and amides, to "fatty amines" (*SZM-II*, 56, 80). Also, the application of the Ritter reaction (3.10.2.2, paragraph 1) to alphaolefins gives rise to α-methylalkylamines [$R-CH(CH_3)-NH_2$], that were not readily available on a large scale.

3. The DSM process to produce succinonitrile from HCN and acrylonitrile en route to tetramethylenediamine and nylon 4/6 was mentioned above.

4. The application of the Strecker synthesis (ONR-87) to propionaldehyde gives α-aminobutyric acid, a component of the new Caridex painless caries removal system.

5. α-Acetamido ketones ($CH_3-CO-CHR-NH-CO-CH_3$) are produced by heating α-amino acids, acetic anhydride, and bases such as pyridine (Dakin–West reaction, ONR-21). This rather complex reaction occurs by way of an azlactone intermediate and involves the acylation of the α-carbon of the initial amino acid and the decarboxylation typical of β-keto carboxylic acids. A simplified mechanism is shown in Scheme 3.41. The use of α-acetamidoketones as building blocks of some heterocyclic systems is illustrated in Scheme 3.42.

# 4

# C₂ BUILDING BLOCKS

## 4.1  THE DOMINANT ROLE OF ETHYLENE AMONG C$_2$ BUILDING BLOCKS: A LESSON IN CHEMICAL ECONOMICS

For several decades ethylene has occupied the most prominent position among all large-volume petrochemicals primarily because of the polymers derived from this C$_2$ building blocks (see Table 1.1). The production capacity of ethylene in the United States peaked at about 41B lb in the late 1970s, but by 1986 it shrank to about 33B lb. The actual output rose from 30B lb in 1979 to the 33B-lb level in 1986 except for a sharp drop in 1982 (see below). The story of the rise and fall of the demand for ethylene is a valuable lesson in the complexities of chemical economics when one deals with a large-volume commodity chemical.

An important determining factor is the supply and cost of the different feedstocks used in the production of ethylene. In 1975 approximately half of the 20B-lb production of ethylene in the United States was obtained by cracking of ethane, while 7B lb was derived from cracking of propane and butanes, LPG, and the the remaining 2B–3B lb were obtained by cracking higher-molecular-weight petroleum fractions, namely, naphtha and gas oil (see Fig. 2.2). In view of the uncertainties in the supply and cost of natural gas before deregulation of its price structure (2.2.1), and because of the increased demand for natural gas as a domestic and industrial fuel, the producers of ethylene began to switch to the use of heavier feedstocks. Thus, by 1979 out of a total of 29B lb of ethylene, 7B lb were derived from naphtha and gas oil, 15B lb from ethane, and 6B lb from propane and the butanes. According to an SRI study (*CMR* 5/11/87), by 1982 the United States was producing only 24.5B lb (out of a nameplate capacity of 38B lb for a very-poor-capacity utilization of only 64.2%), and ethane was the source of 13.6B lb, propane and butanes gave 6.1B lb, and naphtha and gas oil gave only about 4.75B lb of ethylene. The ethylene picture improved by 1985, when about 30.5B lb of ethylene was produced out of a shrunk nameplate capacity of 33.5B lb (for an excellent capacity utilization of 91%) and the sources of ethylene were about 16B, 7B and 7B lb for ethane, propane–butane, and naphtha–gas oil, respectively. By January 1988 the nameplate capacity of 36B lb matched the demand, and contract prices reached the 23¢/lb level. A reasonably strong demand caused contract prices to increase to 24¢ by April 1988, and by June 1988 Exxon and Chevron had made plans to import liquid ethylene by tankers (*CW* 6/22/88).

The prognosis for the ethylene production in the United States is rather favorable since the production in 1995 is projected at over 37B lb (with a 97% capacity utilization) and a significant shift to the utilization of heavier petroleum fractions in view of an increasing demand for ethane and LPGs in the United States. The total free-world production in 1985 of 84.5B lb of ethylene rose to 110B lb in 1987 and is expected to increase to 143B lb by 1994 with ethane, the LPGs, and naphtha–gas oil contributing about 32B, 8B, and 60B lb, respectively, for an approximately 92% capacity utilization. By comparison with these statistics, the production of ethylene in 1986 by countries with planned economies is reported (*CW* 5/27/87) to be 13.9B lb.

In retrospect, the above-mentioned situation in 1982 resulted from an increase in the U.S. production capacity between 1976 and 1981 by about 15B lb. This financial commitment was based on promising growth rates of many large-volume chemicals derived from ethylene but, unfortunately, suffered from several judgment errors (see C. J. Verbanic, "A Make–Do Decade for Petrochemicals," *CB* 3/87, 12–15). To wit, it failed to consider the following factors:

1. The progress in separation technology applied to the most straightforward raw material for the production of ethylene (viz., ethane). The recovery of ethane by cryogenic processes from natural gas and LPG caused its supply to increase from 8B lb in 1982 to 22B lb in 1985. Low-temperature fractionation displaced the traditional technology of absorption–desorption of the gaseous components from heavy petroleum oil. Consequently, the expansion of ethylene production became based on the relatively low temperature cracking of ethane rather than the high-temperature (550–750°C) cracking of the heavier petroleum fractions, which produces only about 25–30% yields of ethylene (with the accompanying production of propylene, 1,3-butadiene, and other products (see Fig. 2.2).

2. The technological progress in the manufacture of the most important of ethylene derivatives, namely, PE. The development and popularity of linear low-density polyethylene (LLDPE) (4.3.1), at the expense of traditional low-density polyethylene (LDPE), actually decreased at first the demand for ethylene since each pound of the final product contains about 5–8% of a higher-molecular-weight olefinic copolymer. Furthermore, the superior mechanical properties of LLDPE allows the film market to manufacture threefold thinner films, and this, of course, decreases the consumption of the

nonprocessed polymer since it is purchased on the basis of mass units but sold on the basis of area units (see A. S. Brown, "LLDPE Stampede: A Rush to Unprofitability," *CB* 2/8/82, 25–31).

3. The maturing and other adverse effects on the growth of the ultimate ethylene-derived consumer products, such as:

(a) The displacement of glass and waxed paper beverage containers by high-density polyethylene (HDPE).

(b) The shrinkage of the ethylene glycol (EG) antifreeze market because of the popularity of smaller-sized (compact and subcompact) automobiles and the introduction of aluminum radiators that are more efficient than the traditional ones manufactured from steel. Also, the domestic EG market destined for the manufacture of poly(ethylene terephthalate)(PET) textiles shrunk by about 20% since 1979 because of foreign competition, while an increase in the use of PET beverage containers (2-L, and more recently still, 3-L soft-drink bottles) did not quite compensate for the former. Finally, the adverse image of ethylene glycol derivatives promulgated (2.4.4.6, paragraph 2) on the basis of the toxicity of EG itself has induced manufacturers of monoethylene, diethylene, and higher-ethylene glycol ethers and derivatives—commonly used industrial solvents for a variety of coating applications, to switch to analogous derivatives of propylene oxide (PO).

(c) The nearly maturing of the PVC tubing market—the winner in the interproduct competition with copper- and iron-based piping, has slowed down the growth of this end use of ethylene in the United States (4.3.2.1).

(d) The significant decrease (by 800MM lb between 1979 and 1986) in the production of industrial ethanol (obtained by the hydration of ethylene; see 4.4.1) because of politically motivated subsidies of ethanol obtained by fermentation of surplus corn and other grains. This was worsened by massive imports ( > 200MM lb in 1986) of Brazilian ethanol (4.4.2) and contributed to a somewhat weaker demand for ethylene. Along the same lines, the EPA accelerated removal of "lead" (tetraethyl lead, TEL, 4.3.3.2) from gasoline has decreased the production of ethyl chloride derived from either ethylene or ethanol and had a similar effect on the demand for ethylene dichloride and ethylene dibromide.

(e) The maturing of the vinyl acetate monomer (VAM) market reflected in the shrinkage of its annual growth rate of 9% during the 1970s to 2.5% in 1986.

4. The development of the methanol–acetic acid–acetic anhydride process (3.3.2.4) that eclipsed the traditional ethylene–acetaldehyde–acetic acid route (4.5.1).

5. The rising production of commodity chemicals in hydrocarbon-rich, LDCs located in the Middle East (2.4.4.4) and elsewhere that have displaced some of the U.S. exports to Europe and other markets (see D. A. O'Sullivan, "After 50 Years, Polyethylene Producers Face Grim Prospects," *CEN* 6/27/83, 17–19).

6. Finally, the decision-makers of the CPIs and the chemical plant construction companies continued the "bad habits nurtured during post World War II era" according to Richard E. Heckert, Chairman of Du Pont (as quoted by C. J. Verbanic, *CB* 3/87) by overpricing construction costs and running "fat" operations that escalated the price of many 1B-lb/yr production facilities to a minimum cost of $600 million per plant during the expansion period of 1976–1981. Actually, in the opinion of this author, the "bad habits" originated with the padding of cost-plus contracts awarded by the Federal Government in preparation for and during World War II and were further cultivated by

fiscally careless industrial management that passed on all sorts of higher operational costs to government and civilian clients. The prevalent ambience was exemplified by the leadership of the United Auto Workers Union that promoted a fully subsidized $2\frac{1}{2}$-day workweek for all of its membership.

The excessively high capital investments in the production of ethylene from heavy-petroleum fractions created fiscal burdens that were magnified by low prices of ethylene and its coproducts propylene, 1,3-butadiene, pyrolysis gasoline, and other fuels. Thus, the prices of propylene and 1,3-butadiene dropped from the 20¢/lb range to below 10¢/lb. As ethane became more abundant, the producers chose to retrofit crackers designed originally for heavy-petroleum feedstocks to function with LPGs, and this increased the supply of ethylene coproducts further with a detrimental effect on prices. In the meantime, the cost of crude oil dropped sharply near the end of 1985, and Middle East production units went onstream and began to flood the world market with second- and third-generation petrochemicals.

This illustrates the complexity of the chemical commodities market and the dangers of billion-dollar capital investments unless all technological, sociopolitical (i.e., government-mandated), and geopolitical factors are considered. As of this writing, the dust has not yet settled on the "bruised and shaken giant" (see C. J. Verbanic, "A Make–Do Decade for Petrochemicals" *CB* 3/87, 12–15) of the U.S. petroleum and petrochemicals industry. The confusing situation in this industrial sector is not limited to the shores of the United States, although it is magnified in this country by a rash of "restructuring" decisions, plant closures, intercompany and leveraged buyouts (LBOs), mergers, wild speculations on the stock market, and other unsettling events. The number of worldwide acquisitions in the chemical and allied industries rose from 914 in 1985 to 1359 in 1986, and dropped somewhat to 1153 during 1987 (according to Kline & Co. as reported in *CW* 6/15/88) and most of these changes occurred in the U.S. marketplace (see E. Goldbaum, *CW* 1/20/88, for a list of the larger ones).

The hydrocarbon-rich Middle East countries were able to move ahead very quickly with the conversion of their petroleum resources into marketable, higher-value-added products and they have the European and Asiatic markets within reach, although their economies have recently suffered from extensive fiscal commitments based on ex-pectations of continued income from high-priced petroleum and its derivatives. Equally successful have been the chemistry-based economic growth efforts of hydrocarbon-poor countries of Southeast Asia such as Singapore, South Korea, and Taiwan (see H. H. Szmant, *International Lab.*, May–June 1987, pp. 6–9). By contrast, the natural gas–petroleum rich countries of the Americas are experiencing great difficulties: most were easily seduced to incur large debts by reckless lending policies to finance gigantic petrochemical projects in expectations that profits from future exports would finance the commitments to the international banking community. Unfortunately, in most cases, slow, inefficient construction and local management, and poorly monitored progress of the projects have resulted in delayed acceptable productivity. In some apparently incurable situations, prospective exporters are either being forced to import chemicals needed to operate their present-day economies or to restrict such imports in order to save scarce hard currency (expected to pay interests on the huge foreign debts) with detrimental effects on existing industrial activities. Discussion of the causes and effects of the serious economic predicament commonly referred to as the "Third World Debt Problem" and the apparent inability of political and banking leadership to find suitable solutions is beyond the scope of this book. However, the author feels justified to mention these issues here because the production and utilization of industrial organic chemicals is crucial to the economic growth of all nonprimitive societies, and the awareness of chemical professionals of these problems is in their self-interest and may stimulate their contributions to the long-awaited solutions.

## 4.2   AN OVERVIEW OF THE PRODUCTION AND UTILIZATION OF C$_2$ HYDROCARBON BUILDING BLOCKS

There can be little doubt that of the three C$_2$ hydrocarbon building blocks—ethane, ethylene, and acetylene—the olefin is of greatest economic importance (Table 1.1). Also, it is clear that much of the demand for ethylene stems from the large-volume production of polymers (1.1.1) discussed in the pages that follow. The production of ethylene and its coproducts by cracking hydrocarbon feedstocks at temperatures ranging from 550 to 860°C is summarized in Fig. 2.2. With a price of 50¢/gal for ethane and 58¢/gal for propane, the price of ethylene reached a high of 28.5¢/lb during the late 1970s, but by mid-1982 it decreased to 18.5¢ and to 15¢/lb in 10/87 but then began to rise to the April 1987 and January 1989 contract price of 24¢ and 32¢, respectively. By comparison, the current price of propylene—a competitor of ethylene for the production of some of the end products—is 17¢ for chemical grade and 20.5¢ for polymer grade.

The possibilities of new sources of ethylene are mentioned elsewhere (3.3.2.8.A). The MTO process operated by BASF in Ludwigshafen, West Germany produces 33,000 lb of ethylene per month. The yields of ethylene and propylene are 70–80%, while the cracking of naphtha yields only 44–50%. Also, the temperature requirement of the MTO process is 400°C lower than the conventional cracking of naphtha.

Because of the special circumstances of Brazil (4.4.2), the dehydration of ethanol supplements the demand for ethylene produced by cracking of mostly imported petroleum (H. H. Szmant, CW 4/9/87). This is exactly what Union Carbide was doing in the United States until the advent of large-scale petrochemical industry in the late 1930s and ethanol obtained from black-strap molasses—the residue obtained from the refining of cane-derived sucrose (SZM-II, 24, 35–38)—was the starting material for some 70 different products marketed by Union Carbide. A return to an economically sound production of ethanol from renewable resources awaits the development of a practical cellulose to glucose to ethanol conversion technology (SZM-II, 89).

All the hydrogen atoms of ethane are relatively unreactive primary ones, and thus high-energy reaction conditions are required in order to achieve chemical transformations in reasonably good yields. Thus cracking reactions are carried out at temperatures that also lead to fragmentation of the C$_2$ structure and recombination of some of the fragments to give propylene and other by-products. Chlorination of ethane to ethyl chloride is feasible at about 400°C when an excess of ethane is employed, but this process must compete with the reaction of ethanol and gaseous or liquid HCl (4.4.3.6) and the addition of HCl to ethylene (4.3.3.2). The traditional way to produce tetrachloroethylene, perc, is the chlorination–chlorinolysis of propane at 550–700°C, which converts readily available propane into two valuable but currently threatened (2.4.4.6, paragraph 2) products: perc and carbon tetrachloride:

$$C_3H_8 + 8Cl_2 \rightarrow Cl_2C{=}CCl_2 + CCl_4 + 8HCl$$

By analogy with the vapor-phase nitration of propane (5.2), the nitration of ethane could be the commercial source of nitroethane and nitromethane; however, again for reasons of long-standing availability and tradition, the production of nitroparaffins, thus far employs propane.

A sign of the current abundance of ethane is the announcement of Phillips 66 Natural Gas Company (*CMR* 7/13/87) that for the moment the current price of ethane (14.25¢–14.75¢/gal) does not justify the separation of ethane from natural gas; thus ethane remains in the gas stream, and hence its heating value will increase. The latest inventory of the American Petroleum Institute (API) reports the April 1987 level to be 20.6MM bbl—a value 30.6% higher than that of the previous year.

Assuming the density of liquified ethane to be 0.446g/mL and considering its 14.5¢/gal price, the cost of ethane in cents per pound is calculated as follows:

$$?¢ = 1 \text{ lb} \times \frac{454 \text{ g}}{1 \text{ lb}} \times \frac{1 \text{ mL}}{0.446 \text{ g}} \times \frac{1 \text{ gal}}{3785 \text{ mL}} \times \frac{14.5¢}{1 \text{ gal}} = 3.9¢$$

The current price ($\sim$ 3.9¢/lb) is certainly attractive for use of ethane as a starting material for transformations to higher-value-added products. One of these value-added transformations could be a catalytic dehydrogenation of ethane to ethylene and hydrogen or an oxidative dehydrogenation of ethane to ethylene and some acetic acid. The latter process is at a pilot-plant stage by Union Carbide (*CE* 2/17/86) and is reported to require 300–400°C and a propietary catalyst to give a 90% conversion of ethane to the above mentioned products.

The use of acetylene as a building block of industrial organic chemicals predates the large-scale production of ethylene. However, the energy intensity of the traditional sources of acetylene, namely, calcium carbide and an oxidative cracking of natural gas or higher hydrocarbons at $\geqslant 1000°$C, caused a partial decline in the use of acetylene as a building block of industrial organic chemicals. The chemical uses of acetylene declined from about 800MM lb in the 1950s and 1960s to a 285MM-lb level maintained for over a decade. Wherever feasible, ethylene has replaced acetylene as a starting material, and except for some use of the calcium carbide route, most of the acetylene in the United States is currently isolated from olefin cracking operations. The higher polarizability of acetylene as compared to simple olefins makes its selective absorption in dipolar aprotic solvents such as DMF and its cyclic analog NMP (3.5.3.1) possible, and most of it is utilized for the synthesis of a group of so-called acetylene chemicals described below (4.7). The unit price of acetylene is not normally found in price lists because of contractual arrangements between olefin cracking operators and their acetylene clients or because of captive use of acetylene by calcium carbide producers. Estimates of acetylene prices based on the pricing of some of its simple derivatives suggest a unit cost of 40¢–50¢/lb.

A potential source of acetylene that may signal a renaissance of the large-volume use of acetylene as an organic building block of industrial organic chemicals is described in Section 4.7.

## 4.3   INDUSTRIAL USES OF ETHYLENE: POLYMERS

### 4.3.1   Polyethylenes (PEs) and Ethylene-Containing Polymers

Of the family of polyolefins (POs), the polymers that contain the ethylene mononmer are of greatest importance. The three most common varieties of polyethylenes are currently known as low-density PE (LDPE), high-density PE (HDPE), and linear low-density PE (LLDPE), and they appeared in the marketplace in that chronological order.

Soon after the termination of World War II the technology of LDPE was developed by ICI in Great Britain. The difficulties of polymerizing ethylene were overcome by means of high pressure ($\leqslant 40{,}000$ psi) and relatively high temperature ($180–200°C$) conditions and the use of oxygen (a diradical) or peroxide catalysts (which also generate free radicals). The free-radical catalyzed chain growth becomes complicated by intermolecular and intramolecular radical transfer reactions illustrated in Scheme 4.1. The resulting radical centers located at nonterminal carbon centers can initiate chain growth, and the result is an aggregate of branched polyethylene chains that do not pack efficiently in space. Hence, there is formed a rather noncrystalline, waxy, *low*-density solid, LDPE, priced at about 43¢. It is processed by blow molding to fabricate bottles (for milk and other consumer products), by extrusion to produce films and sheets, and by injection molding to yield a variety of products. Some major applications include packaging of foods by shrink-wrapping, coating of paper cartons by means of extruded films, and coating of electric wires and cables.

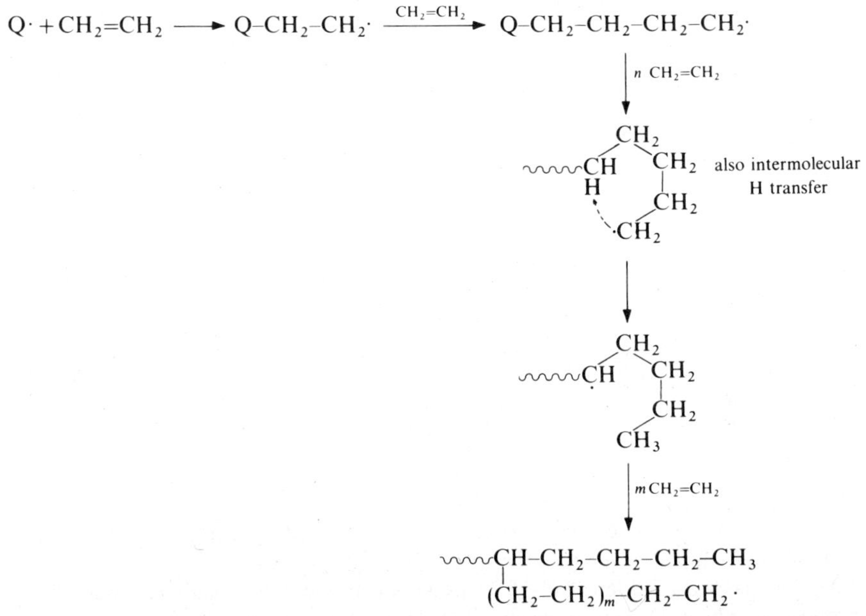

$$Q\cdot + CH_2{=}CH_2 \longrightarrow Q{-}CH_2{-}CH_2\cdot \xrightarrow{\ CH_2{=}CH_2\ } Q{-}CH_2{-}CH_2{-}CH_2{-}CH_2\cdot$$

$\Big\downarrow n\ CH_2{=}CH_2$

also intermolecular H transfer

$\Big\downarrow m\ CH_2{=}CH_2$

**Scheme 4.1**

High-density polyethylene was introduced as a result of the development of Ziegler–Natta transition-metal catalysis, which induces the polymerization of ethylene at low temperatures ($\sim 60°C$) and pressures (1–2 bar). Most importantly, it gives rise to linear, that is, nonbranched ethylene chains. The latter fill space rather efficiently, pack well, and hence produce a *high*-density product in which the intermolecular attractive forces are relatively strong even though they involve only London dispersion forces between $CH_2$ groups of adjacent macromolecules.

A simplistic representation of the original Ziegler catalytic system starts with the activation of $TiCl_3$ or $TiCl_4$ (30¢) by means of a trialkylaluminum compound, $R_3Al$

(4.3.7.2) to give Cl$_4$TiR. Since titanium is a $d^4$ metal, the last-mentioned titanium complex still contains a vacant orbital capable of coordinating the $\pi$-electrons of an ethylene molecule. This is followed by the insertion of the terminal carbon of the coordinated ethylene between titanium and R, and this is the first step in a chain-assembly process that at the same time regenerates a vacant orbital capable of receiving the next ethylene molecule. The alternating steps of insertion and coordination of additional ethylene molecules are reminiscent of a knitting operation in which the needle moves back and forth, "picking up" new loops of thread. Since the catalyst is insoluble, the growing chains tend to reach out into the hydrocarbon solvent. Consequently, the polymeric chains do not clutter the catalyst surface and do not inhibit the diffusion of ethylene molecules toward the metallic sites where the chemical activity occurs. The polymerization mechanism is depicted in Scheme 4.2.

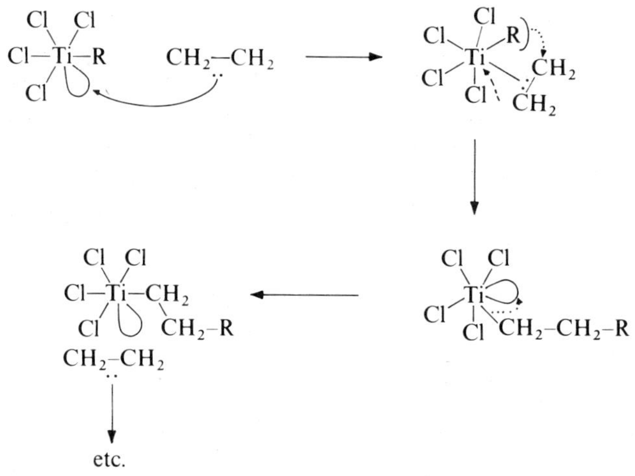

Scheme 4.2

An even more spectacular achievement of Ziegler–Natta catalysis concerns the stereochemical control of the polymerization of propylene (5.4.1).

As soon as the formation of HDPE was demonstrated, other catalytic systems were developed to achieve the same objective. Thus, for example, Phillips Petroleum announced a heterogeneous chromium oxide–silica–alumina catalyst that functions at 70–200°C and 20–50 atm in cyclohexane.

High-density polyethylene is priced higher than LDPE; until recently, grades suitable for blow molding and injection molding cost 44¢, while the extrusion and wire and cable coating grades cost 47¢ and 54¢, respectively. Currently (January 1989), demand boosted the price of all grades to 52–53¢/lb. The last-mentioned grade is also produced with a high carbon-black content for wire and cable coating applications and costs 65¢. As implied by the listing of these unit prices, HDPE is employed in the fabrication of all types of final products, but blow molding of containers and extrusion of sheets, pipes, coated wires, and cables predominate.

A comparison of some of the properties of LDPE and HDPE is of interest. While the rather amorphous LDPE (0.91–0.94 g/cm$^3$) softens at 85–87°C, melts completely at

112°C, and has a tensile strength of 1800–2000 psi, the rather crystalline HDPE, with a density as high as 0.96 g/cm$^3$, softens above 100°C and melts completely in the 125–135°C range, and its tensile strength is about double of that of LDPE (3500–5500 psi). Of course, these differences affect different processing requirements and utilization possibilities.

The third commodity PE (viz., LLDPE), evolved, when relatively small quantities ($\sim$5–8%) of linear and nonlinear but always terminal $C_4$–$C_8$ olefins (e.g., 4-methyl-pentene-1, 4MP-1) were copolymerized by means of the catalytic system that gave rise to HDPE. Now, the basically linear polyethylene molecules contains small, randomly introduced $C_2$–$C_6$ side chains, which inhibit the formation of highly compact aggregates, as was the case with HDPE. The result is a material that combines the presence of crystalline segments of linear PE that provide intermolecular attractive forces (but also tend to induce brittleness characteristic of all crystalline materials), with regions that are noncrystalline because of the disturbing presence of the on-purpose (intentionally) introduced side chains. The latter perform the same function that plasticizers achieve in the case of highly crystalline and hence intrinsically brittle polymers such as PVC. Thus, the still linear, but now low-density polymer, or LLDPE, can be considered to be "internally plasticized," and the remarkable property that made LLDPE so successful is the extraordinary mechanical strength of very thin films, illustrated best by the worldwide popularity of LLDPE shopping bags. About two thirds of all LLDPE is processed into films fabricated as packaging materials, waterproof liners for diapers and heavy-duty packaging, and so on. LLDPE is currently (January 1989) priced at 52¢–53¢ and is very competitive with other materials because of the tensile strength of its thin films (see 4.1).

The density of LLDPE depends obviously on the amount and nature of the terminal (or alpha) olefin comonomer, but most often it is in the range 0.916–0.930 g/cm$^3$. As stated above, it is the mechanical strength (resistance to tear of very thin films) that constitutes its principal virtue. Discussion of how this property is affected by the concentration and molecular size of the alphaolefin comonomer is beyond the scope of this book.

The current production of the three commodity PEs in the United States is nearly 11B lb for the combined demand for LDPE and LLDPE and about 8B lb for HDPE. In terms of ethylene utilization in the United States, LFDPE and LLDPE consume about 28%, while HDPE consumes about 24%. The production capacity for all of these polymers is undergoing expansion (*CW* 10/21/87). A trend to retrofit the original high-pressure equipment installed for the production of LDPE to supply the fastest growing demand for LLDPE (and/or HDPE) is noteworthy. Also, both of these polymers can now be produced by means of more economical gas-phase polymerization methods (*CE* 2/18/85) that employ a fluidized-bed catalyst (Union Carbide's Unipol technology; see 2.4.4.3.D, *CEN* 12/5/77; and F. J. Karol, *CT* 4/83, 222–228; and related developments by BP, BASF, Phillips, etc.). The presence of terminal olefin copolymer and the choice of catalyst determine whether the reactors deliver LLDPE or HDPE. Similar trends apply to the production of PEs in other parts of the world (*CEN* 4/28/86). The popularity of the family of PEs has even spread to many LDCs, where water required for domestic needs that used to be carried in tin or galvanized iron containers, is now transported in colorful PE buckets.

In addition of the above-mentioned ethylene-based commodity PEs there exist several more specialized ethylene-based homopolymers and copolymers.

The most astounding material is the ultra-high-strength, ultra-high-molecular weight ($\geqslant$1M) HDPE (UHMWPE) produced by Allied as its Spectra 900 fiber by means of a gel-spinning process [U.S. Patent 4,413,110, issued in 1983, but the development was not revealed until 1984 (*CMR* 9/10/84) and 1985 (*CW* 2/27/85; *CE* 3/4/85)]. Pound-for-pound, the fiber is claimed to be "10 times stronger than steel and at least 30% stronger than the aramid fiber Kevlar" (9.2.2, 9.3.2.2, Scheme 9.6). The surprisingly durable material again clearly demonstrates the importance of an optimum alignment of macromolecules in order to optimize the intermolecular attractive forces—a lesson learned first in the case of nylon 6/6. However, since we deal here with the lowly London dispersion forces (and not hydrogen bonds, dipole–induced dipole, dipole–dipole, or ionic forces), the Spectra 900 material demonstrates the traditional tendency to underestimate the potential of London dispersion forces operating between C–H systems of neighboring macromolecules when the latter are linear, large, and well aligned. The preceding patent (U.S. Patent 4,413,110) focuses precisely on how HDPE molecules are coaxed to align themselves as the polymer is processed by shifting from one solvent system to another. Allied's UHMWPE melts at 147°C or higher and is extruded as a solution in a relatively nonvolatile solvent (such as mineral oil), the extrudate is then cooled to form a gel, and the latter is extracted with a relatively volatile solvent (such as a fluorocarbon) to give a second gel that is dried and repeatedly stretched. The price of the resulting fiber is estimated at about $22/lb. This development resulted from a cooperative effort of Allied and DSM, and Toyobo of Japan plans to build a plant in Osaka. The gel spinning technique serves to disentangle the high-molecular-weight macromolecules.

Another UHMWPE of molecular weight $3–6 \times 10^6$, 0.93 g/cm$^3$, 1900 UHMW polymer ($1.06), is produced in powder form by Himont, a joint venture of Hercules and Montedison, by means of a Ziegler-like catalytic system. This product cannot be processed by conventional extrusion or injection molding techniques because the length of the macromolecules inhibits their flow even above the melting point of the crystalline form (132°C). At that temperature or above, the opaque solid becomes clear without significant changes in shape of the sample. However, the UHMWPE powder can be compression-molded forged into complex objects or ram-extruded into rods, tubes, and so on. Also, the desired objects can be stamped out of sheets or fabricated by machining semifinished stock. The final products stand out for their exceptionally high resistance to abrasion, self-lubrication, noise-damping properties, and nonstick surfaces.

In 1984 Union Carbide introduced (*CW* 9/19/84) a very-low-density PE (VLDPE) with densities below the traditional 0.915-g/cm$^3$ value and priced at 62¢/lb. It is assumed that VLDPE is obtained by Union Carbide's Unipol technology and, most likely, arises from a higher concentration of $\alpha$-olefins than that employed in the conventional LLDPE process. VLDPE is expected to compete with other film-producing materials such as EVA (see below) because it too is sold on a weight basis.

Also in 1984, ARCO initiated the production of a PE expandable foam (*CW* 5/30/84). The following ethylene copolymers have become rather significant.

### 4.3.1.1    The Ethylene–Propylene–Diene Monomer (EPDM) Elastomer

This elastomer is obtained from ethylene, propylene, and a small amount ($\sim 3\%$) of a diene present in order to create elastic memory in the final product. This material is also referred to as *ethylene*–propylene–rubber (EPR) and is currently produced at a level of 350MM lb. The initial polymerization cannot be carried out homolytically because of the susceptibility of propylene to oxygen-containing radicals, and thus Ziegler–Natta-type catalysts are employed. The ethylene:propylene ratio is adjusted to suit the desired properties, and the ideal diene is one that contains two double bonds of different reactivities: the first is readily copolymerized during the transition-metal catalyzed polymerization process, and the second participates in the curing or vulcanizing phase catalyzed by free-radical polymerization catalysts. The most common dienes employed in the production of EPDM elastomers are 1,4-hexadiene (see below), 1,3-butadiene (6.8.1), ethylidenenorbornene (Chapter 8), and dicyclopentadiene (6.8.3).

The above-mentioned 1,4-hexadiene is a product derived from the addition of ethylene to 1,3-butadiene catalyzed by a nickel complex in which the ligands are triethyl phosphite [$(EtO)_3P$:] molecules, as illustrated in Scheme 4.3.

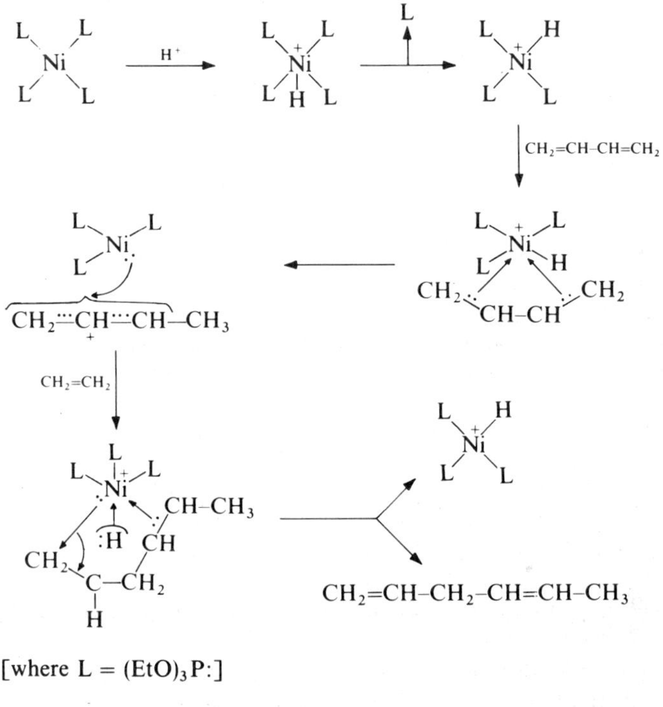

[where $L = (EtO)_3P$:]

**Scheme 4.3**

## 4.3.1.2  Ethylene–Vinyl Acetate (EVA) and Vinyl Acetate–Ethylene (VAE)

These materials are obtained by incorporating 5–50%, or a quantity higher than 50%, respectively, of vinyl acetate (40¢), (4.3.4) into a LDPE material during a typical homolytic, high-pressure polymerization. The presence of vinyl acetate in the copolymer reduces crystallinity, opacity, and heat-seal temperature of the product, and this explains that much of the 1B-lb demand is channeled into the fabrication of film used for shrink-wrapping and heat-sealing of numerous consumer products. EVA is also coextruded with PP or PET films, or combined with aluminum foil, for readily heat-sealed packaging of food products. An increasing presence of acetate groups promotes flexibility, resistance to hydrophobic materials, adhesive properties to polar surfaces, and ease of cross-linking by means of peroxides or radiation. The copolymer that contains 28% vinyl acetate is compatible with both waxes and polybutylene tackifiers and consequently is widely used as a component of adhesives employed for hot-melt bonding of nonwoven fabrics and paper products. EVA that contains about 5–20% vinyl acetate monomer is used for the production of flexible toys, tubing, gaskets, and other items. Much of VAE is used in the form of emulsions for the formulation of industrial adhesives.

### 4.3.1.3 Ethylene–Acrylic or Methacrylic Acid Systems

These systems include ethylene–acrylic acid (EAA), ethylene–methacrylic acid (EMAA), ethylene–methyl acrylate (EMA), and ethylene–ethyl acrylate (EEA) polymers. EAA and EMAA are normally used in the form of their sodium or zinc salts and they represent the general class of *ionomers*, that is, polymers in which intermolecular attractive forces are constituted by ionic or Coulombic forces. Thermal energy can overcome these forces and thus, whereas ionomers can be processed by conventional means, the presence of strong ionic forces is responsible for excellent mechanical strength and resistance to abrasion, as well as excellent adhesion to metallic surfaces. The carboxylic acid content of these ethylene–acid copolymers affects the ultimate properties and is adjusted to fit the intended uses of the polymers. Of particular interest are the polymers that contain as much as 20 and 25% of the carboxylic acids in EAA and EMAA, respectively, the limit established by the FDA for polymers used in packaging of foods. Athletic equipment, footwear, and certain automotive componenets, are examples of other ionomer applications. The unit prices of acrylic and methacrylic acids are 60¢ and 78¢, respectively, and the price of Du Pont's Surlyn plastics is in the $0.92–$1.87 range.

Allied Signal offers a series of Aclyn *low*-molecular-weight ionomers priced at about $2.00, in which the carboxylic acid functions of the EAA copolymer (priced at $1.13–$1.40) are neutralized by magnesium, sodium, calcium, or zinc ions. These metallic salts exist in the form of macromolecular aggregates that exhibit some interesting properties:

- Small quantities (<1% by weight) added to PET induce crystallization of the polymeric materials, an effect that speeds up the fabrication of thermoformed end products.
- They function as thickeners that counteract the commonly observed decrease in viscosity of solutions with rising temperatures.
- They improve the adhesion of coatings, sealants, rust-proofing materials applied to surfaces of HDPE, nylon, aluminum, glass, and steel.
- They promote the dispersion of pigments and other additives in polymers.
- They promote the formation of PE gels employed in some polishes, inks, paints, lacquers, cosmetics, and so on.

The presence of 15–30% methyl or ethyl carboxylate groups in the copolymers of EMA or EEA reduces the crystallinity of the PE chains and increases the flexibility and toughness of the resulting thermoplastics. About 50MM lb of these materials are used for packaging of foods, as gaskets, disposable gloves, and so on. The compatability of these plastics with polyolefins and other classes of plastics accounts for their use as *impact modifiers*. The unit price of the acrylic ester monomers is 66¢, and the copolymers are priced at about 72¢/lb.

### 4.3.1.4 Ethylene–Vinyl Alcohol Polymers (EVOH)

These substances are of major importance as *barrier resins*, that is, plastics in which the presence of hydroxylic groups impedes the diffusion of gases such as oxygen and carbon dioxide across the material. On the other hand, the diffusion of moisture increases with a rising content of hydroxyl groups; hence, in most packaging applications, EVOH films are combined with hydrophobic, hydrocarbon plastics in multilayer arrangements when the packaged product is to be protected from $O_2$, $CO_2$, and moisture. The hydrophilic, polar nature of EVOH explains applications where resistance to mineral and other oils is critical. Food products are the foremost materials packaged in EVOH-containing plastics, and the ease of processing favors such applications. The shelf life of products particularly sensitive to atmospheric oxygen can be prolonged by packaging in EVOH-containing plastic containers under an atmosphere of nitrogen or carbon dioxide. EVOH is obtained from the corresponding EVA polymer (see above) by hydrolytic removal of the acetate groups.

### 4.3.1.5  Ethylene–Propylene–Copolymer Block Polymers

Uniroyal Chemical produces a series of relatively low molecular weight (MW 6500–8000) liquid, block polymers of ethylene–propylene–dicyclopentadiene (DCPD; 6.8.3) and ethylene–propylene–norbornadiene (8.1.2). They are represented as follows:

Trilene CP 80        $-(CHMe-CH_2)_m-(CH_2-CH_2)_n-$

Trilene 65        $-(CHMe-CH_2)_m-(CH_2-CH_2)_n-$

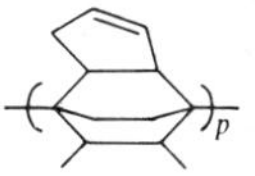

Trilene 66 and 67    $-(CHMe-CH_2)_m-(CH_2-CH_2)_n-$

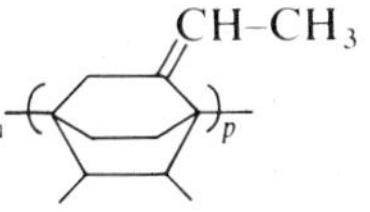

These block polymers are most likely obtained by anionic polymerization. The two last-mentioned products differ mainly in the degree of unsaturation, that is, the content of ethylidenenorbornadiene. These interesting materials can be used as reactive plasticizers of other polymeric systems, and they improve various processing and application properties.

### 4.3.2  Ethylene Oligomers: alphaolefins

There are two kinds of alphaolefins (AOs) to be found in the marketplace. The first is a narrow-boiling mixture of odd- and even-numbered linear olefins that contain a terminal double bond; these are obtained by a propietary (Chevron) cracking of the paraffinic fraction of petroleum (7.3). The second kind of AOs consists of even-numbered, linear olefins with a terminal double bond; these are obtained by oligomerization of ethylene or dehydration of fatty alcohols derived from naturally occurring fatty acids (SZM-II, 8, 62, 80).

The oligomerization of ethylene is catalyzed by aluminum triethyl and is reminescent of Ziegler–Natta catalysis (4.3.1) in that the coordination of ethylene by the electron-deficient metal is followed by insertion of the ethylene molecule between one of the aluminum–carbon bonds. This process is repeated until ethylene is consumed and then thermal decomposition regenerates the catalyst while the AO is liberated as shown in Scheme 4.4.

The lower olefins, 1-butene (26¢), 1-hexene, and 1-octene (the latter two are priced at about 40¢–46¢), are the comonomers in the assembly of LLDPE (4.3.1). The utilization of

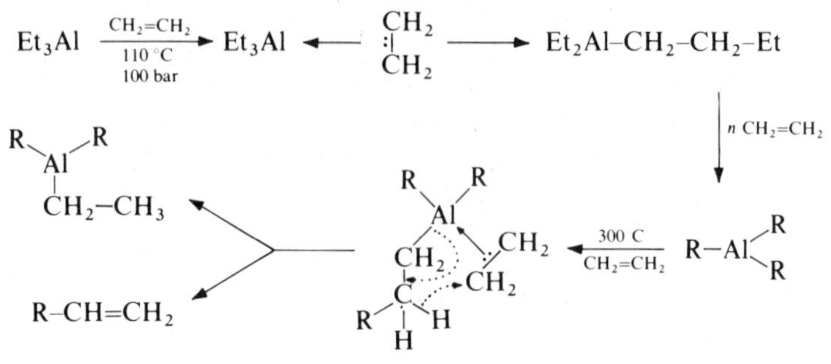

**Scheme 4.4**

the higher AOs is discussed elsewhere (7.4.1). The total demand for all kinds of AOs is in excess of 1.2B lb.

### 4.3.3   PVC, CPVC, and Other Halogenated Derivatives of Ethylene

#### 4.3.3.1   PVC, CPVC, PVdC, and Chlorinated and Chlorosulfonated PE

The production of ethylene dichloride (EDC), the precursor of vinyl chloride monomer (VCM) (24¢), consumes over 12% of the demand for ethylene. The magnitude of the demand for PVC (nearly 7.5B lb) attests to its usefulness. Unlike a "mature" material after decades of use, the demand for PVC remains strong (see A. Agoos, "PVC Acts Like a Growth Market" *CW* 11/4/87; B. F. Greek, "Demand for PVC Outstrips Supply," *CEN* 12/14/87). The resin is available in different forms, ranging in price (January 1989) from 42¢ for general-purpose suspension-polymerized grade, 42¢ for film or pipe grade, and 68¢ for homopolymer dispersion.

Vinyl chloride monomer is produced by dehydrohalogenation of EDC or more effectively by oxychlorination (see 3.2.3) of ethylene. While the overall reaction appears rather simple

$$CH_2=CH_2 + HCl + \tfrac{1}{2}O_2 \rightarrow CH_2=CH-Cl + H_2O$$

the use of chlorine (rather than hydrogen chloride) at about 250°C, and of CuCl$_2$ deposited on KCl, alumina, and silica suggests a greater complexity. Actually, oxychlorination is designed to fully utilize chlorine (10¢), and to avoid its partial conversion to hydrogen chloride that, for practical operational reasons, would be isolated as hydrochloric acid valued as low as 2.75¢ and used commonly as pickling acid or steelmakers-grade muriatic acid for about 2¢/lb. The hydrogen chloride generated *in situ* is converted to chlorine by a sequence of the following reactions:

$$2HCl + 2Cu^{2+} \rightarrow Cl_2 + Cu_2^{2+} + 2H^+$$

$$Cu_2^{2+} + \tfrac{1}{2}O_2 + 2H^+ \rightarrow 2Cu^{2+} + H_2O$$

Vinyl chloride monomer (b.p. $-13.3°C$) is rather insoluble in water and is polymerized under pressure in the presence of peroxides at about 50°C. It has been classified as a carcinogen since 1981, and the EPA regulates its permissible concentration in the workplace. Currently, the EPA is considering a "hazardous" classification for solid PVC waste when a newly developed "toxicity characteristic leaching procedure" (TCLP) finds evidence for a VCM concentration above 50 ppb. Once such ruling is officially adopted, the cost of disposal of PVC wastes is expected to rise from about $10/yd$^3$ to $100–$200/yd$^3$, and the cost of incineration may rise to $1000/yd$^3$.

*Polymerization Methods*

As suggested above by the pricing information, most of the PVC is obtained by means of *suspension polymerization* that involves a mechanical dispersion of VCM in water in the presence of *protective colloids* [e.g., gelatin, PVA, modified cellulose and gums (*SZM-II*, 96)] and peroxide catalysts (6.3.1.7). A small concentration of protective colloids suffices to promote the formation of solid polymer particles that remain suspended because of partial agglomeration until the reaction is completed and until the solid product is separated by centrifugation. Some PVC is produced by *mass polymerization* carried out in two stages in order to avoid a disastrous filling of the reaction vessel with a solid mass of polymer. The solid PVC suspension product obtained as described above is mixed with VCM in a powerful blender and agitated until the polymerization is complete, and PVC is obtained in powder form. Finally, some PVC is obtained as a latex by *emulsion polymerization*: VCM and a peroxide catalyst are emulsified in water by the addition of an appropriate surface-active agent, and this produces finely dispersed PVC used primarily for the production of plastisols or organosols (see below). Similarly, an organic solvent, rather than water, can be employed to produce PVC in a *solution polymerization*.

Polyvinyl chloride per se is a brittle solid and must be plasticized in order to be useful. The most common plasticizers are the esters of phthalic acid (9.3.2) such as dioctyl phthalate (DOP). Also, PVC readily suffers thermal decomposition (dehydrohalogenation), and the initial decomposition product is even more susceptible to additional decomposition because of the formation of highly reactive allylic chloride moieties:

$$-CH_2-CHCl-CH_2-CHCl-CH_2-CHCl- \rightarrow -CH_2-CHCl-CH=CH-CH_2-CHCl-+HCl$$
$$\uparrow$$
$$\text{allylic chloride system}$$

Thus, on heating PVC decomposes to a black mass unless it is appropriately stabilized by means of a variety of metallic and organometallic compounds of tin, zinc, calcium, cadmium, barium, lead, and other materials that include epoxidized soybean oil (*SZM-II*, 59). Other stabilizers are used to protect PVC from degradation by light in case of outdoor applications.

The preceding comments suggest that, for most applications, PVC must be compounded with plasticizers and stabilizers as well as pigments and fillers. The resulting PVC formulations can then be extruded or calendered in the form of sheets, films, pipes, garden hose, weatherstripping, roofing, sidings, flooring, leatherlike upholstery materials, and so on or injection molded to give a variety of desired objects. The amount of plasticizer determines the rigidity of the final products. In addition to plasticizers, the brittle PVC may be blended with *impact modifiers* such as EVA (see 4.3.1), methyl methacrylate–butadiene–styrene (MBS), and other polymers (ABS, MBS, etc.). Large metallic objects can be given a protective and decorative coating by locating the object preheated at about 200°C above a turbulent bed of PVC powder.

*Plastisols* and *organosols* are fluid pastes of properly formulated PVC (obtained as a dispersion resin by means of emulsion polymerization; see above) that contain either water or volatile organic solvents. The pastes are heated to fuse the PVC particles

( > 150°C in the case of a homopolymer or to lower temperatures in case of copolymers), and eventually a solid mass forms that can be deposited, for example, by coating an aluminum panel for use as a house siding, by dipping an irregular shaped object used as a mold and stripping the PVC coating on cooling (in the manufacture of rubberlike gloves), spraying the paste to give drum linings, and so on.

Polyvinyl chloride is chlorinated to give CPVC, which is employed for the manufacture of plumbing supplies for hot-water installations. At least in name, a somewhat related product is chlorosulfonated PE (CSM rubber), produced for many years by Du Pont as Hypalon. In fact, in 1985 Du Pont announced (*CMR* 12/16/85) the production of the billionth pound of its Hypalon and an expansion of the family of chlorinated PEs. The product is obtained by means of a reaction of PE with sulfur dioxide (11.5¢) and chlorine (10¢) under homolytic conditions that give a product that contains a random distribution of about 27% chlorine and slightly over 1% sulfur. The presence of sulfur (as sulfonyl groups, $-SO_2-$) explains the elastomeric properties of Hypalon as it is indicative of a small degree of cross-linking responsible for elastic memory of the material. CSM rubber produces stable coatings for wires, cables, and similar.

To a minor extent (2%), VCM is involved in the production of poly(vinylidene dichloride) or poly(vinylidene chloride) (PVdC). Addition of chlorine to VCM yields 1,1,2-trichloroethane (42¢), and dehydrohalogenation of the latter produces vinylidene dichloride, 1,1-dichloroethylene (VDC; $CH_2=CCl_2$). About 300MM lb of VDC are produced in the United States, and its homopolymer and copolymer—mainly with VCM—constitute the family of Dow's Saran, best represented by the well-known food-wrapping film. The outstanding property of PVdC films is the resistance to diffusion of oxygen, moisture, and other gases and liquids. Because of these barrier-resin properties, PVdC is often processed by coextrusion with cellophane and other films.

The presence of VDC (as well as vinyl acetate, VAM; see above) in copolymers with VCM creates a family of vinyl resins that have been employed for several decades as upholstery and artificial leather materials, protective and decorative coatings, insulating coatings for electric wires, floor tiles, materials for unbreakable phonograph records, and so on. Both VDC and VAM function as internal plasticizers of the otherwise brittle PVC, and their concentration profoundly affects the physical properties of the resins. However, external plasticizers such as the above-mentioned phthalates and esters such as di-*n*-butyl sebacate (6.4.2, 7.4.3) ($1.72) and diisobutyl adipate (6.4.2, 7.4.3), are also employed, depending on the projected uses of the plastic.

### 4.3.3.2 Miscellaneous Halogenated Derivatives of Ethylene

Ethylene dichloride, 1,2-dichloroethane (EDC; 17¢) was mentioned above as a likely transient intermediate in the high-temperature chlorination of ethylene to produce VCM. However, EDC is, in its own right, a useful building block of a variety of industrial organic chemicals (4.3.3.3), and, indeed, it can be employed for the production of VCM by means of thermal decomposition carried out at 450–550°C.

An interesting derivative of EDC is 2-aminoethylsulfonic acid, or taurine, present in bile and lungs of mammals and in other living systems. It is prepared by treatment of EDC with sodium bisulfite (21¢), followed by reaction of sodium 2-chloroethylsulfonate with ammonia. A recent report (Associated Press, 8/14/87) indicates that low levels of taurine in prepared cat foods can cause a heart ailment among the 56MM pet population in the United States. Sodium 2-chloroethylsulfonate is the most likely precursor of another heterodifunctional monomer that has become available recently, namely, 3- and 4-

vinylbenzyl 2-chloroethyl sulfone. The preceding reactions are summarized as follows:

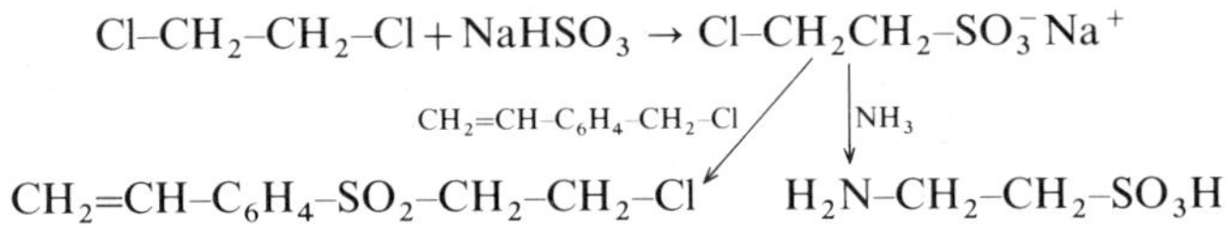

$$Cl-CH_2-CH_2-Cl + NaHSO_3 \rightarrow Cl-CH_2CH_2-SO_3^- \, Na^+$$

$$CH_2=CH-C_6H_4-CH_2-Cl \diagup \qquad \Big\vert NH_3$$

$$CH_2=CH-C_6H_4-SO_2-CH_2-CH_2-Cl^\diagup \quad H_2N-CH_2-CH_2-SO_3H$$

The addition of hydrogen fluoride (anhydrous, 69¢) to VCM followed by thermal dehydrohalogenation, gives vinyl fluoride:

$$CH_2=CH-Cl + HF \rightarrow CH_3-CHClF \rightarrow CH_2=CH-F + HCl$$

in accord with the greater stability of C–F bonds as compared to that of C–Cl bonds. The polymerization of vinyl fluoride yields poly(vinyl fluoride) (PVF), a highly weather resistant thermoplastic used to protect more susceptible polymers (PE, PP, etc.) from photochemically induced degradation.

Ethylene dichloride can be the source not only of VCM, but in the presence of chlorine, high-temperature conditions give rise to the above-mentioned VDC, and three other important chlorinated derivatives of ethylene, namely, 1,1,2-trichloroethane (tri; 42¢), trichloroethylene (TCE; 38.5¢), and tetrachloroethylene, perchloroethylene (perc, dry-cleaning grade 28.5¢, industrial grade 31¢, USP grade 30.25¢). All the chlorine addition, substitution, and dehydrohalogenation reactions occur simultaneously and sequentially, together with oxychlorination in the presence of appropriate catalysts (*FKC*, 604–611, 4.3.3), and give rise to the above-mentioned compounds as shown in Scheme 4.5.

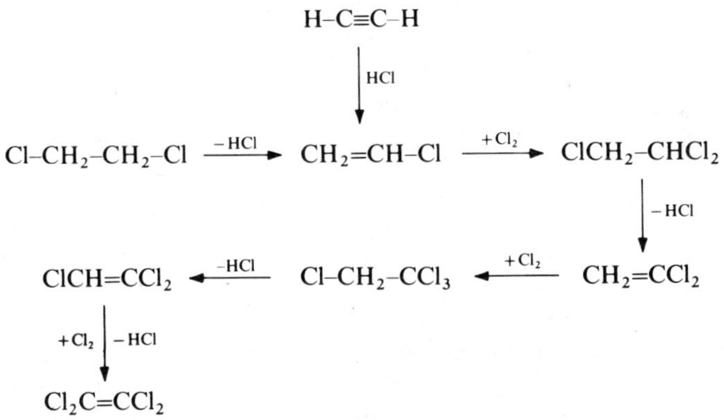

$$H-C\equiv C-H$$

$$\Big\downarrow HCl$$

$$Cl-CH_2-CH_2-Cl \xrightarrow{-HCl} CH_2=CH-Cl \xrightarrow{+Cl_2} ClCH_2-CHCl_2$$

$$\Big\vert -HCl$$

$$ClCH=CCl_2 \xleftarrow{-HCl} Cl-CH_2-CCl_3 \xleftarrow{+Cl_2} CH_2=CCl_2$$

$$+Cl_2 \Big\vert -HCl$$

$$Cl_2C=CCl_2$$

**Scheme 4.5**

As is the case with other chlorinated hydrocarbons (3.2.3), the production of TCE, previously a most popular dry-cleaning fluid, has dropped precipitously from 1.1B lb in 1974 to 300 MM lb in 1986 (see C. J. Verbanic, "Hard Times for Halogenated Hydrocarbons," *CB* 11/86, 32–24), not only because of increasing imports, but mostly because of the "chlorinated hydrocarbon scare." TCE is still used for degreasing machined metallic products before electroplating and other finishing operations. Also, it is the precursor of two important fluorocarbons, 1,1,2-trichloro-1,2,2-trifluoroethane, or Fluorocarbon 113

(FC 113, 89¢), and trifluorochloroethylene (TFCE). The first transformation is a typical fluorination by means of HF–CoF$_3$, and this is followed by a thermal elimination of two adjacent chlorine substituents again demonstrating the superior stability of C–F bonds:

$$ClCH{=}CCl_2 \rightarrow F_2ClC{-}CCl_2F \rightarrow F_2C{=}CClF + Cl_2$$

FC 113             TFCE

Trifluorochloroethylene polymerizes to a high-melting (218°C), clear, solid, poly(chlorotrifluoroethylene) (PCTFE) that is resitant to almost all chemicals except active metals such as molten sodium. The presence of one bulky chlorine substituent in each repeating structural unit, in what would otherwise be a polymer of tetrafluoro-ethylene—Du Pont's Teflon (3.2.3), inhibits the formation of crystalline arrays of macromolecules. This results in the clear appearance of sheets of PCTFE and the lowering of the softening temperature, better expressed as the *glass transition temperature* or $T_g$, from 126°C for Teflon to 45°C for PCTFE. The $T_g$ is the temperature at which a mostly glasslike polymer that behaves like a supercooled liquid (even though it may still be a partially crystalline material) begins to exhibit plastic properties; that is, it flows under the effect of heat and pressure. A lower $T_g$ obviously facilitates the molding and other processing methods of all thermoplastics.

A copolymer of TFCE and ethylene, namely, poly(ethylene–chlorotrifluoroethylene) (ECTFE), is a linear thermoplastic distributed in the United States by Ausimont as Halar, in which the ethylene and TFCE units are arranged mostly in an alternating fashion. The same tendency of alternating arrangements of different structural moieties in copolymers is observed in all vinyl plastics in which the two comonomers differ greatly in terms of electron-donating and electron-accepting character. This phenomenon is related to transient charge-transfer complexes formed between the two vinyl systems (symbolized here by R–CH=CH$_2$ and R'–CH=CH$_2$) and the two different terminal radical species involved in the propagation of the polymerization and derived from the comonomers (symbolized here in a simplified manner by R–CH$_2$· and R'–CH$_2$·). Without going into specific details, it is logical to expect that the formation of charge-transfer complexes by species of opposite electronic character is preferred.

A similar situation applies in the case of the ethylene–TFE copolymer (ETFE) produced in the United States as Du Pont's Tefzel and also distributed as Ausimont's Halon ET.

The domestic production of perc has decreased from 1.9B lb in 1974 to about 450MM lb in 1986, and imports of about 150MM lb supplement the demand for dry cleaning and degreasing of metals and textiles, and about 30% is converted to FC 113. About 100MM lb of FC 113, together with related chlorofluorocarbons, are consumed by the high-tech electronic industry in the fabrication of integrated circuits and circuit boards.

The ban on the use of TCE stimulated the production of 1,1,1-trichloroethane or methylchloroform (40.5¢), but, again, the adverse sociopolitical climate for the use of chlorinated hydrocarbons has caused a drop in its demand from 2.4B lb in 1974 to 1.3B lb in 1986. Much of this demand is satisfied by imports even though, so far, it is free of carcinogenic suspicions. Not even the California Air Resources Board, (CARB) intends for the moment, to characterize 1,1,1-trichloroethane as a toxic air contaminant (*CMR* 6/8/87). This major survivor of the "chlorinated hydrocarbon scare" is used extensively in organic solvent-based paints, adhesives, and other consumer products.

One method of producing methylchloroform involves a Lewis acid-catalyzed addition of HCl to VCM at 40°C in the presence of FeCl$_3$ that follows the Markownikoff rule

(*ONR*-57), and this is followed by a high-temperature (400°C) chlorination of the 1,1-dichloroethane intermediate:

$$CH_2\!=\!CHCl \xrightarrow[HCl]{} CH_3\!-\!CHCl_2 \xrightarrow[Cl_2]{} CH_3\!-\!CCl_3 \quad (\sim 95\%)$$

The high selectivity of the chlorination step can again (3.2.3) be attributed to the stabilization of the intermediate radical by the chlorine substituents.

Other syntheses of methylchloroform include the acid-catalyzed addition of HCl to vinylidene dichloride that occurs in very high yields (as high as 98%) and the chlorination of ethane at about 350°C in which the desired product is accompanied by several by-products.

Ethyl chloride (24¢) can be obtained from either ethane, ethylene, or ethanol. An excess of ethane is chlorinated satisfactorily to ethyl chloride at about 400°C, ethylene adds hydrogen chloride readily in the presence of Lewis acid catalysts (e.g., ferric or aluminum chlorides), and now that ethane is readily available, the two processes can be operated in a sequential manner in order to fully utilize the HCl by-product of the high-temperature chlorination. For that purpose, since addition of chlorine to ethylene is not favored at high temperatures, a mixture of both gases can be chlorinated first, and the products of this reaction can then be cooled and submitted to the HCl–ethylene reaction.

The reaction of ethanol with either hydrogen chloride or concentrated hydrochloric acid occurs smoothly and may be more convenient when carried out on a relatively small scale.

The demand for ethyl chloride has been decreasing rapidly from the 650MM lb level of the mid-1970s because of the EPA-mandated phasing out of TEL to a current demand of about 150MM lb. By 1986 only one producer of TEL remained operational in the United States, namely, Du Pont, with Ethyl, PPG, and Nalco having departed from the ranks of domestic TEL producers. On the other hand, the Ethyl Corporation continues to operate TEL production in Canada and Greece (*CW* 6/26/86).

As in the case of the production of tetramethyl lead (TML), TEL (b.p. 202°C) is obtained by reaction of ethyl chloride with an alloy of lead and sodium. TEL is used jointly with some EDC and the analogous ethylene dibromide that prevent the deposition of lead in engines by producing volatile halides during the combustion process.

Besides the production of TEL (increasingly for export only), uses of ethyl chloride include the production of ethyl cellulose, EC, (*SZM-II*, 87) as a local anesthetic (when applied topically in liquid form and allowed to evaporate, b.p. 12°C), and in other miscellaneous ethylation reactions.

Ethyl bromide (98%, 82¢) is obtained conveniently from ethanol and HBr, although this operation is best carried out at the site of HBr production (by "burning" hydrogen with bromine gas) since the corrosive nature of anhydrous HBr causes the cost of HBr to rise because of the cylinders in which it is transported ($7/lb in 30,000-lb quantities). The production of ethyl iodide ($6.25) by a reaction of ethanol and HI results in complications because of the reducing action of HI, and thus the most convenient method for generating ethyl iodide is by means of the Finkelstein reaction (ONR-30), which takes advantage of the solubility of sodium iodide ($10.15) in acetone while the corresponding chloride and bromide are, for all practical purposes, insoluble:

$$Et\!-\!X + Na^+ \, I^- \rightarrow Et\!-\!I + NaX \downarrow$$

(where X = chloro or bromo substituent).

The reactivity of the ethyl halides as ethylating agents increases with an increase in the atomic weight of the halogens, but so does their unit cost (especially in the case of the intrisically costly derivatives of iodine (crude \$8.65, USP \$17.00), and thus large-scale ethylation reactions, such as the preparation of ethyl cellulose, are carried out by means of ethyl chloride. Also, ethyl magnesium chloride and bromide are available commercially for use in the variety of Grignard reactions (ONR-37).

### 4.3.3.3  Sulfur-, Nitrogen-, and Phosphorus-Containing Derivatives of Ethylene and EDC

Some of the heteroatom derivatives of ethylene are obtained directly from the olefin, and others must be prepared from a more highly functionalized ethylene derivative, namely, EDC. Hydroxyl-group-containing compounds are dealt with elsewhere (4.3.6) since they are usually obtained from ethylene oxide or ethylene chlorohydrin.

Hydrogen sulfide and mercaptans add to ethylene in the presence of oxygen or peroxide catalysts, but the resulting products are of minor industrial importance. For example, ethylene and $H_2S$ give ethyl mercaptan or ethanethiol and diethyl sulfide or, simply, ethyl sulfide. The former can be converted to 2-hydroxyethyl ethyl sulfide ($HO–CH_2CH_2–S–C_2H_5$) by means of EO or ethylene chlorohydrin, and it is used as an additive in electroplating operations. The latter (b.p. 92°C) is added to natural gas as an odorant as a warning against disastrous gas leaks. Oxidation of the preceding hydroxyethyl sulfide gives the corresponding sulfone, $HO–CH_2CH_2–SO_2–C_2H_5$, 2-hydroxyethyl ethyl sulfone, 2-(ethylsulfonyl)ethanol, which finds limited use as a solvent, an intermediate in the preparation of plasticizers, antistatic additives for textiles, and so on.

Ethylene also reacts with sulfur dichloride to give the infamous mustard gas, a chemical warfare agent of World War I (*KS*, 51–52), 2,2'-dichlorodiethyl sulfide:

$$2CH_2{=}CH_2 + SCl_2 \rightarrow Cl–CH_2CH_2–S–CH_2CH_2–Cl$$

Mustard gas is also obtained from di-$\beta$-hydroxyethyl sulfide and hydrogen chloride. In passing, one must acknowledge the role that mustard gas played during the World War II era as a stimulus of mechanistic studies dealing with intramolecular nucleophilic substitution reactions and the development of one class of chemotherapeutic (antineoplastic) agents.

A relatively small amount of EDC is converted by means of sodium polysulfide to specialty rubbers that belong to the polysulfide or Thiokol family:

$$Cl–CH_2CH_2–Cl + Na_2S_4 \rightarrow \left(–CH_2CH_2–S-\overset{\displaystyle S}{\underset{\displaystyle S}{\overset{\uparrow}{\underset{\downarrow}{S}}}}-S\right)_n \qquad \text{Thiokol A}$$

Analogous representatives of such elastomers are

$$\text{Thiokol ST} \qquad \left(–CH_2CH_2–O–CH_2–O–CH_2CH_2-\overset{\displaystyle S}{\underset{\displaystyle S}{\overset{\uparrow}{\underset{\downarrow}{S}}}}-S\right)_n$$

$$\text{Thiokol B} \qquad \left(–CH_2CH_2–O–CH_2CH_2-\overset{\displaystyle S}{\underset{\displaystyle S}{\overset{\uparrow}{\underset{\downarrow}{S}}}}-S\right)_n$$

prepared for ethylene glycol formal and di-(2-chloroethyl) ether (DCEE), respectively (4.3.6.2).

Thiokol rubbers are vulcanized by heating with sulfur in the presence of vulcanizing accelerators in order to render them highly resistant to gasoline, lubricating oils, and most common solvents.

The reaction of EDC with aqueous sodium sulfite produces sodium vinyl sulfonate (SVS) or sodium ethenesulfonate (Air Products, $1.01, 25% aqueous solution):

$$Cl-CH_2CH_2-Cl + Na_2SO_3 \rightarrow CH_2{=}CH-SO_3^- \ Na^+ + NaCl$$

Sodium vinyl sulfonate is employed as a comonomer with other vinyl monomers such as VAM and VDC to assemble ionomers, and it polymerizes to poly(sodium vinylsulfonate) (PSVS; 25% aqueous solution $1.19, Air Products). These polymeric materials are obtained in latex form when potassium persulfate, $K_2S_2O_8$ catalyst is added to an aqueous solution of the monomer(s), and the resulting latex is used for treatment of paper and paperboard.

The plant growth regulator ethepon, 2-chloroethylphosphonic acid, is obtained by means of the Michaelis–Arbuzov reaction (ONR-61) of EDC and trimethyl phosphite (prepared, in turn, from methanol and phosphorus trichloride, (32¢):

$$Cl-CH_2CH_2-Cl + P(OMe)_3 \rightarrow Cl-CH_2CH_2-P(O)(OMe)_2 + MeCl$$

Ethylene dichloride is employed in the traditional production of ethylenediamine (EDA; $1.30) and some of its derivatives:

$$Cl-CH_2CH_2-Cl + NH_3 \quad \text{(followed by NaOH)} \rightarrow H_2N-CH_2CH_2-NH_2$$

As may well be expected, excess ammonia functions as a base and generates some free EDA that, in turn, gives rise to diethylenetriamine ($H_2NCH_2CH_2NHCH_2CH_2NH_2$, $1.60), and the reaction can continue to give triethylenetetramine ($1.43), tetraethylenepentamine ($1.70), and so on. A recent announcement (*CE* 9/28/87) by Air Products reveals that a new facility to produce the higher analogs of EDA avoids the formation of NaCl by-product by utilizing ethanolamine and ethylenediamine as starting materials (also see 4.3.6.2). Most likely, this transformation is a *reductive amination* (see 3.10.3.2) by means of hydrogen and dehydrogenation–hydrogenation catalysts that convert the alcohol to a transient aldehyde and then hydrogenate the resulting imine to the disubstituted amine:

$$R-CH_2-OH \xrightarrow{-H_2} R-CH{=}O \xrightarrow[-H_2O]{+H_2N-R'} R-CH{=}N-R' \xrightarrow{+H_2} R-CH_2-NH-R'$$

This standard industrial alcohol amination process has been employed by Union Carbide and Texaco (*CMR* 2/11/85) under similar circumstances and follows the "do not use chlorine-containing intermediates in vain" rule. The ethylenediamine oligomers are used as corrosion inhibiting lubricating oil additives, epoxy curing agents ("hardeners"), intermediates for the preparation of paper wet-strength polymers, PUR "extenders," and so on.

Ethylenediamine and its oligomers are precursors of a series of chelating agents that can be prepared by alkylation of the amino group with sodium chloroacetate (solid chloroacetic acid, 99%, is priced at 56¢):

$$H_2N-CH_2CH_2-NH_2 + 4Cl-CH_2CO_2Na \rightarrow [-CH_2N(CH_2CO_2Na)_2]_2$$

The simplest member of this family of chelating agents, ethylenediaminetetraacetic acid tetrasodium salt (EDTA), is priced at 36.5¢, and this suggests a more economical route, the use of formaldehyde and sodium cyanide followed by alkaline hydrolysis of the intermediate nitrile. The chelation of heavy metals by EDTA (and its analogs) occurs by virtue of the conveniently located electron-donating nitrogen functions, as shown in Scheme 4.6.

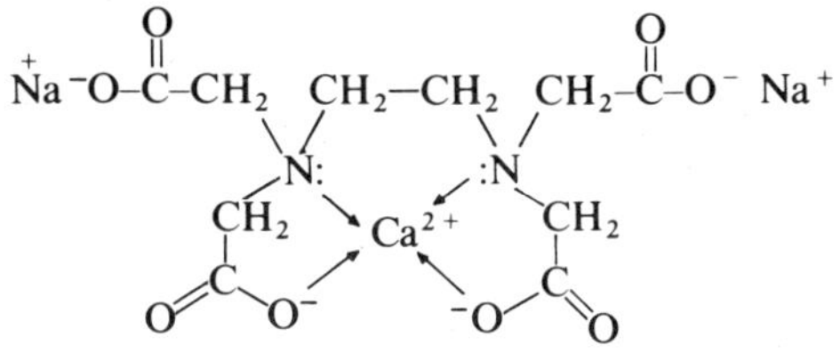

**Scheme 4.6**

All EDTA complexes with heavy-metal ions provide useful catalysts in a number of chemical reactions (J. R. Hart, *CT* 5/87, 313–315). For example, the iron–EDTA complex catalyzes the direct hydroxylation of phenolic compounds (9.3.4) such as salicylic acid to 2,3-dihydroxybenzoic acid by means of air or the conversion of phenol to catechol by means of hydrogen peroxide, while copper–EDTA catalyzes oxidative coupling reactions such as the conversion of phenol to 4,4′-dihydroxydiphenyl or bisphenol by means of oxygen, and so on.

The use of EDA as PUR extender and as a building block of carbamate fungicides are mentioned elsewhere. The reaction of EDA with carboxylic acids produces a variety of imidazolines (10.2.2), but long-chain fatty acids give, for example, ethylene bisoleamide and bisstearamide used as lubricants and mold-release agents.

Ethylenediamine and 2 mol of EDC produce the tricylic, bird-cage-like, sterically nonhindered triethylenediamine commonly known by the name "DABCO" ($10.50), which stands for 1,4-diazobicyclo[2.2.2]octane (originally reported in British Patent 871, 754, 1958 assigned to Houdry Process) but now a registered trademark of Air Products. This original DABCO is probably the most common PUR catalyst (3.8, *Polyurethanes* interlude, p. 147), used especially in the production of foams, although a number of additional amine catalysts designed to promote specific PUR structures. Examples of some of the commonly employed PUR catalysts are shown in Scheme 4.7.

Ethylenediamine dihydroiodide (EDDI) ($\sim$\$8) is a feed additive for cattle. About 0.8MM lb are used in the United States to supply a few milligrams per day of iodine to lactating cows and to prevent foot rot during the winter season.

The demand for EDA in the United States is about 80MM lb, and nearly 30% is exported, another 30% is converted to chelates, and 15% is used by the polymer industry. The remaining EDA is converted with EO to the corresponding hydroxyethyl derivatives, fungicides of the carbamate family, ethylene urea and its dimethylol derivative (3.9.3), and so on. N,N'-Dimethylolethylene urea is employed to convert cellulosic garments to permanent-press or "wrinkleproof" clothing.

(a) $Me_3\overset{+}{N}-CH_2-CH(OH)-CH_3$    $CH_3-(CH_2)_3-CH(Et)-CO_2^-$

(b) $(MeO-CH_2-CH_2-O)_2A\bar{l}H_2$  $Na^+$

(c)

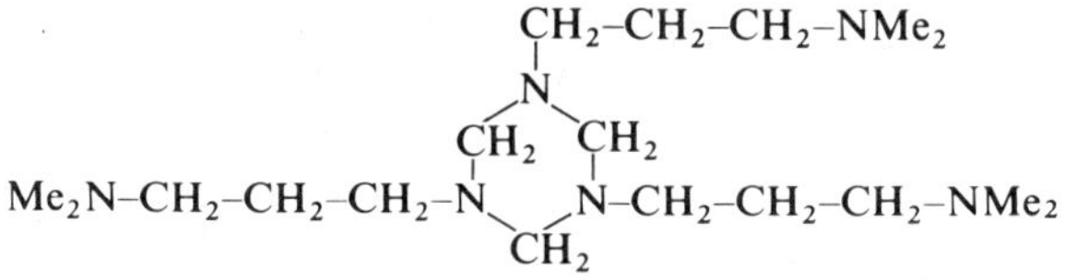

$$CH_2-CH_2-CH_2-NMe_2$$

$$Me_2N-CH_2-CH_2-CH_2-N \underset{CH_2}{\overset{N}{\bigvee}} N-CH_2-CH_2-CH_2-NMe_2$$

(d)

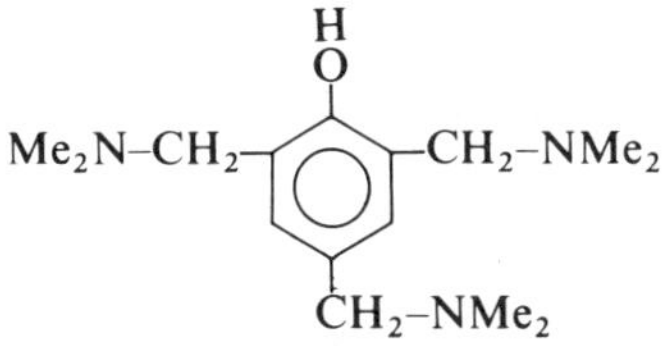

$$Me_2N-CH_2-\underset{CH_2-NMe_2}{\overset{\overset{H}{O}}{\bigcirc}}-CH_2-NMe_2$$

(e)

$$\underset{CH_2}{\overset{CH_2-CH_2}{\bigvee}} \longrightarrow \underset{CH_2}{\overset{CH_2-CH_2}{\bigvee}} \quad ^-O-Q$$

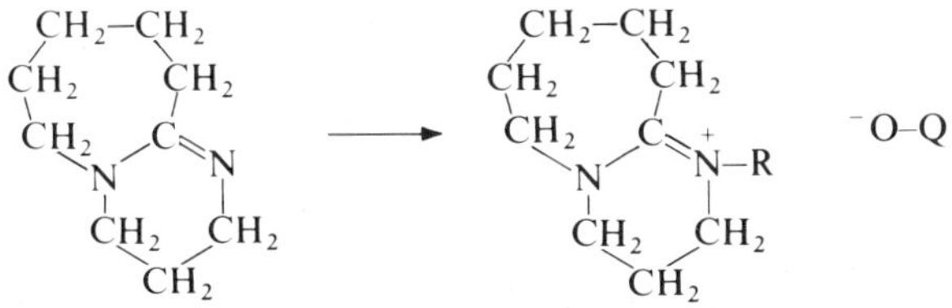

where R = H, or a residue of an alkylating agent, and Q = residue of phenolic, carboxylic, or sulfonic acids

**Scheme 4.7.** (a) Trimethyl 2-hydroxypropyl ammonium 2-ethylhexanoate; (b) sodium dihydro-bis(2-methoxyethoxy)aluminum; (c) N,N',N''-tris(-3-dimethylaminopropyl)-sym-hexahydrotriazine; (d) 2,4,6-tris(dimethylaminomethyl)phenol; (e) 1,8-diaza-bicylo(5,4,0) undecene-7 [Abbott's DBU, derived from caprolactam and acrylonitrile (see Scheme 5.18), is converted to a family of Abbott's Polycat SA catalysts].

#### 4.3.4   Styrene, Polystyrene (PS), and Related Polymers

About 8% of ethylene is channeled into the production of several polymers based on styrene. As discussed elsewhere (9.2.1), the alkylation of benzene with ethylene produces ethylbenzene as well as some diethylbenzene and dehydrogenation of these major alkylation products give styrene and divinylbenzene (DVB) monomers, respectively.

The demand for styrene in the United States is approaching 8B lb, and the present demand pressure, as well as rising costs of benzene, have raised its unit (as of April 1988) cost to 45¢ from 34¢ in 1983, and by January 1989 the price rose to 47¢. The current price of DVB is $3.00 on a 100% basis, although it is sold as a solution in styrene because of difficulties with explosively, exothermic polymerization of neat DVB, or, for that matter, DVB in concentrations higher than 55%. Even styrene alone tends to polymerize on standing unless small amounts of inhibitors such as t-butylcatechol (TBC) are added. Before use, TBC is usually removed from the monomer by extraction with an alkaline solution.

About 52% of styrene is converted to polystyrene (PS; 60–63¢), available commercially for over 50 years. Addition of impact modifiers (4.3.3.1) does not change significantly the price of PS, and the impact modifiers of choice are poly(butadiene), (6.8.1.2, and poly(styrene–butadiene) rubber (SBR), mentioned below. SBR particles dispersed in styrene give, on polymerization, high-impact PS (HIPS), used to fabricate objects more resistant to mechanical shock than all-purpose, plain PS. The production of

HIPS amounts currently to about 2B lb. The addition of small amounts of comonomers or unsaturated, polymerizable impact modifiers developed more recently gives "modified PS" priced at 75¢. Among the different comonomers used for this purpose are vinyltoluene (67¢), a mixture of *m*- and *p*-methylstyrenes, and the more recently developed *p*-methylstyrene (PMS) (9.2.2.4).

Another common form of PS is expandable polystyrene (EPS; 74¢), which produces packaging materials such as the well-known PS popcorn pellets and Dow's Styrofoam on heating. The latter is obtained by processing small beads into final products in appropriately shaped molds (drinking cups, insulating slabs, etc.) and "expanding" and sintering the small hollow PS beads obtained when the polymer prepared by suspension polymerization contains dissolved gases, or a volatile solvent such as pentane (b.p. 36°C) that function as a blowing agent. Larger beads, (e.g., 2–3 mm in diameter) give the PS popcorn.

Styrene is a very versatile monomer, as witnessed by the variety of polymers that are based on its presence. In part, this is due to the fact that it can be polymerized by both homolytic and ionic mechanisms and is compatible with a variety of more or less polar monomers. The latter point explains the relative ease with which PS and many styrene-containing polymers form blends or alloys with other macromolecular families. An excellent example of such polymer alloys is the family of PS–PPE—poly(phenylene ether) (PPE), or PS–GE's PPO—poly(phenylene oxide). These blends can be taylored to suit the given cost–performance situation (see Preface).

The anionic polymerization of styrene is catalyzed by alkyl lithium compounds, usually *n*-butyl lithium, that give rise to "living" polymer chains:

$$R:^- Li^+ + n\, PhCH{=}CH_2 \rightarrow R{-}CH_2CHPh{-}(CH_2CHPh)_{n-2}CH_2\bar{C}HPh\; Li^+$$

The term "living polymer" refers to the fact that the macromolecular anion can be stored and subjected to subsequent polymerization by addition of new quantities of monomer. As we shall see below, this facilitates the production of block polymers already encountered in the case of Uniroyal's Trilene materials, and also in the production of the rapidly expanding (>10% per year) thermoplastic elastomers (TPEs) (also see 3.8.1, PUR).

Another reason for the popularity of styrene monomer is the fact that PS, and most styrene-containing copolymers (except highly cross-linked ones), are thermoplastics that can be obtained by either suspension, emulsion, solution and mass polymerization (4.3.3.1). Large amounts of PS are produced by continuous mass polymerization methods designed to dissipate the heat liberated during the highly exothermic polymerization. Finally, PS and most styrene-derived polymers (again, except highly cross-linked ones) can be conveniently processed by extrusion, coextrusion, injection, blow, sheet and foam molding, and thermoforming, and biaxially oriented films can be fabricated with an improvement of thermomechanical properties.

A slight content of DVB responsible for cross-linking of PS raises the softening point of the resulting "high-temperature–high-impact PS" (58¢), while a larger PVB content (e.g., 2–20%) creates extensively cross-linked, insoluble, and infusible PS particles suitable for the production of ion-exchange resins of different ion-exchange capacities and pore sizes for specific applications (3.4.3.4).

About 13% of styrene monomer is converted to the most common synthetic rubber, namely, SBR produced either as an elastic mass, or in the form of a latex used in the manufacture of water-based paints or for carpet backing. Actually, styrene is a minority component of SBR (~25%) and serves to raise the softening temperature and mechanical strength of the predominant poly(butadiene). About 80% of SBR is transformed into conventional rubber tires on addition of significant amounts of carbon black (about half of the final mass), antioxidants, antiozonants, mold-release agents, and other additives, not to mention sulfur and vulcanization agents (accelerators and/or retardants). The price of SBR is in the 60¢ range and is sensitive to the fluctuating cost of the monomers.

About 10% of styrene monomer is consumed in the production of the terpolymer acrylonitrile–butadiene–styrene (ABS) (pipe grade $0.82; high impact $1.17) and the co-polymer of styrene and acrylonitrile (SAN; 88¢). In ABS the monomer composition is about 18 : 40 : 42, and the material is prepared by copolymerizing "A" and "S" and *grafting* the resulting polymer onto a prepolymerized, tacky polybutadiene. Similarly, there exists the olefin-modified SAN in which the grafting occurs on a polyolefin elastomer. ABS is a tough polymer used extensively in the manufacture of components of household appliances, business machines, personal computers, automotive parts, and so on. The addition of poly(methyl methacrylate) (PMM) produces a transparent grade of modified ABS.

Acrylonitrile-containing polymers, or, for that matter, the acrylonitrile monomer itself, are very susceptible to grafting because of the relative stability of intermediate

$$\cdot \overset{|}{\underset{|}{C}} -C\equiv N$$

radicals formed by the addition of acrylonitrile monomer to a radical center introduced on the substrate by some appropriate means (5.4.2.2).

Variations on the ABS theme are represented by ASA—a terpolymer of acrylate ester–styrene–acrylonitrile, and ACS—a terpolymer of acrylonitrile–chlorinated PE–styrene.

About 6% of styrene monomer is consumed in the production of unsaturated polyester resins (UPRs). The evolution of these materials from glyptal and alkyd resins and their preparation and uses are discussed elsewhere (6.6.2).

The above-mentioned anionic polymerization of styrene gives various styrenic TPEs in which the "hard" PS block is, most commonly, linked to a soft polybutadiene, polyisoprene, poly(ethylene/propylene), or poly(ethylene–butylene) block, and the macro-molecule terminates with another hard PS block. This class of polymers has proliferated since the introduction of spandex (3.8.1, PUR) because they function as elastomers without need of vulcanization, the materials can be processed conveniently as any other thermoplastic, and, finally, they can be readily recycled and thus avoid waste. The styrene–butadiene block polymer (SB) is constituted simply by three S–B–S blocks, with styrene contributing most of the mass.

### 4.3.5  Vinyl Acetate (VAM) and Its Derivatives

The current production of VAM (36¢) is about 1.9B lb, of which approximately 400MM lb is exported. The process that consumes about 2% of the huge production of ethylene, and which has eclipsed all previous sources of VAM, is represented by the following stoichiometry:

$$CH_2{=}CH_2 + CH_3CO_2H + \tfrac{1}{2}O_2 \rightarrow CH_2{=}CH{-}O{-}CO{-}CH_3 + H_2O$$

and parallels the Wacker–Hoechst process, which converts ethylene to acetaldehyde (2.4.4.3.E). The mechanism of the reaction is not as simple as the balanced equation. It depends on (1) the coordinating capability of Pd(II) and its ability to participate in a conventional redox reaction, (2) the ability of Cu(II) to reoxidize Pd(0) to Pd(II), and, finally, (3) the ability of molecular oxygen to reoxidize Cu(I) to Cu(II). All three reactions occur in an integrated fashion as shown in Scheme 4.8.

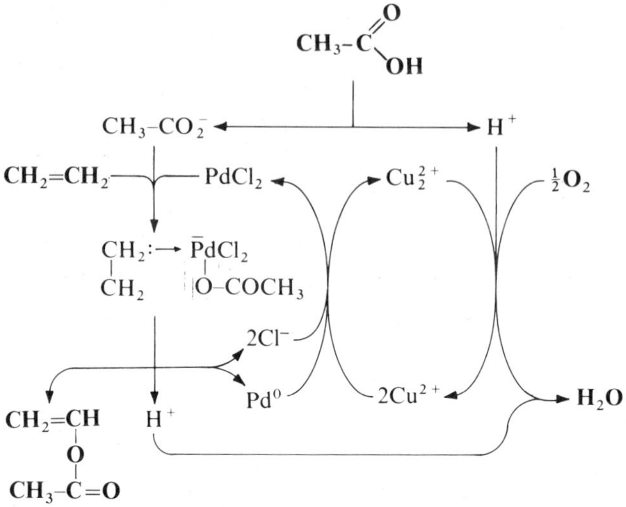

**Scheme 4.8**

For the record, the obsolete VAM processes (since 1980) depend on acetylene and catalysis by mercuric acetate:

$$HC \equiv CH + Ac\dot{O}H \rightarrow CH_2 = CH-O-Ac$$

$$HC \equiv CH + AcOH \ (at \ 200°C) \rightarrow CH_3-CH(O-Ac)_2$$

$$CH_3-CH(O-Ac)_2 \rightarrow VAM + (CH_3-CO)_2O$$

The second option shown here, the thermal decomposition of vinylidene diacetate to vinyl acetate and acetic anhydride, is available to Tennessee–Eastman by modifying the acetic acid process described in 3.3.2.4, with the significant difference that the starting materials are methanol and carbon monoxide.

About 57% of VAM is polymerized to PVA, and the latter is used per se as an emulsion for the manufacture of water-based paints, as an ingredient of adhesives, and for the treatment of textiles and paper products. Another 22% of VAM is polymerized and then fully hydrolyzed to poly(vinyl alcohol) (PVOH; $1.00) for conversion to acetals of formaldehyde (3.4.3.2) or isobutyraldehyde (6.5)—both employed in the manufacture of shatterproof (safety) glass. The formal resin serves as a coating of magnetic wires. The partially hydrolyzed PVA ($1.05) is also used in some adhesive formulations. Finally, about 5% of VAM becomes the minority comonomer in EVA and the majority comonomer in VAE (4.3.3.1). Both of these copolymers are becoming popular as hot-melt adhesives. An interesting but limited use of vinyl acetate is the transvinylation reaction that serves to prepare vinyl esters of relatively high-molecular-weight carboxylic acids:

$$R-CO-OH + CH_2 = CH-O-CO-CH_3 \rightarrow R-CO-O-CH = CH_2 + AcOH$$

The reaction is catalyzed by mercury salts and it is particularly useful for the conversion of naturally occurring fatty acids (*SZM-II*, 66–69) to the corresponding vinyl esters that are then polymerized to produce antistick, hydrophobic coatings, oil additives, and so on.

### 4.3.6  Ethylene Oxide (EO): Production and an Overview of Its Uses

In spite of the 5.8B-lb EO consumption level, the U.S. production of EO is suffering (C. J. Verbanic, "Ethylene and Propylene Oxide: Gloom and Glamor," *CB* 2/88, 24–29) mainly because of the nongrowth of the EG market as a result of

(a)  Increased production of PET in newly industrialized countries.

(b)  Stagnant automotive antifreeze and coolant market traced to the use of smaller and imported automobiles.

(c)  Unfavorable political atmosphere with respect to the use of EG-based ethers and other EG-based solvents (see 2.4.4.6.2).

(d)  Domestic overcapacity and increasing imports.

About 15% of ethylene is converted to about 5.8B lb of EO (35¢), and since 1937 the time-honored catalyst for this purpose is elementary silver (15%) deposited on alumina. The reaction is carried out at 230–290°C, uses either oxygen or air, and is accompanied by some formation of $CO_2$ (*FKC*, 408):

$$CH_2{=}CH_2 + O_2 \rightarrow CH_2{-}CH_2 \atop \diagdown O \diagup$$

The catalytic role of a cluster of silver atoms located at the surface of the heterogeneous catalyst can be rationalized as shown in Scheme 4.9.

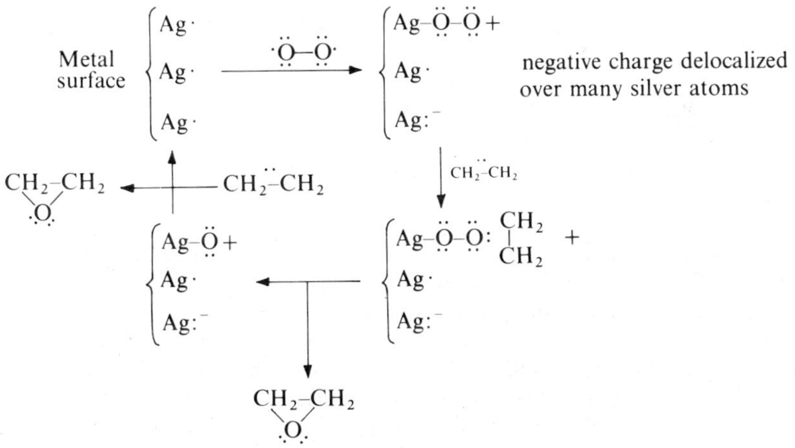

**Scheme 4.9**

The older, two-stage EO process consists of, first, the formation of ethylene chlorohydrin

$$CH_2{=}CH_2 + HOCl \rightarrow HO{-}CH_2CH_2{-}Cl$$

followed by the conversion of the latter to EO by means of the cheapest industrial base, namely, hydrated lime, Ca(OH)$_2$ 2.3¢), which produces the economically unattractive calcium chloride (5¢, liquid, 100% basis; 7.6¢, 77–80% flakes; 10.8¢, anhydrous, 94–97% flakes; refined, USP grade, 79¢).

Hypochlorous is generated *in situ* by the disproportionation of chlorine in water:

$$Cl_2 + H_2O \rightarrow HOCl + HCl$$

Other problems with the chlorohydrin process arise because of the formation of EDC and 2,2′-dichlorodiethyl ether (DCEE) or chlorex [(Cl–CH$_2$CH$_2$)$_2$O], once a common industrial by-product of EO manufacture used in the production of Thiokol B rubber (4.3.3.3), and now a more costly, on-purpose-prepared chemical. All of these reasons illustrate, once more, the "do not use chlorine-containing intermediates in vain" rule, especially when the same objective can be reached by means of air and water, and explain why the chlorohydrin process became obsolete in 1972.

The major uses of EO are as follows:

60% for the production of EG consumed mostly in the production of PET and as an antifreeze liquid

15% for the production of nonionic surfactants by ethoxylation of a variety of precursors that contain an active hydrogen

10% for the production of ethanolamines consumed mostly for the removal of acidic components of gases that contain CO$_2$ and H$_2$S, in the synthesis of ethylenediamine and higher ethyleneamines (4.3.3), alkanolamides of fatty acids (*SZM-II*, 64) used as foam stabilizers and thickeners in shampoos, cellulose derivatives (*SZM-II*, 87) that function as anion-exchange resins or as finishing agents for cotton textiles, and so on

8% for the production of diethylene and triethylene glycols

7% for the production of ethylene glycol ethers (3.3.2.2), including PEG utilized by the PUR industry

Noteworthy by its absence from the preceding listing is the use of EO as a sterilizing gas for germ-free, biological operations. This practice was abandoned because of the hazardous nature of this highly reactive gas.

Since ethylene glycol is the major product derived from EO, it is not surprising that attempts to develop direct routes for the conversion of ethylene to EG, that is, methods that avoid the utilization of EO. An interesting attempt (R. Hughes, *CT* 9/74, 516, *CEN* 7/29/74) consisted of a reaction of ethylene with acetic acid at 170°C by means of tellurium dioxide, HBr, and LiBr to give a mixture of mono- and diacetates of EG that was to be hydrolyzed to EG while acetic acid was recycled. The fate of this large-scale experiment is mentioned elsewhere (2.3.3).

Other methods for the production of EG without utilization of EO and based on syngas derivatives are reported here (3.3.2.8.C) and elsewhere (*WH*, 75–77, *EM*, 49–50). It remains to be seen whether the technology for obtaining high-purity EG required for the formation of high-quality PET textile fibers becomes competitive while ethylene is abundant and relatively cheap.

### 4.3.6.1  Ethoxylation: Ethylene Glycol and Its Oligomers and Polymers, Including PET, and Miscellaneous Products of Ethoxylation

The strained oxirane system represented, by its simplest compound, namely EO, adds to water, alcohols, mono- and diglycerides (*HHS-II*, 56, 61), phenols, mercaptans, primary and secondary amines, and practically any substrate that contains an active-hydrogen function. This ethoxylation reaction is catalyzed by either acids or bases but the latter are more commonly used. For the attachment of multi(ethylene oxide) chains to such substrates, one adds ethylene oxide (b.p. 11°C), with caution because of the highly exothermic reaction. Explosions have resulted in the past from an overenthusiastic addition of EO, and good agitation assures a narrow molecular-weight distribution of multi(ethylene oxide) derivatives.

The demand for ethylene glycol (EG), diethylene glycol (DEG), and triethylene glycol (TEG), is about 4.8B, 0.5B and 0.1B lb, respectively, and these chemicals are priced at 31¢, 31.5¢, and 41¢, respectively, reflecting the increasingly more energy-intensive separation requirements, and a decreasing market.

About 45% of EG is utilized in automotive antifreeze, or more correctly said, all-year-around radiator fluids that contain some water, anti-rust, and other additives such as water-pump lubricants and coloring matter. These traditional specialty materials are often formulated and distributed by EG producers themselves. Thus, for example, for several decades Du Pont distributed the radiator fluid under its trademark Zerex, but recently the highly recognizable trademark and this particular business were purchased by BASF Wyandotte. In the same category of usage are the heat transfer fluids based on aqueous solutions of EG that also contain about 7% inhibitors and a fluorescent orange dye (Dow's DOWTHERM 4000).

Except for some exports (10–12%), practically all of the rest of EG is channeled into the production of about 4.7B lb of poly(ethylene terephthalate) (PET).

This highly versatile thermoplastic polyester is obtained by a reaction of EG with either terephthalic acid (TPA) or its dimethyl ester (DMT) (9.3.2). The mechanical and thermal properties of PET depend on the linear, "hard," and somewhat polar terephthalate moieties alternating with short, flexible, and "soft" ethylene glycol residues. Hence, high-grade PET plastic requires the use of pure monomers, and this is especially true for the production of high-performance textile fibers and films. Considering the origin of terephthalic acid (9.2.2), it is not surprising, therefore, that purified terephthalic acid (PTA) is the grade consumed for the preparation of most PET plastics and that many PET producers have propietary technology for the assembly of their products. Furthermore, in order to optimize the performance of PET fibers and films, the degree of alignment of neighboring macromolecules is manipulated by heating and stretching (see below) of the initial, mostly amorphous product above its $T_g$ of about 74°C. Naturally, the molecular weight distribution also affects the performance of PET fibers and films.

Poly(ethylene terephthalate) is the largest source of textile fibers in our economy as can be appreciated from the following tabulation of the approximate *production of all types of textiles*:

|                                      | B lb |
|--------------------------------------|------|
| Natural and semisynthetic fibers     |      |
| Cotton and wool                      | 3.5  |
| Rayon and cellulose acetate          | 0.5  |
|                                      |      |
| Subtotal                             | 4.0  |

Synthetic fibers

| | |
|---|---|
| PET | 3.2 |
| Nylons (mostly 6/6 and 6) | 2.5 |
| Polyolefins (mostly PP) | 1.5 |
| Acrylics | 0.5 |
| Subtotal | 7.7 |
| Grand Total | 11.7 |

The major uses of PET are as follows:

- Textile fiber converted into polyester clothing and staple fiber for pillows
- Film and sheets used to manufacture photographic and X-ray films as well as transparencies, magnetic tapes for sound and video recordings, food packaging
- Containers for cooking in conventional and microwave ovens as well as boil-in-bag pouches
- Strapping tapes and ropes
- All types of molded objects, including blow-molded bottles for soft drinks that have, for all practical reasons, displaced traditional glass containers

Crystallized PET resin (cPET) is used in the fabrication of trays and similar firm objects and carries a higher price tag (65¢–75¢), and specialty copolymer grade PET used for structural purposes may be priced as high as $1.50, while bottle-grade PET costs 52¢–55¢/lb. Partially oriented PET yarn (POY), used for weaving and knitting, like Celanese's Fortrel 150-denier yarn, costs 75¢–80¢/lb.

As implied above, the processing of high-performance films and bottles includes uni- and biaxial orientation of the macromolecules. In the case of the now common 2-L bottles, the preformed container is subjected to additional blow molding at a temperature above the $T_g$ to produce the lightweight, tough container.

Recycling of PET plastic objects is another advantage of this family of materials except that the recovered plastic cannot be used for food packaging ends.

About one-third of the diethylene glycol consumption is utilized in the assembly of PURs (as a chain extender; 3.8.1, PUR) and as a substitute for glycerol in the formulation of unsaturated polyester resins, UPRs (6.6.2). Reductive amination converts about 8% of DEG into morpholine:

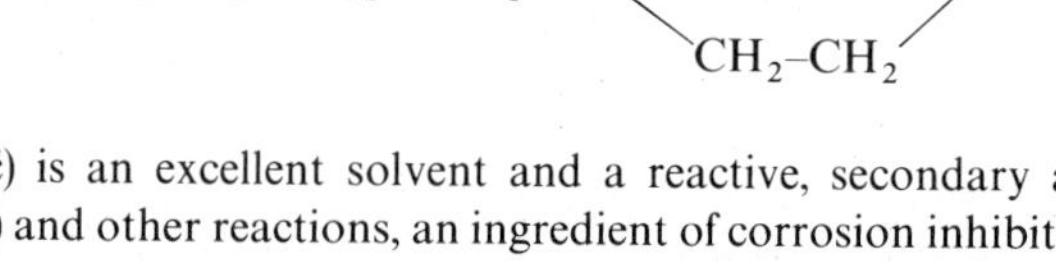

$$O(CH_2CH_2-OH)_2 + NH_3 \rightarrow O{\Large\diagdown}{\overset{CH_2-CH_2}{\underset{CH_2-CH_2}{}}}{\Large\diagdown}NH$$

Morpholine (94¢) is an excellent solvent and a reactive, secondary amine utilized in Mannich (3.4.3.7) and other reactions, an ingredient of corrosion inhibitors, antioxidants, and so on, and its purified, USP-grade sulfate ($387.00) is a building block of various medicinals. The reaction of morpholine with sulfur monochloride (16.5¢) produces the vulcanizing agent 4,4'-dithiodimorpholine ($3.15) $R_2N-S-S-NR_2$, where $R_2N$ represents the morpholino residue.

Diethylene glycol is used to absorb water from natural gas, and this is even more true for triethylene glycol, which shares with its smaller analog the role of a diol in PURs and UPRs and functions as a humectant. Both diols, and even tetraethylene glycol (67¢) are incorporated into functional fluids such as transmission, hydraulic, brake, and other liquid formulations, and these relatively high molecular weight diols are also converted to esters. An example of such a transformation is the preparation of a synthetic lubricating oil (for jet engines), from the $C_9$ pelargonic acid (70¢; *SZM-II*, 70), namely, triethylene glycol dipelargonate (30¢). Pelargonic acid is obtained (together with azelaic acid, 7.3.1) by ozonolysis of oleic acid (see p. 385) or by the oxo reaction (3.5.3.4) of 1-hexene. A somewhat different objective is served by tetraethylene glycol diacrylate ($1.50) (5.5.2), used as a special multifunctional cross-linking agent in the production of unsaturated polyester resins (UPRs).

Higher-molecular weight PEGs are used to replace the water content of wood products and to prevent the cracking of such objects of value on drying. An interesting application of this use of PEG is the restoration and preservation of *Mary Rose*, a ship built in 1509–1510, sunk in 1545, and raised from the sea in 1982 (P. L. Layman, *CEN* 6/1/87, 19–21).

Still higher, oligomeric ethylene glycols (or ethylene glycol ethers) are obtained in part as by-products of the production of EG, DEG, and TEG, but this source is supplemented by the, on-purpose reaction of EO with EG. The polymeric ethylene glycols, PEGs, are available in (average) molecular weights ranging from 100,000 to 500,000, Union Carbide's Polyox ($2.80–$4.40, depending on MW distribution). These water-dispersable polymers are used as thickeners, lubricants, hydrodynamic drag reducing agents for water and aqueous solutions (i.e., increasing the fluidity of water as it moves through pipes or hoses, of special interest for fire-fighting), flocculants, and so on. At the high end of the molecular-weight distribution we have Union Carbide's Carbowaxes—waxy solids available as Polyethylene glycols 4000, 6000, and so on, where the number indicate the average molecular weight of the materials. Thus, the number 600 indicates the presence of about 13 $-CH_2CH_2-O-$ units in the macromolecules. Unlike traditional waxes (*SZM-II*, 81–83), these synthetic PEG waxes are self-dispersable in water and thus more convenient vehicles for some pharmaceutical and cosmetic products such as suppositories, beauty creams, hair-conditioning products, and so on. Poly(ethylene glycols) of $MW = 10^3$ play the role of the "soft" polyol blocks in elastomeric PURs such as Spandex (3.8.1, PUR).

In order to appreciate the physicochemical basis for the affinity of PEGs for water, one must compare the difference between their behavior and that of the homologous poly(propylene glycols) (PPGs), derived from PO. To start with, the absence of ether linkages in a linear PE does not provide any opportunity for hydrogen bonding, and thus it is not surprising that the PE macromolecules are hydrophobic or, in the reverse sense, lipophilic. What may be surprising, on the other hand, is that the presence of methyl substituents in PPGs exerts a steric effect that, to a considerable extent, inhibits hydrogen bonding with water molecules to the point where these polyethers are also lipophilic.

*Nonionic Surface-Active Agents and HLB*

The above-mentioned difference between PEGs and PPGs gives rise to a family of *surface-active agents*, or detergents, syndets ("synthetic detergents"), emulsifying agents, or wetting agents—all rather synonymous terms but indicative of subtle functional differences—in which blocks of PEG and PPG are assembled to provide the appropriate hydrophilic–lipophilic balance (HLB) for a given use. For example, a wide selection of such materials is available from BASF Wyandotte under the tradenames Pluronic

($0.89–$1.04) and Tetronic ($0.94–$1.09). In the last-mentioned family of surfactants the EO–PO blocks are built around a centrally located EDA molecule. The HLB aspect of molecular engineering of surfactants is discussed further in 5.4.3.2, but the importance of the HLB concept in the design of literally thousands of surface-active agents by ethoxylation of lipophilic substrates needs to be illustrated here.

Thus, long-chain, linear alcohols containing 30–50 or more carbon atoms, obtained by oligomerization of ethylene followed by propietary conversion to alcohols, are ethoxylated to produce a series of Petrolite's Unithox Ethoxylates that have HLB values ranging from 4 to 16 (the values *increase* with an *increasing* ratio of EO units attached to paraffinic chains and, for a given average MW of lipophilic hydrocarbon, correspond to about 10, 20, and 50 mol% of EO). These products, valued at about $2/lb, are readily dispersed in water-based systems as well as in organic media. They function as "slip" and lubricating agents, antiblocking, or release agents; promote the formation of films by codispersed polymers; emulsify vegetable oils; and so on.

Shorter paraffinic alcohols are represented by fatty alcohols (*SZM-II*, 78–79) symbolized here by R$_f$–OH and derived from naturally occurring fats and oils, and also by oxo alcohols derived from α-olefins (7.3). These alcohols are ethoxylated to give materials that contain four to seven EO residues, and the products per se are used as nonionic, low-foaming detergents in dishwashing and laundry formulations. Also, they are sulfated to give anionic detergents

$$R_f-O-(CH_2CH_2-O-)_{4-7}-SO_2-OH$$

after they are neutralized by soda ash, ethanolamines, and other bases. The worldwide use of such detergents is estimated at 1.5B lb.

Another common class of surfactants are derived from *p*-alkylated phenols such as octyl, actually 2-ethylhexyl (75¢); nonyl, derived from the propylene trimer (51¢); or dodecyl, derived from the trimer of isobutylene (48¢). These are ethoxylated to introduce 5–10 EO residues to give nonionic detergents of somewhat lower unit price because of the intrinsically low price of EO.

More complex ethoxylated surface-active agents are obtained from lipophilic diglycerides (*SZM-II*, 63) and another complex lipophilic substrate, namely, lanolin or "wool grease" (*SZM-II*, 82–83), priced at $1.13–$1.18 depending on grade. Ethoxylated lanolin products are available as cosmetic, pharmaceutical, and "extra-deodorized" grades priced at about 77¢, 79¢ and 90¢, respectively. In the same category of ethoxylated surface-active agents we find the derivatives of sorbitol (*SZM-II*, 50, 106–109), 68¢–70¢ for solid and 30¢ for a 70% aqueous solution. Sorbitol is a readily obtained reduction product of glucose. It is dehydrated while subjected to esterification with stearic acid (26¢–32¢) to give either the expected sorbitan monostearate or tristearate (76¢ and 80¢, respectively). Each of these esters is then ethoxylated with about 20 EO units to give the corresponding poly(ethylene oxide) derivatives valued at 73¢/lb as shown in Scheme 4.10.

Elsewhere (3.8.2, PUR) it is mentioned that EO is utilized for the capping or tipping of polyol components of PURs in order to convert secondary hydroxylic or carboxylic functions to primary alcohols. Similar chemistry with a different purpose is illustrated by the announced (*CMR* 5/11/87) production of ethoxylated derivatives of bisphenol A (see 5.5.3) by AKZO as modifiers of UPRs and PURs in order to improve the flexibility, shock-absorbing, metal adhesive properties of the final products.

The mono- and diethoxylation products of alcohols such as methanol, namely, Union Carbide's Methyl Cellosolve (2-ME) and Methyl Carbitol, are mentioned in 3.3.2.2.

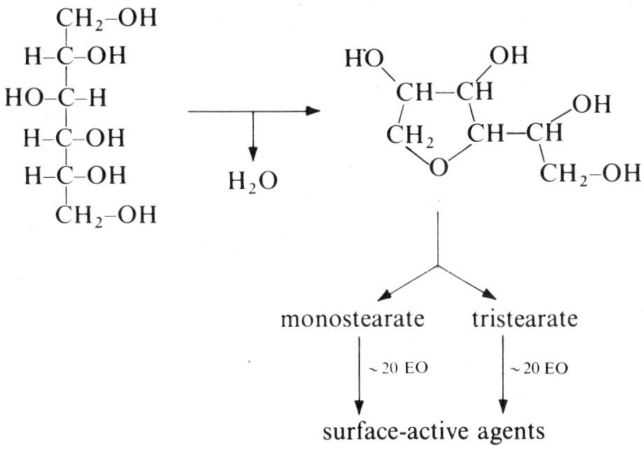

**Scheme 4.10**

Actually, this family of half-century-old Union Carbide solvents also includes the analogous derivatives of other simple alcohols with those derived from ethanol being the original Cellosolve, or 2-ethoxyethanol (2-EE; 51¢) and Carbitol, or diethylene glycol monoethyl ether (56¢). Similarly, *n*-butanol gives rise to Butyl Cellosolve (64.5¢) and Butyl Carbitol (51¢). The family of the glycol ethers was later enlarged by the production of the corresponding acetates that are priced about 15¢–16¢ higher than the alcohol precursors and identified by the addition of an "A" to the acronym of the alcohol (thus 2-EE and 2-ME become 2-EEA and 2-MEA, respectively). The availability from the same producer of a relatively large number of rather similar solvents is explained on the basis of their differences in solubility, volatility, viscosity, and other physical properties. These properties determine which member of the homologous family is most suitable, for example, for the formulation of a given paint, lacquer, or varnish so that all of the components remain properly dispersed and display the best flow, spreading, and drying properties, and so on. The same structure–useful properties rationale applies to the formulation of paint removers and other traditional specialty products.

The preceding EG-derived solvents are currently under scrutiny by OSHA because of suspected health hazards to workers who apply them, and EPA is also concerned with "volatile organic compound" (VOC) emissions that cause environmental pollution (*CW* 10/17/84, 14–15; J. F. Dunphy, *CW* 6/4/86, 8/13/86; J. Chowdhury, *CE* 1/19/87, 14–19). Consequently, EPA established a limit on VOC emissions by restricting the use of organic solvents in metal finishing operations to 3.5 lb of solvent per gallon of coating formulations. Industry is reacting to these restrictions by the installation of complex and costly vapor-recovery systems, and by switching to water-based, radiation-cured powder, or high-solids coatings. Another consequence is the replacement of solvents derived from EO and EG to similar compounds derived from PO and PG since the latter do not carry the stigma of EG toxicity. Also, the "hazardous" methyl groups are being replaced by isopropyl or *t*-butyl groups at the same time, incidentally, that methanol is being promoted as a gasoline extender and substitute. Considering the magnitude of the demand for original equipment coatings (OEC) by the automotive and machine-construction industries, the OSHA and EPA rulings have a significant effect on the chemical industry and the glycol ether demand of over 300MM lb is likely to decrease in the future. One can always, of course, find a solution to boost the demand for ethylene glycols if one tries hard enough. For example, one can follow the example of the government of Austria that was confronted (Associated Press, 12/22/86) with the problem of a safe disposal of millions of gallons of wine sweetend with DEG. Even though no cases of illness or death could be attributed to the consumption of this spiked wine, the government decided to use the mixture of the product with salt for the deicing of hazardous, icy highways. But then, one runs into the possibility that our more alert watchdog organizations will declare the ethanol content of similar mixtures hazardous in view of DUI statistics and, in particular, the fatal experience of Darrell Shepherd who was found dead shortly upon drinking two fifths of vodka (Associated Press 11/4/87).

2-Methoxyethanol is used (about 30MM lb) as an anti-icing additive in aviation fuel. Also, its aluminum hydride complex, Hexcel's Vitride, is a powerful and convenient reducing agent because the coordinated structure renders it soluble in hydrocarbon and ether solvents:

$$
\begin{array}{c}
\text{CH}_2\text{—CH}_2 \\
\text{CH}_3\text{—O} \qquad \text{O} \qquad \text{H} \\
\text{Na}^+ \quad \text{Al}^- \\
\text{CH}_3\text{—O} \qquad \text{O} \qquad \text{H} \\
\text{CH}_2\text{—CH}_2
\end{array}
$$

Vitride is distributed as a 70% solution in toluene at about $2.75, and, unlike pyrophoric sodium aluminum hydride, $Na^+ \ AlH_4^-$, it is safe to handle.

*Glymes, Crown Ethers, Cryptates, and Phase-Transfer Catalysis*

The dimethyl ethers of the ethylene glycol family, the *glymes*, are introduced in 3.3.2.3 and mentioned again in 3.4.3.2. The optimum coordination of the alkali metal cations by ethereal oxygen functions depends on the size of the cation and increases as the appropriate number of oxygen centers creates a well-fitting, tight cyclic complex. Thus, a lithium cation forms a strong complex with *diglyme*, or diethylene glycol dimethyl ether, while the sodium cation is complexed best by *triglyme*, or triethyleneglycol dimethyl ether. The relatively small terminal methyl groups interfere with complex formation less than other alkyl goups, but a completely cyclic structure of appropriate size leads to an even stronger coordination. The first such macrocyclic *crown ether*, dibenzo-18-crown-6, was synthesized from catechol (9.3.4) and chlorex or DCEE (4.3.6) and described by Charles J. Pedersen of Du Pont [*JACS* **89**, 2495, 7017 (1967)]. It became the prototype of numerous analogous structures designed to coordinate metallic species (known as *cryptates* when other than ether functions are responsible for the formation of metallic complexes), and led to a better understanding of many physical, chemical, and biological phenomena. Pedersen's pioneering contribution was recognized in 1987 by a Nobel prize in chemistry shared with Donald J. Cram, who, in recent years, has been elucidating the broader aspects of "host–guest" relationships in enzymatic systems. Even after two decades, dibenzo-18-crown-6 ($341/lb) and the analogous 18-crown-6 ($227/lb) are rather costly, and large-scale industrial applications of crown ethers tend to depend on "poor man's" crown ethers, namely the appropriate glymes. The structures of the above-mentioned compounds are shown in Scheme 4.11.

The original industrial source of glymes, Ansul, was acquired more recently by SpecialtyChem Products (SPC), and the number of chemicals of this type was greatly

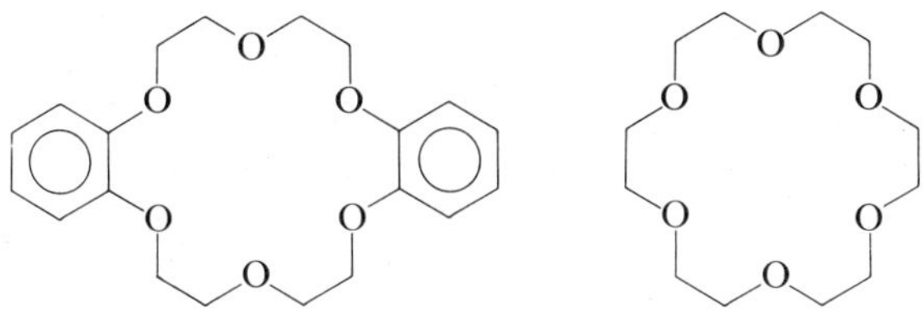

**Scheme 4.11**

expanded. Some of these products available from SPC are priced as follows:

Glyme (SPC E-121), $2.21
Diglyme (SPC E-141), $2.06
Triglyme (SPC E-161), $3.40
Tetraglyme (SPC E-181), $3.50
Diethylene glycol diethyl ether (SPC E-242), $3.50
Diethylene glycol di-n-butyl ether (SPC E-444), $4.65

The complexing of cations by glymes causes, as mentioned elsewhere (3.3.2.3), an increased nucleophilic reactivity of negatively charged counterions, and this accelerates substitution reactions carried out in these solvents and catalysts under discussion and/or allows the reactions to be carried out at temperatures lower than required when the same reactions are carried out in their absence. Either way, this effect is important in industrial processes because it promotes a more efficient utilization of costly production facilities and an energy saving. The complexing of metallic cations by glymes also provides phase-transfer catalysis when a given process is carried out in an aqueous and water-immiscible, two-phase system as described in 3.8.3 for the reaction of the sodium salt of bisphenol A with phosgene (and elsewhere with epichlorohydrin; see 5.4.8). Since glymes are soluble in both phases, they distribute an ionic reactant between the two phases and function in a manner analogous to that described for quaternary ammonium salts. Also, solid ionic reactants that are insoluble in inert organic media and are decomposed in aqueous and other polar media can be solubilized in the former by complexes with glymes. An example of such a situation is the reaction of sodium naphthalide with polychlorinated biphenyls, PCB, (9.2.3), or the solubilization of potassium permanganate in organic media. The discussion of many other industrial applications of glymes is beyond the scope of this book.

### 4.3.6.2 Ethanolamines

The production of ethanolamines by means of ethoxylation occurs in a rather similar fashion as the ethoxylation of hydroxylic compounds, but this topic is dealt with separately because of the different chemistry of the resulting products.

Of the U.S. demand of about 550MM lb of ethanolamines, nearly 40% is utilized in the production of surface-active agents, about 30% is consumed for the removal of moisture and acidic constitutents (i.e., "sweetening" of "sour" natural gas and petroleum), as well as in the separation of syngas (see 3.5.1.6), and the remainder is used to produce textile treatment and anticorrosion products, ingredients of PUR and epoxy resins, and other derivatives. An interesting and possibly unexpected use of ethanolamines is their role as cement grinding aids.

Let us start with the unique situation of a tertiary amine, namely, trimethylamine, that is transformed to the food and feed additive choline—a component of the physiologically important phospholipids (*SZM-II*, 8). The formation of this quaternary ammonium compound can be achieved by treatment of trimethylamine with either ethylene chlorohydrin or by ethoxylation under pressure at about 100°C:

$$Me_3N + EO \rightarrow HO-CH_2CH_2-NMe_3^+ \quad HO^-$$

$$\downarrow \text{HCl}$$

$$Me_3N + Cl-CH_2-CH_2OH \rightarrow HO-CH_2CH_2-NMe_3^+ \quad Cl^-$$

Choline salts are produced at a level of some 60MM lb and are used most extensively by the poultry and swine industry. Feed-grade choline chloride is priced at 28¢ (70% solution) or 39¢ (60% dry solid). Pharmaceutical-grade choline is priced at $2.28, and the corresponding dihydrogen citrate and bitartrate cost $2.73 and $3.14, respectively.

Triethanolamine (35¢) and diethanolamine (34¢) are the products of the ethoxylation of ammonia and are the favored bases for the neutralization of anionic surfactants destined for use in shampoos and other personal-care products:

$$n\text{-}C_{12}H_{25}-O-SO_3^- \quad H\overset{+}{N}(CH_2CH_2OH)_3, \text{ triethanolamine lauryl sulfate, } 27.5¢$$

$$n\text{-}C_{12}H_{25}-O-SO_3^- \quad H_2\overset{+}{N}(CH_2CH_2OH)_2, \text{ diethanolamine lauryl sulfate, } 41¢$$

Lauryl alcohol is a coconut oil-derived fatty alcohol (*SZM*–II, 79).

Reductive amination of triethanolamine is the probable source of tris(2-aminoethyl)amine, Hampshire's TREN—a developmental product that can be used as a chelating agent, lubricant additive, catalyst, and so on.

A hybrid between the structures of a glyme and triethanolamine, namely, tris(2,6-dioxaheptyl)amine, or TDA-1 $[N(CH_2CH_2OCH_2CH_2OCH_3)_3]$, is available from Rhône–Poulenc. This specialty solvent (b.p. 339°C) is also a solid–liquid-phase-transfer catalyst and is priced at $19/lb. It can be assembled by several routes, of which alcohol amination of the diethoxylation product of methanol seems least problematic.

While diethanolamine can be dehydrated to morpholine, the alcohol amination of diethylene glycol shown above (4.3.6.1) seems to be a more practical route to morpholine (*CEN* 1/28/80, 2/25/80). However, the acid-catalyzed cyclodehydration route is the likely process for the preparation of N-substituted morpholines from N-substituted diethanolamines such as the conversion of N-ethyldiethanolamine to N-ethylmorpholine ($1.92):

$$Et-N(CH_2CH_2-OH)_2 \rightarrow Et-N(CH_2CH_2)_2O$$

The use of N-methyldiethanolamine as a building block and catalyst in the assembly of PURs is mentioned elsewhere (3.8.2.1, PUR)

A commonly employed treatment of brain cancer utilizes the antineoplastic N,N′-bis(2-chloroethyl)-N-nitrosourea or carmustine:

$$\overset{\displaystyle O}{\overset{\displaystyle \|}{Cl-CH_2-CH_2-N(N=O)-C-NH-CH_2-CH_2-Cl}}$$

This nitrogen mustard is assembled from monoethanolamine, an organic carbonate such as DMC, followed by the replacement of the hydroxyl groups with thionyl chloride and nitrosation. A recent report (*CW* 10/28/87) indicates an improved delivery system developed by Nora Pharmaceutical by which the antineoplastic reaches the brain tumor from a surgical implant obtained by using a biodegradable polymer derived by condensation of sebacic acid and carbophenoxypropane ($p$-$HO_2C$–$C_6H_4$–$O$–$C_3H_7$) in a 80 : 20 ratio.

Another antineoplastic in common use is cyclophosphonamide (CP), in which the nitrogen mustard moiety is attached to a cyclic phosphonamide structure:

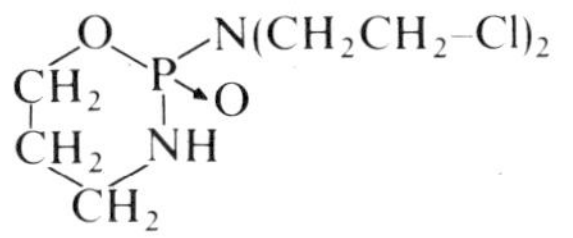

The diethanolamine moiety is incorporated in the recently introduced hetero-multifunctional monomer $N,N$-bis(2-hydroxyethyl)-$N'$($\alpha,\alpha$-dimethyl-3-isopropenyl-benzyl)urea:

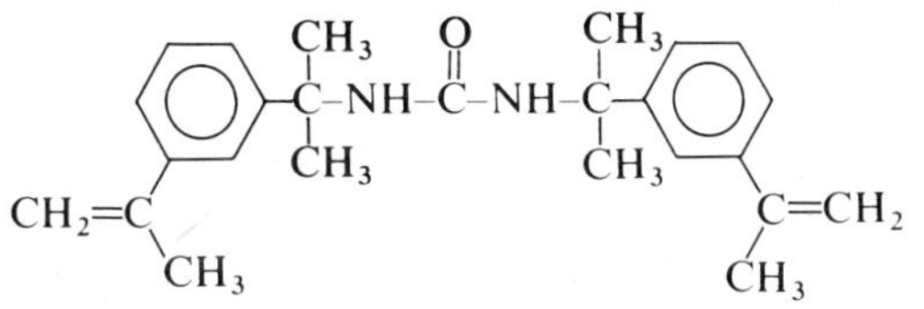

This material is obtained by subjecting *m*-diisopropenylbenzene to a Ritter reaction, followed by a reaction, of the resulting amine with DEA and DMC or phosgene.

Monoethanolamine (43¢) and propionic acid (33¢) produce a cyclic condensation product, 2-ethyl-2-oxazoline (Dow XAS-1454) that can serve as an intermediate in numerous synthetic possibilities [T. Saegusa, *Angew. Chem., Int. Ed. Engl.* **16**, 826 (1977); *Makrom. Chem. Suppl.* **3**, 157 (1979)]. Dow has extended this reaction of ethanolamine to include the use of methacrylic acid (78¢), followed by hydrolysis of the intermediate oxazoline to 2-aminoethyl methacrylate, which is then converted to isocyanatoethyl methacrylate (IEM), the heterodifunctional PUR monomer mentioned in 3.8.1, PUR; this is illustrated in Scheme 4.12. In 1983 (*CE* 4/4/83) Dow was projecting a 0.25MM-lb/yr production of IEM for use as a heterodifunctional monomer (3.8.1, PUR).

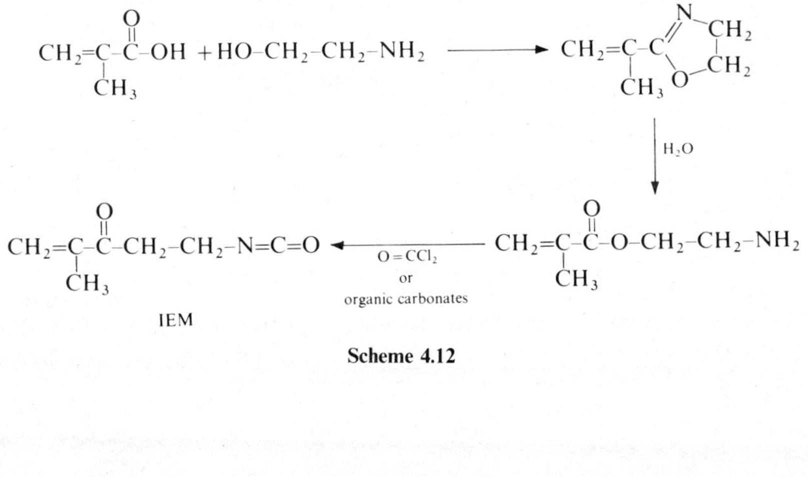

**Scheme 4.12**

Monoethanolamine also undergoes an interesting transformation to ethyleneimine, the aziridine analog of oxirane. First, the ethanolamine is converted to the corresponding sulfate bisulfate

$$H_2N-CH_2CH_2-OH + 2H_2SO_4 \rightarrow HSO_4^- \ H_3N^+-CH_2CH_2-OSO_3H$$

and then the resulting product is added to caustic soda, whereupon an intramolecular nucleophilic substitution (with the bisulfate ion being the leaving group) generates the low-boiling ethyleneimine

$$H_2N-CH_2CH_2-OSO_3^- \rightarrow \underset{\underset{H}{\overset{|}{N}}}{CH_2-CH_2} + HSO_4^-$$

Like EO, ethyleneimine polymerizes readily to poly(ethyleneamines):

$$H_2O + n(CH_2CH_2)NH \rightarrow HO-CH_2CH_2(NH-CH_2CH_2)_{n-1}NH_2$$

$$NH_3 + n(CH_2CH_2)NH \rightarrow H_2N-CH_2CH_2(NH-CH_2CH_2)_{n-1}NH_2$$

However, the strained aziridine ring can be preserved when ethyleneimine participates in substitution or addition reactions. Thus, with cyanuric chloride (3.10.2.3) and phosphorus oxychloride (43¢), ethyleneimine gives triethylenemelamine, and triethylenephosphoramide, as shown in Scheme 4.13. Both of these ethyleneimine derivatives are textile finishing agents, and ethyleneimines, in general, are specialty chemicals used in paper treatments, coatings, and adhesives. Recently, Virginia Chemicals acquired the ethyleneimine technology from the previous and only domestic manufacturer, Cordova Chemical (*CW* 7/24/85, *CMR* 7/29/85).

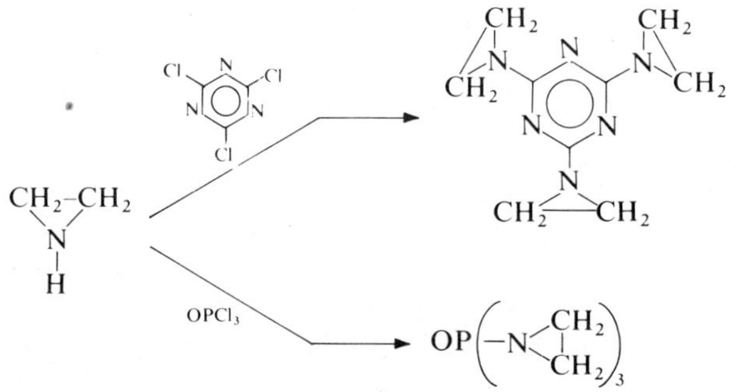

Scheme 4.13

Although ethyleneimine can be used to prepare ethylenediamine, diethylenetriamine and higher oligomers, it has been customary to use the rather inexpensive ethylene dichloride (17¢) starting material (see 4.3.3.3). However, the recent announcement (*CE* 9/28/87) by Air Products (already mentioned in 4.3.3.3) of the new facility in St. Gabriel, Louisiana to produce 20MM lb of diethylenetriamine, triethylenetetramine, and higher ethyleneamines by means of "a propietary process ... [that is, a] ... cost-effective alternative to traditional routes that use ethylene dichloride" suggests that alcohol amination, with a likely propietary catalyst, involving monoethanolamine and ethylenediamine is replacing the unnecessary use of a chlorine-containing intermediate. The family of ethyleneamines "can be used as lube-oil additives, ... [and as a component of] ... paper wet-strength resins, polyamides, epoxy curing agents, and for making other fatty chemical derivatives."

The chelating agents derived from ethylenediamine and other aspects of ethylenediamine chemistry are discussed in connection with EDC (4.3.3.3).

Two important cyclic derivatives of ethylenediamine are piperazine ($1.80, anhydrous), hexahydrate (44%, $1.60), dihydrochloride (%2.00), phosphate ($1.80), citrate ($2.25), and the previously mentioned (4.3.3.3) triethylenediamine or Air Product's original DABCO. Most likely, the preparation of these compounds still involves the use of ethylene dichloride, but piperazine could also be obtained by cyclization of $H_2N-CH_2-CH_2-NH-CH_2-CH_2-OH$, ($1.335), as shown in Scheme 4.14.

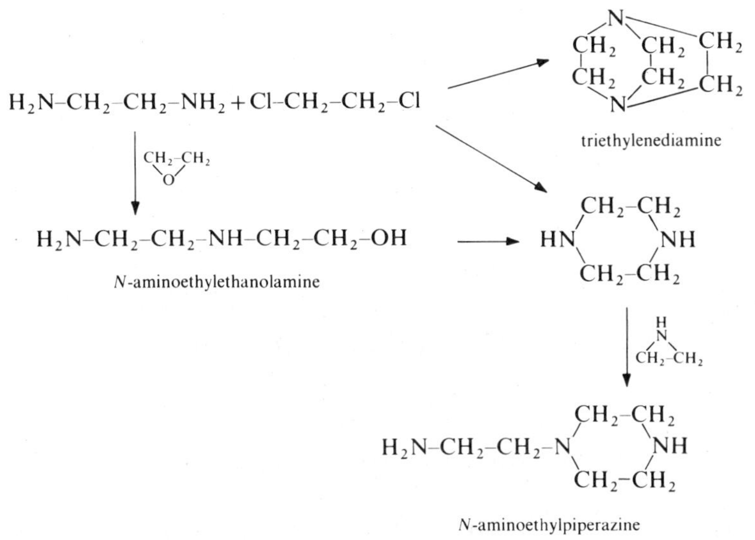

**Scheme 4.14**

Piperazine and its various salts are highly effective anthelmintics.

### 4.3.6.3  Sulfur-Containing Derivatives of EO

The reaction of EO with hydrogen sulfide produces 2-mercaptoethanol and di(2-hydroxyethyl) sulfide or thiodiglycol (TDG):

$$HS-CH_2-CH_2-OH \quad \text{and} \quad HO-CH_2-CH_2-S-CH_2-CH_2-OH$$

The mercaptan function of 2-mercaptoethanol is readily oxidized to the corresponding disulfide, while TDG enhances the point density of photographic developers, improves the performance of electroplating baths, functions as a solvent and cosolvent in ink formulations, and so on. Also, TDG can be subjected to ethoxylation, propoxylation, esterification, and converted to a polyformal

$$\text{HS--(CH}_2\text{--CH}_2\text{--S--CH}_2\text{--CH}_2\text{--O--CH}_2\text{--O)}_n\text{--CH}_2\text{--CH}_2\text{--S--CH}_2\text{--CH}_2\text{--OH}$$

that, in turn, can be used as an antioxidant and heat stabilizer of natural rubber, PAN, and other polymers. TDG is converted to polycarbonates that can function as polyols in the assembly of PURs and treatment with HCl produces the original mustard gas. Finally, the sulfide function of TDG and its derivatives can be oxidized to the corresponding sulfoxides and sulfones by means of peracids and other oxidizing agents.

### 4.3.7   Miscellaneous Derivatives of Ethylene

#### 4.3.7.1   Triethylaluminum (TEA)

This important catalyst in the Ziegler oligomerization of ethylene to α-olefins (7.4.1) is prepared by reductive addition of the olefin to pure aluminum (free of surface oxide):

$$\text{Al (76¢, 99.5\% minimum)} + 3\text{CH}_2\text{=CH}_2 + \tfrac{1}{2}\text{H}_2 \rightarrow \text{Et}_3\text{Al}$$

#### 4.3.7.2   Ethyl Mercaptan and Some of Its Derivatives

Hydrogen sulfide and mercaptans add to ethylene in the presence of peroxides or oxygen to give the expected products. For example:

$$\text{CH}_2\text{=CH}_2 + \text{H}_2\text{S} \rightarrow \text{CH}_3\text{CH}_2\text{--SH}$$

ethyl mercaptan or ethanethiol

$$+\,\text{H--S--CH}_2\text{--CH}_2\text{--OH} \rightarrow \text{CH}_3\text{CH}_2\text{--S--CH}_2\text{--CH}_2\text{--OH}$$

(see 4.3.6.3)        2-ethylmercaptoethanol, or<br>ethyl 2-hydroxyethyl sulfide

Ethyl mercaptan is incorporated into the structure of the preplanting herbicide butylate, Et–S–CO·N($i$-Bu)$_2$.

#### 4.3.7.3   Ethylene Dibromide (EDB)

Ethylene dibromide, together with the corresponding dichloride, is an additive to TEL–TML-containing gasoline that prevents the deposition of lead on pistons, valves, and other parts of the internal combustion engine.

The other major market for EDB was its use as a soil and grain fumigant, but in 1983/84 the EPA canceled its pesticidal applications and, according to current law, EPA was required to dispose 328,000 gal of the material in stock at a cost (to the taxpayers, of course) of about \$2.8 million (*CMR* 3/7/88). Disposal by burning was found by EPA to be most practical (*CMR* 11/30/87). One wonders whether EDB could not have been purchased by some enterprising chemical group to convert it into useful and acceptable derivatives and thus utilize the investment in bromine that, after all, is not so cheap (40¢/lb). The loss of these two markets for EDB depressed the price of bromine, even though its use as an ingredient of flame-retardants (see 9.3.1b) is increasing.

### 4.3.7.4.  The Participation of Ethylene in Metathesis

The unsymmetrical structure of propylene facilitates the presentation of the reaction known as metathesis, and for this reason this topic is introduced in Section 5.4.9. As described there and elsewhere in this book, however, ethylene plays an important role in metathesis and its participation in the process even carries a distinct name, "ethenylation."

## 4.4  ETHANOL AS A FUEL AND AS AN INDUSTRIAL FEEDSTOCK

### 4.4.1  Introduction I: Synthetic and Fermentation Ethanol

The conversion of ethylene to synthetic or industrial ethanol was one of the notable shifts in raw material utilization that accompanied the development of the petrochemical industry. Until the 1940s, most of the ethanol was produced by fermentation of black-strap molasses—the residue from the crystallization of canesugar, and shiploads of molasses used to arrive, mostly from Cuba, to Philadelphia and other ports. Gradually these fermentation plants (not to be confused with the manufacture of alcoholic beverages) were scrapped, and hardly any remained when the 1973 OPEC embargo and high petroleum prices revived interest in fermentation ethanol as a fuel extender. President Carter's administration and politicians of certain agricultural states promoted the fermentation of corn-starch hydrolysate to ethanol, and 10% of the latter and gasoline became the well-known "gasohol." Current technology yields about 2.5 gal of ethanol per bushel of corn. The Carter Administration was projecting a 4B-gal ethanol production by 1990—an insufficient quantity in light of a national gasoline consumption of about 100B gal. By 1984 the sale of gasohol reached 5.3B gal, and by January 1988 the domestic production of fermentation ethanol reached the 1.1B-gal level while imports (from Brazil and other sugarcane-growing countries) supplemented local sources. As the price of crude petroleum decreased in the postembargo era, emphasis on the use of fermentation ethanol shifted to its role as an octane enhancer rather than as a source of energy because EPA was accelerating the elimination of TEL and TML (4.7.1) from gasoline. Today, ethanol must compete with MTBE and similar oxygen-containing antiknock agents such as TAME (3.3.2.2), and, of course, methanol is also being promoted as a high-octane fuel (3.3.1). However, oxygenated fuels such as ethanol are now recognized to diminish CO exhaust emissions by as much as 20–25%, and Colorado, one state that is currently violating EPA air-quality standards, is mandating a minimum oxygen content in marketed gasoline (*CMR* 3/30/87, 7/6/87). Consumer resistance to the use of ethanol-enhanced gasoline has developed lately, and several major gasoline distributors (Amoco, Texaco, and Tenneco) has ceased sales of gasohol in certain regions of the country (P. Savage, *CW* 4/22/87, 35–41).

The production of fermentation ethanol in the United States is promoted through tax incentives (C. J. Verbanic, *CB* 6/87, 27–29) at federal and local levels, and its 1987 market price was $1.06/gal fob works for 95% and $1.18 for 100% (absolute) grade fob works. Federal subsidies reduce the cost to 80¢/gal to blenders of gasoline. This compares with the 95%, tax-free "synthetic ethanol" ($1.80/gal, or 23¢/lb), sold for industrial purposes and produced by an acid-catalyzed hydration of ethylene (32¢). This reaction can be carried out in the gas phase at 300°C and 82 atm over a phosphoric acid-impregnated solid matrix. At present, about 3% of ethylene is used to produce about 215MM gal or about 1.6B lb of synthetic ethanol.

It is clear that the hydrolysis of corn starch ($\sim 10 \text{¢}$) obtained by wet milling of corn, followed by fermentation of the resulting glucose, is not economically sound as long as corn is grown primarily as a food and feed product, and that this practice is motivated by politics rather than by good economics or a wise utilization of renewable resources. This situation is further complicated by short-range and unpredictable government policies (A. Slakter, *CW* 8/19/87) and EPA's application of the same volatility rules to ethanol-enhanced gasoline as it is considering for traditional, all-hydrocarbon gasoline (P. Savage, *CW* 4/22/87, 35–41). Excess agricultural production in western Europe is now being channeled into the production of ethanol and even ethylene (*CW* 11/19/86, 25, 54; 1/7–14/87, 7/22/87), but there, a two-tier price structure serves to satisfy the economic requirements of both food and fuel–chemical sectors (H. H. Szmant, "Proposed: An 'AVA' Policy" *CW*, 1/7–14/87). The United States government pays $11 billion to farmers for their surplus corn (*CMR* 3/30/87), provides disincentives to grow more corn, guarantees $3 per bushel regardless of what the commodity market price of food–feed corn may be, stores whatever surplus is produced in spite of the disincentives, and does not discriminate between its ultimate usage. What results is a series of defaulted loans (that eventually come out of the tax payers' pockets) by fermentation alcohol producers, and DOE's role as a sales agent of bankrupt plants (C. J. Verbanic, *CW* 6/87, 27–29; *CMR* 9/14/87). The latter include fermentation ethanol plants based on molasses—the by-product of a highly subsidized sugarcane industry that, apparently, cannot operate without imported farm workers and, so far, has not utilized Brazilian technology (see below) to produce fermentation ethanol.

A breakthrough in the efficient separation of cellulose from lignin, regardless of the origin of such biomass (*SZM-II*, 6–11) and progress in enzymatic hydrolysis of cellulose to its glucose units will provide new insights into on the production of fuel ethanol without the need of subsidies or "economic encouragements."

### 4.4.2   Introduction II: Brazil's Ethanol Initiative

With mounting imports of crude petroleum and the accompanying drain on its hard-earned and scarce hard-currency reserves, in November 1975 Brazil launched a sugarcane-derived ethanol initiative promoted by the Brazilian physicist Jose Goldemberg (currently the rector of the University of São Paulo), backed personally by the then president-general Ernesto Geisel [A. L. Hammond, *SC* **195**, 564–567 (1977)] and against the advise of some consultants from Western industrialized countries. The program involved an expansion of sugarcane cultivation and the installation of several dozens of distillation facilities fed by sugarcane juice (not molasses) and fueled by bagasse. At the same time, the government imposed high taxes on the use of traditional gasoline fuel and forced a transformation of Brazilian vehicles to employ either gasoline containing 20% (of 95%!) ethanol or 100% (of 95%!) ethanol. Interestingly, Brazilian subsidiaries of American automobile manufacturers were able to comply with this mandate. The program has saved Brazil over $1 billion per year on petroleum imports and created direct and indirect employment for 800,000 and 1.5MM workers, respectively (paid in local, soft currency). The production of ethanol rose from 0.6MM m³ in 1975/76 to 9.2MM m³ in 1985 and was expected to reach 14.3MM m³ in 1988 [G. Hamer, *Trends Biotechn.* **3**, 73 (1985)]. The quantity of 4B gal/yr was sufficient for some exports (*CW* 3/12/86) that produce hard-currency income and for the use of some fermentation ethanol as a chemical feedstock (E. Anderson, *CEN* 11/9/81; H. H. Szmant, *CW* 4/9/86). By the end of 1984 Brazil had a surplus of about 250MM gal, part of which was imported by the United

States. Brazil has become the world's largest producer of ethanol and exports its ethanol production technology to other developing, petroleum-poor countries. Also, the ethanol initiative became a powerful stimulus to improve technology, in general, and specifically an increase in the yield of ethanol from 70 to 75 L/t of cane to over 80 L/t, and the manufacture of distillation plants. In August 1986, the government of President Sarney published a reexamination of the ethanol project: "Energia da Biomassa: Alavanca de uma Nova Politica Industrial" ("Energy from Biomass: In Praise of a New Industrial Policy"), and measures are currently under way to correct some of the undesirable economic and social side effects of the massive program. Foremost among them is the decentralization of the distillation facilities to reduce transportation costs and the prevention of abandoning food crops in favour of energy ethanol.

### 4.4.3  Industrial Uses of Ethanol

#### 4.4.3.1  Solvent Uses

About 50–60% of industrial ethanol is used as a solvent in the formulation of numerous commercial products. In order to prevent human consumption and circumvent losses of heavy taxes imposed by the federal government on all liquors, industrial alcohol is "denatured" by admixture of benzene, toluene, pyridines, the extremely bitter sucrose octaacetate, brucine, and other rather unappetizing materials. The price of the different denatured ethanol formulations ranges between $1.765 and $1.89/gal depending on the particular formula. Unadulterated ethanol is sold for special uses under tight control by the government.

Chemically pure, absolute ethanol is obtained by subjecting the 95.6% ethanol–4.4% water azeotrope (produced by distillation of aqueous ethanol at 1 atm of pressure) to distillation on addition of benzene. Water is removed from this ternary mixture as a benzene–water azeotrope, and fractionation yields pure, anhydrous ethanol. This operation adds about 12¢/gal to the basic price of 95% ethanol, and progress is being made to reduce such costs by employing extractive distillation and membrane technologies.

For historic reasons, ethanol is a common solvent in many chemical operations, but, because of complications due to government regulations, there is a tendency to replace it by isopropyl or methyl alcohols. Fermentation ethanol is preferred for use in the formulation of medicinal products such as tincture of iodine, merthiolate, cough syrups, and elixirs.

#### 4.4.3.2  Ethyl Esters

Approximately 15% of industrial ethanol is converted to esters, and most of those listed below are of sufficient commercial importance to be listed in the weekly *CMR* price lists. Among them we recognize materials that are used as solvents and/or building blocks of more complex structures obtained by Claisen condensation (4.6.4) and other reactions:

Ethyl acetate, 41.5¢, 250MM lb
Ethyl acetoacetate, $1.05
Ethyl butyrate, $1.35
Ethyl benzoate, $1.35
Ethyl carbamate, 32.5¢
Diethyl carbonate, $1.40
Diethyl oxalate, $1.80

Then, we recognize ethyl esters that function as solvent–plasticizers:

Ethyl hexanoate, $4.25, derived from caproic acid
Diethyl phthalate, 69¢, and an odorless, cosmetic grade, 97.5¢
Triethyl citrate, $1.82
Triethyl phosphate, $1.15

Building blocks of polymers are represented by the following monomers:

Ethyl acrylate, 59¢
Ethyl methacrylate, $1.06

A fragrance and flavor ingredient is represented by:

Ethyl cinnamate, $18.60
(also the ethyl esters of benzoic, citric, and phthalic acids)

Inorganic reactants are represented by:

Diethyl sulfate, 59¢
Diethyl thiophosphoryl chloride, $(EtO)_2PS \cdot Cl$
Tetraethyl silicate, technical grade $1.39; distilled tetraethyl orthosilicate, $1.53

Ethyl esters with pronounced biological activity are represented by:

Ethyl p-aminobenzoate, benzocaine, local anesthetic USP $4.25 (9.3.2)
Ethyl parathion, insecticide, $1.75 (9.3.6)
Pyrazithion, insecticide, (10.2.2)
Dimpylate, insecticide, (10.3.2)
Coumithoate, insecticide, (9.3.2)
Malathion, insecticide, $1.62 (6.6.8)

The preparation of these esters is mentioned in connection with the discussion of the corresponding carboxylic acid systems. The preparation of diethyl thiophosphoryl chloride, or diethylchlorothiophosphonate (DECTP) (Ethyl Corp., $1.30)—the building block of several insecticides, some of which are mentioned above—can employ phosphorus trichloride (32¢) and sulfur (14–20¢, depending on grade), followed by a reaction of ethanol with the intermediate thiophosphoric trichloride ($SPCl_3$); or, preferably, it can start with phosphorus pentasulfide (52¢) and ethanol to yield first $O,O$-diethyl dithiophosphoric acid that is converted to the desired product by means of chlorine, as shown in Scheme 4.15.

### 4.4.3.3  Ethyl Ethers

Ethyl ether (50MM lb, 46¢) is the traditional general anesthetic, replaced for human use because of undesirable side effects by more recently developed general anesthetics. Most of the ethyl ether is consumed as solvent in delicate extraction and synthetic processes, although, again, it tends to be replaced by higher-boiling solvents that are less volatile, less inflammable, and less prone to form explosive peroxides on storage.

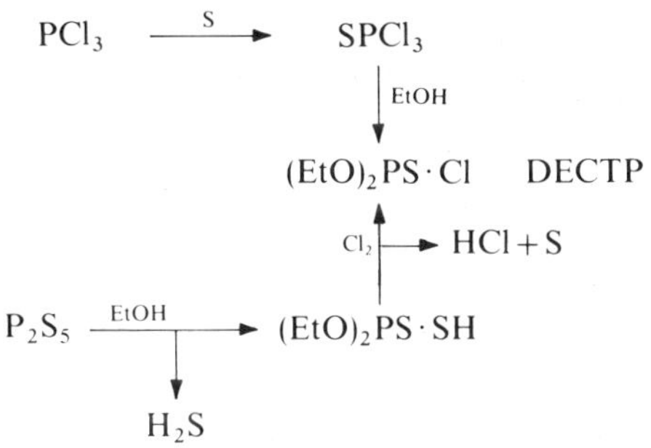

$$PCl_3 \xrightarrow{\text{S}} SPCl_3$$

$$\downarrow \text{EtOH}$$

$$(EtO)_2PS \cdot Cl \quad DECTP$$

$$Cl_2 \longrightarrow HCl + S$$

$$P_2S_5 \xrightarrow{\text{EtOH}} (EtO)_2PS \cdot SH$$

$$\downarrow H_2S$$

**Scheme 4.15**

On an industrial scale, ethyl ether is produced in the gas phase by dehydration of ethanol at 120–130°C, under pressure as high as 100 atm, and in the presence of $WO_3$ catalyst. It can also be obtained at lower temperatures by means of phosphoric or sulfuric acids. The ethoxylation of ethanol is mentioned elsewhere (4.3.6.1).

Most other ethyl ethers are usually produced by ethylation of the appropriate alcohol or phenol with ethyl chloride (250MM lb, 24¢) derived either from ethane (4.2) or more likely from ethanol or ethylene:

$$CH_3CH_2\text{–}OH + HCl \text{ (ZnCl}_2 \text{ at } 140°C) \rightarrow CH_3CH_2\text{–}Cl$$

$$CH_2{=}CH_2 + HCl \text{ (ZnCl}_2 \text{ at } 150\text{–}250°C) \rightarrow CH_3CH_2\text{–}Cl$$

Since the reactivity of ethyl halides in nucleophilic substitutions increases with the application of higher halogen derivatives, delicate substrates are ethylated by means of ethyl bromide (76¢) prepared from ethanol and hydrobromic acid (48% HBr, 38.5¢). The price of ethyl iodide is still higher ($6.25); ethyl iodide is prepared from ethyl chloride by means of the Finkelstein reaction (4.3.3.2).

Another ethylating agent of alcohols, and especially phenols, is diethyl sulfate listed in Section 4.4.3.2 among the ethyl esters. In spite of its bifunctional structure, diethyl sulfate usually functions only as a monofunctional ethylating agent in nucleophilic displacement reactions of alkoxide or phenoxide ions because the sulfate ion is a relatively poor leaving group:

$$R\text{–}O^- \ Na^+ + (EtO)_2SO_2 \rightarrow R\text{–}O\text{–}Et + Na^+ \ {}^-OSO_2(OEt)$$

Ethyl cellulose (EC, 48% ethoxyl content $4.69–5.41, depending on grade and viscosity, reflecting the molecular weight of the product), is obtained in a manner analogous to the preparation of methyl cellulose (3.3.2.3). Both EC and MC and other cellulose ethers are used extensively as water-soluble gums in paints, adhesives, textile sizing formulations, and so on (*SZM-II*, 87).

The formation of ethyl ethers of phenols is illustrated by means of the preparation of phenetol—the precursor of phenetidine ($2.00) (an intermediate in the synthesis of dyes) and in the conversion of piperonal or heliotropine (*SZM-II*, 115) to ethyl vanillin, or bourbonal ($13.50) (a valuable, a flavoring and perfume ingredient); see Scheme 4.16.

### 4.4.3.4 Ethylamines

Reductive amination of ethanol is the source of ethylamine, monoethylamine (anhydrous, 72¢), diethylamine ($1.02), and triethylamine ($1.20). These amines are the building blocks

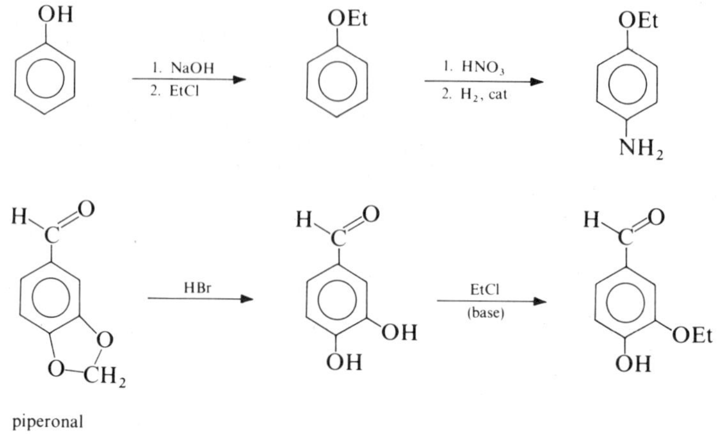

piperonal

**Scheme 4.16**

for a series of industrially important compounds such as the herbicide cynazine (3.10.3.4), the rubber processing agent *N,N'*-diethylthiourea ($2.48; see 3.7.3.3), and the common mosquito repellent *N,N*-diethyl-*m*-toluamide ($2.75), the product of diethylamine and the acid chloride of *m*-toluic acid (9.3.2).

### 4.4.3.5   Ethylation of Carbon Structures by Means of Ethanol-Derived Reactants

The ethylation of carbon structures by means of ethyl halides or diethyl sulfate requires the *in situ* formation of carbanions. This presents no problem in the case of sodium cyanide and acetylene structures. The preparation of 1-butyne from sodium acetylide and ethyl chloride is of little industrial significance. More practical examples are provided by the alkylation of diethyl malonate ($4.35, Kodak), as shown in Scheme 4.17.

### 4.4.3.6   Miscellaneous Ethanol-Derived Materials: DDT, Chloral, and Iodoform

The apparent incongruity between the three substances mentioned in the heading of this section is clarified by the fact that ethanol is the preferred starting material for their origin.

Probably no other single chemical is more responsible for the turmoil of environmental concerns and the spread of chemophobia (H. H. Szmant, *CEN* 11/16/81) than DDT, the acronym for dichlorodiphenyltrichloroethane or 1,1,1-trichloro-2,2-bis(*p*-chlorophenyl)ethane, the centerpiece of Rachel Carson's *Silent Spring*, published in 1962 and popularized by the *New Yorker* magazine. The study of the insecticidal virtues of DDT earned the Swiss chemist P. Mueller the Nobel prize in 1948, in recognition of the fact that DDT is credited with having prevented millions of human deaths attributable to mosquito-transmitted malaria, and with the prevention of typhoid epidemics during World War II through massive delousing of civilian and military populations as the Allies began to invade the Mediterranean theater of war. DDT continues to be used in many tropical countries that are prone to malaria, encephelatis, and other diseases transmitted by biting insects (*KCB*, 98–100). In November 1987 the Pan American Health Organization revealed that 950,000 cases of malaria were registered during 1986 in Latin America, of which 600,000 occurred in the eight countries that border the Amazon area. The popular use of DDT in the United States was banned in 1972 by EPA's William

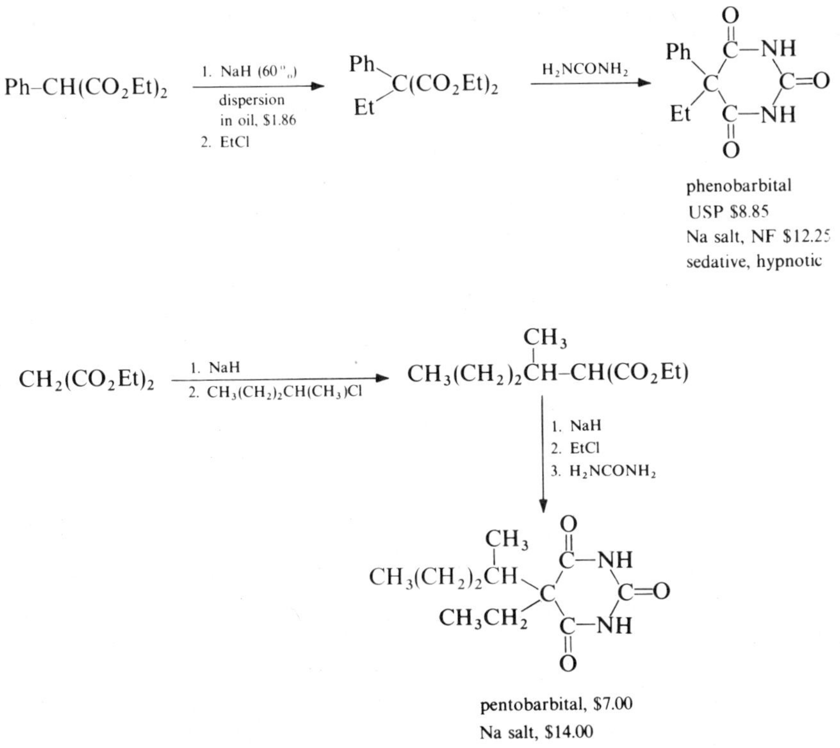

$Ph-CH(CO_2Et)_2$  →(1. NaH (60"$_o$) dispersion in oil, $1.86  2. EtCl)→  $\underset{Et}{\overset{Ph}{>}}C(CO_2Et)_2$  →($H_2NCONH_2$)→  [phenobarbital structure]

phenobarbital
USP $8.85
Na salt, NF $12.25
sedative, hypnotic

$CH_2(CO_2Et)_2$  →(1. NaH  2. $CH_3(CH_2)_2CH(CH_3)Cl$)→  $CH_3(CH_2)_2\overset{\overset{CH_3}{|}}{C}H-CH(CO_2Et)$

1. NaH
2. EtCl
3. $H_2NCONH_2$

[pentobarbital structure with $CH_3(CH_2)_2\overset{CH_3}{CH}$ and $CH_3CH_2$ groups]

pentobarbital, $7.00
Na salt, $14.00

**Scheme 4.17**

Ruckelhaus (see J. G. Edwards, "A Decision on DDT Called an Opinion," *National Pest Control Operator News*, 8/72, 7–12) after an emotional and politicized anti-DDT campaign launched by environmental activists, including Stanford University's Paul Ehrlich, Barry Commoner of Washington University in St. Louis, and others. (For an in-depth analysis of this movement, see Edith Efron's "The Apocalyptics," Simon & Schuster, New York, 1984, and T. H. Jukes, *Farm Chemicals*, 12/84). DDT is still used on a limited scale by the U.S. government to control outbreaks of equine encephelatis and other mosquito-transmitted diseases. The theory that DDT is responsible for the decline of the population of eagles and other birds of prey because of a thinning of their eggshells (*KCB*, 176) is accepted by the general population as gospel truth in spite of some evidence to the contrary (J. G. Edwards, "Pesticides and Politics," in *Pollution and Water Resources*, G. J. Halasi-Kun, Ed., Columbia University Seminar Series, Vol. 16, Pergamon Press, New York, 1984, pp. 57–68; J. G. Edwards, "The Myth of Food-Chaim Biomagnification," *Agrichemical Age*, 4/84, 10, 32–33). This argument will probably be settled when changes in bird population continue to be monitored in the absence of DDT. While the nature of Rachel Carson's book is more poetic than scientific, there are numerous scientific publications that contradict the allegation that DDT was responsible for the scarcity of the American eagle population. (A bibliography of 250 references is available from J. G. Edwards, in *Nature*, 322, 8/28/86. For a balanced analysis of Rachel Carson's claims, see G. J. Marco, R. M. Hollingworth, and W. Durham, Eds., *Silent Spring*

*Revisited*, American Chemical Society, Washington, DC, 1986 and the summary article by the same title in *CT* 6/88, 350–353).

The spread of the Asian "tiger mosquito," *Aedes albopictus*, to 17 states by the end of 1987 since its first appearance in Houston, Texas in 1985 may require the renewed use of DDT. This mosquito species is responsible for the spread of dengue fever from Africa to Asia and to Brazil, and during 1986 88,750 cases of dengue fever were reported throughout Latin America (United Press International, 3/5/88).

Dichlorodiphenyltrichloroethane is obtained by the acid-catalyzed condensation of chloral and chlorobenzene (42.5¢), and chloral, in turn, is obtained by an oxidative chlorination of ethanol:

$$CH_3CH_2-OH + NaOCl \rightarrow Cl_3C-CH=O + NaCl$$

$$Cl_3C-CH=O + 2C_6H_5-Cl \rightarrow Cl_3C-CH(C_6H_4-4-Cl)_2 + H_2O$$

During the era of DDT's popularity, this preparation of DDT was a routine sophomore organic laboratory preparation [Bailes, *JCE* **22**, 122 (1944)].

Another objection to DDT is its slow biodegradability, and this objection is somewhat overcome by the analogous condensation product of chloral with anisole (9.3.4.1), namely, methoxychlor ($2.05), sold to dealers as 50% wettable powder, $Cl_3C-CH(C_6H_4-4OCH_3)_2$.

Also structurally related to DDT, but containing a "handle" for facile biodegradation, is dicofol, $Cl_3C-C(OH)(C_6H_4-4Cl)_2$, a powerful miticide or acaricide, well known to horticulturists as Rohm & Hass' Kelthane. Its toxicity to rats is reported to be rather low (1.495 g/kg body weight when administered orally), but the EPA objects to its use when it contains more than 2.5% of DDT impurity (*CMR* 9/29/86, 10/13/86).

The high reactivity of the aldehyde function in chloral is attributed to the electron-withdrawing effect of the trichloromethyl group. This accounts for its condensation with dimethyl phosphite to give the insecticide trichlorfon:

$$(MeO)_2P(=O)H + Cl_3C-CHO \rightarrow (MeO)_2P(=O)-CH(OH)-CCl_3$$

Trichlorfon is particularly effective for the control of flies, roaches, and foliage-chewing insects. It has been used for 30 years and is toxic to insects because of the slow degradation to the high-boiling liquid dichlorvos, DDVP [$(MeO)_2P(=O)-O-CH=CCl_2$; vapor pressure 0.012 mmHg at 20°C]. The decomposition of trichlorfon to DDVP is triggered by the tendency of phosphites to rearrange to the more stable phosphates in which the central phosphorus atom attains the full complement of oxygen substituents. In any case, DDVP, the common active ingredient of hanging pest-strips and flea collars for dogs and cats, has been in use without mishaps since 1948 at the rate of 2MM lb/yr, but it is now subject of extensive scrutiny by the EPA (*CMR* 2/22/88, 3/21/88; *CW* 2/24/88).

## 4.5 ACETALDEHYDE

### 4.5.1 Production and an Overview of Its uses

Acetaldehyde is a versatile building block because of the reactivity of the aldehyde function and its activating effect on the three neighboring hydrogen atoms. The latter can participate in the classical aldol condensation (ONR-2) reaction.

The production of acetaldehyde took a revolutionary turn with the advent of the Wacker process (*CEN* 8/24/59, 4/17/61) analogous to the conversion of ethylene to vinyl acetate (4.3.4) except that water plays the role of acetic acid. Thus the stoichiometry of acetaldehyde production is simply

$$CH_2=CH_2 + \tfrac{1}{2}O_2 \rightarrow CH_3\text{–}CHO$$

The process requires temperatures of 50–130°C, pressures between 3 and 10 atm, and the presence of copper and palladium chloride catalysts. The abundance of ethylene provides an inexpensive source of acetaldehyde (45.5¢) and explains the displacement of the older processes such as

(a) The dehydrogenation or oxidation of ethanol ($\sim$26¢) by means of copper chromite catalyst at 280°C or oxygen over silver gauze at 450–550°C, respectively.

(b) The addition of methanol or acetic acid to acetylene ($\sim$45¢) to give either methyl vinyl ether (at $\sim$200°C) or vinyl acetate (in the presence of mercuric or zinc acetates), respectively, followed by hydrolytic removal of the methoxyl or acetate groups.

(c) The direct hydration of acetylene to acetaldehyde at about 100°C and 15 psi in the presence of mercuric sulfate.

(d) The addition of two moles of acetic acid to acetylene in the presence of mercury catalysts to give the vinylidene diacetate itermediate that is thermally decomposed to acetaldehyde and acetic anhydride (see the Tennessee Eastman option mentioned in 3.3.2.4).

The revival of acetaldehyde production from acetylene must await the development of less costly sources of this $C_2$ building block (see 4.7).

It is interesting that while the cost of ethylene climbed from 14¢ to 32¢/lb over the past 3 years, the unit cost of acetaldehyde changed only from 37¢ to 45.5¢ during the same period of time.

Nearly half of the production of about 650MM lb of acetaldehyde is converted to acetic acid and its anhydride, although the methanol to acetic acid and anhydride conversion processes (3.3.2.4) are highly competitive in view of the relatively low price of acetic acid (27¢). However, the construction of such production facilities is costly and time-consuming, and the use of the older acetaldehyde process may not be eliminated until the 1990s.

The established Hoechst–Shawinigan process oxidizes acetaldehyde directly to acetic anhydride by means of oxygen and cobalt, manganese, and other metallic catalysts. It involves a complex mechanism in which the key step is the conversion of transient acetyl radicals to the corresponding carbon cations so that the latter can react with acetic acid to give the desired acid anhydride:

$$CH_3\text{–}\dot{C}{=}O \rightarrow CH_3\text{–}\overset{+}{C}{=}O + e^-$$
$$CH_3\text{–}\overset{+}{C}{=}O + CH_3\text{–}CO_2H \rightarrow (CH_3\text{–}CO)_2O + H^+$$

All acetaldehyde–air oxidation processes are hazardous because of the potential accumulation of explosive acetyl peroxide $(CH_3CO)_2O_2$, one of the chain-termination products of the free-radical chain reaction shown below. The Hoechst–Shawinigan process circumvents this by loss of an electron from the acetyl radical in favor of a reducible metal ion. Also, the traditional acetaldehyde to acetic acid conversion process is carried out at relatively low temperatures ($<40°C$) in solvents such as ethyl acetate. This allows the acetaldehyde–peracetic acid complex to form and to decompose to two acetic acid molecules, as illustrated in Scheme 4.18.

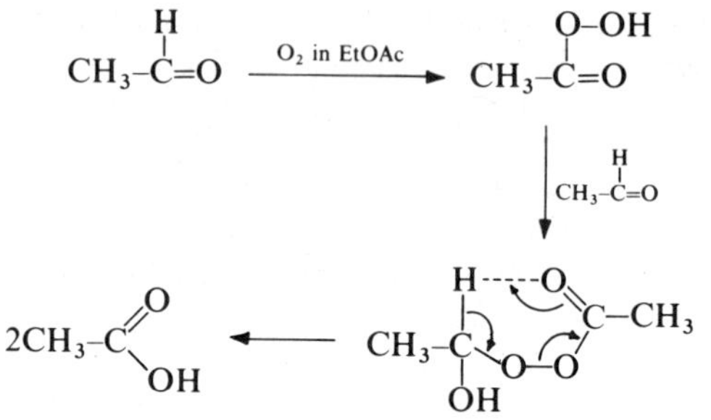

**Scheme 4.18**

The air oxidation of acetaldehyde to peracetic acid is a viable source of the latter, consuming about 6% of the acetaldehyde production. It is a typical free-radical chain reaction:

Initiation steps:   $R-CHO + O_2 \rightarrow R-\dot{C}=O + HOO\cdot$

$HOO\cdot + R-CHO \rightarrow R-\dot{C}=O + HOOH^-$

Propagation steps:   $R-\dot{C}=O + O_2 \rightarrow R-C(=O)-OO\cdot$
$R-C(=O)-OO\cdot + R-CHO \rightarrow R-\dot{C}=O + R-C(=O)-OOH$

(where R– represents the methyl group). This source of peracetic acid is competitive with the reaction between acetic anhydride (43.5¢) and hydrogen peroxide (70%, 45¢; 35%, 23.5¢):

$$(CH_3CO)_2O + HOOH \rightarrow CH_3CO_2H + CH_3C(=O)-OOH$$

especially because of the worldwide expansion of hydrogen peroxide production beyond the current capacity of 2.4B lb (100% basis) (J. Rivoire, *CW* 9/16/87). Peracetic acid is a convenient oxidizing agent in numerous, readily controlled reactions.

Each of the following synthetic uses of acetaldehyde consumes about 6–8% of its total demand.

The aldol–Cannizzaro reaction of acetaldehyde and formaldehyde is the source of pentaerythritol (3.4.3.3), an important building block of thermoset polyesters.

The reaction of acetaldehyde and HCN produces lactic acid (3.10.3.4), and the aldol condensation gives rise to crotonaldehyde and 1,3-butylene glycol (4.5.2). Aldol condensation products and ammonia are the source of synthetic pyridine compounds discussed elsewhere (10.3.1).

The carbonyl group of acetaldehyde is less reactive than that of formaldehyde, and this difference explains the choice of acetaldehyde in the condensation of two *o*-xylene molecules to yield 1,1-bis(3′,4′-dimethylphenyl)ethane, which is subsequently oxidized to 3,3′,4,4′-benzophenonetetracarboxylic dianhydride (BTDA; discussed in 9.3.2.2). The analogous acid-catalyzed condensation of *o*-xylene with formaldehyde is likely to give a mixture of isomeric products in line with the general rule regarding the *inverse relationship between selectivity* and the *vigor* (spontaneous or induced) *of experimental conditions* with which molecules react, *or reactants*.

### 4.5.2 Aldol Condensation Products of Acetaldehyde

At one time, the aldol condensation of acetaldehyde was the source of several large-scale $C_4$ and even $C_8$ chemicals produced by Union Carbide as shown in Scheme 4.19. Some of these preparative routes survive to this day, but the hydroformylation of propylene to *n*-butyraldehyde (3.5.3.4) has undermined the acetaldehyde source of this $C_4$ aldehyde and its derivatives.

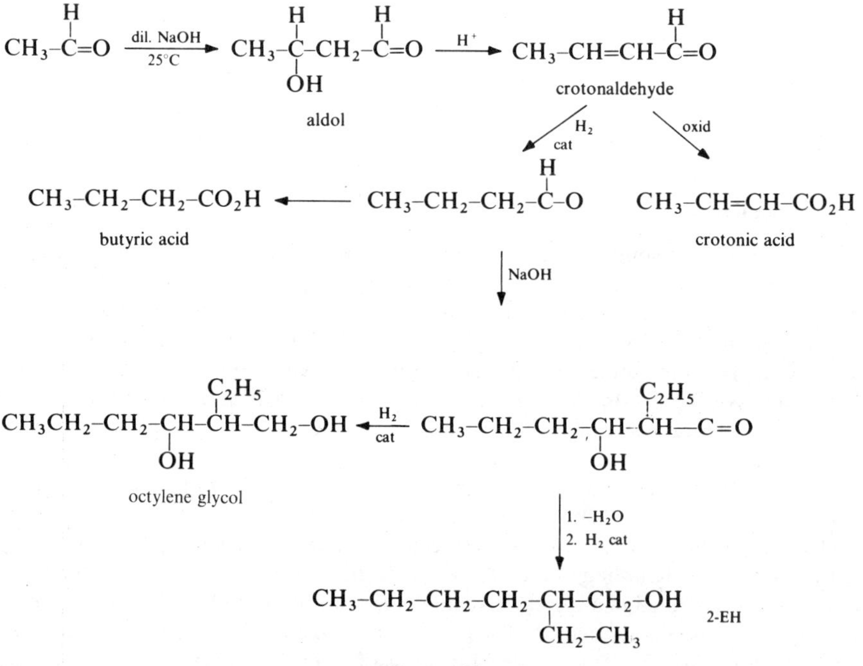

Scheme 4.19

The following fine chemicals are descendants of acetaldehyde:

- Lysine, an essential amino acid in human nutrition can be supplied from renewable resources, and is obtained by fermentation or by a combination of synthetic and biological methods (*SZM-II*, 20, 102, 161, 164). As mentioned elsewhere (3.10.3.3), it can also be synthesized from acetaldehyde and acrylonitrile. This route takes advantage of the carbanion formation from acetaldehyde and the propensity of acrylonitrile for addition of electron-rich species (5.4.2.2).

- Sorbic acid ($2.50), a fungicide used to protect spoilage in cheese and other food products, is obtained by Knoevenagel condensation (ONR-50) of crotonaldehyde, a second-generation acetaldehyde-derived building block, and malonic acid (5.6.4), as shown in Scheme 4.20.

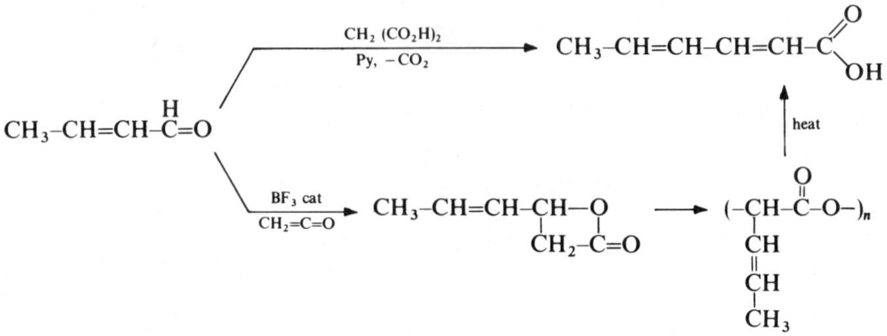

**Scheme 4.20**

This condensation reaction is catalyzed by weak organic bases such as pyridine [Allen & Van Allan, *OS* **III**, 783 (1955)] and takes advantage of the facile decarboxylation of 1,3-dicarboxylic acids. Another route to sorbic acid from crotoaldehyde is shown in 4.6.4.

### 4.5.3  Miscellaneous Derivatives of Acetaldehyde

Acetaldehyde retains some tendency to polymerize that is so prominent in its lower homolog, namely, formaldehyde. At low temperatures and in the presence of mineral acids, a high-melting cyclic tetramer known as metaldehyde (m.p. 247°C) forms in a sealed tube as a result of sublimation that begins at about 112°C. Under similar conditions except for higher temperatures, acetaldehyde forms a cyclic trimer (b.p. ~124°C) known as *paraldehyde* (58.5¢). Metaldehyde is a very effective snail and slug poison used extensively in greenhouses, and it can also serve as a cooking fuel in place of "solid alcohol" (a mixture of ethanol and sodium stearate).

Carefully controlled chlorination of acetaldehyde leads to the formation of chloro-acetaldehyde (Cl–CH$_2$–CHO), dichloroacetaldehyde (Cl$_2$CH–CHO), and trichloro-acetaldehyde or chloral (Cl$_3$C–CHO). These compounds are usually prepared for captive use because of their irritating and corrosive properties. They are readily hydrated and polymerize on standing but are useful building blocks for the assembly of more complex, particularly heterocyclic, structures. The preparation of chloral from ethanol and its use in the synthesis of DDT and methoxychlor is described in 4.4.3.6. Chloral is a hypnotic and the active ingredient of the legendary "Mickey Finn knock-out potion"—an alcoholic beverage to which some chloral is added.

Photochemical chlorination of the chlorinated acetaldehydes produces the corresponding acetyl chlorides that can be hydrolyzed to the chlorinated acetic acids:

$$R-CHO + Cl_2 \xrightarrow{h\nu} R-C(=O)Cl \rightarrow R-C(=O)OH$$

(where R represents the chlorinated methyl groups). However, the chlorinated acetic acids can be obtained more directly by oxidation of the chlorinated aldehydes (e.g., with dilute nitric acid) or by the chlorination of acetic acid. The two more common chlorinated acetic acids are monochloroacetic acid (56¢) and trichloroacetic acid (94¢; USP grade 99.5¢).

Some preparative uses of the above-mentioned chloroacetaldehyde are illustrated in Scheme 4.21.

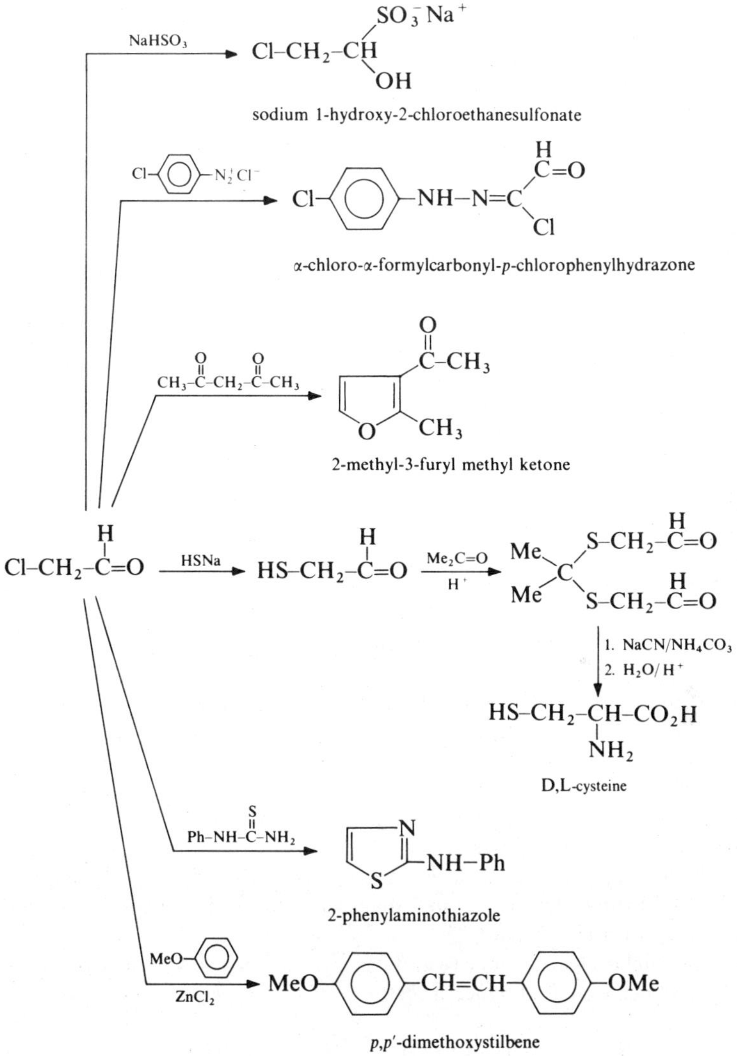

**Scheme 4.21**

## 4.6  ACETIC ACID AND ACETIC ANHYDRIDE

### 4.6.1  Production and an Overview of Their Uses

The impact of the methanol to acetic acid conversion processes (3.3.2.4) is evident from their 44% contribution to the nearly 2.8B-lb production of acetic acid in 1982 and increasing by 1985 to about 84% of the U.S. production capacity of about 3.5B lb. By late 1985 it was reported (*CMR* 9/23/85) that Monsanto had awarded nine licenses in seven countries for use of its acetic acid production technology. The remaining sources of acetic acid are

(a) The oxidation of acetaldehyde and the related production and utilization of peracetic acid (4.5.1).

(b) The recovery of acetic acid from the production of PVOH (4.3.5) cellulose acetate (see below) and from the oxidation of *p*-xylene to terephthalic acid (9.2.2, 9.3.2.2).

(c) The oxidative cracking of butane (6.2.2).

About 20% of acetic acid is destined to be used as acetic anhydride, and the production of both of these materials is, therefore, interconnected as illustrated by the coal-gasification-based production capacity of 0.5B lb of acetic anhydride inaugurated in 1983 by Eastman in Kingsport, Tennessee and described elsewhere (3.3.2.4). The total demand for acetic anhydride is about 1.7B lb, and as stated above, significant amounts of the latter are recovered from the production of cellulose acetate, which consumes about 80% of the total demand for the anhydride.

Both PVA and PVOH, and copolymers of vinyl acetate monomer (4.3.1) either retain the vinyl acetate moiety or release the acetate portion on hydrolysis (once that PVA is safely incorporated in a polymeric chain). These polymers are increasing in popularity, and the current use of nearly half of the acetic acid demand for the production of VAM is expected to grow in the foreseeable future. In the absence of a direct source of acetic anhydride (such as the above-mentioned Tennessee Eastman facility), acetic acid must be converted to the anhydride (for use in the cellulose acetate production), and this consumes about 20% of the demand for the acid. The remaining 30% of the acetic acid consumption is utilized for the preparation of miscellaneous acetates and chlorinated acetic acids (4.5.3), in the treatment of textiles, and as a solvent for oxidation reactions (including the above-mentioned production of terephthalic acid). The current (January 1989) unit prices of acetic acid and acetic anhydride are 29¢ and 44¢, respectively.

The production of vinyl acetate from ethylene and acetic acid is described elsewhere (4.3.5). The formation of the acetates of commonplace alcohols can be based on the classical alcohol–acetic acid equilibration to give the desired ester and water (and the removal of the latter, in one way or another, to displace the equilibrium in favor of the ester). The use of acetic anhydride is a more effective route to esters because the acetate moiety is a superior "leaving group" than water from the transient tetrahedral intermediate, as shown in Scheme 4.22.

Another highly efficient route to acetates is the use of ketene, a highly reactive internal anhydride of acetic acid obtained at 750°C:

$$CH_3\text{--}CO\cdot OH \rightarrow CH_2\text{=}C\text{=}O + H_2O$$

Left to its own resources, ketene tends to dimerize (4.6.4, 6.9), but it can be trapped in acetic acid to give acetic anhydride:

$$CH_2\text{=}C\text{=}O + CH_3\text{--}CO\cdot OH \rightarrow (CH_3\text{--}CO)_2O$$

The energy-intensive conversion of acetic acid to its anhydride by way of ketene explains the difference in their pricing. On the other hand, this is still the most convenient

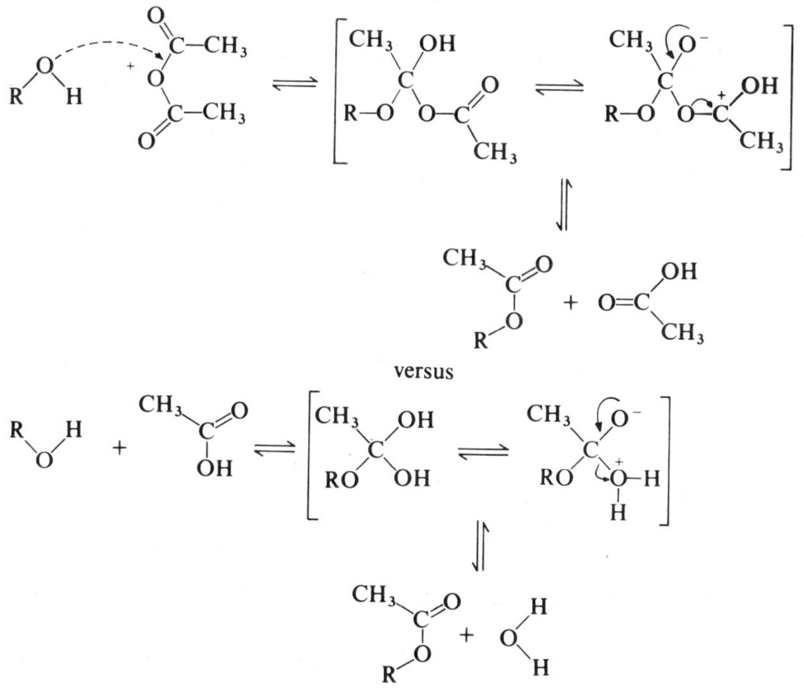

versus

Scheme 4.22

industrial process for the production of acetic anhydride since it requires less capital investment than the methanol to acetic anhydride conversion facility.

The formation of acetates by means of acetyl chloride is economically acceptable only for the preparation of special fine chemicals. Acetyl chloride is relatively costly because its preparation from acetic acid requires the rather wasteful use of thionyl chloride (55¢) or the use of somewhat cheaper but less convenient phosphorus chlorides:

$$CH_3\text{--}CO\cdot OH + OSCl_2 \rightarrow CH_3\text{--}CO\cdot Cl + SO_2 + HCl$$

Also, depending on the nature of materials subjected to acetylation, the hydrogen chloride generated by the preceding reaction may have to be neutralized by addition of tertiary amines, and this represents another complication when one is dealing with large-scale operations.

On the other hand, acetyl chloride is preferred over acetic anhydride in conventional Friedel–Crafts reactions (ONR-33) of aromatic compounds (9.3.5, 10.2.1) because the anhydride forms messy complexes with aluminum chloride.

The conversion of primary and secondary amines to the corresponding acetamides is analogous to the formation of acetates from primary and secondary alcohols. In both cases, the acetylation becomes more difficult as we go from primary to secondary systems.

### 4.6.2  Representative Acetate Esters, Acetamides, Salts, and Acetonitrile

The wide variety of simple and complex organic acetates, inorganic acetates, and acetamides found on the chemical marketplace includes the following chemicals:

1. Cellulose acetate (CA; 850MM lb, $1.30) and the mixed ester of cellulose, namely, cellulose acetate butyrate (CAB; $1.59–$1.81, depending on the butyrate content, which can range between 17 and 55%) are thermoplastic resins prepared from wood pulp and cotton linters (*SZM-II*, 88). The CA usually contains 2.75 acetate groups (out of a maximum of three) per each anhydroglucose unit, or 38–40% acetyl groups by weight. Actually, the production of CA consists of a complete acetylation of all available hydroxyl groups followed by controlled, hydrolytic removal of a fraction of the acetates so that some free hydroxyl groups improve the mechanical properties of the final product. Because of its high softening point, CA is processed into film ("safety film") or sheets by casting; that is, thin layers of solutions of the polymer in mixtures of methylene chloride and methanol are evaporated to give photographic film, optically clear and glossy sheets for use in the assembly of photographic albums, folders, and other consumer goods of this kind.

Cellulose acetate butyrate, as well as cellulose acetate propionate, are obtained by replacing most of the acetic anhydride by butyric and propionic acids (44.5¢ and 33¢, respectively) and the corresponding anhydrides. In addition to the internal plasticizing function of these larger carboxylic ester groups, external plasticizers are also added for ease of thermal processing into countless consumer articles distinguished by their excellent transparency, luster, and colors.

2. Ethyl acetate (290MM lb, 99%, 41.5¢) is a common solvent for coatings, plastics, and so on. Only a small fraction (5%) is used for synthetic purposes. The EPA-induced shift from solvent-to water-based coatings is decreasing the demand for ethyl acetate.

3. *n*-Propyl acetate (53.5¢) and *n*-butyl acetate (52.5¢), as well as isopropyl and isobutyl acetates (47¢ and 45¢, respectively), are also used primarily as solvents. Their relatively low water-solubility, as compared to that of ethyl acetate, is advantageous in extractions of materials of interest (e.g., botanicals) from aqueous systems, but their higher boiling points raise the energy requirements during solvent recovery operations.

4. Phenyl acetate ($1.04) is a special solvent because of its relatively high unit cost and high boiling point (195°C).

5. Phenethyl acetate (C$_6$H$_5$–CH$_2$CH$_2$–O–CO·CH$_3$; $3.35), like its precursor, 2-phenethyl alcohol (NF $2.10), is an ingredient of cosmetic and personal care formulations because of its roselike odor.

6. Triacetin (75¢), the high-boiling (258°C) triacetate of glycerol, is used as a solvent–plasticizer for cellulose nitrate (celluloid), a fixative in perfumes, a solvent for certain dyes, and a topical antifungal agent.

Two acetylated *sucrose derivatives* (*SZM-II*, 104) are sucrose octaacetate ($5.70), mentioned elsewhere (4.3.3.1) as an extremely bitter denaturant of ethanol, and sucrose acetate isobutyrate (SAIB; $1.10), used as a plasticizer with an approximate 2:1 ratio of isobutyrate/acetate groups.

Aspirin (33MM lb, USP $2.05) is the acetate of salicylic acid (9.3.2.1) and still occupies the first place among the over-the-counter (OTC) pain-killers or analgesics.

A few examples of acetates of more complex and costly organic structures that function as pharmaceuticals or perfume ingredients are presented in Scheme 4.23.

Among the more prominent acetamides, one can mention *N*,*N*-dimethylacetamide (DMA; 88¢, b.p. 163°C), the homolog of the more popular dipolar aprotic solvent DMF (b.p. 153°C). It boils lower than the usual 30°C difference between successive members of the same chemical family because the presence of the relatively large methyl groups interferes somewhat with the dipole-dipole attractive forces.

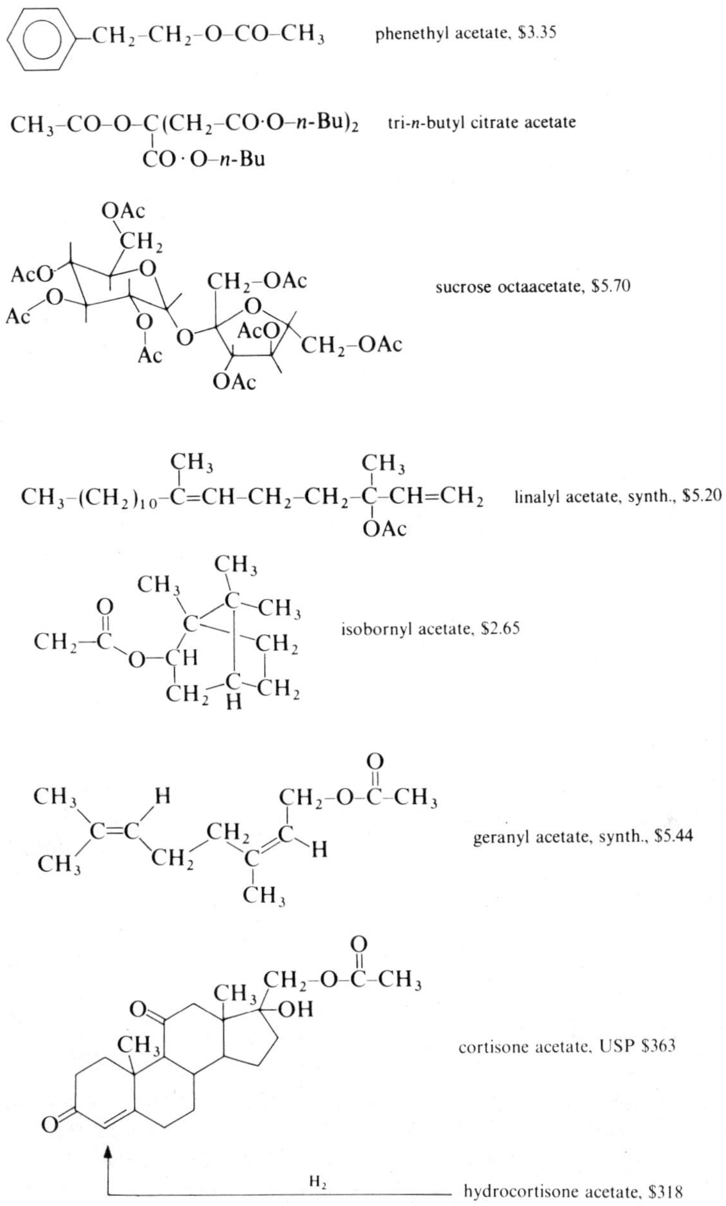

$\text{C}_6\text{H}_5$–$\text{CH}_2$–$\text{CH}_2$–$\text{O}$–$\text{CO}$–$\text{CH}_3$   phenethyl acetate, $3.35

$\text{CH}_3$–$\text{CO}$–$\text{O}$–$\text{C}(\text{CH}_2$–$\text{CO·O}$–$n\text{-Bu})_2$   tri-$n$-butyl citrate acetate
$\text{CO·O}$–$n\text{-Bu}$

sucrose octaacetate, $5.70

$\text{CH}_3$–$(\text{CH}_2)_{10}$–$\overset{\text{CH}_3}{\text{C}}$=$\text{CH}$–$\text{CH}_2$–$\text{CH}_2$–$\overset{\text{CH}_3}{\underset{\text{OAc}}{\text{C}}}$–$\text{CH}$=$\text{CH}_2$   linalyl acetate, synth., $5.20

isobornyl acetate, $2.65

geranyl acetate, synth., $5.44

cortisone acetate, USP $363

$\text{H}_2$   hydrocortisone acetate, $318

**Scheme 4.23**

Three antipyretics and analgesics derived from aniline are

- Acetanilide (9.3.2), ($1.29), used primarily as a chemical intermediate.
- Acetaminophen, $N$-acetyl-$p$-aminophenol (APAP), also distributed under the tradename Tylenol (see 9.3.2) ($2.71), a strong competitor of aspirin in the OTC analgesic marketplace.

• Phenacetin, *p*-ethoxyacetanilide (USP $2.20), classified as a carcinogen by the EPA and current use probably limited to veterinary applications.

One member of the sulfa family of antimicrobials is sulfacetamide, *p*-aminobenzene-sulfonoacetamide, (9.3.3) ($9.10).

Inorganic acetates are soluble in water and are used in electroplating and for the preparation of catalysts, pigments, pharmaceuticals, textile treatments and dyeing formulations:

Sodium acetate, NaOAc, anhydrous, 54¢, USP 60%, 57¢

Potassium acetate, KOAc, NF 90¢

Zinc acetate, $Zn(OAc)_2$, NF $1.00, techn. dihydrate $1.60

Nickel diacetate, $Ni(OAc)_2$, $1.82;

Lead diacetate, techn. 37¢, purif. 46¢; an FDA-approved hair dye

Copper diacetate, $Cu(OAc)_2 \cdot H_2O$, 71¢

Zirconium acetate, $Zr(OAc)_4$, solution = 25% $ZrO_2$

Cobalt acetate, $Co(OAc)_2$, $3.61

Amonium acetate, $NH_4^+ \, ^-OAc$, 78¢, nutrient in fermentations (production of L-lysine, L-glutamic acid, antibiotics, etc.; see *SZM-II*, 35–52); ammonium acetate may also be an intermediate in the Dupont–Dixie Chemical joint venture for the production of acetonitrile (*CMR* 9/21/87):

$$CH_3-CO_2^- NH_4^+ \rightarrow CH_3-C\equiv N + 2H_2O$$

Acetonitrile is a relatively low boiling (b.p. 81°C) member of the dipolar aprotic solvent family (3.5.3.1). It is an inert, nonaqueous solvent for many inorganic salts and functions as an extractant of polar materials from saturated hydrocarbons, as a dehydrating agent in azeotropic distillation by forming a minimum constant-boiling mixture (CBM), with water (b.p. 76°C, 16% water content) and also as an acylating agent in the Houben–Hoesch reaction (ONR-46) by way of the formation of an intermediate iminium chloride:

$$Ar-H + CH_3C\equiv N \; (+ ZnCl_2-HCl) \rightarrow Ar-C(CH_3)=NH_2^+ Cl^-$$

(where Ar–H represents an aromatic structure susceptible to electrophilic substitution reactions). Gentle hydrolysis of the imidinium chloride provides a convenient route to substituted acetophenones and related compounds (9.3.5). Acetonitrile also functions as a building block by virtue of its activated α-hydrogens that generate transient carbanions in the presence of bases. The reaction of acetonitrile, first, with HCl in absolute ethanol, followed by ammonia, produces acetamidine hydrochloride:

$$CH_3-C\equiv N \rightarrow CH_3-C(=NH_2)^+ OEt \; Cl^- \rightarrow CH_3-C(=NH_2^+)NH_2Cl^-$$

The latter is a building block of certain heterocyclic systems, as illustrated by the assembly of thiamine hydrochloride, USP $17.75, or vitamin B$_1$ (see Scheme 10.58, p. 624).

### 4.6.3  Halogenated Acetic Acid Compounds and Glycine

Chlorine-containing acetic acids can be prepared by chlorination of acetaldehyde (4.5.3) or direct chlorination of acetic acid. The latter synthesis is carried out ($FKC$, 254–257) in the presence of sulfur or red phosphorus catalysts and gives excellent (90%) yields of chloroacetic acid. The price differential between acetaldehyde (37¢) and acetic acid (27¢) favors acetic acid as a starting material unless one is interested in multichlorinated products such as trichloroacetyl chloride obtained by way of chloral ($SMW$ 10/14/85, 81; see also 4.5.3).

Chloroacetic acid (120MM lb, 56¢) is used on a relatively large scale in the production of carboxymethyl cellulose (CMC, techn. low or medium viscosity, $1.25; detergent grade, 64¢). This material is prepared by subjecting "alkali cellulose" (see 3.7.3.3.A) to a typical Williamson ether synthesis ($ONR$-96) with sodium monochloroacetate (SMCA; 53.5¢):

$$\text{Cell–O}^-\,\text{Na}^+ + \text{Cl–CH}_2\text{–CO}_2^-\,\text{Na}^+ \rightarrow \text{Cell–O–CH}_2\text{–CO}_2^-\,\text{Na}^+ + \text{NaCl}$$

CMC acts as a soil suspecting agent in laundry detergent formulations; that is, it prevents a redeposition of soil particles loosened during washing. For that purpose, the molecular weight of the cellulose can be relatively low. On the other hand, CMC derived from higher-molecular-weight cellulose, near the 80,000–350,000 range as obtained from the usual pulping operation ($SZM$-$II$, 83–87), is known as "cellulose gum." It produces aqueous solutions of relatively high viscosity and is priced at $1.60 (purified grade). The demand for all grades of CMC amounts to about 750MM lb.

The other relatively large use of chloroacetic acid is the production of a family of phenoxy herbicides

$$\text{Ar–O}^-\,\text{Na}^+ + \text{Cl–CH}_2\text{–CO}_2^-\,\text{Na}^+ \rightarrow \text{Ar–O–CH}_2\text{–CO}_2^-\,\text{Na}^+ + \text{NaCl}$$

represented here by

- Sodium 2,4-dichlorophenoxyacetate (2,4-D; techn. $1.10), the corresponding 2,4-D butyl ester ($1.25), and the corresponding 2,4-D dimethylamine salt (aqueous solution $8.05/gal)
- Sodium 2,4,5-trichlorophenoxyacetate (2,4,5-T; $2.50) and the corresponding iso-octyl ester (91¢)

The latter, the "Agent Orange" herbicide used during the Vietnam war to defoliate areas hiding enemy forces, became the cause of a long-standing damage suit by Vietnam veterans who claim impaired health due to the presence of dioxin contaminants, especially 2,3,7,8-tetrachlorodibenzo-$p$-dioxin (TCDD). The unintentional formation of TCDD during the preparation of 2,4,5-trichlorophenol and other relevant aspects of dioxin hazards are discussed elsewhere (9.3.1.1).

Another herbicide that suffers from the unfavorable publicity created by Agent Orange is silvex, or 2-(2,4,5-trichlorophenoxy)propionic acid, because its preparation also involves the use of 2,4,5-trichlorophenol. On the other hand, 2-methyl-4-chloro-phenoxyacetic acid (MCPA) is free from suspicions because the phenoxy moiety is derived by chlorination of $o$-cresol (9.3.4.1).

The following are other derivatives of chloroacetic acid of industrial interest. The herbicide alachlor, well known as Monsanto's Lasso, is a preemergence herbicide obtained from chloroacetyl chloride and *N*-methoxymethyl-2,6-diethylaniline (9.3.6). Similarly, the herbicide propachlor (Monsanto's Ramrod) is derived from chloroacetyl chloride and *N*-isopropylaniline.

The following compounds are products of nucleophilic substitution reactions of the chlorine group in chloroacetic acid or its derivatives:

1. Thioglycolic acid (TGA; $2.07, refined), obtained by means of sodium hydrogen sulfide

$$Na^+SH^- + Cl\text{-}CH_2\text{-}CO_2H \rightarrow HS\text{-}CH_2\text{-}CO_2H + NaCl$$

and its ammonium salt are employed by the cosmetic industry as an ingredient of hair-waving formulations.

2. Isooctyl mercaptoacetate, obtained when TGA is esterified with isooctyl alcohol (44¢; see 7.3), is the building block of tin- and antimony-based stabilizers of PVC. Two common examples of the latter are $n\text{-}Bu_2Sn(S\text{-}CH_2CO_2R)_2$, derived from di-*n*-butyltin oxide, and $Sb(S\text{-}CH_2CO_2R)_3$ derived from antimony oxide [$Sb_2O_3$, $1.35 (where R represents the isooctyl group)]. On the average, the unit price of tin- and antimony-based heat stabilizers depends on the metal content and ranges from $1.50 to $4.85 and $1.10 to $2.25, respectively.

In the company of other primary mercaptans or thiols, isooctyl mercaptoacetate is also employed as a chain-transfer agent to regulate the molecular-weight distribution during free-radical polymerizations of vinyl monomers. Small amounts of such chain-transfer agents intercept carbon free radicals and thus terminate chain growth because of the preferential coupling of two thiyl radicals to form a stable disulfide:

$$Q\cdot + HS\text{-}R \rightarrow Q\text{-}H + \cdot S\text{-}R$$

$$2\cdot S\text{-}R \rightarrow R\text{-}S\text{-}S\text{-}R$$

where the $Q\cdot$ radical represents a growing polymer radical and HS–R represent the chain-transfer agent.

3. Fluoroacetic acid, known as "1080" in the form of the sodium salt, is a powerful rodenticide but is also highly toxic to humans ($LD_{50}$ 2–5 mg/kg body weight). The poor nucleophilicity of the fluoride ion requires thermal activation during the preparation of 1080 from methyl chloroacetate (b.p. 130–132°C) and potassium fluoride. Methyl fluoroacetate (b.p. 104.5°C) is separated and the ester is hydrolyzed to the desired acid or its sodium salt. This preparative route takes advantage of the well-known (*SZM-I*, 205–212; Section 3.2.3) decrease of intermolecular attractive forces, and hence boiling points, caused by the presence of fluorine substituents.

4. Methyl and higher esters of cyanoacetic acid, are products of the highly nucleophilic cyanide ion (3.10.3.5.B), and they constitute the building blocks of cyanoacrylate polymers.

5. Aminoacetic acid or glycine (techn. $1.86, USP $2.12) can be obtained from ammonia and either chloroacetic or glycolic acids. The relative pricing of these two acids (56¢ and 49.5¢, respectively), the difference in the number of equivalent weight of the $-CH_2CO_2H$ moiety in the two starting materials, and the matter of salt impurities favor the use of glycolic acid (4.3.3.3, 4.3.6). The same arguments apply to the alternative routes to the preparation of EDTA (from EDC and glycine or from EDA and glycolic acid) and,

for that matter, to the preparation of EDA (from EDC and ammonia or EG and ammonia).

6. Glycine is used by the diet-beverage and food industry, which employs saccharine as a sweetener because it masks the bitter aftertaste of saccharine. As stated above, glycine and EDC are used, in part, to assemble the most comon chelating agent, namely, EDTA, and Grace's current advertisements of "Ideas for Creative Chemists" suggest the conversion of glycine to:

(a) Amidoketones such as $CH_3$–CO–NH–$CH_2$–CO–$CH_3$, N-acetonylacetamide (by heating glycine with acetic acid to produce an acetylation of the amino function and, at the same time, a classical decomposition of two carboxylic acids to ketones with the elimination of $CO_2$ and $H_2O$.

(b) The salt of carboxymethyl taurine, $Na^+$ $^-O_2C$–$CH_2$–NH–$CH_2CH_2$–$CH_2$–$Na^+$ from glycine and Cl–$CH_2$–$CH_2$–$SO_3^-$ $Na^+$, sodium 2-chloroethanesulfonate, the intermediate in the production of sodium vinylsulfonate.

(c) N-Methylene amido glycine such as $C_6H_5$–CO–$CH_2$–NH–$CH_2$–$CO_2H$, N-phenacylglycine (by the reaction of glycine and phenacyl chloride, $C_6H_5$–CO–$CH_2$–Cl, the chlorination product of acetophenone, 9.3.5).

(d) Pyrrolidinedione (10.2.1) by heating glycine with succinic anhydride (6.6.5).

(e) Oxazolones (10.2.2) by heating glycine with formic acid to give the oxazolone ring and subjecting the latter to an aldol-like condensation reaction with aldehydes.

(f) Pyrrolidones (10.2.1) by means of a reaction of glycine with a butyrolactone system.

7. Iminodiacetic acid [$HN(CH_2CH_2CO_2H)_2$; $3.00] can be assembled from glycine and either chloroacetic or glycolic acids. Its unit cost argues against the role of iminodiacetic acid as a precursor to EDTA.

8. Sarcosine (3.3.2.6) is a derivative of methylamine, and it can be prepared by a Mannich-like reaction (3.4.3.7) of the cyanide ion or by the reaction of methylamine with either chloroacetic or glycolic acids. In either case, sarcosine (tech. 50¢, purified $1.03) is transformed into surfactants of structure R–CO–N($CH_3$)$CH_2CO_2H$ by means of a reaction with fatty acids or with the more reactive methyl esters (SZM-II, 64), and the resulting N-acylsarcosinate salts, such as the one obtained with triethanolamine,

$$R-CO-N(CH_3)CH_2CO_2^- \qquad ^+HN(CH_2CH_2-OH)_3$$

serve as active ingredients of liquid soaps, shampoos, skin cleansers, and other personal-care products. Sarcosine is also converted into an enzyme-inhibiting toothpaste additive.

9. Dichloroacetic acid can be used in place of glyoxylic (or glyoxalic) acid (4.6.4) in the assembly of the medicinal allontoin with the participation of two urea molecules, as shown in Scheme 4.24. Allontoin, a member of the imidazolidine family of heterocycles

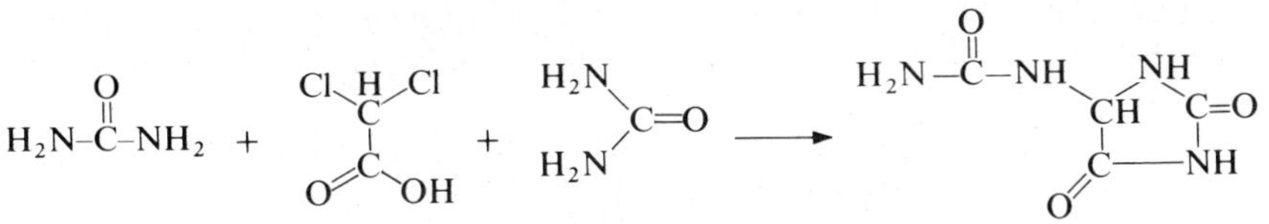

Scheme 4.24

(10.2.2), and also a substituted hydantoin such as Dalantin, described elsewhere, is a vulnerary, that is, a material used for topical treatment of wounds.

10. Trichloroacetic acid (see 4.5.3 and above) is a rather strong acid and forms salts that facilitate the isolation of alkaloids and proteins. It is employed as a herbicide and is converted by conventional means to esters. The latter serve as intermediates for the preparation of trichloroacetamide that on dehydration with phosphorus pentoxide (82¢) yields the insecticide trichloroacetonitrile. The latter can also be manufactured by chlorination of acetonitrile (see above).

### 4.6.4    Condensation Products of Acetic Acid Esters, Anhydride, and Ketene

The presence of active hydrogens in methyl and ethyl acetate (85¢ and 41.5¢, respectively) and acetic anhydride (43.5¢) opens the way for several traditional condensation reactions. The first, the Claisen (acetoacetic ester) condensation ($ONR$-18) occurs with acetic esters themselves:

$$2CH_3-CO-OR \rightarrow CH_3-CO-CH_2-CO\cdot OR + ROH$$

The condensation requires the presence of a base strong enough to generate a transient carbanion, and this is provided by either sodium methoxide (dry, 97¢) or sodium ethoxide to match the R group in the initial acetic acid ester in order to avoid a confusing exchange of one alkoxide by another. Also, the condensation reaction must be carried out under anhydrous conditions since moisture decomposes the sodium alkoxides and forms sodium hydroxide, and the latter saponifies the starting materials and final products. It of historical interest to note that the traditional use of ethyl esters dates back to the greater availability of that alcohol as compared to methanol, but this situation is, obviously, no longer true. In either case, the anhydrous sodium alkoxide can be generated by the reaction of the dry alcohol with metallic sodium (70¢). In order to accelerate the formation of the alkoxides, chunks of sodium are converted to "sodium sand" by dispersing the molten metal in an inert medium (e.g., toluene) and by cooling the mixture before the molten sodium particles coalesce. Methyl acetoacetate (85¢) and ethyl acetoacetate ($1.05) are versatile downstream building blocks for numerous more complex compounds of industrial importance (4.6.4, 6.9).

A variety of other $\beta$-keto esters can be assembled by means of Claisen condensation reactions that employ two unlike esters as long as one contains an $\alpha$-methylene group (site of the active hydrogens) and the other provides an ester function (susceptible to attack by the transient carbanion) and an alkoxide leaving group (that stabilizes the final product). Several examples of such Claisen condensations limited to a situation in which acetic acid esters are condensed with partners that lack an $\alpha$-methylene group are given in Scheme 4.25. It is noteworthy that the condensation products usually offer a more acidic hydrogen than the initial reactant and that the product is protected against further condensation reactions by forming the salt of the most stable carbanion (until the reaction mixture is worked up by decomposing the carbanion and alkoxide salts with water, and extraction of the final product).

Most of the products of these and other condensation reaction shown below are fine chemicals that are intermediates for the synthesis of even more complex structures and are either prepared for captive use by chemical businesses engaged in the production of flavors, fragrances, pharmaceuticals, dyes, and other specialty materials or supplied through contractual arrangements with companies engaged in custom syntheses.

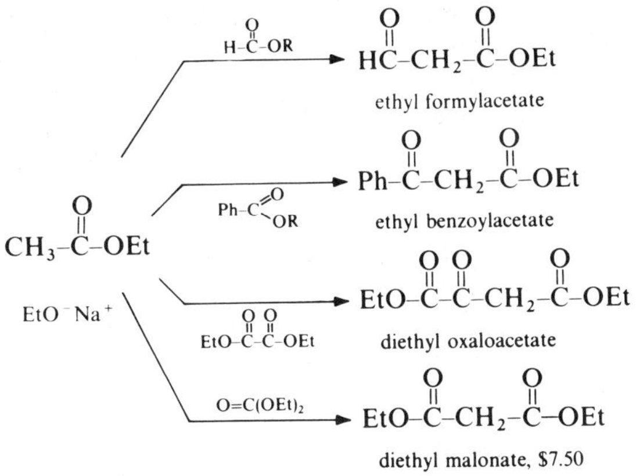

**Scheme 4.25**

Things become more complicated when the initial condensation product is prone to undergo further reactions such as the intramolecular Claisen-like Dieckmann (*ONR*-23) condensation that gives five- or six-membered cyclic structures are intermolecular condensations that lead to intractable, oligomeric products. These possibilities are suggested in the case of a condensation reaction, shown in Scheme 4.26, that involves esters of acetic and succinnic acids:

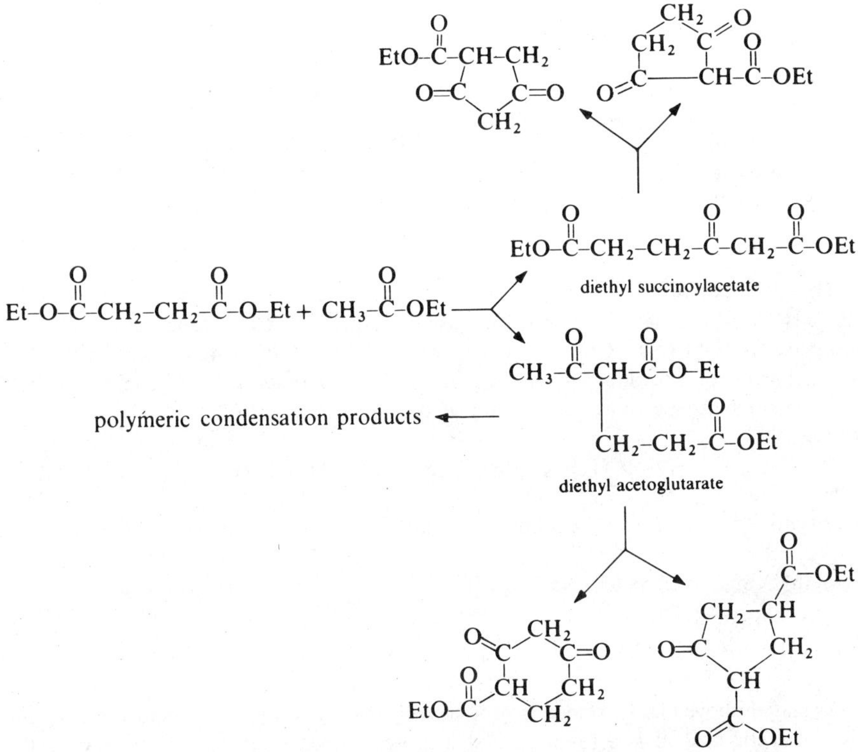

**Scheme 4.26**

For these reasons the reader is advised to consult the literature when interested in the preparation of more complex, specific condensation products.

The Claisen condensation can be extended to higher homologs of acetic esters, but this may require base catalysts that are stronger than the above-mentioned alkoxides because the acidity of the α-hydrogens decreases with substitution at the α-carbon of esters. Among such strong bases we find the carbanion of triphenylmethane, $(C_6H_5)_3C:^-$ (generated by the reaction of sodium amide and triphenylmethane), lithium diisopropylamide, $LiN(i-Pr)_2$, and others. This topic and some of the modifications of the Claisen condensation are beyond the scope of this book.

The acetoacetic esters are classical building blocks of complex carbon skeletons because they can be alkylated at the doubly activated methylene group, and once the assembly of the desired carbon skeleton is complete, the intermediate substituted acetoacetic acids, as is characteristic of all β-keto carboxylic acids, are readily decarboxylated. An excellent example of an industrially important compound prepared in this fashion is the anticonvulsant and, specifically, the antiepileptic agent valproic acid, 2-propylvaleric, or 2-propylpentanoic acid. It is administered as the sodium salt and is commonly known as Abbott's Depakene. Its medicinal virtues were recognized only in 1964, and it was commercialized in 1967. The synthesis of valproic acid is shown in Scheme 4.27.

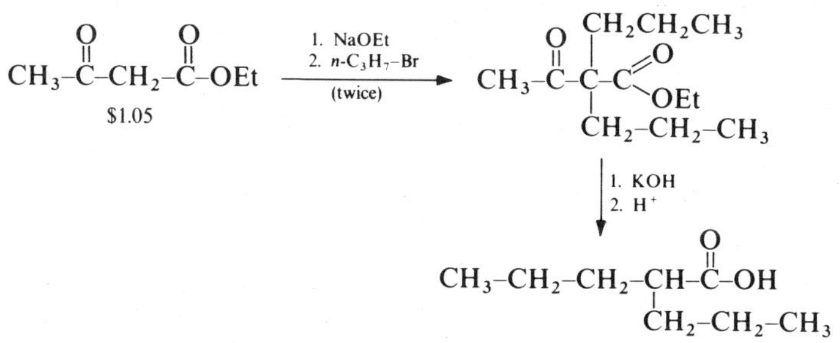

$$CH_3\text{-}\overset{O}{\overset{\|}{C}}\text{-}CH_2\text{-}\overset{O}{\overset{\|}{C}}\text{-}OEt \quad \xrightarrow[\substack{\text{2. }n\text{-}C_3H_7\text{-}Br \\ \text{(twice)}}]{\text{1. NaOEt}} \quad CH_3\text{-}\overset{O}{\overset{\|}{C}}\text{-}\underset{CH_2\text{-}CH_2\text{-}CH_3}{\overset{CH_2CH_2CH_3}{\overset{|}{C}}}\text{-}C\overset{O}{\underset{OEt}{\diagdown}}$$

$1.05

$$\xrightarrow[\text{2. }H^+]{\text{1. KOH}}$$

$$CH_3\text{-}CH_2\text{-}CH_2\text{-}\underset{CH_2\text{-}CH_2\text{-}CH_3}{\overset{|}{CH}}\text{-}\overset{O}{\overset{\|}{C}}\text{-}OH$$

**Scheme 4.27**

The condensation of acetic anhydride with aromatic aldehydes is known as the *Perkin reaction (ONR-67)*. It is catalyzed by dry, fused sodium acetate that generates a transient carbanion $^-:CH_2\text{-}CO\text{-}O\text{-}CO\text{-}CH_3$ at about 100°C. The addition of the carbanion to the carbonyl group of benzaldehyde (9.3.5.1), followed by elimination of acetic acid, provides a convenient synthetic route to cinnamic acid:

$$C_6H_5\text{-}CH\text{=}O+(CH_3\text{-}CO)_2O \rightarrow C_6H_5\text{-}CH\text{=}CH\text{-}CO_2H+CH_3\text{-}CO_2H$$

Esterification of cinnamic acid produces valuable perfume ingredients such as

Methyl cinnamate, $4.65
Ethyl cinnamate, $18.65
Benzyl cinnamate, $8.50

As stated above (4.6.1), ketene is produced by means of a high-temperature dehydration of acetic acid. It is a toxic gas that can be trapped in acetic acid to give acetic anhydride, but otherwise it tends to dimerize spontaneously to diketene (b.p. 127°C):

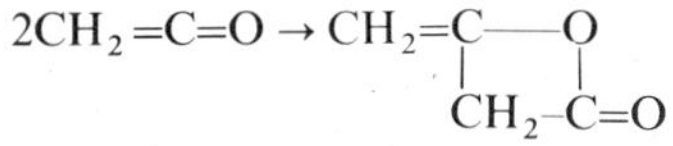

$$2CH_2=C=O \rightarrow CH_2=C\underset{\underset{\displaystyle CH_2-C=O}{|}}{\overset{\overset{\displaystyle O}{|}}{}}$$

The strained, four-membered $\beta$-lactone structure of diketene suggests the presence of a highly reactive system, and, indeed, the reaction of diketene with alcohols or amines is a facile route to the preparation of the above-mentioned acetoacetic esters and acetoacetanilides ($CH_3-CO-CH_2-CO-NH-Ar$) derived from aniline and substituted anilines (9.3.6). The acetoacetanilides, in turn, are the building blocks of commercially important dyes described in 9.3.6. Addition of HCl and alcohols to diketene provides $\gamma$-chloroacetoacetic esters, which are useful building blocks of more complex structures.

Ketene itself reacts with formaldehyde to form the highly reactive $\beta$-propiolactone (3.4.3.8), and with crotonaldehyde (4.5.2) it provides a convenient route to sorbic acid, $2.50.

### 4.6.5 Glyoxal, Glyoxylic and Oxalic Acids, and Related Structures

This section deals with difunctional $C_2$ oxygen-containing structures at oxidation levels higher than ethylene glycol. They are listed in increasing order of oxidation as follows:

| | | | |
|---|---|---|---|
| $HO-CH_2-CHO$ | $H-CO-CO-H$ | $H-CO-CO_2H$ | $HO_2C-CO_2H$ |
| Hydroxyacetaldehyde | Glyoxal | Glyoxylic acid | Oxalic acid |

$$HO-CH_2-CO_2H$$
Glycolic acid

Hydroxyacetaldehyde does not appear to be of industrial importance. Alkaline hydrolysis of chloroacetaldehyde produces a mixture of 2,4-dichloroacetaldol [$Cl-CH_2-CH(OH)-CHCl-CH=O$], cyclic hemiacetals such as 2,5-dihydroxy-1,4-dioxane, and oligomeric hemiacetals.

The two highly reactive aldehyde functions of glyoxal lend themselves for effective cross-linking of hydroxyl- or amino-group-containing structures, and glyoxal is inexpensive (40% aqueous solution, 44.5¢). Hence the main use of glyoxal is for treatment of cellulosic textiles, and the chemical effect can be visualized as a formation of acetal bridges between neighboring cellulose fibrils with the resulting formation of "permanent-pressed" fabrics, as shown in Scheme 4.28.

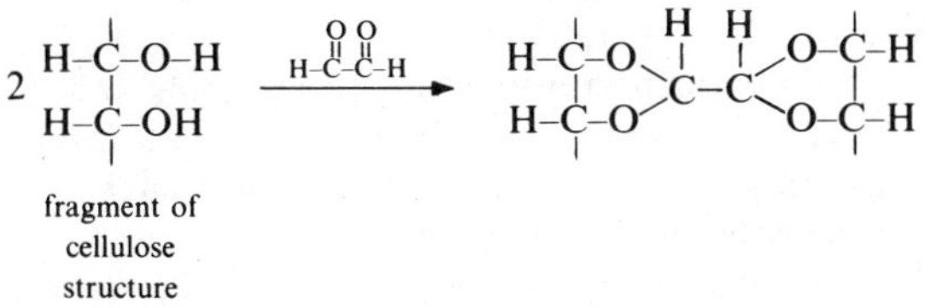

fragment of
cellulose
structure

**Scheme 4.28**

Glyoxal can be obtained by oxidation of EG in the gas phase at about 300°C over silver or copper catalysts or of acetaldehyde with nitric acid at about 40°C. In either case, it is difficult to avoid the formation of the higher oxidation by-products as well as degradation products. Glyoxal exists as a dihydrate in the presence of water, and such a solution can be purified by removing the acidic by-products by means of ion-exchange resins. The reaction of glyoxal with urea to give acetylene diureine, followed by reaction of the latter with formaldehyde to give the trimethylol derivative TMDAU, is mentioned elsewhere (3.9.3). The chemistry of TMDAU formation is another example of the high reactivity of the vicinal dialdehyde system of glyoxal.

Glycolic or hydroxyacetic acid (70% solution, 49.5¢) can be obtained in several ways the least attractive of which is the hydrolysis of chloroacetic acid (4.6.3). It is a by-product of the above-mentioned oxidation of EG by means of nitric acid and can be assembled from formaldehyde and HCN (3.4.3.7), but the most elegant route to the acid or its methyl ester is the acid-catalyzed carbonylation of formaldehyde in the presence of water or methanol, respectively (3.5.3.7). The last mentioned synthetic approach has impacted the production of EDTA and other chelating agents that were previously obtained by means of chloroacetic acid or from formaldehyde–HCN.

Glycolic acid serves as a convenient precursor for glyoxylic acid but is also used in cleaning of metals and electroplating, the processing of leather and textiles, and so on.

Glyoxylic acid inherits the high reactivity of the aldehyde function (characteristic of all aldehyde groups attached to electron-withdrawing moieties—note the behavior of chloral; see 4.4.3.6). Thus the availability on the marketplace of hemiacetals such as 2-methoxy-2-hydroxyacetic acid methyl ester, for example [HO(MeO)CH–CO–OMe], is not surprising. Most likely, this material is obtained by air oxidation of methyl glycolic ester in the presence of methanol, or by ozonolysis of dimethyl maleate, and it is a convenient form of glyoxylic acid when the latter is used as a building block of organic structures of industrial importance. An example of such synthetic use is the preparation of allontoin from urea and glyoxylic acid (rather than dichloroacetic acid; see 4.3.6).

Oxalic acid, a stable and highest oxidation product (except for oxidative degradation) of EG, has become a chemical supplied mostly by newly industrialized developing countries such as the People's Republic of China, followed by Brazil, Taiwan, and others (O. Axtell, "Oxalic: Looming Third-World Monopoly?" *CB* 3/87, 38–40) that may be employing the traditional route of oxidative alkaline degradation of sawdust, or the oxidation of cellulosic feedstocks with dilute nitric acid (*SZM-II*, 90). A modification of this approach to oxalic acid is the nitric acid oxidation of starch. A more modern route is the oxidative fusion of sodium formate (3.5.3.1), but the former U.S. producers of oxalic acid who used this process (the original Hooker and Stauffer companies, which became victims of the current "restructuring" and buyout rage (the former was bought out by Occidental Petroleum, and parts of the latter are now owned by Imperial Chemicals, Akzo, and Rhône-Poulenc), as well as Allied, which used the above-mentioned cornstarch process), all dropped out of the U.S. market, which amounts to about 16MM lb. Even Pfizer, which obtains oxalic acid as the by-product of its citric acid fermentation process, closed down the oxalic acid separation facility.

Oxalic acid per se is widely employed in the bleaching and dyeing operations of textiles, in formulations to remove rust and ink stains (oxalic acid reduces brown ferric to ferrous ions and gives nearly colorless and somewhat water-soluble ferrous oxalate), in the production of pigments (where it can act as a reducing agent because of its thermal decomposition to CO as well as CO$_2$ and water), and so on. Oxalic acid (44¢) is converted to diethyl oxalate ($1.80), which participates in Claisen condensation reactions (4.6.4), and, it and the ethyl ester of the resulting ethyl oxaloacetate are building blocks of other of fine chemicals.


## 4.7    ACETYLENE

### 4.7.1    Introduction: The Past, Present, and Future Status of Acetylene

Acetylene reminds one of the once glamorous and popular queen of organic synthesis whose brilliant star faded because of economic hardships (escalating energy costs) and the appearance of a more competitive rival (ethylene). The future of acetylene is not necessarily forever doomed, but it depends on a rescue by new production technology. One such possibility is mentioned below.

In 1962 the demand for "chemical acetylene," acetylene utilized as a versatile $C_2$ building block of many interesting chemicals and excluding acetylene used in metal welding and cutting operations, was a whopping 1.2B lb (see Fig. 4.1). This demand shrank with the energy crisis to about 290MM lb in 1981 and remained at about that level until 1985, when it began to rise rather slowly to a current demand of about 340MM lb. This demand matches comfortably the production capacity of about 370MM lb. Most of the acetylene is recovered from the cracking operations designed to produce ethylene (the more polarizable acetylene is absorbed in dipolar aprotic solvents such as DMF or NMP), but cracking of ethylene can also be used as an on-purpose source of acetylene. Most consumers of acetylene have confidential arrangements with operators of natural gas and/or petroleum cracking facilities and for this reason the market price of acetylene is difficult to ascertain. For example, GAF—a source of acetylene chemicals of long standing—signed a 10-year agreement with Union Carbide (*CMR* 10/13/86) that provides for the supply of acetylene by the latter and, in addition, stipulates a peaceful coexistence between these two corporations by an abstinence of GAF from the acquisition of additional Union Carbide stocks. Also, a joint venture between Du Pont and Idemitsu Petrochemical at Chiba, Japan designed to produce 1,4-butanediol (see below), will depend on the latter's ethylene production for a source of acetylene (*CW* 11/28/84). In view of the intrinsic instability of acetylene and difficulties in its transportation, all producer–consumer arrangements depend on close proximity of the parties involved.

The only producer of acetylene by way of the traditional calcium carbide route is Air Products, which operates a 75MM-lb acetylene production-level facility in coal-rich Kentucky. The production of $CaC_2$ is highly energy intensive: it consumes about 3000 kWh per ton of product.

On the basis of the pricing of impure calcium carbide (\$402/t or ~20¢/lb), and assuming a 93% yield of acetylene (*FKC*, 26–27), the unit price of acetylene is estimated at about 45¢. A similar result is obtained when the estimates are based on the unit costs of rather simple acetylene derivatives (see below) such as acetylene tetrabromide ($Br_2CH–CHBr_2$; 97¢) and 1,4-butanediol (80¢).

In any case, the estimated unit price of acetylene is higher than that of ethylene, and acetylene could become truly competitive when the merchant price in the open market is appreciably reduced.

A promising possibility for this to happen in the near future (1991) is the result of development work on the electric-arc process by Acetylene Canada. An engineering calculation based on the use of natural-gas feedstock (priced at \$3.20 per $10^6$ BTU) and relatively cheap electricity available in Quebec (2¢/kWh) suggests a unit price for

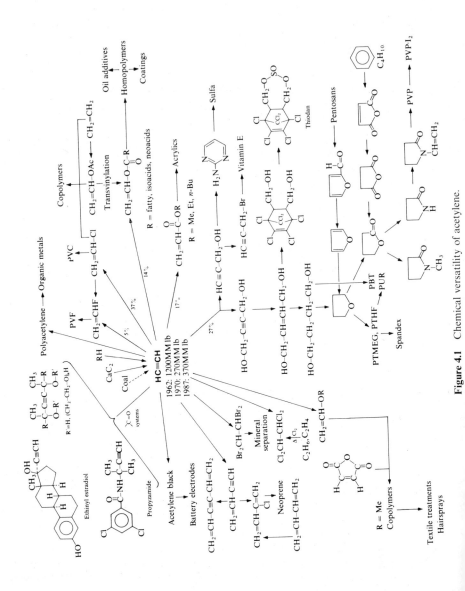

**Figure 4.1** Chemical versatility of acetylene.

acetylene of 32.5¢ at a plant capacity level of 300MM lb/yr. A larger plant capacity (450–500MM lb) is expected to reduce the unit cost of acetylene to 28¢. The essence of the electric-arc process is the employment of very high temperatures (1200–1800°C) and very short contact times (5–20 msec) so that the acetylene generated by fragmentation of natural gas does not suffer various decomposition reactions before the gas stream is quenched. In West Germany Huels produces as much as 265MM lb of acetylene by means of 19 electric-arc facilities. BASF pioneered this technology in 1939, but attempts in 1940 to employ the electric-arc route for the production of acetylene in the United States were abandoned as late as in 1973.

The production of vinyl chloride still consumes about 40% of the demand for acetylene, but this situation is not expected to persist for long. The moderately rising demand for acetylene is attributed mostly to the increasing use of 1,4-butanediol for the production of PBT (6.7) and by the relentlessly growing PURs that consume the diol as well as its oligomeric polyether glycols. Currently, the production of 1,4-butanediol consumes nearly 30% of the acetylene demand in the United States. Acetylene black (95¢), the raw material for the manufacture of electrodes for batteries, consumes about 5% of the demand for acetylene, and the remaining 25% is converted into a series of interesting fine chemicals mentioned below.

The role of acetylene as a building block of industrially important materials can be attributed to various reaction modes:

(a) The carbanion character of one or both of its hydrogen atoms
(b) The capacity to add one or two reactants across its triple bond
(c) The ability to function as a Diels–Alder (ONR-24) dienophile
(d) The posibility of dimerization to vinylacetylene, cyclization to $C_8$ and higher alicyclic systems, and polymerization to give polyacetylenes capable of conducting electricity.

### 4.7.2  Carbanion Behavior of Acetylene

The carbanion character of acetylene is manifest by its reactions with carbonyl groups and participation in nucleophilic substitution reactions. Of these reactions, the former is currently of greatest industrial importance.

With formaldehyde, in the presence of copper acetylide catalyst at 100°C and 5 bar pressure, acetylene first forms propargyl alcohol (HC≡C–CH₂–OH; $1.79), a fine chemical isolated for use as a reagent in the syntheses of valuable terpenes (SZM-II, 125–127) such as vitamin A (feed grade $12.60, pharmaceutical grade $15.00) and vitamin E or d,l-α-tocopherol ($12.50). The role of propargyl alcohol in the preparation of 2-aminopyrimidine, the building block of sulfadiazine, is shown elsewhere (10.3.2).

The addition of a second formaldehyde molecule converts acetylene to 1,4-butynediol ($1.54). It is isolated for special synthetic purposes such as the conversion by catalytic hydrogenation to cis-1,4-butenediol, which, in turn, can be used to prepare the insecticide endosulfan by starting with a Diels–Alder reaction with hexachlorocyclopentadiene (see 6.8.3). Another interesting use of cis-1,4-butenediol is the BASF process for the synthesis of vitamin A acetate. It involves the reaction of the butenediol with two molecules of acetic acid to give the diacetate of 1-butene-3,4-diol (one esterification occurs in a "normal" fashion, while the other occurs with an allylic shift), followed by a hydroformylation reaction that is accompanied by an elimination of acetic acid, as shown in Scheme 4.29.

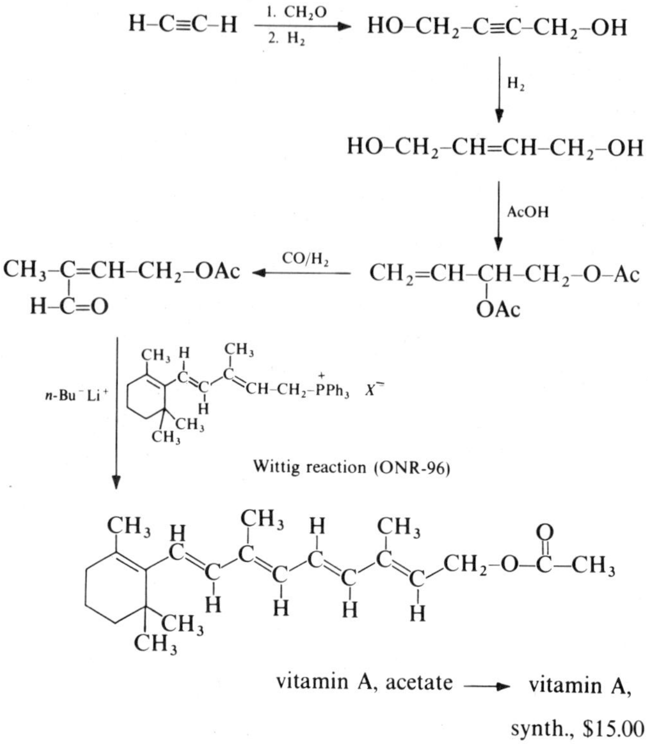

$$H-C\equiv C-H \xrightarrow[\text{2. } H_2]{\text{1. } CH_2O} HO-CH_2-C\equiv C-CH_2-OH$$

$$\downarrow H_2$$

$$HO-CH_2-CH=CH-CH_2-OH$$

$$\downarrow \text{AcOH}$$

$$\underset{H-C=O}{CH_3-C=CH-CH_2-OAc} \xleftarrow{CO/H_2} \underset{\underset{OAc}{|}}{CH_2=CH-CH-CH_2-O-Ac}$$

Wittig reaction (ONR-96)

vitamin A, acetate ⟶ vitamin A,

synth., $15.00

**Scheme 4.29**

The bulk of 1,4-butynediol is hydrogenated directly to 1,4-butanediol (80¢), the key pseudocommodity chemical that is the copolymer of poly(butylene terephthalate) (PBT) and a chain extender in some PURs as well as a steppingstone in the production of poly(tetramethylene etherglycol) or PTMEG (3.8.1, PUR). A competitive source of 1,4-butanediol is the carbonylation of 1,3-butadiene (see the account by Mitsubishi Kasei Corporation in *CT* 12/88, 759–703).

The conversion of 1,4-butanediol to PTMEG involves, first, a phosphoric- or sulfuric acid-catalyzed cyclization to tetrahydrofuran, THF, 96¢, and then, the cationic polymerization of THF. These two separate processes are depicted in Scheme 4.30.

Another role of 1,4-butanediol as a key acetylene chemical is its conversion to γ-butyrolactone ($1.20) by means of a dehydration–dehydrogenation reaction catalyzed by copper at about 200°C:

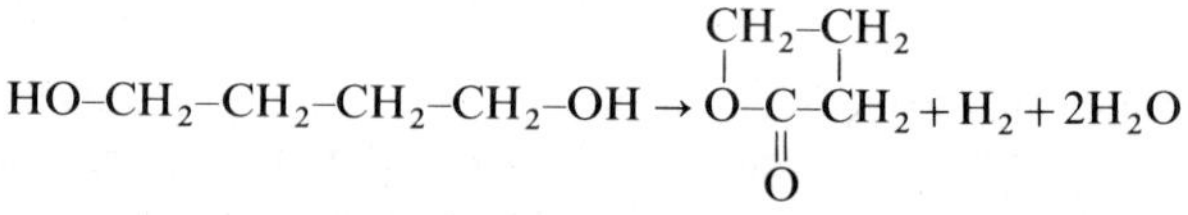

$$HO-CH_2-CH_2-CH_2-CH_2-OH \rightarrow \underset{\underset{O}{\|}}{O-C-CH_2} \overset{CH_2-CH_2}{\overset{|\quad\;\;|}{}} + H_2 + 2H_2O$$

Now γ-butyrolactone becomes the parent of a series of traditional acetylene chemicals, as shown in Scheme 4.31.

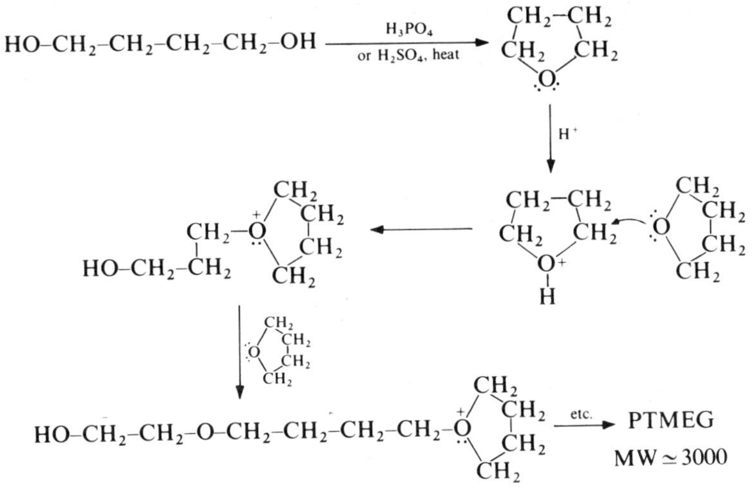

**Scheme 4.30**

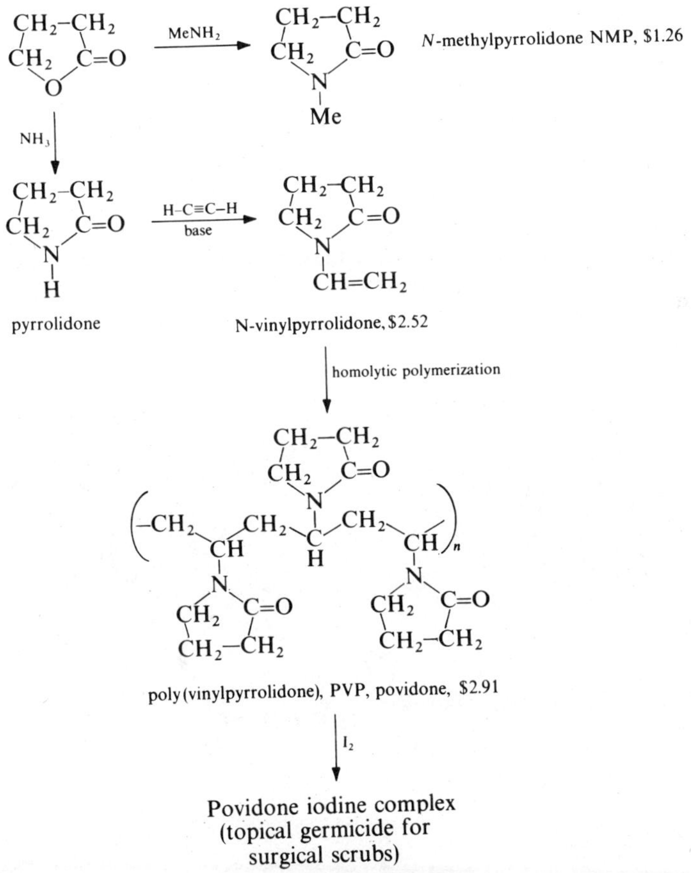

Povidone iodine complex
(topical germicide for
surgical scrubs)

**Scheme 4.31**

Recently Air Products has utilized some of its coal-derived acetylene (4.7.1) to produce a series of acetylene–ketone addition products that are used as defoamers, or, on ethoxylation, as surface-active agents; see Scheme 4.32.

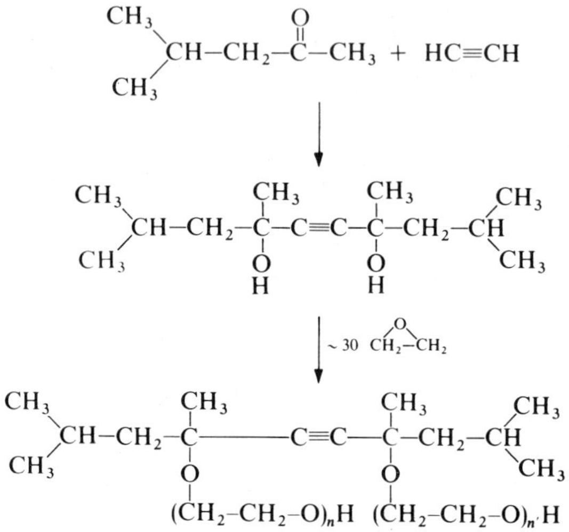

**Scheme 4.32**

Finally, one can illustrate the synthetic applications of the addition of acetylene to carbonyl-group-containing compounds with the preparation of ethynylestradiol, one of a series of related oral contraceptives that contain the acetylenic addendum (*HHS-II*, 51, 141). Another illustration is the preparation of a selective, preemergence herbicide propyzamide, shown in Scheme 4.33.

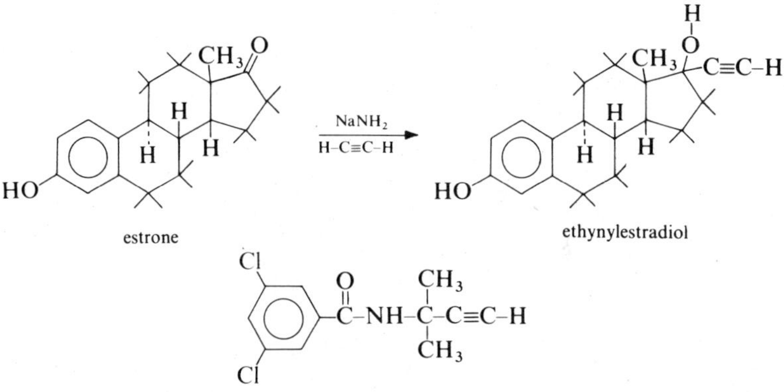

estrone

ethynylestradiol

pyopyzamide

**Scheme 4.33**

Sodium acetylide ($HC\equiv C^-Na^+$) is prepared either in liquid ammonia, (b.p. $-33°C$), by means of sodium amide, or by means of the reaction of acetylene with molten sodium at $110°C$. Sodium acetylide can then be involved in nucleophilic substitutions to produce specialty alkynes such as 1-hexyne, for example, for use in various synthetic operations:

$$HC\equiv C^-Na^+ + n\text{-}Bu\text{-}Cl \rightarrow CH_3(CH_2)_3\text{-}C\equiv CH + NaCl$$

### 4.7.3  Addition Reaction Products of Acetylene

The addition of hydrogen chloride to acetylene to yield vinyl chloride monomer (VCM) occurs smoothly at about $200°C$ over activated carbon impregnated with mercurous chloride catalyst. As stated above (4.7.1), this technology still survives, but its prognosis for the future is bleak unless the unit costs of the two hydrocarbons, ethylene and acetylene, become more competitive. The current use of this route to VCM is justified on the basis of the utilization of acetylene by-product generated during cracking operations designed to produce ethylene.

The addition of hydrogen fluoride to acetylene (in the gas phase over a catalysts such as aluminium fluoride) yields vinyl fluoride monomer (VFM), which is polymerized to the weather-resistant PVF, known as Du Pont's Teflar. VFM can be also obtained by an addition–elimination reaction of HF with VCM, thus demonstrating the relative strength of C–F and C–Cl bonds.

The base-catalyzed (MeONa) addition of methanol to acetylene yields methyl vinyl ether ($CH_2=CH\text{-}OMe$), which is co-polymerized with maleic anhydride (MA), to produce GFA's Gantrez:

$$(\text{-}CH_2\text{-}CH(OMe)\text{-}CH\text{-}CH\text{-})_n$$

The MA residue in poly(vinyl ether–maleic anhydride) can be hydrolyzed to the corresponding dicarboxylic acid, and the latter polymer is available in the form of various salts. Among many useful application of the anhydride or certain ionomeric members of the Gantrez family is the formulation of hair sprays.

The addition of acetylene to a solution of bromine in an inert solvent, say, carbon tetrachloride, yields acetylene tetrabromide, or *sym*-tetrabromoethane. This is one of the most dense (nearly 3 g/mL, but still reasonably inexpensive (97¢) liquids available for flotation–separation of pulverized, valuable materials on the basis of density differences. Methylene diiodide has a density of about 3.33 g/mL but is more costly than acetylene tetrabromide.

Elimination of two vicinal bromines by means of zinc yields 1,2-dibromoethylene ($Br\text{-}CH=CH\text{-}Br$), used as a narcotic.

The analogous preparation of acetylene tetrachloride is possible but very hazardous because of the potential formation of explosive chloroacetylene ($Cl\text{-}C\equiv CH$). For this reason *sym*-tetrachloroethane is prepared by vigorous chlorination of ethane or ethylene. It is an excellent solvent for lipids, sulfur, various resins, and so on, and is also used as a soil fumigant.

The Reppe hydrocarbonylation of acetylene (3.5.3.5) is a convenient preparation of higher esters of acrylic acids.

### 4.7.4   Dimerization and Polymerization of Acetylene

The dimerization of acetylene to vinyl acetylene in the presence of cuprous chloride–ammonium chloride catalyst is of historical importance because it led to the development of one of the early synthetic rubbers, namely, neoprene (Carothers at Du Pont in the 1930s). The 1,4 addition of HCl to vinylacetylene yielded 2-chloro-1,3-butadiene, or chloroprene, by way of an allene intermediate, followed by its allylic rearrangement:

$$2HC\equiv CH \rightarrow HC\equiv CH-CH=CH_2 \rightarrow CH_2=C=CH-CH_2-Cl \rightarrow CH_2=C(Cl)-CH=CH_2$$

A free-radical catalyzed polymerization of chloroprene yields neoprene, or Duprene. Neoprene is still an important specialty polymer because, unlike SBR or natural rubber, it is not affected by gasoline and other petroleum products and hence neoprene hose is used under the hood of automobiles and at the gasoline pump. The current source of chloroprene is a high-temperature chlorination of abundant and less costly 1,3-butadiene (24.75¢).

*Organic Metals*

The first success in the search for an organic conductor of electricity ("organic metals") occurred in 1972 with the discovery of the conducting properties of tetrathiafulvalene–tetracyanoquinodimethane (TTF–TCNQ) complexes. In 1977 A. G. MacDiarmid discovered the conducting properties of properly "doped" polyacetylene obtained by polymerization of acetylene by means of Ziegler–Natta catalyst. The desired polyacetylene is an assembly of conjugated, alternating trans double and single carbon–carbon bonds. The doping of polyacetylene consists of the addition of small quantities of either oxidizing or reducing impurities. The discussion of this fascinating topic is beyond the scope of this book, but the interested reader is referred to a brief review by D. O. Cowan and F. M. Wiygul, "The Organic Solid State," *CEN* 7/21/86, 28–45 and a 656-page monograph entitled *Polyacetylene* by J. Chien, Academic Press, Orlando, FL, 1984 (see also J. R. Reynolds, "Electrically Conductive Polymers," *CT* 7/88, 440–447). The possibility of compact batteries and other devices based on polyacetylene is the reason for a significant R&D effort by the chemical industry and other laboratories as witnessed by such recent revelations as

- The Los Alamos National Laboratory (Los Alamos, NM) announces (*CE* 3/4/85) the formation of polyacetylene in the presence of $AsF_3$–$AsF_5$ at $-90°C$ and casting films and otherwise-shaped polymer from solvents such as acetone.
- Sandia National Laboratory (Albuquerque, NM) develops a neutron detector based on polyacetylene (*CEN* 2/9/87).
- BASF in West Germany develops an iodine-laced polyacetylene with conductivity approaching that of copper (*CEN* 6/22/87, *CMR* 9/7/87).
- Air Products is awarded U.S. Patent 4,657,564, which deals with improved gas separation membranes based on polyacetylene treated with fluorine (*CW* 9/2/87).

Another approach to conducting polymers derived from acetylene is the polydiacetylenes obtained by polymerization of disubstituted diacetylenes, $R-C\equiv C-C\equiv C-R$. These monomers form conjugated chains of alternating single, triple, and double carbon–carbon bonds with a trans orientation of the R groups attached at the double bonds in the resulting polymer (*CEN* 5/5/86, 33–37). An industrial application of these materials are crystals obtained by GTE Laboratories of Waltham, Massachusetts, which are alleged to outperform silicon- and germanium arsenide-based crystals in electronic devices (*CE* 8/18/86).

The preparation of high-performance materials by polymerization of an acetylene "handle" attached to the main aromatic building block of the polymer is mentioned elsewhere (9.3.6), and the cyclization of acetylene to unsaturated derivatives of cyclooctane and cyclododecane is discussed in 10.4.

### 4.7.5  An Obituary or a Prognosis for a Successful Future of Acetylene-Based Processes?

In addition to the above-mentioned acetylene-based production of chloroprene, which is threatened by the competitive chlorination of butadiene, the following are other processes that have become obsolete as a result of the competitive position of ethylene, propylene, and current technology:

1. Acrylonitrile can be prepared by the addition of HCN to acetylene in the presence of cuprous chloride and ammonium chloride or to EO followed by dehydration of the cyanohydrin. These routes cannot compete with the ammoxidation of propylene (5.4.2.1).

2. Acetaldehyde, vinyl acetate, and acetic anhydride (4.5.1) can be prepared by the mercury-salt-catalyzed hydration of acetylene at $100°C$ and 15 psi and the addition of acetic acid to acetylene under somewhat more strenuous conditions to produce acetaldehyde or vinyl acetate, respectively. At this time, these processes cannot compete with the Wacker–Hoechst processes, which use ethylene. The same is true in the case of acetic anhydride that used to be prepared by a zinc acetate-catalyzed addition of acetic acid to vinyl acetate to yield vinylidene diacetate, followed by the thermal decomposition of the latter to yield both acetic anhydride and acetaldehyde. It is difficult to foresee how these processes can compete successfully with carbonylation of methyl acetate prepared *in situ* from methanol and carbon monoxide.

3. At one time, isoprene was prepared on a large scale by addition of acetylene to acetone followed by hydrogenation of the residual triple bond and dehydration of the adduct:

$$HC\equiv CH + CH_3-CO-CH_3 \rightarrow CH_3-\underset{\underset{\displaystyle CH_3}{|}}{C}(OH)-C\equiv CH \xrightarrow[\text{2. } H_2O]{\text{1. } H_2,\text{ cat}} CH_2=C(CH_3)-CH=CH_2$$

Currently isoprene is isolated from the cracking of higher petroleum fractions destined to produce mainly ethylene and propylene, and, when necessary, it is supplemented by the reaction of isobutylene and formaldehyde.

4. In the past, acetylene was also the feedstock for the production of chlorinated two-carbon compounds such as vinyl chloride, 1,1,2-trichloroethane, vinylidene dichloride, 1,1,1,2-tetrachloroethane, tetrachloroethylene, and trichloroethylene. The genealogical relationship between all of these materials is depicted elsewhere (4.3.3.2).

In conclusion, it is fair to say that the future of acetylene as a large-scale source of materials of industrial importance depends on the efficiency of the electric-arc process mentioned in 4.7.1 or on the appearance of some other significant technological breaktroughs.

# 5

# C₃ BUILDING BLOCKS

**TABLE 5.1    Coproduct Yields from Severe Cracking**

| Feed | Products (lb/bbl of Feed) | | | |
|------|------|------|------|------|
| | $CH_2=CH_2$ | $CH_2=CHCH_3$ | $CH_2=CHCH=CH_2$ | $C_6H_6$ |
| Ethane | 97 | 4 | 1 | 1 |
| Propane | 73 | 30 | 4 | 4 |
| Gas oil | 67 | 43 | 11 | 119 |
| Paraffinic naphtha | 78 | 33 | 9 | 21 |

## 5.1  PROPANE AND THE DOMINANT ROLE OF PROPYLENE AMONG THE $C_3$ BUILDING BLOCKS

Among the $C_3$ hydrocarbon building blocks, propane is of relatively minor synthetic importance except as the main source of propylene. Propyne or methylacetylene is a specialty fine chemical prepared from acetylene (4.7) on a small scale. The production of propylene in the United States is about 18B lb (see Table 1.1). It is the runner-up among the large-volume organic building blocks, and most of this chapter is concerned with its transformations to industrially important chemicals.

Propane is the major component of LPG (2.2.1), and the United States is subject to increasing imports of LPG stimulated by the growing use of heavy feedstocks by producers of ethylene. This trend produced a propane buying panic in 1983 ($CW$ 6/22/83) that raised the propane prices along the Gulf Coast as high as 55¢–60¢/gal (or ~ 11¢–12¢/lb). As shown in Fig. 2.2, the distribution of ethylene, propylene, and other cracking products is very sensitive to the nature of the feedstock. A very instructive and more extensive summary of this relationship is reproduced from $CW$ 6/22/83 in Table 5.1.

Recently (E. S. Kische, $CW$ 9/2/87) the New York Mercantile Exchange added propane to other energy commodities traded there, and the 726 contracts for December 1987 were settled at 25.62¢/gal (or about 5¢/lb). When delivered to private consumers located in remote areas in liquefied form (b.p. 79°C at 30 atm), the retail price of propane is about 20¢/lb. Approximately 40% of all propane consumption is used as a fuel for domestic and commercial buildings. Ford Motor Company is experimenting with propane-powered automobiles, even though this fuel delivers a 10–15% lower mileage than does conventional gasoline.

Catalytic dehydrogenation of propane to propylene is becoming a reality, but so far most of the demand for propylene is being satisfied by means of energy-intensive cracking processes. Facilities to dehydrogenate propane are currently being installed in Thailand and Mexico to produce 220MM and 770MM lb of propylene, respectively.

The oxidation of propane–butane to a series of useful products (3.3.1, 6.2.2) is still being practiced, but the explosion at the Celanese facility at Pampa, Texas during November 1987 forced the company to purchase acetic acid elsewhere in order to resume the production of acetate and acrylate esters ($CW$ 12/9/87). One wonders whether the oxidative fragmentation process of LPG will survive the competitive sources of acetic acid and coproducts based on methanol carbonylation.

Other significant chemical uses of propane include the production of nitroparaffins (5.3) and its reaction with chlorine.

The future of high-temperature chlorination–chlorinolysis of propane (4.2) depends greatly on the fate of the major products, carbon tetrachloride and perchloroethylene, and the demand for these chemicals is a function of the growth of fluorine derivatives in the case of carbon tetrachloride and the EPA-imposed restrictions on the use of perc—the current replacement for outlawed trichloroethylene—a dry-cleaning and metal degreasing solvent of the past.

The fate of the propane–butane oxidation to give a complex mixture of methanol, formaldehyde, formic acid, acetaldehyde, acetic acid, and so on, is discussed elsewhere (6.6.2).

Undoubtedly, attempts to use propane directly for the production of acrylonitrile (5.4.3) and acrylic acid (5.4.5) will be forthcoming, and such successful developments will increase the demand for chemical propane.

As long as there is a great demand for ethylene and this material is obtained by severe cracking of petroleum and LPG, the supply of propylene is assured (see Table 5.1 and Fig. 2.2). On the other hand, when ethylene is obtained from ethane, the yield of propylene drops to 2–4%, and the best source of propylene becomes the steam reforming of propane.

Actually, the manufacture of gasoline generates a great deal of propylene together with other $C_4$ and $C_5$ olefins, but practically all of these compounds are subjected to dimerization and trimerization (5.4.1) reactions, as well as alkylation processes involving isobutane, and the resulting mixture of highly branched alkanes of high octane number ($\sim 95$) becomes a valuable component of the marketed gasoline pool. A cash margin analysis (P. Savage, $CW$ 7/22/87, 26–29) of chemical propylene to petroleum refiners, whose primary interest is the production of alkylate gasoline, oscillated between 1¢ and 7¢/lb during the last 4 years and depended greatly on the demand for chemical propylene. In the past, the price of ethylene was traditionally greater than that of propylene, but this situation has changed dramatically in recent years because of the strong international growth ($> 10\%$ per year) in the demand for polypropylene (PP) (5.4.1). There exists the danger (P. Savage, $CW$ 5/13/87, 6–8) that the worldwide expansion of PP production facilities will create a "bandwagon effect" (2.4.4.4). Next to PP, there is a strong demand for propylene oxide (PO).

## 5.2   THE COMPETITIVE EDGE OF PROPYLENE RELATIVE TO THAT OF ETHYLENE

The presence of a methyl group in the homolog of ethylene has both favorable and unfavorable consequences, but the former seem to outweigh the latter.

The most detrimental effect of the presence of three allylic hydrogens in propylene is the difficulty to convert the olefin to PO in a manner as simple as that practiced in the ethylene to ethylene oxide conversion (4.3.6). Instead, the molecular oxygen diradical attacks one of the allylic hydrogen atoms to yield transient acrolein, and this is oxidized further to acrylic acid. While acrylic acid is a useful derivative of propylene, PO is, at this time, even more useful, and hence we may consider this to be a weakness on the part of propylene, except that, as we shall see, ingenious chemistry manages to convert the olefin into the desired PO at the same time that $t$-butyl alcohol is obtained from isobutane. Another detrimental aspect of propylene vis-à-vis ethylene concerns the respective polymers. The tertiary hydrogen of PP is very susceptible to photochemically induced air oxidation and, more so than PE, PP requires larger concentrations of effective antioxidants for outdoor use. On the other hand, the susceptibility to oxygen diradicals of the same tertiary hydrogen inherited from propylene in isopropyl alkylation products such as cumene becomes an asset when one considers the importance of the Hock process (5.4.6). Also, the presence of three allylic hydrogens in propylene opens the way for the ammoxidation route to acrylonitrile (5.4.3). One allylic hydrogen suffices to yield allyl chloride (5.4.8), and, unlike vinyl chloride, because of the presence of the double bond, allyl chloride is a very reactive intermediate for the preparation of equally reactive allylic ethers, esters, and so on.

To summarize, the competitive olefins under discussion provide different practical opportunities as the result of an apparently minor difference in their respective structures. They are indicative of the need to analyze industrially significant consequences of differences between any two rather similar structures and to extract from such analyses advantageous applications for each material. Normally, the economic advantages and disadvantages between two competitive materials must also be taken into consideration, but, in the case of the two olefins under discussion, the price differential is minimal and elastic, in January 1989 for instance, ethylene was priced at 32¢/lb and propylene cost 7¢ and 20.5¢/lb for chemical and polymer grades, respectively.

## 5.3  PRODUCTION AND UTILIZATION OF NITROPARAFFINS

The vapor-phase nitration of alkanes at about 420°C was studied extensively by Henry B. Hass (see Preface) and collaborators at Purdue University, and the research culminated in the establishment of the Commercial Solvents company in 1940 at nearby Terre Haute, Indiana. The research led to the promulgation of the Hass nitration rules (*ONR*-40); the most important rule was the formation of mononitroalkanes by substitution of all available hydrogens in the feedstock and the formation of mononitroalkanes at the site of all possible C–C bond fragments generated by cracking the original alkane. Thus, when propane is employed in vapor-phase nitration, one obtains 1- and 2-nitropropanes from the substitution of the two kinds of hydrogens in propane and nitromethane and nitroethane from the fragmentation of the starting material. Indeed, these four nitro-alkanes are obtained in yields of about 25, 40, 10, and 25%, respectively, with a total conversion of about 40% of propane to nitroparaffins. It is clear that the vigorous nitration–oxidation conditions cause much of the propane to be oxidized, but the low cost of the feedstock justifies such waste. It is also clear, that the use of higher alkanes gives rise to more complex mixtures of nitroparaffins with separation problems more challenging than the separation of the currently manufactured products:

|  | b.p. (°C) |
|---|---|
| Nitromethane | 101 |
| Nitroethane | 114 |
| 2-Nitropropane | 120 |
| 1-Nitropropane | 131 |

Thus, after nearly half a century of industrial nitroparaffin production, propane is still the feedstock of choice, even though the pioneering Commercial Solvents operation was acquired in 1975 by IMC (International Minerals & Chemicals), and the latter became Angus Chemical in 1982. The domestic production of nitroparaffins in the United States was expanded in 1983/84 by the entry of W. R. Grace (see *CW* 3/10/86).

The production level of nitroparaffins in the United States is about 80MM lb. The presence of oxygen in the gaseous nitration mixture favors the formation of the lower-molecular-weight compounds. A modification of the nitration process employs nitrogen dioxide ($NO_2$; 46¢) in place of nitric acid (94.5–98%, 14¢) and the higher cost of the former is compensated somewhat by a greater yield of nitration products and a decreased formation of 2-nitropropane, a suspect carcinogen.

The current prices of the industrially available nitroalkanes: nitromethane ($2.37), nitroethane ($2.50), 2-nitropropane and (55¢) reflect the relative demand for these materials as additives in gasoline used for automobile races, as solvents, as stabilizers of chlorinated solvents, and, of course, as chemical building blocks. In particular, nitromethane is an excellent solvent of poly(cyanoacrylates) and some other polymers.

The chlorination of nitromethane with chlorine or sodium hypochlorite leads to the formation of trichloronitromethane, or chloropicrin ($Cl_3C–NO_2$; $1.25), a grain fumigant and soil insecticide. It also functions as an oxidizing reactant in the synthesis of the dye Basic Violet I or Methyl Violet ($6.80) used commonly in inks, typewriter ribbons, permanent markers, and so on. As $N,N$-dimethylaniline ($1.75) is subjected to the oxidative condensation with chloropicrin, the partial formation of $N$-methylaniline, allows the two anilines to form a resonance-stabilized structure typical of a triarylmethane dye, as shown in Scheme 5.1.

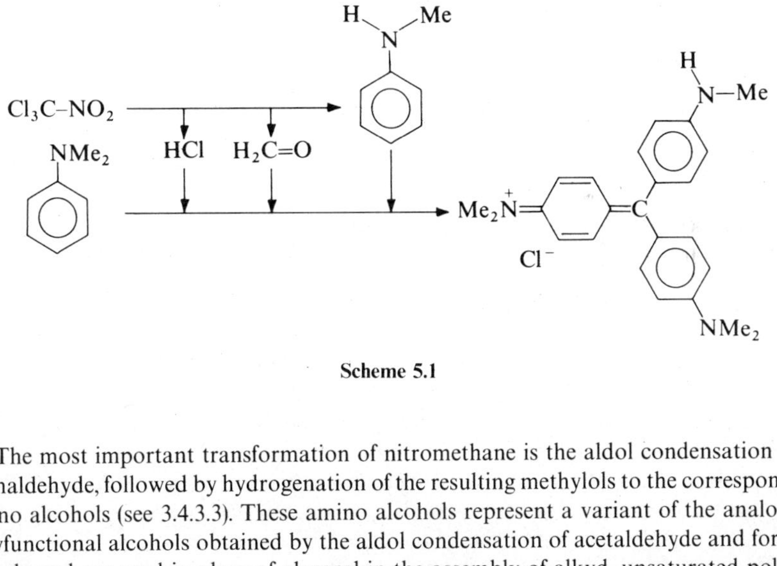

**Scheme 5.1**

The most important transformation of nitromethane is the aldol condensation with formaldehyde, followed by hydrogenation of the resulting methylols to the corresponding amino alcohols (see 3.4.3.3). These amino alcohols represent a variant of the analogous polyfunctional alcohols obtained by the aldol condensation of acetaldehyde and formaldehyde and are used in place of glycerol in the assembly of alkyd, unsaturated polester resins (UPRs), or epoxy resins. Tris(hydroxymethyl)aminomethane is used as a scavenger of formaldehyde during curing of formaldehyde-derived resins, and in pure form ($3.35), it is a buffer of biological systems, while tris(hydroxymethyl)nitromethane itself is an antimicrobial used to prevent odors and clogging of aqueous systems due to microbial action.

The reaction of the preceding amino alcohol with carboxylic acids or aldehydes leads to the formation of oxazolines (A) and oxazolidines (B), respectively (10.2.2), as shown in Scheme 5.2. As shown in Scheme 5.2, ethoxylation of A gives a series of surface-active

Oxazolines:

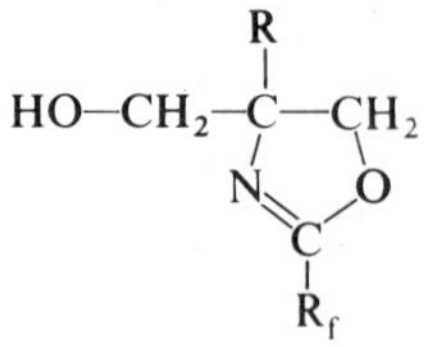

$R_f$ = fatty acids,
R = C$_2$H$_5$,     CH$_2$–OH

R = –C$_2$H$_5$     surface-active agents,
corrosion inhibitors,
antifoam agents

R = –CH$_2$OH   emulsifiers,
corrosion inhibitors,
dispersants of pigments

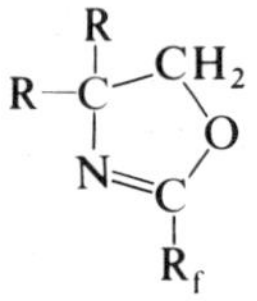

R = –CH$_2$–(OCH$_2$CH$_2$)$_2$–OH
emulsifier,
corrosion inhibitor

Oxazolidines:

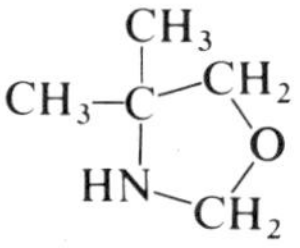

formaldehyde-release agent

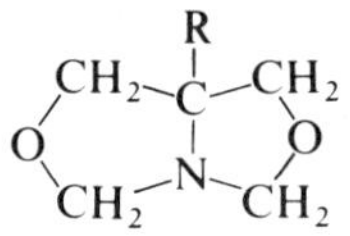

R = –C$_2$H$_5$
corrosion inhibitor,
tanning agent,
formaldehyde-release agent

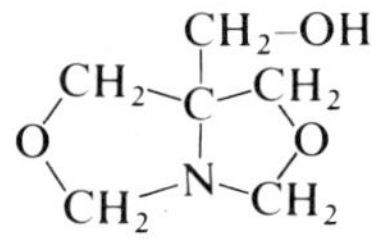

formaldehyde-release agent
(resorcinol–phenol–formaldehyde resins),
protein cross-linking agent in
hair-care products

**Scheme 5.2**

agents, and B is a potential source of formaldehyde and hence functions as a cross-linking agent in phenolic and proteinaceous materials. Analogous derivatives can be obtained from the aldol condensation products of the higher nitroparaffins.

Bromosol [(HO–CH$_2$)$_2$CBr–NO$_2$] is an antibacterial and antifungal used in cosmetic and other personal-care products.

The aldol condensation of *nitroethane* with benzaldehyde is a convenient stepping stone to the synthesis of three related pharmaceuticals:

1. C$_6$H$_5$–CH(OH)–CH(CH$_3$)–NH–CH$_3$, ephedrine (synth. USP $1.25/oz, NF hydro-chloride ∼ $22/lb, USP sulfate ∼ $24) is used at the level of about 1.5MM lb/yr as a bronchodilator and as an active ingredient in nasal decongestants. With two chiral

centers, there exist four stereoisomers of ephedrine: a pair of enantiomers with an erythro configuration and a pair of pseudoephedrine enantiomers with a threo configuration. Strictly speaking, the term "ephedrine" refers to the levo-rotating isomer with the R configuration at both chiral centers that is isolated from *Ephedra vulgaris*—a Chinese herb known as *Ma Huang*. The synthetic pseudoephedrine is a dextro-rotating mixture of isomers and is the ingredient of many OTC formulations.

2. $C_6H_5$–$CH_2CH(CH_3)$–$NH_2$ (amphetamine; racemic sulfate $4.00, *d*-sulfate $12.00) is a prescription drug because of abusive uses. It is a habit-forming stimulant of the central nervous system.

3. $C_6H_5$–$CH(OH)$–$CH(CH_3)$–$NH_2$ [phenylpropanolamine (PPA) HCl $10.90] is an OTC bronchodilator but it is better known as an appetite-suppressant. Because of the common synthetic pathway of PPA and amphetamine, the FDA restrict the content of the latter to 400 ppm.

All three compounds are derivatives of $C_6H_5$–$CH_2$–$CH_2$–$NH_2$, $\beta$-phenethylamine, the hypothetical parent of a numerous family of *sympathomimetic or pressor amines*. The most prominent member of the family is adrenaline or epinephrine—the analog of ephedrine except for 3,4-dihydroxy substituents on the aromatic ring. The synthesis of epinephrine requires that nitromethane be condensed with protocatechualdehyde or, better yet, with piperonal or heliotropine ($9.75; see 9.3.4.2) in which the phenolic ring is momentarily protected by a formal group, $-O-CH_2-O-$. Adrenaline or epinephrine is a hormone secreted by the adrenal gland. The naturally occurring compound is the levo-rotating isomer, while synthetic USP-grade epinephrine is a racemic mixture priced at 60¢/g or $27.75/lb. Structurally related to adrenaline is the antihyperintensive drug methyldopa, 3,4–$(HO)_2$–$C_6H_3$–$CH_2$–$C(NH_2)CH_3$–$CO_2H$, which is also synthesized from piperonal; see Scheme 5.3. Since both protocatechualdehyde and piperonal are derivatives of catechol, the last-mentioned pressor amines are also referred to as *catecholamines*.

The monomethylol derivative of 1-*nitropropane*, on hydrogantion in the presence of EG, or by means of a reaction of the corresponding amino alcohol with EDC, gives rise to the tuberculostatic ethambutol, $[CH_3-CH_2-CH(CH_2-OH)-NH-CH_2-]_2$. While tuberculosis may no longer be a significant medical problem in the United States, it is still a serious health threat in many less industrialized countries.

The monomethylol derivative of 2-*nitropropane* is hydrogenated to the corresponding amino alcohol, and the latter is methylated to give 2-dimethylamino-2-methyl-1-propanol, $Me_2N-C(Me)_2-CH_2-OH$. This tertiary amino alcohol turns out to be an emulsifying agent alone, or in combination with fatty acids, and it is also a corrosion inhibitor and a catalyst in the manufacture of PUR foams (interlude in Section 3.8.1).

The preceding survey of industrially significant nitroparaffins and their derivatives represents only the proverbial tip of the iceberg of chemical possibilities. Note, for example, the recent announcement (*CE* 11/86) that chlorinated and brominated imidazolidinones are effective biocides for the control of algae in swimming pools, spas, cooling towers, heat exchangers, and air-conditioning installations. These compounds are relatively stable, and their uses offer several advantages when compared with the traditional use of chlorine or cyanuric chloride. Their synthesis starts with 2-*nitropropane* by taking advantage of the somewhat acidic nature of its alpha hydrogen, as shown in Scheme 5.4.

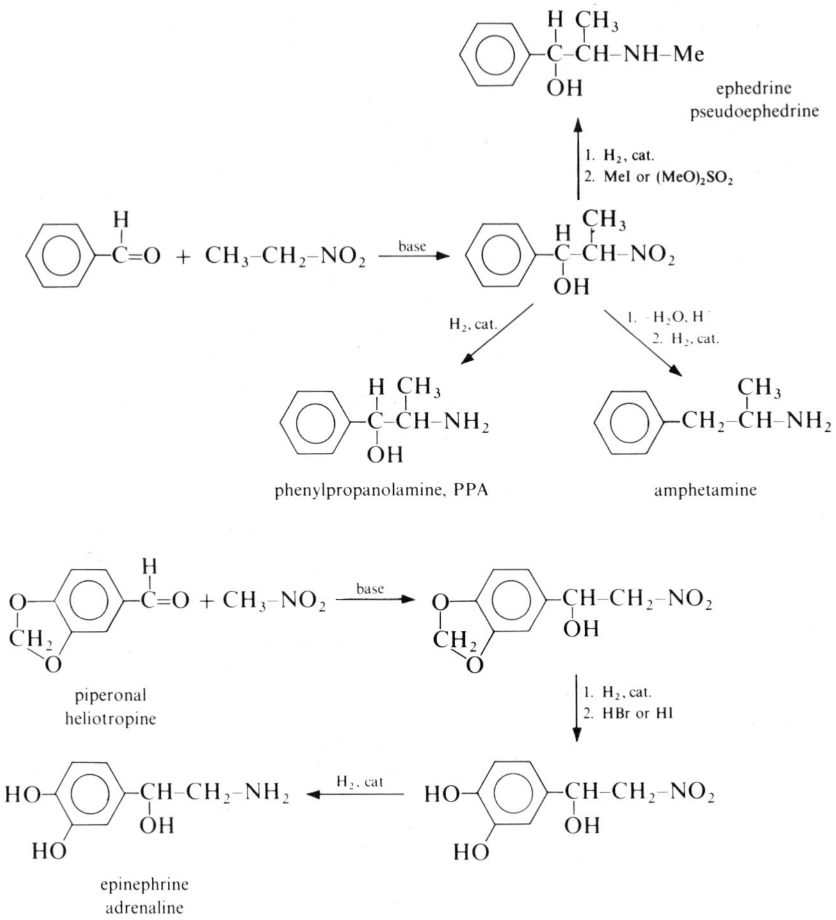

Scheme 5.3 contains chemical structures and reactions. The labeled compounds and reaction conditions are:

ephedrine
pseudoephedrine

$CH_3-CH_2-NO_2$ + benzaldehyde, base →

1. $H_2$, cat.
2. MeI or $(MeO)_2SO_2$

1. $-H_2O$, $H^+$
2. $H_2$, cat.

$H_2$, cat.

phenylpropanolamine, PPA

amphetamine

piperonal
heliotropine

$CH_3-NO_2$, base →

1. $H_2$, cat.
2. HBr or HI

$H_2$, cat.

epinephrine
adrenaline

**Scheme 5.3**

## 5.4  AN OVERVIEW OF THE CHEMICAL USES OF PROPYLENE

As is also true in the case of ethylene, polymers are directly and indirectly responsible for much of the 18B-lb demand for propylene, and this demand is driven by the PP boom (P. Savage, *CW* 4/13/88, 36–38).

About 36% of propylene is converted into the homopolymer PP, most of which becomes the stereoregulated rather than the amorphous product. Another 6% is used in ethylene–propylene copolymers and for the preparation of dimers, trimers, and oligomers of propylene.

About 16% of propylene is subjected to ammoxidation, and the bulk of the resulting AN either becomes polyacrylonitrile (PAN) or is incorporated in copolymers of AN such as ABS, SAN, and NR.

About 11% of propylene is epoxidized to PO, and a large portion of PO is converted to PPG—an important constituent of PURs.

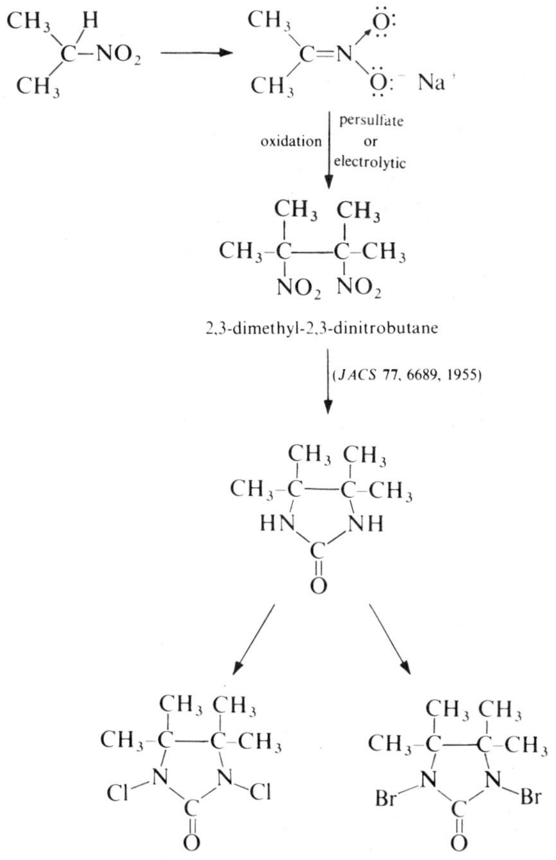

$$\underset{CH_3}{\overset{CH_3}{\underset{|}{C}}}\!\!-\!\!NO_2 \quad \longrightarrow \quad \underset{CH_3}{\overset{CH_3}{C}}\!\!=\!\!N\!\!\underset{\ddot{O}:}{\overset{\ddot{O}:}{}}\ Na^+$$

oxidation | persulfate or electrolytic

CH₃ CH₃

$$CH_3\!-\!\underset{NO_2}{\overset{CH_3}{\underset{|}{C}}}\!\!-\!\!\underset{NO_2}{\overset{CH_3}{\underset{|}{C}}}\!\!-\!CH_3$$

2,3-dimethyl-2,3-dinitrobutane

(*JACS* 77, 6689, 1955)

$$CH_3\!-\!\underset{HN}{\overset{CH_3}{\underset{|}{C}}}\!\!-\!\!\underset{NH}{\overset{CH_3}{\underset{|}{C}}}\!\!-\!CH_3$$
C=O

$$CH_3\!-\!\underset{Cl-N}{\overset{CH_3}{\underset{|}{C}}}\!\!-\!\!\underset{N-Cl}{\overset{CH_3}{\underset{|}{C}}}\!\!-\!CH_3 \qquad CH_3\!-\!\underset{Br-N}{\overset{CH_3}{\underset{|}{C}}}\!\!-\!\!\underset{N-Br}{\overset{CH_3}{\underset{|}{C}}}\!\!-\!CH_3$$
C=O                              C=O

**Scheme 5.4**

About 9% of propylene is used to alkylate benzene to yield cumene, and the latter becomes a source of phenol—the building block of PF—and acetone—the building block of methyl methacrylate monomer and hence of the polymer PMM, as well as IPDI—another building block of PURs.

About 7% of propylene is subjected to hydroformylation, and the main end products of this transformation are *n*-butyraldehydes and isobutyraldehydes and their respective alcohols and 2-ethylhexanol (2-EH).

The solvent and chemical intermediate isopropyl alcohol (IPA) occupies the next position among the industrial derivatives of propylene with 6%, and IPA is a supplementary source of the above-mentioned acetone.

The conversion of propylene to allyl alcohol, acrolein, acrylic acid, and epichlorohydrin, (the last-mentioned compound by way of allyl chloride) consume another 6% of propylene.

These major uses account for about 91% of the total demand for propylene, and this still leaves about 9% for miscellaneous uses. The detailed discussion of the industrial role of propylene follows, more or less, the order of decreasing demand except that some interesting, and significant derivatives of the major products are also included.

### 5.4.1 Production of Polypropylene (PP) and Copolymers, Dimers, Trimers, and Oligomers of Propylene

The production of PP is increasing worldwide at an impressive rate (10%; see P. L. Layman, *CEN* 2/1/88, 15–18), and the same is true of high-impact propylene copolymers (*CMR* 7/27/87). The latter may contain 8–28% ethylene, and they are rather suitable for structural applications in the automotive industry (e.g., body trim panels and interior trim, battery components, bumper covers, airducts), manufacture of furniture parts and household appliances such as washing machines and fabrication of food containers suitable for microwave oven use. This high-impact propylene is actually a block polymer of propylene and ethylene, and its production requires a 20–30% higher capital investment than does the production of a random propylene–ethylene copolymer or a PP homopolymer because of the obvious need to control the feed of the monomers. Consequently, while the cost of general-purpose molding grade PP homopolymer is priced (12/87) at 45¢, high-impact PP copolymer costs 53¢. The advances in PP technology are reminiscent of the preceding evolution of PEs—from LDPE, to HDPE, to LLDPE, to HMWPE, to UHMWPE, and so on, as discussed in 4.3.1. One might say that the PP technology is catching up and meeting the challenge of PE materials.

Because of the above-mentioned susceptibility of allylic hydrogens to oxygen-containing reagents that include polymerization catalysts, the large-scale polymerization of propylene had to await the development in 1950 of Ziegler–Natta catalysis (4.3.1). For this achievement and the elucidation of the stereochemistry of the resulting polypropylene structure, Karl Ziegler and Giulio Natta shared the Nobel prize.

First, we must clarify that the common polymerization pattern of propylene tends to produce a head-to-tail polymer, but then, there can exist stereoregulated or *tactic PP* in two distinct isomeric forms. In *isotactic PP* the methyl groups located on alternating carbons of the polymer chain (viewed along an elongated, linear, but zigzagging chain) are all oriented in the *same* direction. In *syndiotactic PP*, the same methyl groups are oriented in *opposite* directions on alternating carbons. Then, of course, one can also have nonstereoregulated or *atactic PP*, in which the methyl groups are oriented in a random fashion. Tactic PP tends to have a high degree of crystallinity and, consequently, is rather rigid, while atactic PP is soft and tacky. The differences in the thermal and mechanical properties of commercially available stereoregulated (isotactic) and atactic PP is enormous and reflect the difference between success and failure in some marketplace applications. Thus, while atactic PP begins to soften and lose tensile strength below 75°C, isotactic PP does not melt until the temperature approaches about 165°C, and it can be used safely up to 120°C (its "heat-distortion temperature") before the shape of the solid object becomes distorted. The presence of small amounts (2–7%) of ethylene during the polymerization of propylene suffices to destroy crystallinity in the random copolymer and produces an atactic PP useful as a packaging material and for the fabrication of a variety of rather flexible, optically clear, consumer products destined for use at relatively low temperatures. When the molecules of isotactic PP become aligned by subjecting the semimolten ("plastic") material to unidirectional flow accompanied by some stretching, its tensile strength increases to as much as 4300–5500 psi. The properties of isotactic PP can be further improved by means of *biaxial orientation* of the polymer molecules, and this, obviously, is done most easily in the case of films referred to as "BOPP". As one may well expect, by 1970 the Ziegler–Natta catalysts were challenged by the appearance of competitive quasi-Ziegler–Natta catalysts that offer better yields and improved stereospecificity.

In addition to these structural applications, which use either injection or extrusion molding, PP is processed to give filaments employed in the manufacture of carpets, indoor–outdoor floor coverings, upholstery, automobile seat covers, sportswear, geo-textiles (i.e., textiles used to stabilize soil), and so on. PP film is used for food packaging in competition with cellophane and other materials, in disposable diapers (1.2), and in other products. The use of a low-cost PP sheet for "blister packaging" of pharmaceutical products was announced recently by Exxon (*CMR* 8/10/87). This technology consists of thermoforming, filling (e.g., with medicinal capsules), and sealing the contents in a single operation that produces an attractive package highly resistant to moisture.

The domestic production of PP is projected to grow beyond the 5.8B-lb level indicated in 1.1.1, and most of it consists of the tactic homopolymer with atactic PP contributing only about 6% of the total. The high-impact PP block polymer is a relatively new commercial development, and this material finds increasing applications in the manufac-ture of automobile components, luggage, and other items exposed to mechanical abuse.

The worldwide demand for PP is currently estimated at nearly 20B lb, and the two leading production processes are the previously mentioned (4.3) Union Carbide Unipol process and Himont's Spheripol process, which also uses a gas-phase reactor but depends on a high-activity catalyst (SHAC) developed by Shell about 10 years ago. These two PP processes are competing head-to-head on the international market (P. Savage, *CW* 11/25/87, 99–101), but they do not stand alone: a pact announced (*CW* 10/2/87) between three PP producers—BASF in West Germany, ICI of Great Britain, and National Distillers and Chemical of the United States—is designed to pool their experience and technical resources for the improvement of PP production technology.

The oxidative degradation of PP films can be inhibited by the incorporation of antioxidants such as butylated hydroxytoluene (BHT), which functions as a scavanger of free radicals (represented by $Q\cdot$) generated from PP and molecular oxygen. The chemistry of this photosensitized process can be represented in the fashion illustrated in Scheme 5.5.

Polypropylene films can be also stabilized by means of UV light-absorbing com-pounds that convert photochemical energy into vibrational–rotational energy released as heat. Hindered-amine light stabilizers, or HALS, and other types of stabilizers of polymers against UV-induced degradation are mentioned elsewhere (10.2).

The stabilization of PP products, against oxidative degradation while they are exposed to outdoor uses reminds us of the worldwide environmental pollution problem caused by mishandling of polymer wastes. Since human behavior is difficult to control, chemistry can contribute to the solution of this societal problem by creating *photo-degradable polymers*. This can be achieved by incorporating.in the macromolecules comonomers which become active sites for photochemically induced degradation of PP, PEs, and other POs, as well as PVC, PS, and other polymers. One such class of comonomers is vinyl ketones such as methyl vinyl ketone ($Me–CO–CH=CH_2$) and methyl isopropenyl ketone ($Me–CO–C(Me)=CH_2$). These compounds can be prepared by means of the well-known (*SZM-I*, 330) thermal decarboxylation of a mixture of acetic and acrylic (5.4.5) or methacrylic (6.3.1.4) acids:

$$Me–CO_2H + CH_2=CH–CO_2H \rightarrow Me–CO–CH=CH_2 + CO_2 + H_2O$$

$$Me–CO_2H + CH_2=C(Me)–CO_2H \rightarrow Me–CO–C(Me)=CH_2 + CO_2 + H_2O$$

The desired reaction is catalyzed by the oxides of calcium, barium, and thorium (see *OS-I*, 192; *OS-II*, 389). Pharmaglobe Labs Ltd. of Rexdale, Ontario announced (*CMR* 2/15/88

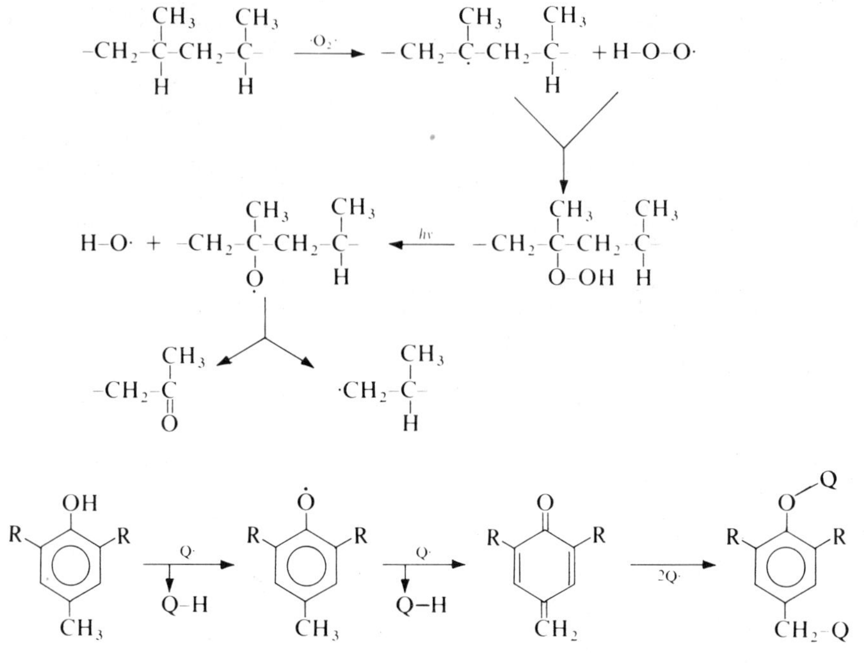

where R = *t*-Bu–
Q· = H–O–O·, H–O·, ·$O_2$·

**Scheme 5.5**

and 3/28/88) the production of methyl isopropenyl ketone (MIPK) on a 6000 kg/month scale. MIPK is priced at \$11.80/lb.

The *dimerization* of propylene to yield isoprene is of potential industrial importance, although currently most of the latter is isolated as a by-product of cracking operations (Fig. 2.2). Trialkyl aluminum catalyzes the formation of 2-methyl-1-pentene, and the latter is pyrolyzed with an interesting elimination of methane, as shown in Scheme 5.6.

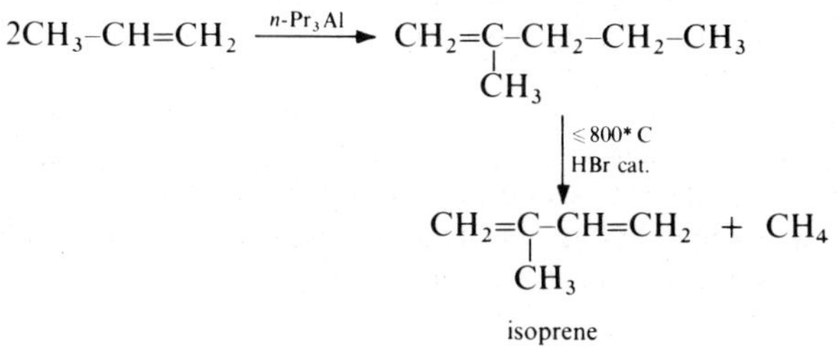

**Scheme 5.6**

Of great practical importance in connection with the production of alkylate gasoline (5.1) is the acid-catalyzed dimerization, trimerization, and so on, of propylene (also in the company of isobutylene; see 6.3.1.2). This process is accompanied by alkylation of alkanes that possess a secondary or preferably tertiary hydrogen (6.3.1.1). The dimerization and alkylation processes are complicated by rearrangements of intermediate carbocations illustrated by means of the simplest pair of reactants, namely, propylene and propane, in Scheme 5.7. Needless to add, things become more complex when higher than $C_3$ building blocks are involved. The olefinic products are hydrogenated before use as gasoline components in order to avoid formation of gums and tars on long-range exposure to oxygen. Hydrogen fluoride and sulfuric acid are the most common alkylation catalysts.

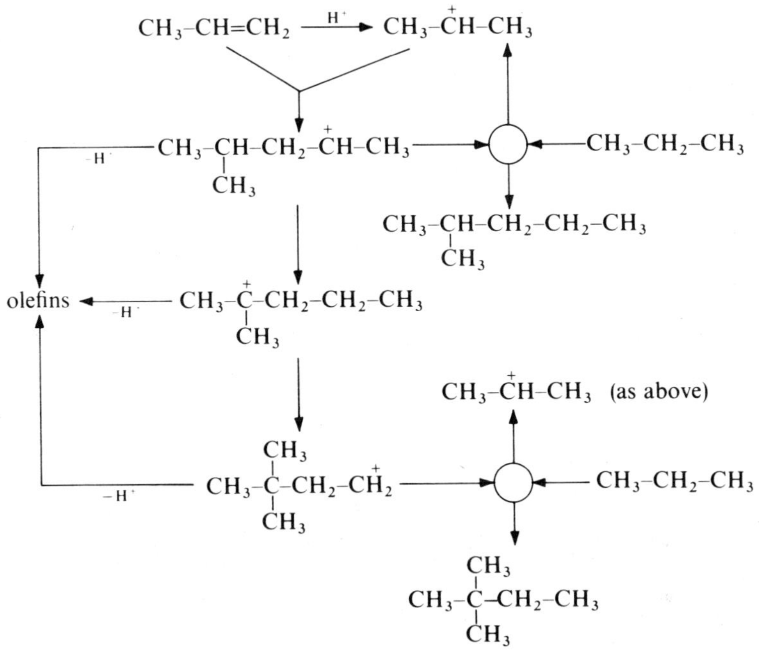

$$CH_3-CH=CH_2 \xrightarrow{\;H\cdot\;} CH_3-\overset{+}{C}H-CH_3$$

$$CH_3-\underset{\underset{CH_3}{|}}{CH}-CH_2-\overset{+}{C}H-CH_3 \longrightarrow \bigcirc \longleftarrow CH_3-CH_2-CH_3$$

$$CH_3-\underset{\underset{CH_3}{|}}{CH}-CH_2-CH_2-CH_3$$

$$\text{olefins} \xleftarrow{\;-H\cdot\;} CH_3-\underset{\underset{CH_3}{|}}{\overset{+}{C}}-CH_2-CH_2-CH_3$$

$$CH_3-\overset{+}{C}H-CH_3 \quad \text{(as above)}$$

$$CH_3-\underset{\underset{CH_3}{|}}{\overset{\overset{CH_3}{|}}{C}}-CH_2-\overset{+}{C}H_2 \longrightarrow \bigcirc \longleftarrow CH_3-CH_2-CH_3$$

$$CH_3-\underset{\underset{CH_3}{|}}{\overset{\overset{CH_3}{|}}{C}}-CH_2-CH_3$$

**Scheme 5.7**

In the absence of propane, the oligomerization of propylene in the presence of phosphoric acid and other acidic catalysts produces propylene dimers consisting mainly of 4-methyl-1-pentene (4-MPI) and 4-methyl-2-pentene:

$$(CH_3)_2CH-CH_2-CH=CH_2 \quad \text{and} \quad (CH_3)_2CH-CH=CH-CH_3$$

4-MPI is currently in demand as a copolymer of ethylene in LLDPE and gives a high-melting (240°C) homopolymer useful in the fabrication of microwave-oven utensils, medical equipment, and so on. This poly(methylpentene) (PMP) material—Mitsui's TPX—is claimed to have the lowest density (0.83 g/mL) of all commercially available thermoplastics, and the fiberglass-reinforced product, Mitsui's FRTPX, sells for $3.00. Phillips also produces PMP under the trade name "Crystalor."

The oxidative dimerization of propylene by means of trialkylaluminum produces 2-methyl-1,4-pentadiene, a terminal diolefin that is a homolog of isoprene. This material is the building block used by Amoco to prepare one of its aromatic dicarboxylic and tricarboxylic acids (9.3.2.2), namely, 1,1,3-trimethyl-3-phenylindan-4′,5-dicarboxylic acids (PIDA). The formation of the diolefin and PIDA is shown in Scheme 5.8.

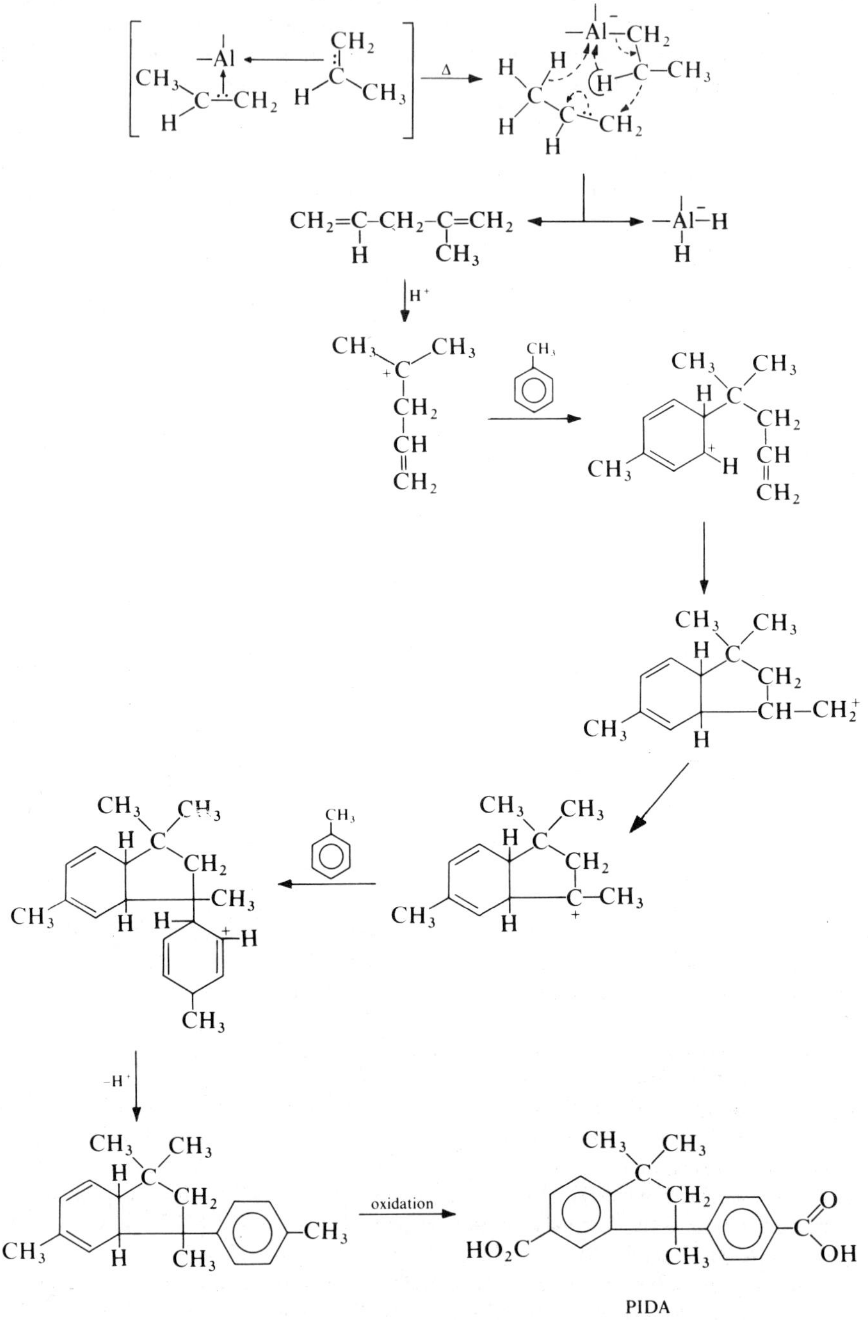

PIDA

Scheme 5.8

Propylene trimers are also known as isononenes and are produced at a 430 MM-lb level to be transformed by hydration to isononyl alcohol (48¢) or by Friedel–Crafts alkylation of phenol to nonylphenol (53¢)—a building block of surfactants obtained by ethoxylation or sulfation; see Scheme 5.9. The demand for nonylphenol is about 200 MM lb, and, besides surfactants, it is also used an antioxidant for rubber, some plastics, and lubricating oils. The ethoxylated product with an average number of nine EO units is the spermatocide nonoxynol or nonoxynol-9 present in contraceptive sponges.

Another category of propylene trimer derivatives is Pennwalt's di-$t$-nonyl polysulfide (TNPS), a mixture of isomeric $(C_9H_{19})_2S_5$ (87¢) that serves as a lubricating oil additive and, most likely, is prepared by addition of hydrogen sulfide to produce $t$-mercaptans that are then treated with sulfur monochloride and sulfur. In an analogous fashion propylene is oligomerized to tetramers, or isododecenes, that give rise to dodecylphenol (48¢).

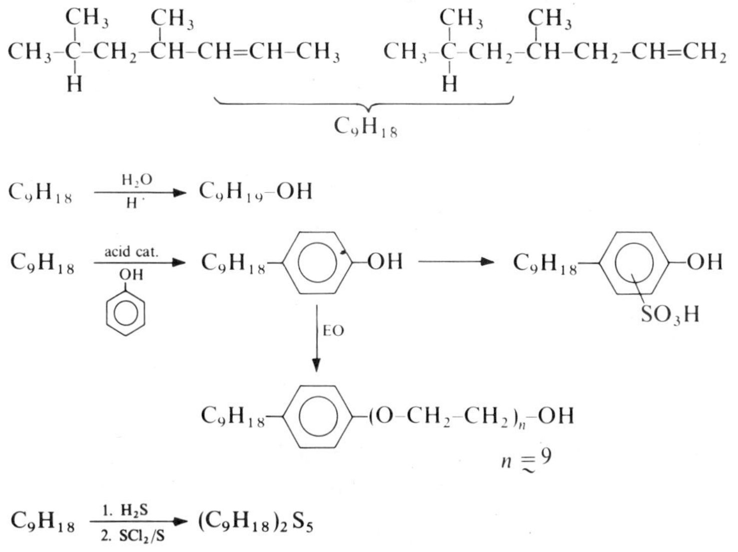

Scheme 5.9

Amoco supplies several polypropenes with an average degree of polymerization of 11–20 propylene units.

### 5.4.2  Propylene Oxide (PO)

5.4.2.1  Production and Prospects

The gloomy image of EO derivatives is contrasted by the glamorous prospects of the analogous structures derived from PO (C. J. Verbanic, $CB$ 2/88, 24–29). The latter include propylene glycol (PG), PO-based glycol ethers and their acetates, amines, and so on. Thus, propoxylation, the reaction of PO with substrates that contain an active hydrogen, is challenging the long-established ethoxylation reactions. Also, PO derivatives are filling the breach caused by health hazards attributed to methylene chloride.

The advantage of a direct oxidation of ethylene to EO by $O_2$ is compensated by a formation of valuable coproducts in the case of PO production. As we shall see below, these coproducts currently are either styrene, $t$-butyl alcohol, or methyl ethyl ketone.

Thus, for example, in the case of the evolving chemical economy of South Korea, there are plans (*CW* 9/17/86, *CMR* 1/19/87) to produce simultaneously 100,000 t$^m$ of PO and 225,000 t$^m$ of styrene based on propietary ARCO technology that starts with oxygen, ethylbenzene, and propylene, as shown in Scheme 5.10. Naturally, other convenient hydroperoxides, such as *t*-butyl hydroperoxide (6.2.5), may also be used to convert propylene to PO:

$$t\text{-Bu}-\text{O}-\text{OH} + \text{CH}_3-\text{CH}=\text{CH}_2 \rightarrow \text{CH}_3-\underset{\backslash_{\text{O}}/}{\text{CH}-\text{CH}_2} + t\text{-Bu}-\text{OH}$$

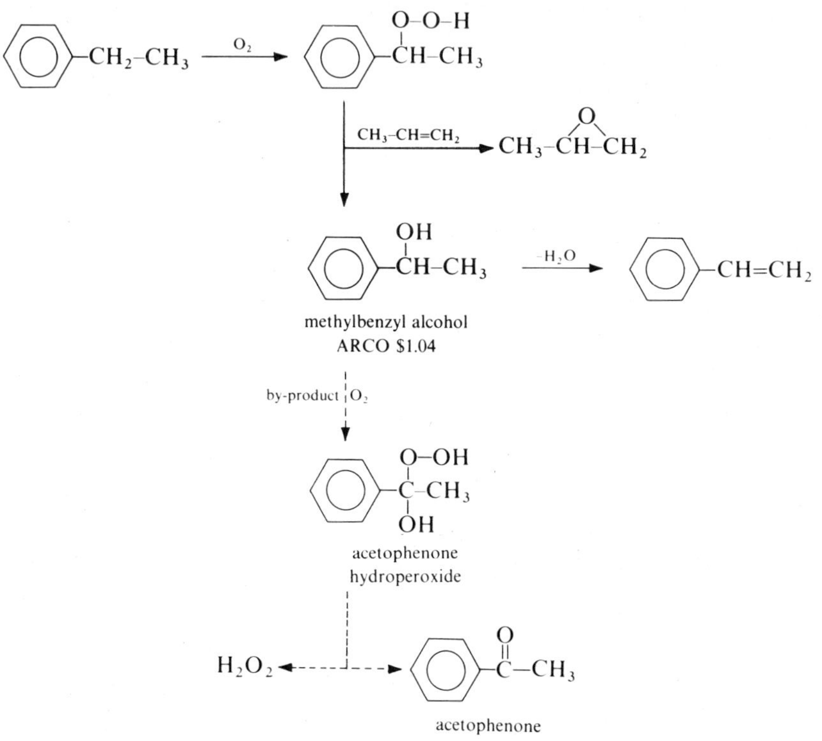

**Scheme 5.10**

This process starts with isobutane and oxygen, and in addition to PO, (47.5¢), it gives *t*-butyl alcohol (TBA; 70¢). The latter can be used as a gasoline enhancer and as a source of chemically pure (99%) isobutylene, (32¢), suitable for polymerization to polyisobutylene (unlike the crude isobutylene obtained by cracking of the butane fraction of LPG).

Again, ARCO's propietary technology is responsible for a $340 million plant in France (*CMR* 6/6/88) designed to produce 330MM lb of PO together with 790MM lb of TBA, as well as 110MM lb of the downstream product propylene glycol. In fact, this process can start with a mixture of isobutane and *n*-butane since only the former is oxygenated and the latter is recovered (as shown in Fig. 2.2).

Another candidate for the generation of a hydroperoxide needed to produce PO is *sec*-butyl alcohol (36.5¢). It is first transformed to a rather stable methyl ethyl ketone hydroperoxide (6.3.2.4) and eventually leads to methyl ethyl ketone (MEK; 28.5¢); see Scheme 5.11.

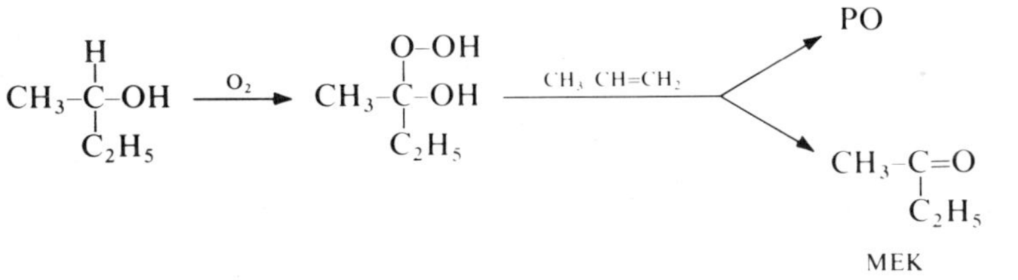

**Scheme 5.11**

The analogous process, which starts with IPA ($1.51/gal or ~23¢/lb), yields acetone (24¢), and the use of the rather stable hydroperoxide of cumene as an oxidizing agent produces the expected alcohol, which, when subjected to dehydration, yields α-methylstyrene (9.2.1), although the latter is also obtained by dehydrogenation of cumene (20¢). The hydroperoxide processes leading to PO were developed by Oxirane (a subsidiary of Halcon International and ARCO). After the technological failure to convert ethylene directly to EG by means of diacetoxylation, Oxirane was acquired completely by ARCO, and the latter alone has the capacity to produce about half of the 3B-lb demand for PO in the United States. Not accidentally, ARCO is also producing *t*-butyl alcohol, acetophenone, $C_6H_5-CO-CH_3$, and methylbenzyl alcohol or α-hydroxyethylbenzene $[C_6H_5-CH(OH)CH_3]$.

The traditional method of converting propylene to PO involves the addition of hypochlorous acid (generated *in situ* from chlorine and water) followed by the treatment of the isomeric chlorohydrins with base, as shown in Scheme 5.12. In spite of many technological variations of the hypochlorous acid route, it is being replaced by the hydroperoxide process except for Dow, which remains faithful to the chlorohydrin intermediates and announced recently (*CW* 7/2/86) an increase of its PO production capacity in Stade, West Germany to 420,000 $t^m$/yr.

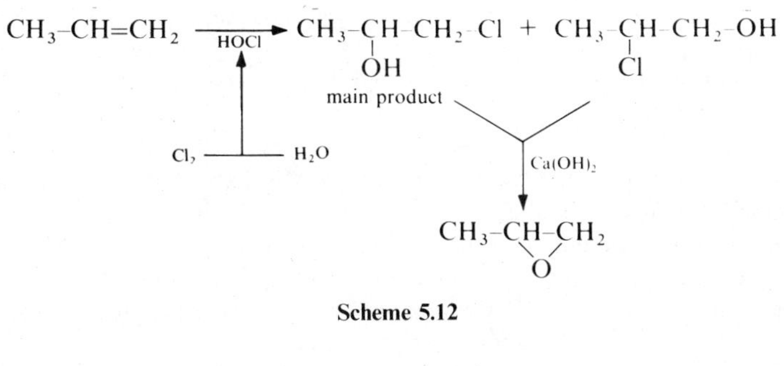

**Scheme 5.12**

The multitude of routes to PO, and the acceptance of its derivatives in place of the traditional EO derivatives, assures a prosperous future for this chemical. At this time the United States PO production capacity is 2.9B lb, shared almost equally by ARCO and Dow.

5.4.2.2   The PO Building Block

**A. ADDITION REACTIONS OF THE PO OXIRANE SYSTEM.** As mentioned above, some of the uses of PO parallel those of EO, except that the PO products differ in three important aspects:

1. By no stretch of imagination can the derivatives of PO be converted to those of EO that carry the stigma of EG toxicity under physiological conditions, and hence PO derivatives can be employed in the food industry.

2. The presence of the additional methyl group in PO derivatives decreases hydrogen bonding at the residual ethereal oxygen of the $-CH_2-CH(CH_3)-O-$, moiety and oligomers and polymers of PO are hydrophobic; also, methyl branching reduces rotational freedom along the $-C-O-C-$ bonds and causes the PO chains to be more "stiff" than the analogous EO chains.

3. The unsymmetrical nature of PO can give rise to either primary or secondary hydroxyl-containing derivatives depending on the nature of the reactant and the pH of reaction conditions. Thus, nucleophilic reactants attack preferentially the unsubstituted terminal of the oxirane ring with the resultant formation of secondary alcohol products. On the other hand, acidic conditions proceed by way of protonation of the oxirane and tend to induce the formation of primary hydroxyl terminals and derivatives of the more stable transient, secondary carbon cations. Ambiguous reaction conditions give a mixture of both types of products. The primary or secondary nature of the hydroxyl groups may be crucial to further chemical transformations of the PO derivatives [see, for example, the reason for tipping or capping of PO oligomers when these are used as polyols in the assembly of PURs (3.8.1, PUR)].

About 55% of PO is utilized for the production poly(propylene glycol ethers)—the favored polyol ingredients of flexible and semirigid PUR foams (interlude in Section 3.8) used in the manufacture of upholstery, bedding, automobile components, and so on:

$$HO-CH_2CH(CH_3)-[O-CH_2-CH(CH_3)]_n-OH \qquad (MW = 10^3-10^4)$$

The improvements in RIM and RRIM technology promise to increase future demand of this class of PO derivatives.

The production of PG and its simple derivatives consumes about 25% of PO:

$CH_3CH(OH)CH_2OH$, PG, 40¢, USP 43¢, b.p. 187°C, versus 196–198°C for EG, 31¢ (no USP grade!)

The demand for PG in the United States is nearly 600MM lb, and PG is employed in the formulation of UPRs (45%); in pharmaceuticals and food products, including semimoist pet foods ($\sim$ 15%), and about 3–5% each as a humectant of tobacco, plasticizer, coatings, and brake, hydraulic, and other functional fluids. At this point in time, the United States still exports about 18% of its PG production, but this market is likely to shrink as hydrocarbon-rich developing countries enter the ranks of producers.

Propylene glycol is converted to the corresponding carbonate by means of phosgene (3.5.3.6), but poly(propylene carbonate) is produced from PO and $CO_2$ (3.7.2.5).

Propylene carbonate is an excellent dipolar aprotic solvent (3.5.3.1), ARCO's slightly yellow (due to traces of catalyst) product Arconate costs 52.5¢, while Texaco's colorless product is priced at 77¢ (*CMR* 5/18/87) propylene carbonate is used, among other things, for the removal of $CO_2$ from the water-gas-shift reaction mixture, and as a reactive diluent of PMDI—the isocyanate-based wood binder used in the manufacture of reconstituted wood boards in competition with PF-based binders. Also, up to 10 parts of propylene carbonate can be allowed to react with 100 parts of polyol to give improved PUR foams because the open-cell structure promotes air circulation.

Propylene oxide is converted to PG ethers:

$$CH_3\text{--}CH\text{--}CH_2 + HO\text{--}CH_2CH(CH_3)OH \rightarrow HO\text{--}CH(CH_3)CH_2\text{--}O\text{--}CH_2CH(CH_3)\text{--}OH$$

dipropylene glycol, DPG, 40.5¢

and

$$HO\text{--}CH(CH_3)CH_2\text{--}[O\text{--}CH_2CH(CH_3)]_2OH$$

tripropylene glycol, TPG, 64¢

The demand for DPG is only about 60MM lb, and both DPG and TPG are obtained as by-products of PG production. They are used primarily in the formulation of unsaturated polyester and alkyd resins as plasticizers and solvents.

Propylene glycol methyl ethers are formed in a routine manner:

$CH_3\text{--}O\text{--}CH_2\text{--}CH(CH_3)\text{--}OH$, propylene glycol monomethyl ether, 49¢, and the corresponding acetate, PMA, Dow 63.5¢.

$CH_3\text{--}O\text{--}CH_2\text{--}CH(CH_3)\text{--}O\text{--}CH_2\text{--}CH(CH_3)\text{--}OH$, dipropylene glycol monomethyl ether (DPMG; 46¢). DPMG has recently been introduced in solvent-based coating systems to replace the environmentally suspect analog based on EG. The same is true of the analogous *n*-butyl ethers that compete with butyl cellosolve and butyl carbitol (4.3.6).

*t*-Bu$\text{--}O\text{--}CH_2\text{--}CH(CH_3)\text{--}OH$, propylene glycol mono-*t*-butyl ether, ARCO's Arcosolv PTB, 65¢.

Another ARCO solvent available on the marketplace is a mixture of the above-mentioned propylene glycol methyl ether and the corresponding acetate and is proposed as a replacement for methylene chloride in the cleanup of PU residues obtained during the manufacture of low-density, flexible foams (*CMR* 4/27/87). Similarly, residues of high-density spray foam are said to be cleaned by a mixture of PMA and DPMG.

Also of interest are the amino alcohols derived from PO:

Monoisopropanolamine, $CH_3\text{--}CH(OH)\text{--}CH_2\text{--}NH_2$, 66¢

*N,N*-Dimethylisopropanolamine, $CH_3\text{--}CH(OH)\text{--}CH_2\text{--}NMe_2$ (DMIPA), 88¢ (77%)

Diisopropanolamine, $[CH_3\text{--}CH(OH)\text{--}CH_2]_2NH$, 58.5¢

Triisopropanolamine, $[CH_3\text{--}CH(OH)\text{--}CH_2]_3N$, 57.5¢

Texaco produces numerous poly(oxypropylene)di-, tri-, and higher diamines ("Jeff-amines") priced mostly in the $1.40–$1.90 range:

$$H_2N\text{--}CH(CH_3)CH_2[O\text{--}CH_2CH(CH_3)]_nNH_2, \text{ where } n = 2 - 68$$

Also available are

- Analogous diamines except that they are derived from both EO and PO.
- Monoamines built up from monoalkyl ethers of ethylene glycol.
- Hydroxypropyl derivatives of diamines obtained by treatment of the above-mentioned Jeffamines with PO.
- Urea derivatives obtained by condensation of the Jeffamines with urea.
- Triamines built up from either glycerol or 1,1,1-trimethylolpropane.

Most of the amines are likely prepared by alcohol amination of the appropriate propylene glycol ethers. They are used as curing agents of epoxy resins in the formulation of PURs, especially in RIM applications, where the introduction of urea moieties favors the formation of structurally firm products. Also, these compounds serve as building blocks of emulsifiers, lubricants, fuel additives, corrosion inhibitors, in textile treatment, and so on.

As mentioned above, a poly(propylene glycol) block is not hydrophilic as the analogous poly(ethylene glycol) block because the methyl substituent inhibits hydrogen bonding with water. This difference is the basis of a series of surface-active agents developed by BASF Wyandotte and known as Pluronics. Actually, imaginative molecular engineering has produced four major classes of surfactants that can be represented by indicating the manner in which the EO and the PO blocks are assembled:

Pluronic surfactants            EO–PO–EO
Pluronic R surfactants          PO–EO–PO

and the Tetronic family of surfactants in which the EO and PO blocks are attached to a central ethylenediamine, EDA, molecule:

Tetronic surfactants            (EO–PO)$_2$–EDA–(PO–EO)$_2$
Tetronic R surfactants          (PO–EO)$_2$–EDA–(EO–PO)$_2$

Since the length of each block and the ratio of the hydrophilic and lipophilic blocks are also subject to manipulation, there results a large number of surface-active agents that exhibit optimum properties under different circumstances.

**B. REARRANGEMENT REACTION OF PO AND SOME USES OF ALLYL ALCOHOL.**
Allyl alcohol (90¢) is formed by a rearrangement of PO (47.5¢) brought about at about 280°C in the presence of lithium dihydrogen phosphate, LiH$_2$PO$_4$:

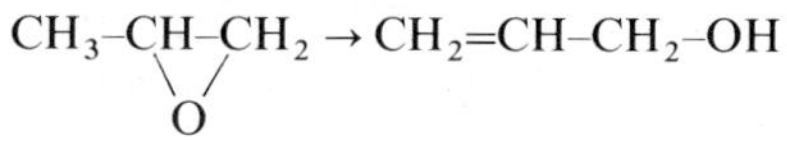

$$CH_3-CH-CH_2 \rightarrow CH_2=CH-CH_2-OH$$
$$\diagdown_{O}\diagup$$

This route to allyl alcohol seems preferable to a selective hydrogenation of acrolein (62¢) (5.4.4) or the hydrolysis of allyl chloride (65¢) (5.4.8).

Allyl alcohol is used primarily for the synthesis of diallyl and triallyl esters that serve as comonomers in radiation-induced polymerizations. These multifunctional monomers are of increasing industrial importance in the formulation of UV-curable inks for high-speed printing operations or for inclusion in composites that require a higher-energy electron-beam (EB) activation in order to penetrate bulky objects. Examples of such allyl esters are

$(CH_2=CH-CH_2-O-CO-O-CH_2CH_2)_2O$, diethylene glycol bis(allyl carbonate), Akzo's Nouryset 200 ($1.90), 6MM lb, used to cast ophthalmic lenses and thus referred to as "optical monomer." Most likely, the monomer is obtained from allyl alcohol and ethylene glycol bis(chloroformate), $O(CH_2-CH_2-O-CO-Cl)_2$.

$cis$-$(=CH-CO\cdot O-CH_2-CH=CH_2)_2$, diallyl maleate, DAM, Alcolac's Sipomer, used for second-stage cross-linking of vinyl polymers.

$o$-$C_6H_4(CO\cdot O-CH_2-CH=CH_2)_2$, diallyl phthalate.

$HO-C(CH_2-CO\cdot O-CH_2-CH=CH_2)_2$, triallyl citrate.
$\quad\quad\;\; |$
$\quad\quad C(=O)-O-CH_2-CH=CH_2$

Allyl ethers are included here with the esters because they serve the same function even though they may be synthesized by means of allyl chloride (5.4.8):

$CH_3CH_2C(CH_2OH)(CH_2-O-CH_2-CH=CH_2)_2$, 2-ethyl-2-hydroxymethyl-1,3-di-allyloxypropanediol, National Starch's trimethylolpropane diallyl ether, $3.22.

This heteromultifunctional monomer is used to incorporate a hydroxyl function into a vinyl copolymer or introduce allyl groups into a polymer assembled first by virtue of the alcohol function.

The reaction of allyl alcohol with cyanuric chloride (10.3.3), gives triallyl cyanurate (TAC), also available from National Starch ($4.40) for cross-linking of vinyl polymers.

Monofunctional allylic esters are incorporated into unsaturated polyester formulations especially when the latter are meant to be used in coatings:

$n$-$C_5H_{13}-CO-O-CH_2-CH=CH_2$, allyl caproate, $3.90
$CH_2=C(CH_3)CO\cdot O-CH_2-CH=CH_2$, allyl methacrylate

A different heterodifunctional monomer is the reaction product of allyl alcohol and epichlorohydrin (5.4.8):

Allyl glycidyl ether, $CH_2=CH-CH_2-O-CH_2-CH-CH_2$
$\quad\quad\quad\quad\quad\quad\quad\quad\quad\quad\quad\quad\quad\quad\;\; \backslash\;/$
$\quad\quad\quad\quad\quad\quad\quad\quad\quad\quad\quad\quad\quad\quad\;\;\; O$

This compound can be used to prepare the corresponding amino alcohols, glycols and other products by way of addition reactions of the epoxide function, or it can be incorporated as is in epoxy resins.

The use of allyl alcohol in the production of synthetic glycerol is described below (5.4.8).

Allyl bromide ($5.50) is a specialty alkylating agent and is obtained from the alcohol and hydrogen bromide ($7.00).

A free-radical catalyzed addition of sodium bisulfite to allyl alcohol leads to the formation of propane sultone:

$$\text{HO-CH}_2\text{-CH=CH}_2 \xrightarrow{\text{NaHSO}_3} \text{HO-CH}_2\text{-CH}_2\text{-CH}_2\text{-SO}_3\text{Na}$$

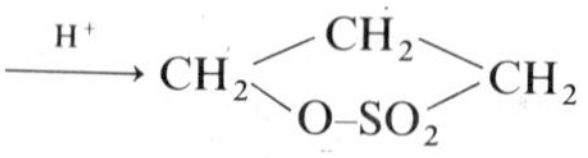

$$\xrightarrow{\text{H}^+} \text{CH}_2 \begin{array}{c} \diagup \text{CH}_2 \diagdown \\ \diagdown \text{O-SO}_2 \diagup \end{array} \text{CH}_2$$

Propane sultone is the intermediate for the production of some amphoteric or acidic surfactants available from Raschig:

$$\text{CH}_3(\text{CH}_2)_{11}\text{-NMe}_2 \rightarrow \text{CH}_3(\text{CH}_2)_{11}\text{-}\overset{+}{\text{N}}\text{Me}_2\text{-CH}_2\text{CH}_2\text{CH}_2\text{-SO}_3^-$$

Ralufon DCH, $1.50 as a 50% solution

$$\text{iso-C}_{13}\text{H}_{27}\text{-O-(CH}_2\text{CH}_2\text{O)}_4\text{H} \rightarrow \text{iso-C}_{13}\text{H}_{27}\text{-O-(CH}_2\text{CH}_2\text{O)}_4\text{-O(CH}_2)_3\text{SO}_3\text{H}$$

K salt is Raulfon F4-1

also used as a leveling agent in electroplating of nickel:

$$\text{CH}_3(\text{CH}_2)_{10}\text{CO-NH-(CH}_2)_3\text{NMe}_2 \rightarrow$$

$$\text{CH}_3(\text{CH}_2)_{10}\text{CO-NH-(CH}_2)_3\overset{+}{\text{N}}\text{Me}_2^-(\text{CH}_2)_3\text{SO}_3^-$$

Ralufon CA, $2.50 as a 50% solution

An interesting utilization of allyl alcohol is the preparation of a fire-retarding monomer that can be incorporated into epoxy resins or, after transformations of the epoxide function, into other types of polymers. The first step represents a typical free-radical catalyzed *chain reaction*: an addition of carbon tetrachloride to the terminal double bond of allyl alcohol. The free radicals are generated by thermal decomposition of benzoyl peroxide.

*Initiation step*:

$$\text{R-CO-OO-CO-R} \rightarrow 2\text{R}\cdot + 2\text{CO}_2$$

$$\text{R}\cdot + \text{CCl}_4 \rightarrow \text{R-Cl} + \text{Cl}_3\text{C}\cdot$$

where R represents the phenyl radical, C$_6$H$_5$·.

*Propagation step*:

$$\text{Cl}_3\text{C}\cdot + \text{CH}_2\text{=CH-CH}_2\text{-OH} \rightarrow \text{Cl}_3\text{C-CH}_2\text{-}\overset{\cdot}{\text{C}}\text{H-CH}_2\text{-OH}$$

$$\text{Cl}_3\text{C-CH}_2\text{-}\overset{\cdot}{\text{C}}\text{H-CH}_2\text{-OH} + \text{CCl}_4 \rightarrow \text{Cl}_3\text{C-CH}_2\text{-CHCl-CH}_2\text{-OH} + \text{Cl}_3\text{C}\cdot$$

The product of this step is now subjected to an aqueous base that converts the chlorohydrin moiety into an epoxide:

$$\text{Cl}_3\text{C-CH}_2\text{-CHCl-CH}_2\text{-OH} \rightarrow \text{Cl}_3\text{C-CH}_2\text{-CH-CH}_2 \begin{array}{c} \diagdown \diagup \\ \text{O} \end{array}$$

The resulting trichlorobutylene oxide, Olin's TCBO, is a beautiful example of industrial chemistry that combines homolytic and heterolytic reactions to achieve the desired purpose of assembling a fire-retardant with a reactive structural moiety that allows the molecule to be introduced into a variety of products.

For sake of completion, the *termination step* of the chain reaction can be the formation of a covalent bond between any of the intermediate radicals shown above.

Another example of imaginative industrial organic chemistry is offered by the recent announcement (*CW* 4/15/87) that ARCO plans the production of 75MM lb of 1,4-butanediol by means of Japan's Kuraray technology, which subjects allyl alcohol to hydroformylation:

$$HO-CH_2-CH=CH_2 + CO/2H_2 \rightarrow HO-CH_2-CH_2-CH_2-CH_2-OH$$

The conversion of allyl alcohol priced at 90¢ to 1,4-butanediol, priced at 80¢, does not seem to be an economically sound idea, but the picture becomes more favorable when we consider PO (47.5¢) to be the feed stock of an integrated process.

### 5.4.3    Acrylonitrile (AN)

5.4.3.1    An Overview of Ammoxidation

The ammoxidation of propylene to acrylonitrile was introduced in 1961 by Sohio:

$$CH_2=CH-CH_3 + NH_3/O_2 \rightarrow CH_2=CH-C\equiv N + HCN + CH_3-C\equiv N$$

The reaction is carried out at about 400°C in the presence of phosphomolybdate and other catalysts. Acrylonitrile is obtained in better than 80% yields, but the coproducts, acetonitrile (53¢) (4.6.2) and HCN (50¢) (3.10) are also valuable. This revolutionary process has eclipsed all previous tedious routes to AN. It currently satisfies the 2.5B-lb demand for AN (40¢) and has had a profound effect on many areas of industrial organic chemistry:

- It provided a relatively inexpensive source of acrylonitrile that, in turn, became an attractive component of a variety of polymers.
- It stimulated the industrial utilization of the cyanoethylation reaction that allows a substrate that contains an active hydrogen, symbolized by Q–H, to be transformed easily into Q–CH_2–CH_2–C≡N or its hydrolysis or hydrogenation products.
- It became a convenient source of HCN and catalyzed an expansion of its uses (3.10).
- It became a model for analogous ammoxidations of other allylic as well as benzylic systems.
- It provided an easy access to two dipolar, aprotic solvents: acetonitrile and benzonitrile.

Ammoxidation replaced the tedious preparations of AN based on acetylene, acetaldehyde, and ethylene oxide. It is reminiscent of the Andrussov process (*ONR*-3) in which methane is converted to HCN and can also be applied to propane but understandably, with much lower yields of AN.

The ammoxidation reaction has been extended to other aliphatic systems:

Isobutylene gives methacrylonitrile.

2-Methyl-2-butene is converted to 3-methyl-2-butenonitrile.

Also, other amylenes (7.4) and a variety of methyl-substituted aromatic systems such as the xylenes (which give the corresponding dinitriles of isophthalic and terephthalic acids; 9.2.2), and 2- and 4-methylpyridines or α- and γ-picolines (which yield 2- and 4-cyanopyridine, 9.2.1), are obtained by means of ammoxidation. Furthermore, partial hydrogenation of the aromatic nitriles produces aminomethyl derivatives of the nitrile precursors, and complete hydrogenation gives rise to the corresponding aminomethyl-cyclohexanes (8.3). Both types of cyclic amines have become significant building blocks of PURs after the amino groups are converted to isocyanato functions.

### 5.4.3.2   The Acrylonitrile Building Block

The telltale involvement of AN in the assembly of rather complex organic structures is the presence of a linear three-carbon moiety with either a terminal nitrogen group or a transformation product of the original cyano function. Also, AN is a vinyl monomer with a polar and chemically rather inert cyano group, and for these reasons alone, it is an important and valuable component of many polymers.

The worldwide interest in AN is reflected by the 7B-lb demand and the extraordinary level of exports: about 40% of the U.S. production of 2.5B lb. Of the remaining 60%, 28% is consumed in the domestic production of poly(acrylonitrile) (PAN) present in the so-called acrylic and modified acrylic (modacrylic) textile fibers that are converted to bulky yarns known to consumers as Chemstrand's Acrilan and Du Pont's Orlon. These acrylic wool-substitutes constitute about 20% of all synthetic noncellulosic textile fibers. Because of the chemical inertia of the cyano function, and in order to facilitate dyeing of the fiber, small amounts of either acidic or basic monomers, or monomers that provide polarizable structural features, are copolymerized with AN. Typical comonomers used for that purpose are some substituted styrenes, vinyl pyridines, acrylamide, vinyl acetate, iso-butylene, and vinyl chloride. It is customary to refer to the fiber as "modacrylic" when the concentration of such comonomers exceeds 15%.

Another 15% of the AN production is incorporated into various polymers in order to increase intermolecular attractive forces and thus improve thermal and mechanical properties of the final products. This is the role of AN in SAN (the styrene–acrylonitrile resin, 88¢) produced at a 125MM-lb level, NBR (the nitrile rubber composed of AN and butadiene present in a ratio of 20–40:60–80), and ABS (the acrylonitrile–butadiene–styrene resin, $1.07–$1.17, depending on its impact resistance). The last-mentioned resin is obtained by grafting AN and styrene onto a polybutadiene matrix. Also, ABS forms useful alloys with polycarbonate (PC), nylons, SMA, polysulfones, and others.

About 10% and 4% of AN demand is channeled into the production of adiponitrile and acrylamide, respectively, both of which are discussed below.

The electrohydrodimerization (EHD) of acrylonitrile to adiponitirle (7.2.1) was developed at Monsanto by M. M. Baizer, recipient of the 1976 ACS Award for Creative Invention. It is carried out at a lead cathode in an aqueous solution that contains tetrabutylammonium tosylate [$(n\text{-Bu})_4\text{N}^+ 4\text{-MeC}_6\text{H}_4\text{SO}_3^-$] and ammonia. The latter is present in order to suppress the reduction of the cyano group to imines and amines. The tail-to-tail dimerization of AN can be visualized to occur as shown in Scheme 5.13.

Acid hydrolysis of AN produces about 155MM lb of acrylamide ($1.00 or aqueous solution 69¢; $CH_2=CH\text{-}CO\text{-}NH_2$), which is polymerized under carefully controlled conditions to polymers of very high molecular weight ($10–15 \times 10^6$), and these, on partial alkaline hydrolysis, give flocculants that facilitate the clarification and filtration of

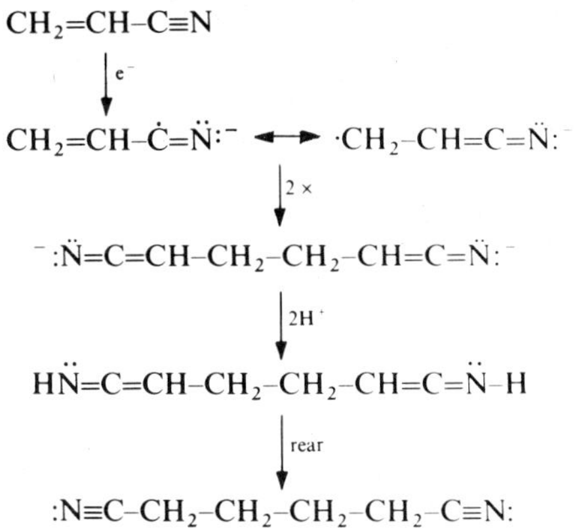

$$CH_2=CH-C\equiv N$$

$$\downarrow e^-$$

$$CH_2=CH-\dot{C}=\ddot{N}{:}^- \longleftrightarrow \cdot CH_2-CH=C=\ddot{N}{:}^-$$

$$\downarrow 2\times$$

$${}^-{:}\ddot{N}=C=CH-CH_2-CH_2-CH=C=\ddot{N}{:}^-$$

$$\downarrow 2H\cdot$$

$$H\ddot{N}=C=CH-CH_2-CH_2-CH=C=\ddot{N}-H$$

$$\downarrow \text{rear}$$

$${:}N\equiv C-CH_2-CH_2-CH_2-CH_2-C\equiv N{:}$$

Scheme 5.13

wastewater produced during the disposal of municipal sewage, paper pulping and mineral processing liquors, and so on. These ionomers contain a random distribution of sodium carboxylate and amide groups.

Acrylamide (AM) is also polymerized by means of a basic catalyst such as sodium methoxide in sulfolane. This type of polymerization occurs by hydrogen transfer [L. W. Bush and D. S. Breslow, *Macromolecules* **1**, 189 (1968)]:

$$MeO^- + CH_2=CH-CO\cdot NH_2 \rightarrow MeO-CH_2-\bar{C}H-CO\cdot NH_2$$

$$\downarrow$$

$$MeOCH_2CH_2-(C=O)NH^-$$

$$\downarrow AM$$

$$MeOCH_2CH_2-C(=O)NH-CH_2\bar{C}H-CO\cdot NH_2$$

$$\downarrow$$

$$MeOCH_2CH_2-C(=O)NH-CH_2CH_2-CO\cdot NH^-$$

$$\downarrow n AM$$

$$MeO[CH_2CH_2-(C=O)NH]_nCH_2CH_2CO\cdot NH^-$$

$$\downarrow$$

etc.

Chains of AN are easily grafted onto different polymeric backbones because of the stabilization of the radical center by the AN molecule. Thus, if a radical center is created by the use of $Ce^{4+}$ or by means of gamma radiation, the polymeric radical symbolized by $Q\cdot$ interacts with AN molecules because of resonance stabilization of the adducts:

$$-\dot{C}H-C\equiv N{:} \leftrightarrow -CH=C=\ddot{N}\cdot$$

$$Q\cdot + CH_2=CH-C\equiv N \rightarrow Q-CH_2-\dot{C}H-C\equiv N \leftrightarrow Q-CH_2-CH=C=\dot{N}{:}$$

$$\xrightarrow{n CH_2=CH-C\equiv N} Q-[CH_2-CH(CN)]_n-CH_2-\dot{C}H-C\equiv N, \text{ etc.}$$

The grafting of AN chains onto starch, followed by partial alkaline hydrolysis of some of the cyano groups, gave the highly absorbent material nicknamed "superslurper"

(*SZM-II*, 93). As mentioned elsewhere, competitive materials of extraordinary absorption capacity for aqueous fluids are now manufactured by cross-linking acrylate salts.

A radically different use of AN is its conversion to "carbon fibers" used for the reinforcement of plastic products, especially composites employed in the manufacture of objects that must meet very stringent strength requirements. The manufacture of such fibers starts with PAN fibers that are subjected to graphitization, that is, pyrolysis in the absence of oxygen. Under carefully controlled conditions, the PAN chains cyclize to give linear arrays of fused pyridines that form even more highly fused polymeric chains on dehydrogenation. If a polymer that is held together by two internally cross-linked chains of atoms is known as a "ladder polymer," then the ultimate fusion of PAN molecules is recognized to be a double-ladder polymer, as shown in Scheme 5.14. It is easy to understand why PAN-derived "carbon or graphite fibers" are rather costly ($15–$45/lb) and why, therefore, they are high-tech specialty materials used mainly by the aerospace industry (nozzles of the space shuttle *Columbia* that must deliver a thrust of at least 1MM lb from solid-fuel rockets; also brakes, cargo-bay doors, and other components), and in military applications (>10% of the Boeing F-18 fighter's structural weight is contributed by PAN-derived fibers), and so on. In 1979 (*CEN* 1/22/79) Union Carbide announced the construction of its first, 0.8MM-lb, PAN-based fiber production facility under a license from Torey of Japan, and since then the domestic capacity rose to 7.0MM lb. Recently (*CMR* 4/4/88) DOD prompted Congress to pass a law that mandates DOD contractors to employ domestic rather than imported PAN-derived graphite fibers. The latter must not be confused with cheaper materials obtained by controlled carbonization of coal tar, petroleum pitch, or rayon fibers, even though these may also be called "carbon fibers" (*CW* 4/29/87). The worldwide demand for PAN-based carbon fibers in 1987 was about 10MM lb, for use in the manufacture of golf clubs, fishing rods, and other athletic equipment. The producers besides the United States, Japan, and western Europe, include Taiwan and South Korea.

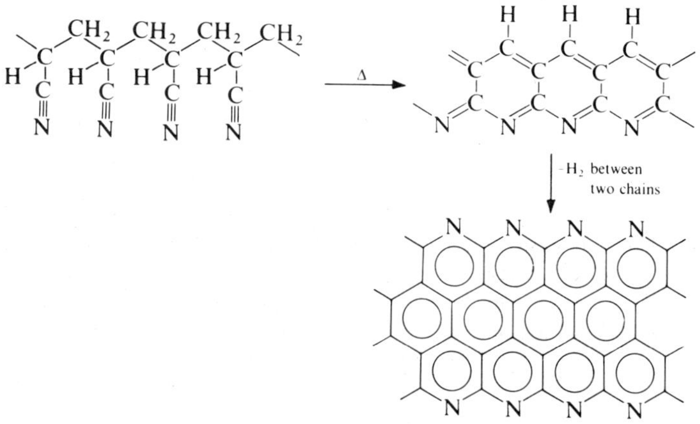

**Scheme 5.14**

To continue with the role of AN in the realm of polymers, we must mention some specialty monomers that are used to modify other polymeric materials. Thus, there is *N*-methylolacrylamide used to modify starch (*SZM-II*, 94) and in the manufacture of nonwoven textiles, and *N,N'*-methylene*bis*acrylamide, mentioned above in connection with cross-linked, high-aqueous-fluids-absorbing (AFA, p. 14) poly(acrylates) which constitute the principal class of superabsorbent polymers (SAPs, p. 14). These chemicals are obtained directly from AN by means of a Ritter reaction with formaldehyde. The Ritter reaction of AN with the trimethylcarbocations derived from either isobutylene or t-butyl alcohol gives rise to N-t-butylacrylamide, which provides a hydrophobic environment in copolymers. The formation of this and analogous t-octylacrylamide is shown in Scheme 5.15. More complex Ritter reaction products of AN are 2-acrylamido-2-methylpropanesulfonic acid, Lubrizol's AMPS ($1.15–$1.30, depending on grade), which provides anionic character to the resulting copolymers and thus enhances their receptivity to basic dyes, and diacetone acrylamide derived from mesityl oxide (5.5.4). The formation of the former and the structure of the latter monomers are shown in Scheme 5.16.

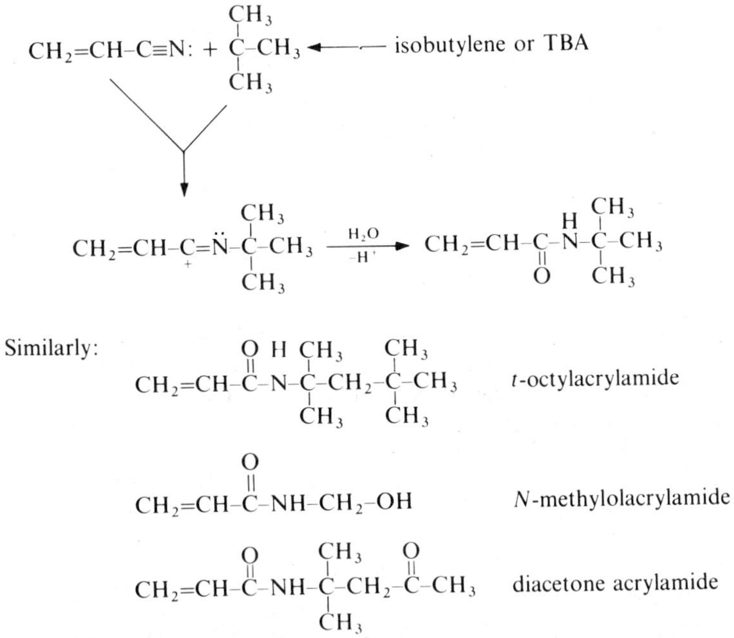

$$CH_2=CH-C\equiv N: + \underset{\underset{CH_3}{|}}{\overset{\overset{CH_3}{|}}{C}}-CH_3 \longleftarrow \text{ isobutylene or TBA}$$

$$CH_2=CH-C=\overset{+}{\underset{}{\overset{..}{N}}}-\underset{\underset{CH_3}{|}}{\overset{\overset{CH_3}{|}}{C}}-CH_3 \xrightarrow[-H^+]{H_2O} CH_2=CH-\underset{\underset{O}{\|}}{C}-\underset{}{\overset{H}{N}}-\underset{\underset{CH_3}{|}}{\overset{\overset{CH_3}{|}}{C}}-CH_3$$

Similarly:

$$CH_2=CH-\underset{\underset{O}{\|}}{C}-\overset{H}{N}-\underset{\underset{CH_3}{|}}{\overset{\overset{CH_3}{|}}{C}}-CH_2-\underset{\underset{CH_3}{|}}{\overset{\overset{CH_3}{|}}{C}}-CH_3 \qquad \textit{t-octylacrylamide}$$

$$CH_2=CH-\underset{\underset{O}{\|}}{C}-NH-CH_2-OH \qquad \textit{N-methylolacrylamide}$$

$$CH_2=CH-\underset{\underset{O}{\|}}{C}-NH-\underset{\underset{CH_3}{|}}{\overset{\overset{CH_3}{|}}{C}}-CH_2-\underset{\underset{O}{\|}}{C}-CH_3 \qquad \text{diacetone acrylamide}$$

**Scheme 5.15**

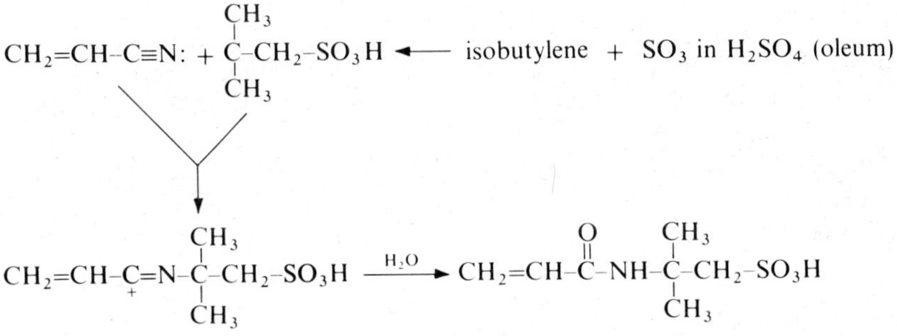

$$CH_2=CH-C\equiv N: + \underset{\underset{CH_3}{|}}{\overset{\overset{CH_3}{|}}{C}}-CH_2-SO_3H \longleftarrow \text{ isobutylene } + SO_3 \text{ in } H_2SO_4 \text{ (oleum)}$$

$$CH_2=CH-C=\overset{+}{N}-\underset{\underset{CH_3}{|}}{\overset{\overset{CH_3}{|}}{C}}-CH_2-SO_3H \xrightarrow{H_2O} CH_2=CH-\underset{\underset{O}{\|}}{C}-NH-\underset{\underset{CH_3}{|}}{\overset{\overset{CH_3}{|}}{C}}-CH_2-SO_3H$$

**Scheme 5.16**

We come now to the potentially countless cyanoethylation products obtained when almost any compound that contains an active hydrogen substituent (symbolized here by Q–H) is allowed to react with AN in the presence of a base catalyst. Traditionally, Rohm and Haas's Triton B, or benzyltrimethylammonium hydroxide, is used for that purpose:

$$Q\text{–}H + CH_2\text{=}CH\text{–}C\text{≡}N \rightarrow Q\text{–}CH_2\text{–}CH_2\text{–}C\text{≡}N$$

This special case of the Michael reaction (*ONR*-60) was studied extensively by H. A. Bruson of Rohm & Haas (see *Organic Reactions*, Vol. 5, Wiley, New York, 1949 p. 79), ironically, before ammoxidation opened the floodgates of AN. Some industrially significant cyanoethylation products are as follows:

CH$_3$–O–CH$_2$CH$_2$CH$_2$–NH$_2$, 3-methoxypropylamine BASF $1.01

Me$_2$N–CH$_2$CH$_2$CH$_2$–NH$_2$, 3-*N*,*N*-dimethylaminopropylamine, DMAPA, BASF $1.18, corrosion inhibitor in aviation gasoline, epoxy resin catalyst, and precursor of three compounds:

CH$_2$=C(CH$_3$)–CO·NH–CH$_2$CH$_2$CH$_2$–NMe$_2$, *N*-dimethylaminopropyl methacrylamide, Rohm Tech.'s DMAPMA

*N*,*N'*,*N''*-Tris(3-dimethylaminopropyl)-*sym*-hexahydrotriazine, or tris-*N*,*N'*,*N''*-dimethylaminopropylhexahydro-1,3,5-triazine, PUR catalyst (see Scheme 3.15)

CH$_2$=C(CH$_3$)–C(=O)–NH–CH$_2$CH$_2$CH$_2$–NMe$_3^+$ Cl$^-$    methacrylamidopropyltrimethylammonium chloride, Rohm Tech.'s MAPTAC

*N*,*N'*-Isophthaloyl-bis(trimethylene urea) is a material registered by Upjohn with the EPA in November 1981. It can be assembled by hydrogenation of the cyanoethylated derivative of the diamide of isophthalic acid, followed by reaction with phosgene or a carbonate ester, as shown in Scheme 5.17.

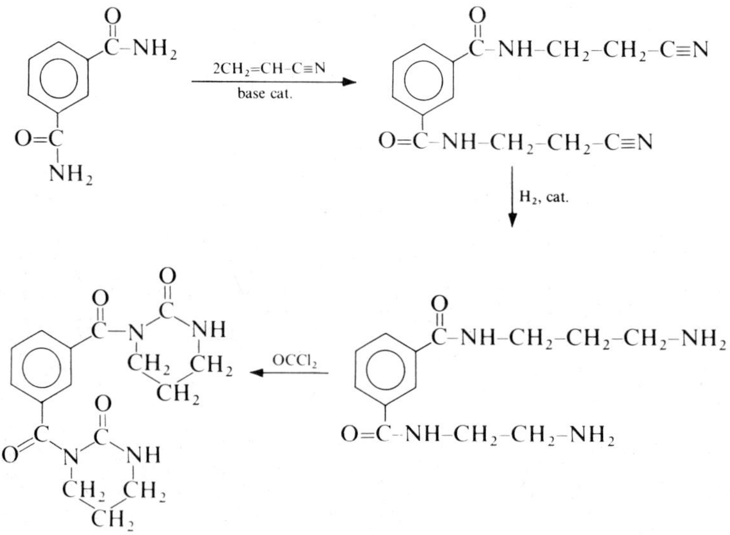

Scheme 5.17

1,5-Diazabicyclo(5.4.0)undec-5-ene, Abbott's Polycat DBU, is a PUR catalyst assembled by cyanoethylation of caprolactam, hydrogenation of the cyano group, and an acid-catalyzed thermal cyclization. DBU forms resonance-stabilized amidinium salts with protic acids, symbolized here by H–X, that function as PUR catalysts and are known as Abbott's Polycat SA; see Scheme 5.18.

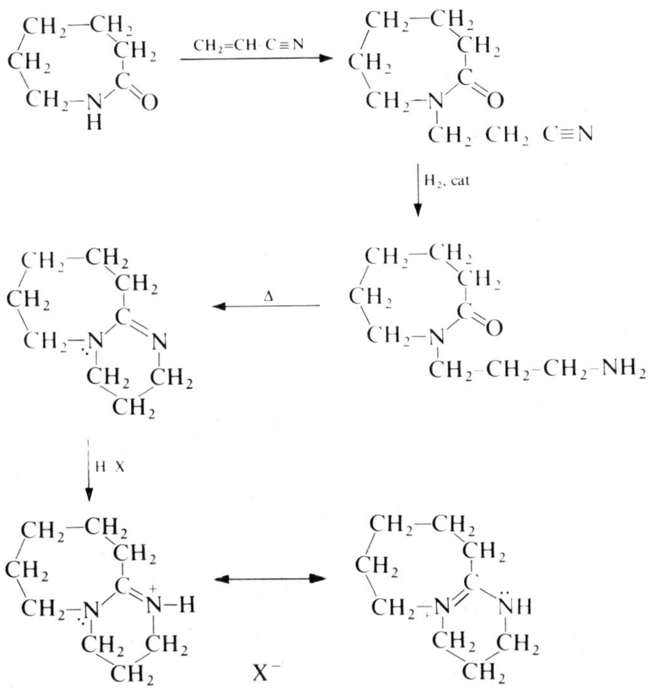

**Scheme 5.18**

$C_6H_5$–$N(CH_3)$–$CH_2CH_2$–C≡N    *N*-cyanoethyl-*N*-methylaniline, 0.2MM lb

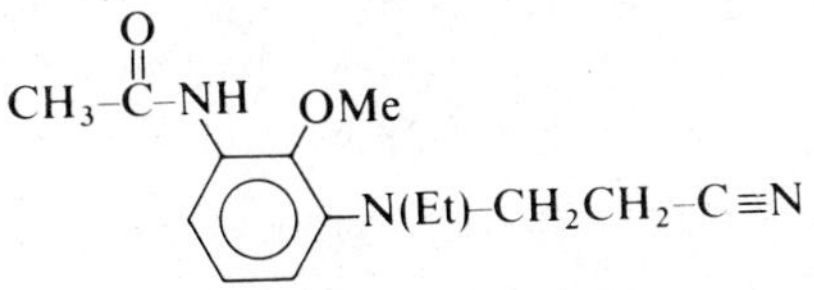

*N*-cyanoethyl-*N*-ethyl-3-acetamido-*o*-anisidine,

Fatty amines, symbolized here by $R_f$–$NH_2$ (*SZM-II*, 56, 80), are cyanoethylated and converted by hydrolysis or hydrogenation to the corresponding carboxyethyl and aminopropyl derivatives:

$R_f$–NH–$CH_2CH_2$–$CO_2H$    and    $R_f$–NH–$CH_2CH_2CH_2$–$NH_2$

Salts of the former are surfactants and the latter are used as corrosion inhibitors and as gasoline detergents.

The synthesis of *d,l*-lysine, which begins with the cyanoethylation of acetaldehyde, is mentioned elsewhere. The analogous cyanoethylation of acetone gives $CH_3-CO-CH_2-CH_2-CH_2-C\equiv N$, 5-ketocapronitrile, which is converted to 2-methylpyridine, or α-picoline, and 2-methylpiperidine (10.2.1) on hydrogenation and cyclization.

A broad-spectrum biocide announced recently (*CW* 2/1/82) for use in industrial aqueous systems such as paints, inks, and variety of household products is 1,2-dibromo-2,4-dicyanobutane, Merck–Calgon's Tektamer 38. Most likely, it is synthesized by cyanoethylation of the dibromo adduct of AN:

$$Br-CH_2-CHBr-C\equiv N + CH_2=CH-C\equiv N \rightarrow Br-CH_2-CHBr(C\equiv N)-CH_2CH_2-C\equiv N$$

For this reaction, the base catalyst must be of a kind that abstracts a proton but is a poor nucleophile, and a fluoride may serve this purpose.

A somewhat similar reaction between *n*-propionitrile and AN:

$$N\equiv C-CH_2CH_3 + CH_2=CH-C\equiv N \rightarrow N\equiv C-CH(CH_3)CH_2CH_2-C=N$$

is the likely source of the low-cost, aliphatic diamine (see 2.4.5) offered since 1986 by Du Pont as Dytek A (89¢). Most likely, it is obtained by hydrogenation of the preceding 2-methylglutaronitrile. 2-Methylpentamethylenediamine can be used as a curing agent for epoxy resins and in place of the more costly hexamethylenediamine when the nonlinear structure of this diamine does not matter. The propionitrile may well be a by-product of the EHD process used by Du Pont to prepare adiponitrile.

*An Interlude: Silane Coupling Agents*

Silane coupling agents are chemicals that modify the surface between two incompatible materials such as (1) polymers and (2) silicious minerals, glass fibers, limestone, gypsum, metallic oxides, and other reinforcing agents or fillers. The surface-modifying agents under discussion bridge the a priori incompatible materials, enhance adhesion, and consequently have a favorable effect on thermomechanical and electrical properties of the composite, as well as on the rheology (flow properties) of the composite during processing operations.

The most common surface-modifying agents are products of the hydrosilylation reaction, that is, the addition of silicon-hydrogen-containing reactants to olefinic (or acetylenic) systems. Trialkoxysilanes $[H-Si(OR)_3]$ are frequently used, although titanates $[Ti(OR)_4]$ and zirconates $[Zr(OR)_4]$ are also employed to react with inorganic surfaces. The addition of trialkoxysilanes to three-carbon, unsaturated systems in the presence of catalytic amounts of chloroplatinic acid $(H_2PtCl_6)$ yields typical silane coupling agents:

$$Q-CH_2-CH=CH_2 + H-Si(OR)_3 \rightarrow Q-CH_2-CH_2-CH_2-Si(OR)_3$$

where Q represents a variety of functional groups capable of interacting by chemical and physical forces with the polymer to be used in the composite. On the other hand, the trialkoxy silicon terminal of the coupling agent or for that matter other hydrolyzable silicon substituents such as chloro and acetoxy groups, react at the surface of the dispersed mineral particle, glass fiber, and so on, symbolized in Scheme 5.19 by a hydroxy silicate structure. Thus, a pretreatment of the otherwise inert reinforcing agent and/or filler with a coupling agent provides a surface that may even copolymerize with the polymer in question by virtue of the appropriately chosen functional group Q, and the otherwise incompatible phases are bound, or coupled, to each other.

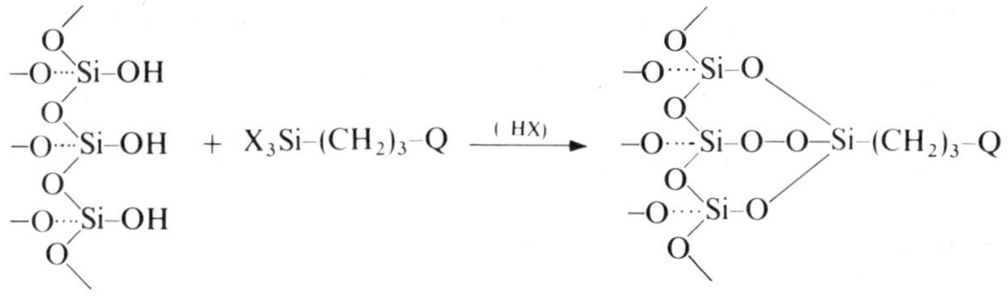

X = Cl, –OAc, –OR
Q = desired structural moiety

**Scheme 5.19**

The three-carbon chain that bridges the two phases is usually long enough to also couple enzymes to glass support systems so that the active site of the enzyme is not obstructed and can perform its catalytic function in the "bound enzyme." The same can be said of valuable transition metal catalysts that are bound to mineral support systems. Another important function of the silane coupling agents is the modification of chromatographic packings in order to obtain systems capable of delicate separations on a large scale that are of increasing industrial importance. Related to the last-mentioned development is the use of polymeric silicones as a chromatographic stationary phase deposited on an appropriate support.

Representative silane coupling agents are shown as follows, and it is clear that most of them are obtained by hydrosilation of the $C_3$ building blocks discussed in this chapter:

$H_2N–CH_2CH_2CH_2–Si(OEt)_3$    $\gamma$-aminopropyltriethoxysilane, \$6.75

$HS–CH_2CH_2CH_2–Si(OMe)_3$    $\gamma$-mercaptopropyltrimethoxysilane, \$10.00

$CH_2=C(Me)–CO–O–CH_2CH_2CH_2–Si(OMe)_3$    $\gamma$-methacroyloxypropyltrimethoxysilane, \$11.00

$CH_2–CH–CH_2–O–CH_2CH_2CH_2–Si(OMe)_3$    $\gamma$-glycidyloxypropyltrimethoxysilane, \$10.35
$\quad\backslash_O/$

$CH_2=CH–C_6H_4–CH_2–\overset{+}{N}H_2–CH_2CH_2CH_2–Si(OMe)_3\ Cl^-$    vinylbenzylammonium-propyltrimethoxysilane chloride

$CH_2=CH–Si(OMe)_3$    vinyltrimethoxysilane (obtained from acetylene)

The use of these relatively costly high-tech specialty chemicals is justified by small concentrations required to profoundly improve the desired properties of the final products. For additional information in this area of chemistry, see E. P. Plueddemann, *Silane Coupling Agents*, Plenum Press, New York, 1982 and *Silicon Compounds—Register and Review*, Petrarch Systems, Bristol, Pennsylvania, 1987.

### 5.4.4  Oxidation of Propylene to Acrolein and Some of Its Uses

The homolytic oxidation of propylene by the molecular oxygen diradical occurs in the presence of catalysts such as bismuth phosphomolybdate (the same catalysts used in ammoxidation), and the intermediate hydroperoxide decomposes spontaneously to acrolein (62¢):

$$CH_2=CH-CH_3 \rightarrow CH_2=CH-CH_2-O-OH \rightarrow CH_2-CH-CH=O+H_2O$$

The vigorous conditions (temperatures $\leqslant 400°C$) cause the formation of by-products such as acetic and acrylic acids.

Acrolein can be also formed by means of the aldol-type reaction of formaldehyde and acetaldehyde carried out at about 320°C over sodium silicate:

$$CH_2O+CH_3-CHO \rightarrow CH_2=CH-CHO+H_2O$$

The traditional route to acrolein is a thermal decomposition of glycerol over magnesium sulfate at about 330°C (or the involuntary thermal decomposition of glycerides during pyrolytic reactions in frying pans):

$$HO-CH_2-CH(OH)-CH_2-OH \rightarrow CH_2=CH-CH=O+2H_2O$$

Even at the relatively low cost of synthetic glycerine (56¢), the stoichiometry does not favor this reaction except under special circumstances.

Acrolein is a low-boiling (52.5°C) lachrymatory liquid generally used as an aqueous solution to cross-link cellulose fibers and in the synthesis of methionine and lysine. In connection with the production of methionine from acrolein, Degussa recently announced (*CW* 2/15/88) the construction of a 80MM-lb acrolein plant in Mobile, Alabama that would expand its methionine production capacity by 35MM lb to a total capacity of 110MM lb.

A "hot tube reaction" of acrolein and ammonia at pressures of up to 200 atm and temperatures of up to 200°C produces $\beta$-picoline (10.3.1):

$$2CH_2=CH-CH=O+NH_3 \rightarrow$$

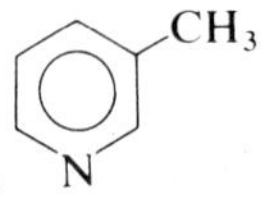

This reaction is important since $\beta$-picoline is the intermediate in the synthesis of the vitamins nicotinic acid or niacin (NF $3.40, feed grade $2.77) and its amide or niacinamide (NF $3.40). In place of the building blocks $C_3+C_3+N_1=C_6N_1$, one may replace acrolein by acetaldehyde and formaldehyde, in which case the desired $\beta$-picoline is assembled by building blocks $2C_2+2C_1+N_1=C_6N_1$.

An example of the synthetic use of acrolein in the preparation of another fine chemical is its Diels–Alder reaction (*ONR*-24; see 6.8.1.4) with myrcenol—the hydration product of myrcene—obtained by the pyrolysis of $\beta$-pinene or from isoprene. This yields the perfume ingredient lyral (see *SZM-II*, 122).

An ingenious modification of the classical Meerwein–Ponndorf reaction (*ONR*-58) converts acrolein to allyl alcohol (90¢). Instead of aluminum alkoxide prepared from aluminum metal ($1.24/lb for 50-lb "pigs" that must be processed into powder valued at about $4.00), one employs on an industrial scale the more amenable magnesium metal ($1.63/lb) and zinc oxide (82¢) to produce magnesium alkoxide from a secondary alcohol. Thus, one can employ *sec*-butyl alcohol (36.5¢) capable of a hydride transfer from the alcohol moiety of the magnesium complex to acrolein; as shown in Scheme 5.20. On hydrolysis of the magnesium complex, one obtains allyl alcohol and methyl ethyl ketone (MEK; 50¢).

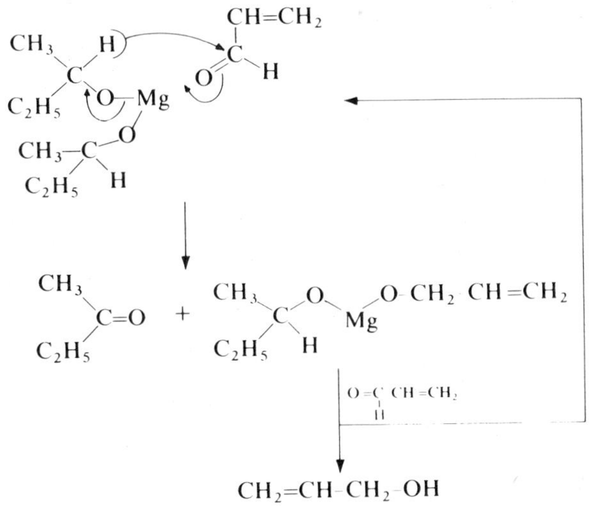

**Scheme 5.20**

Allyl alcohol is a valuable building block of multifunctional esters mentioned in connection with propylene oxide (5.4.2.2.B) since it is the preferred industrial precursor to allyl chloride. The hydrolysis of allyl chloride (5.4.8) to allyl alcohol is not particularly practical because of the unfavorable stoichiometry (46% of allyl chloride is accounted for by chlorine) and the formation of undesirable by-products.

### 5.4.5  Oxidation of Propylene to Acrylic Acid and Some of Its Uses

The U.S. production capacity of acrylic acid (techn. 54¢, glacial 67¢) for 1987 is estimated at about 1.1B lb and is fully utilized because of the growing demand as a component of carboxylated vinyl polymers used in a variety of coatings, for the synthesis single-function and multifunctional acrylate ester monomers, and because of the exceptional growth of polyacrylate superabsorbing polymers (SAP), the key materials in the manufacture of the most modern version of disposable diapers and adult incontinence and feminine hygiene products. Multifunctional acrylate esters are in demand in radiation-cured coatings and high-speed printing operations, and polyacrylates are also in demand as a substitute for phosphate builders in detergent formulations and as dispersants in drilling muds.

There are several approaches to the production of acrylic acid and its derivatives. The oxidation of propylene to acrylic acid occurs by way of acrolein (5.4.4) and is carried out by means of oxygen at 260–300°C. These vigorous conditions are responsible for the formation of by-products such as acetaldehyde, acetic acid, acetone, propionic acid, and even a $C_4$ by-product, maleic anhydride. A direct conversion of propylene to acrylic acid at 200–500°C by means of oxygen in the presence of molybdate and propietary catalysts is practiced by BASF in West Germany and by Badische in Freeport, Texas (*CEN* 7/14/80).

Even the readily available acrylonitrile (5.4.3) (39.5¢) is hydrolyzed by means of sulfuric acid to acrylamide (69¢) and then, under basic conditions, to acrylic acid salts:

$$CH_2=CH-C\equiv N \rightarrow CH_2=CH-CO\cdot NH_2 \rightarrow CH_2=CH-CO_2^- \cdot Na^+$$

Acrylonitrile can also serve as a precursor of acrylate esters:

$$CH_2=CH-C\equiv N + R-OH \rightarrow CH_2=CH-CO-O-R$$

and this transformation is usually catalyzed by sulfuric acid.

Older methods to prepare acrylic acid or its esters include the Reppe hydrocarbonylation of acetylene (4.7) in the presence of nickel carbonyl. This route is particularly useful for the synthesis of higher-molecular-weight esters such as lauryl acrylate, which, together with lauryl methacrylate ($1.72), is copolymerized to give coating materials:

$$HC\equiv CH + CO + n\text{-}C_{12}H_{25}-OH \rightarrow CH_2=CH-CO\cdot O-n\text{-}C_{12}H_{25}$$

The captive use of propiolactone leads to the same result, and another route to acrylic esters involves a transesterification of methyl acrylate (66¢).

Some additional monoacrylic esters of commercial importance are

Ethyl acrylate, 52.5¢

n-Butyl acrylate, 54.5¢

Isobutyl acrylate, 71¢

2-Ethylhexyl acrylate, 79.5¢

2-Hydroxyethyl acrylate (HAA), Dow, $1.40

Hydroxypropyl acrylate, Dow, $1.46

Benzyl acrylate

2-Phenoxyethyl acrylate

Tetrahydrofurfuryl acrylate

Ethylene glycol acrylate phthalate

Coatings derived from the higher-molecular-weight acrylate esters are internally plasticized, rather flexible, and hence quite impact-resistant.

Multifunctional acrylates available for radiation-cured polymeric products include the following:

1,1,1-Trimethylolpropane triacrylate (TMPTA), $1.50

Pentaerythritol triacrylate (PETA), $1.50

Tetraethylene glycol diacrylate (TEGDA), $1.50

Tripropylene glycol diacrylate (TPGDA)

1,6-Hexanediol diacrylate (HDODA)

Glyceryl propoxy triacrylate, Celanese's GPTA, combines three different building blocks:

$$CH_2=CH-CO-O-C_3H_6O-CH(CH_2-O-C_3H_6O-CO-CH=CH_2)_2$$

Acrylates that contain unsubstituted hydroxy groups are being introduced by BP Chemicals as components of thermoset resins and radiation-cured coatings for metal, paper, and other products. The above-mentioned pentaerythritol ester, for example, functions as a cross-linking ingredient in unsaturated polyester resins (UPRs).

As in the case of the higher acrylic esters, methyl acrylate can also serve as an intermediate in the preparation of amides. Thus, for example, N,N-dimethylacrylamide (Alcolac, $3.04) can be prepared from the ester by means of dimethylamine.

An interesting situation that illustrates interproduct competition of different chemical entities arose as a consequence of the November 1987 explosion of the Celanese plant in Pampa, Texas (5.1), which also affected the production of 2-ethylhexyl acrylate. Now, acrylate ester-based adhesives have replaced to some extent the adhesive uses of SBR latex—a development of interest to all producers of 2-ethylhexyl acrylate. To avoid losing the ground conquered by the acrylate-based adhesives, Celanese competitors (BASF, Rohm & Haas, Union Carbide) agreed to convert Celanese acrylic acid to the ester of interest in their own production facilities (*CMR* 12/7/87) while the Pampa facility was being repaired.

The homopolymer of acrylic acid is utilized in the form of its alkali metal and ammonium salts as a thickening and dispersing agent, a protective colloid, and a flocculant in the disposal of wastewater from metallurgical and pulping operations and municipal waste treatments.

A slight degree of cross-linking of polyacrylic acid salts with ethylenediamine bisacrylamide is the basis of gels with an outstanding capacity to absorb aqueous solutions (see above). The annual production of disposable diapers in the United States is about 17 billion units, only half of which contain superabsorbent polyacrylic acid, sold at about $1.45. Thus, the current demand of about 70–80 MM lb of these polyacrylate salts could easily double in the near future. In view of this and other growing acrylic acid markets, all three domestic producers—BASF, Hoechst Celanese, and Rohm & Haas— are expanding their production capacity to 1.350B lb (*CMR* 5/4/87, *CW* 5/6/87).

The vinylidene dichloride–methyl acrylate copolymer was introduced by Dow in 1986 under the tradename Saran MA as a food packaging material superior to the traditional PVDC film because not only does it exhibit excellent barrier properties to water and oxygen but also only half as much material is used in a coextruded "sandwich" between two layers of poly(vinyl acetate) (K. Pinckney, *CW* 5/20/87).

The presence of acrylic acid in copolymers increases the adhesive properties of the material. Dow announced (*CW* 11/3/82) the construction of a 60MM-lb/yr plant to produce the ethylene–acrylic acid copolymer (EAA) in Freeport, Texas.

Acrylic acid and some of its simple derivatives undergo 1,4-addition reactions with nucleophilic reactants. Some examples of this behavior are shown below.

With hydrogen sulfide acrylic acid gives 3,3-thiodipropionic acid [$S(CH_2CH_2CO_2H)_2$], and similarly, lauryl acrylate (see above) and $H_2S$ give dilauryl 3,3-thiodipropionate ($1.53). The analogous dimyristyl and distearyl thiodipropionates, as well as the 3-mercaptoacrylates, and the corresponding disulfides are available on the marketplace;

$$S(CH_2CH_2CO_2R)_2, \quad HS{-}CH_2CH_2CO_2R, \quad ({-}S{-}CH_2CH_2CO_2R)_2$$

Some of these compounds are used as plasticizers, but more importantly, they are incorporated in a variety of products as antioxidants because they decompose peroxides that may accumulate with time. Actually, a combination of hindered phenolic anti-oxidants such as BHT (9.3.4) and thiodipropionates exerts a synergestic effect on the retardation of discoloration and embrittlement of PE and PP films.

A more complex $H_2S$ adduct is obtained from pentaerythritol tetraacrylate:

$$C(CH_2-O-CO-CH_2CH_2-SH)_4$$

The quadruple thiol functionality of pentaerythritol tetrakis-$\beta$-mercaptopropionate, Cincinnati Milacron's Mercaptate Q-43 ester and its star-shaped carbon skeleton, imply a powerful effect of this additive on mechanical and thermal properties of polymers.

A valuable constituent of flavors and fragrances is methyl $\beta$-methylmercaptopropion-ate ($CH_3-S-CH_2CH_2-CO-OCH_3$; $48), the product of the addition of methyl mercaptan (3.3.2.6.C) to methyl acrylate.

The rather similar adduct of ethanol to ethyl acrylate is ethyl 3-ethoxypropionate (EEP), produced by Eastman in a 70MM-lb facilty as the solvent Ektapro (*CMR* 6/15/87) to replace the endangered solvent (4.3.6.1) ethylene glycol ethyl ether (EE).

The addition of aqueous acrylate to acrylic acid gives rise to a novel $\beta$-carboxy-ethylacrylate:

$$CH_2=CH-CO_2^- + CH_2=CH-CO\cdot OH \rightarrow CH_2=CH-CO-O-CH_2CH_2-CO\cdot O^-$$

The corresponding carboxylic acid ($1.05) is available from Alcolac as Sipomer beta-CEA and, when used as a vinyl copolymer, provides pendant carboxyethyl groups that induce excellent adhesion of the polymer to metallic surfaces.

Another novel product is Texaco's MA-300 amphoteric surfactant:

$$R-O-CH_2-CHMe-O-CH_2-CHMe-NH_2^+-CH_2CH_2-CO_2^-$$

(where R represents $C_{10}H_{21}-C_{12}H_{25}$). The dipropoxylated alcohol R–OH is subjected to reductive amination (4.3.3.3), and the resulting amine undergoes a 1,4- addition with acrylic acid to give the betaine. The surfactant is said to have antistatic and anticorrosive properties over a wide pH range.

A Diels–Alder (*ONR*-24) reaction product of acrylic acid and the conjugated isomer of linoleic or dehydrated ricinoleic acid (*SZM-II*, 54, 76–77) is Westvaco's Diacid 1550:

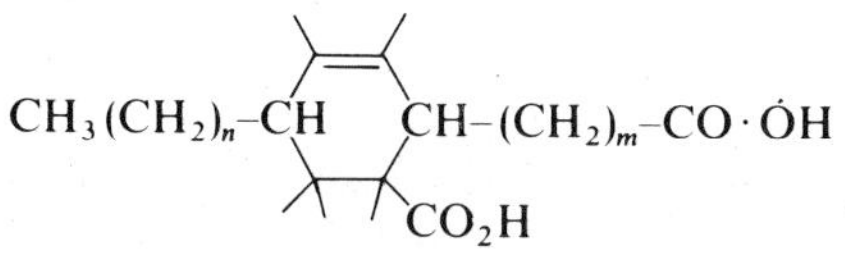

$$CH_3(CH_2)_n-CH \qquad CH-(CH_2)_m-CO\cdot OH$$
$$CO_2H$$

(where $n + m = 12$, most likely 5 and 7, respectively). A partially neutralized potassium salt is sold as Wastvaco-H-240 and functions as a hydrotrope and as a corrosion inhibitor.

### 5.4.6 Alkylation of Aromatic Systems with Propylene and Some of the Resulting Products

The use of propylene in the Friedel–Crafts (*ONR*-33) reaction of benzene to give cumene (18¢) and the subsequent Hock process (9.2.1) to yield phenol (34¢) and acetone (24¢) had a revolutionary impact (2.4.4.3) on the traditional routes to the two last-mentioned chemicals. Currently, these transformations consume about 9% of the propylene demand, as shown in Scheme 5.21. Almost all of the cumene is utilized for the production of phenol

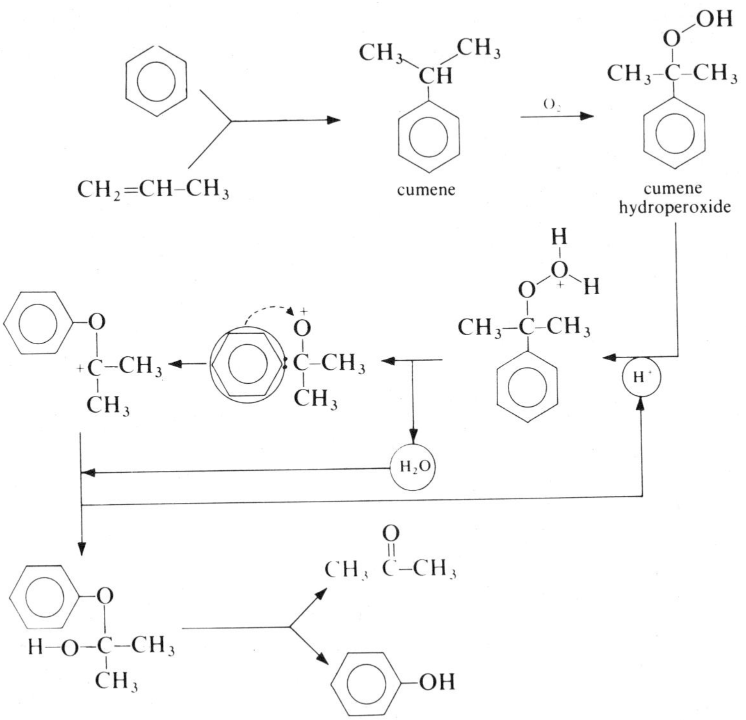

**Scheme 5.21**

and acetone except for a few percent that is dehydrogenated to isopropenylbenzene, α-methylstyrene (see below) or converted to cumene hydroperoxide, a relatively stable hydroperoxide that is permitted to be transported and is employed as a vinyl polymerization catalyst (see 6.3.1.7) and as an oxidizing agent.

More recently, the by-products of the cumene production, namely, *m*- and *p*-diisopropylbenzenes, have also been subjected to the Hock process to give resorcinol and hydroquinone, respectively, in addition, of course, to acetone. In 1987 Cyanamid introduced on the marketplace *m*-diisopropylbenzene (DIPEB), as well as 2-isopropylnaphthalene and 2,6-diisopropylnaphthalene. At about the same time, the announcement (*CMR* 5/11/87) by PMC Specialties of the availability of *p*-cumylphenol or 4-(4'-isopropylphenyl)phenol suggests that the compound may be obtained by means of a single Hock process carried out on doubly alkylated biphenyl.

Isopropenylbenzene or α-methylstyrene (AMS; 44¢) is produced at a level of 50MM lb and is used as a comonomer in unsaturated polyester resins, UPRs, ABS, and PVC in order to improve their thermal and flow properties. The use of cumene hydroperoxide as an oxidizing agent results in the formation of the corresponding alcohol 2-phenyl-2-propanol, which can then be dehydrated to AMS, but AMS can also be obtained simply by dehydrogenation of cumene if the producer is not interested in taking advantage of the Hock process. It is not clear at this time which policy is being followed by Cyanamid in connection with its production of DIPEB and the isopropylated naphthalenes.

Propylene or its hydration product, IPA (5.4.7) is employed in the ortho alkylation of phenols and anilines, and, for example, *o*-isopropylphenol (OIP) and 2,6-diisopropyl-aniline (DIPA) are available from Ethyl Corporation.

### 5.4.7  Hydroformylation and Hydration of Propylene

The hydroformylation of propylene serves as the primary source of the *n*-C$_4$ and iso-C$_4$ building blocks produced as described in 3.5.3.4 and subject of numerous transformations as described in Chapter 6.

The hydration of propylene to isopropyl alcohol (IPA; \$1.51/gal or 22¢/lb) is carried out by absorbing propylene in 85% sulfuric acid at about 25°C in a countercurrent manner (*FKC*, 496–501), followed by dilution with water of the intermediate isopropyl sulfate and distillation of the resulting IPA to give a 91% alcohol-containing water azeotrope. This is converted to 99% ("absolute") IPA by addition of diisopropyl ether (b.p. 68–69°C) and the mixture is subjected to additional fractionation. This results in a ternary azeotrope of composition 91.1% diisopropyl ether, 5.8% IPA, and 3.1% water at 61.4°C, and there remains the almost dry IPA (b.p. 82–83°C).

Other acidic catalysts, including phosphoric acid and acidic ion exchange resins, can be employed for the hydration of propylene.

Unlike ethanol, IPA is not subject to taxation and "rubbing alcohol" is an over-the-counter (OTC) 70% solution with water as a diluent that can be found in many domestic medicine cabinets. Currently, EPA proposes (*CEN* 3/28/88) that the manufacturers of the 1B-lb quantity of IPA test the material for "subchronic toxicity, oncogenicity, mutagenicity, reproductive toxicity, developmental toxicity, and pharmacokinetics"!

Isopropyl alcohol is the logical precursor of acetone, but this use has decreased from 40% of the acetone demand in 1980 to the current source of about 25% because of competition from the expanding use of the Hock process. However, since the production of phenol is sensitive to the construction level, the IPA route to acetone is a backup source.

Some of the chemicals derived from IPA are enumerated as follows.

- Diisopropyl ether, or simply, isopropyl ether (37–44¢, depending on purity) is obtained by means of a relatively low temperature, acid-catalyzed dehydration of the alcohol. It forms explosive peroxides even more readily than does diethyl ether and should be tested for the presence of peroxides (starch–iodide) before use. Otherwise, it may be a more convenient extraction solvent than diethyl ether because of the higher boiling point (66–68°C vs. 34.6°C) and because of a significantly lower solubility in water.

- Isopropyl acetate (47¢) is used as a solvent for derivatives of cellulose.

- Isopropyl myristate (\$1.19) is an ingredient of cosmetic formulations, in which it functions as an emollient and lubricant, and it can be also used as a blending agent in inks, plasticizer, and so on (*SZM-II*, 78).

- *t*-Butylperoxy isopropyl carbonate (*t*-Bu–OO–CO–O–iso-Pr) is a typical peroxy polymerization catalyst (see 6.3.1.7) and curing agent of unsaturated polyester resins available from Lucidol as Lupersol TBIC-M75 and from Akzo as Trigonox BPIC.

- Monoisopropylamine (76¢) is a conveniently low-boiling (b.p. 34°C) primary amine.

- Diisopropylamine (\$1.07) is treated with lithium metal (\$24.45) to give lithium diisopropylamide [LiN(*i*-Pr)$_2$], a strong base soluble in cyclohexane or THF. It is available from LITHCO for less than $12/lb as a 2 M solution when purchased in large quantities.

### 5.4.8   Chlorination of Propylene and Some Uses of Allyl Chloride

The chlorination of propylene to yield allyl chloride is carried out at high temperatures ($\sim 500°$C) in order to avoid addition reactions. Allyl chloride ($CH_2$=CH–$CH_2$–Cl; 65¢) is, as expected, a reactive alkylating agent, and the following are some examples of its industrially important transformation products. The introduction of an allyl group into a given substrate makes the product capable of vinyl polymerization:

- Diallyl ether of bisphenol A (9.3.4) is prepared by means of an alkaline solution of the phenol, as shown in Scheme 5.22.

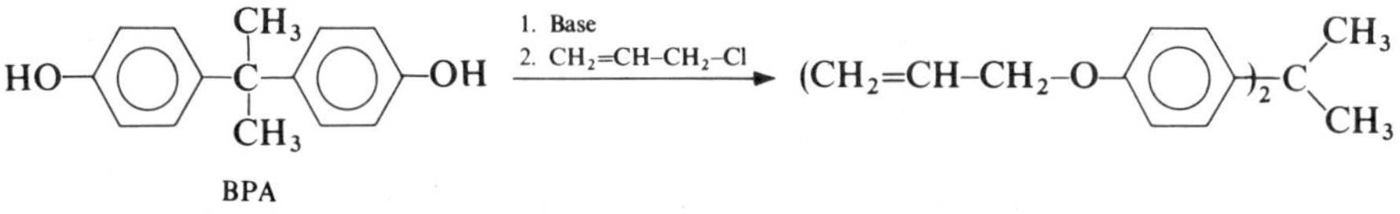

Scheme 5.22

- Sodium allylsulfonate ($CH_2$=CH–$CH_2$–$SO_3^-$ $Na^+$; Ritchem, 42¢) is the product of the reaction of allyl chloride with sodium sulfite. It can be copolymerized in order to introduce an ionic function into the polymeric product.
- Dimethyldiallylammonium chloride [$Me_2N^+(CH_2$–CH=$CH_2)_2$ $Cl^-$, National Starch's DMDAAC] illustrates the ease with which allyl chloride can produce a quaternary ammonium salt and thus introduce a cationic function when used as a comonomer. Cationic polymers are desirable for use in some flocculants, the production of electrically conducting paper, and the formulation of conditioners for damaged hair. The free-radical polymerization of DMDAAC takes an interesting, although not necessarily unexpected, course since the initial radical intermediate of DMDAAC tends to react first intramolecularly rather than intermolecularly; see Scheme 5.23.
- Recent newcomers on the scene of allyl compounds are the products of Virginia Chemicals, currently priced at about $3.00:

  Mono-, di-, and triallylamines

  N-Cyclohexylallyl- and diallylamines

  N-Methyldiallylamine

- The reaction of allyl chloride with sodium cyanate ($Na^+$ $^-OC\equiv N$; 85¢) does not stop with the formation of allyl cyanate because of its trimerization to triallyl cyanurate, TAC; see Scheme 5.24. This trifunctional allylic monomer is available from National Starch ($4.40). It is of interest to note that the preceding reaction does not give allyl isocyanate, $CH_2$=CH–$CH_2$–N=C=O, because the cyanate ambident

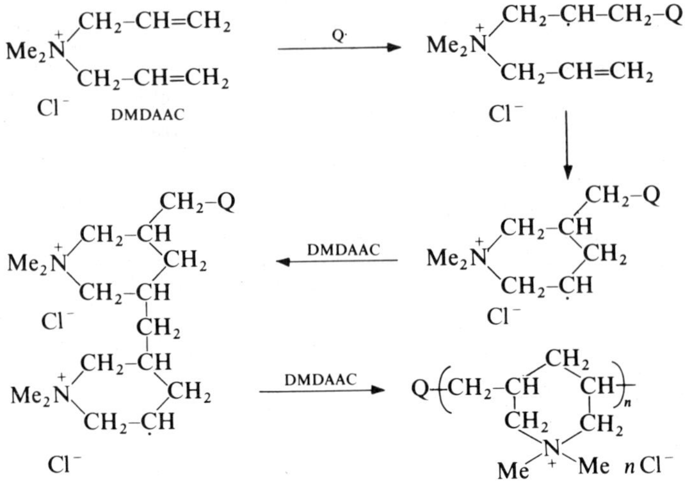

$$Me_2\overset{+}{N} \begin{cases} CH_2-CH=CH_2 \\ CH_2-CH=CH_2 \end{cases} \quad \xrightarrow{Q^\cdot} \quad Me_2\overset{+}{N} \begin{cases} CH_2-\overset{\cdot}{C}H-CH_2-Q \\ CH_2-CH=CH_2 \end{cases}$$

$$Cl^- \quad \text{DMDAAC} \qquad\qquad Cl^-$$

$$Me_2\overset{+}{N}\begin{cases} CH_2-\overset{\overset{\displaystyle CH_2-Q}{|}}{CH} \\ \ \ \ \ \ \ \ \ \ CH_2 \\ CH_2-\overset{\cdot}{C}H \\ \ \ \ \ \ \ \ \ CH_2 \end{cases} \quad\xleftarrow{\text{DMDAAC}}\quad Me_2\overset{+}{N}\begin{cases} CH_2-\overset{\overset{\displaystyle CH_2-Q}{|}}{CH} \\ \ \ \ \ \ \ \ \ \ CH_2 \\ CH_2-\overset{\cdot}{C}H \end{cases}$$

$$Cl^- \qquad\qquad\qquad Cl^-$$

$$Me_2\overset{+}{N}\begin{cases} CH_2-\overset{\cdot}{C}H \\ \ \ \ \ \ \ \ \ CH_2 \\ CH_2-\overset{\cdot}{C}H \end{cases} \quad\xrightarrow{\text{DMDAAC}}\quad Q\left(CH_2-\overset{\overset{\displaystyle CH_2}{|}}{CH}\ \ \ \overset{\overset{\displaystyle CH_2}{|}}{CH}\right)_n$$

$$Cl^- \qquad\qquad\qquad\qquad \underset{Me}{\overset{}{}}\overset{+}{N}\underset{Me}{\ \ }\ n\,Cl^-$$

**Scheme 5.23**

$$\left.\begin{array}{l} CH_2=CH-CH_2-Cl \\[8pt] Na^+\ {}^-O-C\equiv N \end{array}\right\} \longrightarrow CH_2=CH-CH_2-O-C\equiv N \xrightarrow{3\times}$$

$$R-O\overset{\displaystyle N}{\underset{\displaystyle N}{\bigcirc}}\overset{\ \ }{\underset{\displaystyle N}{}}O-R$$
$$O-R$$

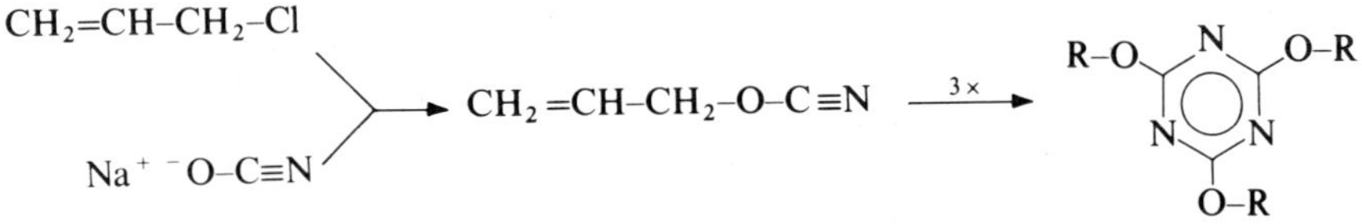

(where R = allyl)

**Scheme 5.24**

nucleophile reacts at its "hard" terminal apparently in response to the highly polarized carbon–chlorine bond in allyl chloride.

• Epichlorohydrin

$$\overset{\displaystyle O}{\overset{\displaystyle \diagup\ \diagdown}{CH_2-CH-CH_2-Cl}},\ 86\cent$$

may, volumewise, be the most important derivative of allyl chloride in view of the United States production capacity of 640MM lb. It is obtained by the addition of hypochlorous acid generated, as usual, from chlorine and water, followed by treatment of the isomeric dichlorohydrins with caustic soda (*FKC* 335–338) or limewater, and it is isolated by distillation (b.p. 116°C), as shown in Scheme 5.25.

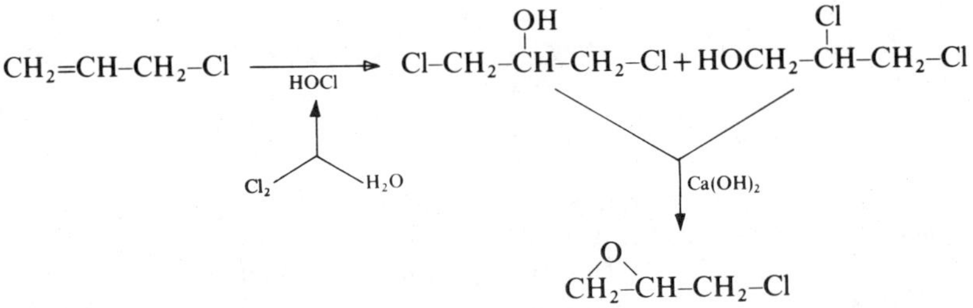

$$CH_2=CH-CH_2-Cl \xrightarrow[\text{HOCl}]{} Cl-CH_2-\overset{\overset{\displaystyle OH}{|}}{CH}-CH_2-Cl + HOCH_2-\overset{\overset{\displaystyle Cl}{|}}{CH}-CH_2-Cl$$

$$Cl_2 \diagup\ \diagdown H_2O \qquad\qquad Ca(OH)_2$$

$$\overset{\displaystyle O}{\overset{\displaystyle \diagup\ \diagdown}{CH_2-CH-CH_2-Cl}}$$

**Scheme 5.25**

Epichlorohydrin can be also obtained by the treatment of allyl chloride with hydrogen peroxide (35% 23.25¢, 50% 32.25¢, or 70% 45¢), and this route may become more popular in view of the rapidly expanding production of hydrogen peroxide (J. Rivoire, *CW* 9/16/87, 38–39). Epichlorohydrin is the building block of glycidyl ethers, which are the backbone of epoxy resins. The most prominent monomer for the production of epoxy resins is the diglycidyl ether of bisphenol A (DGEBA):

$$CH_2\text{-}CH\text{-}CH_2\text{-}O\text{-}\langle\bigcirc\rangle\text{-}C(Me)_2\text{-}\langle\bigcirc\rangle\text{-}O\text{-}CH_2\text{-}CH\text{-}CH_2$$

prepared from epichlorohydrin and an alkaline solution of bisphenol A (epoxy grade, 67¢). It is of interest to note that this and other nucleophilic substitution reactions of epichlorohydrin occur at its epoxide terminal but that the opening of the oxirane ring relays the nucleophilic reactivity to the chlorine-containing terminal. This is represented as follows with Nu:⁻ symbolizing a nucleophilic reactant:

$$Nu\text{:}^- + CH_2\text{-}CH\text{-}CH_2\text{-}Cl \rightarrow Nu\text{-}CH_2\text{-}CH\text{-}CH_2\text{-}Cl \rightarrow Nu\text{-}CH_2\text{-}CH\text{-}CH_2 + Cl^-$$

In the case of the reaction of the bisphenol A anions with epichlorohydrin, the nucleophilic substitution process does not stop with the formation of the above-mentioned diglycidyl ether but continues in an intermolecular fashion to give an amorphous prepolymer. The formation of this liquid, epoxy resinous prepolymer ($1.31) is represented here in a simplified fashion by limiting the reaction to one of the two glycidyl terminals, and then by showing, in Scheme 5.26, a typical structure of the linear assembly of the epoxy prepolymer. Three or more repeating units give solid epoxy prepolymers priced at $1.285. When the ultimate epoxy resin is formed, the prepolymer is "set" by addition of multifunctional "hardeners" such amines, thiols, alcohols, and carboxylic acids, and now the surviving epoxy groups produce a cross-linked, thermoset, insoluble structure. It is clear that even if we limit ourselves to the bisphenol A-derived epoxy prepolymer, there are countless possibilities for the design of different epoxy resins, and the choice depends on the proposed use of these materials. Thus, while coating resins are priced at $1.00–$1.20, epoxy resins for electronic applications may be priced as high as $7.00/lb. Epoxy-coating resins are used primarily for protection and decoration of steel products, boats, and so on. In fiberglass-reinforced epoxy laminates the resin acts as a binder, and these structural applications of epoxies give rise to inert pipes, vessels, and other large objects. Modified epoxies are obtained by incorporation of polyesters or polyacrylates, and these materials are used, among other things, for the coating of beer cans and the manufacture of flooring products.

In addition to the epoxy structures assembled from bisphenol A, there are in use also aromatic structures in which the glycidyl group is attached to an aggregate of phenol molecules held together by formaldehyde-derived methylene bridges (epoxy novolac resins), or glycidyl ethers of tris-4-hydroxyphenylmethane (9.3.4). The last-mentioned trifunctional epoxy building block is one of the materials used for encapsulation of electronic devices. In order to introduce fire-retardant properties, the diglycidyl ether can be assembled from tetrabromo–bisphenol A, which is obtained by virtue of the facile bromination of all phenols (in this case only in the ortho positions in relation to the oxygen function). Another trifunctional epoxide, triglycidyl isocyanurate (TGIC), can be

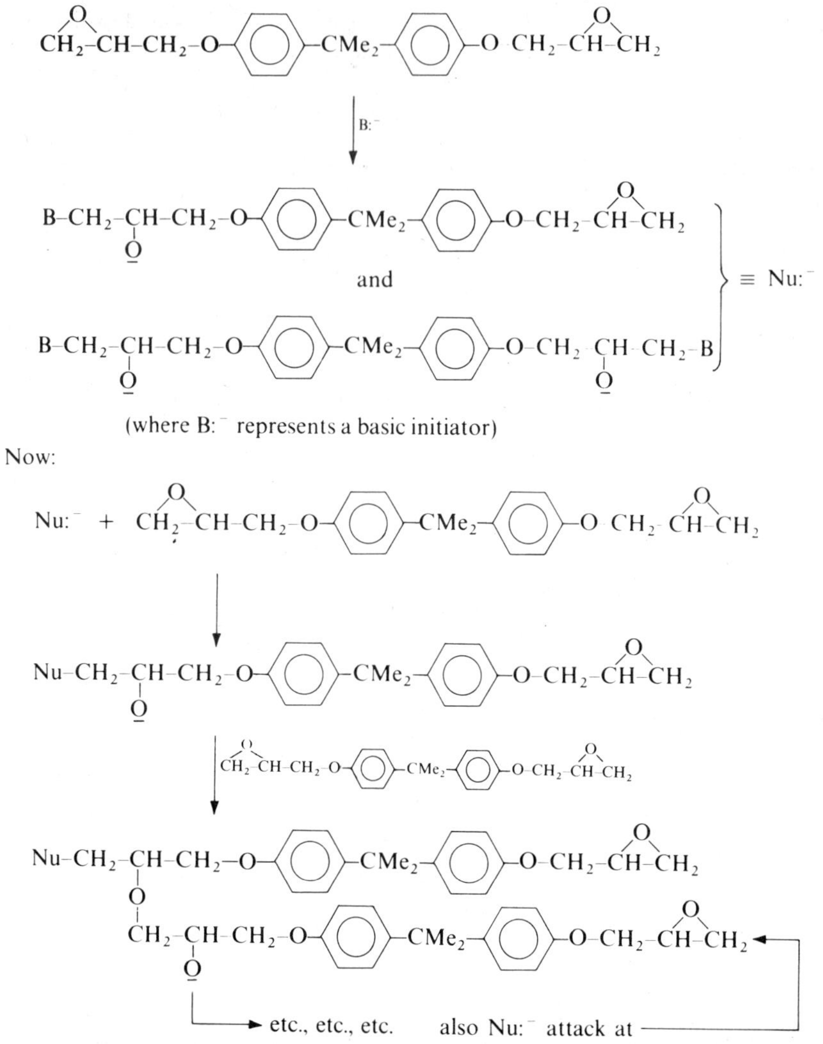

**Scheme 5.26**

derived from triallyl isocyanurate. It is employed in the formation of thermoset powder coatings that are applied electrostatically on metallic surfaces of automotive and appliance components and cured by heat. The structures of the above-mentioned epoxy building blocks are shown in Scheme 5.27.

The reaction of epichlorohydrin with numerous aliphatic and alicyclic hydroxylic compounds further expands the choice of epoxy building blocks. Some examples of such glycidyl compounds are shown as follows:

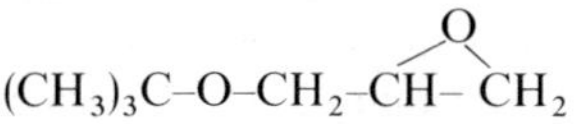

$(CH_3)_3C-O-CH_2-CH-CH_2$   *t*-butyl glycidyl ether

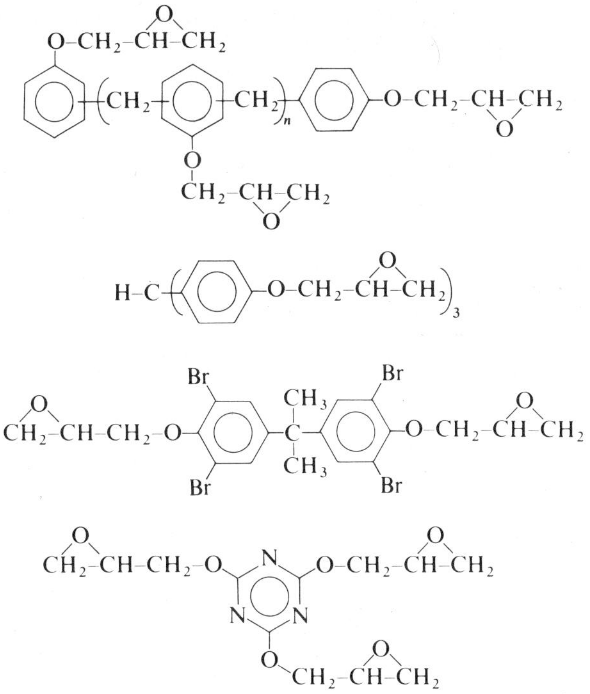

$$HO-CH_2CH(OH)CH_2-O-CH_2-CH-CH_2 \quad \text{1-glyceryl glycidyl ether}$$

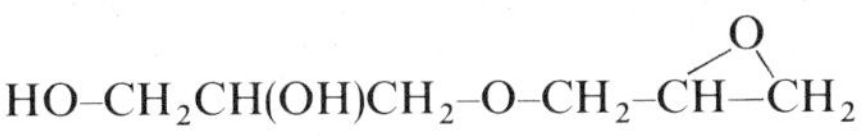

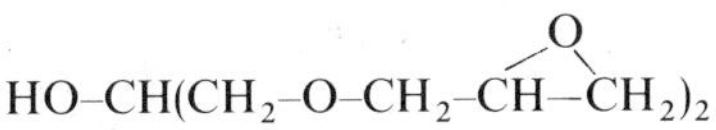

$$HO-CH(CH_2-O-CH_2-CH-CH_2)_2 \quad \text{1,3-glyceryl diglycidyl ether, Howard Hall's glycidyl ether-100, \$2.73}$$

An acid-catalyzed hydrolysis of epichlorohydrin produces glycerol monochloro-hydrin, 3-chloro-1,2-propanediol, or $\alpha$-chlorohydrin [$Cl-CH_2-CH(OH)CH_2-OH$]. This chemical is the Epibloc rodenticide from Pestcon Systems, and its hydrolysis leads to synthetic glycerol. However, synthetic glycerol is obtained preferentially from allyl alcohol and 50% hydrogen peroxide (in the presence of sodium tungstate catalyst at temperatures below 50°C according to Degussa technology), and this synthetic glycerol or glycerine (96% 56¢, 99.5% 78¢) supplements the glycerine obtained from natural sources (*SZM-II*, p. 65 and elsewhere) (natural refined 99.5%, USP, 78¢, natural 96%, USP, CP, 76.25¢). About one third of the current demand of about 375MM lb of glycerine is satisfied by the synthetic product. Hydrogenolysis of sorbitol (*SZM-II*, 106) and mannitol is a source of glycerine not used in the United States. About 35% of glycerol is used in the formulation of pharmaceutical and personal-care products, and about 20% functions as a humefactant in tobacco. Alkyd resins and PURs consume about 15% of the glycerol demand, and the rest includes plastification of cellophane, the preparation of explosives (e.g., nitroglycerine or glycerol trinitrate), the production of glycerol triacetate or triacetin, and so on.

Epichlorohydrin is used to cross-link starch before this intermediate is converted to the metal-scavenging xanthate.

The reaction of epichlorohydrin with diethylamine gives

$$Et_2N-CH_2-CH(OH)-CH_2-OH,$$

3-diethylaminopropane-1,2-diol. Similarly, the reaction of epichlorohydrin with iso-propylamine is the first step in the assembly of propranol—one of the most widely prescribed drugs against heart disease; see Scheme 5.28. Propranolol is an effective "calcium blocker"; that is, it inhibits the contraction of the smooth muscle of coronary arteries and allows blood flow to the heart. Its sale in 1981 reached a value of $225 million.

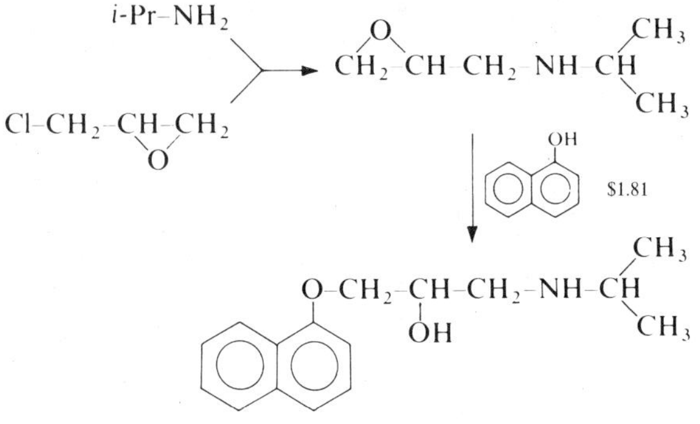

**Scheme 5.28**

Either epichlorohydrin or the above-mentioned dichlorohydrin and glycidol is converted by means of hydrogen sulfide (13¢), sodium hydrogen sulfide, or sodium sulfhydrate (25¢) to monothioglycerol [HS–CH$_2$CH(OH)CH$_2$–OH], a water-soluble antioxidant stabilizer of pharmaceutical and cosmetic formulations, a free-radical catalyzed polymerization modifier, and a promoter of wound healing.

The reaction of epichlorohydrin with pyridine, followed by the addition of sodium sulfite, produces 3-pyridinium-2-hydroxypropanesulfonate, available from Rit-Chem:

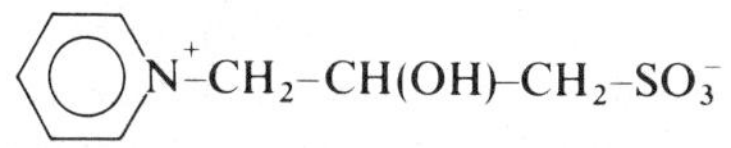

Oce Andeno of the Netherlands recently introduced chiral building blocks derived from epichlorohydrin, namely, R-glycidyl butyrate and the tosylate, which provide an opportunity for the synthesis of enantiomeric fine chemicals such as the above-mentioned beta blockers. A likely preparation of these compounds involves either the resolution of an optically active derivative of glycidol say, d-camphorsulfonic acid, followed by the replacement of the sulfonate group:

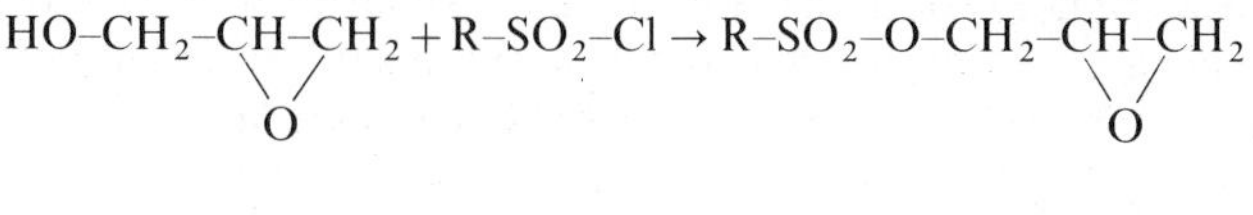

$$HO-CH_2-CH-CH_2 + R-SO_2-Cl \rightarrow R-SO_2-O-CH_2-CH-CH_2$$
$$\underset{O}{\diagdown\diagup} \qquad\qquad\qquad\qquad \underset{O}{\diagdown\diagup}$$

(where $R-SO_2-$ represents the moiety of a convenient optically active sulfonic acid). An alternative source of chiral reagents is the resolution of the racemic tosylates of glycidol by means of a chiral chromatographic packing to give

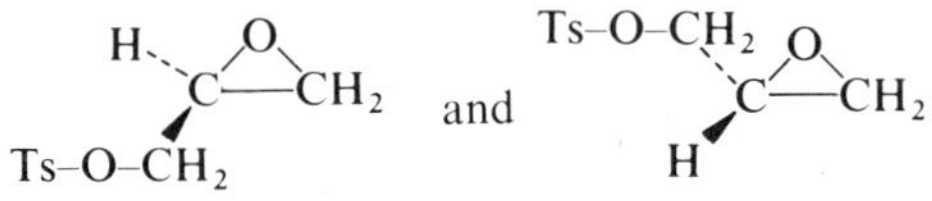

R-Glycidyl tosylate          S-Glysidyl tosylate

The bromine adduct of allyl chloride, $Br-CH_2-CHBr-CH_2-Cl$, namely, 1,2-dibromo-3-chloropropane (DBCP), was employed for the control of nematodes until it was banned by the EPA because of its carcinogenicity and its effect on male sterility. Such powerful physiological activity is not surprising when one considers the alkylation behavior of the dehydrohalogenation product that involves the 2-bromo substituent. Because of the presence of one allylic chlorine substituent, a similar fate may be awaiting 1,3-dichloro-propene ($Cl-CH_2-CH=CH-Cl$, Dow's Telone II), a soil fumigant allowed by the EPA in the hands of "certified applicators for restricted use" (as should be the practice for large uses of most industrial chemicals). On the other hand, Dow's 1,2,3-trichloropropane, the addition product of chlorine to allyl chloride, seems to be safe from EPA restrictions since its dehydrochlorination to 1,3-dichloropropene does not occur with great ease as is the case with the analogous bromine compound.

### 5.4.9  Metathesis of Propylene and Metathesis in General

The unsymmetrical nature of propylene is a convenient feature for the introduction of metathesis (Farona, $CT$ 1/78) even though, as mentioned in 4.3.7.4, the reaction occurs also with the simplest of all olefins, namely, ethylene. Metathesis is a disproportionation that involves two olefinic molecules and results in a redistribution of the moieties that arise from a cleavage of the double bond. Naturally, this is of little significance in the case of symmetrical olefins (unless one half should contain an isotopic marker), except that in the case of symmetrical cyclic olefins, the reaction produces noncyclic products.

The simplest system to illustrate metathesis is the reaction of propylene that gives ethylene and 2-butene:

$$2CH_3-CH=CH_2 \rightarrow CH_2=CH_2 + CH_3-CH=CH-CH_3$$

The reverse reaction is actually used as a source of propylene (see below). Metathesis is catalyzed by transition metals such as tungsten (e.g., $WCl_6$, $EtAlCl_2$, and Et–OH for a homogeneous reaction at room temperature), or $WO_3$, $Re_2O_7$, CoO, and $MoO_3$ on aluminum silicate (for a heterogeneous reaction at 400–500°C). A highly active homo-geneous catalyst has been reported recently ($CEN$ 4/28/86), and a great deal of research continues in this field in view of the many opportunities for novel preparative routes to industrial organic chemicals offered by metathesis. Metathesis in the presence of ethylene, or ethenylation, of long-chain olefins obtained by cracking of petroleum wax and with a random distribution of double bonds, is one source of alphaolefins (AOs) and is dealt with elsewhere (7.4). Other examples of practical applications of metathesis are shown below.

A facility that combines the dimerization of ethylene to 2-butene and the metathesis of this olefin with additional ethylene provides a new source of propylene, and a 300MM-lb plant designed to produce polymerization-grade propylene has been established (*CMR* 11/18, 12/23/85) at Channelview, Texas by Lyondell Petrochemical (a division of Atlantic Richfield) using technology developed by Phillips Petroleum. A pilot plant with a capacity of 1 t$^m$ of propylene has been installed in Taiwan (*CE* 9/28/87).

Phillips produces neohexene [(CH$_3$)$_3$C–CH=CH$_2$] by ethenylation of diisobutylene [(CH$_3$)$_3$C–CH=C(CH$_3$)$_2$; see 6.3.1.2].

Interesting materials are obtained by the application of metathesis to unsaturated fatty acids and derivatives (*SZM-II*, 75).

Multifunctional olefins such as 1,5-hexadiene ($\sim$\$3.80) and 1,9-decadiene ($\sim$\$5.60) are being produced (*CEN* 12/16/85, 5/25/87; *CMR* 6/8/87; H. C. Short, *CW* 6/3/87, 30–33; *CE* 8/17/87, 22–25) by Shell using ethenylation of 1,5-cyclooctadiene and cyclooctene, respectively, at a \$4.8-million facility in Marseilles, France by means of a propietary rhenium catalyst: see Scheme 5.29. The 6.6 MM-lb production will provide intermediates for the synthesis of pharmaceuticals, aroma chemicals, and other specialty products.

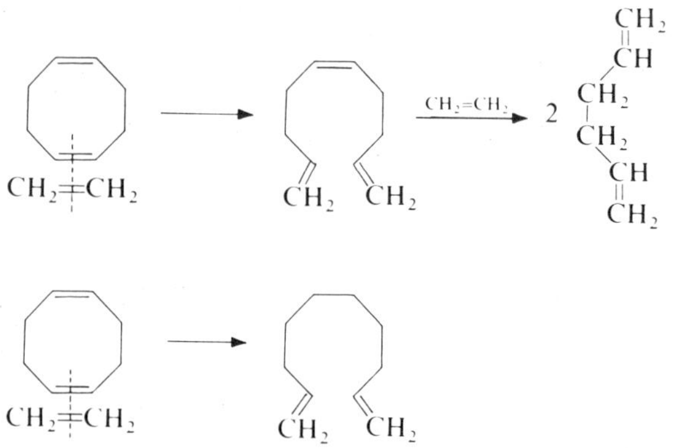

**Scheme 5.29**

Metathesis oligomerization–polymerization has been employed by CdF since 1967 to convert norbornene—the Diels–Alder adduct of acetylene (4.7) and cyclopentadiene (6.8.3)—to polynorbornene, and in 1980 Huels launched the production of the polymer derived from cyclooctene. Currently, Hercules is commercializing the polymer Metton obtained by metathesis of dicyclopentadiene. It is not surprising that metathesis polymerization is evoking great interest in this and analogous reactions (S. C. Stinson, *CEN* 4/27/87). The formation of these polymers is shown in Scheme 5.30.

## 5.5  ACETONE AND ITS ROLE AS A C$_3$ BUILDING BLOCK

### 5.5.1  Sources of Acetone and an Overview of Its Uses

Acetone is an "ancient" industrial chemical because its original source, namely, the decomposition of calcium acetate (obtained by treating pyroligneous acid with lime) even precedes the discovery of petroleum. Pyroligneous acid is the liquid product of the destructive distillation of wood and it contains, in addition to water, about 6% acetic acid,

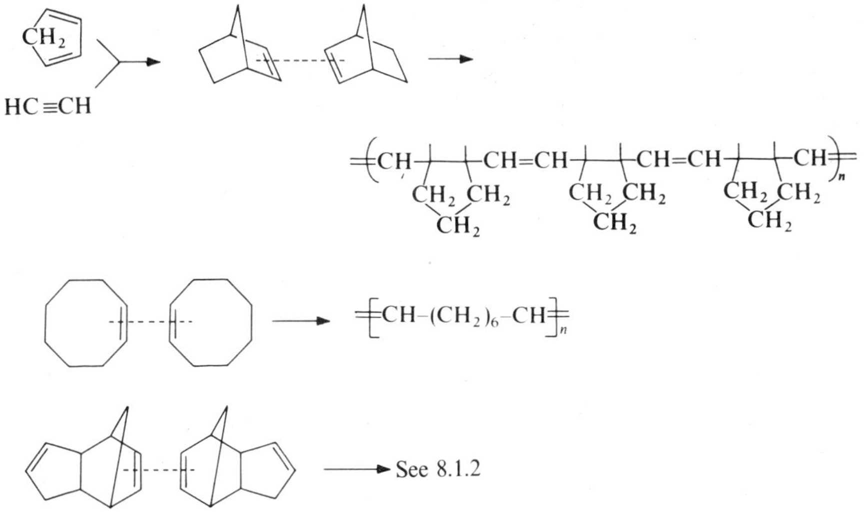

$$+CH+ +CH=CH+ +CH=CH+ +CH+_n$$
$$\quad CH_2\ CH_2 \qquad CH_2\ CH_2 \qquad CH_2\ CH_2$$
$$\qquad CH_2 \qquad\qquad CH_2 \qquad\qquad CH_2$$

$$\longrightarrow \quad +CH-(CH_2)_6-CH+_n$$

$$\longrightarrow \text{See 8.1.2}$$

**Scheme 5.30**

methanol (wood alcohol), and numerous other compounds.

$$Ca(O-CO-CH_3)_2 \rightarrow CaCO_3 + CH_3-CO-CH_3$$

The demand for acetone in the United States is about 2B lb, and while most of the demand is satisfied by the Hock process applied to cumene (5.4.6), the remainder is derived from the oxidation of IPA (5.4.7).

Minor additional sources of acetone are as follows:

1. Hock process applied to isopropylaromatics other than cumene (e.g., the preparation of hydroquinone; see 9.2.2)
2. Pyrolytic oxidation of propane–butane (6.2)
3. Reduction of acrolein to allyl alcohol (5.4.4)
4. Production of hydrogen peroxide from IPA (5.4.7)
6. By-product of PO manufacture (5.4.2)

The current unit price of acetone is 27¢ in the eastern United States and 30¢ west of the Rockies. About 15% of the acetone consumed in the United States is used as solvent. For example, tanks of acetylene utilized in metal welding and cutting operations contain acetone and a porous mineral (kieselguhr). Acetylene is dissolved under moderate pressure in this heterogeneous system. Methyl methacrylate monomer (MMA) and other members of the methacrylic acid system occupy the first place among the derivatives of acetone and consume about 35% of the total demand. About 12% of the demand for acetone is channeled into the production of bisphenol A, and another 10% becomes methyl isobutyl ketone (MIBK) and its reduction product methyl isobutyl carbinol. This still leaves a considerable amount of acetone to be converted into numerous other interesting compounds, some of which are mentioned in the pages that follow.

The continued growth of MMA as the building block of the homopolymer and of some copolymers mentioned elsewhere, because of their impact-improving properties, is stimulating the development of processes that compete with the traditional acetone to MMA conversion route. On the other hand, the production of bisphenol A depends, at this time, exclusively on the employment of acetone, and the demand for bisphenol A is growing at a respectable rate. The appearance and increasing use of IPDI for the production of light-stable PURs is bound to boost the demand for acetone. Finally, the shift from organic solvent-based coatings to water-based systems threatens the future use of acetone and some of its derivatives (e.g., MIBK) as an industrial solvent. Thus, we may conclude that the future of acetone is subject to several counteracting factors and is difficult to predict at this time.

### 5.5.2  Methyl Methacrylate and the Methacrylic Acid System

The time-honored use of acetone, HCN, methanol, and sulfuric acid to yield methyl methacrylate (62¢) (*FKC*, 547–551) is currently being challenged by oxidation and ammoxidation of isobutylene to methacrylic acid (78¢) and methacrylonitrile, respectively (see 6.3.1.4), and this challenge must be taken seriously in view of the availability and low cost of isobutylene (32¢) and companion reactants. It is reported (*CE* 6/22/87) that ARCO has acquired worldwide rights to the two-stage oxidation of isobutylene to methacrylic acid that is esterified to MMA. This technology was developed by Halcon, and the key to success is the use of two different types of catalyst: (1) oxides of Mo, Bi, Co, Fe, Ni, and other metallic elements in the formation of the aldehyde and heteropolyacids of either Mo or phosphomolybdates and (2) oxides of Bi, Sb, and Th in the formation of methacrylic acid.

Another competitive process scheduled for production of MMA in 1986 is the cooperative venture of CdF Chimie of France and Roehm of West Germany based on a process developed by Ashland Chemical of Columbus, Ohio. Presumably it starts with the Reppe hydrocarbonylation (3.5.3.5) of propylene in the presence of methanol and the dehydrogenation of the intermediate methyl isobutyrate.

The most recent development in this competitive world is the BASF process inaugurated in West Germany (*CE* 11/23/87; *CW* 1/20/88, 64–65) based on the hydro-formylation of ethylene to propionaldehyde, the condensation of the latter with formaldehyde to give methacrolein, and a conventional oxidation and esterification of the latter to MMA.

We note that the case of MMA illustrates beautifully the application of the building-block principle to the design of the methacrylic moiety of MMA: it is assembled by the following combinations of building blocks, $C_4$, $C_3 + C_1$, $C_2 + C_1 + C_1'$.

The different approaches to the production of MMA are pictured in Scheme 5.31.

Poly(methyl methacrylate) (PMM) is an attractive structural material and is recognized for exceptional optical clarity. This property accounted for its use during World War II for the fabrication of fighterplane canopies, and more recently it has been used as a structural material for the manufacture of windowglass replacements in buses, schools, and other hazardous places; decorative components of furniture and lighting fixtures; and so on. Rods of PMM can "bend" (deflect) light and are used to illuminate objects that are otherwise not easily accessible to traditional illumination. PMM can be processed by extrusion, injection molding, and other methods typical of thermoplastics, and protective coatings have been developed that protect PMM surfaces against damage by scratching and hence make this material more competitive with the tougher polycarbonate surfaces.

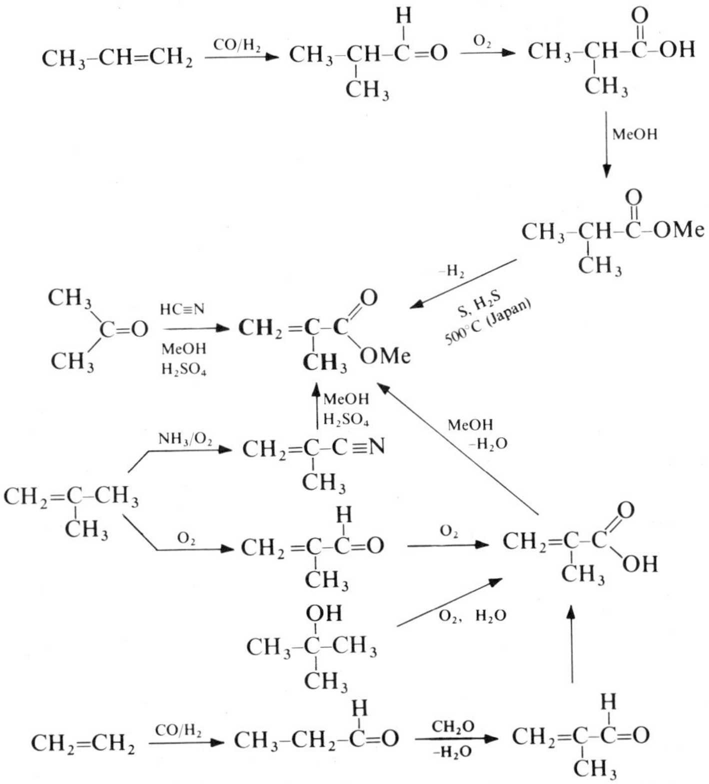

**Scheme 5.31**

Over the years, the public has become familiar with the two common tradenames for PMM: Rohm & Haas's "Plexiglass" and Du Pont's "Lucite."

Besides methyl methacrylate monomer, there are other methacrylate systems available for the modification of polymer properties. These multifunctional monomers include compounds such as 1,1,1-trimethylolpropane trimethacrylate (TMPTMA) and heterodifunctional monomers such as

Allyl methacrylate (5.4.2)

2-Hydroxyethyl methacrylate (HEMA)

2-Dimethylaminoethyl methacrylate (DMAEMA)

2-Trimethylammoniumethyl methacrylate (TMAEMC)

2-Acetoacetoxyethyl methacrylate (AAEM),
$CH_3-CO-CH_2-CO-O-CH_2-CH_2-O-CO-C(CH_3)=CH_2$

2-Aminoethyl methacrylate hydrochloride (AEM),
$Cl^- H_3N^+-CH_2-CH_2-O-CO-C(CH_3)=CH_2$

$N$-(3-Aminopropyl) methacrylamide hydrochlride (APMA),
$Cl^- H_3N^+-CH_2-CH_2-CH_2-NH-CO-C(CH_3)=CH_2$

N-(3-Dimethylaminopropyl) methacrylamide (DMAPMA),
$(CH_3)_2N-CH_2-CH_2-CH_2-NH-CO-C(CH_3)=CH_2$

N-(Trimethylammoniumpropyl) methacrylamide,
$Cl^- (CH_3)_3N^+-CH_2-CH_2-CH_2-NH-CO-C(CH_3)=CH_2$

### 5.5.3    Bisphenol A and Its Uses

The importance of bisphenol A (BPA)—the acid-catalyzed condensation product of acetone and phenol—in the assembly of epoxy resins is stressed in 5.5.1. A possibly even more important reason that explains the growing demand for this building block are the polycarbonate resins (PCs), which are currently registering an annual growth rate of 7–8% (a growth rate threefold that of epoxies). The pricing of PBA distinguishes between epoxy grade (67¢) and polycarbonate grade (71¢) because, while the amorphous, cross-linked nature of epoxy resins does not depends on the presence of linear BPA molecules and can tolerate the presence of $o,p'$- and even $o,o'$-condensation products, optimum properties of the PCs require the linear, $p,p'$-BPA building block. The purification of BPA adds 4¢ to the cost of the polycarbonate grade. The production of PCs from BPA and either phosgene or its esters is described elsewhere. We can appreciate the effect on crystallinity and intermolecular attractive forces of the linear array of polymer molecules when we also consider the relatively limited freedom of rotation around the dimethyl-substituted central carbon.

Of the total demand of about 800MM lb of BPA, nearly 40% is consumed in the production of epoxy resins while practically all of the remaining 60% is converted into the 400MM lb of PCs and other, specialty high-performance thermoplastics referred to below. The recently expanding market for compact disks (CDs; > $930-million retail market in 1986 to supply the 225,000 CD players purchased in 1986 alone) is one of the driving forces behind the growth of PCs, but for this purpose the resin must be "superclean." Until recently, when Mobay began to produce this grade of PCs in the United States, most of this material was imported from West Germany and Japan. The unit price of PC ranges from $1.90 for general-purpose grade to about $2.15 for UV- and flame-resistant grades, and still higher for special grades.

Polycarbonates blend with ABS, PET, PBT, PPE, and other polymer families, and this is another important reason for their growth on the marketplace.

Apart of the central role of BPA in the production of epoxy and PC polymers, this difunctional phenol is the building block of high-performance specialty thermoplastics such as the poly(ether sulfones), polyimides, polyarylates, poly(ether ketones), and other families described in Chapter 9. Also, as mentioned previously, the phenolic nature of BPA facilitates its bromination of give the tetrabromo derivative that is incorporated in flame retarding materials.

### 5.5.4    Methyl Isobutyl Ketone (MIBK) and Other Aldol Condensation Products of Acetone

Like acetaldehyde (4.5.2), acetone possesses active hydrogens conducive to aldol conden-sations ($ONR$-2) and a carbonyl group that is not excessively hindered and is receptive to an attack by carbanions. In addition, the reactivity of acetone is manifested on both sides of the central carbonyl function. On the other hand, steric inhibition in acetone does prevent the formation of oligomers analogous to metaldehyde and paraldehyde (4.5.1).

The simplest aldol condensation product of acetone is diacetone alcohol [$(CH_3)_2C(OH)-CH_2-CO-CH_3$; 52¢]. The formation of this material is a classical illustration of "too much is *not* better." Thus, while alkaline conditions are needed to bring about the aldol condensation that generates diacetone alcohol, at the same time the alkaline conditions also catalyze the retro-aldol decomposition of diacetone alcohol to the monomeric molecules. Consequently, acetone is exposed in a fleeting manner to solid basic catalysts (such as oxides of barium or calcium or solid KOH), and the aldol condensation product is allowed to accumulate away from the catalyst that promoted its birth (see *OS* **I**, 193). Diacetone alcohol (b.p. 168°C) is stable under neutral conditions and can be used as a solvent, as a component of hydraulic fluids, and so on, but in the presence of acids it is dehydrated to mesityl oxide (46¢):

$$2CH_3-CO-CH_3 \rightleftharpoons (CH_3)_2C(OH)-CH_2-CO-CH_3 \rightarrow (CH_3)_2C=CH-CO-CH_3$$

Diacetone alcohol can be trapped and made to react under acidic conditions with appropriate substrates as illustrated by the synthesis of ethoquin, Monsanto's Santoquin ($2.25), an antioxidant added to animal feeds and also used as a rubber additive, as shown in Scheme 5.32. Another way to prevent a retro-aldol decomposition of diacetone alcohol is to remove its active hydrogens. This is achieved by hydrogenation to hexylene glycol (50¢):

$$(CH_3)_2C(OH)-CH_2-CH(OH)-CH_3, \text{ b.p. } 198°C$$

Hexylene glycol is used as a solvent and as an ingredient of brake and hydraulic fluids, particularly when the latter are based on castor oil (*SZM-II*, 26, 54–57, 71).

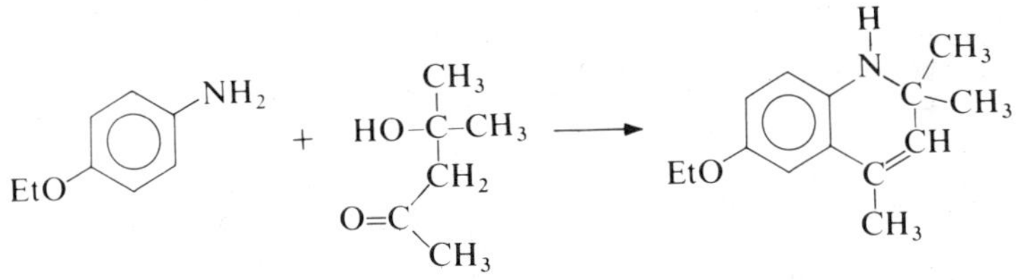

**Scheme 5.32**

Selective, mild hydrogenation of the carbon–carbon double bond of mesityl oxide produces (*FKC*, 543–546) about 150MM lb of the rather industrially important solvent MIBK (42¢ and 48¢, east and west of the Rockies, respectively), and further hydrogenation of the carbonyl group leads to another solvent, "methyl amyl alcohol," "methyl isobutyl carbinol," or 4-methyl-2-pentanol (55¢):

$$(CH_3)_2CH-CH_2-CO-CH_3 \quad \text{and} \quad (CH_3)_2CH-CH_2-CH(OH)-CH_3$$

MIBK is an excellent solvent for cellulose derivatives and is also used an extractant of niobium, zirconium, and hafnium compounds and is the building block of one of the acetylenic diols mentioned in 4.7.

The aldol condensation of mesityl oxide with an additional molecule of acetone (accompanied by spontaneous dehydration driven by sufficient resonance stabilization of the product) leads to the formation of *phorone*, and hydrogenation of phorone results in another useful solvent, diisobutyl ketone, (60¢):

$$(CH_3)_2C=CH-CO-CH=C(CH_3)_2 \xrightarrow[\text{cat}]{H_2} (CH_3)_2CHCH_2-CO-CH_2CH(CH_3)_2$$

Mesityl oxide also undergoes a base-catalyzed, intramolecular, 1,4-addition reaction that creates another building block of great current interest, isophorone (81¢; see Scheme 5.33. Isophorone has been used as a solvent for vinyl and cellulosic polymers for some time, but its importance increased by its conversion to isophoronediamine (IPD; Huels, $1.85) and the conversion of this diamine to the corresponding isophorone diisocyanate (IPDI; Huels, $2.70), mentioned (3.8.1, PUR) in connection with the design of light-stable PURs. The first step in the conversion of IPD to IPDI is an expected 1,4 addition of HCN to the conjugated system (3.10.2.3), and this is followed by an exhaustive hydrogenation of the HCN adduct in the presence of ammonia, which results in the reductive amination of the ketone function and a hydrogenation of the cyano function. The final step is, of course, a conversion of the diamine to the diisocyanate (Scheme 5.34):

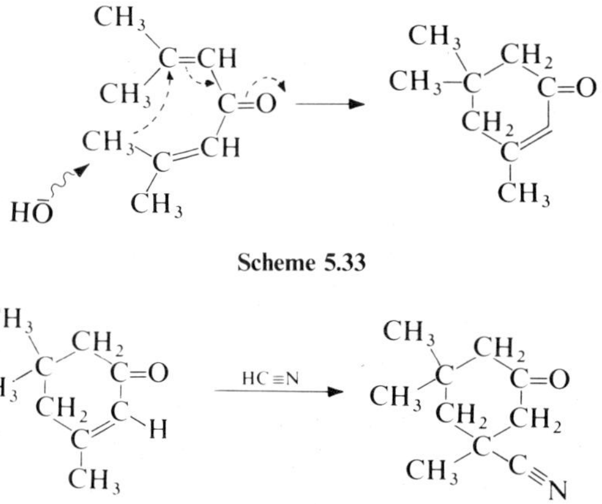

**Scheme 5.33**

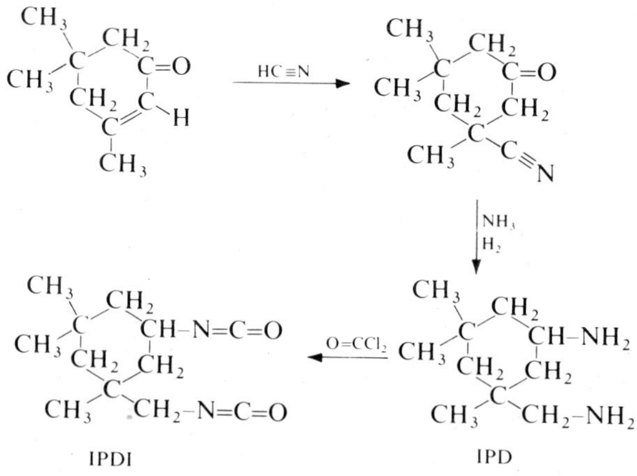

**Scheme 5.34**

The conversion of isophorone to 3,5-*xylenol* is also of interest but is discussed elsewhere (9.3.4). Another transformation of the fundamental acetone building block to an aromatic derivative is the formation, alas in poor yield, of 1,3,5-trimethylbenzene or mesitylene by means of sulfuric acid (*OS* II, 41). Along the same line, the cyanoethylation of acetone, followed by hydrogenation over nickel or palladium catalyst of the intermediate 5-keto-capronitrile, produces 2-methylpiperidine that can be dehydrogenated to α-picoline (10.3.1):

$$CH_3COCH_3 + CH_2{=}CH{-}C{\equiv}N \longrightarrow CH_3{-}CO{-}CH_2CH_2CH_2{-}C{\equiv}N$$

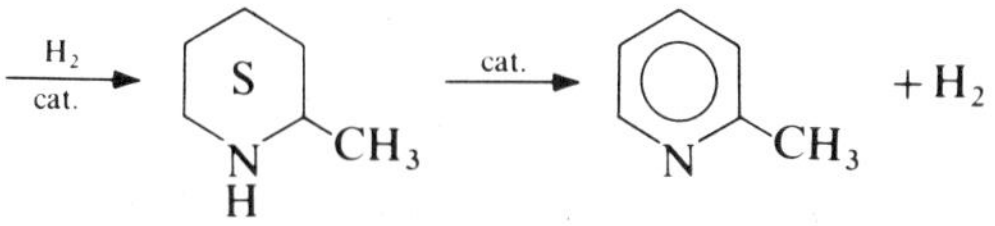

An amusing, but economically significant, utilization of isophorone is its role as a woodpecker repellent since these birds are responsible for significant destruction of electric and telephone poles.

### 5.5.5 Isopropenyl Acetate and Acetylacetone

The formation of *isopropenyl acetate* takes advantage of the almost negligible enol formation by acetone and the high reactivity of ketene that traps the enol in spite of the otherwise one-sided keto–enol equilibrium:

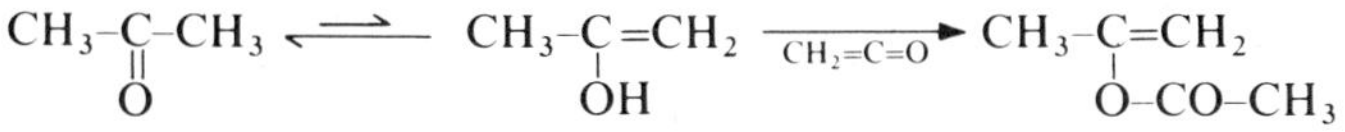

On a laboratory scale, isopropenyl acetate is formed when ketene is generated by the thermal decomposition of acetone:

$$CH_3{-}CO{-}CH_3 \rightarrow CH_3{-}\dot{C}O + \dot{C}H_3 \rightarrow CH_2{=}C{=}O + CH_4$$

and the gaseous products are passed into dry acetone, which contains a trace of acid catalyst (to promote the generation of the enol).

Isopropenyl acetate is a very effective acetylating agent. It functions under gentle conditions because of the presence of an excellent leaving group, namely, the acetone anion, which is immediately converted back to acetone. Thus, it is used to prepare acetates of secondary and even tertiary alcohols and to convert delicate carboxylic acid (e.g., unsaturated and hydroxylated ones, represented here by R–CO$_2$H) to mixed anhydrides, R–CO–O–CO–CH$_3$.

Isopropenyl acetate is the industrial precursor to acetylacetone generated by a rearrangement at about 450°C:

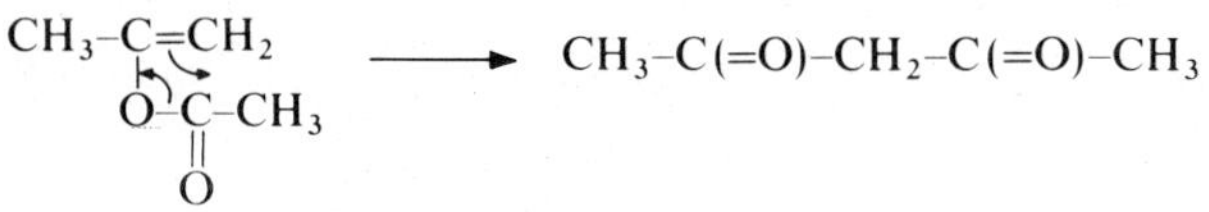

Acetylacetone (2,4-pentanedione, or acac) is best known for its ability to form covalent chelates of a variety of metals; see Scheme 5.35.

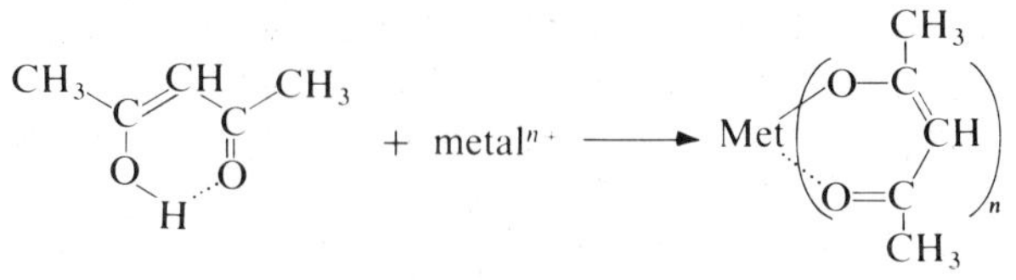

**Scheme 5.35**

These materials are soluble in organic solvents and rather volatile:

b.p., °C

| | |
|---|---|
| Ni(acac)$_2$ | 230 |
| Cr(acac)$_3$ | 340, Harshaw, $7.90 |
| Co(acac)$_3$ | Decomposes, m.p. 214–215°C, Harshaw, $13.00 |
| Fe(acac)$_3$ | Decomposes, m.p. 181°C, Harshaw, $7.80 |
| Al(acac)$_3$ | 315 |
| $n$-Bu$_2$Sn(acac)$_3$ | Liquid |
| Ru(acac)$_3$ | Decomposes, m.p. 235–237°C, Harshaw, $8.00 |

The cost of these metallic complexes reflects primarily the cost of the metal. The metallic chelates are ideal starting materials for the preparation of catalysts and metal-containing organic pigments.

Acetylacetone is used to extract valuable metals into organic solvents and to remove trace metals during the waste water treatments.

### 5.5.6    Miscellaneous Preparative Uses of Acetone

5.5.6.1    Dimethylhydantoin

The hydantoin synthesis involving acetone results in the formation of 5,5-dimethyl-hydantoin. The 1-$N$-methylol derivative, monomethyloldimethylhydantoin (MDMH):

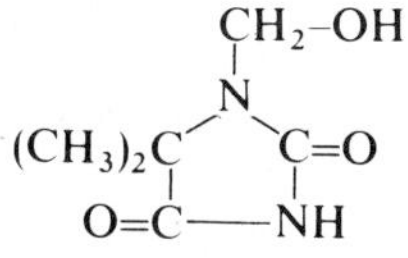

is converted on heating into a resin employed in the formulation of hair lacquers.

5.5.6.2    Ketals or Acetonides

The ketal of acetone and glycerol is distributed by Polysciences as Solketal ($\sim$ $22.00):

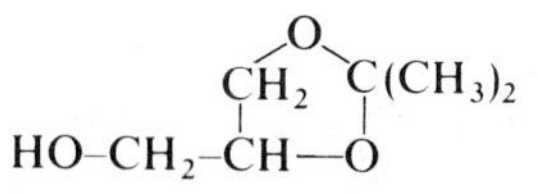

This glycerol dimethylketal, or glycerol acetonide, is claimed to be a solvent and plasticizer that, like most ketals, is stable under alkaline conditions.

A recent newcomer among acetone ketals is 2,2-dimethoxypropane, available from BFC Chemicals and required for the synthesis of the BFC insecticide Bendiocarb (9.2.4). However, 2,2-dimethoxypropane can also be used in transketalization and trans-alkoxylation reactions (to prepare, for example, methyl isopropenyl ether) and the formation of dimethylimines from primary amines. These possibilities are indicated in Scheme 5.36.

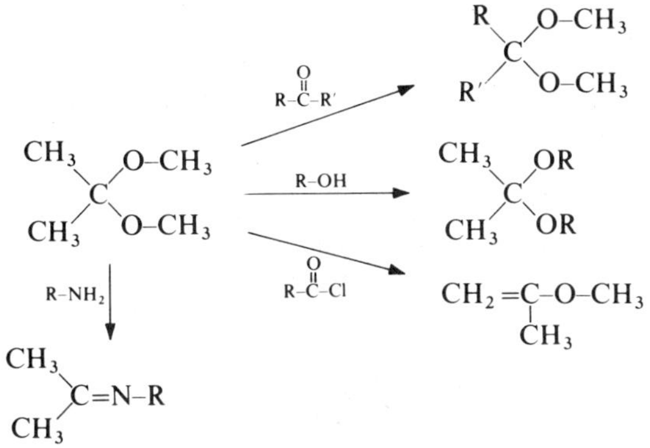

**Scheme 5.36**

### 5.5.6.3   Haloform Reaction Products

Acetone is a reasonably cheap starting material for the preparation of iodoform (NF $24.00) [*OS* **I**, 52; *JCE* **36**, 572 (1959)] as well as bromoform ($2.70) by means of the haloform reaction (*ONR*-56):

$$CH_3-CO-CH_3 + I_2 + NaOH \rightarrow HCI_3 + CH_3-CO_2Na + NaI$$

The high unit cost of iodoform is dependent on the equally high cost of iodine (crude $8.60, USP $17.00). Iodoform is used as a topical antiseptic. It can be also prepared from a mixture of iodine and sodium hydroxide and the organic substrates ethanol or acetaldehyde. However, these materials consume additional halogen because of the oxidative nature of the haloform reaction, and if the use of elementary iodine is to be avoided, one can employ sodium hypochlorite as the oxidizing agent and sodium iodide (USP $10.15) or potassium iodide (USP $9.15) as the source of iodine. Iodoform is reduced to methylene diiodide by means of sodium arsenite.

Bromoform (see above) is a rather dense liquid (2.90 g/mL) used in the separation of materials by difference in their specific gravity as mentioned in the case of methylene bromide (3.3.2.3). Bromoform, like its chlorine-containing analog chloroform, is a sedative but its use is restricted to veterinary medicine.

## 5.6   DOWNSTREAM C$_3$ BUILDING BLOCKS DERIVED NONPROPANE SOURCES FROM SOURCES OTHER THAN THE PROPANE SYSTEM

### 5.6.1   Propionic Acid

The origin of propionic acid and its derivatives is traced to the Reppe hydrocarbonylation reaction of ethylene (3.5.3.5). Propionic acid (33¢) is used for the preservation of fodder and grain, and it, as well as its sodium and calcium salts (54¢ and 50¢, respectively), are important food additives that inhibit the growth of molds in bakery products. The demand for propionic acid and its salts amounts to about 110MM lb. About half of this quantity is consumed as a food and feed additive, another 30% is used as a herbicide, and the remaining 20% becomes cellulose propionate and other propionic acid derivatives. Among the latter, one may mention N-acetopropionamide:

$$CH_3CH_2CO \cdot NH_2 + (CH_3CO)_2O \rightarrow CH_3CH_2CO\text{-}NH\text{-}COCH_3$$

which is the precursor of a substituted imidazole shown elsewhere (9.2.1).

### 5.6.2  Propionaldehyde

Propionaldehyde is the product of the hydrocarbonylation of ethylene. A recent use of propionaldehyde is the novel route to MMA mentioned in 5.5.2.

### 5.6.3  1,3-Propanediol

1,3-Propanediol is more difficult to obtain than either its smaller or larger glycol analogs, or, for that matter, than the isomeric 1,2-diol. Accordingly, 1,3-propanediol carries a rather high price tag ($6.00). It is produced by Shell in England and Degussa in West Germany and is distributed in the United States by Biddle Sawyer. Its origin may be the hydrogenation of malonic acid esters (see below), but it could also be obtained by hydrogenation of esters of 3-hydroxypropionic or hydracrylic acid. The latter can be derived from the corresponding nitrile (that turns out to be ethylene cyanohydrin, the product of HCN and EO), or from the Michael addition of water to the acrylic acid system.

### 5.6.4  Propargyl Alcohol and Its Derivatives

Propargyl alcohol is derived from acetylene, and some of its preparative applications are also shown elsewhere (4.7.2). An additional synthetic use of propargyl alcohol is its reaction with guanidinium chloride, which yields 2-aminopyrimidine—the building block of sulfapyrimidine, as shown in Scheme 5.37.

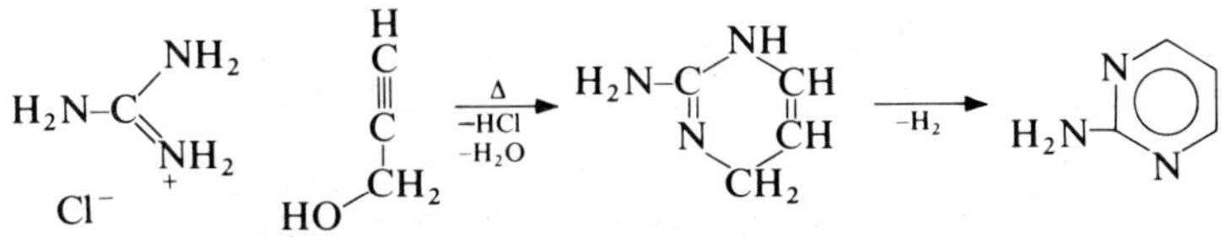

Scheme 5.37

### 5.6.5  Malonic Acid and Its Derivatives

Malonic acid and its derivatives can be derived from malonitrile [CH₂(CN)₂, MDN, Lonza, ~$8.00, m.p. 30–31°C, b.p. 218–219°C], in view of the large-scale production of the latter from acetonitrile and cyanogen chloride (3.10.3.4). Prior to this industrial development, the malonic acid system was assembled most conveniently by the reaction of chloroacetic acid (56¢) and sodium cyanide (71¢) in the presence of soda ash to give cyanoacetic acid, or by a Claisen condensation of methyl acetate with dimethyl carbonate (90¢) in the presence of sodium methylate to yield dimethyl malonate.

The methylation of dimethyl malonate yields the expected ester of 2-carboxypropionic acid, $CH_3-CH(CO \cdot OMe)_2$. However, in place of the malonic ester it is more convenient to employ the acetone ketal of malonic acid known as *Meldrum's acid*. It is obtained by means of the reaction of malonic acid with acetone in the presence of acetic anhydride:

$$CH_2(CO_2H)_2 + O{=}C(CH_3)_2 + Ac_2O \longrightarrow$$

The use of Meldrum's acid in the assembly of ibiprofen—the most recent of the three common OTC pain-killers—is shown elsewhere (9.3.2).

Another malonic acid derivative that is a building block of special $\beta$-keto esters is the magnesium salt of *p*-nitrobenzyl malonate, Orsynex's Mg-PNBM ($250/kg in 1–5-kg quantities). The prospective $\beta$-keto moiety is attached to the malonic ester reagent by means of the corresponding acylimidazole that is prepared from the desired carboxylic acid and carbonyldiimidazole:

$$R{-}CO_2H + Im{-}CO{-}Im \rightarrow R{-}CO{-}Im$$

$$\downarrow (p\text{-}O_2N{-}C_6H_4{-}CH_2{-}O{-}CO{-}CH_2{-}CO_2)_2Mg$$

$$(p\text{-}O_2N{-}C_6H_4{-}CH_2{-}O{-}CO{-}CH(CO{-}R){-}CO_2)_2Mg$$

where Im represents the imidazole moiety

On treatment with dilute acid, the acylated product is converted to the *p*-nitrobenzyl ester of the desired $\beta$-keto system. The blocking *p*-nitrobenzyl ester group is then removed by hydrogenolysis catalyzed by Pd–C to give the desired $\beta$-keto acid $R{-}CO{-}CH_2{-}CO_2H$.

Malononitrile (MDN) is a versatile building block of fine chemicals and plays a particularly important role in the synthesis of heterocyclic compounds (Chapter 10). This versatility can be attributed to

- Its multifunctional nature since it contains two cyano functions and two active hydrogens (its $pK_a$ of 11 is of the magnitude as the $pK_a$ 10 of phenol).
- The ability to form a dimer, namely, 2-amino-1,1,3-tricyano-1-propene, by the addition of the hydrogen of one molecule to the cyano function of another molecule, with a simultaneous formation of a new C–C bond and a tautomeric shift of a hydrogen to create a more highly conjugated system:

$$2CH_2(C{=}N)_2 \rightarrow (N{\equiv}C)_2{-}C({=}NH){-}CH_2{-}C{\equiv}N \rightarrow (N{\equiv}C)_2{-}C({-}NH_2){=}CH{-}C{\equiv}N$$

[This dimer (and the analogous trimer) can participate in a variety of subsequent reactions.]

- The elimination of HCN from initial reaction products when such elimination is conducive to the formation of resonance-stabilized products.

Some of the rather simple products derived from MDN are shown first.

1. The formation of another significant building block, *tetracyanoethylene* (TCNE) by halogenation of MDN:

$$Cl_2 + 2CH_2(C{\equiv}N)_2 \rightarrow (N{=}C)_2C{=}C(C{=}N)_2 + HCl$$

TCNE is a potent electron acceptor in the formation of charge-transfer complexes (CTCs) (R. Foster, *Organic Charge-Transfer Complexes*, Academic Press, New York, 1969). Also, it is an active participant in the formation of Diels–Alder adducts (6.8.1.4) such as the product of the reaction with 1,3-butadiene, namely, 3,3,4,4-tetracyanocyclohexene:

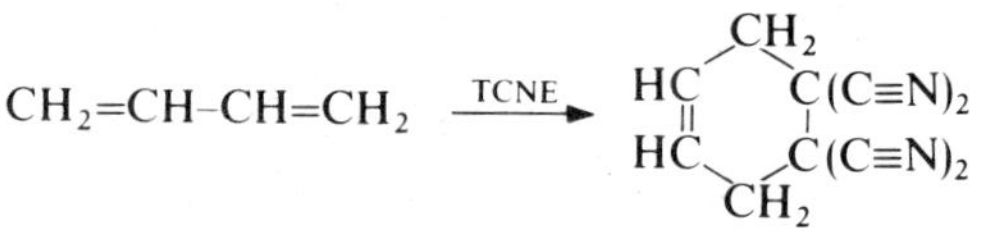

$$CH_2{=}CH{-}CH{=}CH_2 \xrightarrow{\text{TCNE}} \begin{array}{c} CH_2 \\ HC \overbrace{\phantom{xx}} C(C{\equiv}N)_2 \\ HC \underbrace{\phantom{xx}} C(C{\equiv}N)_2 \\ CH_2 \end{array}$$

In the presence of aqueous base TCNE and MDN form 1,1,2,3,3-pentacyanopropene, thus illustrating the base-catalyzed addition of MDN to the unsaturated system of TCNE, followed by the elimination of one HCN molecule:

$$(N{\equiv}C)_2CH_2 + (N{\equiv}C)_2C{=}C(C{\equiv}N)_2 \rightarrow (N{\equiv}C)_2C{=}C(C{\equiv}N){-}CH(C{\equiv}N)_2 + HC{\equiv}N$$

The highly electrophilic nature of TCNE is also demonstrated by its reactions with alcohols and ammonia that result in displacements of two cyano groups to give what at first sight appear to be condensation products of MDN and carbonate esters and urea, respectively:

$$(N{\equiv}C)_2C{=}C(OR)_2 \leftarrow (N{\equiv}C)_2C{=}C(C{\equiv}N)_2 \rightarrow (N{\equiv}C)_2C{=}C(NH_2)_2$$

2. The condensation products of MDN and aldehydes or ketones are formed as expected from the presence of two active hydrogens in MDN and the tendency to give the most resonance-stabilized product. Thus, for example, benzaldehyde and MDN gives benzalmalononitrile:

$$C_6H_5{-}CH{=}O + CH_2(C{\equiv}N)_2 \rightarrow C_6H_5{-}CH{=}C(C{\equiv}N)_2$$

However, the reaction products with formaldehyde demonstrate the methylol formation and methylene-bridging capacity of the latter, which lead to the formation of 2,2-dimethylolmalononitrile and 1,1,3,3-tetracyanopropane, respectively:

$$(HO{-}CH_2)_2C(C{\equiv}N)_2 \quad \text{and} \quad (N{\equiv}C)_2CH{-}CH_2{-}CH(C{\equiv}N)_2$$

3. The presence of active hydrogens in MDN also explains the ease of Michael addition reactions (*ONR*-60) illustrated here by the formation of the adduct with

benzalacetone or chalcone $(C_6H_5\text{-}CO\text{-}CH\text{=}CH\text{-}C_6H_5)$, namely, 1,1-dicyano-2,4-diphenyl-4-ketobutane:

$$C_6H_5\text{-}CO\text{-}CH_2\text{-}CH(C_6H_5)\text{-}CH(C\text{≡}N)_2$$

4. The reduction of the nitrile functions by means of lithium aluminum hydride (LiAlH$_4$) gives malonodialdehyde [CH$_2$(CH=O)$_2$] and partial hydrolysis with sulfuric acid produces malonamide [CH$_2$(CO·NH$_2$)$_2$].

With this glimpse at some of the typical behavior patterns of MDN, and with the forewarning of the ease of the formation of resonance-stabilized five- and six-membered heterocyclic rings, we can now offer a miniscule sample of the use of MDN in the preparation of materials of industrial importance. Also, the preceding list illustrates how the industrial production of a new raw material (in this case MDN) opens the gates to a newly accessible fine chemicals of known, or as yet undetermined, practical value. See Scheme 5.38.

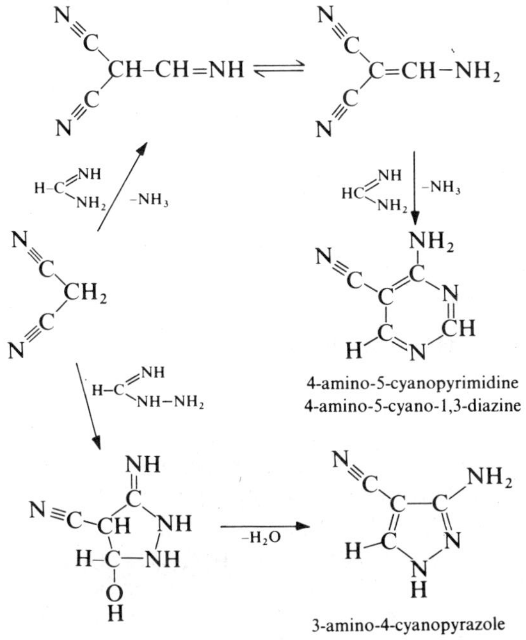

**Scheme 5.38**

### 5.6.6 Lactic Acid

Lactic acid can be obtained by fermentation of renewable feed stocks such as whey with *Lactobacillus*. *d,l*-Lactic acid (88%, $1.03; food grade $1.06; 50%, 62¢) is also obtained by fermentation of glucose, sucrose, and lactose (*SZM-II*, 30, 41). Lactic acid is used as an acidulant in foods, in the production of cheese, in printing and dyeing operations, in dehairing and decalcification of hides during the production of leather products, and in the production of solvents such as ethyl lactate and *n*-butyl lactate ($1.58). Lactic acid is also converted to the food additive calcium lactate (NF pentahydrate, $2.00; NF trihydrate, $2.10; NF dried, $2.80). In spite of the large quantity of whey obtained as a by-product of cheese manufacture and the significant lactic acid and lactose content of whey (most of which, by the way, is wasted), some lactic acid is also synthesized by the acid-catalyzed reaction of acetaldehyde and CO at 900 atm and at a temperature above 130°C (3.5.3.7). An alternative route to synthetic lactic acid is the reaction of acetaldehyde and HCN and the hydrolysis of the intermediate lactonitrile, but this hydrolysis would have to be carried out under alkaline conditions since acids would cause dehydration to give acrylic acid or its amide.

### 5.6.7 Pyruvic Acid and Aldehyde

Pyruvic acid ($CH_3$–CO–CO·OH) can be prepared by decomposition of tartaric acid, (NF $1.50), obtained as a sediment in the manufacture of wine or by oxidation of cellulose with $NO_2$ (*SZM-II*, 39 and 90, respectively). Also, pyruvic acid can be obtained by mild oxidation of propylene glycol (German Patent DE 3,012,004, 1981) or glycerol [*Angew. Chem., Int. Ed. Eng.* **21**, 540 (1982)].

Pyruvaldehyde ($CH_3$–CO–CH=O), in the form of the dimethyl acetal [$CH_3$–CO–CH(OMe)$_2$], is an important intermediate in the synthesis of C-substituted imidazoles (see 9.2.2). This dimethyl acetal is obtained by the acid-catalyzed methanolysis of dihydroxyacetone (*JOC* **41**, 2642 (1976)), a reaction that involves a synchronized shift of a hydride ion toward the transient carbon cation:

$$HO–CH_2–CO–CH_2–OH + H^+ \rightarrow HO–CH–CO–CH_2^+ + H_2O$$

$$\downarrow$$

$$(MeO)_2CH–CO–CH_3 \leftarrow 2MeOH + O=CH–CO–CH_3$$

The dimethylacetal of pyruvaldehyde can also be synthesized (U.S. Patent 4,158,019, 1979) by nitrosation of acetone with an alkyl nitrite (the *n*-butyl nitrite is commonly employed for such nitrosation reactions), a spontaneous tautomeric shift to give the corresponding oxime, followed by methanolysis of the oxime:

$$CH_3–CO–CH_3 + RO–N=O \rightarrow CH_3–CO–CH_2–N=O \rightarrow CH_3–CO–CH=N–OH$$

and

$$CH_3–CO–CH=N–OH + MeOH(H^+) \rightarrow CH_3–CO–CH(OMe)_2 + H_3N^+–OH$$

### 5.6.8 1,3-Dichloroacetone

1,3-Dichloroacetone (b.p. 173°C, m.p. 42–45°C) is obtained by direct chlorination of acetone, together with the lower-boiling monochloroacetone (b.p. 120°C), the isomeric 1,1-dichloroacetone (b.p. 120°C), and the higher chlorinated acetone derivatives such as 1,1,3-trichloroacetone (b.p. 90°C). These building blocks are produced by Wacker in West

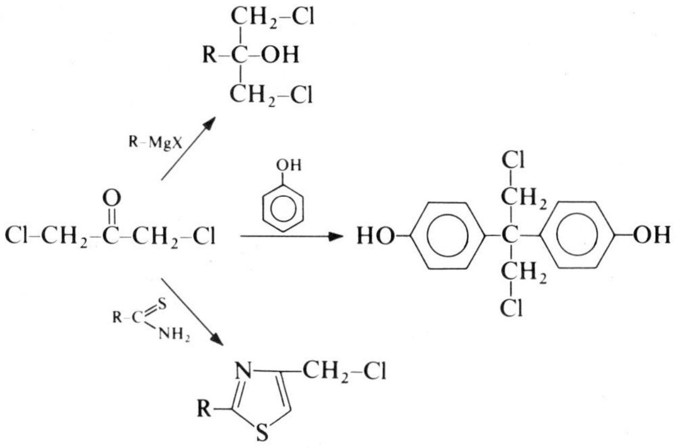

CH$_2$–Cl
|
R–C–OH
|
CH$_2$–Cl

R–MgX

O
||
Cl–CH$_2$–C–CH$_2$–Cl

OH

HO–⬡–C–⬡–OH

Cl
|
CH$_2$
|
|
CH$_2$
|
Cl

R–C⟨$_{NH_2}^{S}$

N——CH$_2$–Cl

R——S

**Scheme 5.39**

Germany and also by Ruetgers represented in the United States by Ruetgers–Nease. A small sample of the synthetic capability of the first-mentioned chlorinated acetone is presented in Scheme 5.39.

# 6

# C$_4$ BUILDING BLOCKS, INCLUDING ISOPRENE AND CYCLOPENTADIENE

## 6.1   OVERVIEW OF THE C$_4$ HYDROCARBON BUILDING BLOCKS

The butanes constitute 65–80% of LPG, with the remainder being propane (5.1). Until recently, most of the approximately 100B lb of available *n*- and iso-butanes were utilized by petroleum refiners to manufacture the alkylate component of marketed gasoline (6.3.1.1), and some of the butanes were added to gasoline in order to increase its vapor pressure. However, recent EPA rulings mandate a reduction of hydrocarbon emissions by lowering the vapor pressure of gasoline, and this creates a large surplus of C$_4$ hydrocarbons (P. Savage, "EPA Gasoline Rules Yield a Butane Bonanza," *CW* 8/12/87, 6–10). Another contribution to the current abundance of the butanes are LPG imports from Canada, the Middle East, and other countries. These circumstances are predicted to cause a drop in the cost of the butanes and a "renaissance in C$_4$ chemistry" (according to K. Brooks, *CW* 9/9/87, 36–39). So far, only about 10B lb of the available butanes were employed for chemical transformations (see Fig. 2.2), but this is bound to increase as a function of the strong demand for those C$_4$ derivatives that are part of the rapidly expanding 55B-lb polymer sector and those that function as fuel additives in the huge 100B-t gasoline market. Foremost among these materials are 1-butene (6.3.2), maleic anhydride (6.2.4, 6.6), and MTBE (6.3.1.4). The abundance of the butanes is also likely to benefit the producers of ethylene and propylene.

Some of the new outlets for the LPG hydrocarbons include

- British Petroleum's Cyclar process for the formation of benzene and other aromatics based on of a special zeolite catalyst regenerated by means of UOP technology. The formation of aromatics is accompanied by the production of valuable hydrogen required, among other things, for the hydrodesulfurization of petroleum and for a decrease in the aromatic content of diesel fuels that is under consideration by the EPA.
- The production of propylene by ethenylation (5.4.9) of butylenes.

Much of the successful expansion of the chemical uses of the butanes and the related butylenes is the result of progress in separation technology. Several routes are indicated in

Fig. 2.2 and are enumerated as follows:

1. The separation of the *n*-butenes and isobutylene by conversion of the latter to either MTBE or *t*-butyl alcohol (TBA). This separation occurs because of the relative ease with which isobutylene becomes protonated to give the *t*-butyl carbocation that reacts with either methanol or water, respectively. The increasing importance of MTBE as an octane-enhancing gasoline additive is mentioned elsewhere (3.3.2.2). TBA can be used for the same purpose, and it can also serve as the intermediate for the production of polymerization-grade isobutylene and the feedstock for pure methacrylic acid and its derivatives (6.3.1.4). Chemically pure isobutylene can also be prepared by thermal decomposition of MTBE, and the first plant of this type is being built by Shamprogetti in Hungary (K. Brooks, *CW* 9/9/87, 36–39).

2. A more delicate separation of the isomeric *n*-butenes from the saturated C$_4$ hydrocarbons (than the preceding case involving isobutylene) is the hydration of the former to *sec*-butyl alcohol that is oxidized to methyl ethyl ketone (MEK), a traditional solvent but now also considered to be an additive for methanol-based gasoline. The role of MEK in this new application is to decrease the volatility and hygroscopic properties of methanol. The formation of a secondary carbocation is achieved by Deutsche Texaco by means of acidic ion-exchange resins (3.4.3.4) under supercritical conditions.

3. The separation of 1,3-butadiene, isoprene, and aromatics from the less polarizable monoolefins or saturated hydrocarbons by means of extraction dipolar aprotic solvents (3.5.3.1) has been a common practice for some time, but a recent development (*CW* 8/26/87) allows the separation of butane and butenes and, furthermore, the separation of 1- and 2-butenes (b.p. −6.5 and +0.35°C, respectively) from the isomerization equilibrium mixture by means of a Krupp–Koppers solvent called "Butenex" (K. Brooks, *CW* 9/9/87, 36–39); the solvent is a morpholine derivative (most likely *N*-formylmorpholine, p. 426), and it affects differently the vapor pressure of each component of the mixture and thus provides the basis for an extractive distillation.

4. The acid-catalyzed formation of MTBE (in order to separate isobutylene from the less reactive components) can be followed by selective hydrogenation of the remaining C$_4$ compounds carried out under conditions that do not affect 1-butene, and a 100MM-lb plant based on this technology produces isobutylene (by cracking MTBE) and 1-butene for Sumitomo Chemical at Chiba, Japan.

The current unit prices of the C$_4$ hydrocarbons are as follows:

C$_4$H$_{10}$, 30¢/gal or 5.4¢/lb

*n*-Butylenes, 26¢/lb;

Isobutylene, 32¢/lb

The higher unit cost of isobutylene relative to that of the *n*-butenes can be attributed to its established uses for the (1) alkylation of benzene derivatives such as BHT (6.3.1.6); (2) production of TBA (6.3.1.3), the diisobutylenes, and butyl rubber (6.3.1.2); and (3) the formation of other compounds discussed in 6.3.1.

## 6.2  CHEMICAL USES OF THE BUTANES

Of the two butanes, *n*-butane and isobutane, the former is used in the majority of transformations in which the linear carbon skeleton is preserved. On the other hand, the presence of the tertiary hydrogen in isobutane is advantageous in either homolytic substitution reactions such as the oxidation to TBHP (6.2.5) and heterolytic reactions such as the alkylation of olefins (6.3.1.1).

### 6.2.1  1,3-Butadiene

The on-purpose cracking of $n$-butane supplements the 1,3-butadiene, or simply butadiene, that is isolated as a by-product of the cracking of higher petroleum fractions (see Fig. 2.2) during the manufacture of ethylene and propylene:

$$CH_3CH_2CH_2CH_3 \rightarrow CH_2{=}CH{-}CH{=}CH_2 + H_2$$

This dehydrogenation reaction is carried out at about 620°C in the presence of $Cr_2O_3$–$Al_2O_3$ catalyst (*FKC* 169). The chemistry of butadiene is discussed in Section 6.8.1, and since it provides the first opportunity to examine the Diels–Alder reaction (*ONR*-24), the related diene systems of isoprene and cyclopentadiene are included in this presentation of cycloaddition chemistry.

### 6.2.2  Oxidative Cracking of the Butanes

The oxidative cracking of the butanes (in the company of propane) is a remarkable feat of industrial separation technology since the mixture of products may include nearly a dozen organic products depeding on the condition of the process (*FKC*, 4–5, 11–12).

The reaction can be carried out in the absence of catalysts in the gas phase at 375–455°C with air at about 100 psi to give mainly formaldehyde, methanol, and acetaldehyde, together with significant amounts of $n$-propanol, isobutyl, and $n$-butyl alcohols, MEK, $C_4$–$C_7$ ketones, $C_5$–$C_7$ alcohols, and oxides of ethylene, propylene, and butylene. In order to simplify the separation, one can isolate formaldehyde, methanol, acetaldehyde, and acetone and hydrogenate the remaining components of the mixture to alcohols and hydrocarbons.

In the presence of steam, the reaction mixture contains mostly formaldehyde, methanol, acetaldehyde, and acetone, minor quantities of $n$-propanol and butanols, and a significant amount of a mixture of carboxylic acids.

The oxidation in the liquid phase is carried out in the presence of cobalt, manganese, or chromium acetates at 150–250°C, while air is bubbled through the reaction mixture to remove methanol, formic, acetic and propionic acids, acetone, MEK, and minor reaction products. The reaction conditions can be varied to yield mostly acetic or formic acids.

The major practitioners of the above-mentioned oxidation processes are Union Carbide and Celanese (now Hoechst-Celanese), and the recent explosion of the Celanese plant in Pampa, Texas, where the liquid-phase process was employed to produce over 500MM lb of acetic acid, is mentioned elsewhere (5.1). It is doubtful that these complex and inherently dangerous operations will survive the passing of time in view of the development of more direct routes to most of the products of interest.

### 6.2.3  Thiophene

The high-temperature reaction at about 600–700°C of $n$-butane with sulfur yields thiophene and hydrogen sulfide as a result of sulfur-induced dehydrogenation followed by addition of some $H_2S$ to unsaturated intermediates:

$$CH_3CH_2CH_2CH_3 + S_8 \rightarrow (C_4H_6, C_4H_8) + H_2S \rightarrow \langle\!\!\underset{S}{\bigcirc}\!\!\rangle$$

Among all heterocyclic compounds, thiophene resembles benzene most, and its chemistry is briefly described in 10.2.1.

### 6.2.4  Maleic Anhydride

The production of maleic anhydride (MA) has evolved from the traditional wasteful oxidative degradation of benzene (*FKC*, 514–518, *HP* 11/80, *ICN* 7/85), which employs vanadium pentoxide, to the more direct oxidation of *n*-butane (see 6.6.1). The success of the latter process depends, in part, on the fluidized-bed technology contributed among others by Standard Oil Chemicals (Cleveland) and The Davy Corporation, PLC of London, the joint effort of Monsanto and Du Pont (*CW* 1/28/87; *CMR* 1/26/87; J. Haggin, *CEN* 2/9/87; *CE* 2/16/87). Since much of the current interest in MA depends on its role as a precursor to 1,4-butanediol and THF (6.7), as well as other acetylene chemicals (4.7), a stimulus to the expansion and simplification of MA production was also provided by its low-pressure hydrogenation technology developed by Davy McKee and Union Carbide (*CE* 1/20/86, 11/24/86, 2/16/87; *CW* 12/10/86; *CB* 1/87).

The utilization of MA is discussed in Section 6.6.

### 6.2.5  Oxidation to *t*-Butyl Hydroperoxide (TBHP)

As stated above (see 5.4.2.1), the tertiary hydrogen of isobutane is susceptible to a selective, free-radical chain reaction with molecular oxygen. The process is carried out in the presence of appropriate initiators at relatively low temperatures (as close to 100°C as possible) and at about 35 bar. Also, for sake of safety, the reaction is *not* carried out to completion:

$$(CH_3)_3C-H + O_2 \rightarrow (CH_3)_3C-O-O-H$$

The desirable product, TBHP, can also be produced by the acid-catalyzed addition of hydrogen peroxide (70% 45¢) to isobutylene and by the acid-catalyzed exchange reaction of TBA with hydrogen peroxide. In order to avoid the accumulation of TBHP, ARCO utilizes the first mentioned route for the *in situ* conversion of propylene to PO and for the production of pure isobutylene by way of TBA:

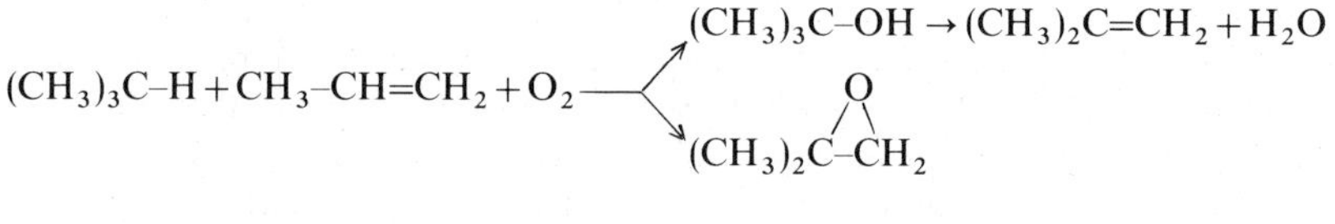

$$(CH_3)_3C-H + CH_3-CH=CH_2 + O_2 \left\langle \begin{array}{l} (CH_3)_3C-OH \rightarrow (CH_3)_2C=CH_2 + H_2O \\[2mm] \overset{\displaystyle O}{\underset{\displaystyle (CH_3)_2C-CH_2}{\bigwedge}} \end{array} \right.$$

The epoxidation of olefins by TBHP is catalyzed by naphthenates (8.1) of Mo, Ti, V, and other transition metals.

*t*-Butyl hydroperoxide and related hydroperoxides, R–O–OH, and organic peroxides, R–O–O–R′, are important free-radical polymerization catalysts and vulcanization agents that cure prepolymers to final polymeric materials. They are discussed in greater detail below (6.3.1.7).

Generally, hydroperoxides that do not contain an α-hydrogen are relatively stable and are approved for shipment by land and over water by the Interstate Commerce Commission.

## 6.3    CHEMICAL USES OF ISOBUTYLENE

### 6.3.1    Isobutylene

#### 6.3.1.1    Alkylation of Isobutylene in the Presence of Isobutane

Alkylation of isobutylene in the presence of isobutane and the analogous reactions involving the $C_3$ building blocks represent one of the major processes utilized by the petroleum refining industry in the manufacture of high-octane gasoline. These operations consume a significant amount of sulfuric acid (virgin 100%, 3.407¢–4.25¢, depending on location) and anhydrous hydrogen fluoride (68.75¢) (G. Parkinson, "Which Catalyst for Alkylation?" $CW$ 2/24/88). The overall reaction is deceptively simple as illustrated by the formation of 2,2,4-trimethylpentane—commonly known as "isooctane," the original standard for determination of octane numbers with an arbitrarily assigned ON of 100:

$$(CH_3)_3C\text{–}H + CH_2{=}C(CH_3)_2 \rightarrow (CH_3)_3C\text{–}CH_2\text{–}CH(CH_3)_2$$

The octane numbers reported on gasoline pumps are an algebraic average of a research octane number (RON) and a motor octane number (MON), determined under specified operating conditions of an automotive engine. The effect of branching of alkanes on the performance of an internal combustion engine operating at relatively high internal pressures is illustrated by the fact that the nearly isomeric $n$-heptane has an assigned ON of zero. Octane numbers are not applicable to diesel engines that operate on fuel evaluated by "cetane numbers."

The production of isooctane is triggered by the acid-catalyzed formation of the relatively stable $t$-butyl carbocation. The latter adds to isobutylene to yield a nearly equally stable carbocation $(CH_3)_3C\text{–}CH_2\text{–}C^+(CH_3)_2$, and this species is capable of abstracting a hydride ion from isobutane to give the desired product, and, at the same time, it generates a new $t$-butyl carbocation. Thus the ionic chain reaction continues until the reactants are depleted. Propylene can participate in the alkylation reaction by forming the isopropyl carbocation that abstracts a hydride ion from isobutane to give a "propylene alkylate" (Fig. 2.2) constituted mainly by 2,2-dimethylpentane (5.4.1). The alkylation process is actually more complex because of the ease with which carbocations rearrange by virtue of a shift of hydride and methyl carbanions to the electron-deficient carbocation centers. Thus, about 60% of the alkylation products obtained from the acid-catalyzed reaction of propylene and isobutane result from initial rearrangements rationalized as shown in Scheme 6.1. However, since a $t$-butyl carbocation is also generated by a hydride ion transfer from isobutane to the isopropyl carbocation

$$(CH_3)_3C\text{–}H + (CH_3)_2CH^+ \rightarrow (CH_3)_3C^+ + (CH_3)_2CH_2$$

and some of the resulting carbocations can lose a proton to give propylene and isobutylene, it is not surprising that the alkylation products obtained from propylene and isobutane contain about 10% of the above-mentioned isooctane.

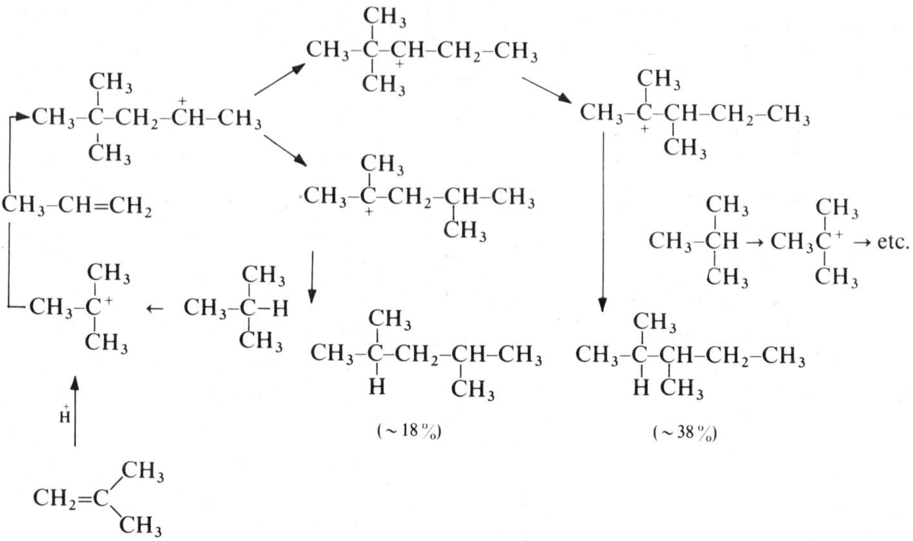

$$CH_3-\overset{\overset{\displaystyle CH_3}{|}}{\underset{\underset{\displaystyle CH_3}{|}}{C}}-\overset{+}{CH}-CH_2-CH_3$$

$$CH_3-\overset{\overset{\displaystyle CH_3}{|}}{\underset{\underset{\displaystyle CH_3}{|}}{C}}-CH_2-\overset{+}{CH}-CH_3$$

$$CH_3-\overset{\overset{\displaystyle CH_3}{|}}{\underset{\underset{\displaystyle CH_3}{|}}{C}}-\overset{+}{CH}-CH_2-CH_3$$

$$CH_3-\overset{\overset{\displaystyle CH_3}{|}}{\underset{}{CH}} \rightarrow CH_3\overset{\overset{\displaystyle CH_3}{|}}{\underset{}{C^+}} \rightarrow \text{etc.}$$

$$\overset{+}{CH_3}-\overset{\overset{\displaystyle CH_3}{|}}{\underset{\underset{\displaystyle CH_3}{|}}{C}}-CH_2-\overset{+}{CH}-CH_3$$

$$CH_3-CH=CH_2$$

$$CH_3-\overset{\overset{\displaystyle CH_3}{|}}{\underset{\underset{\displaystyle CH_3}{|}}{C^+}} \leftarrow CH_3-\overset{\overset{\displaystyle CH_3}{|}}{\underset{\underset{\displaystyle CH_3}{|}}{C}}-H$$

$$CH_3-\overset{\overset{\displaystyle CH_3}{|}}{\underset{\underset{\displaystyle H}{|}}{C}}-CH_2-\overset{\overset{\displaystyle }{}}{\underset{\underset{\displaystyle CH_3}{|}}{CH}}-CH_3$$

$$(\sim 18\%)$$

$$CH_3-\overset{\overset{\displaystyle CH_3}{|}}{\underset{\underset{\displaystyle H}{|}}{C}}-\overset{}{\underset{\underset{\displaystyle CH_3}{}}{CH}}-CH_2-CH_3$$

$$(\sim 38\%)$$

$$\overset{+}{H}$$

$$CH_2=C\overset{\diagup CH_3}{\diagdown CH_3}$$

**Scheme 6.1**

With the dynamic picture of carbon skeleton rearrangements and shifts of hydrogen species in mind, we can now picture the major components that the "butylene alkylate" (Fig. 2.2) that contains, in addition to isooctane, also some 2,3,4-trimethylpentane, 2,3,3-trimethylpentane, and 2,2,3-trimethylpentane, as shown in Scheme 6.2. It is clear that the actual composition of alkylation processes based on $C_3$ and $C_4$ feedstocks depends on the composition of reactants and reaction conditions.

It is also noteworthy that the selective behavior of isobutane as a source of hydride ions vis-à-vis carbocations provides a chemical method for the isolation of the relatively inert n-butane that becomes the modern source of MA (6.2.4) and, on dehydrogenation to the olefins, also the source of 1- and 2-butenes (see 6.3.2 and Fig. 2.2).

### 6.3.1.2   Dimers, Oligomers, and Polymers of Isobutylene

Dimers, oligomers, and polymers of isobutylene are obtained by means of acidic catalysts such as those mentioned in the preceding section, as well as by acidic ion-exchange resins, phosphoric acid-impregnated solid supports, and so on. The final products result from the loss of a proton to give, usually, the most substituted olefins such as:

$$(CH_3)_3C-CH=C(CH_3)_2 \qquad (CH_3)_3CCH_2C(CH_3)_2CH=C(CH_3)_2$$

Diisobutylene, 37¢                  Triisobutylene, 45¢

Hydrogenation of the $C_8$ olefin yields the same isooctane as shown above to be formed in the isobutylene–isobutane alkylation reaction, and this, indeed, can be a method for enriching gasoline with a high-octane additive. The higher oligomerization products of isobutylene provide, on hydrogenation, solvents for the production of HDPE, LLDPE, PP, and other polyolefins since the polymers tend to precipitate from the initially homogeneous reaction mixtures. Exxon's Isopar solvents are an example of such solvents that range in boiling points from 98°C to over 250°C.

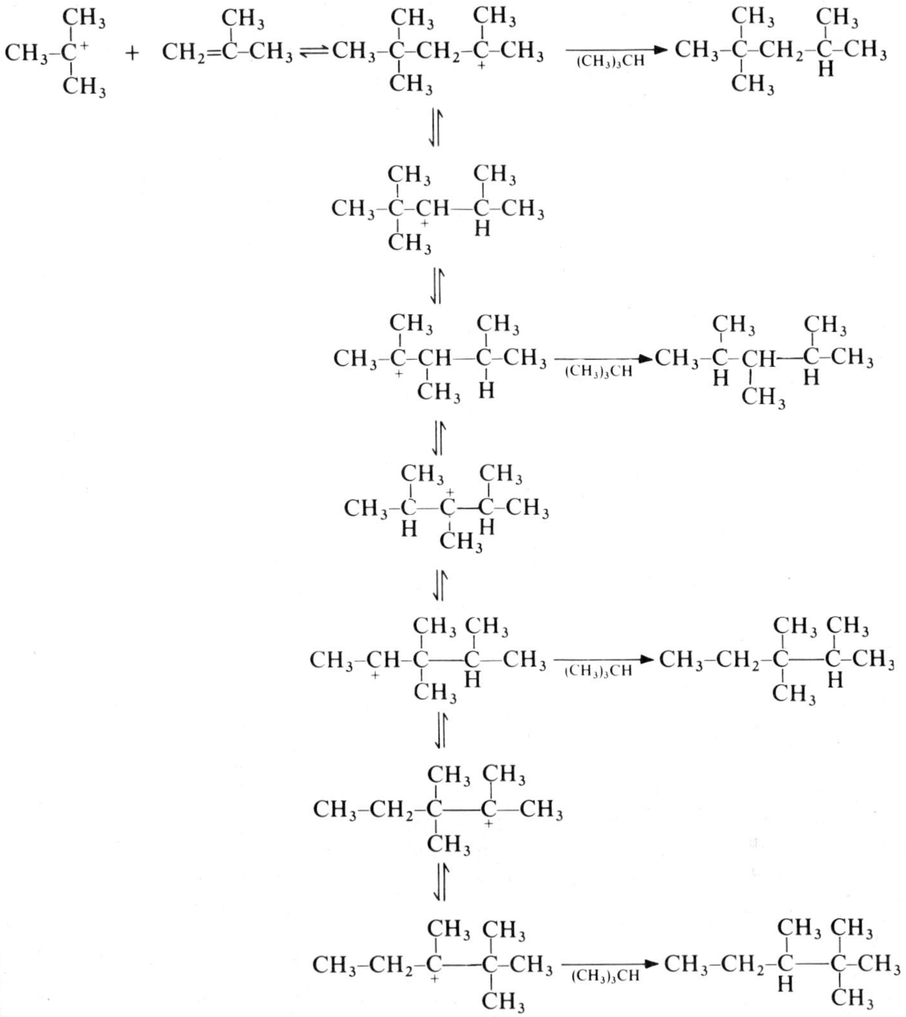

$$\underset{\underset{CH_3}{|}}{\overset{\overset{CH_3}{|}}{CH_3\text{-}C^+}} + \underset{}{\overset{\overset{CH_3}{|}}{CH_2\text{=}C\text{-}CH_3}} \rightleftharpoons \underset{\underset{CH_3}{|}}{\overset{\overset{CH_3}{|}}{CH_3\text{-}C}}\text{-}CH_2\text{-}\overset{\overset{CH_3}{|}}{\underset{+}{C}}\text{-}CH_3 \xrightarrow{(CH_3)_3CH} \underset{\underset{CH_3}{|}}{\overset{\overset{CH_3}{|}}{CH_3\text{-}C}}\text{-}CH_2\text{-}\overset{\overset{CH_3}{|}}{\underset{H}{C}}\text{-}CH_3$$

$$\Updownarrow$$

$$\underset{\underset{+}{\overset{|}{CH_3}}}{\overset{\overset{CH_3}{|}}{CH_3\text{-}C}}\text{-}CH\text{-}\overset{\overset{CH_3}{|}}{\underset{H}{C}}\text{-}CH_3$$

$$\Updownarrow$$

$$\underset{\underset{CH_3}{|}}{\overset{\overset{CH_3}{|}}{CH_3\text{-}\underset{+}{C}}}\text{-}CH\text{-}\overset{\overset{CH_3}{|}}{\underset{H}{C}}\text{-}CH_3 \xrightarrow{(CH_3)_3CH} CH_3\text{-}\overset{\overset{CH_3}{|}}{\underset{H}{C}}\text{-}CH\text{-}\overset{\overset{|}{CH_3}}{\underset{H}{C}}\text{-}CH_3$$

$$\Updownarrow$$

$$CH_3\text{-}\overset{\overset{CH_3}{|}}{\underset{H}{C}}\text{-}\overset{+}{C}\text{-}\overset{\overset{CH_3}{|}}{\underset{H}{C}}\text{-}CH_3$$

$$\Updownarrow$$

$$CH_3\text{-}\underset{+}{CH}\text{-}\overset{\overset{CH_3}{|}}{\underset{CH_3}{|}}{C}\text{—}\overset{\overset{CH_3}{|}}{\underset{H}{C}}\text{-}CH_3 \xrightarrow{(CH_3)_3CH} CH_3\text{-}CH_2\text{-}\overset{\overset{CH_3}{|}}{\underset{CH_3}{|}}{C}\text{—}\overset{\overset{CH_3}{|}}{\underset{H}{C}}\text{-}CH_3$$

$$\Updownarrow$$

$$CH_3\text{-}CH_2\text{-}\overset{\overset{CH_3}{|}}{\underset{CH_3}{|}}{C}\text{—}\overset{\overset{CH_3}{|}}{\underset{+}{C}}\text{-}CH_3$$

$$\Updownarrow$$

$$CH_3\text{-}CH_2\text{-}\overset{\overset{CH_3}{|}}{\underset{CH_3}{|}}{\underset{+}{C}}\text{—}\overset{CH_3}{C}\text{-}CH_3 \xrightarrow{(CH_3)_3CH} CH_3\text{-}CH_2\text{-}\overset{\overset{CH_3}{|}}{\underset{H}{C}}\text{—}\overset{\overset{CH_3}{|}}{\underset{CH_3}{|}}{C}\text{-}CH_3$$

**Scheme 6.2**

The amination (see 6.3.1.5) of diisobutylene yields the highly hindered primary amine, 2,2,4,4-tetramethylbutylamine, or $t$-octylamine ($2.60).

The addition of hydrogen sulfide to triisobutylene gives $t$-dodecyl mercaptan, $(CH_3)_3CCH_2C(CH_3)_2CH_2C(CH_3)_2\text{-}SH$, a typical chain-transfer agent employed to control the molecular-weight distribution in free-radical catalyzed polymerizations.

During the early years of synthetic detergents, triisobutylene was also the main source of alkylbenzenesulfonates, but the absence of an $\alpha$-hydrogen in $(CH_3)_3CCH_2(CH_3)_2CH_2C(CH_3)_2\text{-}C_6H_4\text{-}SO_3^-\,Na^+$ (mostly para and meta) inhibited biodegradability of this branched sodium alkylbenzenesulfonate. This caused an accumulation of foam in rivers and lakes and indicated to the surfactant industry the need to switch to, more or less, linear alkylbenzenesulfonates (LAB) or "detergent alkylates" mentioned in connection with the α-olefins (AOs) (7.3), which do contain an α-hydrogen. In the case of alkylated phenols, the phenolic moiety does provide an opportunity for biodegradability, and thus branched alkyl groups derived from di- and triisobutylene can be employed to contribute the hydrophobic (or lipophilic) component of the detergent. Thus, for example, acid-catalyzed alkylation of phenol (34¢) gives

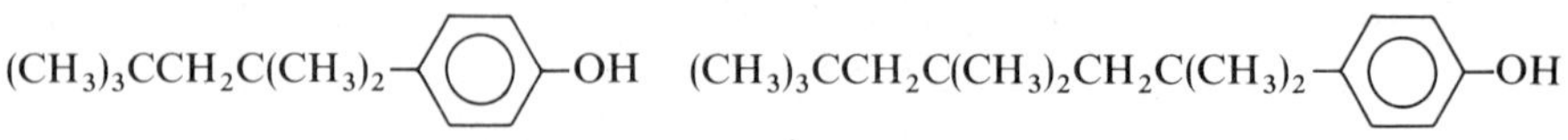

$(CH_3)_3CCH_2C(CH_3)_2$—⟨O⟩—OH    $(CH_3)_3CCH_2C(CH_3)_2CH_2C(CH_3)_2$—⟨O⟩—OH

*p-t*-octylphenol, 75¢                                    *p-t*-dodecylphenol, 48¢

The polymerization of isobutylene in the presence of about 3–5% of isoprene (6.8.2) gives rise to *butyl rubber* (current demand ∼ 300 MM 1b):

$$[-CH_2C(CH_3)_2]_n-CH_2-C(CH_3)=CH-CH_2-$$

The inclusion of some isoprene in the otherwise saturated polybutylene chain allows cross-linking in order to provide elastic memory (pp. 154, 198, 204). Vulcanization of butyl rubber is brought about by means of peroxides or by addition of chlorine or bromine and treatment with metallic oxides. Halogenated butyl rubber, such as Exxon's Chlorobutyl and Goodrich's Hycar 2202, have been known for several decades.

Butyl rubber is used in the manufacture of rubber tires in the form of an inner layer that is self-sealing on accidental puncture. The demand for this and other types of rubber is not keeping up with the improvement of the GNP in the United States because of continuing large imports of fully equipped vehicles, down-sizing of automobiles, and the improved performance of rubber tires.

Polyisobutylenes of relatively low molecular weight (< 3000) are gummy products used in caulking and sealant formulations as well as lubricants and adhesives. Also, they are added to such polymers as PS in order to improve the impact resistance of the final products and the flow properties during processing.

Amoco Chemicals is a supplier of liquid polybutenes (38¢–43¢, depending on viscosity range), obtained by polymerization of an isobutylene-rich butene feed that contains a small amount of linear butylenes. The molecular-weight distribution of the different products peaks between 330 and 4000, and, depending on the average molecular weight, the materials can be used in the formulation of pressure-sensitive or hot-melt adhesives, cutting and other metalworking oils, as lubricants, plasticizers, and so on. The poly-butenes can be chemically modified because of the reactivity of the allylic hydrogen near the residual, terminal double bond. Thus, the reaction with phosphorus pentasulfide or maleic anhydride gives the following derivatives:

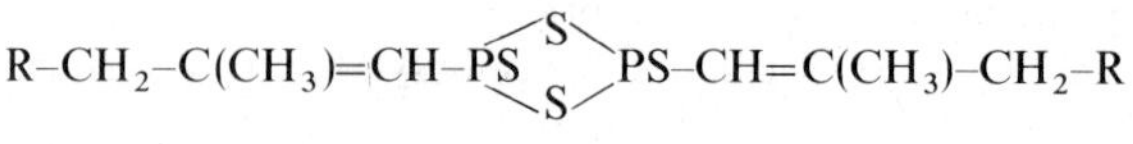

$$R-CH_2-C(CH_3)=CH-\overset{S}{\underset{S}{PS}}PS-CH=C(CH_3)-CH_2-R$$

and

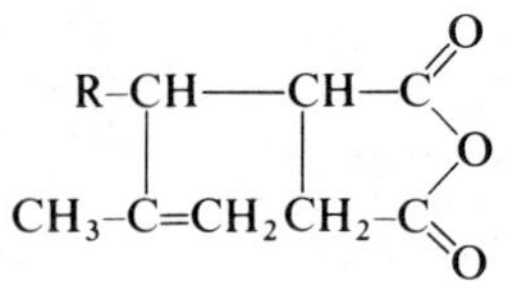

where R represents the residual chain of the polybutene, namely

$$(CH_3)_3C-[CH_2-C(CH_3)_2]_n-$$

and the "activated polybutenes" available as Amoco's Actipol contain a single epoxide function at one of the terminals.

### 6.3.1.3   MTBE, TBA, and Related Compounds

MTBE, TBA, and related compounds are either direct or indirect derivatives of isobutylene. The two first-mentioned materials are obtained by an acid-catalyzed reaction of isobutylene with methanol (3.3.2.2) or water, respectively, and, as stated before, they are of increasing importance as octane-enhancing components of gasoline in place of TEL and TML (p. 207). In November 1986 the unit price of MTBE, 64¢/gal; compared favorably with that of competitive octane enhancers: government-subsidized fuel ethanol, 79¢/gal; and toluene, 67¢/gal. The RON and MON of MTBE are reported as 115–135 and 98–115, respectively, and the demand in the United States for MTBE was about 10B lb in 1987. With a current gasoline demand of about 3MM bbl/day, the demand for MTBE in the United States could theoretically rise to a level of about 23B lb/yr. On a global scale, the demand for MTBE for the decade of the 1990s is expected (R. Remirez, *CE* 5/25/87) to reach about 12MM $t^m$, or 26.5B lb, with the United States and Europe consuming about 50 and 33%, respectively, of this huge amount. Such a demand would be expected to cause a shortage of isobutylene and stimulate the use of isobutane dehydrogenation—the Catofina process—developed by Air Products (*CW* 6/4/86). After subjecting the $C_4$ olefins to an acid-catalyzed reaction with methanol, which traps isobutylene as MTBE, fractionation of the residual *n*-butenes [the boiling points of 1-butene and 2-butene are $-6.5$ and about $+0.4$ (trans) and $+3.7$ (cis) °C, respectively] allows the separation of the isomeric *n*-butenes.

The projected use of TBA (70¢) serves to decrease the moisture-absorbing properties of methanol fuel (3.3.1), and the relatively high cost of TBA is partially offset by the high octane number of methanol (130). The current production of about 400MM lb of TBA is employed in part as a source of "polymer-grade" isobutylene used for the production of well-characterized polymers.

The recently aroused interest in the homolog of MTBE, namely ethyl *t*-butyl ether (ETBE), is reflected by the following reports: C. Verbanic, "ETBE: Ethanol's Motor Fuel Hope?," *CB* 10/88, 38–39; E. Anderson, "Ethyl *tert*-Butyl Ether Shows Promise as Octane Enhancer," *CEN* 10/24/88, 11–12.

From a viewpoint of the national interest, the shift from MTBE to ETBE is logical in view of a greater utilization of domestic renewable resources, and lower volatility and hygroscopic properties, as well as a somewhat higher octane rating of the higher ether. The outcome of this interproduct competition should be decided on the basis of the economic picture, which must take into account taxes imposed on gasoline and subsidies legislated on the production of corn and fermentation ethanol and a potential reduction of imported petroleum.

### Synthesis of Industrially Important Peptides

A discussion of the industrial importance of TBA cannot fail to mention its role in the synthesis of industrially important peptides. Peptides are assemblies of α-amino acids illustrated here by a *tri*peptide, that is, a structure assembled from *three* amino acids:

$$H_2N–CHR''–CO–NH–CHR'–CO–NH–CHR–CO·OH$$

Peptides are, of course, miniproteins, and the entire field of $\alpha$-amino acids and its biopolymers has now become a high-tech segment of the chemical industry growing at an annual rate of about 10%. The consumption of peptides in the United States is expected (*CMR* 11/23/87) to reach a level of $4.3 billion by 1996, and this estimate includes human health-care products—a $700-million business in 1986 and predicted to grow to $2.4 billion by 1996. Other major growth areas of the peptide market are food products, a $720-million business in 1986 and expected to reach $1 billion by 1996; agricultural uses, a $200-million business in 1986 and expected to double by 1996; and industrial applications, consuming only $19 million in 1986 but expected to reach $100 million in 1996. Finally, veterinary applications of peptides are also expected to expand rapidly from the currently small level.

The challenge of assembling peptides from different $\alpha$-amino acids (AAs) was met by two important developments:

1. the successful design of AA-protecting groups, that is, groups that prevent one of the two reactive terminals (usually the amino terminal in the $H_2N–CHR''–CO–OH$ moiety shown above) from undergoing undesirable side reactions while the other reactive group, usually the carboxylic acid function, is being attached to the next AA partner.

2. The simplification of the procedure for the isolation and purification of desired products between successive operations that attach one AA after another until the desired peptide is assembled.

The latter problem was solved by the R. B. Merrifield, who introduced in 1964 the principle of solid phase synthesis of peptides for which he received the Nobel prize in 1984. The technique consists of the attachment of one terminal of the AA to a polymeric matrix, say, for example, a somewhat cross-linked, substituted PS. Now, the insolubilized AA is subjected to coupling with the next AA, thus forming a dipeptide. Successive treatments with appropriate AA reactants are alternated by washing and removal of impurities and by-products from the growing peptide. Eventually, when the desired peptide is assembled, it is detached from the polymeric matrix without causing structural damage to the rather delicate peptide molecules, and also while racemization of the chiral centers present in most AAs at the $\alpha$-carbon is avoided.

Over the decades, since Emil Fischer (1852–1919) elucidated the nature of amino acids, peptides, proteins (and various other areas of organic chemistry for which he became the second awardee of a Nobel prize in 1902), much progress has been made in the design of AA-protecting groups and the modernization of the solid-phase synthesis principle. The latter technology is now automated by computer-controlled sequential cycles of addition of properly protected AAs and the removal of undesired by-products. Also, this technique has been expanded to the synthesis of nucleic acids—structures responsible for genetic characteristics of all living species (see 10.3.2).

One of the early groups introduced to protect the amino terminal of AAs from participation in undesirable reactions is the $t$-butoxycarbonyl moiety $[(CH_3)_3C–O–CO–]$, usually abbreviated as Boc, BOC, or when attached to the AA amino terminal, as $N$-$t$-BOC. It would seem that the Boc group could be introduced by means of the chloroformate of TBA but, unfortunately, this materials decomposes under ordinary conditions:

$$(CH_3)_3C-O-CO-Cl \rightarrow (CH_3)_3C-Cl + CO_2$$

Hence, the original procedure for the attachment of the Boc group to AAs involves the preparation of an intermediate $t$-butoxycarboxazide by way of $t$-butyl $S$-methyl thio-carbonate and $t$-butoxycarbohydrazide:

$$t\text{-BuO}^-\ Na^+ + O{=}C{=}S \rightarrow t\text{-BuO-CO-S}^-\ Na^+ \xrightarrow[\text{MeI}]{\text{MeI}} t\text{-BuO-CO-S-Me}$$

$$t\text{-BuO-CO-S-Me} \xrightarrow{H_2N-NH_2} t\text{-BuO-CO-NH-NH}_2 \xrightarrow{\text{HONO}} t\text{-BuO-CO-N}_3$$

The use of these costly reagents and a three-step synthesis account for the high price of Boc derivatives of AAs. In the case of relatively simple AAs such as alanine $[CH_3-CH(NH_2)-CO_2H]$ and phenylalanine $[Ph-CH_2-CH(NH_2)-CO_2H]$, the unit prices compare as follows (Sigma, 1987):

| | | | |
|---|---|---|---|
| DL-Alanine | $9.40/lb | DL-Ph-alanine | $63.75/lb |
| L-alanine | $60.00/lb | L-Ph-alanine | $56.70/lb |
| D-alanine | $440.00/lb | D-Ph-alanine | $460.00/lb |
| N-t-BOC-L-alanine | $459.00/lb | N-t-Boc-L-Ph-alanine | $683.00/lb |
| N-t-BOC-D-alanine | $2540.00/lb | N-t-Boc-D-Ph-alanine | $2180.00/lb |

Another traditional AA-protecting group is the benzyloxycarbonyl, $Ph-CH_2-O-CO-$, moiety, Cbz, or N–Cbz when attached to the AA amino terminal. It is introduced as the chloroformate of benzyl alcohol ($Ph-CH_2-O-CO-Cl$; Aldrich, $102.00/lb).

The two above-mentioned protecting moieties are removed by different procedures that do not interfere with the partner groups: Boc is removed by mild treatment with HCl–acetic acid, while Cbz is cleaved by mild hydrogenation in the presence of palladium catalysts. A protecting group similar to Cbz is the 9-fluorenylmethyloxycarbonyl moiety, Fmoc or FMOC, introduced as the 9-fluorenylmethyl chloroformate (see 9.3.5), (Sigma, $2,310.00), which differs from the somewhat analogous benzyl compound by being readily cleaved under mildly basic conditions.

Once the protecting group is cleaved (or "deprotected"), and the freed amino terminal is coupled with the next AA through its carboxylic acid terminal (in a manner indicated elsewhere), we now have a dipeptide attached at one end to the polymeric matrix. It still contains a protecting group at the newly introduced AA moiety, but the deprotection, coupling, deprotection, coupling cycles can now be repeated ad infinitum until the desired oligopeptide or polypeptide is assembled. At that point the product is removed from the polymeric support (in a manner also indicated elsewhere).

In the context of industrial peptide synthesis, we must examine the industrial consumption of the amino acid building blocks ($SZM$-$II$, 161–165). The value of the worldwide production of AAs is reported ($CMR$ 5/11/87) to have passed in 1986 the 1B-lb quantity, valued at $1.5 billion. The breakdown of this total AA market showed $90 million consumed in research and for analytical purposes, $120 million used for medicinal and pharmaceutical purposes, $565 million channeled into the animal feed market, about $578 million used as flavoring agents and food supplements, and $190 employed as chemical intermediates. Specifically, cysteine, consumed at a 0.1MM-lb level and valued at about $8–$11/lb, is used as a dough conditioner (to decrease fermentation time because of its ability to break down proteins), as a meat flavoring agent, in cosmetic formulations, as a health food additive, and so on. Methionine (pp. 182, 296) and lysine (*l*-isomer hydrochloride, $1.40, pp. 182, 419) are important food and feed additives. Tryptophan (*l*-form, $28.20) is used mostly as a food supplement and is imported from Japan at a 100,000-lb level. Glycine is a somewhat atypical AA because of its commonplace industrial uses, which include its taste-masking function in saccharine-based foods and beverages. Volumewise, phenylalanine is the current star among AAs because of its role in the production of aspartame—a dipeptide known to consumers in the United States as the sweetener Equal or NutraSweet—manufactured by Monsanto, who bought G. D. Searle presumably in order to acquire this product. The worldwide sales of NutraSweet were reported (*CMR* 4/21/86) to have increased to $700 million since its introduction in 1981, and it has replaced saccharin in many low-calorie beverages and foods.

Aspartame is the methyl ester of the dipeptide composed of L-phenylalanine and L-aspartic acid, and its full name is *N*-L-α-aspartyl-L-phenylalanine 1-methyl ester; see Scheme 6.3. About 3MM lb of L-phenylalanine is required for the production of aspartame. Its market price varies greatly because of imports (6/24/85), but it is in the proximity of $20/lb, while L-aspartic acid is priced at only about $1.50/lb.

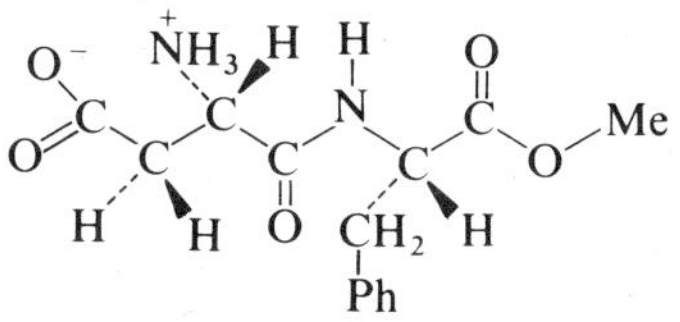

**Scheme 6.3**

It is of interest to note that while the structure in which both AAs possess the L configuration is about 160–180 times more sweet than an equivalent weight of sucrose, the isomer in which the phenylalanine portion has the D configuration is bitter. L-Phenylalanine can be synthesized by stereospecific amination of cinnamic acid (9.3.2) by means of a yeast organism or by condensation of hydantoin with benzaldehyde followed by enzymatic reduction; see Scheme 6.4.

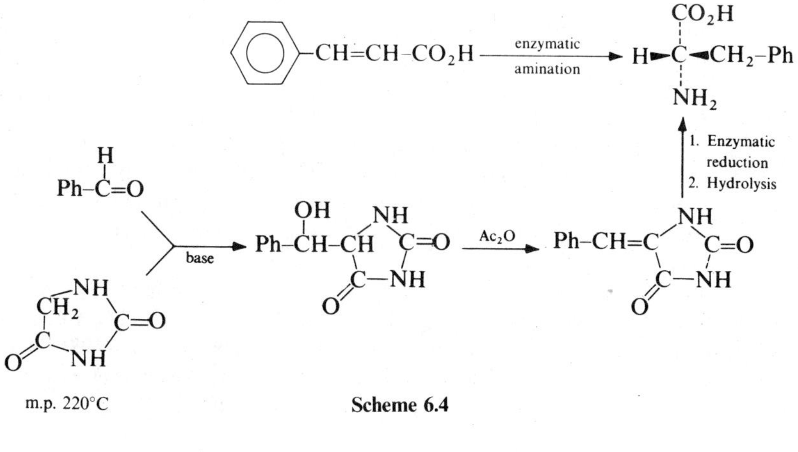

m.p. 220°C                    **Scheme 6.4**

The aspartame patents held by Monsanto have already expired in Canada (a market of about $30 million with a demand of about 700,000 lb; see *CW* 7/15/87), but the U.S. patents do not expire until 1992. It will be interesting to see how the low-calorie sweetener market will fare in the future in view of several competitive sweeteners looming on the industrial horizon (*CW* 7/25/84); S. J. Ainsworth, "New Sweeteners Crowd Sugar out of the Bowl," (*CW* 8/10/88).

### 6.3.1.4  Oxidation and Ammoxidation of Isobutylene

Oxidation and ammoxidation of isobutylene give rise to members of the methacrylic acid family.

Oxidation of isobutylene (32¢) to methacrylic acid (78¢) produces a significant value-added incentive. It can be brought about by means of nitric oxides and nitric acid in the presence of $V_2O_5$ catalyst in the liquid phase, although the yield is only 50–60%. Interest in polymers derived from methacrylic acid and its derivatives undoubtedly stimulates developmental efforts to improve this transformation.

A process that starts with TBA and leads to methyl methacrylate (MMA) may represent an improvement on the direct oxidation of isobutylene and such as route was announced (*CW* 7/24/85) by Mitsubishi Rayon for a 120MM-lb/yr capacity plant to be completed by 1988.

The traditional route to methacrylic esters (from acetone and HCN) is described elsewhere (5.5.2). Other competitive developments (*CW* 10/14/81, 10/23/85; *CMR* 10/21/85) include the acid-catalyzed hydrocarbonylation of propylene (3.5.3.5), in the presence of water or methanol, that yields either methacrylic acid or the methyl ester, MMA, on dehydrogenation of the isobutyric intermediates.

A route to methyl methacrylate that starts with even smaller building blocks, namely, ethylene and CO (see 5.5.2), was announced recently (*CE* 11/23/87) by BASF in West Germany. A plant capable of 80 MM-lb/yr production is expected to be operational in 1988.

The presence of allylic hydrogens in isobutylene allows the application of the ammoxidation process (5.4.3.1), which provides an entry to methacrylonitrile, methacrylamide, and the corresponding acid or esters.

All members of the methacrylic system are valuable monomers in the production of homopolymers and copolymers because of the ease with which these structures participate in vinyl polymerizations.

### 6.3.1.5  Amination and the Ritter Reaction

Amination and the Ritter reaction of isobutylene takes advantage of the relative stability of the *t*-butyl carbocation that reacts with ammonia, amino compounds, or hydrazine, as shown in Scheme 6.5. The *t*-butylhydrazinium chloride is oxidized at room temperature by ferric chloride to the corresponding *t*-butyldiazine, and since the latter decomposes to nitrogen and isobutane, a mixture of the *t*-butylhydrazinium chloride and ferric chloride functions as a blowing agent:

$$(CH_3)_3C-NHNH_3^+ \ Cl^- + FeCl_3 \rightarrow (CH_3)_3C-N{=}NH \rightarrow (CH_3)_3CH + N_2$$

The success in the amination and Ritter reactions depends on a careful control of the acidic catalyst concentration so that the electron pair of the nitrogen-containing reactants is not deactivated by formation of the corresponding -onium compounds.

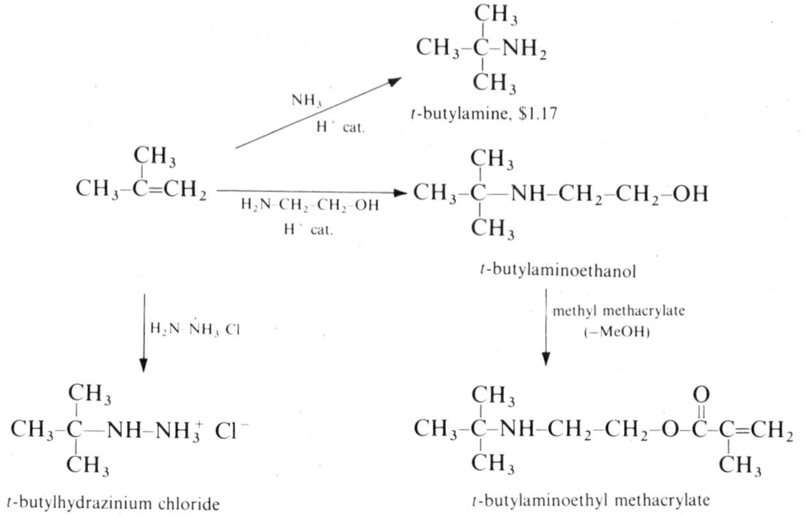

**Scheme 6.5**

### 6.3.1.6 Friedel–Crafts Reaction Products

Friedel–Crafts reaction (*ONR*-33) products of isobutylene are readily available because of the facile generation of the *t*-butyl carbocation, but, because of more convenient material handling reasons, TBA may be the preferred starting material.

The *t*-butyl group may be introduced on aromatic rings for one or more of the following reasons:

1. To stabilize the aromatic compound by protection of the ortho and para positions of phenols and anilines against oxidation and consequent coupling reactions (9.3.4)
2. To increase the solubility of the products in lipophilic media

Among the most common *t*-butylated compounds we find the antioxidants butylated hydroxytoluene (BHT; $1.26–$1.28 technical and food grades, respectively), butylated hydroxyanisole (BHA, food grade $8.80), the styrene and other liquid vinyl polymerization inhibitor butyl catechol ($2.85), and *p-t*-butylphenol (78¢), which is a comonomer in phenolic resins (PF, 3.4.3.5) when the latter are formulated to give oil-soluble varnishes and similar coatings; see Scheme 6.6. Other examples of *t*-butylated aromatics are Shell's Ionox 75, which stands for 2,6-di-*t*-butylphenol, which is coupled with formaldehyde to form Shells's Ionox WTE; see Scheme 6.7. *p-t*-Butyltoluene is the precursor of *p-t*-

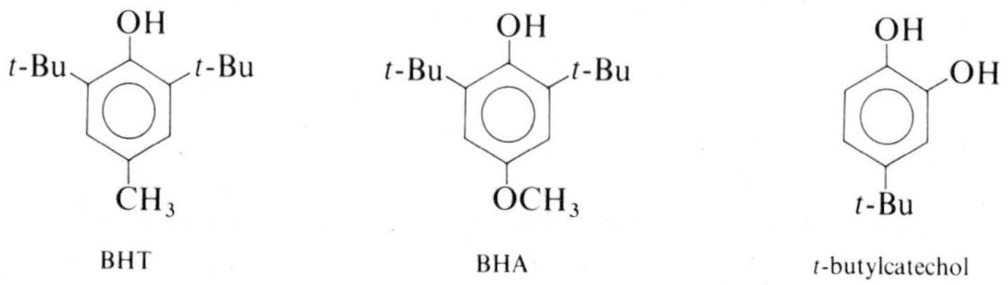

BHT        BHA        *t*-butylcatechol

**Scheme 6.6**

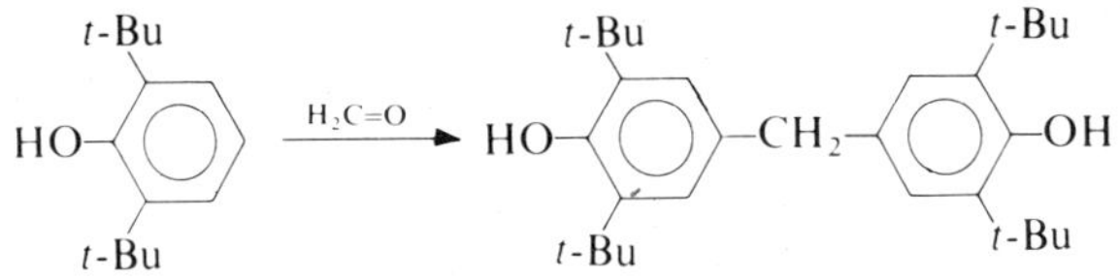

4,4'-methylene-bis-2,6-*t*-butylphenol

**Scheme 6.7**

butylbenzoic acid since the *t*-butyl group resists oxidation by chromic acid or permanganate:

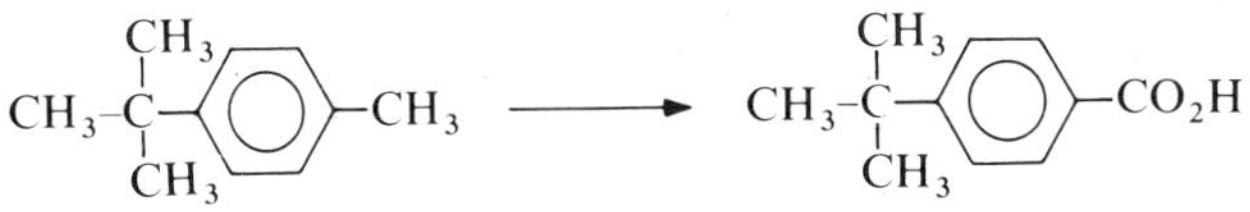

Finally, an example of a more complex butylated aromatic compound is Ciba–Geigy's antioxidant Iraganox 1076, octadecyl 3-(3',5'-di-*t*-butyl-4'-hydroxyphenyl)propionate:

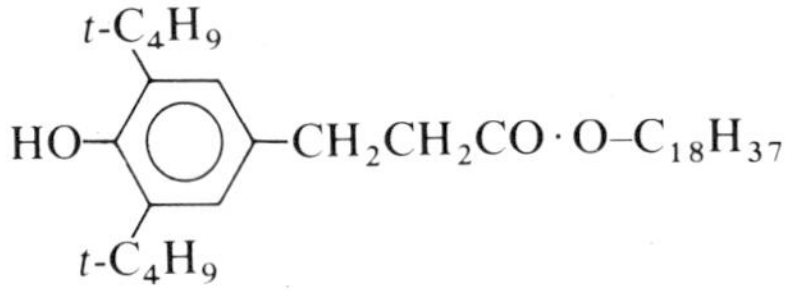

### 6.3.1.7  *t*-Butyl Peroxide and Related Polymerization Catalysts

*t*-Butyl peroxide and related polymerization catalysts serve to either initiate or complete (cure) unsaturated (vinyl and acetylenic) polymerization products and also to cross-link saturated prepolymers that contain C–H bonds vulnerable to attack by oxygen-containing free radicals. In the United States, the use of organic peroxides for the reasons mentioned above amounted to only 42MM lb during 1986 (*CMR* 9/14/87), but the 50-odd compounds valued at about $100 million (for an average price of nearly $2.50) were responsible for the production of over 18B lb of final polymeric products. This obviously important topic of industrial organic chemistry is introduced at this point because the *t*-butyl group is a frequent component of peroxide structures, and, actually, di-*t*-butyl peroxide (*t*-Bu–O–O–*t*-Bu) is the prototype for the great variety of peroxide systems. This is so because the absence of $\alpha$-hydrogen atoms assures the relative stability of the *t*-butyloxy, or simply *t*-butoxy, radicals formed upon controlled decomposition of the peroxide:

$$(CH_3)_3C-O-O-C(CH_3)_3 \rightarrow 2(CH_3)_3C-O\cdot$$

In the presence of $\alpha$-hydrogens, a rearrangement tends to convert an oxygen into a carbon radical center:

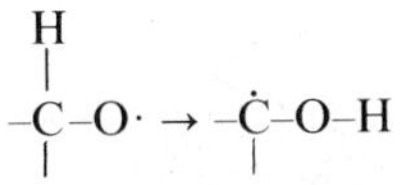

which is less aggressive in bringing about the desired catalytic changes. Closely related to the $t$-butoxy radical by virtue of the absence of $\alpha$-hydrogens are the $t$-amyloxy ($t$-Am–O·) and cumyloxy (Cum–O·) radicals:

$$CH_3CH_2C(CH_3)_2\text{–}O\cdot \quad \text{and} \quad C_6H_5\text{–}C(CH_3)_2\text{–}O\cdot$$

and the frequent presence of these structural moieties is noteworthy among organic peroxides.

There are nine principal categories of industrial peroxy systems, referred to here as A–H, and they are listed as follows, together with representative information concerning the rate of thermal decomposition reported as a function of temperature (in Celsius) of their half-lives ($t_{1/2}$) over a period of 10 hr, and unit prices.

| | °C for $t_{1/2}=10$ hr |
|---|---|
| A. Organic hydroperoxides (R–O–O–H) (see 6.2.5) | |
| $t$-Bu hydrogen peroxide (TBHP), 70% in H$_2$O, Akzo $1.58 | 171–172 |
| $t$-Am hydrogen peroxide | 172 |
| Cum hydroperoxide, 83¢ | 158 |
| Dicumyl-$\alpha,\alpha'$-dihydroperoxide, $2.25 | |
| 1,1,3,3-Tetramethylbutyl 2-hydroperoxide, Akzo $7.00 | |
| 2,5-Dihydroperoxy-2,5-dimethylhexane | 154 |
| B. Dialkyl peroxides | |
| Di-$t$-Bu peroxide (DTBP), $1.80 | 125–126 |
| $t$-Bu Cum peroxide | 121 |
| Di-Cum peroxide, Akzo $4.15 | 115–118 |
| Bis($t$-Bu peroxy)-$\alpha,\alpha'$-diisopropylbenzene, Akzo $5.20 | 117–122 |
| 2,5-Di($t$-Bu peroxy)-2,5-dimethylhexane | 119 |
| 2,5-Di($t$-Bu peroxy)-2,5-dimethyl-3-hexyne | 128 |
| C. Peroxy esters (R–CO–O–O–R′) | |
| $t$-Bu peroxy acetate, 50%, Akzo $2.40 | 102–103 |
| $t$-Bu peroxy octoate (2-Et hexanoate), 50% in DOP, Akzo $2.30 | 74 |
| $t$-Bu peroxy pivalate | 54 |
| $t$-Bu peroxy neodecanoate | 49 |
| $t$-Bu peroxy benzoate, Akzo $2.17 | 104–105 |
| $t$-Bu peroxy $o$-toluate, 75% Akzo $5.21 | 97 |
| Di-$t$-Bu peroxy phthalate | 105 |
| $t$-Bu peroxy maleate (mono) | 87 |
| $t$-Am peroxy acetate | 100 |
| $t$-Am peroxy pivalate | |
| $t$-Am peroxy neodecanoate | |
| $t$-Am peroxy benzoate | 100 |
| Cum peroxy pivalate | 47 |
| Cum peroxy neoheptanoate | 43 |
| Cum peroxy neodecanoate | 38 |
| Bis-2,5-(2-ethylhexanoylperoxy)-2,5-dimethylhexane | |
| Bis-2,5-benzoylperoxy-2,5-dimethylhexane | 100 |

| | $^{\circ}$C for $t_{1/2} = 10$ hr |
|---|---|
| D. Diacyl peroxides (R–CO–O–O–CO–R′) | |
| Diacetyl peroxide | 69 |
| Bis (3,5,5-trimethylhexanoyl) peroxide | 60 |
| Dilauroyl peroxide (7.3.1) | 62 |
| Benzoyl peroxide (BP), \$2.10 (70% wetted), \$5.15 (98%, granular) | 71–73 |
| Di-(p-chlorobenzoyl) peroxide, Akzo 50% paste in silicone oil \$17.42 | 74 |
| Bis-(2,4-dichlorobenzoyl) peroxide, Akzo \$12.85 | 54 |
| Acetyl cyclohexanesulfonyl peroxide | 31–38 |
| E. Peroxycarbonates (R–O–O–CO–O–R′) | |
| t-Bu peroxyisopropyl carbonate, Akzo 75%, \$7.32 | 97–99 |
| t-Bu peroxy 2-ethylhexyl carbonate | 100 |
| t-Am peroxy 2-ethylhexyl carbonate | 98 |
| F. Peroxydicarbonates (R–O–O–CO–O–O–R′) | |
| Di(sec-Bu peroxy) dicarbonate | 44–50 |
| Di(t-Am peroxy) dicarbonate | |
| Bis(2-ethylhexylperoxy) dicarbonate, Akzo 70% in DOP \$3.01 | 40–49 |
| Bis(4-t-Bu cyclohexylperoxy) dicarbonate, Akzo 93%, \$5.18 | 44 |
| G. Peroxyketals [RR′C(O–O–R″)$_2$] | |
| 2,2-Di-(t-Bu peroxy) butane | 104 |
| 2,2-Di-(t-Am peroxy) propane | 108 |
| 1,1-Di-(t-Bu peroxy) cyclohexane | 93 |
| 1,1-Di-(t-Am peroxy) cyclohexane | 93 |
| Ethyl 3,3-di-(t-Bu peroxy) butyrate | 115 |
| Ethyl 3,3-di-(t-Am peroxy) butyrate | 112 |
| n-Bu 4,4-di-(t-Bu peroxy) valerate | 107–109 |
| 1,1-Di-(t-Bu peroxy) 3,3,5-trimethyl-cyclohexane | 92–95 |
| H. Peroxyketones [RR′(O–O–H)$_2$ and/or RR′C(O–O–H)–O–O–CRR′(O–O–H)] | |
| Methyl ethyl ketone peroxide, Akzo 9% active oxygen, \$1.59 | 105 |
| Methyl isobutyl ketone peroxide, Akzo 8.5% active oxygen, \$3.60 | |
| Cyclohexane peroxide, Akzo \$3.85 | |
| Acetylacetone peroxide, Akzo \$2.74 | |
| I. Miscellaneous | |
| Ethyl O-benzoyl laurohydroximate [in $n$-$C_{11}H_{23}$C(OEt)=N–O–CO–$C_6H_5$ the electronegative nitrogen functions as if it were an oxygen, and the homolytic bond scission occurs at the N–O bond] | |
| Acyl aryl sulfonyl peroxide (R–SO$_2$–O–O–CO–R) | 135 |

Noteworthy by their absence from the preceding listing are the organic peracids (R–CO–O–O–H), and the reason for this apparent omission is their behavior as electrophilic heterolytic reactants. Thus, under usual experimental conditions, the reactions of organic peracids with olefins, sulfides, and tertiary amines depend on the nucleophilic nature of these substrates that are converted to epoxides or oxiranes, sulfoxides, and amine oxides, respectively; see Scheme 6.8.

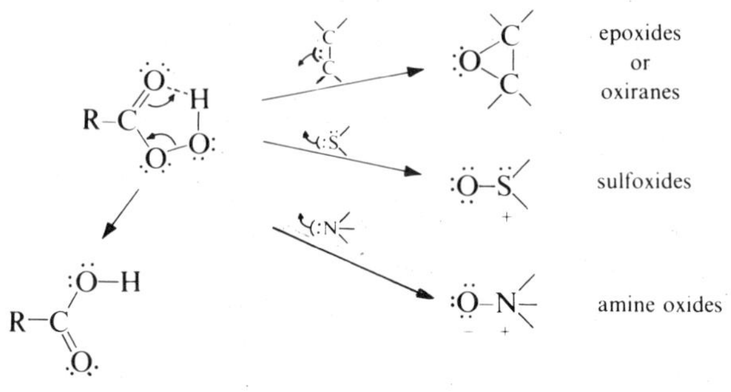

**Scheme 6.8**

The wide selection of organic peroxides as polymerization initiators, as well as curing and processing agents, can be attributed to the multiplicity of situations in which they are expected to function satisfactorily. We not only have a wide variety of monomers and prepolymers but must also consider the different polymerization conditions (emulsion, suspension, solution, bulk, etc.), the effect of the presence of residual decomposition products on the final products, different curing or vulcanization conditions, and the speed and temperature requirements during processing to give the fabricated end products. It is almost impossible to recite which circumstances call for the use of a given organic peroxide, especially since the usefulness of some of the peroxides overlap. However, in order to illustrate certain preferred choices of organic peroxides, we can cite the following:

1. Polyolefins are usually polymerized by means of peroxy esters and peroxydicarbonates with a half-life (as defined above) between 40 and 70°C, while olefin copolymer cross-linking usually employs monoperoxy carbonate esters (R–O–O–CO–O–R′).

2. Styrene polymers and copolymers produced by suspension polymerization may utilize diacyl and dialkyl peroxides or peroxy esters with a half-life of 60–120°C, while bulk polymerization may require a "finishing catalyst" of the peroxy ketal family with a half-life near 100°C. Finally, the relatively low temperature emulsion polymerization of styrene can employ an organic hydroperoxide in the presence of, say, ferrous ion that induces the formation of alkoxy radicals by chemical rather than thermal means:

$$R\text{–}O\text{–}O\text{–}H + Fe^{2+} \rightarrow R\text{–}O\cdot + Fe(OH)^{2+}$$

A similar reaction of redox metals is produced by hydrogen peroxide when R represents H.

3. Unsaturated polyester resins (UPRs) can be cured at ambient or elevated temperatures by means of aromatic diacyl peroxides, peroxy ketones, peroxy carbonates, peroxy dicarbonates, or peroxy esters.

4. Polyacrylates can be obtained by means of alkyl hydroperoxides, peroxy esters,

and peroxy dicarbonate catalysts that dissociate at low temperatures, say, $\sim 40°C$ for a half-life of 10 hr.

5. PVC production utilizes peroxy dicarbonates, diacyl peroxides, and acyl arylsulfonyl peroxides.

6. The cross-linking (vulcanization) of elastomers such a EPDM, SBR, NR, and silicones can be induced by means of diacyl peroxides for which the 10-hr half-life lies between 55 and 135°C (e.g., benzoyl and chlorinated benzoyl peroxides at the lower end of that range).

7. Peroxy ketones are often used during radiation curing by means of electron beams and $\gamma$-rays, and benzoin ethers such as Akzo's Trigonal 14 (Ph–CO–CH(Ph)–O–R), may be used as initiators during UV-catalyzed curing of polyester coatings and for high-speed printing operations.

The catalytic objective of these peroxy catalysts is to produce free radicals derive from the initiator (Q·) and to generate free radicals from either monomer or prepolymer by either addition to an unsaturated system:

$$Q· + CH_2 = CH–R \rightarrow Q–CH_2–\dot{C}H–R$$

or abstraction of a susceptible hydrogen from the substrate:

$$Q· + H–\overset{\overset{\displaystyle CH_3}{|}}{\underset{\underset{\displaystyle CH_3}{|}}{C}}–CH_2–CH_2– \longrightarrow Q–H + ·\overset{\overset{\displaystyle CH_3}{|}}{\underset{\underset{\displaystyle CH_3}{|}}{C}}–CH_2–CH_2–$$

$$Q· + CH_3–\overset{\overset{\displaystyle O-}{|}}{\underset{\underset{\displaystyle O-}{|}}{Si}}–O– \longrightarrow Q–H + ·CH_2–\overset{\overset{\displaystyle O-}{|}}{\underset{\underset{\displaystyle O-}{|}}{Si}}–O–$$

Chain growth or cross-linking is then carried on by the newly generated radical. In the case of carboxylate or cabonate radicals, R–CO·O· or R–O–CO·O·, respectively, the initial species may suffer a loss of $CO_2$ before engaging in the formation of the desired carbon radical that propagates the vinyl polymerization. In the case of saturated systems, cross-linking is produced by linkup of two carbon radicals generated by the free-radical initiator.

The production of TBHP involves an acid-catalyzed addition of hydrogen peroxide to isobutylene (cf. 6.2.5), and a similar addition of TBHP to isobutylene produces DTBP.

$t$-Butyl perbenzoate is obtained by means of a Schotten–Baumann reaction (*ONR*-82) of TBHP and benzoyl chloride (9.3.2), and other peroxy esters are obtained in an analogous manner:

$$R–O–O–H + R'–CO–Cl \text{ (dilute NaOH)} \rightarrow R–O–O–CO–R'$$

where R–O–O–H represents other organic hydroperoxides obtained by either

- Addition of hydrogen peroxide to the corresponding olefin.
- Oxidation of a tertiary and especially tertiary and also benzylic hydrogen.
- An acid-catalyzed replacement of a hydoxyl by a hydroperoxy function.

In the last-mentioned case, the alcohol may be obtained by reduction of a readily available ketone (e.g., cyclohexanone and isophorone).

Peroxy carbonates and peroxy dicarbonates are assembled from organic hydroperoxides and chloroformates or by the treatment of peroxy chloroformates with sodium peroxide:

$$R–O–OH + Cl–CO–O–R' \rightarrow R–O–O–CO–O–R'$$

$$R–O–O–CO–Cl + Na_2O_2 \rightarrow R–O–O–CO–O–O–CO–O–O–R$$

Peroxy ketals are derived by addtion of organic hydroperoxids, usually TBHP, to the keto functions of simple ketones (such as MEK or acetone) or to more complex keto precursors such as acetoacetic (4.6.4) or levulinic esters ($CH_3–CO–CH_2–CH_2–CO·OR$) (*SZM-II*, 90–91).

Benzoyl peroxide not only serves as a common polymerization catalyst but also plays a unique role in the bleaching of flour (see *BP* 3/88) and as a keratolytic ingredient in topical formulations.

### 6.3.1.8   Miscellaneous Derivatives of Isobutylene

1. Neoacids are formed by the acid-catalyzed addition of CO to branched olefins (3.5.3.7.C) and in the case of isoutylene one obtains neopentanoic or pivalic acid (30¢). Thionyl chloride (55¢) converts the acid to pivaloyl chloride [$(CH_3)_3C–CO·Cl$], which becomes the point of entry to pivaloyl esters and amides. For additional information, see Exxon's bulletin *Neo Acids: Properties, Chemistry and Applications*, 1982.

2. Metathesis of isobutylene and 2-butene is a source of isoprene (6.8.2) and propylene (Scheme 6.9). The use of the Prins reaction (*ONR*-72) in the formation of isoprene is mentioned in 3.4.3.8.

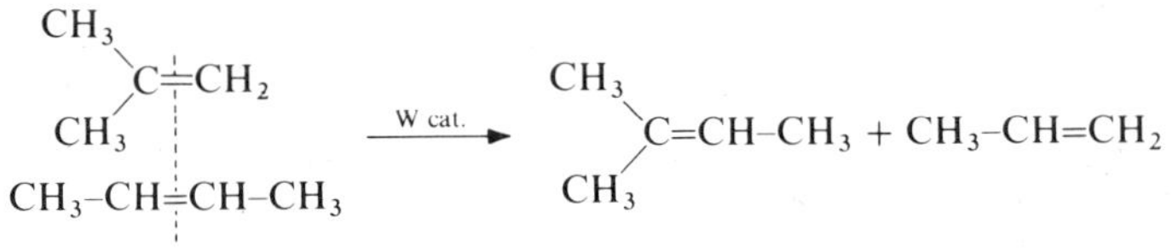

**Scheme 6.9**

3. Methallyl chloride is obtained by the chlorination of isobutylene at relatively elevated temperatures in a manner similar to the chlorination of propylene to allyl chloride (5.4.8). The alkylation of benzene with methallyl chloride in the presence of sulfuric acid gives an interesting synthetic intermediate known as neophyl chloride:

$$C_6H_6 + CH_2=C(CH_3)CH_2–Cl \rightarrow C_6H_5–C(CH_3)_2–CH_2–Cl$$

4. Hydroformylation products (3.5.3.4) derived from isobutylene include isoamyl alcohol [$(CH_3)_2CHCH_2CH_2–OH$, 95%, $1.44], which also happens to be one the principal components of fusel oil (*SZM-II*, 37–38). The corresponding carboxylic acid is isovaleric acid.

The dehydration of isoamyl alcohol to 3-methyl-1-butene is being promoted by Phillips (*CEN* 4/22/85) as the source of another comonomer in the production of LLDPE.

Various esters of isoamyl alcohol are flavoring agents and are used in the food and perfume industries:

Isoamyl formate—fruit flavor

Isoamyl acetate—"banana oil" (also provides a pear flavor) and is used as a solvent

Isoamyl n-butyrate—pear, rum, and fruit flavors

Isoamyl isovalerate—liqueur and candy flavoring agent

Isoamyl benzoate, diisoamyl phthalate, and isoamyl salicylate—high-boiling perfume fixatives and are used as soap scents

### 6.3.2  n-Butylenes and Their Uses

As indicated in Fig. 2.2 and mentioned above (6.1), the isomeric n-butenes can be separated by fractional distillation and equilibrated by the use of appropriate catalysts. The formation of 1-butene by a selective dimerization of ethylene carried out in the presence of a homogeneous catalyst was announced recently (K. Brooks $CW$ 9/9/87 and 12/23/87) by the Institute Français du Petrole (IFP), which installed this so-called Alphabutol process in Thailand and Saudi Arabia and has licensed its use by Indian Petrochemicals. It is argued that developing countries that do not possess the complex extraction facilities to isolate 1-butene from the $C_4$ petroleum cracking fraction can obtain the desirable 1-butene by starting with the $C_2$ building block.

Before the appearance of LLDPE (4.3.1) in the mid-1970s, the demand for 1-butene remained at a 10MM-lb level, but by 1988 it grew to nearly 490MM lb. About one-third of the 1-butene demand is satisfied by dimerization of ethylene, while the rest is obtained by cracking. Over 40% of 1-butene is now consumed in the production of LLDPE since the latter variation on PE eclipsed the 8B-lb market for LDPE and has also encroached on the demand for HDPE. About 8–10% by weight of 1-butene is incoporated in the feed during the production of LLDPE, and its price (26¢) is competitive with its rival comonomers 1-hexene (46¢) and 1-octene (47¢), although the higher olefinic comonomers produce more stretchable films.

Another relatively new market for 1-butene is its linear polymer, poly(1-butene), often referred to simply as polybutylene (PB; Shell, $1.30 pipe grade, 83¢ film grade). It is prepared by means of Ziegler–Natta catalysts as a homopolymer and as a copolymer with ethylene. As suggested by the pricing information, these polymers are used in the production of films and pipes, and low-molecular weight materials can be employed as lubricating oil additives.

Carefully controlled alkylation of phenol with 1-butene gives a mixture of isomeric sec-butylphenols, and treatment with hydrogen peroxide or an organic hydroperoxide converts 1-butene to the epoxide, 1,2-butylene oxide (b.p. 63°C), which is utilized in a similar manner as its lower homolog, PO (b.p. 34°C). 1,2-Butylene oxide can be handled more easily because of the higher boiling point and its preferential reaction with nucleophilic compounds at the unsubstituted carbon terminal. The glycol derived from this epoxide, namely, 1,2-butanediol, is now available on the marketplace.

The analogous epoxide of 2-butene and the corresponding glycol, 2,3-butanediol, can be oxidized to diacetyl ($10–$15) found in butter and hence used to flavor oleomargarine and other dairy products, as well as by the perfume industry. Previous to the availability of 2,3-butanediol, diacetyl was prepared by nitrosation of MEK and the hydrolysis of the resulting oxime; as shown in Scheme 6.10. A flavoring agent related to diacetyl and

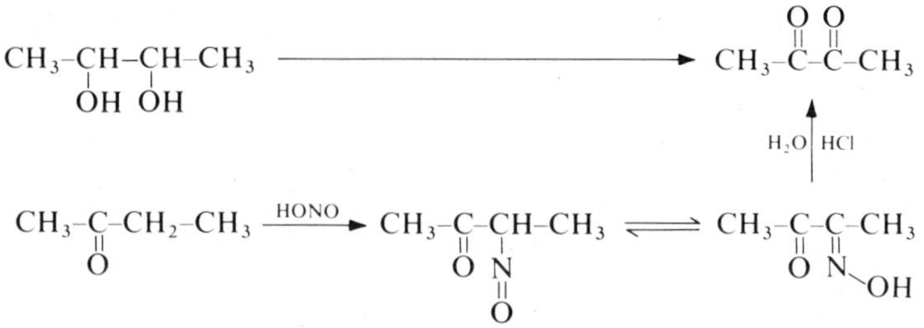

$$\begin{array}{c} CH_3\text{–}CH\text{–}CH\text{–}CH_3 \\ \quad\;\; | \quad | \\ \quad\;\; OH \;\; OH \end{array} \longrightarrow \begin{array}{c} O \;\; O \\ \parallel \;\; \parallel \\ CH_3\text{–}C\text{–}C\text{–}CH_3 \end{array}$$

$$\uparrow \quad H_2O \;\big|\; HCl$$

$$\begin{array}{c} CH_3\text{–}C\text{–}CH_2\text{–}CH_3 \\ \quad\; \parallel \\ \quad\; O \end{array} \xrightarrow{\;HONO\;} \begin{array}{c} CH_3\text{–}C\text{–}CH\text{–}CH_3 \\ \quad\; \parallel \quad \parallel \\ \quad\; O \quad N \\ \qquad\quad \parallel \\ \qquad\quad O \end{array} \rightleftharpoons \begin{array}{c} CH_3\text{–}C\text{–}C\text{–}CH_3 \\ \quad\; \parallel \;\; \parallel \\ \quad\; O \;\; N \\ \qquad\quad\;\; \diagdown OH \end{array}$$

**Scheme 6.10**

copresent in butter is its partial reduction product, acetoin [$CH_3$–$CO$–$CH(OH)$–$CH_3$]. Acetoin can be prepared by mild reduction of diacetyl by means of zinc in acetic acid or by means of several microorganisms such as sorbose bacteria, *Mycoderma aceti*, and some families of fungi. Also, as the prototype of the acyloin family of organic compounds,

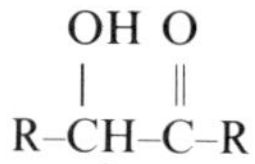

$$\begin{array}{c} OH \quad O \\ | \qquad \parallel \\ R\text{–}CH\text{–}C\text{–}R \end{array}$$

it can be assembled by an acyloin condensation (*ONR*-1) of an acetate ester. For example, in the reaction of ethyl acetate with metallic sodium carried out in an inert medium such as toluene, a sodium atom first adds to each ester molecule to give a carbon radical. Two of the latter couple to give the new carbon–carbon bond, and this is followed by elimination of sodium ethoxide to yield an enol salt of the desired acetoin. Treatment of the enol salt with water liberates the desired acyloin, as illustrated in Scheme 6.11.

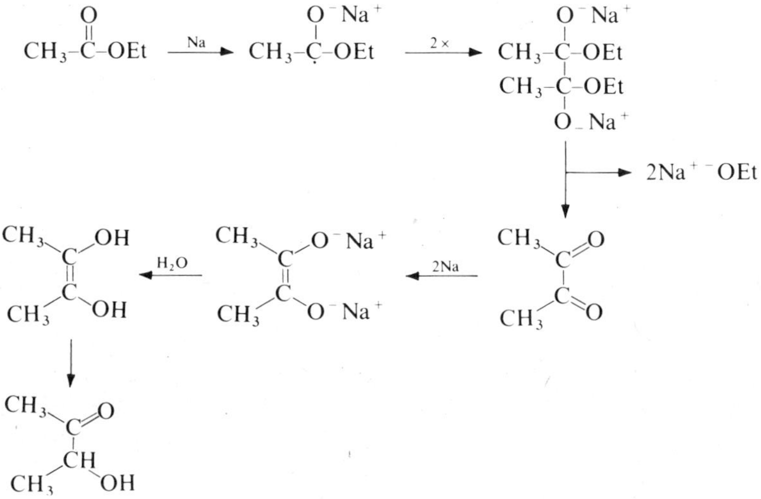

$$\begin{array}{c} O \\ \parallel \\ CH_3\text{–}C\text{–}OEt \end{array} \xrightarrow{\;Na\;} \begin{array}{c} O^-\,Na^+ \\ | \\ CH_3\text{–}C\text{–}OEt \end{array} \xrightarrow{\;2\times\;} \begin{array}{c} O^-\,Na^+ \\ | \\ CH_3\text{–}C\text{–}OEt \\ CH_3\text{–}C\text{–}OEt \\ | \\ O^-\,Na^+ \end{array}$$

$$\big|\!\longrightarrow 2Na^+\,{}^-OEt$$

$$\begin{array}{c} CH_3 \diagdown \quad OH \\ \qquad C \\ \qquad \parallel \\ \qquad C \\ CH_3 \diagup \quad \diagdown OH \end{array} \xleftarrow{\;H_2O\;} \begin{array}{c} CH_3 \diagdown \quad O^-Na^+ \\ \qquad C \\ \qquad \parallel \\ \qquad C \\ CH_3 \diagup \quad \diagdown O^-Na^+ \end{array} \xleftarrow{\;2Na\;} \begin{array}{c} CH_3 \diagdown \quad \diagup O \\ \qquad C \\ \qquad C \\ CH_3 \diagup \quad \diagdown O \end{array}$$

$$\big\downarrow$$

$$\begin{array}{c} CH_3 \diagdown \quad \diagup O \\ \qquad C \\ \qquad \diagdown CH \\ CH_3 \qquad \diagdown OH \end{array}$$

**Scheme 6.11**

The use of nickel catalysts promotes the dimerization of 1-butene to 2-ethyl-1-hexene, and the latter—a relatively nonbranched olefin, as well as 1-butene itself—give rise to a series of oxo chemicals by way of the hydroformylation reaction (3.5.3.4).

The hydroformylation of 1-butene gives valeraldehyde and 2-methylbutyl alcohol, also known as "active amyl alcohol" since the levo-rotating isomer is isolated from fusel oil (*SZM-II*, 38). The mixture of *n*-amyl and 2-methylbutyl alcohols obtained by hydroformylation of *n*-butylenes is marketed as "primary amyl alcohols" (46.5¢).

Of historical interest is the fact that during the 1920s the pentane fraction of petroleum was chlorinated and the resulting mixture of amyl chlorides was hydrolyzed to the corresponding alcohols. The latter were then converted to the acetates (Sharpless process, *SZM-I*, 170, 190, 224 and *FKC*, 2nd ed., 1957) that were used as solvents for lacquers and in other industrial applications; see Scheme 6.12.

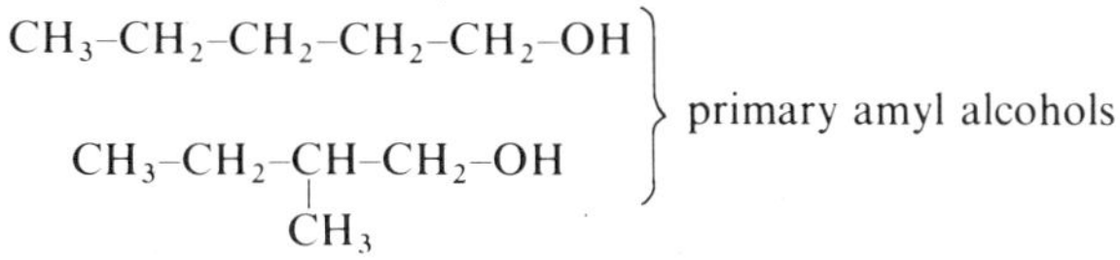

$$CH_3-CH_2-CH_2-CH_2-CH_2-OH$$
$$CH_3-CH_2-CH-CH_2-OH$$
$$\overset{|}{CH_3}$$

primary amyl alcohols

**Scheme 6.12**

Both *n*-butenes are hydrated to *sec*-butyl alcohol (36.5¢), most of which is oxidized to methyl ethyl ketone (MEK; 28.5¢) by means of oxygen at above 400°C and 2–4 atm as the gases are passed over solid packings such as ZnO and brass spheres. The price differential between the related alcohol and ketone suggests that the relatively high cost of the alcohol is attributed to its small-volume market, while the unit cost of MEK reflects its high ($\sim$600MM lb) use as a solvent that must compete with acetone (27¢) and other industrial solvents used in coatings, adhesives, nitrocellulose solutions, printing inks, and other products. In the future, the two-step route from the *n*-butylenes to MEK may be shortened by means of a process announced by Catalytica (*CMR* 7/28/86) that converts *n*-butane directly to MEK by means of molecular oxygen in the presence of a propietary catalyst.

Methyl ethyl ketone and hydrogen peroxide give about 12MM lb of the so-called MEK hydroperoxide ($1.70) (see 6.3.1.7) that is used widely for the high-temperature curing of about 1.3B lb of UPR. MEK hydroperoxide is a mixture in which about 50% consists of a dimeric hydrogen peroxide adduct [(H–O–OC(Me)(Et)–O)$_2$] and another 25% is a trimeric, cyclic peroxide:

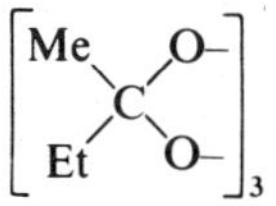

$$\begin{bmatrix} Me & O- \\ & \diagdown C \diagup \\ Et & O- \end{bmatrix}_3$$

The ammoxidation of 2-butene yields the expected (5.4.3.1) crotononitrile (CH$_3$–CH=CH–C≡N), which on acid-catalyzed hydrolysis, forms crotonamide and, finally, crotonic acid ($1.50). Actually, 1-butene can also be subjected to ammoxidation because of an *in situ* isomerization to 2-butene. Crotonic acid is also obtained by oxidation of crotonaldehyde (4.5.2) and is used as a copolymer in vinyl polymerization of VAM. The last-mentioned application is of special interest for the introduction of carboxylic acid functions into the final polymer in order to improve its adhesion to paper and other surfaces. Crotonic acid is also a building block of some fine chemicals such as the four-carbon amino acid threonine:

$$CH_3-CH(OH)-CH(NH_2)-CO_2H, \; dl\text{-}, \; \$58.25.$$

Tiglic acid, *trans*-$CH_3$–CH=C($CH_3$)–$CO_2$H, is a branched homolog of crotonic acid and can be prepared by dehydration of MEK cyanohydrin, followed by hydrolysis of the nitrile function. This compound (Ald, $19/lb) is used to break up emulsions. The corresponding cis isomer is known as "angelic acid" and various esters of both isomers, but in particular esters of geraniol (*SZM-II*, 119, 130) are valuable ingredients of perfumes and flavoring agents. Methyl tiglate is valued at about $64/lb.

## 6.4  *n*-BUTYL AND ISOBUTYL ALCOHOLS AND DERIVATIVES

Of the four isomeric butyl alcohols, TBA and *sec*-butyl alcohol are mentioned in Sections 6.3.1.3 and 6.3.2, respectively, and this leaves only *n*-butyl and isobutyl alcohols to be discussed here.

### 6.4.1  Sources of *n*-Butyl Alcohol and an Overview of Its Uses

Thus far there is no practical way to obtain *n*-butyl alcohol (34¢), from the C₄ hydrocarbons. Previously, this alcohol was obtained exclusively by fermentation of glucose (derived from molasses, cornstarch, and the like; see *SZM-II*, 38) by means of *Clostridium acetobutylicum* (sometimes referred to as "Weizmann's bacterium" in honor of C. Weizmann, a chemist and the first president of Israel). The fermentation yields *n*-butyl alcohol, acetone, and ethanol in a ratio of 70:25:5, and this source of *n*-butyl alcohol is still viable today (R. T. Myers, *CEN* 3/23/87, 3). However, the hydroformylation of propylene (5.4.7) dominates the production of *n*-butyl and isobutyl alcohols, probably because of the industrial importance of their precursors, namely *n*-butyraldehydes and isobutyraldehydes (6.5). The hydroformylation reaction is carried out at about 1000 atm and 150–200°C in the presence of phosphine-modified cobalt carbonyl catalysts, or under considerably milder conditions (10 atm and 110° C) when the ironpentacarbonyl catalyst is used. Shell carries out the hydroformylation of propylene by means of rhodium catalysts that yield directly a mixture of the butyl alcohols and 2-ethylhexyl alcohol (2-EH), the major transformation product of *n*-butyraldehyde. The formation of either aldehyde or alcohol is a function of the $CO:H_2$ ratio. In any case, the catalysts are chosen to favor the production of the linear products that happen to be more valuable as manifested by the traditionally somewhat higher price of the *n*-acohol (38¢) as compared to that of isobutyl alcohol (37¢), while the opposite is true in the case of *n*-butyraldehyde (29.5¢) as compared to isobutyraldehyde (35¢).

The demand for *n*-butyl alcohol in the United States is about 900MM lb. It provides access to a medium-sized alkyl group that borders on water insolubility and offers a functional group capable of a variety of chemical transformations. Consequently, about 40% is employed in the assembly of a variety of hydrophobic esters, about 25% is converted to ethers, about 5% is subjected to alcohol amination to yield amines, and about 14% is employed as a solvent that bridges aqueous and nonaqueous phases and serves to disperse water-borne adhesives, paints, and other coating materials. The last-mentioned uses are likely to increase in the future in view of the shift away from coatings applied in strictly organic solvents. Then, there are several interesting derivatives of the *n*-butyl moiety that are incorporated in phase-transfer catalysts (see 6.4.4) and as partners of lithium alkyls that are soluble in hydrocarbon solvents.

### 6.4.2  Esters of *n*-Butyl Alcohol

Among the more significant derivatives of *n*-butyl alcohol are the following esters:

*n*-Butyl acetate, 110MM lb, 52.5¢, used extensively as a solvent for lacquers, inks, and adhesives

*n*-Butyl acrylate, 69¢, monomer for surface-coating poducts

*n*-Butyl methacrylate, 88¢, monomer and comonomer

*n*-Butyl *p*-hydroxybenzoate, or butylparaben, food preservative

*n*-Butyl octyl phthalate, 40¢, plasticizer

*n*-Butyl benzyl phthalate, 53¢, plasticizer

*n*-Butyl cyclohexyl phthalate, 74¢, plasticizer

*n*-Butyl isodecyl phthalate, 35¢, plasticizer

*n*-Butyl lactate, $1.58, plasticizer

*n*-Butyl oleate, 80¢, plasticizer

*n*-Butyl stearate, 55¢, contributes waxlike character in cosmetics; 55¢, cosmetic grade 91¢

Di-*n*-Butyl fumarate, 77¢, comonomer that contributes flexibility

Di-*n*-butyl maleate, 63¢, comonomer that contributes flexibility

Di-*n*-butyl phthalate, 54¢, plasticizer, insect repellant

Di-*n*-butyl sebacate, $1.72, plasticizer

### 6.4.3  *n*-Butyl Chloride, a Valuable Intermediate

The conversion of *n*-butyl alcohol (b.p. 117°C) to *n*-butyl chloride (b.p. 78°C) by means of anhydrous HCl (62¢) is the key to the preparation of organometallic compounds and other butylated compounds.

*n*-Butyllithium (15% solution in hydrocarbon solvents, $15.95) is obtained from the chloride and lithium metal ($27.95) extruded in the form of a wire:

$$n\text{-Bu}-\text{Cl} + 2\text{Li} \rightarrow n\text{-Bu}-\text{Li} + \text{Li}^+ \ \text{Cl}^-$$

*n*-Butyllithium is a catalyst in anionic polymerizations. For example, 1,3-butadiene or isoprene an be polymerized in the manner illustrated in Scheme 6.13.

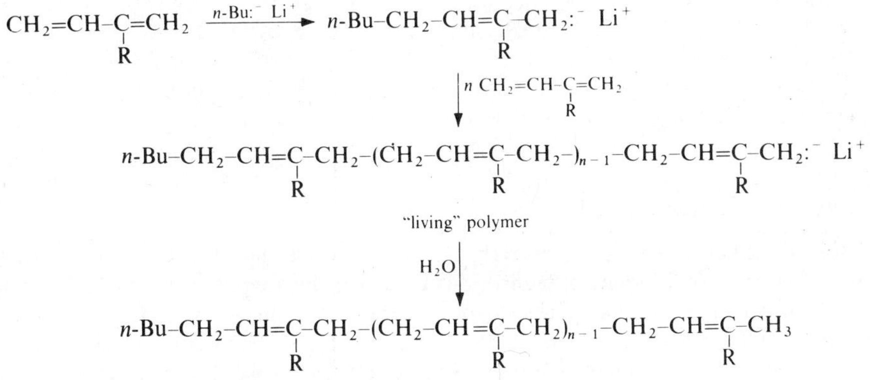

(where R = $H_1$ CH$_3$)

**Scheme 6.13**

*n*-Butyl chloride is converted to the corresponding Grignard reagent, *n*-BuMgCl (Orsynex, $3.70 as 2.5 *M* solution in THF) by treating the chloride with magnesium ($1.53) in the form of freshly cut shavings suspended in suitable solvents (THF, diethyl ether, etc.). This reagent becomes a versatile building block in countless Grignard reactions (*ONR*-37). The solvent-induced shift of the equilibrium:

$$2R-Mg-Cl \rightleftharpoons R_2Mg + MgCl_2$$

gives di-*n*-butylmagnesium that is used to the tune of 250,000 lb as a cocatalyst in the formation of LLDPE.

The raction of the butyl Grignard with tin tetrachloride (SnCl$_4$; ∼ $3.00) provides an entry to the most commonly employed family of organotin compounds:

1. *n*-Butyltin trichloride (*n*-BuSnCl$_3$) is used in the waterproofing of fabrics.

2. Di-*n*-Butyltin dichloride (*n*-Bu$_2$SnCl$_2$) is hydrolyzed to a polymeric oxide [(*n*-Bu$_2$SnO)$_x$], but, more importantly, the chlorine substituents can be replaced to yield a series of stannous esters, *n*-Bu$_2$Sn(O–CO–R)$_2$, where the carboxylate moiety may stand for acetate (∼ $8.00; octoate, ∼ $7.00; laurate, ∼ $6.00; etc.). The dibutylstannous esters are used as PVC stabilizers, as PUR catalysts, and for room-temperature vulcanization of silicone rubber.

3. Tri-*n*-butyltin chloride (*n*-Bu$_3$SnCl) gives, on hydrolysis, the oxide, (*n*-Bu$_3$Sn)$_2$O, TBTO, which equilibrates in aqueous media with tri-*n*-butyltin hydroxide (*n*-Bu$_3$Sn–OH). This material ($8.60), commonly referred to as "tributyl tin" (TBT), is an effective inhibitor of marine organisms (barnacles) that grow on hulls and increase the friction (drag) of a moving vessel. Also, it inhibits other organisms such as marine worms that destroy wooden boats by boring into the hull. The frequent scraping and painting of the vessels can be reduced by means of TBT incorporated in marine paints. Also, TBT derivatives such as tributyltin methacrylate and its polymer have been incorporated in marine paints that slowly release tributyltin hydroxide to repel the attacking organisms. About 5MM lb of TBT have been employed annually for this purpose, mostly by ocean-going ships and about 30% by recreational vehicles, with some antifouling paints based also on copper compounds. However, recent evidence suggests that low concentrations (in ppt range) of tin compounds are toxic to oysters, clams, crabs, and other nontarget marine organisms, and, at the suggestion of EPA, Congress is currently (*CMR* 8/10/87, 10/12/87, 11/16/87, 12/28/87) establishing restrictions on the use of TBT.

Tributyl tin is also an ingredient of paints formulated for the protection of wooden land-based structures against mildew and rotting.

### 6.4.4   *n*-Butylamines

Next to the families of methyl- and ethylamines, the *n*-butylamines occupy an important position among this family of industrial chemicals. They are prepared preferentially by amination of the corresponding alcohol to yield mono-*n*-butylamine (96¢), di-*n*-butyl-amine ($1.06), and tri-*n*-butylamine ($1.33). A more recent application of the *n*-butyl-amines is the preparation of phase-transfer catalysts (PTCs) (3.8.4). The presence of three *n*-butyl groups suffices to create a hydrophobic or lipophilic behavior of quaternary ammonium salts, and the butyl group can be paired up with other hydrocarbon substituents:

$n$-Bu$_3$N + Me–Cl → $n$-Bu$_3$NMe$^+$ Cl$^-$, tri-$n$-butylmethylammonium chloride, Ethyl's TBMAC, \$1.85

$n$-Bu$_3$N + $n$-Bu–Cl → $n$-Bu$_4$N$^+$ Cl$^-$, tetra-$n$-butylammonium chloride

PhCH$_2$–NMe$_2$ + $n$-Bu–Cl → Ph–NMe$_2$($n$-Bu)$^+$ Cl$^-$, benzylbutyltrimethylammonium chloride

The partial hydrophobic–lipophilic nature of the $n$-butyl residues establishes an equilibrium between the concentrations of the organic cation, represented here by R$_4$N$^+$, in the aqueous and nonaqueous phases of a heterogeneous reaction mixture, and the intrinsically water-soluble anionic nucleophilic reactant, symbolized here by Nu$^-$, accompanies the organic cation into the organic phase where it encounters the water-insoluble substrate symbolized here by Q–X. This situation is pictured as follows:

$$Na^+ Nu^- + R_4N^+ Cl^- \rightleftharpoons Na^+Cl^- + R_4N^+ Nu^-$$

| water-soluble reactant | in water and nonaqueous phases | water-soluble counterions | nonaqueous phase where the reaction occurs between Nu$^-$ and Q X |
|---|---|---|---|

In this diagram the counterions are the nonreacting partners of the ions that are actually involved in the phase-transfer catalysis phenomenon. In other words, the counterions are the "innocent bystanders" of the equilibrium that brings together the reacting species.

The $n$-butylamines are also building blocks of some dyes, rubber-processing agents, pesticides, and so on. Examples of fine chemicals that contain $n$-butyl residues are

- The local anesthetic dibucaine hydrochloride derived from 2-hydroxyquinoline-4-carboxylic acid (10.3.1) (A in Scheme 6.14).

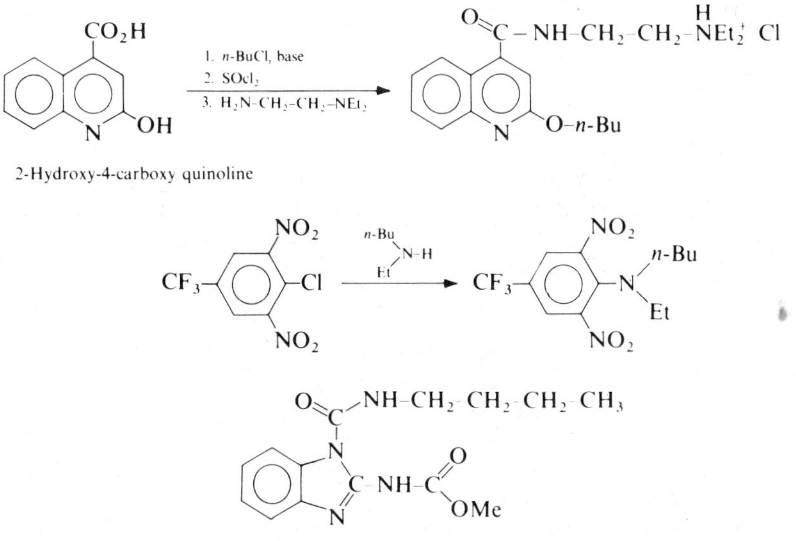

(a)

2-Hydroxy-4-carboxy quinoline

(b)

(c)

**Scheme 6.14**

- The herbicide benfluralin (Lilly's Balan) (B in Scheme 6.14).
- The fungicide and ascaricide benomyl (C in Scheme 6.14).

The structures of these compounds are shown in Scheme 6.14.

### 6.4.5   *n*-Butyl Ethers

Di-*n*-butyl ether, or simply butyl ether ($1.85, b.p. 142°C) is formed by acid-catalyzed dehydration of *n*-butyl alcohol and is a nearly water-insoluble solvent.

Monoethoxylated and diethoxylated derivatives of *n*-butyl alcohol are mentioned in Section 4.3.6.1.

### 6.4.6   Isobutyl Alcohol and Some of Its Derivatives

Unlike its linear isomer, isobutyl alcohol (100MM lb, 32¢) is obtained exclusively by hydroformylation of propylene. It serves as a building block of a number of esters:

Isobutyl acetate, 45¢, solvent

Isobutyl acrylate, 71¢, monomer

Isobutyl isobutyrate, 42.5¢, low-cost solvent [b.p. 147°C, obtained by one-step Tischenko reaction (*ONR*-90) of isobutyraldehyde (35¢)], catalyzed by aluminum or sodium alkoxide catalysts:

$$2(CH_3)_2CH-CH=O \rightarrow (CH_3)_2CH-C(=O)-OCH_2CH(CH_3)_2$$

Isobutyl methacrylate, 87¢, monomer

Isobutyl phenylacetate, $3.10, perfume ingredient

Isobutyl salicylate, $3.45, perfume ingredient

Diisobutyl phthalate, 55¢, plasticizer

### 6.4.7   Guerbet Alcohols

Guerbet alcohols, in general, are obtained (*ONR*-38) by subjecting a primary alcohol to a high-temperature treatment in the presence of the corresponding alkoxide or copper catalyst. In the case of *n*-butyl alcohol, this one-step dehydrogenation–hydrogenation reaction, accompanied by an aldol condensation of the transient aldehyde, produces 2-EH—the identical alcohol obtained from *n*-butyraldehyde when the latter is subjected in a sequential fashion to an aldol condensation, dehydration, and hydrogenation.

## 6.5   THE BUTYRALDEHYDES AND THE CORRESPONDING CARBOXYLIC ACIDS

The hydroformylation reaction of propylene (5.4.7) is the source of *n*-butyraldehyde (29.5¢) and isobutyraldehyde (35¢) and the corresponding oxidation products, butyric (44.5¢) and isobutyric (75¢) acids. The price differentials between the isomeric pairs reflect differences in demand and economy of scale.

The industrially most significant derivatives of *n*-butyraldehyde results from its aldol condensation, which gives rise to the common "octyl alcohol" of commerce, referred to as

"2-EH"—the acronym for 2-ethylhexanol (36¢). The demand for 2-EH in the United States is about 720MM lb, and its production and that of the related 2-ethylhexanoic acid are shown as follows:

$$2CH_3CH_2CH_2CH=O \ (NaOH, -H_2O) \rightarrow \ CH_3CH_3CH_2CH=C(Et)CH=O$$

$$\downarrow H_2, \text{ cat.}$$

$$CH_3CH_2CH_2CH_2CH(Et)CO_2H \qquad CH_3CH_2CH_2CH_2CH(Et)CH_2OH$$

A typical ester derived from 2-EH is the monomer 2-ethylhexyl acrylate (79.5¢), but most other esters are popular plasticizers:

- Di-(2-ethylhexyl) or simply dioctyl phthalate (DOP), 200MM lb, 43¢
- Dioctyl adipate, 66¢
- Dioctyl azelate, 99¢
- Dioctyl sebacate, $1.47

The two last-mentioned $C_9$ and $C_{10}$ dicarboxylic acids that give rise to the octyl esters are obtained from renewable resources (*SZM-II*, 70, 72): the first either from oleic or ricinoleic acids and the second exclusively from ricinoleic acid. This situation illustrates once more that in the "real world" of industrial chemistry it is not advisable to mentally divorce biomass-derived industrial chemicals from petrochemicals.

Before the advent of oxo chemicals, butyraldehyde and 2-EH were derived from acetaldehyde (4.5.2), but when we compare the pricing of acetaldehyde (45.5¢) and butyraldehyde (29.5¢), it becomes obvious that this traditional route is no longer competitive.

An analogous sequence of transformations gives rise to 2-EH, except that acetaldehyde is used in place of one butyraldehyde, yielding 2-ethyl-1-butanol (2.05).

A combination of the Tishchenko and aldol reactions [*JOC*, **8**, 256 (1943)] that uses magnesium aluminum ethoxide gives a common insect repellent 2-ethyl-1,3-hexanediol, ethohexadiol, or octylene glycol, as shown in Scheme 6.15.

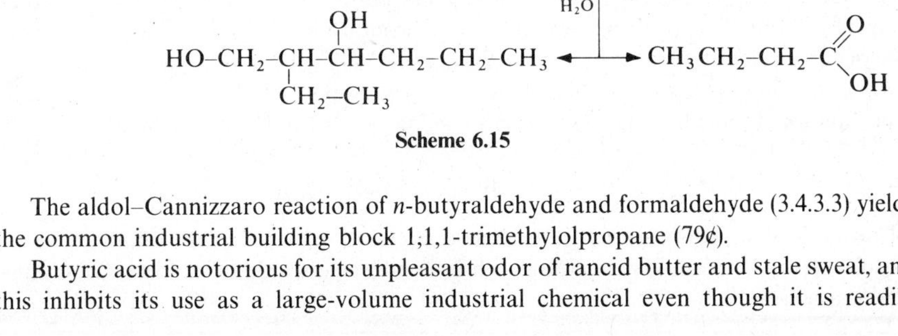

$$3CH_3-CH_2-CH_2-\overset{H}{\underset{}{C}}=O \ \xrightarrow{MgAl(OEt)_n} \ CH_3-CH_2-CH_2-\overset{O}{\overset{\|}{C}}-O-CH_2-CH-CH_2-CH_3$$

with substituents $CH-OH$, $CH_2$, $CH_2$, $CH_3$

$$HO-CH_2-\overset{OH}{\underset{}{CH}}-CH-CH_2-CH_2-CH_3 \ \xleftarrow{H_2O} \ CH_3CH_2-CH_2-C\overset{O}{\underset{OH}{}}$$

$$\overset{}{\underset{CH_2-CH_3}{}}$$

**Scheme 6.15**

The aldol–Cannizzaro reaction of *n*-butyraldehyde and formaldehyde (3.4.3.3) yields the common industrial building block 1,1,1-trimethylolpropane (79¢).

Butyric acid is notorious for its unpleasant odor of rancid butter and stale sweat, and this inhibits its use as a large-volume industrial chemical even though it is readily accessible by means of the Reppe hydrocarbonylation of *n*-propyl alcohol. However, as is true for various foul-smelling compounds, they are appreciated in small doses as ingredients of perfumes and flavors. Thus, more than a dozen esters of butyric acid are listed among Aldrich's 1988 *Flavors and Fragrances* (prices for 25-kg-quantity lots), including:

Ethyl butyrate, FCC, \$2.20; *n*-propyl butyrate, \$11.50
*n*-Heptyl butyrate, \$14.00; *n*-octyl butyrate, \$11.00

Butyric acid esters are used as flavors of margarine and similar food products.

A derivative of butyric acid, namely, *N*-chloro-α-aminobutyric acid

$$CH_3-CH_2-CH(NHCl)-CO_2H$$

has recently attained some notoriety as the means to dissolve decayed portions of teeth without the need of drilling and anesthesia (*CEN* 12/9/85). A buffered saline solution of this material is claimed to function in a matter of minutes and to be harmless to the patients. As a derivative of an α-amino acid, the synthesis of the compound is likely to start with propionaldehyde rather than butyric acid and involve a treatment of α-aminobutyric acid with hypochlorite.

Isobutyraldehyde can be subjected to most of the preceding reactions of its linear isomer. Thus, with formaldehyde (3.4.3.3) one obtains neopentyl glycol (53¢). The sterically crowded aldol condensation product of two isobutyraldehyde molecules can be stabilized by hydrogenation to 2,2,4-trimethyl-1,3-pentanediol, or trimethylpentanediol (Eastman's TMPD, 48.5¢).

The aldol condensation of isobutyraldehyde with acetone, subjected to dehydration and hydrogenation, yields methyl isoamyl ketone [$CH_3-CO-CH_2-CH_2-CH(CH_3)_2$; 51¢].

A somewhat different use of isobutyraldehyde is the production of the slow-release fertilizer isobutylidene diurea [IBDU; $(CH_3)_2CHCH(NH-CO-NH_2)_2$] by a joint venture of Mitsubishi and Celanese (*CW* 11/26/84).

Some of the more common esters of isobutyric acid are

• Isobutyl isobutyrate, 42.5¢, solvent (see 6.4.6).
• Sucrose acetate isobutyrate, Eastman's SAIB, \$1.33, plasticizer (*SZM-II*, 104).

An illustration of the use of isobutyric acid as a building block of a fine chemical is the synthesis of the antiinflammatory medicinal sulindac known to many arthritic patients as Merck Sharp & Dohme's Clinoril. The starting point of the assembly of this rather complex structure is fluorobenzene (9.3.1) and α-bromoisobutyroyl bromide. The latter is obtained from isobutyric acid by means of the Hell–Vollard–Zelinski reaction (*ONR*-41) in which an aliphatic carboxylic acid is treated with bromine (or chlorine) in the presence of phosphorus to yield the corresponding α-halogen-substituted acid halide:

$$(CH_3)_2CH-CO\cdot OH + Br_2-P_4 \rightarrow (CH_3)_2CBr-CO\cdot Br$$

A Friedel–Crafts reaction of fluorobenzene and the brominated acid bromide gives 5-fluoro-2-methyl-1-indanone (see below), which is subjected to additional transformations shown in Scheme 6.16.

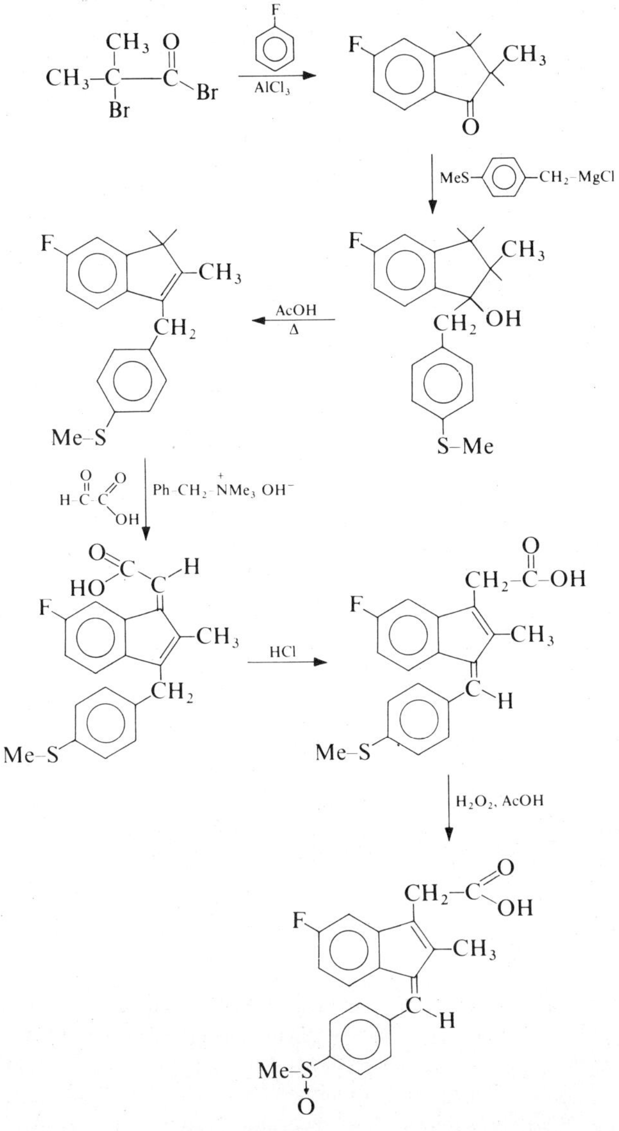

Sulindac, Merck-Sharpe Dohme's Clinoryl; (Z)-5-fluro-2-methyl-
1-(p-methylsufinylbenzylidene)-3-indenylacetic acid,
antiinflammatory [see R. F. Shuman et al.,
JOC **42**, 1914 (1977)]

**Scheme 6.16**

A simpler transformation of isobutyric acid is its conversion to the industrially available diisopropyl ketone (b.p. 124°C), which can be prepared by means of the classical thermal decomposition of the calcium salt or its modern modification in which the carboxylic acid is subjected to a vapor phase decarboxylation at about 400°C over thoria or similar catalysts:

$$2(CH_3)_2CH-CO_2H \rightarrow (CH_3)_2CH-CO-CH(CH_3)_2 + CO_2 + H_2O$$

The reduction product of this ketone is also available as 2,4-dimethyl-3-pentanol. The thermal decarboxylation of carboxylic acids to ketones is not limited to the preceding example. It is also applicable to dicarboxylic acids and to mixtures of monocarboxylic acids (*SZM-I*, 122–123).

## 6.6 MALEIC ANHYDRIDE (MA) AND DERIVATIVES AS A BUILDING BLOCK OF INDUSTRIAL CHEMICALS

### 6.6.1 Production and an Overview of Uses

The production of maleic anhydride (MA) (see 6.2.4) has evolved from a wasteful oxidative degradation of benzene with the loss of two carbon atoms as carbon dioxide:

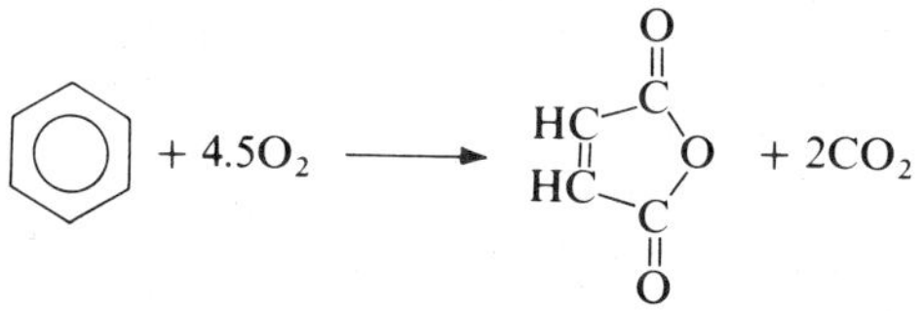

to the use, first, of butene, and then to the current process based on *n*-butane:

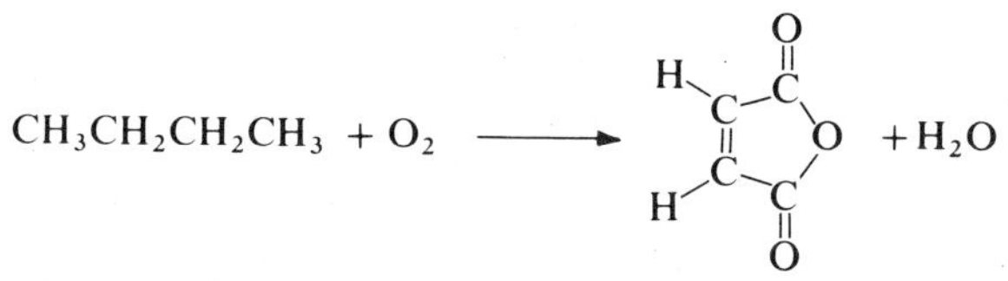

The benzene degradation process was modeled on the analogous transformation of naphthalene to phthalic anhydride (9.2.2, 9.3.2.2) in which the use of costly V$_2$O$_5$ catalyst (prepared from $4.00 raw material) was critical. In 1962 the precursor of the company that became Denka Chemicals Corporation attempted the production of MA starting with *n*-butenes but eventually switched back to the benzene-based process. Finally, in 1983 the production of MA from butane was heralded by the announcement (*CMR* 4/18/83, "Maleic Plant, World's Largest, Starts up in Florida") with reference to Monsanto's facility to produce 110MM lb of MA in Pensacola. The production capacity of this plant was boosted recently (12/14/87) by 79MM lb, and gradually all domestic producers of MA have switched to butane feedstock (*CE* 8/17/87, 29–33).

The current consumption of MA in the United States is about 400MM lb, and its uses cover a broad spectrum of materials, from its traditional role in the production of unsaturated polyester resins (UPRs), absorbing about half of the demand, to some typical acetylene chemicals (4.7), detergents, pesticides, lube oil additives, and fine chemicals such as aspartic acid—the building block of Monsanto's aspartame. The moderate cost of MA

(55¢) and its current origin based on an ample supply of low-cost butane assures a good future for this versatile chemical.

### 6.6.2    Unsaturated Polyester Resins and SMA

The UPRs represent another evolutionary process in the realm of industrial organic chemistry. The original coating material of the 1920s was known as "glyptal," derived from *gly*cerol and *phthal*ic anhydride. These two basic ingredients were dehydrated by heating to form a generally linear polyester that was formulated by addition of pigments and fillers, deposited on metallic objects, and, finally heated (baked) to deposit protective and decorative coatings of the thermoset polyester on appliances and their components. Gradually, as the choice of other industrial organic chemicals expanded and more efficient coating application techniques were developed, some of the glycerol (synthetic 99.5%, 76¢) was replaced by other polyfunctional alcohols such as, initially, pentaerythritol (71¢) and later 1,1,1-trimethylolpropane (79¢) and even cheaper glycols such as propylene glycol) (51¢) and diethylene glycol (43¢). Also, some of the phthalic anhydride (molten 35¢, flake 39¢) was replaced by isophthalic (46¢) and other carboxylic acids such as tall-oil fatty acids (TOFA) (21¢) (*SZM-II*, 134). All prices cited here are as of January 1989. Thus, glyptal resins evolved into "alkyd resins," or alkyds. Finally, the addition of maleic anhydride to the mixture of polyfunctional alcohols and difunctional and higher-functionality carboxylic acids provided the opportunity to combine the stepwise condensation polymerization—a relatively slow process based on elimination of a portion of the reactants (usually water) with a rapid vinyl addition polymerization. MA not only participates in the formation of polyester linkages, but its double bond serves as a site for free-radical-catalyzed chain growth of vinyl monomers such as styrene, α-methylstyrene, or vinyltoluene (9.2.1). Consequently, the modern UPRs are obtained by mixing a selection of the above-mentioned ingredients, heating (or cooking) the mixture to a specified viscosity, diluting this prepolymer with a rather large quantity of low-cost styrene (47¢), and completing the polymerization by additional heating in the presence of a vinyl polymerization catalyst (6.3.1.7). These styrenated alkyd resin coatings are not only less costly to formulate and to apply, but the presence of built-in polyester branches tends to give less brittle and more shockproof, smooth, enamel-like coatings. Fiber-glass-reinforced, and aluminum hydroxide or precipitated calcium carbonate filled polyester formulations are employed for the manufacture of large objects and structural components for the manufacture of transportation equipment, boats, housing, and so on. The total consumption of UPRs in the United States is estimated at about 1.4B lb, and different grades cost between 60¢ and $2.00. Currently, about half of the UPR consumption is channeled into the construction and marine industries (25% each), and newer developments include the fabrication of "cultured marble" objects for kitchen and bathroom, and the use of "polymer concrete" for road construction (see *CMR* 5/23/88).

The other large-scale contribution of MA to polymers is the production of styrene–maleic anhydride (SMA). This polymer tends to form an alternating sequence of the two monomers because, while MA per se does not polymerize under ordinary reaction conditions, it combines with a terminal styrene radical more readily than another styrene monomer molecule, and a terminal maleic anhydride radical combines readily with a styrene monomer molecule. The explanation for such preference in the assembly of two different vinyl monomers is beyond the scope of this book, but its practical result is the formation of coupled styrene–maleic anhydride residues that are subject to chemical modifications by alkali and alcohols; see Scheme 6.17.

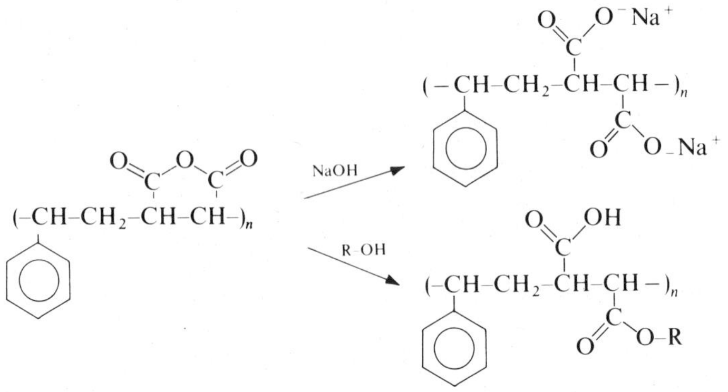

**Scheme 6.17**

The SMA-derived salts are dispersible in water and can be used as thickeners in water-based paints and as sizing for paper and textiles. Otherwise, the straightforward SMA polymer is characterized by a higher tolerance to elevated temperatures than, for example, other styrene-derived thermoplastics such as the ABS resins, thus permitting the processing of SMA above 200°C. The polymerization of styrene and butadiene in the presence of about 10% MA gives impact-resistant SMA terpolymers, and similar impact-resistant materials are obtained by alloying SMA with ABS and other polymers.

### 6.6.3  Sulfosuccinates

The peroxide-catalyzed addition of sodium bisulfite to MA or its esters has been employed for some time by American Cyanamid to produce its Aerosol family of surface-active agents. These materials are useful in emulsion polymerizations and as wetting agents during textile-dyeing operations. The key reaction is the formation of the sulfur free radical derived from sodium bisulfite:

$$Na^{+\,-}O_3S-H + Q\cdot \rightarrow Na^{+\,-}O_3S\cdot + Q-H$$

where Q· represents the peroxide-derived radical and the addition of NaSO$_3$H is propagated by the intermediate carbon radical derived from the MA residue. The incorporation of a series of alkyl groups in maleic esters provides the most suitable balance between the hydrophilic sodium sulfonate groups and the lipophilic moiety of the maleic ester. In this fashion we arrive at Aeorosol OT, in which the sodium sulfosuccinate ester contains two 2-EH residues, Aerosol MA, in which the sulfosuccinate ester contains two 2-methylpentyl residues; Aeroasol AY, with a mixture of *n*-amyl and *sec*-amyl groups; Aerosol IB, with two isobutyl groups; and so on. The maleic esters available for these and other uses include dibutyl maleate (63¢) and dioctyl maleate (62¢). Similar sodium sulfosuccinate surfactants are derived from monoesters and the monoamide of maleic acid.

Sulfosuccinic acid itself has recently become available from Goldschmidt Chem. as a "new building block for creative chemistry."

### 6.6.4  Alkenyl Succinates

Maleic anhydride reacts with terminal olefins in an interesting fashion because the cyclic mechanism of the ene reaction is responsible for the formation of an alkenyl succinic

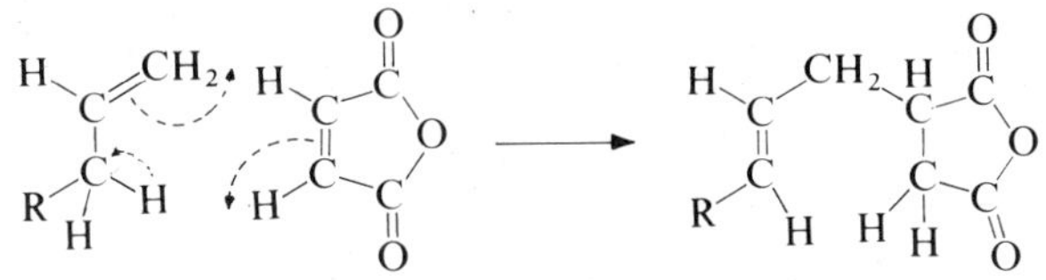

**Scheme 6.18**

anhydride system (see 7.4.1, paragraph C), as shown in Scheme 6.18. This behavior is utilized in the derivatization of alpha-olefins, (AOs; see 7.3) and poly(1-butenes) (6.3.2).

### 6.6.5   Dithiophosphate Derivatives of MA

A highly successful adduct of diethyl dithiophosphate [$HS-PS \cdot (OCH_2CH_3)_2$] to diethyl maleate is the insecticide known best to the general public by American Cyanamid's tradename "Malathion" ($1.62) (Scheme 6.19). Malathion appears to be one of the

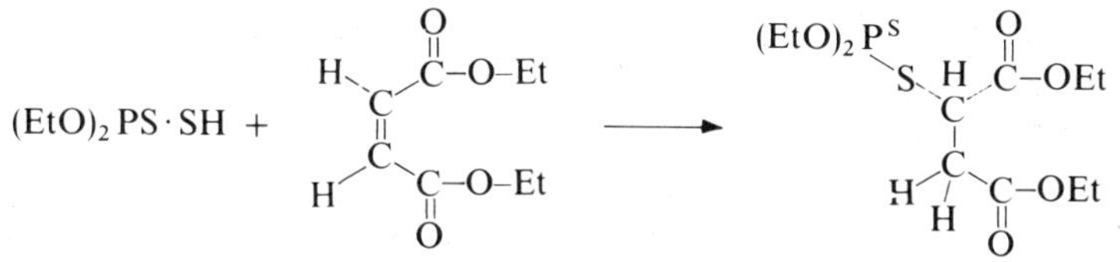

**Scheme 6.19**

pesticides most effective into combating the devastation to fruit crops caused by the Mediterranean fruit fly. It is harmless to humans (*when*, as is true with all chemicals, *it is applied judiciously*), and its use achieved national notoriety during the summer of 1981 when the then governor of California Jerry Brown vetoed the use of Malathion in California while infestation of the Mediterranean fruit fly was spreading across the agricultural section of that state. Finally, after the imposition of an embargo by other states on the entry of California crops and as the danger of infestation in other regions of the United States increased, the federal government mandated the eradication of the pest by means of Malathion. This affair contributed to the eclipse of the political career of Governor Brown, who, at the end of his governorship of California (1975–1981), became a partner of a Los Angeles law firm. The State of Florida routinely controls localized outbreaks of Mediterranean fruit fly infestation by means of Malathion and thus protects its citrus and other fruit crops without risk to the human population.

### 6.6.6   Hydrazide and Bismaleimide Derivatives of MA

The reaction of MA with hydrazine (85% as hydrate $1.25) under rather strongly acidic conditions (to prevent addition of the nitrogen nucleophile to the C=C bond) produces

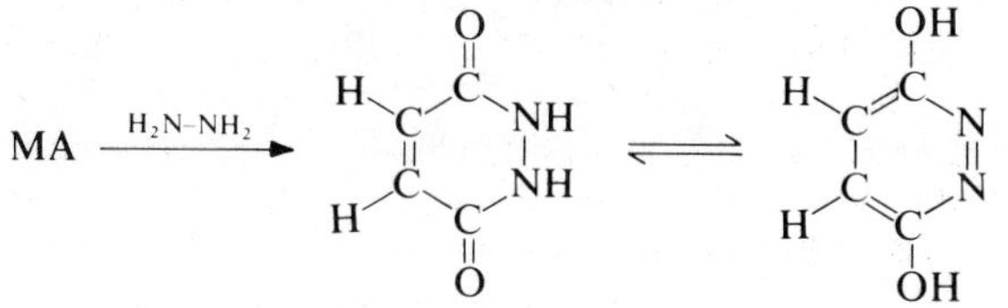

3,6-dihydroxypyridazine

**Scheme 6.20**

maleic hydrazide. It is employed to control the growth of weeds and also serves as an entry to the family of pyridazine compounds (10.3.2); see Scheme 6.20.

A somewhat analogous reaction of MA is the formation of *bismaleimides* from aromatic diamines such as bis-$p,p'$-methylenedianiline and $m$-phenylenediamine (9.3.6), which produce bismethylenedianiline maleimide (Itoh's BMI) and bis-$m$-phenylenediamine maleimide (Itoh's BPI):

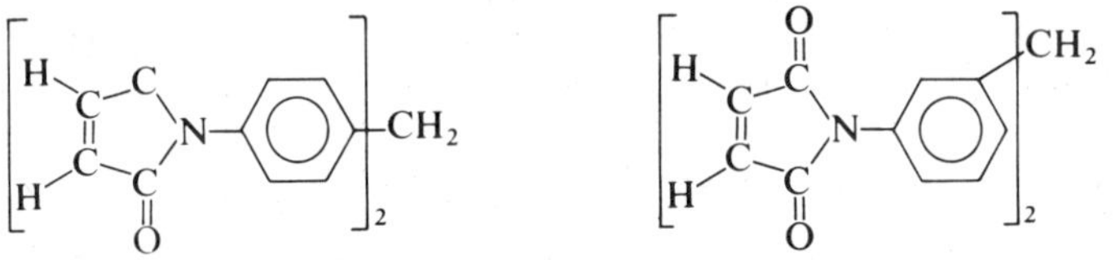

These difunctional maleimides are copolymerized with allylic monomers and cured to produce tough, high-performance materials such as Ciba-Geigy's XU 292, valued at $25/lb.

### 6.6.7 Fumaric, Malic, and Tartaric Acids

The isomerization of maleic to fumaric acid is achieved by heating the former in water, as shown in Scheme 6.21. The greater thermodynamic stability and its poor solubility in water facilitate the isolation of fumaric acid (33MM lb, technical grade 60.5¢, food grade 75.5¢). Fumaric acid is also obtained by fermentation of carbohydrates by means of *Rhizopus* mold (*SZM-II*, 42).

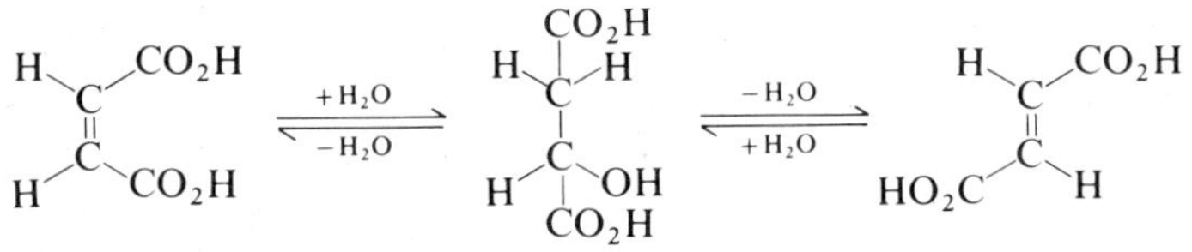

**Scheme 6.21**

Fumaric acid is approved by the FDA as an acidulent and meat tenderizer, and its use in the food industry is expected to grow as asceptic packaging of foods becomes more popular in the United States. Ferrous fumarate ($1.50) is also employed as a food additive. Technical-grade fumaric acid is employed in the production of alkyd resins and UPRs and in the production of polymers for paper-size applications.

The acid-catalyzed hydration of fumaric acid leads to $d,l$-malic acid (20MM lb, 81¢). Malic acid competes with citric acid (see below) as an acidulent in the manufacture of soft drinks, and it is often added to food products that contain aspartame (p. 338) because "it makes aspartame taste more like sugar" (*CMR* 6/29/87). L-Malic acid is obtained by fermentation.

The reaction of MA with a formic acid solution of hydrogen peroxide produces tartaric acid (NF $1.50). Tartaric acid and some of its salts (*SZM-II*, 39–40) are traditional by-products of the wine industry, and about 7MM lb of potassium bitartrate (or "cream of tartar," $1.12) and potassium sodium tartrate (or Rochelle salt, 80¢) are imported from wine-producing countries. Potassium bitartrate is a constituent of some baking powders and is used in dyeing operations based on mordant dyes to reduce $Cr^{6+}$.

Crystals of potassium sodium tartrate are employed in piezoelectric devices, and solutions of this material are used to reduce silver ions in the manufacture of mirrors.

### 6.6.8   Succinic Acid and Other Hydrogenation Products of MA

The catalytic hydrogenation of MA to succinic anhydride ($1.71) and the hydrolysis of the latter to succinic acid ($4.10) are relatively simple because this hydrogenation involves a carbon–carbon double bond rather than a more resistant carboxylic acid function. The relatively high cost of these compounds reflects a small demand rather than chemical complexity of the transformations.

The current (*CE* 8/17/87, 33) joint initiative of Standard Oil of Cleveland, Ohio and Mitsui Toatsu Chemicals to produce (at Osaka, Japan) three typical acetylene chemicals (6.7), namely, 1,4-butanediol (BDO; 88¢), tetrahydrofuran (THF; $1.04), and $\gamma$-butyrolactone (GBL; $1.20) is based on the *in situ* generation of diethyl maleate as the hydrogenation is carried out at 140–220°C and 100–600 psi by means of a copper-promoted catalyst. The esterification–hydrogenation technology is contributed by Davy McKee of London and Union Carbide, and the preference for hydrogenation of an ester, rather than the carboxylic acid or its anhydride, is based on traditionally recognized differences in their susceptibility to hydrogenation. These transformations are summarized in Scheme 6.22.

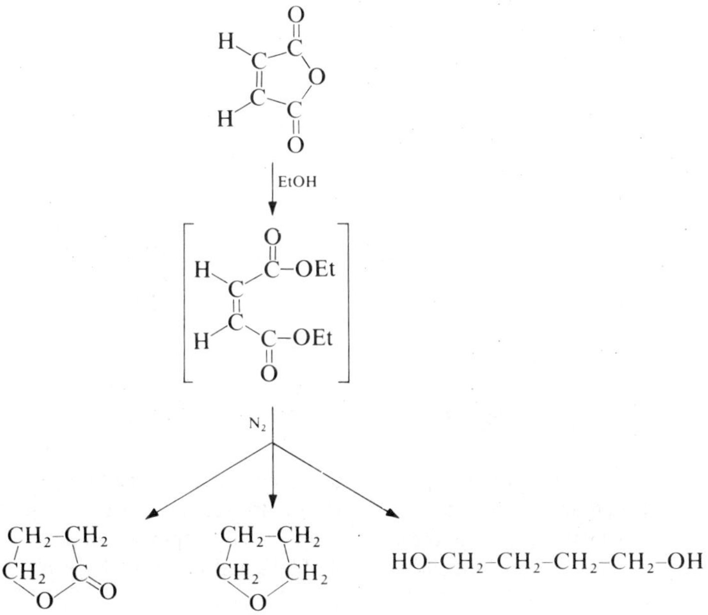

**Scheme 6.22**

### 6.6.9   Diels–Alder Behavior and Ozonolysis of MA Derivatives

The electronegative carboxylic anhydride structure of MA induces a strong dienophile behavior in the Diels–Alder reaction (*ONR*-24). The reaction of MA with 1,3-butadiene (6.8) occurs with such ease that a gravimetric determination of butadiene in a $C_4$ gas stream can be based on the quantitative precipitation of the Diels–Alder adduct as the gas is passed through a solution of MA in an appropriate solvent. Additional examples of Diels–Alder products derived from MA are shown in Section 6.8.

The ozonolysis of C=C systems graduated from academic laboratories to the industrial world when Emery Industries began to produce azelaic and pelargonic acids from oleic acid on a large scale (*SZM-II*, 70). Recently, Chemie Linz of Austria announced (*CE* 10/12/87) ozonolysis facilities for the cleavage of a variety of C=C bond-containing compounds dissolved in methanol. The ozonolysis is coupled with a hydrogenation workup of ozonides to yield the methyl hemiacetals of aldehydes and/or ketones (depending on the number of carbon groups located at the initial double bond). Apparently, the first commercially available ozonolysis–hydrogenation product (advertised in *CMR* 2/23/87 by Chemie Linz and almost simultaneously by Societé Française Hoechst) is the methyl hemiacetal of methyl glyoxylate (see p. 253) derived from dimethyl

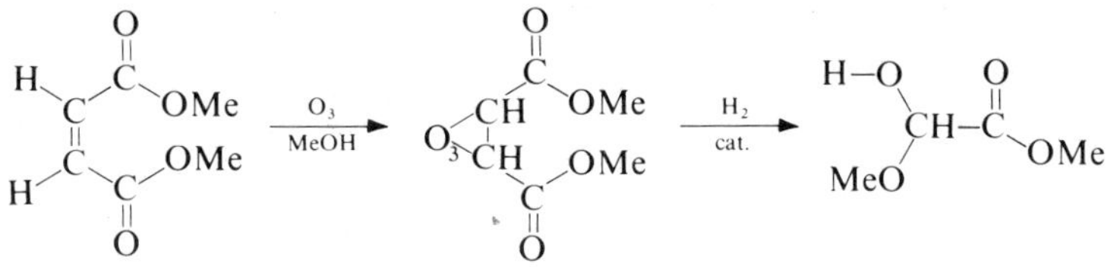

**Scheme 6.23**

maleate (Scheme 6.23). Some other synthetic intermediates accessible by means of the Chemie Linz technology and their precursors are as follows:

$$CH_2=C(CH_3)-CH(OMe)_2 \rightarrow CH_3C(=O)-CH(OMe)_2$$

methacrolein diMe acetal          pyruvic aldehyde diMe acetal
                                  methylglyoxal diMe acetal

Cyclohexene → 1,1,6,6-tetramethoxyhexane

*p*-Nitrostyrene → *p*-nitrobenzaldehyde

Chemie Linz offers custom syntheses of aldehydes, ketones, aldehyde–carboxylic acids, and keto–carboxylic acids from the appropriate precursors.

## 6.7    1,4-BUTANEDIOL (BDO) AND ITS USES

The two major driving forces behind the expanding demand for BDO (88¢) are PBT, poly(1,4-butylene terephthalate), the high-performance version of PET (4.3.6.1), and the direct and indirect uses of BDO in the production of PURs (interlude in Section 3.8), that is, as chain extender and as a precursor of poly(tetramethylene ether glycol) (PTMEG), respectively.

In the latter case, BDO is the precursor of THF, which on cationic polymerization yields PTMEG ($1.70). BASF, one of the producers of PTMEG, supplies a range of polymeric products of average molecular weights 250, 650, 1000, 2000, and 2900 for use in the design of PUR with different performance characteristics.

The alternate name for PTMEG is poly(tetrahydrofuran) (PTHF), and this brings home the possibility that furfuraldehyde—a building block derived from renewable resources (*SZM-II*, 100–103)—may also eventually become the source of BDO and PTHF. In this connection it is encouraging to note that Quaker Oats Chemical—the successor of the furfuraldehyde-producing pioneer, namely, Quaker Oats Company—is expanding the production of furfuraldehyde by about 20MM lb in Memphis, Tennessee, a relatively large increase vis-à-vis the current global production level of about 90MM lb obtained from sugarcane bagasse and grain residues (bran).

The butane-based production of MA (6.6.1) and its conversion to BDO by way of THF is another route (pursued by Standard Oil and Union Carbide) that may gain in importance as that technology is further developed. Not to be left behind, ARCO has licensed from Kuraray of Japan the PO to allyl alcohol–CO route to BDO (5.4.4) and plans to produce 75MM lb of BDO in Texas, while Mitsubishi is producing some BDO by means of a transition metal-catalyzed addition of acetic acid to 1,3-butadiene (Mitsubishi Kasei Corporation, *CT* 12/88, 759–763).

Thus, we find that BDO is at the crossroad of several chemical possibilities and the relationship between the aforementioned materials is summarized in Scheme 6.24.

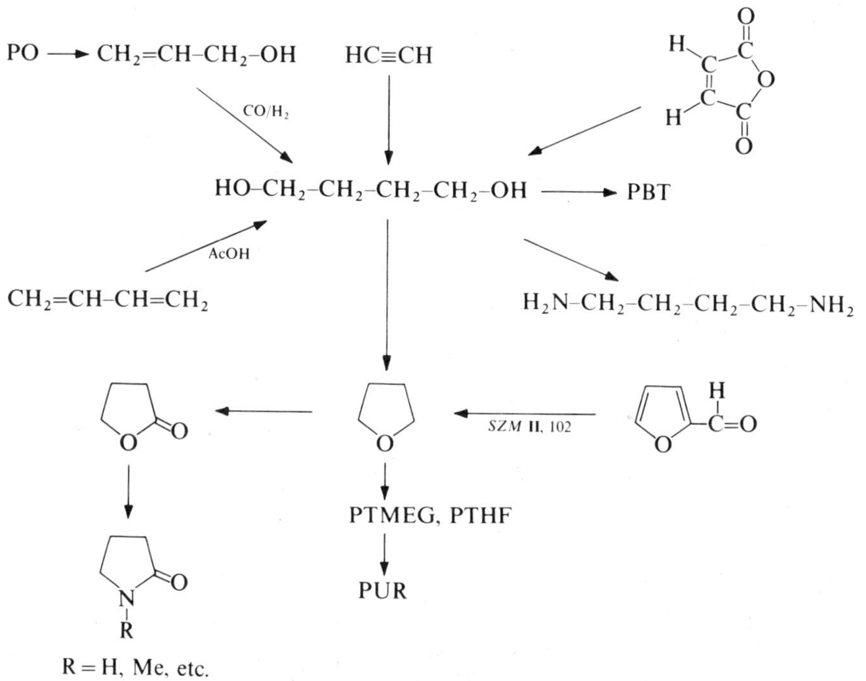

**Scheme 6.24**

For the moment, however, the present major source of BDO is acetylene and formaldehyde (4.7.3) as evidenced by the projected expansion of Du Pont's BDO production in La Porte, Texas (*CMR* 10/12/87) on the basis of a new contract with nearby Exxon as the supplier of acetylene isolated from refinery gases (in addition to the existing supply of acetylene from Rohm and Haas). A similar arrangement with Idemitsu in Japan raises the projected worldwide BDO capacity of Du Pont—the leading producer of PBT—to 275MM lb by 1989. Similarly, BASF is doubling its BDO capacity to 400MM lb in Geismar, Louisiana and in Ludwigshafen, West Germany, and GAF, in conjunction with Huels of West Germany, is doubling its BDO capacity in Marl, West Germany to 155MM lb while it already produces an equal amount in the United States.

The structure of PBT combines the tendency to form crystalline molecular arrays (promoted by the linear terephthalate moiety) and the flexibility of the tetramethylene links; consequently, this semicrystalline thermoplastic exhibits excellent thermal and mechanical properties. Objects fabricated from PBT resist temperatures as high as 140°C and it is an impact- and distortion-resistant, tough engineering material. In addition, the thermoplastic PBT forms useful polymer alloys or blends with PET, PC, and other plastics, and it can be processed by injection, extrusion, and foam molding. Du Pont's Hytrel is a thermoplastic elastomer assembled from "hard blocks" of poly(1,4-butylene glycol terephthalate), "soft blocks" of the terephthalate of PTMEG, and blocks of a random terephthalate of both BDO and PTMEG (*CT* 2/87, 82, 282).

A recent development that involves the use of BDO is its conversion to tetramethylenediamine, the smaller isomer of hexamethylenediamine, and like the latter, a building block of a nylon in conjunction with adipic acid. So far, nylon 4/6, DMS's Stanyl, is being produced in the Netherlands on a relatively small scale (165,000 lb/yr) and exhibits promising properties as an engineering thermoplastic and textile fiber. On the basis of this promising future of nylon 4/6, DSM plans to produce 44MM lb of nylon 4/6 in the United States in collaboration with Allied-Signal in addition to its existing commitments in Japan. Unitika of Japan is currently offering (*CW* 3/11/87) a 30% fiberglass-reinforced nylon 4/6 plastic priced at $3.65–$4.55 suitable for the manufacture of high-performance objects. The material is formulated with DSM-supplied plastic.

Most likely, tetramethylenediamine is obtained by the amination of BDO, although other possibilities of its origin are the reductive amination of butyrolactone beyond the formation of pyrrolidone, or the addition of HCN to AN (the current market prices shown here as well as those shown above in connection with potential sources of BDO may change with an improvement of antiquated technologies); see Scheme 6.25.

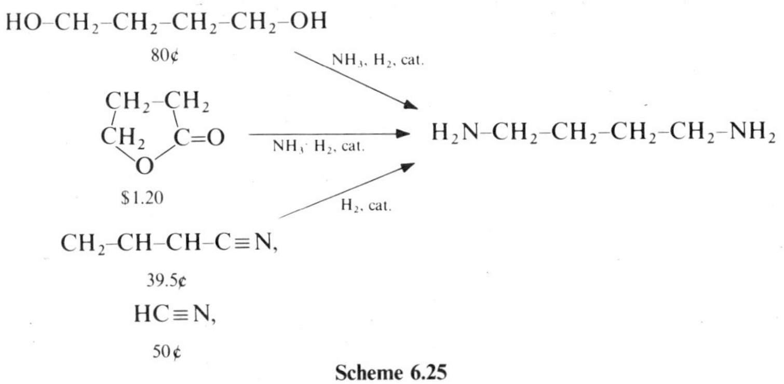

$$HO-CH_2-CH_2-CH_2-CH_2-OH$$

80¢

NH$_3$, H$_2$, cat.

CH$_2$-CH$_2$
CH$_2$   C=O     $\xrightarrow{\text{NH}_3, \text{ H}_2, \text{cat.}}$     $H_2N-CH_2-CH_2-CH_2-CH_2-NH_2$
   O

$1.20

H$_2$, cat.

$$CH_2-CH-CH-C\equiv N,$$

39.5¢

$$HC\equiv N,$$

50¢

**Scheme 6.25**

Some recently announced industrially available, transformation products of BDO are 1,4-dibromobutane, 5-bromovaleronitrile, and 5-bromovaleric acid.

## 6.8   1,3-BUTADIENE AND RELATED CONJUGATED DIENE SYSTEMS

### 6.8.1   1,3-Butadiene

#### 6.8.1.1   Sources and an Overview of Uses

As shown in Fig. 2.2, 1,3-butadiene is a by-product of petroleum cracking operations designed primarily for the production of ethylene. The on-purpose production of butadiene and imports supplement the butadiene by-product, when necessary.

The domestic production of butadiene dropped from the 1978 level of about 3.5B lb to a low of about 1.9B lb in 1982, and since then, it has gradually increased to the current level of about 3.8B lb. The decrease in the demand of butadiene in the United States parallels the decrease in the production of SBR mentioned elsewhere (4.3.4). The worldwide demand for butadiene is about 5.5B lb. In recent years prices of butadiene fluctuated widely between 29¢/lb in January 1986 to a low of 9¢/lb in October 1986 and then recovered to 30¢/lb in 12/87, to drop again to the current, relatively stable price of 22¢/lb (CMR 4/11/88, 1/89).

An important lesson can be learned from the history of butadiene in connection with World War II. At the start of that conflict, when the United States found itself cut off from the supply of natural rubber by the Japanese occupation of Southeast Asia, a crash program directed by an effective War Production Board was instituted to orchestrate butadiene production necessary for the manufacture of tires for military vehicles, airplanes, tank treads, and so on. The standard styrene–butadiene rubber, GRS (for "government rubber styrene") was chosen on the basis of the German BUNA S (butadiene–styrene formula produced by means of sodium—or Natrium in German— catalysis). However, the major bottleneck was the supply of butadiene. Since time was of the essence, in addition to cracking paraffinic hydrocarbons, the War Production Board authorized a simultaneous utilization of the following synthetic pathways:

1. The Lebedev process (ONR-54) that consists of a catalytic cracking of ethanol at about 400°C over the oxides of aluminum and zinc.

2. The formation and hydrogenation of aldol (4.5.2) followed by dehydration of the intermediate 2-buten-1-ol ($CH_3$–CH=CH–$CH_2$–OH) to the desired butadiene.

3. The dehydration of 1,4-butanediol obtained by the classical acetylene– formalde-hyde route (4.7.2).

4. Partial hydrogenation of vinylacetylene (4.7.4) to butadiene. Obviously, the overall butadiene production program was successful since the United States fulfilled its role as the "arsenal of democracy" and supplied all allied forces (on the eastern as well as western fronts of Europe, and also in the Pacific theater of war) with the necessary materials. On termination of the war, GRS became the currently employed SBR.

About 37% of the butadiene demand is channeled into the production of SBR, and about 22% becomes butadiene rubber (BR) (see below). The approximately 1B lb of BR is used mainly (70%) in the construction of rubber tire treads, with the rest being employed as impact modifier of PS and other polymers, and in the manufacture of hoses, gaskets and conveyor belts. The styrene–butadiene latex used in the formulation of water-based paints and for the production of rubber-backed carpeting consumes about 9% of butadiene. A specialty latex obtained by the copolymerization of butadiene with vinylpyridine (10.3.1), gives, in conjunction with resorcinol–formaldehyde (9.3.4.2), an adhesive that binds nylon cords to rubber. The production of adiponitrile (see 3.10.3.6 and below) and HMDA consume about 12% of the butadiene demand, and minor contribu-tions of butadiene go into the production of ABS resins (6%) and nitrile rubber (NR) (2.5%). Still smaller quantities of butadiene are consumed in the production of cyclodi- and cyclotrimerization products and in the Diels–Alder cycloaddition reactions men-tioned below.

The separation of butadiene from the less unsaturated hydrocarbons depends most often on *extractive distillation* with dipolar aprotic solvents such as DMF, DMA, furfural, acetonitrile, and NMP. These solvents (3.5, 3.1) have a relatively high boiling point and dipole moment. Thus, they associate with the rather polarizable butadiene to form molecular complexes that are retained when the less unsaturated hydrocarbons are removed by distillation. Then, at higher temperatures, the molecular complexes break down and butadiene is also distilled while the dipolar aprotic solvents remain behind to be reutilized. Butadiene also forms a molecular complex with cuprous ammonium acetate:

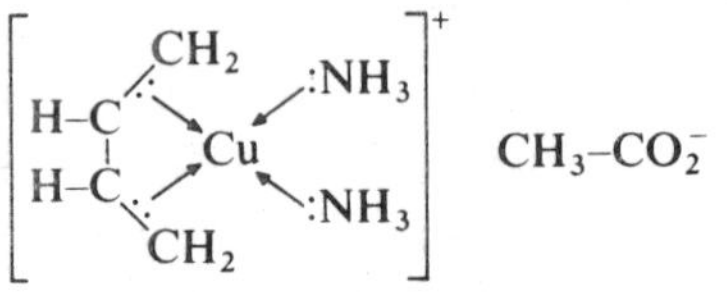

$$\left[ \begin{array}{c} \text{H-C} \overset{\displaystyle CH_2}{\underset{\displaystyle CH_2}{\diagup\diagdown}} \overset{\displaystyle :NH_3}{\underset{\displaystyle :NH_3}{Cu}} \\ \text{H-C} \end{array} \right]^{+} \quad CH_3-CO_2^{-}$$

which can serve to remove butadiene from the gas stream resulting from the cracking operation. However, extractive distillation is more convenient, and the copper complex is mentioned here only because it is reminiscent of the butadiene reactions catalyzed by transition-metal catalysts shown below.

### 6.8.1.2 Polymers Derived from 1,3-Butadiene

Styrene–butadiene rubber is obtained by free-radical-initiated polymerization of about 14% by weight of styrene, with the remainder being butadiene. The butadiene residue, $C_4H_6$, is distributed in the polymer as follows:

About 60% as the 1,4-trans moiety:

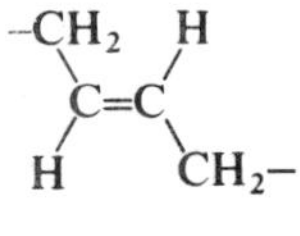

$$\begin{array}{cc} -CH_2 & H \\ \diagdown & \diagup \\ C=C \\ \diagup & \diagdown \\ H & CH_2- \end{array}$$

About equal amounts a the 1,4-cis and 1,2 moieties:

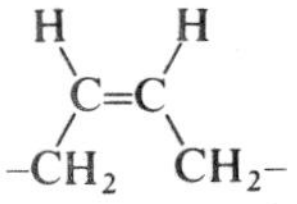

$$\begin{array}{cc} H & H \\ \diagdown & \diagup \\ C=C \\ \diagup & \diagdown \\ -CH_2 & CH_2- \end{array} \quad \text{and} \quad \begin{array}{c} -CH_2-CH- \\ | \\ CH=CH_2 \end{array} \quad \text{respectively}$$

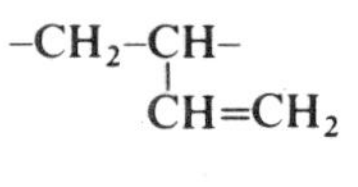

The residual olefinic bonds are cross-linked to a limited extent during the vulcanization process. The reaction of elementary sulfur is controlled by both vulcanization accelerators and delayed vulcanization agents. The actual mix from which a rubber tire is fabricated contains as much as 50% by weight of carbon black, as well as other rubber-processing agents such as antioxidants, antiozonants, lubricants, and pigments.

The structure and hence properties of BR are determined by the choice of polymerization catalysts. For example, the Ziegler–Natta catalytic system of $R_3Al–TiCl_3$ (1 : 1) tends to produce mostly a 1,4-trans polymer, while the addition of some iodine, or the use of $CoX_2$ in place of the titanium chloride, induces the formation of the preferable 1,4-cis polymer. Finally, the use of vanadium-based catalysts such as $VX_3$ and $R_3Al$ tend to give the 1,2 polymer. As pointed out above, most of BR is utilized by milling (blending in a rubber mill) with SBR and other elastomers in order to increase the resiliance and toughness of the final product. Syndiotactic 1,2-polybutadiene (~ $1.00) was produced until recently by Uniroyal and is now marketed by FSR America—a division of Japan Synthetic Rubber.

As mentioned in 6.8.1.1, minor quantities of butadiene are employed as comonomers in ABS polymers and nitrile rubber, NR or NBR. ABS contains 50–75% styrene and the presence of a relatively small amount of butadiene contributes to the impact resistance of this tough engineering plastic. NR contains 20–40% of acrylonitrile and is also a tough, specialty elastomer. A recent addition to the NR family (*CMR* 6/1/87) is a hydrogenated nitrile rubber (HNBR), Polysar's Tornac, which has excellent heat- and oil-resistant properties and is expected to penetrate the under-the-hood automotive market.

Terpolymers of methyl methacrylate–butadiene–styrene (MBS; $1.13–$1.31) are employed as impact modifiers and as processing aids of some polymers.

Butadiene is the current source of chloroprene, the monomer that gives rise to neoprene rubber (CR; 215MM lb, $1.13) and was originally prepared from vinyl acetylene (4.7.4). Neoprene does not swell or soften in the presence of petroleum products and is used as a wire and cable coating and in the manufacture of other industrial products. During the production of chloroprene, the initial product of the substitution–elimination reaction of butadiene and chlorine is a mixture of about 60% $Cl-CH_2-CH=CH-CH_2-Cl$ and about 30% the 1,2-dichloro-3-butene isomer. The latter is dehydrohalogenated to the desired 2-chloro-1,3-butadiene, or chloroprene, while the former is subjected to a $Cu_2Cl_2-FeCl_3$-catalyzed allylic rearrangement that yields additional 1,2-dichloro-3-butene. Fortunately, the two isomeric dichlorobutenes can be readily separated because of a large difference in boiling points:

cis-1,4-Dichloro-2-butene, b.p. 152°C, while the boiling point of the trans isomer is 75°C at 40 mm, and 1,2-Dichloro-3-butene, b.p. 123°C

A specialty hydroxy-terminated polybutadiene is produced by means of a hydroxyl group-generating, radical polymerization initiator:

$$CH_2=CH-CH=CH_2 + \cdot O-H \rightarrow \cdot CH_2-CH=CH-CH_2-O-H \rightarrow etc.$$

(where "etc." signifies the coupling of additional butadiene molecules with the initially formed carbon radical).

Phillips can supply hydroxyl-terminated liquid Butarez (HTL) at a cost of $10–$20/lb with a range of equivalent weights between 800 and 4000 and a choice of primary or secondary alcohol functions located on a $\beta$-carbon relative to the position of the double bond:

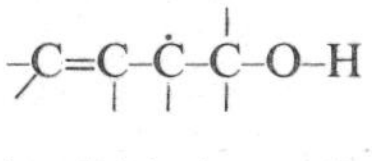

$$\underset{/}{-}C=\underset{|}{C}-\underset{|}{\overset{|}{C}}-\underset{|}{C}-O-H$$

This suggests that the polymerization-initiating radicals are either hydroxymethyl or $\alpha$-hydroxyethyl, $\cdot CH(CH_3)-O-H$, species:

$$CH_2=CH-CH=CH_2 + \cdot CH_2-OH \rightarrow CH_2-CH=CH-CH_2-CH_2-OH \rightarrow etc.$$

$$+ CH_3-\underset{\cdot}{C}H-O-H \rightarrow \cdot CH_2-CH=CH-CH_2-CH(CH_3)-OH \rightarrow etc.$$

Such hydroxyalkyl radicals can be generated *in situ* when the polymerization of butadiene is carried out in an alcohol solution by means of peroxide catalysts (6.3.1.7) that abstract an $\alpha$-hydrogen atom from an alcohol. For example:

$$t-Bu-O\cdot + CH_3-CH_2-O-H \rightarrow t-Bu-O-H + CH_3-\underset{\cdot}{C}H-O-H$$

The α-hydrogen of an alcohol is the most vulnerable hydrogen since the resulting radical is stabilized by resonance with the adjacent nonbonding electrons of the hydroxyl group.

In a somewhat similar fashion, Sartomer produces liquid Poly bd resins, priced at about $1.22/lb, which contain an average of 50 butadiene residues (for an average molecular weight of 2800) and two terminal, primary hydroxyl groups. About 80% of the butadienes are incorporated in a 1,4 manner to give 3:1 ratio of trans:cis residues, while the rest of butadiene monomer undergoes a 1,2-polymerization that gives rise to 20% pendant vinyl groups. This polymerization is likely to be catalyzed by hydroxyl radicals generated by means of a homolytic decomposition of hydrogen peroxide.

### 6.8.1.3  Butadiene as a Source of C$_6$ Monomers

The development of the HCN–butadiene route to adiponitrile (3.10.3.6) that is hydrogenated to hexamethylenediamine (HMDA) contributes not only to nylon 6/6, but also to other two-digit nylons in which the first digit is the number 6, such as nylons 6/8, 6/10, 6/11, and 6/12. HMDA is also converted to the corresponding diisocyanate, HMDI (see interlude in Section 3.8).

The double hydrocarbonylation of butadiene to yield adipic acid is carried out by BASF in Europe. Since November 1984 the hydroformylation of butadiene has also been practiced in the United States by Badische Corp. to produce about 13MM lb of 1,6-hexanediol (HDO). This homolog of 1,4-butanediol is used to assemble high-performance PURs employed in the manufacture of durable products such as rollers for skates and skateboards and outer shells of ski boots.

### 6.8.1.4  Cycloaddition Reactions of 1,3-Butadiene

The Diels–Alder reaction (*ONR*-24) of butadiene (22¢) and MA (55¢) produces tetrahydrophthalic anhydride (65¢). It is noteworthy that the price of the product is less than the sum of the listed prices of the reactants and this suggests that, in order to be competitive, captive use of MA by its producers (2.3.5) converts it to the Diels–Alder product under discussion with some sacrifice of MA profits.

This lower-melting (m.p. 101–102°C) analog of phthalic anhydride (35¢, m.p. 131–133°C) can be used as an ingredient of UPRs, and it is also a building block of an excellent fungicide, namely, captan (Scheme 6.26). The companion component of captan,

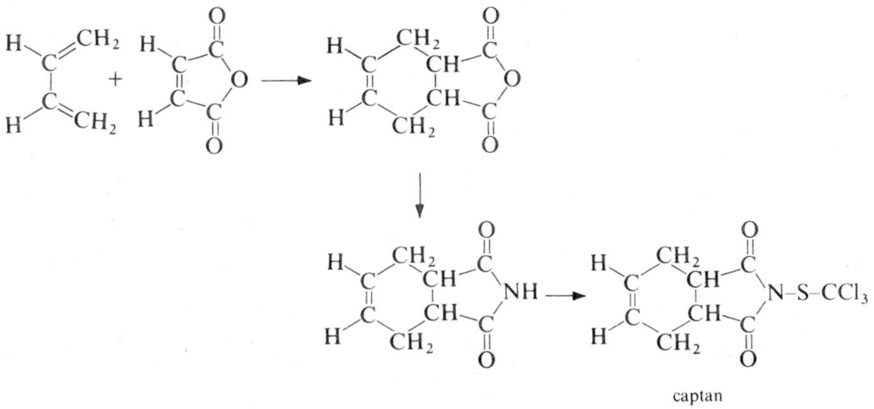

captan

**Scheme 6.26**

namely, trichloromethylsulfenyl chloride, is obtained by the perchlorination of either methyl mercaptan or dimethyl disulfide (see 3.3.2.6.C).

The cycloaddition reaction of butadiene with sulfur dioxide produces the solvent sulfolene (Phillips $1.50), but since this bicyclic product readily undergeoes a retro-Diels–Alder reaction at elevated temperatures, it is stabilized by hydrogenation to another dipolar aprotic solvent (3.5.3.1), namely, sulfolane (b.p. 285°C, m.p. 27.5°C; Phillips $1.82); see Scheme 6.27.

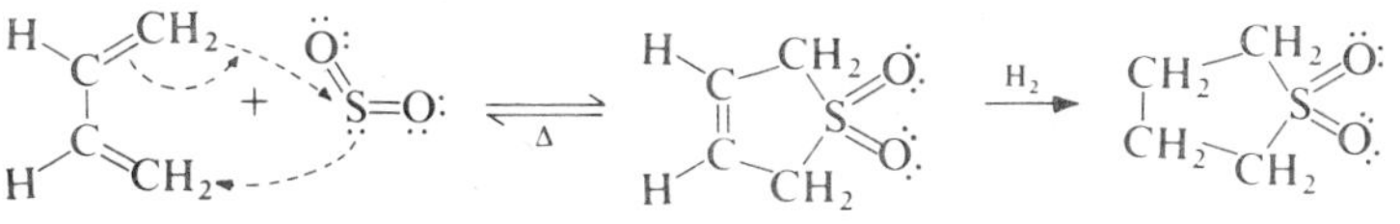

**Scheme 6.27**

The Diels–Alder reaction of butadiene with cyclopentadiene (6.8.3) leads to 5-ethylidenenorbornene, one of the most desirable "D" components of the EPDM (ethylene–propylene–diene–methylene) elastomers (4.3.1, 5.4.1). However, the initial cycloaddition product must be isomerized by treatment with sodium to rearrange the exo-double-bond-containing 5-vinyl-bicyclo-(2.2.1)hept-2-ene to the desirable endo-double-bond-containing isomer (Scheme 6.28). Now, the cyclic diolefin contains double bonds of

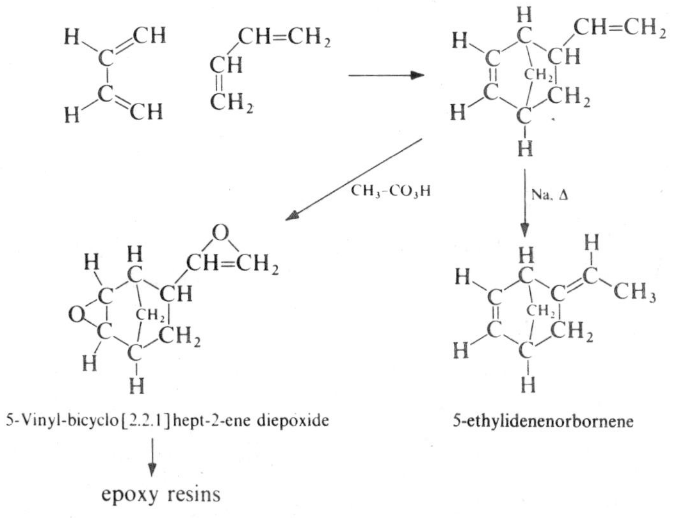

5-Vinyl-bicyclo[2.2.1]hept-2-ene diepoxide

↓

epoxy resins

5-ethylidenenorbornene

**Scheme 6.28**

significantly different reactivity: the endo double bond located in the strained cyclopentene moiety copolymerizes readily with ethylene and/or propylene, and this leaves the less reactive exo double bond for the subsequent peroxide-catalyzed cross-linking or curing process. On the other hand, the intermediate 5-vinyl-bicyclo(2.2.1)hept-2-ene can be treated with peracetic acid to yield the corresponding diepoxide—one of the constituents of epoxide resins derived from epoxides other than glycidyl ethers (5.4.8).

A somewhat different cycloaddition of butadiene is its cyclodimerization and cyclotrimerization to 1,5-cyclooctadiene ($1.08), vinylcyclohexene (another candidate for a "D" in EPDM and for another nonglicidic ether epoxide resin constituent), and two isomeric cyclododecatrienes: 1,5,9-*trans,trans,cis*-cyclododecatriene (CDT; 82¢) and 1,5,9-*trans,trans,trans*-cyclododecatriene (Scheme 6.29). The cyclodimerization and cyclotrimerization reactions of butadiene are catalyzed by Wilke's catalyst introduced in 1955,

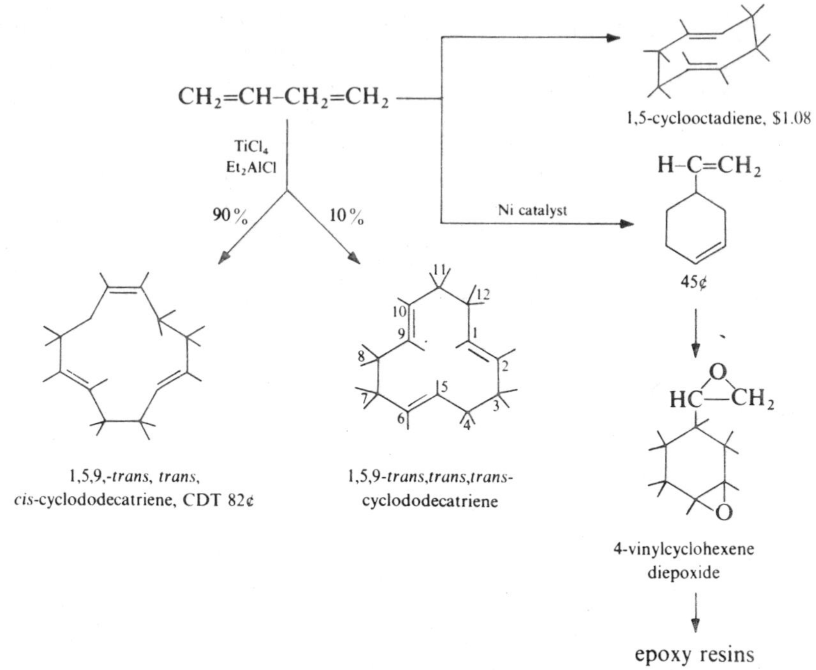

$$CH_2=CH-CH_2=CH_2$$

TiCl$_4$
Et$_2$AlCl

90%      10%

Ni catalyst

1,5-cyclooctadiene, $1.08

$$H-C=CH_2$$

45¢

$$HC—CH_2$$
O

1,5,9,-*trans, trans,*
*cis*-cyclododecatriene, CDT 82¢

1,5,9-*trans,trans,trans*-
cyclododecatriene

4-vinylcyclohexene
diepoxide

epoxy resins

**Scheme 6.29**

which consists of a 1.5:1 mixture of TiCl$_4$ and AlClEt$_2$. On the other hand, vinylcyclohexene is produced by means of a nickel catalyst. The 8- and 12-membered ring structures provide a viable route to both cyclic and open-chain building blocks. The C$_{12}$ triene is

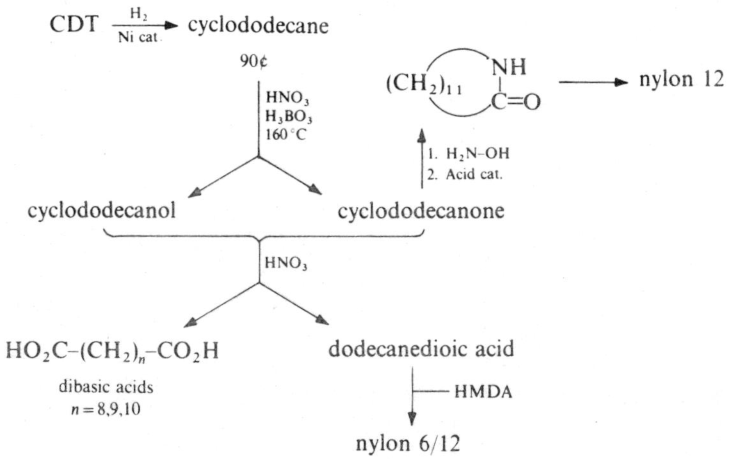

CDT $\xrightarrow{\text{H}_2}{\text{Ni cat}}$ cyclododecane

90¢

HNO$_3$
H$_3$BO$_3$
160 °C

$(CH_2)_{11}$   NH
                C=O      ⟶  nylon 12

1. H$_2$N–OH
2. Acid cat.

cyclododecanol            cyclododecanone

HNO$_3$

HO$_2$C-(CH$_2$)$_n$-CO$_2$H         dodecanedioic acid

dibasic acids
n = 8,9,10

HMDA

nylon 6/12

**Scheme 6.30**

first hydrogenated and then subjected to oxidation with nitric acid to dodecanedioic acid—the buildng blocks of nylon 6/12 and of synthetic lubricants in the form of diesters, while cyclododecanone oxime is subjected to a Beckmann rearrangement (*ONR*-8) to yield lauryl lactam, the building block of nylon 12 (see Scheme 6.30). Similarly, the eight-membered ring structure is a potential source of cyclooctanone and suberic acid, as shown in Scheme 6.31.

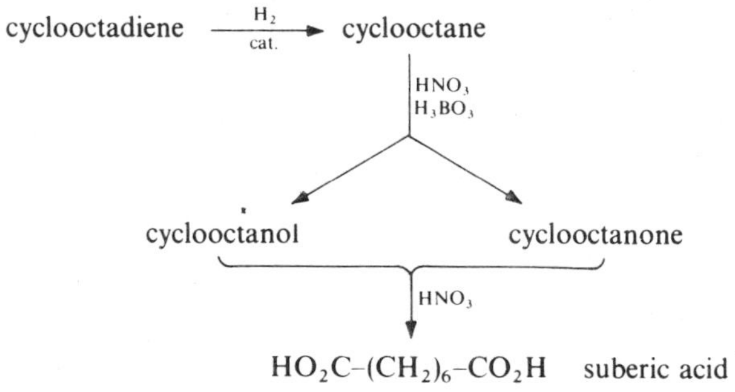

Scheme 6.31

Diels–Alder reactions of butadiene with dienophiles other than MA or $SO_2$ and TCNE (5.6.5) are also of interest. Thus, with crotonaldehyde (4.5.2) one obtains 2-methyl-4-cyclohexene carboxaldehyde, which, on a Tishchenko reaction (*ONR*-90) and epoxidation, provides another epoxide resin ingredient, as shown in Scheme 6.32.

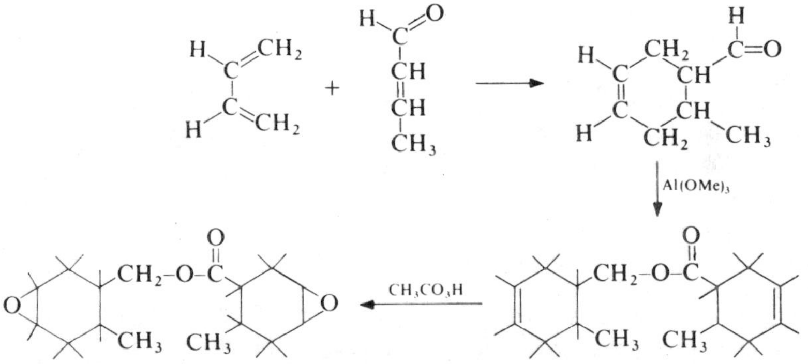

Scheme 6.32

Finally, one must mention the disastrous attempt (2.3.3) to produce tetrahydroanthraquinone (THAQ)—an interesting catalyst that accelerates the pulping of wood. The process consists of the Diels–Alder reaction of butadiene and naphthaquinone (9.3.3d)

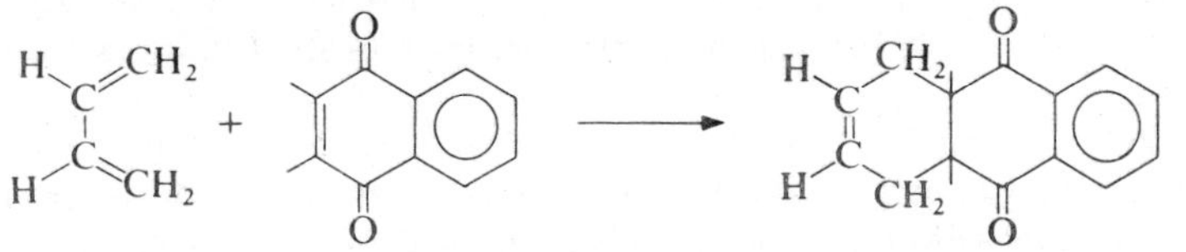

Scheme 6.33

(Scheme 6.33). THAQ continues to be produced by means of the traditional route that starts with the reaction of phthalic anhydride and benzene.

### 6.8.2  Isoprene

As is 1,3-butadiene, isoprene is also isolated as by-product of cracking operations designed to supply ethylene (Fig. 2.2), and any additional demand is supplied by imports. However, companies like Shell, that have idle isoprene plants and depend on a steady supply for the production of its isoprene-derived thermoplastic elastomer Kraton, as well as its lubricating-oil viscosity-index modifier Shell vis, can, if necessary, reactivate these production facilities (*CW* 7/7/82).

Cracking of the higher petroleum fractions may produce a mixture of isoamylenes, 2-methyl-1-butene, and 2-methyl-2-butene, and the linear isomer of isoprene, namely, 1,3-pentadiene, commonly known as piperylene (see below). High-quality isoprene requires the removal of piperylene, and this is achieved by extraction of the isoamylenes with 65% sulfuric acid, separation of the less readily sulfated piperylene, and hydrolysis of the sulfated isoamylenes. Now, the latter are subjected to catalytic dehydrogenation to yield purified isoprene. Some chemical uses of piperylene are mentioned elsewhere (7.2.1).

Several synthetic routes to isoprene are also available and are enumerated as follows:

1. Cracking of the propylene dimer (5.4.1)—actually a mixture of isohexane and isomeric isohexenes. This mixture is treated with an acid at below 300°C to promote the conversion of the isohexenes to 2-methyl-2-pentene, and then the latter and isohexane are thermally dehydrogenated and pyrolyzed to isoprene and methane at 650–800°C.

2. The Prins reaction (*ONR*-72) of isobutylene and formaldehyde (3.4.3.8) and the modernized version by Takeda (*CEN* 10/16/72), which involves a gas-phase, continuous process at about 300°C in the presence of antimony oxide catalyst.

3. Ethenylation of isobutylene (6.3.1) and the analogous metathesis of isobutylene and propylene or 2-butene.

4. The traditional addition of acetylene to acetone (4.7.3), followed by partial hydrogenation and dehydration at about 275°C. This process was employed in Italy in 1965 (*CEN* 5/17/65).

About 95% of isoprene is converted to "synthetic natural rubber," that is, poly-*cis*-1,4-isoprene, which is identical to the material isolated from the rubber tree, *Hevea brasiliensis* (*SZM-II*, 26, 144). This geometric isomer assumes a helical conformation that is most conducive to elastic properties, unlike the trans isomer known as *balata* or *gutta percha* (used in dental root-canal fillings). The desired regio- and stereoselective polymerization of isoprene is achieved by means of carefully chosen transition metal catalysts. Thus, while a relatively low ratio ($\sim 1:1$) of $TiCl_4$ and iso-$Bu_3Al$ produces the trans polymer, a higher ratio of the alkylaluminum component, say, $1:2–3$, gives rise to the desired cis-1,4 product. These results suggest that the catalytic complex functions as a template that orients isoprene molecules in a manner that favors the formation of the cis-1,4 polymer; see Scheme 6.34.

Another method used to obtain poly-*cis*-1,4-isoprene employs anionic polymerization by means of *n*-butyllithium (6.4.3) in a poorly solvating (alkane or cycloalkane) medium, and these conditions favor the formation of a complex between isoprene and the lithium terminal of the catalyst that induces a cis orientation of the monomer.

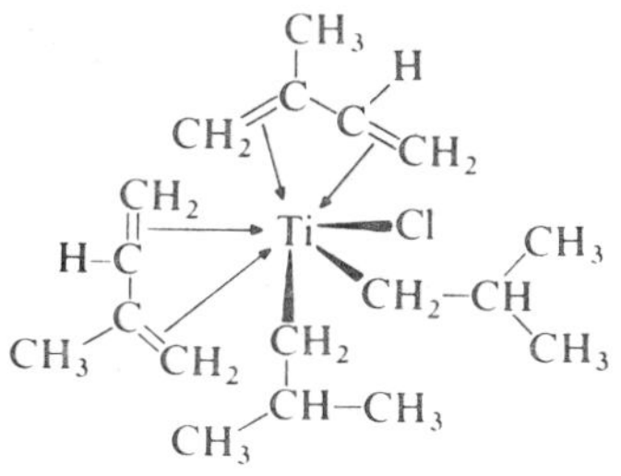

Scheme 6.34

On the Japanese market, synthetic rubber manufactured by Kuraray from isoprene derived from the Prins reaction was selling at about 64¢/lb and was unable to compete with imported natural rubber, priced at about 44¢ (*CW* 4/13/85). A small amount of isoprene is incorporated in BR (6.3.1.2).

Isoprene functions as an active diene in Diels–Alder reactions and yields products that differ from the corresponding butadiene derivatives only by the presence of a methyl side chain.

### 6.8.3   Cyclopentadiene and Its Derivatives

The gradual elimination of leaded gasoline has caused a corresponding increase in the reforming of linear alkanes to BTX and this, in turn, increased the formation of the by-products cyclopentadiene and methylcyclopentadiene. The same effect is produced when the production of ethylene shifts from light petroleum fractions to naphtha and gas oil feedstocks (see Fig. 2.3).

*Cyclopentadiene*, or more properly, its spontaneously formed dimer dicyclopentadiene (DCPD), is priced at 29¢, a remarkably low price for such complex structure. Storage of cyclopentadiene (b.p. 42°C) when freshly distilled converts it to a 4 : 1 mixture of endo and exo isomers of DCPD, and the latter undergo a retro-Diels–Alder reaction on attempted distillation at atmospheric pressure (b.p. ~ 170°C with decomposition). The equilibrium between the three related structures is depicted in Scheme 6.35. Thus, one can obtain

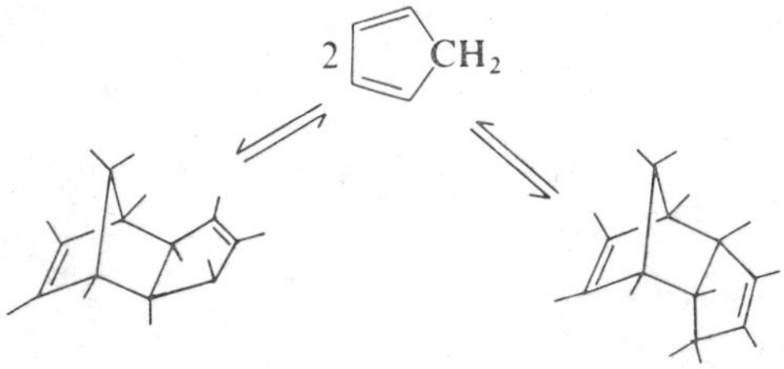

exo                    4 : 1                    endo

Scheme 6.35

derivatives of either the cyclopentadiene monomer or the dimers. For example, Velsicol offers the following (see Scheme 6.36) derivatives of DCPD obtained by hydration and/or epoxidation (hydrogen substituents are omitted for the sake of clarity). Examination of the structures (and pricing) of the preceding chemicals reveals that the more strained double bond located in the bicyclic portion of the molecules is more reactive (cf. 6.8.1.4)

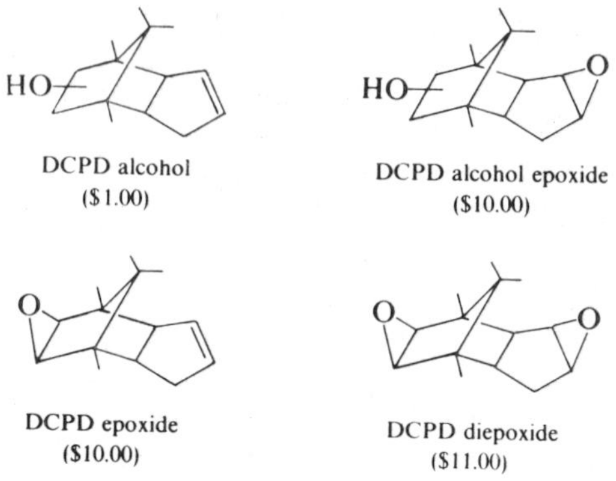

DCPD alcohol
($1.00)

DCPD alcohol epoxide
($10.00)

DCPD epoxide
($10.00)

DCPD diepoxide
($11.00)

**Scheme 6.36**

than the double bond of the fused cyclopentene ring. Alcolac offers the acrylate and methacrylate of the above-mentioned DCPD alcohol.

As mentioned above (6.6.9), the Diels–Alder reaction of cyclopentadiene and MA produces bicyclo(2.2.1)heptene-2,3-dicarboxylic anhydride, or 5-norbornene-2,3-dicarboxylic anhydride. The analogous derivative of methylcyclopentadiene is sold by Buffalo Color Corp. as Nadic methyl anhydride ($1.99), and the completely saturated ring system is also available from ARCHEM (Scheme 6.37).

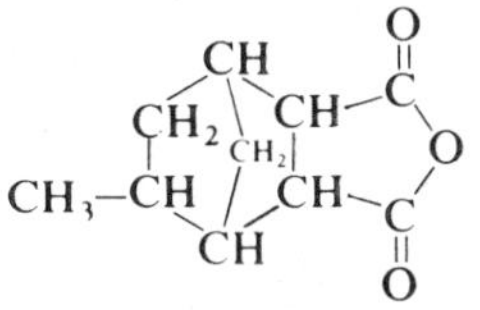

**Scheme 6.37**

The use of DCPD by Hercules in the formation of Metton by metathesis polymerization is shown elsewhere (5.4.9).

Exhaustive chlorination of cyclopentadiene gives hexachlorocyclopentadiene (HCCPD), previously a popular building block of a family of insecticides described elsewhere (8.1.2).

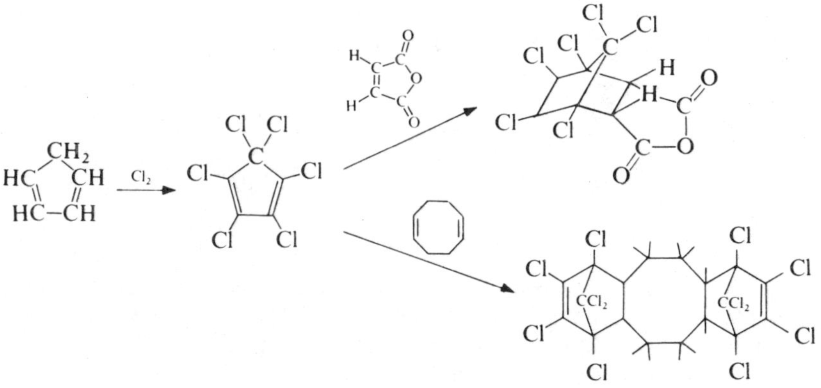

**Scheme 6.38**

Hexachlorocyclopentadiene and its MA adduct, namely, hexachloro-*endo*-methylene-tetrahydrophthalic anhydride, or chlorendic anhydride, as well as the double Diels–Alder adduct with 1,5-cyclooctadiene known as "Dechlorane Plus" (Occidental Chem., ~\$2.40), are available from Velsicol and the subsidiary of Occidental Petroleum, respectively. The latter two materials are used as flame retardants, and their structures are shown in Scheme 6.38. Chlorendic anhydride can be incorporated in polyesters, and Dechlorane Plus can be used as an additive in nylons, EPDM, LDPE, PP, PUR, epoxy resins, and so on.

At this point, a final word about cyclopentadiene pertains to the readily formed carbanion—the cyclopentadienyl anion—which functions as the so-called Cp ligand in a number of transition-metal complexes employed as catalysts. The Cp ligand is a pseudoaromatic structure (it is planar and has a sextet of $\pi$ electrons as in benzene) and forms three types of complexes; as shown in Scheme 6.39. For additional comments concerning the chemistry of DCPD, see 8.1.2.

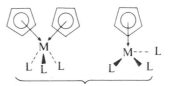

H H
strong base
or
alkali metals
metal $L_n$
p$K_a$ ~20
cyclopentadienyl anion
Fe
M
L L L
M--- L
L L
The number or nature of ligands L may vary depending on the metal and its oxidation state
Ferrocene, typical Cp "sandwich" of many metallocenes, including the methylcyclopentadiene-derived Mn complex originally proposed as an antiknock agent but subsequently outlawed by EPA

**Scheme 6.39**

## 6.9 MISCELLANEOUS DOWNSTREAM $C_4$ DERIVED BUILDING BLOCKS FROM SOURCES OTHER THAN $C_4$ HYDROCARBONS

Diketene and acetoacetic esters are mentioned in 4.6.4. Here we wish to amplify the preparative uses of diketene and its traditional derivative, simple acetoacetic esters.

A heterodifunctional derivative of hydroxyethyl methacrylate and diketene, namely, acetoacetoxyethyl methacrylate, has been produced industrially since 1986:

$$CH_2{=}C(CH_3){-}CO{-}O{-}CH_2CH_2{-}O{-}CO{-}CH_2{-}CO{-}CH_3$$

The reaction product of diketene with EG yields a double acetoacetic ester, and the keto groups of the latter are subjected to an acid-catalyzed condensation with *o*-*t*-

butylphenol to give an antioxidant, American Hoechst's Hostanox O3:

$$CH_3-CAr_2-CH_2-CO-O-CH_2-CH_2-O-CO-CH_2-CAr_2-CH_3$$

(where each Ar group represents the *o-t*-butylphenyl group).

Aniline is allowed to react with an acetoacetic ester to yield 2-methyl-4-hydroxyquinoline, or 4-hydroxyquinaldine (10.3.1):

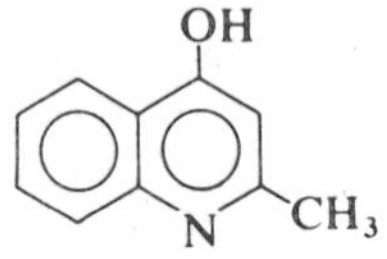

# 7

# C₅ AND HIGHER ACYCLIC BUILDING BLOCKS

## 7.1  INTRODUCTION: PRIMARY AND DOWNSTREAM BUILDING BLOCKS

To provide a systematic and representative coverage of acyclic C₅ and higher compounds of industrial importance, it is convenient to distinguish between primary and downstream building blocks (see Preface).

The primary building blocks are derived in a rather direct fashion from either fossil or biomass sources. A more extensive survey of the latter category is found elsewhere (*SZM-II*), but, since many industrially important chemicals are obtained by combining building blocks derived from both fossil and biomass sources, some of the more important contributions of renewable resources need to be highlighted here.

Many of the downstream building blocks were mentioned in the preceding chapters. The object of this chapter is not to repeat such information, but rather to summarize and broaden the perspective of the usefulness of these "industrial intermediates."

## 7.2  PRIMARY C₅ AND HIGHER BUILDING BLOCKS DERIVED FROM FOSSIL SOURCES

### 7.2.1  Petroleum

Mixtures of isomeric pentanes, hexanes, and heptanes are separated from petroleum refinery streams by fractional distillation. Some of the higher alkanes are isolated during low-temperature fractional crystallization when high-molecular-weight paraffins are "dewaxed" by means of solvents such as MEK, benzene, toluene, ethylene dichloride, methylene dichloride, and propane in order to produce quality lubricating oils. Straight-chain paraffins tend to crystallize as their molecular weights, and hence their softening points, increase.

With rare exceptions, no attempts are made on an industrial scale to isolate specific isomers, and some of the paraffinic materials of varying degree of homogeneity offered on the marketplace are shown as follows:

Hexane, minimum 95%, $1.12/gal, 5.61 lb/gal (all densities at 60°F)

Heptane, 95%, $1.18/gal

$n$-Octane, minimum 97%, $6.25/gal

Cleaners's naphtha, flash point 140°F, $1.40/gal

Lacquer diluent, petroleum fraction b.p. 140–200°F, $1.25/gal; b.p. 200–240°F, $1.12/gal, 6.24 lb/gal

Naphtha, petroleum, VM&P (varnish makers and painters), $1.20/gal, 6.28 lb/gal

Mineral oil, USP $2.18–$2.45/gal, depending on viscosity

Mineral spirits, regular $1.41/gal; odorless $1.78/gal, 6.55 lb/gal

Petrolatum, USP 29¢–38.5¢/lb, depending on color

USP microcrystalline wax, petroleum, coating grades, FDA 36.5¢/lb; laminating grades, FDA 38.5¢/lb

Paraffin, fully refined, 29¢–41.5¢/lb, depending on softening temperature, slack wax, 5–20% oil, 16¢–21¢/lb

The U.S. production capacity of $n$-paraffins ($C_{10}$ and higher) is about 1B lb valued, on the average, 10¢/lb. Certain fractions of paraffinic hydrocarbons are utilized as solvents, lubricating oils, and other functional materials. For example, $n$-hexadecane is a common component of automotive lube oils, and the pure hydrocarbon is chosen ($CW$ 10/21/87) as an indicator of high-temperature antioxidant capability (HTAC) of the paraffinic component in formulated consumer products.

Nearly two-thirds of $n$-paraffins are utilized in the production of linear alkylbenzenes (LAB) (see below); some are converted to the corresponding linear alcohols, and about 5% become chlorinated paraffins.

The chlorination of pentanes to give mixtures of chlorides that were subsequently hydrolyzed and converted to "amyl acetates' for use as a lacquer solvent (Sharples process) is mentioned on p. 349. Currently, one finds on the marketplace:

Amyl alcohols, primary mixed isomers, 46.5¢;

Amyl acetates, primary mixed acetates, 57¢.

Most likely, however, these materials are prepared from amyl alcohols isolated from fusel oil (see 7.3.4). From the chlorination of pentanes one obtains seven chlorides: three primary, three secondary, and one tertiary, but alkaline hydrolysis of the tertiary chloride produces isoamylene, $(CH_3)_2C=CHCH_3$, rather than the corresponding alcohol.

The chlorination of the higher paraffins is practiced to produce about 100MM lb of extensively chlorinated products for use as lubricating oil additives, ingredients of cutting oils, plasticizers and flame retardants for PVC, water repellents for textiles, sealants, and so on:

Chlorinated paraffin, 40% chlorine 45¢
50% chlorine 46¢
60% chlorine 46.5¢
70% chlorine, resinous, 69¢

For the production of LAB, monochlorination of paraffins is achieved by use of an excess of the hydrocarbon and by limiting the reaction to about 30% conversion. The chlorinated products are then subjected to a Friedel–Crafts reaction with benzene (9.2.1) to give linear alkylated benzene (LAB). This mixture, in turn, is sulfonated to about 600MM lb of alkylbenzenesulfonates, linear alkylsulfonates (LAS) for use in biodegradable domestic and industrial cleaning products valued at about 45¢/lb. The production of LAS from alphaolefins is mentioned in 7.4.1.

Other important reactions of paraffinic hydrocarbons include sulfoxidation and sulfochlorination. Both are free-radical chain reactions carried out at room temperature and induced by photochemical means:

$$R-H + SO_2 + O_2 \rightarrow R-SO_3H \text{ petroleum sulfonate, } 60\text{--}62\%$$

sulfonic acid content 48.75¢–49¢/lb available in high-, medium-, and low-molecular-weight grades.

$$R-H + SO_2 + Cl_2 \rightarrow R-SO_2-Cl + HCl$$

Petroleum chlorosulfonate is reminiscent of Du Pont's chlorinated elastomer Hypalon, CSM rubber, mentioned elsewhere (4.3.3.1).

The cracking of paraffinic hydrocarbons is mentioned in connection with the production of ethylene (4.2), propylene (5.1), and alphaolefins (7.4.1). The pyrolysis is carried out at about 500°C in either the presence or absence of steam and catalysts. Depending on the reaction conditions and feedstocks (Fig. 2.2), the distribution of products varies with respect to (a) molecular-weight distribution of olefinic products, (b) location of double bonds in olefins, and (c) relative yields of cyclic products such as cyclopentadiene (6.8.3) and BTX (9.2). There are countless patents, and even a greater number of patent claims and examples, deal with the above-mentioned technology.

The cracking of petroleum also produces, for all practical purposes, inseparable but still useful mixtures of $C_5$ and $C_6$ olefins. An example of utilization of such a mixture is the BP France project ($CW$ 7/29/87) to convert these olefins into the corresponding methyl ethers and thus expand the production of these gasoline additives by 110MM lb/yr at the same time that the production of the companion MTBE (3.3.2.2) is also expanded by the same amount.

Finally, the cracking of petroleum produces, in addition of $C_2$, $C_3$, and $C_4$ olefinic building blocks as described in 4,3,2, 5,4,1, and 6.3, respectively, several dienes such as 1,4-butadiene (6.8.1 and references throughout these pages), isoprene (6.8.2), and piperylene (or 1,3-pentadiene). The two latter hydrocarbons are C$_5$ building blocks per se, while butadiene is shown (6.8.1.4) to form cyclic dimer and trimer derivatives that provide C$_8$ and C$_{12}$ dicarboxylic acids on oxidative cleavage. Also, butadiene adds two molecules of either HCN or CO in response to a search of new routes to C$_6$ building blocks of nylon 6/6—adipic acid and hexamethylenediamine and their derivatives such as adipic and esters, and hexamethylenediisocyanate. Adiponitrile is the product of the acrylonitrile EHD process (5.4.3.2). The same search for nylon 6/6 building blocks produces by-products that become building blocks on their own right: Du Pont's glutaric acid (97¢ as a 50% solution) and Dytek A.

Dai Nippon Ink & Chemicals, represented in the United States by DIC Americas— the recent acquirer of Reichold Chemicals—offers a derivative of piperylene, namely, Epiclon B-4400 ($6.00), which is a specialty hardener of epoxy resins obtained by the reaction of the diene with 2 mol of MA:

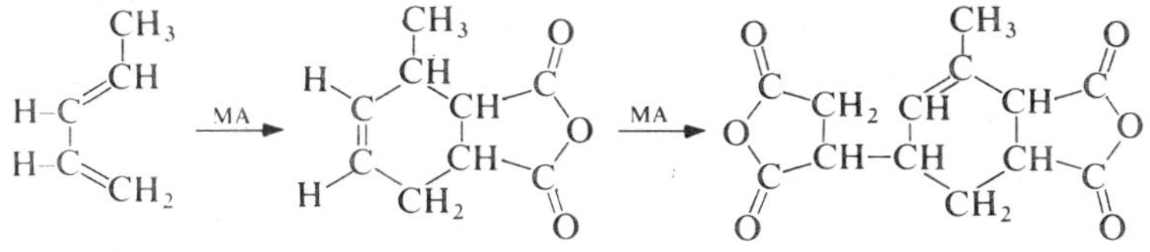

The tetracarboxylic acid dianhydride is formed by an ene reaction (*ONR*-28) analogous to the common behavior of alphaolefins (see 7.4.1) with the Diels–Alder adduct.

Piperylene is also used in the formulation of pressure-sensitive and hot-melt adhesives.

### 7.2.2   Mineral Waxes

Waxes in general are identified according to how they behave and feel to the touch (they are responsible for glossy, water-resistant, slippery coatings) more han their chemical identity. Most waxes in commercial use are derived from renewable sources (*SZM-II*, 81–83), but some are extracted from lignite—geologically speaking, a "young" coal. Montan wax (55–61¢ for West German and domestic material) and ozocerite are the most common mineral waxes on the marketplace. Considering their biological origin, it is not surprising to learn that montan wax, for example, contains the n-C$_{28}$ alcohol and carboxylic acid and probably also the ester of the two components.

### 7.3   PRIMARY C$_5$ AND HIGHER BUILDING BLOCKS DERIVED FROM RENEWABLE SOURCES

#### 7.3.1   Fatty Acids from Fats and Oils

There are probably as many different fats and oils as there are species in the animal and plant kingdom, but only a limited number are of industrial importance (prices, in cents per pound, quoted at the respective commodity markets):

Grease, white 9.5¢ (4/86); 13.75¢ (1/88)
Lard, 10¢ (4/86); 15¢ (1/88);
Tallow, inedible, fancy, 10.25¢ (4/86); 16.5¢ (1/88);
Tallow, inedible, bleached, 10¢ (4/86); 16.25¢ (1/88);

Castor oil, raw, No. 1 Brazil, 46¢;

Castor oil, refined 74–78¢;

Coconut oil, crude, New York, 13¢ (4/86); 28.75¢ (1/86);

Coconut oil, refined, New York, 17.75¢ (4/86); 34.75¢ (1/88);

Corn oil, crude, Midwest( 17.75¢ (4/86); 23.25¢ (1/88);

Corn oil, refined, Midwest, 24.3¢ (8/86); 31.30¢ (1/88);

Cottonseed oil, crude, 18¢ (4/86); 20¢ (1/88);

Cottonseed oil, refined, 27¢ (4/86); 29.5¢ (1/88);

Linseed oil, crude, 30¢ (4/86); 25¢ (1/88);

Palm oil, crude, 14.5¢ (4/86); 24¢ (1/88);

Peanut oil, crude, 29¢ (4/86); 33¢ (1/88);

Peanut oil, refined, 36.5¢ (4/86); 42.6¢ (1/88);

Soybean oil, crude, 17.5¢ (4/86); 20¢ (1/88);

Soybean oil, salad oil, 21.5¢ (4/86); 24.65¢ (1/88)

Some fatty raw materials are utilized chemically without either alkaline hydrolysis (saponification) or acid-catalyzed hydrolysis. Foremost examples of this practice is the production of about 100MM lb of epoxidized soybean oil (61¢) for use as a PVC stabilizer and the sulfation of castor oil to give a surface-active agent for use in textile processing (*SZM-II*, 59). Castor oil is imported at a level of about 200MM lb for this use and for transformations of its predominant fatty acid, namely, ricinoleic acid (see below).

The chemical workup of naturally occurring fats and oils (*SZM-II*, 53–80) produces glycerol or glycerine (natural, USP CP 99.5%, 78¢; synthetic, see 5.4.2.2.B) and a mixture of fatty acids characteristic of a given source. Generally speaking, specific fatty acids, or narrow fractions of fatty acids of similar molecular weight and structure, are normally isolated by fractional distillation of their methyl esters. As pointed out elsewhere (*SZM-II*, 61–65), the methyl esters of fatty acids occupy a cross-road position in the preparation of most specific chemicals derived from fats and oils. Some of the fatty acid methyl esters are marketed:

Methyl laurate, 70%, 66¢;

Methyl laurate, 92%, 69¢;

Coconut 13–18 methyl esters, 50¢;

Palm kernel 13–18 methyl esters, 50¢.

Finally, the hydrolysis of appropriate fractions of the methyl esters gives the corresponding fatty acids. The unit prices, when available, of the more common fatty acids are as follows:

$C_6$ caproic

$C_8$ caprylic, 80¢

$C_{10}$ capric, 60¢

$C_{12}$ lauric, 35¢
   coconut fatty acids, 15–22¢

$C_{14}$ myristic, $1.12

$C_{16}$ palmitic, 51¢

$C_{18}$ stearic, single-pressed, 28¢
stearic, double-pressed, 26¢
stearic, triple-pressed, 32¢
stearic, tallow fatty acids, ~14¢

$C_{18}$ oleic (one double bond), single distilled (red), 35¢
oleic (one double bond), double distilled (white), 38¢
soya acids, ~18¢

$C_{18}$ linoleic and linolenic (two and three double bonds, respectively, marketed as linseed oil fatty acids), 46;

$C_{18}$ ricinoleic (one double bond and one hydroxyl group), 79.5¢
ricinoleic dehydrated $1.10.

These qualifications illustrate the difficulties to obtain pure fractions of fatty acids on an industrial scale.

Fatty acids per se, or in the form of various metallic salts are used (*SZM-II*, 66–77) to reduce friction under a variety of manufacturing conditions. Also, they are subjected to various chemical transformations to produce fatty alcohols, amides, and so on, mentioned below. With respect to chemical transformations, a unique rule is played by ricinoleic acid—the main constituent of castor oil—because, depending on experimental conditions, it becomes the source of sebacic acid (the $C_{10}$ dicarboxylic acid; $1.94), capryl alochol (2-hydroxyoctyl alcohol; 35¢), heptaldehyde; heptanoic acid (65¢), and undecylenic acid (the $C_{11}$ carboxylic acid with a terminal double band; $2.57). Capryl alcohol, in turn, gives rise to methyl hexyl ketone (~50¢) and a mixture of 1- and 2-octenes (37¢). The details of these transformations are discussed elsewhere (*SZM-II*, 71–74).

Naturally derived fatty alcohols are obtained by hydrogenation of the corresponding methyl ester fractions. For example, Sherex lists its NF-grade $C_{14}$ cetyl alcohol at $1.015, its NF grade $C_{18}$ stearyl alcohol (97%) at $1.015, and the $C_{22}$ behenyl alcohol at $1.98. A mixture of cetyl and stearyl alcohols (Aldol 63) is also available at $0.845. According to Henkel, its propietary catalyst allows the direct hydrogenation of triglycerides to glycerol and fatty alcohols (*CW* 6/29/88).

Reductive amination of the fatty alcohols leads to the corresponding fatty amines and to their derivatives. For example, Sherex lists, among others, the following amines:

Lauryl amine, $1.40; distilled $1.57
Coco amine, $1.00; distilled $1.17
Tallow amine, $0.58; distilled $0.70
Oleyl amine, $0.78; distilled $0.90
Methyl dicoconut amine, $1.30
Methyl distearyl amine, $1.11

The formation of quaternary ammonium salts from fatty amines gives antistatic agents priced at about $1.00/lb and useful in a variety of material processing operations. Quaternary ammonium salts of fatty amines are also used in hydrometallurgical operations for the extraction of chromates, vanadates, molybdates, tungstates, and the like.

Fatty esters can be converted to the corresponding fatty amides. Thus, for example, the following materials are available:

Oleic acid amide, Humko's Chemamide U, $1.56

Erucic acid amide, $C_{22}$ one double bond, Humco's Chemamide E, $2.71

Bis(stearamide), Humko's Chemamide W-40, $0.89

Oleylpalmitamide, Humko's Chemamide P-181, $3.25

The treatment of fatty alcohols with HBr gives a series of 1-bromoalkanes such as, for example:

1-Bromotetradecane, Humphrey, 97%, $3.25

1-Bromooctadecane, Humphrey, 97%, $3.52

Fatty acid amides are employed by the plastics industry as mold releasing and internal lubricants during processing (slip and antiblock agents). Thus Witco Corporation's Humko Division lists Chemamide E and Chemamide U, erucamide (the one double-bond-containing $C_{22}$ carboxylic acid obtained from rapeseed) and oleamide, respectively, at $2.71 and $1.56. The secondary fatty acid amides (R–CO–NH–CO–R'), such as oleylpalmitamide (Chemamide P 181) and bis(stearamide) (Chemamide W-40) are listed at $3.25 and 89¢, respectively.

Fatty acid derivatives of diethanolamine are employed as antistatic additives for use with PEs, PS, PP, SAN, and so on:

Bis(hydroxyethyl) tallow amine, Akzo's Armostat 310, $1.50;

Bis(hydroxyethyl) coco amine, Akzo's Armostat 410, $1.80.

Careful oxidation of some fatty alcohols generates the corresponding aldehydes of interest to the perfume industry:

Capric aldehyde, $3.95;

Lauric aldehyde, $7.75.

Emery pioneered the ozonolysis (see 6.6.9) of oleic acid to give the $C_9$ mono- and dicarboxylic acids (pelargonic, 75¢) and azelaic ($1.38) acids, respectively (*SZM-II*, 69–70). Of these two products, the first mentioned is competitive with the synthetic acid (70¢), obtained by hydroformylation of 1-octene.

### 7.3.2 Tall-Oil Fatty Acids (TOFA)

Tall oil is the mixture of lipids isolated as a by-product of pulping operations that convert wood into cellulose fibers that, in turn, become processed into different kinds of paper, paperboard, cartons, and so on. A relatively small amount of cellulose is used for chemical purposes and is then referred to as "chemical cellulose." Tall oil originates from the extractable constituents of the plant cells. The other major structural constituent of most plants, namely, lignin, is not discussed here because its chemical makeup yields predominately phenolic molecules.

The American Paper Institute's estimate of the 1986 production of paper and paperboard exceeded 70MM t (*CW* 3/19/86), and this explains the tremendous quantity of tall oil that is coproduced under current pulping conditions. Only a fraction of this tall oil (crude \$90/t or 4.5¢/lb) is fractionated, thus raising the cost of distilled tall oil (DTO) to 12¢/lb, and some is further refined to a product valued at about 31¢/lb. Even though the distillation capacity of tall oil is about 1.7B lb, most of the tall oil is employed as a fuel in the currently energy-intensive pulping operations. The tall-oil fatty acids still contain some rosin acids (*SZM-II*, 76, 112, 134–135) and are priced accordingly: TOFA 2% or more rosin acids, 13¢; TOFA less than 2% rosin acids, 14¢.

The distilled portion of tall oil consists primarily of oleic and linoleic acids, and this mixture is subjected to thermal and catalytic processes that produce the so-called dimer and trimer acids. These complex products result from Diels–Alder-like cyclization reactions, and their molecular structure can be visualized to consist of a central core that holds together two or three oleic and linoleic residues, as well as hydrocarbon and carboxy-terminated hydrocarbon chains (two or three of each, in the case of dimer and trimer acids, respectively). Emery separates its dimer and trimer acids in a molecular still and offers products of composition that approaches the expected number of carbons:

$$2 \times C_{18} = C_{36} \text{ for dimer acid}$$
$$3 \times C_{18} = C_{54} \text{ for trimer acid}$$

The dimer and trimer acids can be esterified, ethoxylated, converted to amides and polyamides (when subjected to reactions with polyfunctional amines), converted to acid chlorides, hydrogenated to the corresponding alcohols, and so on. These standard transformations of dimer–trimer acids and their derivatives lead to the formation of different types of coatings, lubricants, plasticizers, adhesives, corrosion inhibitors, and the like. Nearly half of the dimer acid production is channeled into the preparation of hot-melt adhesives. Specifically, a "nylon 6/36" is employed in the manufacture of tin cans.

### 7.3.3 Terpenes

Terpenes are even more numerous than the nearly countless kinds of fats and oils. They are isolated from the plant kingdom and also form part of the extractable components of plant cells. Terpenes are usually associated with fragrances and flavors isolated as "essential oils" (*SZM-II*, 28), but tall oil (translated "pine oil" in Swedish) carries along with it the terpene constituents of trees in the form of turpentine (the price of crude sulfate turpentine rose from about 90¢/gal to \$1.80/gal in January 1989). Here, the term sulfate refers to the so-called Kraft pulping process, which yields the "black liquor" by-product. The latter is concentrated mainly to recover inorganic pulping chemicals and to dehydrate it to render it a combustible mixture; in so doing, however, one first obtains the volatile and water-insoluble turpentine (b.p. 150–180°C). Actually, the terpene portion of the black liquor steam distills during the concentration operation, and the crude tall oil that is left behind is either burned (see 7.3.2) or fractionated under vacuum.

The common structural denominator of all terpenes is the isoprene moiety: its presence is recognizable in linear and cyclic terpenes, hydrocarbon and oxygen-containing terpenes, small terpene molecules (referred to as *mono*terpenes that contain 10 carbon atoms; i.e., *two* isoprene units), polyterpenes such as natural rubber, and even some structures that were modified structurally by biochemical processes of the living cells. A more detailed discussion of industrial aspects of terpenes is found elsewhere (*SZM-II*, 110–148). Here we only cite some examples of acyclic terpenes that serve as building blocks of industrial importance.

Lemongrass oil is the volatile oil of *Cymbopogon flexuosus* (India, \$6.13/lb; Guatemala, \$7.50) containing 75–85% of citrals, $Me_2C=CH-CH_2CH_2C(Me)=CH-CH=O$ a mixture of the two geometric isomers:

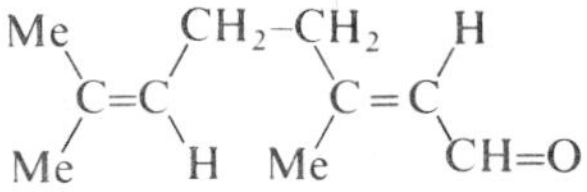

Geranial 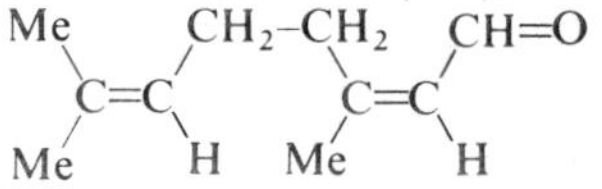                                Neral

The corresponding alcohols of natural origin are geraniol (90–92%, \$10.60) and nerol (technical grade \$5.30, perfume grade \$4.60). Natural geraniol must compete with the synthetic product (96–98%, \$4.00). Geranyl acetate (natural \$10.95, synthetic 5.44) and geranyl formate (natural \$15.95, synthetic \$6.60) are also employed in perfumery. The frequently encountered wide price difference between natural and synthetic materials used in perfumery and as flavoring agents is not surprising because of the delicate nuances produced by trace "impurities" that are appreciated by the senses of sophisticated consumers. It is also noteworthy that synthesis favors the formation of the isomer in which the aldehyde and the largest substituent at the terminal C=C bond are trans to each other.

A common synthesis of geranial starts with the addition product of acetylene and acetone (4.7.2) that is hydrogenated to the corresponding olefin, 2-methyl-3-buten-2-ol,

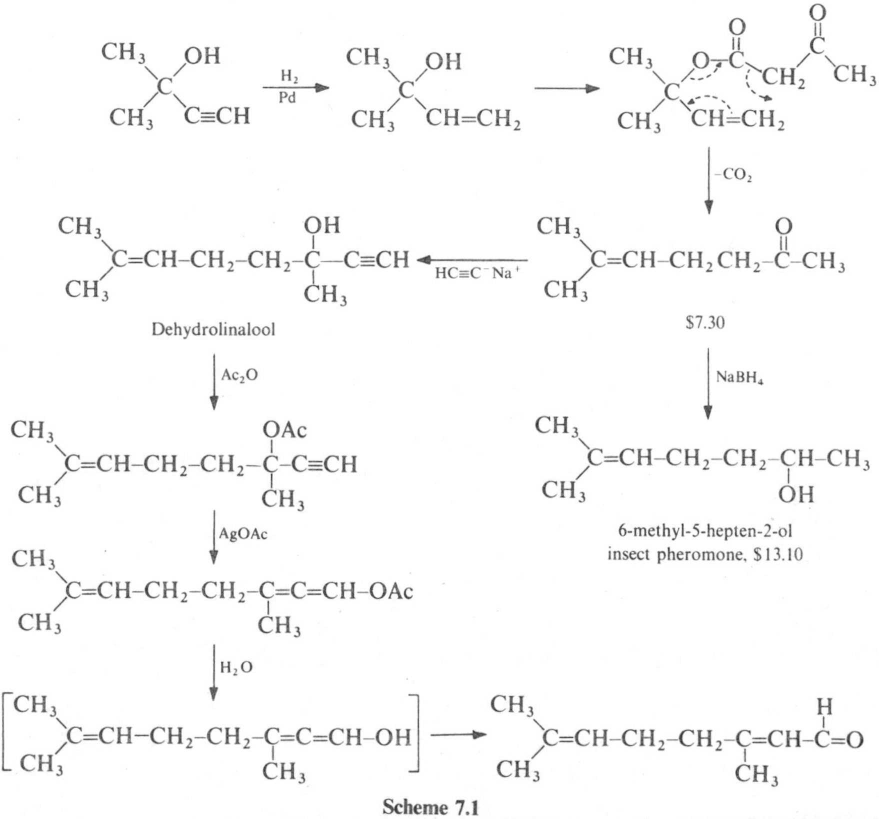

Scheme 7.1

and this intermediate is treated with diketene (6.9.1) to yield the expected acetoacetate. On heating, the latter undergoes a Cope rearrangement (*ONR*-20) to give 6-methyl-5-hepten-2-one or methylheptenone ($7.30), and another addition of acetylene provides the remaining two carbons, except that an allylic rearrangement of an acetate is required to locate the aldehyde group at the terminal carbon atom; see Scheme 7.1. The reduction of methyl heptenone to the corresponding alcohol, methyl heptenol or 6-methyl-5-hepten-2-ol ($14.50), an insect pheromone, is shown in connection with the intermediate ketone.

Geranial can be reduced to geraniol by means of sodium borohydride. The conversion of geranial (or citral a) to other perfume ingredients such as the α- and β-ionones ($7.50 and $10.60, respectively) requires an aldol condensation with acetone to yield the not unexpected pseudoionone. This is followed by an acid treatment that illustrates the facile transformation of acyclic to cyclic terpenes and gives the isomeric ionones (Scheme 7.2). The conversion of β-ionone to vitamin A$_1$, (synthetic grade, 1MM A units/g, in oil; pharmaceutical grade, $18.65/lb; feed grade, 650,000 A units/g, $14.50) is shown elsewhere (*SZM-II*, 125–126).

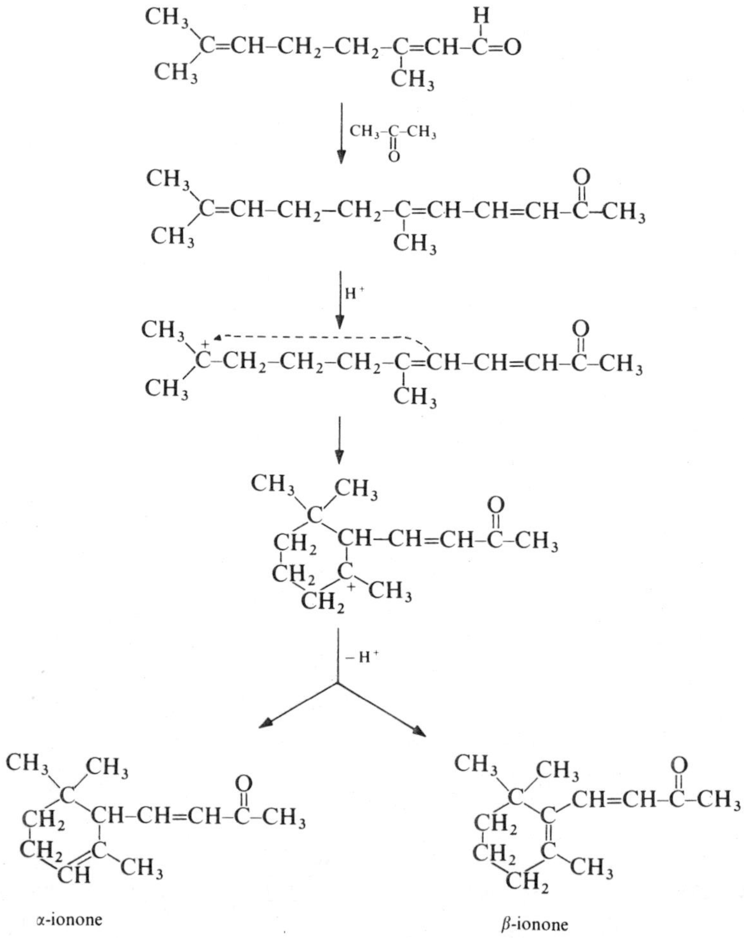

**Scheme 7.2**

Another acyclic building blocks useful in the synthesis of several perfume ingredients (including nerol) are myrcene (85¢), alloocimene, and d-limonene ($1.20), obtained by the pyrolysis at about 550°C of the pinenes—the main constituents of turpentine (Scheme 7.3). Some of the synthetic uses of myrcene are discussed elsewhere (*SZM-II*, 120).

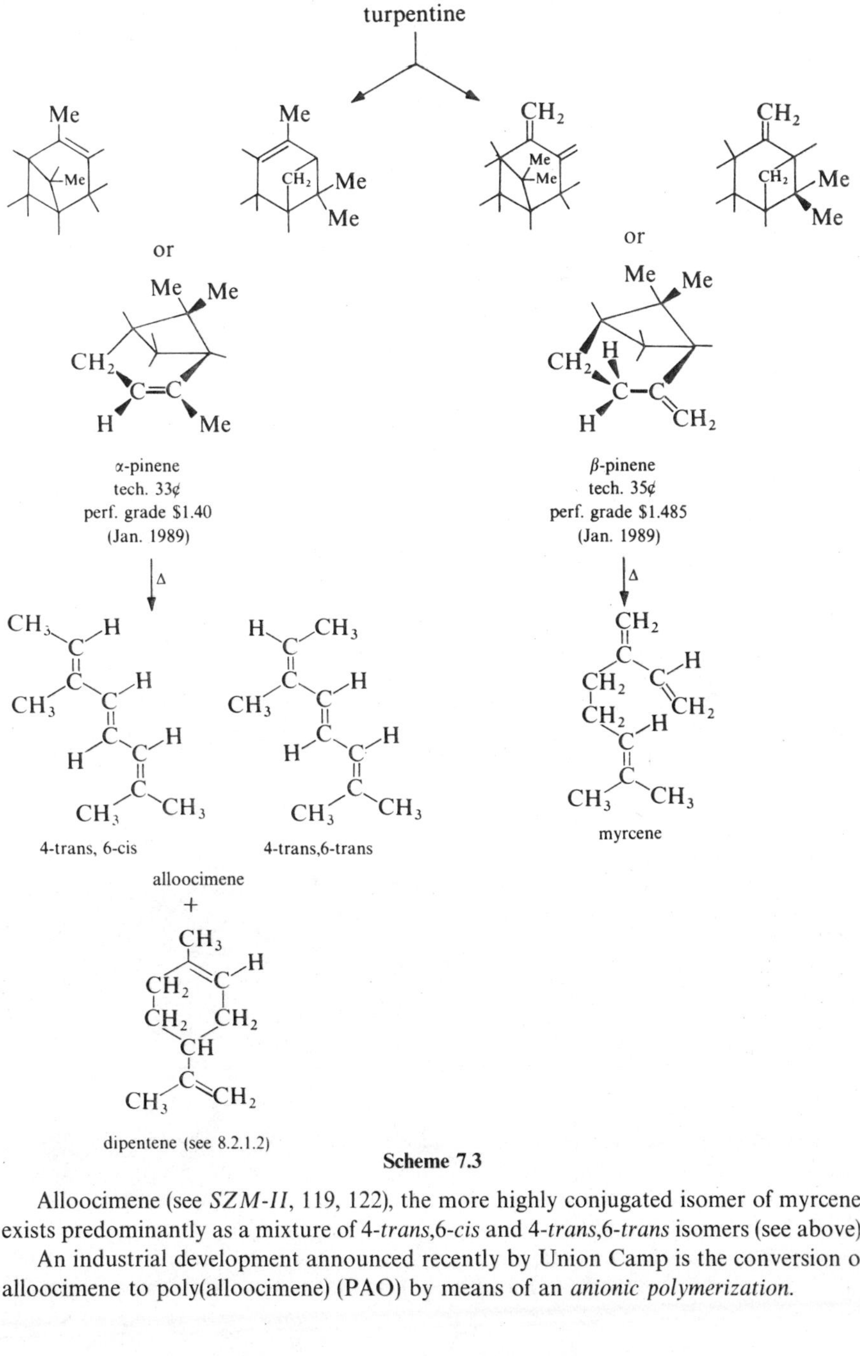

turpentine

or

α-pinene
tech. 33¢
perf. grade $1.40
(Jan. 1989)

β-pinene
tech. 35¢
perf. grade $1.485
(Jan. 1989)

4-trans, 6-cis

4-trans,6-trans

alloocimene

myrcene

+

dipentene (see 8.2.1.2)

**Scheme 7.3**

Alloocimene (see *SZM-II*, 119, 122), the more highly conjugated isomer of myrcene, exists predominantly as a mixture of 4-*trans*,6-*cis* and 4-*trans*,6-*trans* isomers (see above).

An industrial development announced recently by Union Camp is the conversion of alloocimene to poly(alloocimene) (PAO) by means of an *anionic polymerization*.

First, the dry monomer is added to an excess of a finely dispersed active metal such as sodium in the presence of an inert solvent and a cosolvent capable of coordinating metallic cations. THF or glyme serve the latter purpose. The initially formed radical anion reacts with additional active metal to give an anionic complex that functions as a catalyst when more monomer is added. This process is represented as

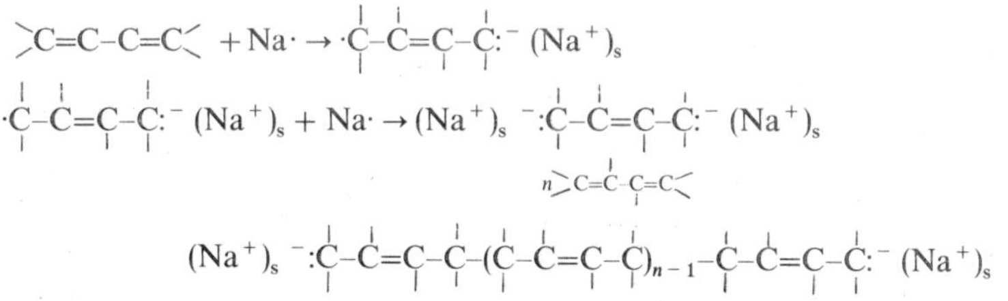

$$>C=C-C=C< + Na\cdot \rightarrow \cdot \overset{|}{C}-\overset{|}{C}=C-\overset{|}{C}:^- (Na^+)_s$$

$$\cdot \overset{|}{\underset{|}{C}}-\overset{|}{\underset{|}{C}}=C-\overset{|}{\underset{|}{C}}:^- (Na^+)_s + Na\cdot \rightarrow (Na^+)_s \ ^-:\overset{|}{\underset{|}{C}}-\overset{|}{\underset{|}{C}}=C-\overset{|}{\underset{|}{C}}:^- (Na^+)_s$$

$$_n{>}C{=}\overset{|}{C}\cdot C{=}C{<}$$

$$(Na^+)_s \ ^-:\overset{|}{\underset{|}{C}}-\overset{|}{\underset{|}{C}}=C-\overset{|}{\underset{|}{C}}-(\overset{|}{\underset{|}{C}}-\overset{|}{\underset{|}{C}}=C-\overset{|}{\underset{|}{C}})_{n-1}-\overset{|}{\underset{|}{C}}-\overset{|}{\underset{|}{C}}=C-\overset{|}{\underset{|}{C}}:^- (Na^+)_s$$

where $(Na^+)_s$ stands for a tightly solvated cation. The molecular weight distribution of the "living polymer" obtained at this stage is readily controlled by the use of appropriate ratios of monomer and the initially formed anionic catalyst, and the living polymer can be utilized in subsequent reactions with electrophilic reagents, or decomposed to the hydrocarbon polymer by addition of protic materials.

PAO obtained according to U.S. Patent 4,694,059 (issued in 1987 to R. Veazey and assigned to Union Camp) consists of a mixture of macromolecules denoted as A, B, and C formed by anionic polymerization of alloocimene at positions 6,7, 2,3, and 4,7, respectively, where A and B together consitute more than half of the product. The structures of the three major PAO components are shown below.

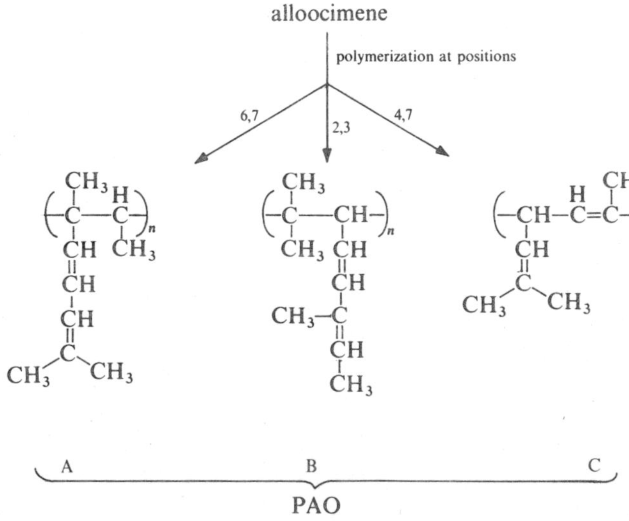

alloocimene

polymerization at positions

6,7    2,3    4,7

A       B       C

PAO

The resulting PAO can be used per se as a binder in inks, as a tackifier in hot-melt adhesives, as a cross-linking agent for vinyl polymers, and so on. It can also be transformed to epoxides (D) by means of peracetic acid or *t*-butyl hydroperoxide and to Diels–Alder adducts (E) of dienophile olefins like MA. The MA adducts in turn can be converted to esters, amides, and other derivatives of the MA residue (see U.S. Patent 4,753,998 issued in 1988 to K. S. Hayes, C. R. Frihart, and R. L. Veazey and assigned to

Union Camp), and PAO can be hydrogenated or subjected to addition of chlorine, and so on. The "living PAO" obtained as described above can be functionalized by reaction with carbon dioxide (F) and other electrophilic reactants such as aldehydes and ketones (U.S. Patent 4, 690,979 issued in 1987 to R. L. Veazey and assigned to Union Camp). Finally, it can be converted to metallic complexes obtained, for example, when it is treated with iron pentacarbonyl, $Fe(CO)_5$. The product (see U.S. Patent 4,690,984 issued in 1987 to R. L. Veazey and M. S. Pavlin, and assigned to Union Camp) of the latter transformation is believed to have the structure G. At this time, the unit price of poly(alloocimene) (PAO) is estimated to be in excess of \$3.00/lb. The formation of some of the above-mentioned derivatives is summarized below.

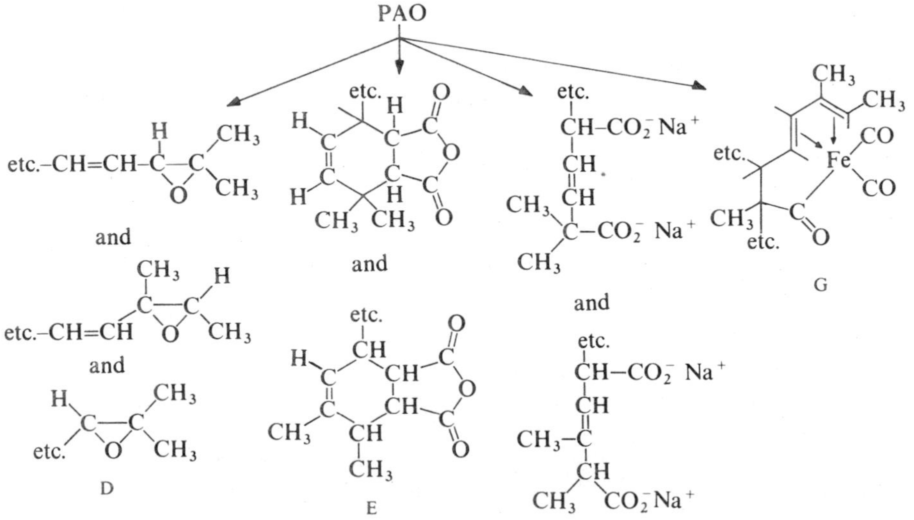

### 7.3.4  Fermentation Products

The synthetic capabilities of microorganisms have been harnessed by humanity since early recorded history, and the evolution of the biotechnological route to industrial organic chemicals is sketched elsewhere (*SZM-II*, 33–52). Here we wish to highlight the production of a few chemicals that are not readily accessible by nonbiological means and that, in turn, serve as building blocks for other meterials of interest.

The formation of two $C_5$ alcohols, namely, isoamyl (95%, \$1.44) and active amyl alcohols the main components of fusel oil

$$(CH_3)_2CH-CH_2-CH_2-OH \quad \text{and} \quad CH_3CH_2-CH(CH_3)-CH_2-OH$$

during the fermentation of glucose or sucrose in the course of ethanol production is mentioned elsewhere (p. 23). These two alcohols occur in a 85:15 ratio and are marketed as a mixture referred to as "mixed primary amyl alcohols" (46.5¢) and are converted to the corresponding esters. It is of interest to note that the workup of fusel oil was described in the second edition of *FKC* published in 1957 (pp. 107–114) but dropped in the subsequent edition as petrochemicals eclipsed materials derived from biomass. However, the escalation of the fermentation ethanol industry based on sugarcane (4.4.2) and the worldwide renewed interest (P. L. Layman, "Sugar Producers Focus Increasingly on Chemical Markets," *CEN* 12/7/87, 9–13) to "sucrochemistry"—a term proposed first by H. B. Hass—should revive industrial interest in its by-products.

The formation of isoamyl alcohol by means of the hydroformylation of 1-butene is competitive with the fermentation route, and the same is true of the hydroformylation of 2-butene to the (potentially) optically active amyl alcohol. The interest of Phillips in 3-methyl-1-butene-obtained by dehydration of the major amyl alcohol component of fusel oil is also noteworthy in this connection.

Among the carboxylic acids produced by fermentation (*FKC*, 275–279) that employs either molds, such as *Aspergillus niger*, or yeast, is the C$_6$ tricarboxylic acid (citric acid; anhydrous powder 89.5¢), USP anhydrous granules 83.5¢ (see Scheme 7.4). The consumption of citric acid in the United States is about 250MM lb, but this demand may increase significantly if a mixture of the acid and its sodium salt (USP 76.5¢) becomes accepted as a method to remove H$_2$S from petroleum refinery gases. Currently, it is a common acidulent in the food industry, a constituent of bubbly (CO$_2$-evolving) pharmaceutical preparations, an ingredient of metal-cleaning formulations, a potentially growing "builder" in laundry products (because of its metal-chelating capability), a source of salts such as

Calcium citrate, $3.82;

Diammonium citrate, $2.79;

Ferric ammonium citrate, NF, $2.00;

Potassium citrate, USP 90.5¢;

and, of course, a building block for the production of other organic chemicals of industrial interest. Among the latter, there are two plasticizers:

Triethyl citrate, $1.82, b.p. 294°C

Tributyl citrate, $1.70

Rapid heating of citric acid at its melting point of 153°C produces a homolog of maleic anhydride, namely, itaconic anhydride, which, on treatment with water, is converted to itaconic acid ($1.45, m.p. 162–164°C with decomposition). When citric acid is heated rapidly at about 175°C, the transient itaconic anhydride isomerizes to citraconic anhydride, which, on treatment with water, gives citraconic acid (m.p. ~90°C with decomposition). The structures of these related compounds are shown in Scheme 7.4. Itaconic acid can also be obtained directly by fermentation that uses *Aspergillus terreus*.

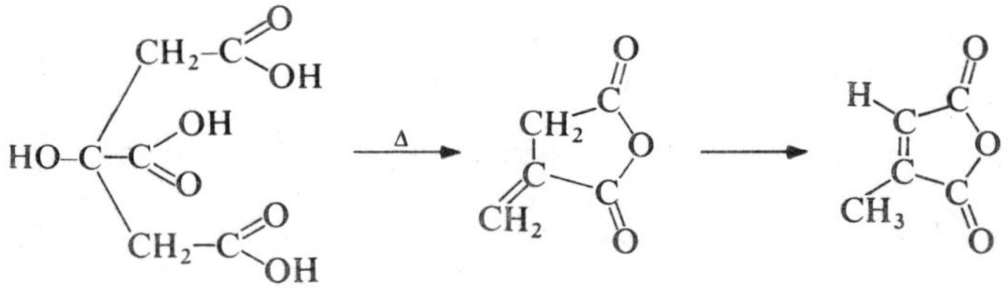

Scheme 7.4

Thus far the use of unsaturated derivatives of citric acid as monomers and/or co-monomers has been relatively little explored.

Another common fermentation product is glutamic acid utilized in the form of monosodium glutamate (MSG; 82¢) as a meat-imitating flavoring agent in soups and other processed food products. Only the dextrorotatory L-isomer is of interest for this purpose since the levorotatory D-isomer is tasteless:

$$
\begin{array}{ccc}
\underset{\text{L-Glutamic acid}}{
\begin{array}{l}
CO_2H \\
H_2N\text{---}C\text{---}H \\
CH_2 \\
CH_2 \\
CO_2H
\end{array}
\equiv
\begin{array}{l}
H \\
HO_2C\text{---}C\text{---}NH_2 \\
CH_2 \\
CH_2 \\
CO_2H
\end{array}
}
& &
\underset{\text{D-Glutamic acid}}{
\begin{array}{l}
CO_2H \\
H\text{---}C\text{---}NH_2 \\
CH_2 \\
CH_2 \\
CO_2H
\end{array}
\equiv
\begin{array}{l}
H \\
H_2N\text{---}C\text{---}CO_2H \\
CH_2 \\
CH_2 \\
CO_2H
\end{array}
}
\end{array}
$$

Glutamic acid can be obtained by fermentation of carbohydrates by means of *Micrococcus glutamicus* or *Bacillus megaterium–cereus*. Also, it can be obtained from fumaric acid by means of *B. pumilus* or α-ketoglutaric acid by means of *Aeromonas* spp. and, finally, can be isolated from the hydrolyzate of casein, gluten, soybean protein, and from sugar beet molasses. In recent years the United States has imported yearly as much as 50MM lb of MSG from Taiwan and South Korea, even though there are several domestic producers.

## 7.4  DOWNSTREAM BUILDING BLOCKS: ALPHAOLEFINS

### 7.4.1  Olefin Oligomers, Including Alphaolefins (AO)

Discussion of the chemistry of ethylene (4.3.1, 4.3.2), propylene (5.4.1), isobutylene (6.3.1.2), and 1-butene (6.3.2.1) unavoidably touches on their dimers, trimers, and oligomers, which, in turn, become the downstream building blocks of a variety of chemicals of industrial importance. Here we wish to summarize the availability and some of the uses of these intermediates.

Specifically, ethylene is converted to 1-butene, 1-hexene, and 1-octene, but with butadiene it gives 1,4-hexadiene. The contributions of these terminal olefins to the production of LLDPE and EPDM elastomer are mentioned in 4.3.1. The same is true of the formation of polybutenes, PBs, from 1-butene (6.3.2). The formation of alphaolefins is elucidated below to show that it occurs not only by oligomerization of ethylene, but also by the cracking of paraffins and by dehydration of fatty alcohols (7.3.1).

Different reaction conditions convert propylene to the dimer 2-methyl-1-pentene, which is cracked to isoprene (6.8.2) or to a mixture of 4-methyl-1-pentene (4MP-1) and 4-methyl-2-pentene. 4MP-1 is obtained by a propietary catalyst developed by Phillips (*CE* 2/16/87) that avoids the formation of 4-methyl-2-pentene. 4MP-1 is a new comonomer in the production of LLDPE already exploited by BP (*CMR* 5/27/87). Also, 4MP-1 is polymerized to poly(methylpentene) and marketed by Mitsuichi as TPXP-C$_6$—a clear plastic of high resistance to chemicals and heat. 2-Methyl-1,4-pentadiene becomes the building block of one of Amoco's aromatic dicarboxylic acids (9.3.2.2), and 1,4-hexadiene—one of the "D"s in EPDM, is obtained by a nickel-catalyzed dimerization of propylene by a mechanism reminiscent of the formation of 2-ethyl-1-hexene from 1-butene. It is noteworthy that the related 1,5-hexadiene is one of Shell's "specialty olefins" obtained by metathesis of a cyclic derivative of 1,3-butadiene (6.8.1.4). The isononenes are the trimers of propylene that are converted to isononyl alcohols, p-nonylphenol or di-t-nonyl polysulfides. Similarly, the isododecenes are the tetramers of propylene and are converted to p-dodecylphenol or t-dodecyl mercaptan (TDM), a chain-transfer agent in free-radical catalyzed polymerizations. Another derivative of propylene tetramer is Milliken's dodecenylsuccinic anhydride (DDSA; \$0.905), obtained by the ene reaction (*ONR*-28) of the oligomer with MA (see paragraph f below, in this section, below for the mechanism). A representative structure of DDSA is

$$CH_3CH_2CH_2CH(CH_3)CH_2CH(CH_3)CH{=}C(CH_3)CH_2{-}CH{-}CO$$

$$CH_2CO \diagdown O$$

Propylene is also converted to oligomers with a degree of polymerization between 11 and 20.

The products of the alkylation reaction of isobutylene, isobutane, and propylene give a mixture of branched octanes and heptanes, and the dimer and trimer of isobutylene give the precursors of highly branched primary amines (such as t-octyl, priced at \$2.60) and di-t-octylamines (\$1.40), mercaptans, and t-octylphenols and t-dodecylphenols—precursors of sulfated and ethoxylated detergents.

It is clear that the propylene tetramer and the isobutylene trimer are both branched olefins that give rise to $C_{12}H_{25}{-}X$ derivatives, but that the two structures differ with respect to branching:

$$CH_3CH(CH_3)CH_2CH(CH_3)CH_2CH(CH_3)CH_2CH(CH_3){-}X$$

and

$$CH_3C(CH_3)_2CH_2C(CH_3)_2CH_2C(CH_3)_2{-}X$$

Isobutylene is the source of liquid polybutenes that can be functionalized, for example, by reactions with phosphorus pentasulfide or maleic anhydride to produce a lubricating oil and a gasoline additive, respectively. In the presence of a nickel-containing catalyst, 1-butene yields a rather unexpected dimer, 2-ethyl-1-hexene, which readily undergoes hydroformylation.

The most common isoamylene is 3-methyl-2-butene (50¢), obtained from C$_5$ refinery fractions and used, for example, for the alkylation of phenol to give p-t-amylphenol, Pennwalt's Pentaphen 61 (91¢), a commonly employed germicidal agent. The 2,4-di-t-amylphenol (~ 97¢) is also commercially available. The methyl ether derived from isoamylene, $(CH_3)_2C(C_2H_5){-}O{-}CH_3$, is t-amyl methyl ether (TAME)—the octane enhancer of gasoline and companion of MTBE.

3-Methyl-1-butene is a valuable (estimated unit price \$3–\$5/lb) building block of synthetic pyrethroids—materials that replace and improve on the naturally occurring insecticide pyrethrin. The latter is isolated from chrysanthenum flowers (0.9% pyrethrin content, \$1.91/lb). The dehydration of isoamyl alcohol to 3-methyl-1-butene (without isomerization of the C=C or of the carbon skeleton) was pursued by Phillips (*CEN* 4/22/85). Similar efforts were revealed by Polysar Limited of Canada (*CMR* 7/25/85, *CE* 8/19/85) in the course of the production of 50MM lb of the desired olefin by means of a catalytic distillation process.

Also of interest in connection with the synthesis of fine chemicals are the members of the C$_5$ prenyl family of compounds, the 4-substituted 2-methyl-2-butenes,

$CH_3$–$C(CH_3)$=$CH$–$CH_2$–X. Prenyl alcohol (X = OH) is obtained by means of the Prins reaction (*ONR*-72) of isobutylene and formaldehyde:

$$CH_3-C(CH_3)=CH_2 + H_2C=O \rightarrow CH_3-C(CH_3)=CH-CH_2-OH$$

Prenyl alcohol is converted to the corresponding halide for use as an organometallic reagent in the synthesis of pyrethrin-like compounds.

The alphaolefins constitute one of the more rapidly growing sectors of industrial organic chemistry with a current demand level approaching 1.5B lb, although the production capacity is estimated at nearly 2B lb and much of the production is subject to captive use. The global consumption of AOs in 1986 was estimated at 1.76B lb and is expected to reach about 3.75B lb by the year 2000 (P. Layman, *CEN* 5/30/88, 9–10). The strongest driving force responsible for the 8–10% yearly growth is the increasing popularity of LLDPE (4.3.1), which consumes about 40% of the demand, mostly in the form of $C_6$–$C_8$ comonomers, and detergents and lubricants consume mostly the $C_{12}$–$C_{18}$ fraction at a rate of 30 and 10%, respectively, of the total demand. Other interesting uses of AO are enumerated below.

The unit prices of AO range from 26¢ for 1-butene to about 40¢ for the $C_6$–$C_{18}$ and higher member of this family. As shown elsewhere (4.3.2), the oligomerization of ethylene is a large source of AO. There are two approaches to the preparation of the higher AO from the still larger alkanes:

1. Thermal degradation at temperatures of up to 600°C in the presence or absence of steam to give mixtures of linear olefins in which the AOs predominate. These are separated into fractions of narrow boiling point ranges and are described as $C_6$–$C_9$, $C_{10}$–$C_{13}$, $C_{14}$–$C_{18}$, and so on, olefins.
2. Catalytic dehydrogenation of paraffins below 600°C to a complex mixture of olefins in which the double bond is randomly distributed. However, these olefins are then converted to AO by ethenylation (5.4.9).

The Shell higher-olefin process (SHOP) utilizes ethenylation in order to taylor the molecular-weight distribution of AO produced by ethylene oligomerization. The latter requires relatively high pressures ($\sim 750$ psi) but relatively low temperatures ($< 120°C$) and nickel catalysts that contain triarylphosphine ligands.

Future production of AO may include the direct hydrogenation at 250°C and 300 bar of fatty acids to fatty alcohols (rather than hydrogenation of intermediate methyl esters) by means of Co–Cu–Mn–Mo–$H_3PO_4$ catalysts followed by the dehydration of the latter (*HP* 4/86, 721). A joint effort of Henkel and Union Carbide to produce AO by dehydration of fatty alcohols was announced several years ago (*CMR* 9/28/81).

In addition to the above-mentioned utilization of AO in LLDPE, detergents, and lubricants, AO is also converted to alcohols for the preparation of plasticizers, and minor amounts become synthetic fatty acids, mercaptans, amine oxides, and other interesting derivatives mentioned below.

The α-olefins are subjected to the following chemical transformations:

(a) *Alkylation* of benzene or phenol under Friedel–Crafts (*ONR*-33) conditions, including the use of $AlCl_3$–HCl equivalent to chloroaluminic acid ($HAlCl_4$), or other acids (sulfuric, $BF_3$, phosphoric adsorbed on solid supports, etc.), gives nearly linear alkylates of the aromatic substrates, $RCH(CH_3)Ar$. The $C_{10}$–$C_{13}$ AOs are used in this way to build up the desirable lipophilic component of the alkylbenzenesulfonates (LAS). This route to LAS products is preferred over the alkylation of benzene by monochlorinated alkanes (7.2.1). The use of AO for the alkylation of phenols also serves to produce internally plasticized and hence less brittle and more impact-resistant PF on some cross-linking with formaldehyde (3.4.3.5). In the same manner bisphenol A can be first (di)alkylated before the introduction of glycidyl ether moieties (5.4.8) to give flexible epoxy or PC resins. The dialkylation of benzene with AO leads to the production of synthetic lubricants and other automotive functional fluids.

(b) The direct *hydration of AO to primary alcohols* under mild reaction conditions by means of *trans*-PtHCl(Me)$_2$ catalyst was described recently by C. M. Jensen and W. C. Trogler [*SC* 9/5/86; see also *CE* 2/16/87 for the use of the *trans*-Pt(OH)$_2$(PMe$_3$)$_2$ catalyst at 60–100°C in the presence of the benzyltrimethylammonium chloride phase-transfer catalyst]. The conversion of AO to primary alcohols can be achieved in a roundabout way by utilizing the anti-Markownikoff (*ONR*-57) addition of HBr in the presence of peroxides or catalyzed by UV light:

$$R-CH=CH_2 + HBr \xrightarrow{\text{cat.}} R-CH_2-CH_2-Br \xrightarrow{\text{H}_2\text{O}} R-CH_2-CH_2-OH$$

Linear primary alcohols are obtained directly from ethylene by oligomerization catalyzed by aqueous hydrogen peroxide. Presumably the reaction is initiated by hydroxyl radicals and chain termination occurs when the intermediate radical picks up a hydrogen from water to form the desired product thus regenerating another hydroxyl radical. In this manner Ethyl Corporation produces a series of pure C$_6$–C$_{18}$ Epal alcohols priced at 86.5¢, while blends of narrow-molecular-weight distribution cost 60¢–70¢/lb. Similarly, Petrolite produces its Unilin alcohols, ($1.46) which contain 15–25 ethylene units. The latter are then ethoxylated to the corresponding Unithox products ($1.85), in which the EO residues constitute about 50% of the total molecular weight, and, finally, some of the ethoxylated materials are also sulfated. The linear primary alcohols obtained by functionalized oligomerization must compete with primary linear alcohols obtained by hydroformylation of AO (see below), but in either case all other structural features being equal, the length of the alkyl chain affects the hydrophilic–lipophilic balance (HLB) (4.3.6.1) and determines the specific function of the surface-active products. Generally, in progressing from a relatively low HLB number (small hydrophilic moiety–large lipophilic moiety) to a high HLB number (large hydrophilic moiety–small lipophilic moiety), the nature of the surface-active agent changes progressively from a water-in-oil emulsifier, to that of a wetting agent, to an oil-in-water emulsifier, to that of a detergent, and finally to that of a solubilizer. While the function of emulsifiers is rather obvious, it should be clarified that solubilizers are employed, for example, by the textile industry to improve the uniformity and efficiency of dyeing operations, while wetting agents improve the dispersion of solids (pigments, ingredients of cement mixtures, etc.) in order to improve the homogeneity and performance of the end products. Similarly, the anti-Markownikoff addition of H$_2$S to AO yields linear primary alkyl mercaptans that are used as chain-transfer agents in vinyl polymerization to control the molecular-weight distribution of the products.

(c) *Hydroformylation* of C$_6$–C$_{14}$ AO (obtained from ethylene oligomers) yields odd-numbered aldehydes, alcohols, and carboxylic acids. At the lower end of molecular-weight distribution, say, C$_6$–C$_{10}$, these alcohols are recognized as building blocks of phthalate plasticizers, while alcohols above C$_{10}$ become building blocks of a variety of surface-active agents. Thus, Shell's series of Neodol $C_9$–$C_{15}$ primary alcohols (71.5¢–80¢) are converted to the corresponding ethoxylates (59¢–71¢), and some of the latter are then sulfated to Neodol ethoxysulfates (45.5¢). The carboxylic acids are esterified with polyfunctional alcohols such as 1,1,1-trimethylolpropane to become the basis of ester-type synthetic lubricants. Hydroformylation of 1-hexene and 1-octene produces *n*-heptaldehyde and *n*-nonaldehyde, respectively, which are oxidized to the corresponding heptanoic and pelargonic acids and compete with the same materials derived from natural sources (7.3.1). When the hydroformylation is carried out with olefin fractions in which the double bonds are distributed in a random fashion, the double bonds tend to migrate along the chain, but since hydroformylation occurs most readily at terminal double bonds, the products tend to be similar to those derived from AO. Primary alcohols obtained by hydroformylation of AO are also converted to amines, and in the case of tertiary amines, the latter can be oxidized with peracids to nonionic detergents of the amine oxide type:

$$R–CH_2–OH + Me_2NH \rightarrow R–CH_2–NMe_2 \rightarrow R–CH_2–N(Me_2) \rightarrow O$$

Other transformation of primary AO alcohols include the preparation of lubricant additives such as zinc dialkyl dithiophosphates (ZDP):

$$R–CH_2–OH + P_2S_5 \rightarrow (R–CH_2–O)_2PS–SH \rightarrow ((R–CH_2–O)_2PS–S)_2Zn$$

(d) *Sulfonation* of AO by means of $SO_3$, followed by quenching the reaction mixture with water, gives a mixture of alkenesulfonic and hydroxyalkanesulfonic acids (Scheme 7.5). These mixtures are neutralized with soda ash and used in detergent formulations.

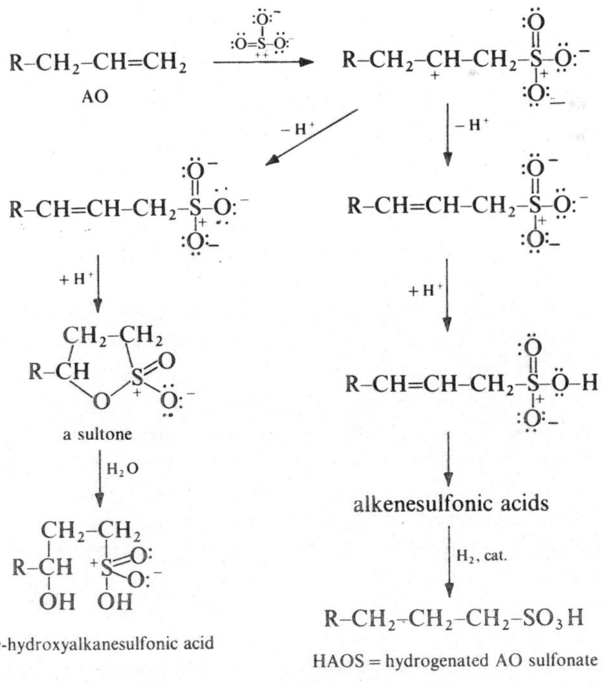

Scheme 7.5

(e) *Oligomerization* of $C_8$–$C_{10}$ AO, followed by hydrogenation, gives poly(alpha-olefin oligomers) (PAOOs), which represent another class of synthetic lubricants. The oligomerization reaction is more complex than a common vinyl polymerization because it is believed to involve not only proton shifts but also the intermediacy of protonated cyclopropanes, which explains the rather unexpected branching in the end products (Scheme 7.6). The first of its kind, the totally synthetic motor oil Mobil 1 contains about

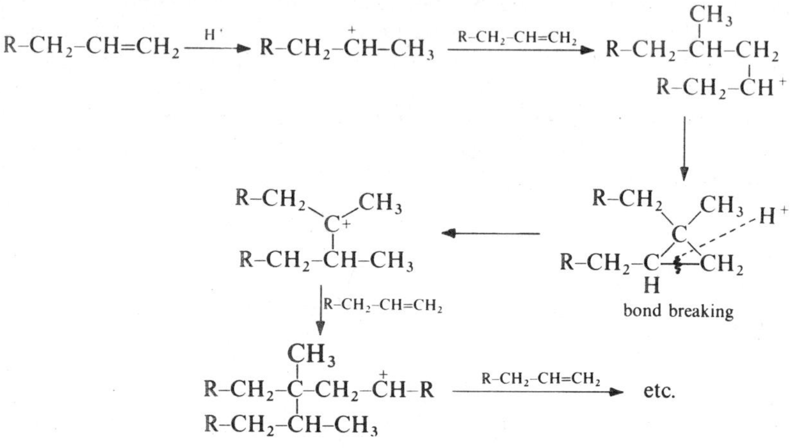

**Scheme 7.6**

70% of PAOO with the remainder contributed by polyol esters (such as the heptanoic acid derivative mentioned in (paragraph b above). While these synthetic lubricating oils represent only a few percent of the 4 billion quarts marketed in the United States, their use is growing because of guaranteed long-range stability.

(f) *Isoacid production* involves the Reppe hydrocarbonylation reaction (3.5.3.5) of AO:

$$R-CH=CH_2 + CO/H_2O \rightarrow R-CH(CH_3)-CO \cdot OH$$

The reaction is carried out in the presence of sulfuric acid, $BF_3$, and other acids, and the products are converted to esters of polyfunctional alcohols such as 1,1,1-trimethylpropane to join other components of synthetic lubricating oils and other functional fluids.

(g) *MA adducts* of AO are better known as the *alkenylsuccinic anhydrides* (ASA), and their formation is best explained by invoking a six-membered transition state in this so-called Alder or ene reaction [R. B. Woodward and R. Hofmann, *JACS* **87**, 2046, 2511 (1965); *ONR*-28; see also 6.6.4]. A whole spectrum of ASA products ranging from *n*-octenyl- through *n*-octadecenyl and isooctadecenyl succinic anhydrides is available from Humphrey Chemical (*CMR* 1/12/87, 15) at unit prices in the $0.98–$1.60 range. Additional developmental ASA are priced higher. ASA products are converted to salts or esters to serve as surface-active agents or plasticizers, respectively. The salts are used in enhanced-oil-recovery (EOR) operations. The alkenylsuccinic anhydrides derived from AO compete with the analogous structures derived from branched oligomers of olefins other than ethylene (see 7.2.1).

## 7.4.2 Aldol-Type and Related Condensation Products

Again, as in the case of the smaller olefins (7.4.1), $C_5$ and further downstream building blocks are inevitably mentioned in preceding sections of this book. They are indicated here in order of an increasing carbon content.

*Among the $C_5$ compounds we find the following:*

- *n*-Amyl alcohol, possibly derived by hydrogenation of a valeric acid ester, is the likely precursor of methyl heptine carbonate, methyl 2-octynoate ($26.75)—a pheromone building block—obtained by reaction of the alcohol with HBr, followed by use of sodium acetylide (4.7.2) and finally a reaction with methyl chloroformate (p. 164), to yield

$$CH_3-(CH_2)_4-C\equiv C-O-C(=O)-CH_3$$

- Valeric acid ester, in the presence of an alkoxide, can be alkylated with a *n*-propyl halide to give 2-propylpentanoic or valproic acid (VPA), the well-known antiepileptic and anticonvulsant. (For additional information concerning pheromones and allelochemicals and their applications in pest control, see M. Jacobsen, *Insect Sex Pheromones*, Academic Press, New York, 1972; G. A. Mathews, *Pest Management*, Wiley, New York, 1984; and A. R. Putnam and C. Tang, Eds., *The Science of Allelopathy*, Wiley, New York, 1986.)

- Pentaerythritol, a tetrafunctional alcohol, is the building block for the assembly of various esters, and ethers, accompanied by similar derivatives of dipentarythritol and tripentarythritol ($1.42 and $1.00, respectively). An example of such pentary-thritol derivative is the thermoplastic ester of rosin acids (*SZM-II*, 135), namely, Hercules' Pental 100, used as a modifier of elastomers, vinyl, acrylic, and other polymers to produce pressure-sensitive adhesives, improved printing inks, and so on.

- Neopentyl glycol (6.5) is a difunctional primary alcohol with restricted rotational freedom and hence a building block of esters or ethers conducive to the production of somewhat rigid polymer structures.

- Acetylacetone (7.4.4).

*Among the $C_6$ compounds we find the following:*

- 1,1,1-Trimethylolpropane (6.5)—a trifunctional primary alcohol with attributes similar to those of the preceding examples; the building block of the triacrylate ($1.50) and other esters of interest.

- 2-Ethyl-1-butanol ($2.05) (6.5).

- Mesityl oxide (5.5.4) (46¢), a likely precursor of pheromones such as 6-methyl-5-hepten-2-one, or heptenone ($7.30) and methyl heptenol, or 6-methyl-5-hepten-2-ol ($13.10):

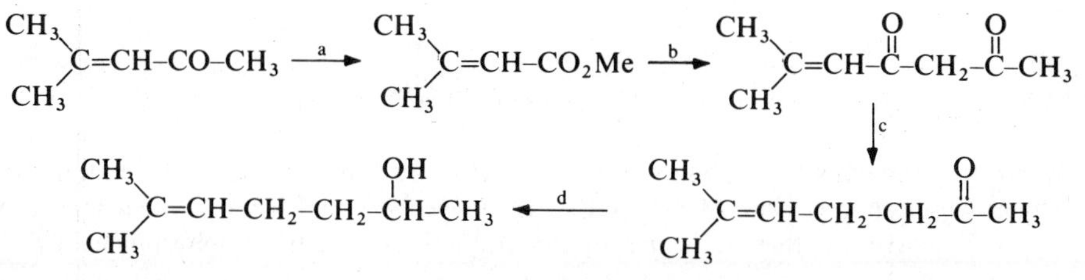

These transformations involve (a) the haloform reaction (*ONR*-39, *ONR*-56) followed by esterification, (b) condensation with acetoacetic ester followed by decarboxylation of the resulting acid, (c) selective reduction of the 4-keto group, and (d) selective reduction of the residual keto group.

- Hexylene glycol (5.5.4) (50¢) is converted to two peroxy esters, 1,1-dimethyl-3-hydroxybutyl peroxyneoheptanoate and 1,1-dimethyl-3-hydroxybutyl 2-ethylhexanoate (CH$_3$–CH(OH)–CH$_2$–CMe$_2$–O–O–CO–R), where R = –CH$_2$–CH$_2$–CMe$_3$ or CH(Et)–CH$_2$–CH$_2$–CH$_2$–CH$_3$, which are effective polymerization catalysts for the production of PVC and other vinyl polymers.
- Diacetone alcohol and the related mesityl oxide and methyl isobutyl ketone, all down-stream C$_6$ building blocks derived from acetone (5.5.4).
- Acetonylacetone (5.5.5, 7.4.3).
- Adipic acid (8.3.1).
- 2-Methyl-1,4-pentadiene obtained by means of an oxidative dimerization of propylene catalyzed by AlR$_3$ (5.4.1).
- The propylene dimer products of the acid-catalyzed reaction of propylene and propane (5.4.1).
- 1,5-Hexadiene obtained by ethenylation of 1,5-cyclooctadiene (5.4.9).
- 1,4-Hexadiene obtained by the nickel complex-catalyzed reaction of ethylene and butadiene (4.3.1.1).

*Among the C$_7$ compounds we find the following*:

- Methyl 3-methyl-1-pentenyl ketone, and its partial reduction product.
- Methyl isoamyl ketone (51¢), the building blocks derived from acetone and isobutyraldehyde (5.5.4, 6.5).

*Among the C$_8$ compounds we find* a whole family of structures derived from an aldol condensation of two *n*-butyraldehydes, specifically (a) 2-ethylhexyl alcohol (2-EH), the industrial octyl alcohol, and (b) the corresponding 2-ethylhexanoic or 2-ethylhexoic acid (57¢). This octyl alcohol is the building block for a series of esters such as dioctyl phthalate (43¢), butyl octyl phthalate (40¢), dioctyl adipate (66¢), dioctyl azelate (99¢), dioctyl sebacate ($1.47), and others that function either as plasticizers and/or synthetic lubricants, while 2-ethylhexyl acrylate (79.5¢) obviously serves as a vinyl comonomer that contributes a hydrophobic moiety to the final polymer. 2-Ethylhexyl alcohol also participates in the assembly of sulfosuccinates, and the nitrate of 2-EH is used as a cetane enhancer in diesel fuels, while the phosphate of 2-EH is used as an extractant of cobalt, zinc, and other metals by virtue of the formation of metallic complexes that are soluble in kerosene. In the case of zinc, the structure of the complex is shown as follows:

$$[\textit{n-}\text{BuCH(Et)CH}_2\text{–O}]_2\,\text{PO·OH} + \text{Zn}^{2+} \rightarrow [(\text{RO})_2\,\text{PO·O}]_2\,\text{Zn}^{2+}\,[\,{}^-\text{O–PO(OR)}_2]_2$$

with an H above the first PO·OH group.

(where R– represents the carbon moiety of 2-EH). The phosphate ester of 2-EH, di(2-ethylhexyl)phosphoric acid, Mobil's DEHPA or Henkel's D$_2$EHPA, is often used in conjunction with another derivative of the alcohol, namely, trioctylphosphine oxide

(TOPO), for the isolation of $U_3O_8$ during the wet phosphoric acid process (treatment of phosphate rock with sulfuric acid). The valuable $U_3O_8$ may be present in concentrations as low as 0.1–0.3 lb/t of phosphoric acid end product, but still the complex of the uranium oxide–DEHPA–TOPO is transfered from the aqueous phosphoric acid stream into an organic phase (*CE* 1/3/77). In 1981 the production of phosphoric acid in the United States amounted to about 20B lb, and hence it was an ample source of uranium.

In a reversal of the ester forming role, 2-ethylhexanoic acid gives rise to esters with various alcohols including 2-EH. Possibly the best illustration of the role of 2-ethylhexanoic acid as a downstream building block is its transformation by means of the acyloin condensation (*ONR*-1) to the corresponding acyloin oxime, Henkel's LIX 63, ($6.43)—an extractant of cupric ions from low-grade ores (Scheme 7.7).

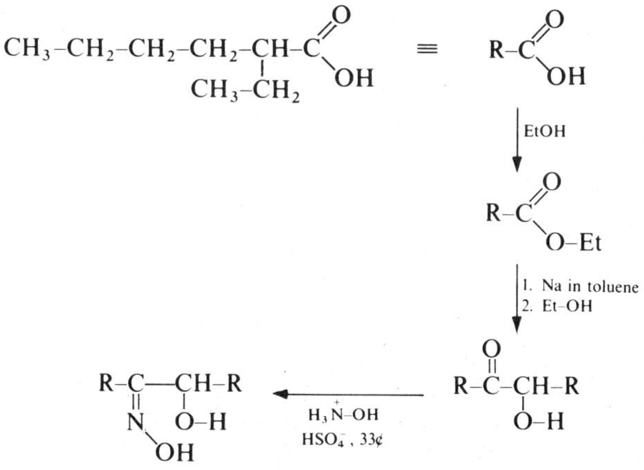

$$CH_3-CH_2-CH_2-CH_2-\underset{\underset{CH_3-CH_2}{|}}{CH}-C\overset{\displaystyle O}{\underset{\displaystyle OH}{}} \quad \equiv \quad R-C\overset{\displaystyle O}{\underset{\displaystyle OH}{}}$$

EtOH

$$R-C\overset{\displaystyle O}{\underset{\displaystyle O-Et}{}}$$

1. Na in toluene
2. Et–OH

$$R-\overset{\displaystyle O}{\overset{\|}{C}}-\underset{\underset{O-H}{|}}{CH}-R$$

$$\underset{HSO_4^-,\ 33¢}{H_3\overset{+}{N}-OH} \quad \longleftarrow$$

$$R-\underset{\underset{\displaystyle OH}{\overset{\displaystyle N}{\|}}}{C}-\underset{\underset{O-H}{|}}{CH}-R$$

**Scheme 7.7**

Another family of $C_8$ building blocks arises from the aldol condensations of methyl ethyl ketone (MEK) (6.3.2): with the addition of another MEK molecule we obtain methyl heptenone (see 7.3.3) and following partial hydrogenation of the latter, methyl heptenol ($14.50) forms.

2-Ethyl-1,3-hexanediol, or octylene glycol, 6.5 apparently has not found much use as a building block of organic compounds of industrial interest even though it offers a not too common 1,3-glycol structure. 2,2,4-Trimethyl-1,3-pentanediol, Eastman's TMPD (48.5¢) is a relatively recent addition to $C_8$ building blocks obtained by an aldol condensation of two isobutyraldehyde molecules (6.5) followed by reduction of the remaining aldehyde group by means of formaldehyde in the presence of alkali (a Cannizzaro reaction).

*Among the $C_9$ compounds we find* phorone, the condensation product of mesityl oxide (see above) with another molecule of acetone (5.5.4) and its cyclic derivative isophorone, as well as its reduction product diisobutyl ketone (60¢). There can be little doubt that phorone is an important building block of industrial organic compounds when we consider its conversion to IPDI and the role of the latter in the assembly of PURs (interlude in Section 3.8).

Also, we recognize the contribution of the phorone building block to the structures of various polymer additives that function as antioxidants and UV-light stabilizers, Ciba's:

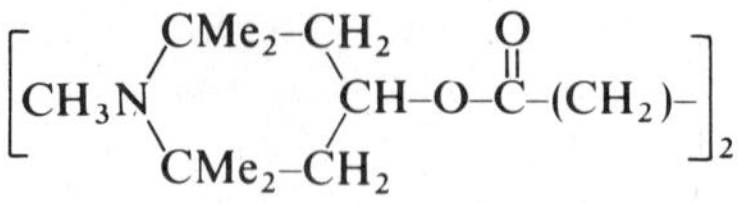

$$\left[ CH_3N \begin{matrix} CMe_2-CH_2 \\ \quad\quad CH-O-\overset{\displaystyle O}{\underset{\displaystyle \|}{C}}-(CH_2)- \\ CMe_2-CH_2 \end{matrix} \right]_2$$

Tinuvin 770

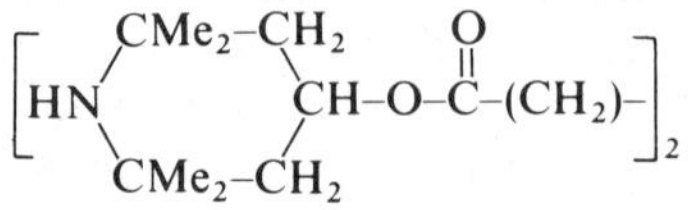

$$\left[ HN \begin{matrix} CMe_2-CH_2 \\ \quad\quad CH-O-\overset{\displaystyle O}{\underset{\displaystyle \|}{C}}-(CH_2)- \\ CMe_2-CH_2 \end{matrix} \right]_2$$

Tinuvin 765

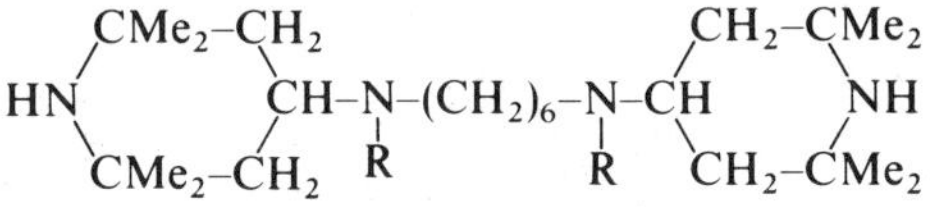

$$HN \begin{matrix} CMe_2-CH_2 \\ \quad\quad CH-N-(CH_2)_6-N-CH \\ CMe_2-CH_2 \;\; R \quad\quad R \end{matrix} \begin{matrix} CH_2-CMe_2 \\ \quad\quad NH \\ CH_2-CMe_2 \end{matrix}$$

Chimasorb 947FL

where

$$R = Me_3C-CH_2-CMe_2-O\text{—}\underset{\underset{N}{N}\;\;\underset{\;}{N}}{\overset{N}{\bigcirc}}$$

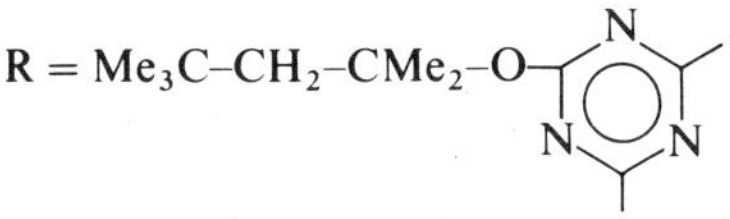

### 7.4.3  Carbonyl and Related Addition Reactions

Under this heading we can list the following downstream building blocks:

• Products of the addition of acetylene to acetone [viz., methyl butynol, $1.79;

$$(CH_3)_2 C(OH)-C\equiv CH],$$

MEK [viz., methyl pentynol, $2.28;

$$CH_3(C_2H_5)C(OH)-C\equiv CH],$$

2–EH aldehyde [viz., ethyl octynol, $3.15;

$$n\text{-}C_4H_9(C_2H_5)C(OH)-C\equiv CH],$$

and MIBK [viz., Air Product's Surfynol 104 ($2.11)]:

$$(CH_3)_2 CHCH_2 C(CH_3)(OH)-C\equiv C-C(CH_3)(OH)CH_2 CH(CH_3)_2$$

and its ethoxylated derivative, Air Product's Surfynol 440 ($2.02):

$$(CH_3)_2 CHCH_2 C(CH_3)(O\text{-}R)-C\equiv C-C(CH_3)(O\text{-}R)CH_2 CH(CH_3)2$$

where each O–R represents, on the average, 3.5 $-(OCH_2 CH_2)_n OH$ groups; the Surfynols are corrosion-inhibiting surfactants.

Products of the addition of $CO-H_2O$ to olefins to give neoacids (3.5.3.5) such as pivalic and neononanoic acids and isoacids from $\alpha$-olefins (see 7.4.1, paragraph e)

An interesting application of the pivalic acid building block is reported (*CW* 4/15/87) to be the thermal decomposition of tetra(neopentyl)titanium at about 250°C, which deposits a corrosion- and wear-resistant titanium carbide coating on metallic surfaces whereas the traditional deposition of titanium carbide requires temperatures of about 1000°C; obviously, the preparation of the titanium derivative requires the reduction of pivalic acid to the corresponding alcohol, possibly by way of pivalyl chloride, the conversion of the alcohol to a Grignard agent by way of a neopentyl halide, and the reaction of the Grignard with $TiCl_4$.

Another product is adipic acid and its derivatives assembled by the double addition of 1,3-butadiene to HCN or CO (6.8.1.3).

### 7.4.4  Alkylation Products of Active-Hydrogen Compounds

Acetylene is alkylated in the form of its sodium derivative (see 7.4.2) prepared by means of sodium (70¢) in liquid ammonia (b.p. $-33$°C), or sodamide, $NaNH_2$ ($\sim \$4.00$):

$$H-C\equiv C^- \ Na^+ + n\text{-}C_5H_{11}-Br \rightarrow n\text{-}C_5H_{11}-C\equiv C-H \qquad \text{1-heptyne, Heico \$9.50}$$

Some of the synthetic possibilities of the downstream building block 1-heptyne are illustrated in Scheme 7.8.

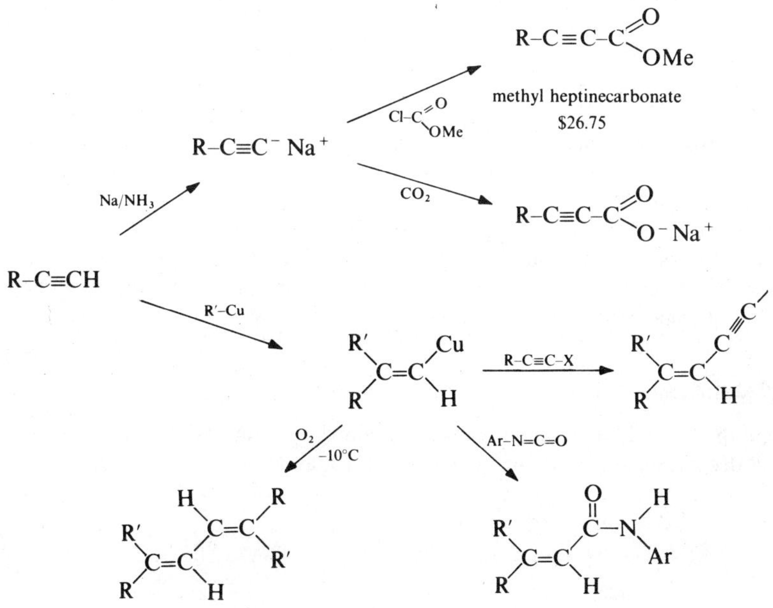

**Scheme 7.8**

Acetoacetic esters are alkylated by alkyl halides and substituted alkyl halides; for example, the synthesis of valproic acid (see Scheme 4.27, 4.6.4) can also involve a double alkylation of an acetoacetic ester with *n*-propyl bromide, followed by a treatment with *concentrated* alkali that produces substituted acetic acids (*ONR*-1).

Acetonylacetone (CH$_3$–CO–CH$_2$–CH$_2$–CO–CH$_3$), a useful C$_6$ building block of many heterocyclic structures (10.2.2), is obtained by the alkylation of an acetoacetic ester with a 2-chloroacetoacetic ester, followed by the decomposition of the resulting product by means of *dilute* alkali or an acid-catalyzed hydrolysis–decarboxylation (*ONR*-1), as shown in Scheme 7.9.

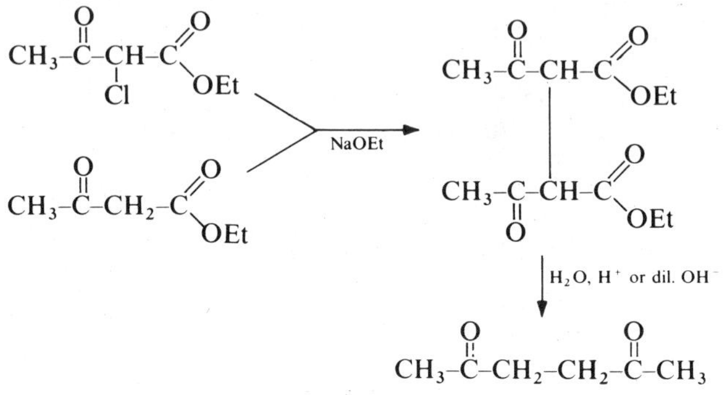

**Scheme 7.9**

The reaction of the sodium salt of acetoacetic ester with either ethylene oxide or ethylene chlorohydrin is the likely source of 2-acetylbutyrolactone ($6–$12):

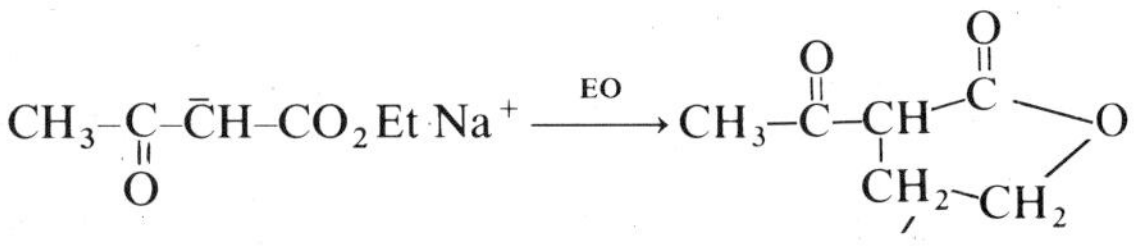

Acetylbutyrolactone is an intermediate in the synthesis of heterocyclic systems such as 5-hydroxymethyl-4-methylthiazole—a structural component of thiamine or vitamin B$_1$ (see Scheme 10.58, 10.2.2).

Acetylacetone (CH$_3$–CO–CH$_2$–CO–CH$_3$), the deceptively similar but smaller homolog of acetonylacetone, is obtained by the reaction of ketene with acetone followed by the rearrangement of isopropenyl acetate shown in 5.5.5; it is a useful building block in the assembly of some heterocyclic structures (10.2.2).

### 7.4.5  Metathesis Products

Some of the C$_5$ and higer building blocks obtained by metathesis, such as 1,5-hexadiene and 1,9-decadiene, are mentioned elsewhere (5.4.9, and 6.8.1.4).

# PART THREE

# CYCLIC BUILDING BLOCKS

For pragmatic reasons this presentation of cyclic building blocks is divided in three parts:

Chapter 8—nonaromatic carbocyclic compounds
Chapter 9—aromatic carbocyclic compounds
Chapter 10—heterocyclic compounds

There is, however, some unavoidable overlap between these three chapters, and, for that matter, between the noncyclic structures and the cyclic ones, because

1. Some aromatic carbocyclic structures are hydrogenated and thus are precursors of the nonaromatic carbocyclics.
2. Many aromatic carbocyclics serve as nuclei from which heterocyclics are constructed.
3. Nonaromatic carbocyclic compounds are often cleaved to give noncyclic derivatives, and vice versa: noncyclic compounds often serve as precursors of both carbocyclic and heterocyclic structures.

Some additional explanations are appropriate. We do not chose to subdivide the heterocyclic compounds into nonaromatic and aromatic because these differences are usually readily bridged by either hydrogenation that destroys the resonance stabilization of the aromatic structures or oxidation of the nonaromatic heterocyclic to the resonance-stabilized, aromatic heterocyclics. For all practical reasons, the last-mentioned transformation often occurs spontaneously in the presence of air. Finally, in the preceding sections of this book, we find a few families of compounds that, strictly speaking, are heterocycles: epoxides or oxiranes, anhydrides of dicarboxylic acids, lactones, lactams, and others. However, since these cyclic structures are readily interconverted to their noncyclic relatives and, indeed, in some cases the cyclic and noncyclic structures exist as partners of a dynamic equilibrium, we do not repeat the discussion of such of cyclic structures here except for incidental reasons. The reader is encouraged to follow up on cross-references and consult the Index to connect related parts of chemistry wherever they happen to occur.

# 8

# NONAROMATIC CARBOCYCLIC COMPOUNDS

## 8.1 PRIMARY BUILDING BLOCKS DERIVED FROM FOSSIL SOURCES

### 8.1.1 Naphthenic Acids and Related Structures

The naphtha fraction of petroleum contains a variety of saturated cyclic hydrocarbons, usually alkyl derivatives of cyclopentane and cyclohexane. These compounds are difficult to separate, and traditional refinery operations convert them into aromatic hydrocarbons and olefinic fragments. Thus, the alicyclic constituents of petroleum account, in part, for the production of BTX during reforming (2.2.1) and also, for cyclopentadiene and methylcyclopentadiene (6.8.3).

    The naphthenic acids are the oxidation products of the above-mentioned alkylcyclopentanes. They are isolated from the fraction of petroleum of boiling range 300–400°C, and their predominant composition can be represented by

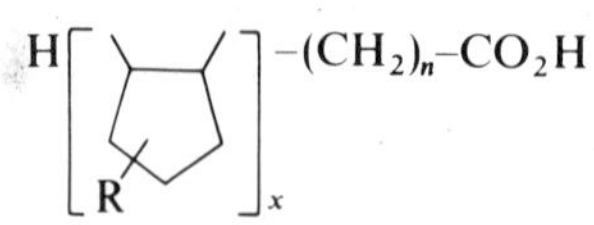

$$H\left[\begin{array}{c}\\ \\ R\end{array}\right]_x -(CH_2)_n-CO_2H$$

where R is usually a methyl group; $x = 1–5$, with the frequency of $1 > 2 > 3 > 4 > 5$; and $n > 1$. The crude naphthenic acids are priced at 30¢ and refined at a price of 80¢, and the demand in the United States is about 32MM lb, of which a portion is exported. Naphthenate salts of industrial importance are Pb ($1.11), Ca (85¢), Co ($2.06), Zn (95¢), and Mn (67¢), used for the "drying" of paints and other coatings, as well as some inks, that solidify on air oxidation of polyunsaturated constituents (*SZM-II*, 57–59, 66). Similarly, the Co salt catalyzes the curing of UPRs, and the Zn and Cu ($1.19) salts are used as fungicides; the latter becomes likely replacement of pentachlorophenol (9.3.4), now classified by the EPA as a "restricted use pesticide" in the treatment of lumber.

Other industrial applications of the naphthenates include the formulation of lubricating, cutting oil, and surfactant additives and as ingredients of corrosion inhibitors.

### 8.1.2 Cyclopentadiene and Analogous Cyclic Olefins and Their Derivatives

In the United States the high-temperature petroleum-refining operations produce large amounts ( > 250MM lb) of cyclopentadiene and methylcyclopentadiene (6.8.3), probably as the result of naphthene fragmentation. The cycloaddition behavior of these compounds is mentioned in connection with the chemistry of 1,3-butadiene (6.8) and is amplified as follows and summarized, in part, in Fig. 8.1.

Dicyclopentadiene (DCPD) is incorporated in unsaturated polyester resins (UPRs), where it joins maleic anhydride and styrene during the cross-linking process, which results in the formation of these thermoset materials produced at a level of nearly 1B lb.

The cyclopentadiene–MA adduct, 5-norbornene-2,3-dicarboxylic anhydride, is converted by Ethyl Corporation to its Saytex BN-451 flame-retardant by reaction with ethylenediamine and then with bromine to yield ethylene bisdibromonorbornane dicarboximide ($4.25).

Recently (*CMR* 3/28/88), Hercules inaugurated a 30MM-lb-capacity plant at Deer Park, Texas to produce its Metton poly(dicyclopentadiene), an injection molding resin based on the exothermic metathesis polymerization that occurs in two stages: (1) at the highly strained double bond of the bicyclic moiety and, (2) at the double bond of the residual cyclopentene:

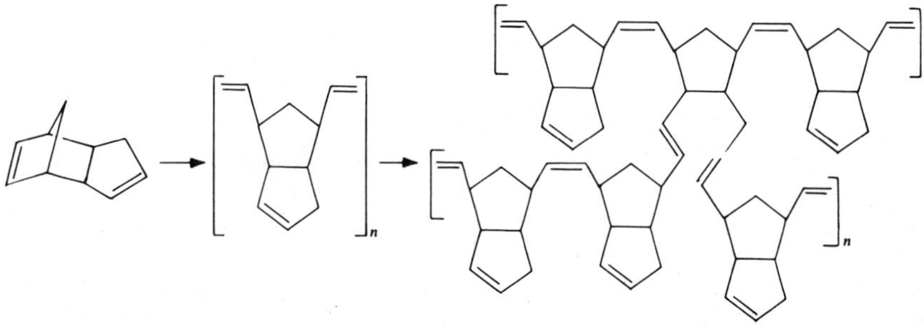

A promising use of poly(dicyclopentadiene) resins formulated with inclusion of reinforcing fillers involves RIM applications for the production of automotive components.

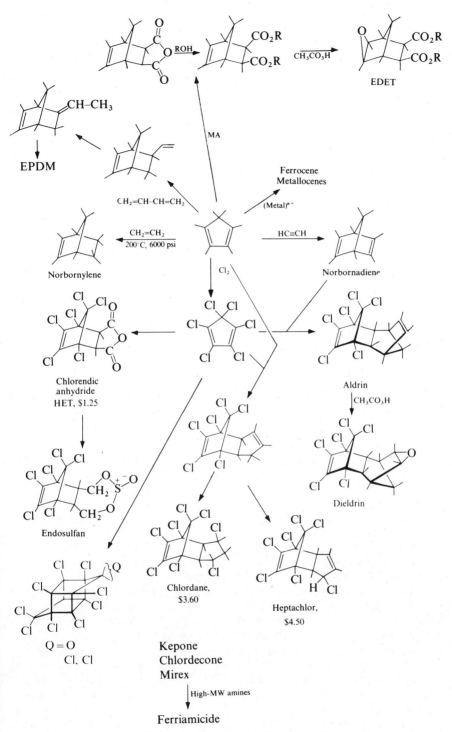

**Figure 8.1** Some cyclopentadiene derivatives.

Another current development is Ciba-Geigy's unsaturated thermosetting polyimide resins (identified as the RD86 series) based on DCPD feedstock (*MP* 6/88, 264–265). The so-called allylnadicimide resins are prepared by alkylating the sodium salt of cyclopentadiene with allyl chloride and subjecting allylcyclopentadiene to a Diels–Alder reaction with MA. The resulting allyl-substituted nadic anhydride can now be treated with variety of diamines such as MDA and HMDA to give bismaleimides that are cured by virtue of the allylic functionality. This chemistry is summarized in Scheme 8.1.

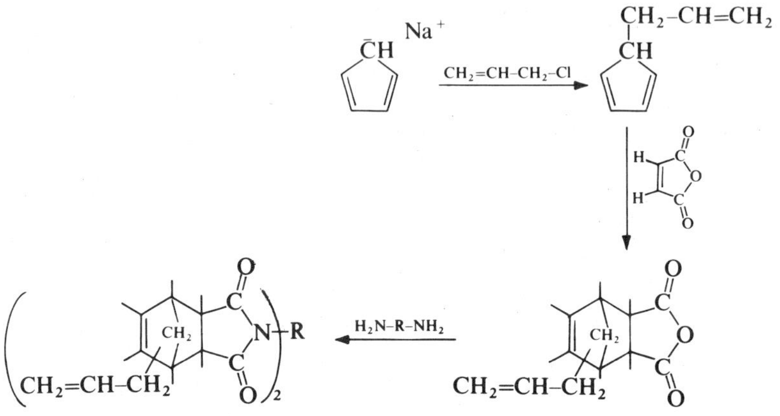

**Scheme 8.1**

### The Case of the Chlorinated Cyclodiene Pesticides

Several decades ago, the product of exhaustive chlorination of cyclopentadiene, namely, hexachlorocyclopentadiene (HCCPD) became the cornerstone of a series of pesticides usually referred to as "chlorinated cyclodienes." Their formation and structures are shown in Fig. 8.1. The omnipresence of chlorine substituents in cyclopentadiene exacerbates the electron-attracting character of the diene system, and HCCPD readily undergoes Diels–Alder reactions (as illustrated by reactions with 1,5-cyclooctadiene and even MA).

The multiple chlorine substituents in the insecticides of the chlorinated cyclodienes also render the resulting molecules rather stable to chemical and biochemical degradation, and as a result of the "DDT scare" incited by Rachel Carson's *Silent Spring*, in 1975 the EPA banned the use of heptachlor ($LD_{50}$ oral administration, male rats, 100 mg/kg body weight; 162 mg/kg female rats) for general pesticide control. In 1983 the EPA allowed subsurface applications for termite control and treatment of roots in the case of nonfood plants on the basis that the chlorinated cyclodienes "appear to cause cancer in lab animals" (*CEN* 8/17/87). No human health problems attributed to chlordane—a water-insoluble material with $LD_{50}$ intraperitoneal application in male rats 343 mg/kg (cf. the $LD_{50}$ of 13.0 mg/kg for oral application in male rats in the case of ethyl alcohol and the well-documented correlation between the tobacco smoking and lung cancer in humans)—have been reported by the chlordane manufacturer Velsicol (*CMR* 6/15/87) after 40 years of production and use by pest-control operators. Nevertheless, headlines such as "Cyclodienes Again Target of Activists" and "Chlordane

Ban Seen Possible by Late Spring or Early Summer" (*CMR* 3/30/87, 2/23/87, respectively) heralded a renewed threat to the survival of this industrial chemical. The defense of chlordane ("Velsicol Executive Rallies Defense for Termicides," *CMR* 6/29/87; "Chlordane, Heptachor Study Said to Indicate Safe Use," *CMR* 7/6/87) was countered by "Chlordane 'Mountain' Feared by Environmental Group" in *CMR* 9/28/87. There followed (*CEN* 8/17/87) the ". . . EPA/Velsicol pact . . . (that) halts chlordane sales" and the announcement (*CMR* 10/12/87) of the "final limitations on the continued distribution, sale and use of existing stocks of heptachlor and chlordane" and the imposition of a complicated time table. Until November 30, 1987 the materials were used as in the past; between December 1, 1987 and April 15, 1988 the sales were "limited to certified applicators or persons under such supervision" and postconstruction use of pesticides was (and continues to be) limited to low-pressure applications. In the meantime, the judicial system in the person of a U.S. District Court judge became involved in the controversy by (*CW* 9/9/87)

> ordering . . . (the EPA) . . . to explain why it is allowing the continued sale of pesticides containing heptachlor and chlordane even though the manufacturer, Velsicol Chemical, has voluntarily agreed to halt all sales of the product pending an evaluation of application methods designed to reduce exposure to the substance.

The EPA's "final" disposition of chlordane and heptachlor leaves two unanswered questions, specifically, whether (1) chlordane is a carcinogen in humans exposed to it under conditions of normal usage and (2) the "final" disposition of chlordane usage is truly final in view of the apparently most recent change in attitude within the leadership of the EPA. The answer to question 1 is of special personal interest to the author, who, starting 31 years ago, while living for 4 years in the Caribbean, used several pounds of chlordane powder on various occasions and without any special precautions to combat ferocious ants capable of stripping overnight every green leaf of shrubs and garden plants in the way of their "warpath": am I carrying the "bad seed" of carcinogenic chlordane in my tissues? For a partial answer to question 2, see "Toxics Called Less Dangerous after Review," *CMR* 1/11/88 and the statement attributed to Dr. John Moore, assistant administrator of the EPA: "this agency has been identified by many as possibly adopting an ultraconservative approach in many previous assessments that were based on 'simplistic' assumptions that 'now strain credulity'." The controversy concerning the use of chlorinated cyclodiene pesticides lingers on, surfacing on occasions such as the locust plague in northern Africa (*CW* 4/13/88): whereas the United Nations Food and Agriculture Organization (FAO) proposes the use of dieldrin, the governments of the afflicted areas and donors of aid are reluctant to use this pesticide because of the presumed environmental hazards. To cite Hugh D. Crone—the author of *Chemicals and Society: A Guide to the New Chemical Age*, Cambridge University Press, Cambridge, UK, 1986—who promises the reader "to tempt you with reason" but then adds "There is little in this world to suggest we are in the Age of Reason, no more than was revealed to Candide in a previous era of supposed light."

## 8.2  PRIMARY BUILDING BLOCKS DERIVED FROM RENEWABLE SOURCES

The countless examples of the use of biomass-derived building blocks for syntheses of, again, countless transformation products of industrial importance can be treated in a book of this size in only a cursory fashion. The frequently cited reference (*SZM-II*) is somewhat more detailed but, again, falls short of the immense literature in this field.

### 8.2.1   Terpenes

The ease of cyclization of an isoprene dimer is illustrated in connection with the synthesis of the ionones. Other examples of cyclic terpenes available for the synthesis of materials of industrial importance are described elsewhere (*SZM-II*, 110–148). Here we simply limit ourselves to a review of certain high-volume building blocks:

1. The isomeric pinenes (see 7.3.3, Scheme 7.3) are an outstanding example of underutilized building blocks even though a number of fine chemicals are derived from them by the fragrance and flavor industries. These include synthetic camphor (technical grade $1.80, refined $3.50; see Figure 10, pp. 130–131, in *SZM-II*) and about 30MM lb of tacky polyterpene resins manufactured by Arizona Chemical Company from the pinenes and dipentene (see below). The exact chemical structure of the cationic polymerization products is not known (see *SZM-II*, 116–117); the pyrolysis of the pinenes to myrcene and its use in the synthesis of geranial and other members of that family of compounds is shown elsewhere.

2. 1-Methyl-4-isopropenylcyclohexene (b.p. 175°C) is known as dipentene (25¢) when isolated by steam distillation of sulfate pulping "black liquor" (7.3.2) or as *d*-limonene (36.5¢) when isolated as a by-product of the concentrated orange juice industry. The presence of allylic hydrogens at both terminals of this molecule, and the presence of the isopropenyl group that can participate in Ritter reactions are two of many potentially interesting transformations of this building block. Other than its use as the above-mentioned polyterpene resin constituent and its use as a solvent, thus far other major chemical transformations of dipentene have been limited to the hydration of the isopropenyl group to give terpineol ($1.10)—the major constituent of "pine oil" cleaning preparations and some of its esters such as terpinyl acetate ($1.35, extra grade $2.40) and terpinyl propionate ($4.50), all used by the perfume industry, and the dihydration product and expectorant *cis*-terpin hydrate (NF $1.35), which is of interesting structure (*SZM-II*, 117).

3. Menthol, which is well known to consumers for its "cooling effect," is obtained from natural sources as the levo-rotatory isomer of 1-methyl-4-isopropyl-3-cyclohexanol ($8.00) (there are three chiral centers; see *SZM-II*, 124), but the racemic synthetic material can be prepared from β-pinene, or still better from *m*-cresol (9.3.4) (95–98%, $1.65). In any case, synthetic menthol exhibits only half of the characteristic cooling effect of the naturally derived levo-rotatory isomer yet is priced at $9.00 (USP grade). Menthol serves as a building block of various odoriferous menthyl esters (e.g., the acetate) and the salicylate used in personal sunscreening formulations.

4. Abietic, pimaric, and other diterpenoid rosin acids (38¢–40¢) are complex tricyclic structures isolated from tall oil in an impure state. Nevertheless, they are subjected to a variety of chemical transformations (*SZM-II*, 134–135) to yield ingredients of alkyd resins, paper-sizing materials, plasticizers, amides that serve as hot-melt adhesives, and so on.

5. Steroids (*SZM-II*, 135–143) are somewhat similar to, but even more complex than, the rosin acids; they are isolated from a range of plant and animal sources and are the building blocks of fine chemicals such as the members of the vitamin D family, sex

hormones, cortisone and related antiinflammatory medicinals, oral contraceptives, and so on.

### 8.2.2  Fermentation Products

The most important cyclic structures produced by fermentation may well be certain antibiotics (*SZM-II*, 43–48). These materials function as building blocks in the sense that they are subjected to chemical modifications, frequently also with the aid of microorganisms. The foremost example of such a procedure is the family of penicillins.

## 8.3  DOWNSTREAM BUILDING BLOCKS

### 8.3.1  Hydrogenation of Aromatic Structures and Related Reactions

The saturation of unsaturated, aromatic structures to yield the related nonaromatic cyclic derivatives is utilized extensively on an industrial scale because of readily available aromatic starting materials. The two chemical routes employed most commonly for this purpose are hydrogenation and halogen addition. In the case of hydrogenation, we can distinguish between processes that produce the nonaromatic cyclic structure without modification of whatever functional group(s) may exist in the aromatic precursor, and other processes in which the aromatic ring *and* the existing functional group(s) are both transformed by hydrogenation.

Cyclohexane (or "cyclo"; 280MM gal or 1.83B lb, $1.095/gal or 16.75¢/lb) is obtained by extensive fractionation of certain naphtha fractions and hydrogenation of benzene at 150–250°C and 25–55 atm in the presence of Ni–$Al_2O_3$ catalysts. The unit price of cyclohexane tends to follow the oscillation of the unit price of benzene (see 2.4 and Figure 2.4). All but 10% of the cyclohexane production is destined for the manufacture of two nylons: 60% for nylon 6/6 and 30% for nylon 6, and for this reason the demand for cyclohexane is sensitive to imports of textile fibers and materials, and to the state of the economy. Imports of competitive textiles from recently industrialized countries, especially polyester materials based on the less costly PET (50–65¢/lb vis-à-vis a minimum of $1.28/lb for nylon 6/6) strongly affect domestic production of the nylons. Thus, while the U.S. capacity for nylon 6/6 filament is about 500MM lb, the annual demand (in millions of pounds) has been decreasing steadily over the years: from 461 in 1979, to 427 in 1980, to 400 in 1981, and so on.

The EPA is currently (*CW* 8/5/87) initiating an investigation of the potential dangers of exposure to cyclohexane by some 42,000 workers with respect to subchronic, reproductive, developmental, and neurotoxic effects; oncogenicity; and dermal absorption and sensitization at an estimated cost of $2.6 million.

Another aromatic hydrocarbon that has a long history of hydrogenated derivatives is naphthalene (9.2.1), which is hydrogenated in two stages, first producing tetrahydronaphthalene, or Du Pont's Tetralin, and then becoming decahydronaphthalene, or Du Pont's Decalin ($2.00). Current interest in tetrahydronaphthalene rests with its use as a hydrogen donor in coal liquefaction (2.2.3) and as a starting material in the synthesis of α-naphthol (9.3.4).

Cyclohexanone (technical grade fob works 59¢, delivered 70¢) and cyclohexanol (63.5¢) are important stepping stones in the transformation of cyclohexane to the nylons. Both these derivatives of cyclohexane are obtained by means of molecular oxygen at about 200°C and 50 atm in the absence of catalysts and under milder reaction conditions when cobalt and other transition-metal catalysts are present. An increased yield of cyclohexanol is promoted by the presence of boric acid, and the alcohol can then be oxidized to the ketone by means of air at about 100°C and atmospheric pressure.

Cyclohexanone is cleaved to adipic acid by means of 60% nitric acid (94.5–98%, 14¢/lb; more dilute, 9.8¢/lb). The same result can be achieved by air in the presence of Cu and Mn catalysts.

*The Nylon Story*

In the fall of 1988 Du Pont celebrated the fiftieth anniversary of the polymerization of adipic acid and hexamethylenediamine (HMDA) to nylon 6/6:

$$H(-NH-(CH_2)_6-NH-CO-(CH_2)_4-CO-)_x OH$$

and 1939 marked the year when women in the United States generally abandoned the use of unattractive cotton stockings. Since then, nylon 6/6 has become also a high-performance thermoplastic used for structural purposes, and its role during World War II is mentioned elsewhere.

The development of improved textile fibers from nylon 6/6 is the classical example of the importance of aligning polymer molecules in order to optimize their thermo-mechanical properties. The chemical structure of nylon 6/6 suggests that intermolecular attractions between neighboring polymer molecules depend on hydrogen bonding between nearby substituted amide linkages, $-CO-NH-$. However, polymer molecules are born in a chaotic three-dimensional array that does not maximize hydrogen bonding. It was the discovery of the importance of stretching the semifluid, (i.e., "plastic") material that converted a mechanically weak filament into a strand of filaments of high tensile strength. This lesson is still being learned, as witnessed by the development of surprisingly high mechanical strength in UHMWPE, in which only relatively weak London dispersion forces can be credited for intermolecular attractive forces between hydrocarbon chains, or by the development of superior mechanical strength in "partially oriented" (PO) and biaxially oriented polyester (BOPET) products. Also, processing of nylon 6/6 to give high-performance, engineering-quality polymer depends on aligning the macromolecules.

Nylon 6/6 paved the way for the development of analogous nylon 6/10, nylon 6/11, nylon 6/12, and the most recent addition to the two-component nylons, nylon 4/6. In these members of the nylon family, the first digit denotes the carbon chain length of the diamine, and the second digit denotes the carbon-chain length of the dicarboxylic acid. The chain length of the two components influences, among other things, the probability of hydrogen-bonded alignment of neighboring polymer molecules even under optimum conditions of unfolding the entangled molecules into a zigzagging "linear" mode and also determines the balance between the hydrophobic nature of the hydrocarbon portion of the material vis-à-vis the hydrophilic nature of the amide linkages.

Nylon 6/6 also paved the way for the development of one-component nylons such as its foremost representative derived from caprolactam, namely, nylon 6:

$$H-[NH-(CH_2)_6-]_x B$$

where B represents a base catalyst that initiates the ring-opening polymerization of the monomer and thus becomes incorporated into one of the terminals in the macromolecule. Caprolactam (molten 85¢, flakes 87¢) is obtained from cyclohexanone by different routes.

First, caprolactam can be produced by means of the classical Beckmann rearrangement (ONR-80) of cyclohexanone oxime, but for this purpose one requires hydroxylamine (H$_2$N–OH), available commercially as hydroxylammonium sulfate [2(H$_3$N–OH)$^+$ SO$_4^{2-}$, 83¢]. At ~85°C and at a pH of ~7, cyclohexanone is converted to the oxime, and the latter undergoes the desired rearrangement under rather acidic reaction conditions (Scheme 8.2). The adjustment of pH 7 for the production of the oxime

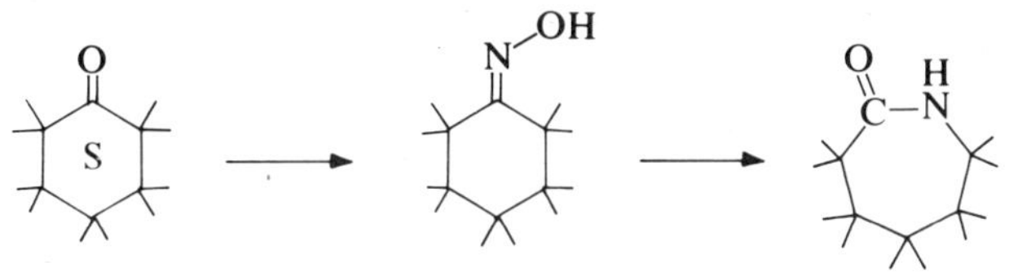

**Scheme 8.2**

is achieved by addition of ammonia. (A higher acidity tends to reverse the equilibrium and the oxime is hydrolyzed to cyclohexanone.) This process generates a great deal of ammonium sulfate (4¢; 4.5 parts of by-product for each part of final caprolactam product), and this circumstance has contributed to the search for alternative routes to caprolactam. A relatively minor modification of the preceding process in the use of hydroxylamine phosphate in the production of the oxime—the HPO process. A radically different alternative is the Italian (Snia Viscosa) process in which the primary building block of caprolactam is toluene rather than benzene and in which nitrosylsulfuric acid is used in place of hydroxylamine (Scheme 8.3).

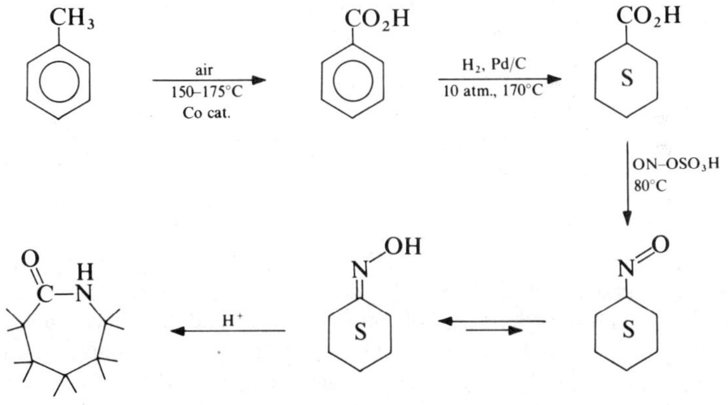

**Scheme 8.3**

Another radically different route converts cyclohexane directly to the oxime by means of a photochemically induced nitrosation by means of nitrosyl chloride, which is generated, in turn, from nitrosylsulfuric acid. The sequence of reactions is shown in Scheme 8.4.

The above-mentioned caprolactam processes require the intervention of somewhat less common inorganic reactants and their origin is also of interest to an industrial chemist. Hydroxylamine is produced by the Rasching process, in which ammonium carbonate (obtained by heating ammonium bicarbonate (26¢) with ammonia (anhydrous

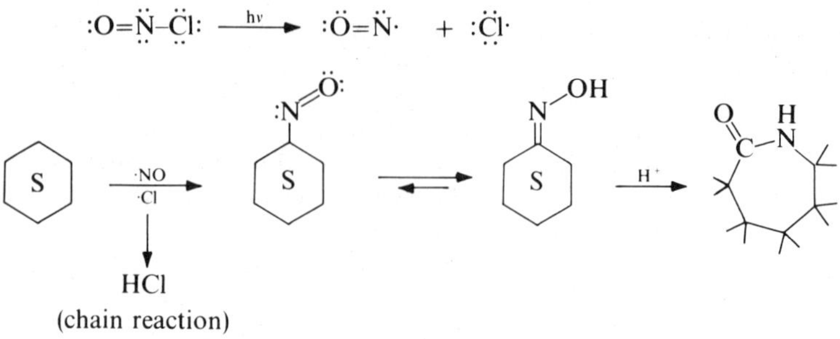

$$:O=\ddot{N}-\ddot{C}l: \xrightarrow{hv} :\ddot{O}=\ddot{N}\cdot \;+\; :\ddot{C}l\cdot$$

(chain reaction)

**Scheme 8.4**

3.75¢) is allowed to react with nitrogen oxides (obtained by oxidation of ammonia) to give, first, ammonium nitrite that, finally, is converted to hydroxylamine sulfate by means of sulfur dioxide (11.5¢) and ammonium hydroxide (aqueous solution 29.4%; $NH_3$ 100% basis, 13¢). The sequence of these critical transformations is

$$(NH_4)_2CO_3 + NO/NO_2 \rightarrow NH_4^+\,NO_2^-$$

$$NH_4^+NO_2^- + SO_2/NH_4^+OH^- \rightarrow HO-NH_3^+HSO_4^- + NH_4^+HSO_4^-$$

The other interesting inorganic reactants mentioned above are nitrosyl chloride and nitrosylsulfuric acid, and their origin is

$$NH_3 + O_2 \text{ (cat.)} \rightarrow NO + NO_2, \text{ or simply } NO_x$$

$$NO_x + H_2SO_4 \rightarrow ON-O-SO_2-OH$$

$$ON-O-SO_2-OH + HCl \rightarrow ON-Cl + H_2SO_4$$

The current demand for caprolactam in the United States is about 1.1B lb, and it is driven by the marketing of newly introduced stain-resistant carpets (*CMR* 9/21/87).

A higher homolog of nylon 6 is nylon 11 (Atochem's Rilsan, $3.45), obtained from undecylenic acid ($2.57), a thermal decomposition product of ricinoleic acid, the main fatty acid of castor oil (7.3.1; see also *SZM-II*, 73–74). The terminal C=C bond of this $C_{11}$ building block is treated with HBr ($7.00) under anti-Markownikoff (*ONR*-57) conditions to give the terminal bromo derivative of undecanoic acid, and this intermediate is treated with ammonia to produce the ω-amino acid. Dehydration of this amino acid leads to the rather costly nylon 11 produced in Europe on the order of 20MM lb. The main virtue of nylon 11 is its relatively hydrophobic character, which renders it highly suitable for the manufacture of water-nonabsorbing end products such as ski boots.

A still higher homolog of nylon 6 is nylon 12. Its origin is traced to the cyclic trimer of butadiene (8.3.2.2). Nylon 12 is a versatile, high-melting (175°C) thermoplastic that can be processed into fibers and films or into other desired objects by extrusion, injection, and rotational and blow molding. It is also used as a hot-melt adhesive, and most of the fiber is used in the manufacture of carpets.

The common nylons (6/6 and 6) represent about 95% of the total 400MM-lb consumption of nylons for nontextile purposes in the United States. They are processed into all sorts of objects including nearly friction free and tough machine components. The latter properties can be enhanced by addition of fillers such as molybdenum disulfide

(MoS$_2$ obtained from the metal, valued at \$13.50/lb), or highly fluorinated polymers such as poly(tetrafluoroethylene) (PTFE). Blends of the common nylons are produced to adjust the thermomechanical properties and to facilitate processing. A variety of reinforcing materials (graphite and PAN-type carbon fibers, and different mineral additives) are used to manufacture durable components for the fabrication of industrial and personal end products.

Caprolactam is the building block of materials other than nylon 6. Thus, the reaction of caprolactam with acrylonitrile is shown elsewhere (5.4.3.2) to yield PUR catalysts, and complete hydrogenation in the hands of Mitsubishi converts it to hexamethyleneimine [(CH$_2$)$_6$NH] (*CW* 10/6/82).

Cyclohexanone is subjected to oxidation by means of peracetic acid [S. L. Friess, *JACS* **71**, 2571 (1949)] to give the corresponding caprolactone that is oligomerized to polycaprolactone (3.8.1, PUR)—an important polyol for the assembly of PURs; see Scheme 8.5.

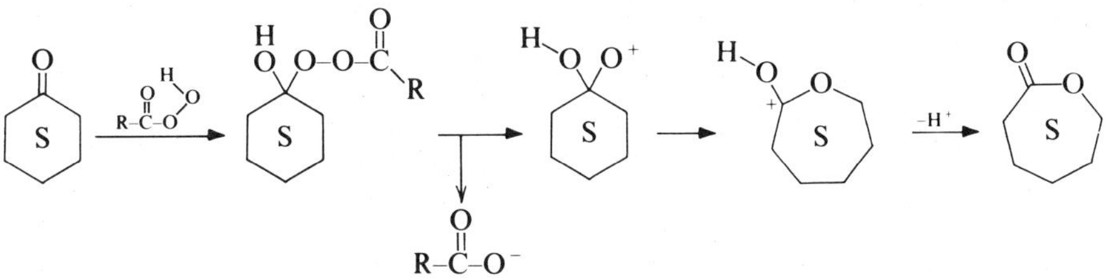

**Scheme 8.5**

Cyclohexanone undergoes an aldol-type condensation to give 2,6-biscyclohexylidene-cyclohexanone, and this cyclohexanone trimer is dehydrogenated by means of oxygen and a copper catalyst to yield 2,6-diphenylphenol. The last-mentioned material can be polymerized to the diphenyl analog of the important 2,6-dimethylphenol-derived polymer (see 9.3.4) known as polyphenylene ether (PPE) or General Electric's polyphenylene oxide, PPO (Scheme 8.6). It is noteworthy that the transformation of cyclohexanone to the aromatic polymer represents a reversal of the synthetic concept embodied by the heading of this section of the book; it takes advantage of the reactivity of the nonaromatic

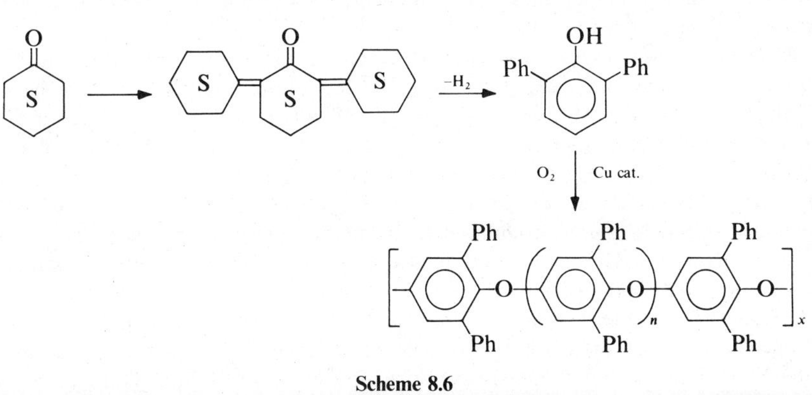

**Scheme 8.6**

building block and then applies a dehydrogenation step to obtain the desired *o,o'*-disubstituted phenol.

Cyclohexanol and/or cyclohexanone serve to produce cyclohexylamine (10MM lb, technical grade 95¢) and dicyclohexylamine ($1.25) by means of the conventional reductive amination process. About 70% of the first-mentioned amine is currently employed for the treatment of water for use in boilers, and the remainder becomes converted into photographic and rubber-processing chemicals. A relatively small amount of cyclohexylamine is used as a chain terminator in radical-catalyzed polymerizations.

At one time, the growth prospects of cyclohexylamine were very bright because of the potential use of its sulfonic acid, cyclamic acid ($C_6H_{13}$–NH–$SO_3H$), in the form of sodium or calcium salts, as synthetic sweeteners. These products became best known to U.S. consumers by Abbott's tradename "Sucaryl," but even though the $LD_{50}$ of sodium cyclamate is reassuring [15.25 g(!)/kg] when this agent is administered orally to mice or rats, the FDA banned its use in 1969 on the basis of one alleged bladder cancer observation in rats administered massive doses of the material. This observation could not be confirmed by subsequent studies, and the ban was eventually rescinded, but the whole irrational experience discouraged the manufacturer from continued production. Sodium cyclamate is now advertized in the United States (e.g., *CMR* 1/19/87) as a product of Eng Fong Chemical Industries of Taiwan and Eng Fung Chemical Company of Thailand.

The conversion of cyclohexylamine to *N,N'*-dicyclohexylthiourea (preferably by means of carbon disulfide rather than thiophosgene) is the first step in the preparation of *N,N'*-dicyclohexylcarbodiimide (DCC; R–N=C=N–R, where R = cyclohexyl). The second step involves the use of mercuric oxide, which eliminates $H_2S$ and generates DCC—an effective, mild reagent used to activate carboxylic acids for the preparation of esters, amides, and other derivatives because of the good leaving-group character of *N,N'*-dicyclohexylurea:

$$R-N=C=N-R + R'-CO-OH \rightarrow R-NH-C=N-R$$
$$|$$
$$O-CO-R'$$

$$Q-H\downarrow$$

$$R'-CO-Q + RNH-CO-NHR$$

(where Q–H is the reagent to be coupled to the original R'–CO–OH). The more costly dicyclohexylamine is employed as a corrosion inhibitor and as an ingredient of paints, coatings, inks, and other products.

The reaction of cyclohexanol with HCl gives cyclohexyl chloride, and the latter becomes the intermediate in the preparation of tricyclohexyltin hydroxide, or cyhexatin [$(C_6H_{11})_3SnOH$]—a powerful miticide introduced by Dow in 1982 as its Plictran. Until recently it was used extensively for the protection of fruit, but in 1987, after the California Department of Food & Agriculture suspended sales of the product (*CEN* 7/6/87), Dow requested the EPA to cancel the registration of this pesticide (*CW* 10/7/87) in view of the potential birth defects to unborn children of pregnant women workers. The preparation of cyhexatin involves a Grignard reaction of cyclohexyl chloride with tin tetrachloride followed by a hydrolytic workup of the reaction mixture.

Cyclohexene is a fine chemical obtained by dehydration of cyclohexanol since partial hydrogenation of benzene is difficult to control. Much of the cyclohexene used in the United States is imported from Japan at a price of about $1.50–$1.80/lb. Uniroyal offers a pure grade at about \$2.07 and converts it to cyclohexene oxide (\$5.01). Toray of Japan uses the reaction of cyclohexene and nitrosyl chloride for the preparation of L-lysine, one of the essential amino acids used as a food additive for people who depend on lysin-poor corn. Fortunately, Purdue University researchers have developed a high-lysin variety of sorghum by chemical mutation of a wild Ethiopian strain, and L-lysine is produced by fermentation and also by at least two synthetic methods (*SZM-II*, 102, 164), one of which is the Toray process, illustrated in Scheme 8.7. Synthetic lysine ester is obtained as a

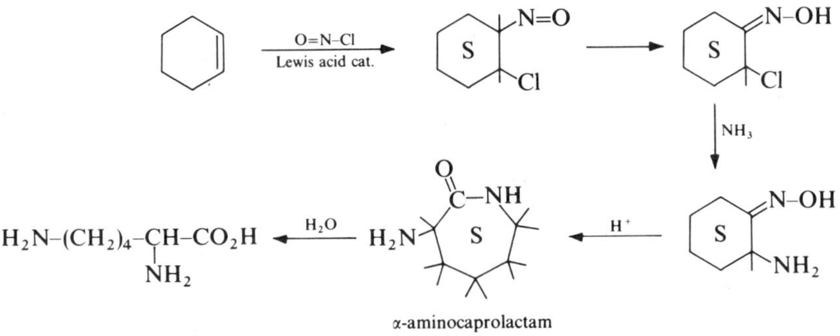

α-aminocaprolactam

**Scheme 8.7**

racemate but the enzyme L-hydrolase (immobilized in order to avoid mechanical losses) allows the desired isomer to be hydrolyzed while the unreacted D-isomer is isolated and recycled for racemization. The global production of L-lysine is about 100MM lb, most of which is obtained by fermentation dominated by Ajinomoto of Japan and of Brazil. Food-grade L-lysine is priced at about \$5.50, and feed-grade L-lysine monohydrochloride is offered at \$1.40.

The Birch reduction (*ONR*-11) consists of the reaction of an aromatic compound with metallic sodium (or lithium) in liquid ammonia in the presence of a proton source such as an alcohol. It yields nonconjugated dihydro derivatives of the aromatic substrate. In the case of phenol, the Birch reduction produces a transient dihydrophenolate salt that gives rise to 2-cyclohexenone on acidification (Scheme 8.8): This material (available from Heico at \$20.00) is a building block in the synthesis of certain pharmaceuticals and fragrances.

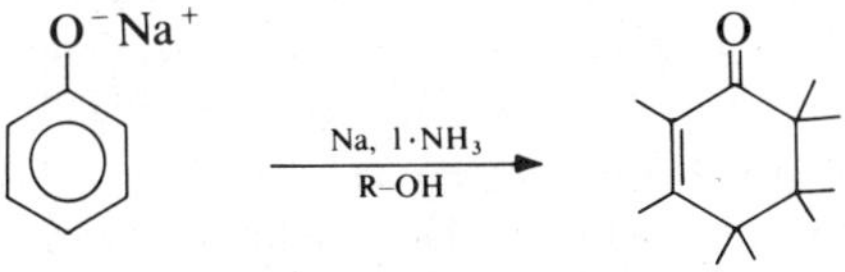

**Scheme 8.8**

Another example of the Birch reduction is the partial hydrogenation of the benzenoid ring of β-estradiol (*SZM-II*, 142).

Imperial Chemical Industries in Manchester, UK produces interesting dihydro-benzene building blocks by a combination of biotechnological and chemical procedures. Thus benzene, toluene, and some other substituted benzenes, are subjected to oxygen in the presence of some ethanol and an aqueous suspension of intact cells of a specific strain (UV 4) of *Pseudomonas putida* on a rather large scale (1000:1) reactor to produce the corresponding *cis*-dihydrocatechols, which are isolated by extraction of the fermentation mixture; see Scheme 8.9.

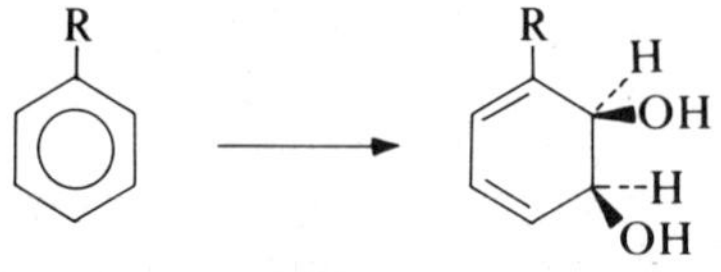

**Scheme 8.9**

(where R $=$ H, CH$_3$, F, CF$_3$, C$_6$H$_5$). It is significant that the substituted benzenes are converted both regio- and enantioselectively to the *cis*-1,2-dihydro-3-substituted 1*S*,2*R* catechol enantiomers valued at about 680/lb, and the availability of these high-tech, chiral building blocks provides an opportunity for numerous synthetic possibilities described by B. Taylor and S. Brown in *Performance Chemicals*, 1986, November 18–23. The role of ethanol in the above-mentioned reaction mixture is to regenerate the NADH cofactor of the benzenedioxygenase enzyme responsible for the desired reduction according to the equation

$$NAD + C_2H_5-OH \rightarrow NADH + CO_2$$

A number of aromatic monomers are converted to the corresponding saturated monomers by means of hydrogenation and generally are utilized either in conjunction with or in place of the aromatic monomers. As compared to the latter, the polymers derived from the hydrogenated monomers exhibit improved stability to light, and, because of decreased crystallinity, they have lower glass-transition temperatures $T_g$. Thus, although they can be processed at lower temperatures than the aromatic counterparts, they are less temperature resistant. Examples of hydrogenated monomers are

- Hydrogenated bisphenol A, Milliken's Millad (HBPA; $2.23).
- Hexahydrophthalic anhydride ($1.42), possibly obtained by the more readily controlled hydrogenation of tetrahydrophthalic anhydride (6.8.1.4), (65¢) rather than by hydrogenation of phthalic anhydride (32¢).
- Methylhexahydrophthalic anhydride (derived from piperylene), Milliken's 626 MHHPA ($1.85).
- Methylene bis-4-cyclohexylamine or bis-(*p*-aminocyclohexyl)methane (PACM; $2.05), obtained from MDA and converted to the MDI analog, namely, methylene bis-4-cyclohexylisocyanate, Mobay's Desmodur W ($3.00).
- Dimethyl *trans*-1,4-cyclohexanedicarboxylate (Eastman's DMCD, $1.00), obtained from DMT and converted, in turn, to 1,4-cyclohexanedicarboxylic acid (1,4-CHDA), also obtained by hydrogenation of terephthalic acid. The product consists of an approximately 6:4 mixture of cis and trans isomers, and each of the latter, in turn, exists as an equilibrium mixture of conformational isomers:

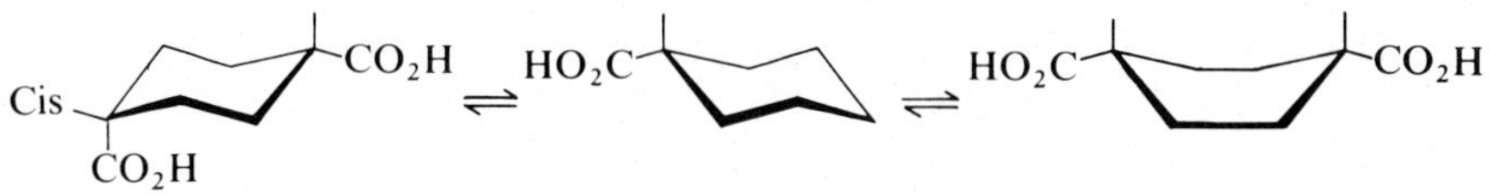

chair-*a,e*          chair-*e, a*          boat-*e, e*

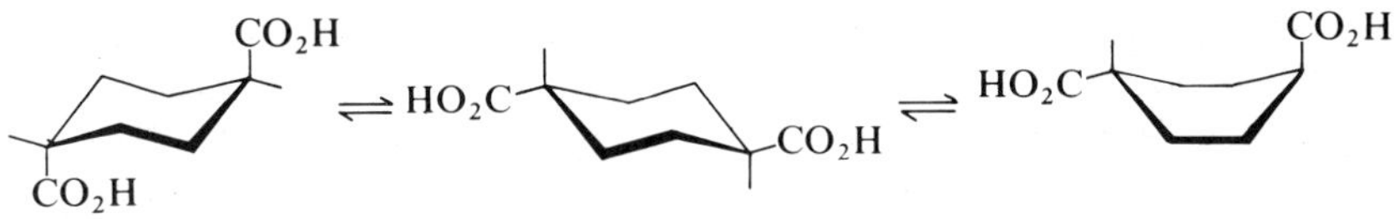

| chair-*a,a* | chair-*e,e* | boat-*e,a* |

The existence of conformational isomers improves the flexibility of polymeric derivates of 1,4-cyclohexanedicarboxylic acids.

- 1,4-Bishydroxymethylcyclohexane, or cyclohexanedimethanol (CHDM; Eastman 68¢). Both CHDM and terephthalic acid or its dimethyl ester DMT (9.3.2.2) are the building blocks of Kodak's Kodar PCTG copolyester 5445 (96¢) which is a clear, amorphous polymer with a $T_g$ of about 190°F and is compatible with PC.
- 5,8-Dihydro-1-naphthol—most likely a product of the Birch reduction of α-naphthol—is available in commercial quantities from Heico.

Nonaromatic diisocyanates are of interest for the assembly of light-resistant PURs (3.8.1, PUR), and hydrogenation of the ammoxidation products of xylenes is a convenient synthetic route (Scheme 8.10). Nitroaromatics can, of course, be hydrogenated to the corresponding aminoalicyclic structures (9.3.6).

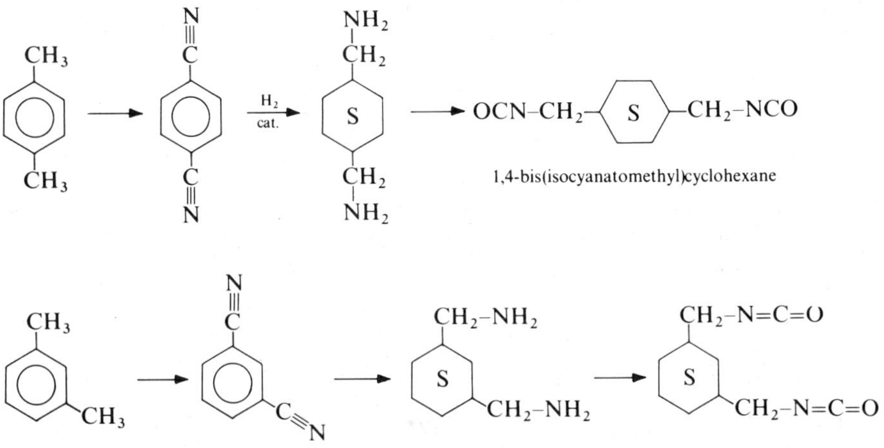

1,4-bis(isocyanatomethyl)cyclohexane

1,3-bis(isocyanatomethyl)cyclohexane

**Scheme 8.10**

Ciba-Geigy produces a cycloaliphatic diepoxy carboxylate under the tradename Araldite CY 179 that appears to originate with the Tishchenko reaction (*ONR*-90) of benzaldehyde, followed by a mild and selective partial hydrogenation of the two aromatic rings before the residual double bonds are epoxidized (Scheme 8.11). This liquid diepoxide

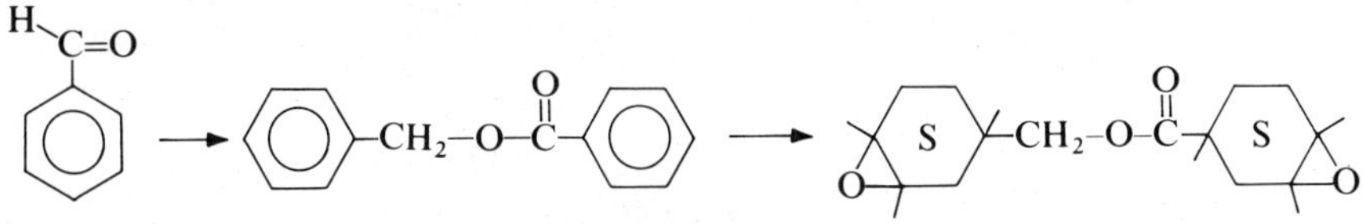

**Scheme 8.11**

is recommended as a building block of epoxy resins used for outdoor electrical applications.

As stated at the beginning of this section, halogen addition can also saturate an unsaturated cyclic chemical. The addition of chlorine to benzene yields eight isomeric benzene hexachlorides, $C_6H_6Cl_6$, of which one isomer (m.p. 112.5°C) is known as $\gamma$-benzene hexachloride or lindane and exhibits insecticidal properties. Even though it has been listed as a carcinogen in 1981, lindane is found on the marketplace. It is priced at $6.50 (99.9%, technical grade). A 20% formulation is imported from Spain, France, and the People's Republic of China and is priced at $16–$19. The structure of lindane is

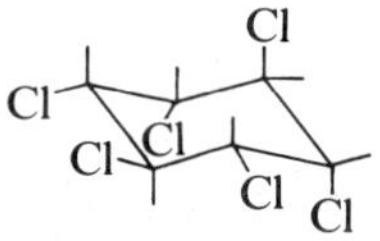

The addition of bromine to cyclododecatriene (6.8.1) produces the flame-retardant hexabromocyclododecane or Ethyl's Saytex ($2.20).

### 8.3.2 Cyclization Products

#### 8.3.2.1 Aldol-Type Cyclization Reactions

Isophorone, the product of an acetone condensation reaction carried out under strongly alkaline conditions, is becoming an increasingly important industrial building block. As shown elsewhere (5.5.4), it has become the source of isophoronediamine and of the light-stable, alicyclic PUR ingredient—isophorone diisocyanate, IPDI (interlude in Section 3.8). Recently, the Nuodex division of Huels began to market a mixture of two open-chain aliphatic diisocyanates—the trimethylhexamethylene diisocyanates (TMDI), currently sold for $4.30. The mixture of the isomeric hexamethylene diisocyanates is composed of the 2,2,4-trimethyl isomer $OCN-CH_2C(Me)_2CH_2CHMeCH_2CH_2-NCO$ and the 2,4,4-

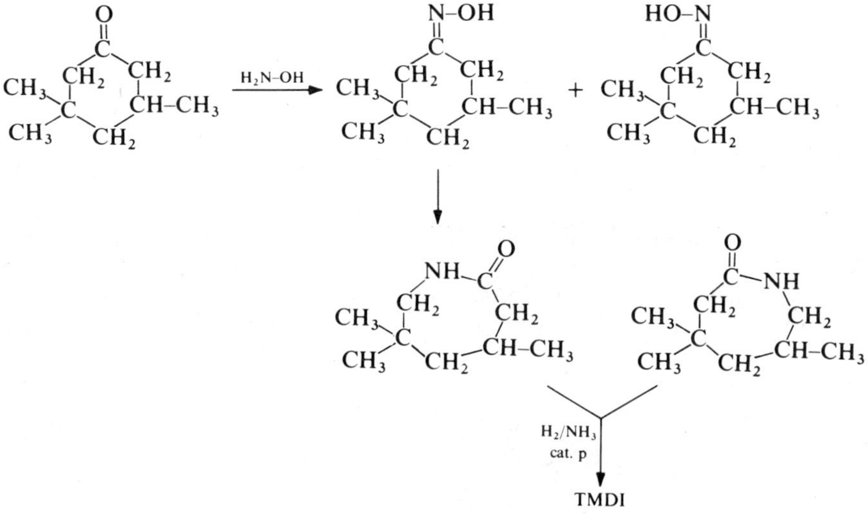

**Scheme 8.12**

trimethyl isomer $OCN-CH_2CHMeCH_2C(Me)_2CH_2CH_2-NCO$. The carbon skeleton of the TMDI twins is derived from the oxidative cleavage of the isophorone ring, and the most likely route to the final products is the hydrogenation of the C=C double bond, the formation of the oxime of the resulting saturated ketone, and a Beckmann rearrangement to obtain two isomeric lactams, as shown in Scheme 8.12. The formation of a lactam mixture is expected since oxime rearrangements occur with the migration of the carbon terminal that is trans or anti to the orientation of the original hydroxyl group of the oxime. There is little reason to expect the formation of only one geometrical isomer of the intermediate oxime. The conversion of the two lactams to the corresponding diamines— the precursors of the two isocyanates—is most likely achieved by forceful hydrogenation in the presence of ammonia.

Another synthetic use of isophorone that occurs with a partial cleavage of its carbon skeleton is the formation of 3,5-dimethylphenol or 3,5-xylenol (9.3.4), shown in Scheme 8.13. The loss of a methane moiety in this transformation illustrates the powerful driving force of aromatization.

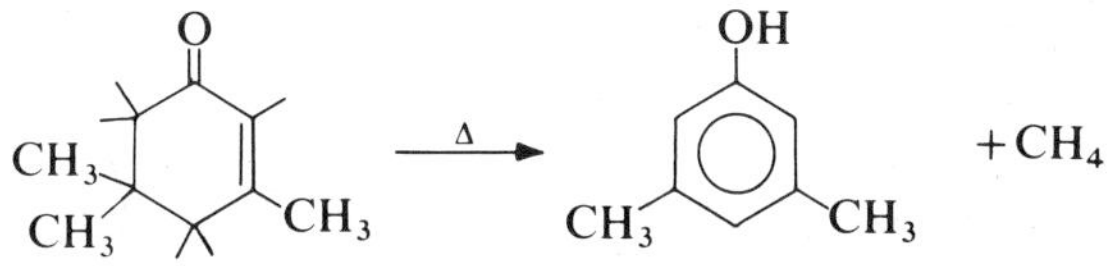

**Scheme 8.13**

The presence of the 3,3,5-trimethylcyclohexanol moiety is recognized in the structures of some peroxy polymerization catalysts (see 6.3.1.7).

### 8.3.2.2 Non-Aldol-Type Cyclization Reactions

A by-product of the hydrogenation of adiponitrile to hexamethylenediamine is the formation of 1,2-diaminocyclohexane, shown in Scheme 8.14. Du Pont and Milliken Chemicals have collaborated to make this diamine available as Milliken's Millamine 5260 ($1.28), for use, among other things, as an ingredient of epoxy resins.

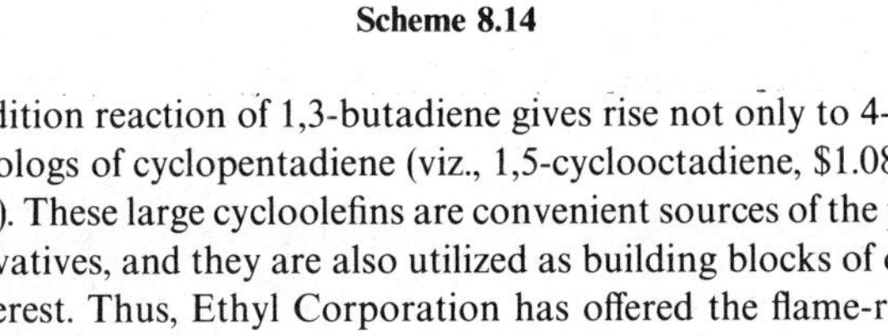

**Scheme 8.14**

The cycloaddition reaction of 1,3-butadiene gives rise not only to 4-vinylcyclohexene but also to homologs of cyclopentadiene (viz., 1,5-cyclooctadiene, $1.08) and cyclododecatrienes (6.8.1.4). These large cycloolefins are convenient sources of the previously shown $C_8$ and $C_{12}$ derivatives, and they are also utilized as building blocks of other compounds of industrial interest. Thus, Ethyl Corporation has offered the flame-retardant additive hexabromocyclododecane (see above), and Huels offers its Vestenamer family of polyoctenylene rubber additives obtained, most likely, by metathesis of 1,5-cyclooctadiene. Cycloaddition reactions of maleic anhydride with 1,3-butadiene and with cyclopentadiene are shown elsewhere (6.6.9).

# 9

# AROMATIC CARBOCYCLIC COMPOUNDS

## 9.1   INTRODUCTION: ORGANIZATION OF THIS CHAPTER

The multitude of aromatic carbocyclic compounds, and their synthetic interdependence, demands an organizational effort designed to avoid repetition and to promote conciseness. These objectives should be achieved by starting with a discussion of sources of the fundamental aromatic hydrocarbons, followed by relatively short presentations of five major chemical transformations that account for the formation of large-volume derivatives (9.2.1–9.2.5). Following this overview of the role that each major synthetic route plays in the production of the principal derivatives of aromatic hydrocarbons, we discuss, with a broader perspective than in Sections 9.2.1 to 9.2.5, the variety of reactions and uses of some additional, industrially significant transformation products.

Numerous diagrams summarize synthetic pathways that radiate from individual upstream building blocks and are offered in order to avoid verbose explanations. Again, as elsewhere throughout this book, the reader is encouraged to consult the cross-references cited in the Index.

## 9.2   SOURCES OF THE FUNDAMENTAL AROMATIC HYDROCARBONS AND THEIR MAJOR CHEMICAL TRANSFORMATIONS

As mentioned elsewhere (see Table 1.1), the components of BTX are among the principal upstream building blocks of the chemical industry and are derived mainly from petroleum, while coal tar makes rather small contributions (Table 9.1). BTX is produced in the course of steam-cracking of naphtha (Fig. 2.3) and by means of reforming. The latter, a dehydrocycloaromatization process, requires dual-function catalysts: one promotes dehydrogenation, and the other functions as an acid. Such catalysts are obtained by incorporating Pt, Pd, Re, and Rh metal clusters into an alumina–silica

**TABLE 9.1  Sources of BTX**

| Source | Production, B gal (B lb) | | |
|---|---|---|---|
| | B | T | X |
| Petroleum | 1.65 | 7.6 | 8.1 |
| Coal tar | 0.053 | 0.008 | 0.010 |

matrix. The redistribution of BTX components obtained from petroleum is discussed below.

Admittedly, the subject of *industrial catalysis* is mentioned in this book only in a sporadic fashion (see Index). For a recent coverage of activities in this important area, see J. M. Winton et al., "Catalysts '88: Restructuring for Technical Clout," *CW* 6/29/88, pp. 20–64. Other sources of information include the following:

L. L. Hegedus et al., *Catalyst Design: Progress and Perspectives*, Wiley, New York, 1987.

T. S. R. P. Rao, Ed., *Advances in Catalysis Science and Technology*, Wiley, New York, 1975.

G. W. Parshall, *Homogeneous Catalysis: The Applications and Chemistry of Catalysis by Soluble Transition Metals*, Wiley, New York, 1980.

G. W. Parshall and W. A. Nugent, a three-article series regarding the use of homogeneous catalysis for the production of pharmaceuticals, agrochemicals, flavors, fragrances, and other fine chemicals, published in *CT* 3/88, pp. 184–190; 5/88, pp. 314–320; and 6/88, pp. 376–383.

J. M. Thomas and R. M. Lambert, *Characterisation of Catalysts*, Wiley, New York, 1980.

The aromatic products are extracted from alkanes and olefins either by means of either dipolar aprotic solvents (sulfolane, DMF, NMP, DMSO, N-formylmorpholine, etc.—consult Index) or strongly hydrogen bonding solvents (mono-, di-, tri-, and tetraethylene glycols; N-methylformamide; benzyl alcohol; phenols; etc.). Extractive distillation techniques are also employed extensively. Naphthalene, methylnaphthalenes, biphenyl, indene, fluorene, and other aromatic hydrocarbons can be isolated in small amounts (except for the naphthalenes) from the higher-boiling fractions of residual petroleum, but coal tar is a better source of these materials (Fig. 2.3).

The coal-tar fraction of b.p. 230–300°C consists of about 10–15% naphthalene, 40–45% methylnaphthalenes, 5–10% diphenyl, and 30–35% dimethyl and higher alkylated naphthalenes. This mixture is used as a solvent for asphalt, an emulsifiable disinfectant, an ore flotation agent, and so on. In this context we must mention the "coal-tar acids," a complex mixture of *o*-, *m*-, and *p*-cresols (9.3.4) and of some higher alkylated phenols consumed in the United States at a level of about 135MM lb. Individual fractions of the coal-tar acids are employed in the manufacture of plasticizers, phenolic resins, and so on. On the marketplace, they are priced according to the *m*- and *p*-cresol content because *o*-cresol is the least reactive, and hence least desirable, tricresyl phosphate plasticizer:

Cresylic acids, *m*- and *p*-, >25%, 58¢

Cresols, 92%, *m*- and *p*-, 75¢

*o*-Cresol 75¢; *m*-cresol, 95–98%, $1.65

Coal tar supplies about 70% of the demand for naphthalene, and while coal tar satisfies the relatively small demand for higher multiring aromatic hydrocarbons (e.g., anthracene, phenanthrene, and acenaphthene), practically all of the latter are imported from West Germany.

Naphthalene (340MM lb, petr. m.p. 80°C, 30¢; crude dom. m.p. 78°C, 22¢; phthalic, anhydrous grade 23.5¢) is the source of a variety of derivatives illustrated by the following examples:

- Tetrahydronaphthalene (8.3.1), used as a solvent, heat-transfer medium, and an intermediate
- α-Naphthol or 1-naphthol ($1.81) (9.3.4), a downstream building block of dyes, antioxidants, medicinals, agrichemicals
- 2,6-Dihydroxynaphthalene, possibly related by origin (9.2.1) to naphthalene-2,6-dicarboxylic acid, an intermediate in poly(ethylene 2,6-naphthalenedicarboxylate), (PEN) synthesis, the analog of PET that, according to Eastman, exhibits fivefold better barrier properties than the latter with respect to oxygen, $CO_2$, and moisture (*MP* 4/88, 90)
- α-Naphthylamine ($2.10), an intermediate in the synthesis of rubber antioxidants, dyes, and so on
- 6-Amino-2-naphthalenesulfonic acid, or Broener's acid ($4.15), a dye intermediate
- 1-Naphthyl-*N*-methyl carbamate or carbaryl insecticide (Union Carbide's Sevin)

The coal-tar source of diphenyl or biphenyl (99.9%, 65¢) is supplemented by the by-product of hydrodealkylation (HDA) of toluene (see below) and by synthesis when benzene is passed over a heated metallic surface maintained at about 750°C. About 1 part by weight (pbw) of diphenyl is formed for each 100 pbw of benzene during the HDA of toluene, and the formation of diphenyl is accompanied by some terphenyl. Diphenyl is used as a fungistat during shipments of oranges and is also alkylated to give heat-transfer media and other functional fluids. Polyphenyls are also used in the manufacture of carbonless copy paper, and Monsanto has recently (*CB* 2/88) raised its production of polyphenyls to 15MM lb.

Previously diphenyl and analogs were converted to chlorinated biphenyls (CBPs) for use as a transformer fluid and heat-transfer medium. The use of the CBPs is now banned and has become a scourge of the chemical industry (9.2.3).

The rather crude mixture of indene and alkylstyrenes is polymerized and used in the production of rubber, caulks, and other polymeric materials.

Probably the most commerically promising aromatic hydrocarbon derived from renewable resources is *p*-cymene, or *p*-isopropyltoluene (~72¢), the product of the cracking of the pinene fraction of turpentine (8.2.1; see also *SZM-II*, 113, 129, 131).

The BTX mixture of aromatic hydrocarbons obtained from steam cracking of naphtha contains about 40% of B, 20% of T, and 5% X, as well as some ethylbenzene and other minor constituents. The reforming process produces a mixture of the BTX hydrocarbons composed of about 3% B, 13% T, and 18% X. Thus, these major refining processes give a high content of toluene (~45%) and xylenes (also 45%) and a relatively small content of benzene (~10%). Approximately 75% of this mixture contributes about 25% of the total gasoline pool in order to raise the octane rating of the end product since the octane numbers of benzene and toluene are about 106 and 110, respectively. The remaining 25% of the BTX mixture is channeled into chemical use, and the current demand for the individual components is highest for benzene, and least for the xylenes. To adjust the supply of individual BTX components to the demand, the mixture is subjected to two additional refining processes: hydrodealkylation (HDA) and redistribution or disproportionation of methyl substituents.

The HDA process employs hydrogen and catalysts such as $Cr_2O_3$ and $Mo_2O_3$, at 500–650°C, and strips off methyl groups in the form of methane.

The redistribution of methyl groups is catalyzed by strong acids in a reaction somewhat reminiscent of the classical Jacobsen rearrangement (*ONR*-48): the protonated electron-rich multisubstituted aromatic hydrocarbons are attacked in a nucleophilic fashion by other electron-rich hydrocarbons with a net transfer of a methyl group from one hydrocarbon to another. This process is depicted in Scheme 9.1.

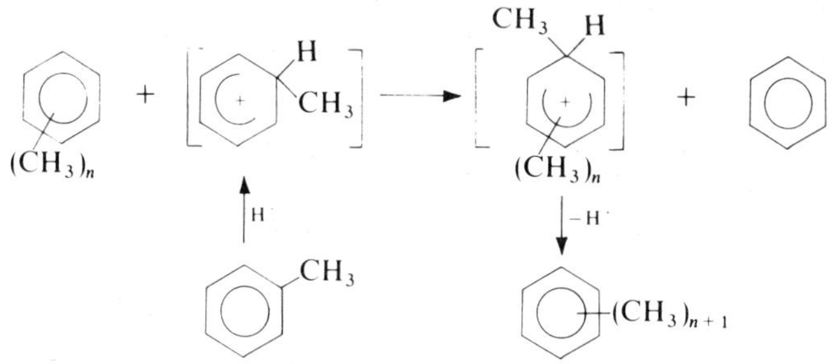

**Scheme 9.1**

The acid-catalyzed mobility of alkyl groups attached to aromatic hydrocarbons increases as a function of the stability of the transient alkyl-derived carbocations. Thus, it is not surprising, for example, that the *t*-butyl group migrates or is displaced by an electrophilic reagent with the greatest of ease. This accounts for the observation that the alkylation of *m*-xylene with a *t*-butyl-generating reactant (e.g., the corresponding alcohol or olefin) yields 3,5-dimethyl-*t*-butylbenzene (rather than an ortho- and/or para-substituted product) because this particular protonated intermediate is least stabilized by the two methyl substituents. Thus, it is the isomer that tends to accumulate in the equilibrium mixture. Another experimental fact that sheds light on the mobility of alkyl groups is the facile displacement of a *t*-butyl group by a bromonium ion (9.3.1.2) during the conversion of *p*-di-*t*-butylbenzene to *p*-bromo-*t*-butylbenzene.

In this manner the original BTX feedstock destined for use by the chemicals industry is transformed to a mixture containing about 15% B, 50% T, and 35% X, with the following distribution of the X isomers: 65% para, 20% meta, and 15% ortho. We note the advantageous high content of *p*-X among the X isomers.

Despite the long-standing practice of subjecting the original BTX mixture to these refining operations, there is always room for significant improvements. Thus, for example, Coseden Chemical Company (a division of Fina Oil & Chemicals Co.) announced (*CE* 11/10/86) that a new, long-lasting zeolite catalyst improves the toluene to benzene (and X) conversion by reducing the consumption of hydrogen by 50% and producing 5–10% fewer high-boiling residues. This "T2BX" process is carried out by passing a preheated mixture of toluene and hydrogen over the catalyst at 600–700°F and 400–900 psig, followed by recycling of hydrogen and fractionation of the liquid products. Similarly, Mobil announced (*CW* 3/5/86) progress in the conversion of olefins and low-molecular-weight paraffins into BTX by means of the HZSM-5 zeolite catalyst with a $SiO_2$–$Al_2O_3$

ratio of 70:1. This so-called M2 process is carried out at 538°C and 1–20 atm, and the yield of BTX is a function of the hydrogen concentration. In the same vein, UOP and BP announced (*CE* 4/27/87) a significant cost reduction (10¢/lb) in the conversion of LPG to valuable *p*-xylene, benzene, and other components of BTX, and this success is attributed to a combination of BP's Cyclar catalyst and UOP's continuous catalyst regeneration process.

Higher methylated benzene derivatives, such as pseudocumene (or 1,2,3-trimethylbenzene; Amoco, 80¢) and 1,2,4,5-tetramethylbenzene (or durene) are valuable by-products of the above-mentioned disproportionations of BTX. Their importance lies in the corresponding oxidation products, namely, trimellitic anhydride (TMA) and pyromellitic anhydride (PMDA), respectively (9.2.2). The interesting uses of these building blocks are discussed in 9.3.2.3. In this context, the unexpected bonanza of appreciable quantities of durene formed as a by-product of the MTG process (3.3.2.7.B) operated on a large scale in New Zealand (*CMR* 2/3/86) is noteworthy. It is estimated that as much as 60MM lb of durene may be isolated from this source—10 times the estimated production in the United States. PMDA currently sells at $6.80–$13.80/lb, and the New Zealand government is contemplating the construction of a $20–$30-million plant with an annual production capacity of 88MM lb of PMDA (*CE* 4/25/88). The potential importance of the durene building block explains the joint venture of Du Pont and Mitsubishi designed to import pseudocumene from China to Japan for conversion into PMDA.

About 50% of "chemical benzene" is alkylated (9.2.1) to ethylbenzene destined to become styrene, 25% becomes isopropylbenzene or cumene for conversion to phenol and acetone, and about 5% is transformed into linear alkylated benzene (LAB), to be converted, in turn, to LAS detergents (9.2.4).

The pricing of the BTX components reflects the cost of crude petroleum and fluctuations in the demand–supply equilibrium of individual chemicals (Table 9.2). The prices cited in Table 9.2 are quotations fob at the lowest-priced location, and in order to convert the traditional gallon pricing to cents per pound, the following density values (at 60°F) for benzene, toluene, and xylene are used: 7.36, 7.26, and 7.25 lb/gal, respectively. Fluctuations in the pricing of these primary commodity chemicals are echoed by changes in the unit pricing of their secondary commodity derivatives. Thus, for example, between May 1987 and mid-June 1987 the prices of styrene and cumene rose from 42¢ to 48¢/lb and from 25¢ to 30.25¢/lb, respectively.

Of the three xylenes, the *o*-isomer, consumed in the United States at a level of about 900MM lb, is converted mainly to phthalic anhydride (9.2.2)—the building block of the largest group of plasticizers (DOP, etc.) and of UPR and alkyd resins, markets that show

**TABLE 9.2    Pricing of BTX Components**

|  | Benzene | | Toluene | | Xylenes | | |
|  | | | | | (Ortho) | (Meta) | (Para) |
|  | ($/gal) | (¢/lb) | $/gal | ¢/lb | (¢/lb) | (¢/lb) | (¢/lb) |
|---|---|---|---|---|---|---|---|
| 1/82 | 1.57 | 21.3 | 1.38 | 19.0 |  | Mixed $1.47/gal | |
| 2/85 | 1.17 | 15.9 | 0.95 | 13.1 |  | Mixed $0.98/gal | |
| 4/87 | 1.55 | 21.1 | 0.80 | 11.0 |  | Mixed $0.795/gal | |
|  |  |  |  |  | 14.75 |  | 19 |
| 1/88 | 0.85 | 11.5 | 0.60 | 8.25 |  | Mixed $0.70/gal | |
|  |  |  |  |  | 13.0 | 36 | 17.5 |

only moderate growth. These three markets consume about 50, 25, and 20% of the phthalic anhydride demand, respectively. The production of phthalic anhydride from o-xylene is displacing gradually the traditional the oxidative degradation of naphthalene. On the other hand, p-xylene production has increased from about 3.5B lb in 1976 to about 5.2B lb in 1987 mainly because of the continued growth of the terephthalate-based polymers, of which PET is the predominant end product (> 1.7B lb in 1987). About 1B lb of p-xylene was exported in 1986 to newly industrialized countries, and much of it was returned to the United States in the form of polyester fabrics and textiles. The export market for p-xylene shrank to 0.9B lb in 1987 and is likely to shrink further as the less developed countries become more industrialized and as exports from the Middle East increase. m-Xylene is oxidized to isophthalic acid that is incorporated into UPRs and alkyd resins.

### 9.2.1  Alkylation, Chloromethylation, Acylation, and Related Reaction Products

Of the three processes that introduce new C–C bonds into the aromatic ring, the first-mentioned one serves to produce the high-volume commodity chemicals ethylbenzene and cumene, while the latter two processes serve to prepare an array of fine chemicals.

The Friedel–Crafts (ONR-33) alkylation of benzene with ethylene (4.3.4) is usually carried out in the presence of a small amount of an acidic (protic and/or Lewis acid) catalyst and yields ethylbenzene (22¢, ~ 9B lb), together with some diethylbenzene (98¢) and higher ethylbenzenes. The latter two products are recycled to maximize the formation of the desired ethylbenzene by transalkylation. Some diethylbenzene-containing fractions, however, are isolated to produce divinylbenzene (DVB), in a manner analogous to the conversion of ethylbenzene to styrene (see below).

The dominant component of diethylbenzene (and hence also of its dehydrogenation product, DVB) is the m-isomer, and this result can also be explained by considering the relative degree of stabilization of the transient protonated species by the two ethyl substituents; the protonated p-isomer tends to accumulate in the reaction mixture (as a result of preferential stabilization of the positive charge by both electron-donating ethyl groups), and hence it becomes the victim of transalkylation. The difference in the behavior of p- and m-isomers is illustrated in Scheme 9.2.

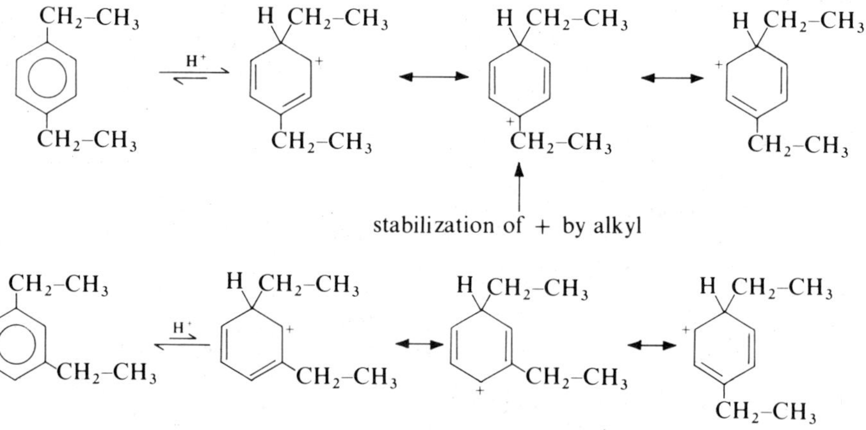

stabilization of + by alkyl

**Scheme 9.2**

The Friedel–Crafts process can be carried out in either liquid or gas phase ($\leqslant 450°C$), and catalysts such as aluminum chloride, ferric chloride, tin tetrachloride, phosphoric acid–boron trifluoride, phosphoric acid on alumina–silica, calcium hydrogen phosphate, and boron trifluoride–alumina can be employed.

Dehydrogenation of ethylbenzene to styrene (9B lb) is achieved most commonly at 550–700°C in the presence of catalysts such as iron oxide–chromium oxide, vanadium pentoxide, magnesium oxide–alumina, and zinc oxide–alumina. Minor amounts of benzene and toluene are formed during this high-temperature process, and the hydrogen by-product is used to fuel the energy-intensive transformation. On dehydrogenation, the resulting styrene must be quickly cooled in order to avoid polymerization. It is fractionated under vacuum in the presence of a polymerization inhibitor such as sulfur and stored for any length of time in the presence of antioxidants such as BHT and TBC.

Fractionation of the dehydrogenation product also yields DVB–styrene mixtures rather than pure DVB because of the danger of spontaneous, exothermic polymerization of the latter. A few years ago the intention to market 100% pure DVB priced at $2.75 never materialized presumably because of its risky storage. The production of somewhat cross-linked PS used to manufacture PS-based ion-exchange resins is carried out by means of solutions of DVB in styrene.

Stabilized styrene, or solutions of DVB in styrene, are washed before use with aqueous alkali to remove the phenolic polymerization inhibitors.

The enduring (see Table 1.11) industrial importance of styrene has stimulated a search for alternative sources. Before we mention those, we must remember that about 7% of the demand is satisfied by extraction from petroleum cracking operations. The most successful alternative route to styrene is mentioned in connection with the production of PO by means of the hydroperoxide of ethylbenzene (5.4.2.1) since the resulting methyl-benzyl alcohol, ARCO's MBA ($1.04), if isolated and marketed, can be easily dehydrated to styrene. Actually, ARCO has launched MBA as a competitor of benzyl alcohol ($C_6H_5CH_2OH$; technical grade $1.26; NF $1.37; photo grade 80¢), and ARCO has also announced the availability of acetophenone—the decomposition product of 1-hydroper-oxyethylbenzene [$C_6H_5CH(OOH)CH_3$]. The economic incentive for such a process is to "kill two birds with one stone" and avoid the energy-intensive dehydrogenation of ethylbenzene as well as the energy-intensive source of hypochlorous acid when PO is obtained by way of the chlorohydrin. This principle is served by a process reported some time ago from the Soviet Union in which the dehydrogenation of ethylbenzene is coupled with the hydrogenation of nitrobenzene to aniline (Scheme 9.3).

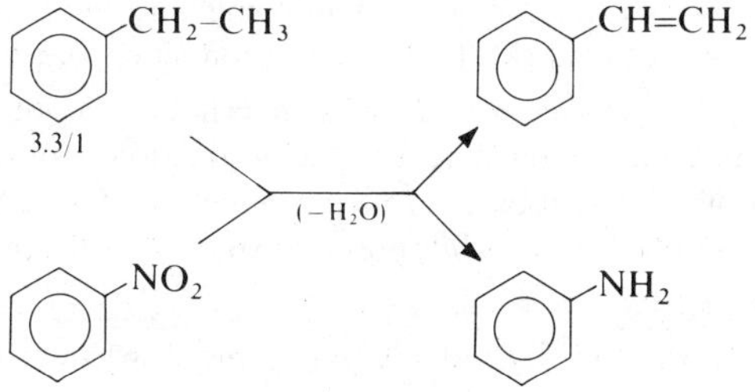

**Scheme 9.3**

Wherever electricity and hence chlorine are costly, the chlorination of ethylbenzene to 1-chloroethylbenzene and its dehydrohalogenation to styrene will remain obsolete. In the

United States the cost of electricity is about 3.5¢/kWh, and the electrolysis of sodium chloride requires 3310 kWh per metric ton of chlorine. In other words, out of the 9.75¢/lb price of chlorine, 5.25¢ is accounted for by the cost of electricity. These circumstances may not be valid in developing countries with untapped hydroelectric resources.

Other potential methods for the production of styrene include

1. A simultaneous alkylation and dehydrogenation of benzene in the presence of a Rh catalyst (Japan):

$$C_6H_6 + CH_2{=}CH_2 + O_2 \rightarrow C_6H_5{-}CH{=}CH_2 + H_2O$$

2. A reaction of toluene and formaldehyde (or a formaldehyde-generating system involving methanol or even methane–$O_2$):

$$C_6H_5{-}CH_3 + H_2C{=}O \rightarrow C_6H_5{-}CH{=}CH_2 + H_2O$$

3. A variation on the preceding route (Monsanto) to convert toluene to stilbene and subject the latter to ethenylation:

$$2C_6H_5{-}CH_3 \rightarrow C_6H_5{-}CH{=}CH{-}C_6H_5 \xrightarrow{CH_2{=}CH_2} 2C_6H_5{-}CH{=}CH_2$$

$$+ H_2(\sim \$3.50/1000 \text{ scf})$$

It is difficult to predict at this time which, if any, of the preceding failing college sophomore organic chemistry answers will turn out to become an industrial reality!

The price of styrene rose from 22¢ in early 1987 to 45¢ in April 1988 and 50¢ in January 1989 (mostly because of a strong demand and record profit-taking in the U.S. chemicals industry) but recently has generally hovered near a 30¢/lb level.

Styrene is a continuously growing building block for the production of thermoplastic, thermosetting, and elastomeric polymers (see 4.3.4). Over 6.2B lb of this monomer are consumed in the following fashion (approximate weight ratios of monomers are shown in parentheses):

- PS (5.5B lb, nearly 100% S, 53–60¢ during 6/87, 58¢–75¢ during 4/88, depending on amount of additives to improve impact resistance; see 4.3.4). Major processing techniques used for the fabrication of end products consumed domestically at a level of 4.3B lb are 43% by extrusion, 37% by molding, and about 12% by formation of expandable PS. About 110MM lb of the PS production is exported.
- SBR (S:B ratio $\sim 6$:1), (see 6.8.1.2), 2.2B lb. About 73% of SBR is used in the production of tires and the remainder in the production of other rubber objects, as well as latex for water-based paints. In the production of tires, SBR is blended with paraffin oil and is priced at 58¢, while non-oil-extended SBR can cost as much as 81¢.
- SMA (S/M $\sim 1$), 2.2B lb (see 6.6.2).
- ABS (A:B:S ratios $\sim 16$–25:6:70–80, plus some dispersed SBR), 1.25B lb (4.3.4, 5.4.3.2, and 6.8.1.2).
- PSB (S:B < 9), 0.6B lb (4.3.4, 6.8.1.2).
- UPR ($\leqslant 30\%$ S), 0.4B lb, 51¢–56¢.
- SAN (S:AN $\sim 3$), 0.12B lb (5.4.3.2), 88¢.

Considerable amounts of benzene ($\sim 3\%$ of 6.2B lb) are alkylated to produce LABs (9.2.4) that are converted to LAS, 800MM lb, the largest class among different categories of surfactants that constitute the main ingredients of a $10-billion annual market in the United States (*CW* 1/20/88, 28–56). These materials are also referred to as "synthetic detergents", "syndets", "detergents", "elusifiers", and "suspending and dispersing agents"—where some of these terms are synonymous while others indicate differences in performance. However, since the common denominator of all of these materials is their activity at the water–oil interface, the terms "surface-active agents" (surfactants for short), are most applicable to all. The importance of hydrophilic–lipophilic balance (HLB) is mentioned elsewhere (4.3.6.1).

*Categories of Surfactants*

Alkylbenzene-derived LAS constitutes part of the largest category of surfactants classified according to the charge carried by the main structural moiety of the product:

1. Anionic surfactants, 3.7B lb (up from about 2.3B lb over the past 10 years), carry a negative charge on the lipophilic portion of the product and are represented by sulfonates, monosubstituted sulfates of different medium-molecular-weight alcohols ($\sim 260$MM lb) and their ethoxylated derivatives ($\sim 500$MM lb), and, of course, the carboxylates. The last-mentioned class includes the sodium and potassium salts of fatty acids—old-fashioned soaps ($900 million). The latter became outdistanced by "soapless soaps" for the first time in 1953, and by 1955 synthetic surface-active agents were already contributing 2B lb vis-à-vis 1.5B lb of soap products (*SZM-I*, 684).

2. Cationic surfactants, carry a positive charge on the lipophilic portion of the product and have been contributing a steady 300MM lb/yr to the total demand for all surfactants. They are represented primarily by quaternary ammonium salts.

3. Nonionic surfactants are the most rapidly growing category of surfactants. Their use expanded over the last 10 years from about 900MM lb to about 1.7B lb, and they are represented by ethoxylates (4.3.6.1) of all sorts of lipophilic substrates, including the block polymers of EO–PO (5.4.2.2.A). The alcohol ethoxylates contribute about 725MM lb to the total consumption of all nonionic surfactants.

4. Amphoteric surfactants represent a relatively small group of surface-active agents that contain oppositely charged ionic moieties. Normally, the positive charge is located at a quaternary ammonium ion center, while the negative charge is carried by either a carboxylate, a sulfonate, or a sulfate terminal. For example, tertiary amines (which provide the lipophilic portion of the surfactant) can be treated with sodium chloroacetate to produce the family of betaines (*SZM-II*, 62–64):

$$R_f{-}NMe_2 + Cl{-}CH_2{-}CO_2^- \, Na^+ \rightarrow R_f{-}\overset{+}{N}Me_2{-}CH_2{-}CO_2^-$$

(where $R_f$ represents a fatty amine residue).

Numerous surfactants are derived from renewable starting materials with or without the incorporation of components derived from fossil resources (see index in *SZM-II*). There are many thousands of different surfactants available on the marketplace designed for optimum performance under countless circumstances.

Nearly 50% of the approximately 6B-lb/yr surfactant consumption is absorbed by the personal consumer market, with the rest destined for institutional and industrial consumption. Gradually, the traditional solid products are now being displaced by liquid formulations that include many additives such as softeners, antistatics, bleaches, anti-deposition and optical screening agents, and proteolytic enzymes (for use in laundry products), as well as emollients, scents, coloring agents, skin- and hair-treatment materials, and other ingredients found in personal-care products.

About 20% of benzene is alkylated with propylene (5.4.1) to yield cumene, 4B lb (see Table 1.1) (20¢), and the Hock process converts the latter to phenol (9.2.1) and acetone (5.5.1). Cumene hydroperoxide (74¢) and peroxide are mentioned elsewhere (6.3.1.7) among polymerization catalysts. Dehydrogenation of cumene produces about 50MM lb of α-methylstyrene (AMS), employed principally as a modifier of ABS and UPR resins. The replacement of styrene by AMS reduces the crystallinity of the initial polymers and is conducive to their improved performance as flexible coatings and adhesives. A double isopropylation of benzene produces a mixture of *m*- and *p*-diisopropylbenzenes (DIPB), and a new plant built by Eastman will provide these starting materials for the preparation of a variety of derivatives (*CMR* 6/30/86).

The alkylation of naphthalene with a variety of alkyl halides or olefins produces mixtures employed as additives to lubricating oils, including "pour-point depressants," which inhibit undesirable crystallization of paraffinic components, and "viscosity index modifiers", which have been used for some time.

Recent activity in the isopropylation of aromatics is noteworthy and appears to have two major objectives:

1. Dehydrogenation to the corresponding isopropenyl derivatives and their use in a Ritter reaction or as copolymers.
2. Utilization of the readily accessible hydroperoxides for the preparation of the corresponding phenols (and acetone by-product) by means of the Hock process, as polymerization catalysts (6.3.1.7), and for the epoxidation of olefins and trans-formation of the benzylic alcohol coproduct of the epoxidation to isopropenyl derivatives for use as indicated in 1 above.

The choice of these alternative possibilities depends on the broader objectives of each manufacturer.

Thus, Eastman has chosen to convert *p*-diisopropylbenzene to hydroquinone ($1.95; photographic grade $2.54). An interesting side effect of this development was the sudden scarcity of manganese sulfate on the marketplace because of the displacement of the traditional route to hydroquinone that involves the oxidation of aniline by means of manganese dioxide.

American Cyanamid has recently launched the production of benzylic diisocyanates derived from diisopropenylbenzenes by means of the Ritter reaction. A $20-million facility began to operate (*CEN* 2/8/88) in Willow Island, West Virginia with the production of *m*-isopropenyldimethylbenzyl isocyanate (TMI; $5.00) and *m*-tetramethylxylidene diiso-cyanate (TMXDI; $3.00). Both materials are products of the Ritter reaction of the isopropenyl substituent:

$$Ar–C(CH_3)=CH_2 \rightarrow Ar–C(CH_3)_2–NH_2 \rightarrow Ar–(CH_3)_2–N=C=O$$

(where Ar represents the residue of the initial xylene). It is not clear at this time whether Cyanamid chose to dehydrogenate *m*-diisopropylbenzene to obtain the isopropenyl intermediates [e.g., *m*-diisopropenylbenzene (*m*-DIPEB); $2.75] or if the latter are obtained by way of the Hock process. These so-called aliphatic isocyanates are useful monomers in the assembly of light-stable nonaromatic PURs (interlude in Section 3.8). Also, since 1981, American Cyanamid has been offering (*CW* 9/16/81) 2- and 2,6-diisopropylnaphthalene (IPN, 93¢; DIPN, 89¢) without a clue as to whether these products may become 2-naphthol and 2,6-dihydroxynaphthalene or the corresponding isocyanates. More recently Virginia Chemicals and Kureha of Japan announced (*CW* 10/14/87) a joint venture to produce diisopropylnaphthalene used currently by the Japanese partner as a solvent for carbonless copy paper.

Goodyear has announced the availability of *m*- and *p*-diisopropenylbenzene, as well as the corresponding hydration product *p*-bis-α-hydroxyisopropylbenzene (p-DIOL).

Eastman Kodak can provide heterotrifunctional monomers obtained by subjecting *m*-diisopropenylbenzene to a single Ritter reaction (most likely with methyl carbamate), followed by reaction with either diethanolamine (34¢) or the analogous amine derived from PO, namely, *N,N*-bis(2-hydroxyethyl)-*N'*-(*a,a*-dimethyl-3-isopropenylbenzyl)urea and *N,N*-bis(2-hydroxypropyl)-*N'*-(*a,a*-dimethyl-3-isopropenylbenzyl)urea:

$$m\text{-CH}_2\text{=C(CH}_3)\text{-C}_6\text{H}_4\text{-C(CH}_3)_2\text{-NH-CO-N(CH}_2\text{-CHR-OH)}_2$$

(where R represents either H or methyl). These new specialty monomers are priced at $842/lb in 5-kg or higher quantities.

Mitsui Petrochemical announced in May 1980 that it had mastered the troublesome conversion of *m*-diisopropylbenzene to resorcinol (technical grade $1.80, USP $4.25 crystals, $4.40) and, for good measure, also offers resorcinol monoacetate ($1.98).

The production of *p*-cumylphenol, or 4-α-methylstyrylphenol, by PMC Specialties Group (*CMR* 5/11/87) may start with a double isopropylation of diphenyl followed by a single application of the Hock reaction, as shown in Scheme 9.4. The material is an intermediate in the synthesis of UV stabilizers and antioxidants.

The alkylation of toluene with ethylene produces a 2:1 mixture of *m*-ethyltoluene and *p*-ethyltoluene, and these intermediates have traditionally been dehydrogenated to a mixture of the corresponding vinyltoluenes. The worldwide demand for vinyltoluenes (67¢) is only about 45MM lb, but they perform a useful function as additives to PS in order to reduce the troublesome shrinkage as the liquid monomer polymerizes (a property of most polymers). The same function, incidentally, is performed by *t*-butylstyrene in the case of the UPRs. The placid situation of the traditional usage of vinyltoluenes was perturbed in 1982 by Mobil's announcement that their HZSM-5 catalyst selectively produces *p*-ethyltoluene that, on dehydrogenation to *p*-vinyltoluene, yields 97% homogenous poly(*p*-methylstyrene) (PMS). The thermomechanical properties of PMS are actually distinctly different and in many ways superior to those of PS (*CT* 9/82). A 35MM-lb/yr plant to produce *p*-vinyltoluene was completed at Baton Rouge, Louisiana in 1986 and is now operated by Hoechst Celanese.

The Friedel–Crafts reaction of benzene with methylene chloride gives diphenylmethane [$(C_6H_5)_2CH_2$, $4.75], employed as a perfume ingredient because of its orangish odor. Diphenylmethane can also be prepared by a Friedel–Crafts reaction of benzene with benzyl chloride (see below); in either case, an excess of benzene functions as a solvent and minimizes multiple alkylations.

A variation on the theme of alkylation of benzenoid structures is the chloromethylation or Blanc reaction (*ONR*-12). The formation of benzyl chloride per se by means of chloromethylation is of little interest because the benzene ring is not particularly reactive

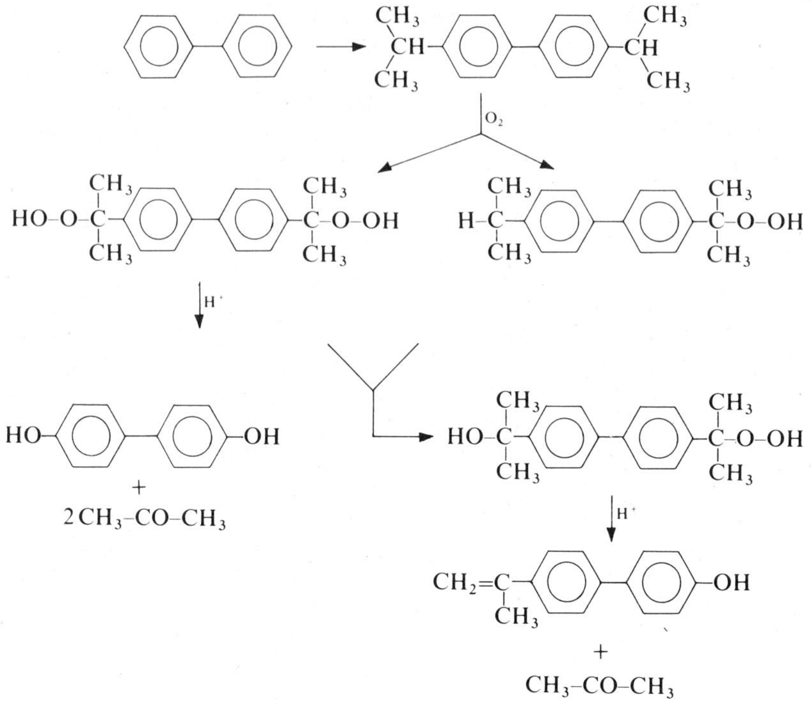

**Scheme 9.4**

under normal chloromethylation conditions ($CH_2O$–$HCl$–$ZnCl_2$) and also benzyl chloride can be readily obtained by side-chain chlorination of toluene (9.2.3). In either case, benzyl acetate—a perfume ingredient priced at about $1.30 and imported at a 1MM-lb level from Italy and Mexico (*CMR* 8/14/87)—can be obtained from benzyl chloride and sodium acetate.

Of greater interest is the chloromethylation of aromatic structures more susceptible than benzene to weakly nucleophilic reactants (such as compounds that contain multiple alkyl groups, methoxy, and other electron-releasing substituents) and the conversion of these chloromethyl derivatives to a variety of substitution products (9.3.1). This sequence of transformation is used, for example, in the synthesis of ion-exchange resins (3.4.3.4). However, chloromethylated styrene polymers can also be assembled by means of Dow's vinylbenzyl chloride, a 40:60 mixture of *m*- and *p*-isomers, which are employed as comonomers. Similarly, toluene or naphthalene are converted to *p*-methylbenzyl chloride and α-chloromethylnaphthalene, respectively. A reaction of the latter with sodium cyanide and the hydrolysis of the resulting nitrile yield α-naphthaleneacetamide and the corresponding carboxylic acid—two of several common plant hormones, growth accelerators, or auxins, used in horticulture. Actually, in view of the susceptibility of the polynuclear aromatic hydrocarbons to attack by free radicals, the same auxin can be obtained from naphthalene in a more direct fashion by means of acetic anhydride and potassium permanganate—a reaction mixture that generates carboxymethyl ($\cdot CH_2$–$CO_2H$) radicals.

Additional synthetic uses of chloromethylated intermediates are described in 9.3.1, and the assembly of some medicinal products derived from anisole is described elsewhere in this book.

The Friedel–Crafts acylation of aromatic carbocycles gives rise to numerous aldehydes and ketones mentioned more extensively in 9.3.5. The reaction conditions are similar to those employed in the analogous alkylation reactions discussed above except for two factors: the $AlCl_3$ (anhydrous ~61¢) or similar Lewis acid catalysts are required in equimolar or higher ratios with respect to the acylating reactant because the resulting carbonyl products inactivate the catalysts through formation of complexes; and acyl chlorides are more costly than alkyl halides, olefins, or alcohols utilized in alkylations.

Two examples of products that reflect the relatively costly acylation reactions are

- Acetophenone, perfume grade $2.15, technical grade 76¢.
- Benzophenone, technical-grade flakes $2.40, NF crystals $2.50, flakes $3.65.

The preparation of these ketones by means of the Friedel–Crafts process requires either the use of acetyl chloride or benzoyl chloride (57¢) or phosgene (55¢), respectively. Acetyl chloride (b.p. 52°C), is a powerful irritant that is available commercially but is best prepared for captive use. Thus, acetophenone is preferably prepared from benzoic acid (technical grade 55¢) and acetic anhydride (43.5¢), a reaction that avoids the use of acetyl chloride and, more importantly, avoids traces of chlorine-containing by-products that are deleterious to the use of acetophenone as a perfume ingredient or intermediate. The latter reaction occurs most likely by way of a thermal decarboxylation of the mixed anhydride:

$$C_6H_5-CO-O-CO-CH_3 \rightarrow C_6H_5-CO-CH_3 + CO_2$$

A third example that serves to illustrate one of the many modifications of the classical Friedel–Crafts process based on the use of aluminum chloride (G. A. Olah, *Friedel–Crafts and Related Reactions*, Vols. 1–3, Interscience, New York, 1963–1965) is the preparation of β-acetylnaphthalene or β-methyl naphthyl ketone (crystals $14.00), which employs polyphosphoric acid (PPA), prepared from phosphoric acid (technical grade 85%, 26.5¢) and phosphorus pentoxide (82¢).

A different type of alkylation reaction occurs with styrene-like systems in the presence of protic acids. It is illustrated in Scheme 9.5 by the formation of 1,3,3-trimethyl-1-phenylindane from two molecules of isopropenylbenzene. The transformation of this hydrocarbon into a building block of polymers is shown elsewhere (9.3.2.1).

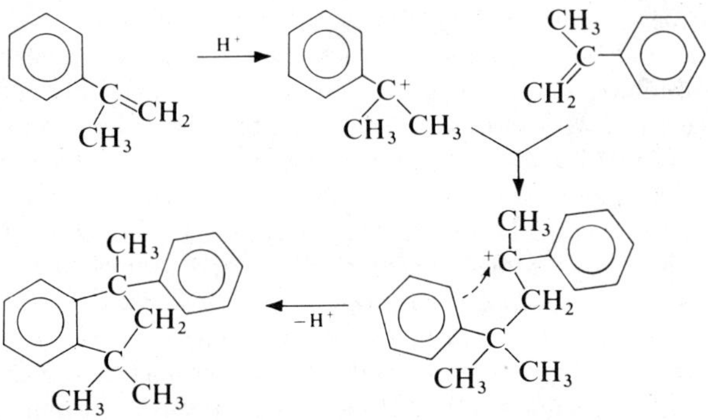

**Scheme 9.5**

A combination of Friedel–Crafts alkylation and condensation reactions is shown next because of its importance as a route to stilbenes that can be employed as starting materials for the preparation of optical brighteners (one of the laundry additives mentioned above in connection with the short survey of surfactants) and sunscreening agents. Thus, in the presence of acid catalysts, chloroacetaldehyde reacts with 2 mol of benzene, or an appropriate benzene derivatives to form the stilbene system:

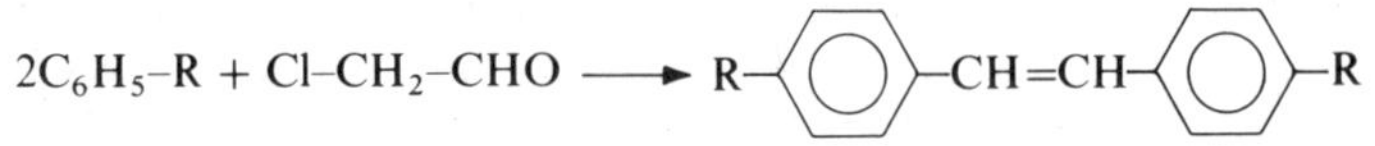

$$2C_6H_5-R + Cl-CH_2-CHO \longrightarrow R-\langle\bigcirc\rangle-CH=CH-\langle\bigcirc\rangle-R$$

The preparation of substituted stilbenes obtained by virtue of the effect of nitro substituents on methyl groups of certain benzene derivatives is shown in 9.3.6.

### 9.2.2  Oxidation Products

The relatively high stability of aromatic rings to oxidation and to drastic thermal conditions (unless electron-donating substituents such as hydroxyl or amino functions are present) provides many opportunities for the conversion of alkyl side chains to carboxylic acids. The formation of oxidation products that contain functional groups at a lower oxidation state is discussed elsewhere (9.3.5).

The most common industrially important oxidation processes of aromatic carbocycles are as follows:

#### 9.2.2.1  Air Oxidation of Methyl Substituents

Xylenes + air at 350°C, atmospheric pressure, $V_2O_5$ catalyst:

o-isomer, 16.5¢ → phthalic anhydride, 1B lb, 21¢ until recently, 41¢ in January 1989

m-isomer, 36¢ → isophthalic acid, 46¢

p-isomer, 19¢ → terephthalic acid, purified, PTA, ~37.5¢

The relative unit prices of these three oxidation products reflect the economy of scale in the case of isophthalic acid, and the operation of competitive forces in the case of the other two benzene dicarboxylic acids. In particular, terephthalic acid is most valuable when purified of nonlinear impurities for use in the assembly of polyester and other polymer families (e.g., the aramids—see below) in which a linear structure optimizes crystallinity, and hence thermomechanical properties. For this reason, terephthalic acid is also produced as the dimethyl ester (DMT), which can be purified by fractional distillation. The production of DMT can be carried out by oxidation of p-xylene in the presence of acetic acid, methanol, and a cobalt catalyst, and the production–demand figures for the terephthalic acid system usually are reported jointly for PTA and DMT. The demand for PTA–DMT in the United States is nearly 7.0B lb, and over 50% of this amount is converted to PET textile fibers (4.3.6.1), while about 20% is used in the fabrication of blow-molded bottles and film. Smaller quantities are used for the production of PBT (6.7) and as building blocks of specialty high-performance thermoplastics such as the *aramids* (9.3.2)—condensation products of DMT and m- or p-phenylenediamines pioneered by Du Pont as its Nomex and Kevlar, respectively (Scheme 9.6). By-products of the large-

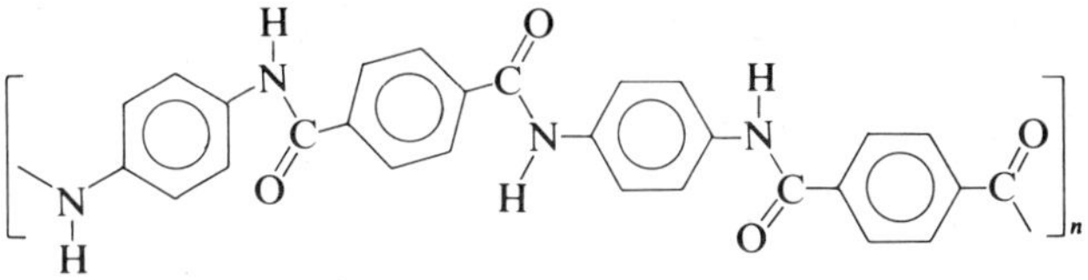

Kevlar

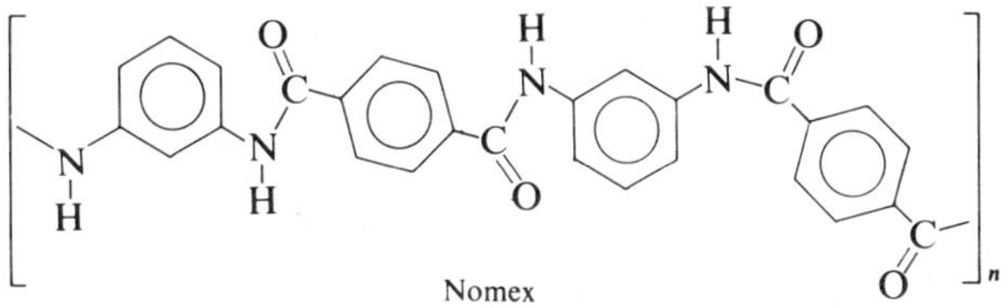

Nomex

**Scheme 9.6**

scale production of terephthalic acid are

- $p$-Toluic acid ($p$-CH$_3$–C$_6$H$_4$–CO$_2$H) and its methyl ester (46.5¢).
- Methyl paraformylbenzoate (MFB; $p$-MeO·CO–C$_6$H$_4$–CH=O).
- $p$-Carboxybenzaldehyde ($p$-CBA, $p$-HO$_2$C–C$_6$H$_4$–CH=O).
- A mixture of methyl benzoate and methyl $p$-tolylate ($p$-Me–C$_6$H$_4$–CO$_2$Me).

The partial oxidation of $m$-xylene yields $m$-toluic acid—the building block of the common mosquito repellent $N,N$-diethyltoluamide, (deet) while the partial oxidation of $o$-xylene with air at 160°C and 14 atm results in the formation of $o$-toluic acid. Phthalic anhydride (PA) originally obtained by oxidation of naphthalene, is currently derived principally from $o$-xylene. The demand for phthalic anhydride in the United States hovers around the 1B-lb mark. About 50% of PA is consumed in the production of plasticizers such as DOP (7.4.2) ($\sim$170MM lb), and about 25% and 20% is incorporated in UPRs and alkyd resins, respectively. Additional minor uses of PA are mentioned in 9.3.2. Isophthalic acid is employed mostly as a modifier of alkyds and UPRs, but it also participates in the assembly of the above-mentioned aramids in order to lower their crystallinity and, consequently, lower the processing temperatures. An oxidation on a scale smaller than that in the preceding cases is the conversion of toluene to about 170MM lb of benzoic acid (55¢, USP $1.73). One-half of this amount is still employed in the production of phenol (9.3.4), although the relatively low price (34¢ in January 1988 but 46¢ in January 1989) of the latter and the advantages of the Hock process are eroding the traditional routes to phenol. The remainder of benzoic acid is converted mainly to sodium benzoate (technical grade 70.5¢, USP grade 83.5¢), employed as an antifungal food and soft-drink preservative. The methylnaphthalenes are oxidized to the corresponding naphthoic acids (see 9.2).

### 9.2.2.2  Oxidative Degradation of Aromatic Hydrocarbons

The oxidative degradation of naphthalene and benzene (in that chronological order) to phthalic anhydride and maleic anhydride, respectively, requires, of course, more vigorous conditions than the preceding side-chain oxidations. In the case of naphthalene, the reaction is carried out at between 350 and 550°C by means of oxygen and the uniquely suitable catalyst V$_2$O$_5$—elaborated from technical-grade oxide priced at $3.35–$4.10

depending on the physical state. With the maturing of the petrochemicals industry, the production routes of first naphthalene and more recently benzene-derived maleic anhydride (6.6.1) began to be phased out by routes based on o-xylene and the $n$-$C_4$ building blocks (6.2.4, 6.3.2), respectively. The oxidative degradation of naphthalene also generates some MA by-product that can be separated from the gaseous effluents.

### 9.2.2.3   Oxidation of Benzylic Systems

The oxidation of benzylic hydrogens by means of molecular oxygen is promoted by metallic oxides such as chromic acid ($CrO_3$, $1.10) and is used extensively in the preparation of hydroperoxides from isopropyl-substituted aromatics such as cumene and diisopropyl compounds (9.2.1). The analogous oxidation of ethyl-substituted aromatics such as ethylbenzene produces rather unstable hydroperoxides that tend to decompose to the corresponding ketones as illustrated by the facil conversion of $m$-diethylbenzene (9.2.1) to the otherwise rather inaccessible $m$-diacetylbenzene; see Scheme 9.7. The partial oxidation of the primary hydrogens of methyl-substituted aromatics can be achieved by means of manganese dioxide [$MnO_2$; crude 84%, 12.5¢; synthetic chemical ferrite grade, 49¢; crystalline battery grade, 70¢), in the presence of acetic anhydride that protects the aldehyde intermediate from further oxidation. Thus, toluene can be converted to benzaldehyde, technical grade 73¢, NF $1.25, by way of benzal diacetate (Scheme 9.8). This procedure avoids traces of chlorinated by-products obtained in alternate routes (9.2.3) that are deleterious to the use of benzaldehyde ("oil of almond") as a fragrance or flavoring agent or as a building block of cinnamic and mandelic acids and derivatives (9.3.5) that may serve the same purpose. A methylene group flanked by two benzoid rings is oxidized to a ketone. Thus, diphenylmethane is oxidized by means of a solution of chromic acid in acetic acid to benzophenone (technical grade $2.40, NF crystals $2.50, NF flakes $3.65) used in perfumery and as a UV-light-absorbing ingredient in personal-care products. A good example of this situation is the oxidation of a structure in which two o-xylene molecules are bridged by a carbon group introduced by means of a condensation reaction of acetaldehyde (rather than formaldehyde; see 3.4.3.5) in order to achieve greater regioselectivity). The oxidation of the resulting hydrocarbon yields 3,3′,4,4′-benzo-phenonetetracarboxylic dianhydride, or Allco's BTDA ($5.5), the building block of polyimides and other specialty high-performance thermoplastics described in 9.3.2.3 (see Scheme 9.9).

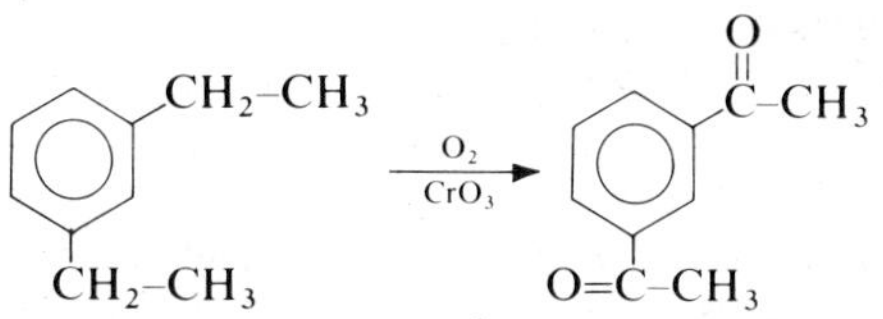

Scheme 9.7

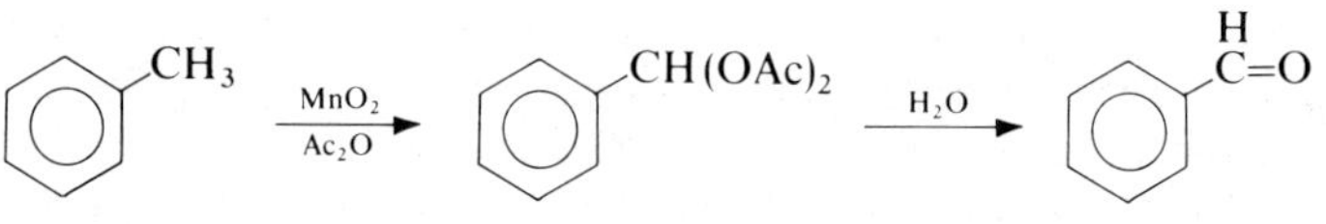

Scheme 9.8

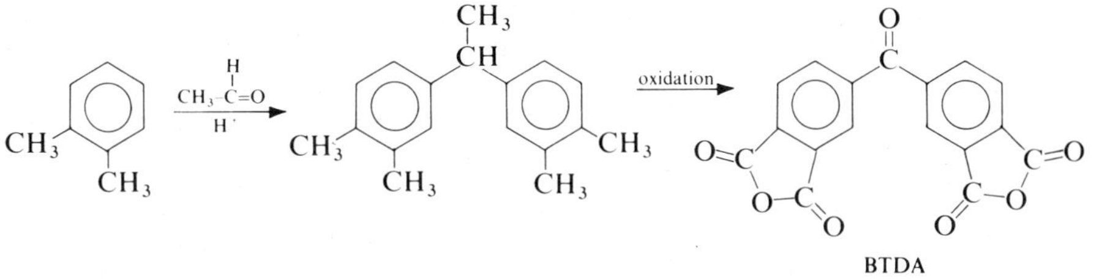

**Scheme 9.9**

### 9.2.2.4  Dehydrogenation of Aromatic Systems

Formally speaking, dehydrogenation is equivalent to oxidation, and the "hot tube reaction" of benzene to yield diphenyl or biphenyl is mentioned elsewhere (9.2). An eutectic mixture of diphenyl (b.p. 255°C) and diphenyl oxide (b.p. 259°C) is used as a heat-transfer medium known to consumers as Dow's Dowtherm. A mixture of isomeric terphenyls, and even higher analogs, is obtained as a by-product of the diphenyl production. Conversions of ethylbenzene to styrene, ethyltoluene to vinylbenzene or $p$-methylstyrene and cumene to $\alpha$-methylstyrene (9.2.1) are other examples of dehydrogenation reactions, as is the fusion of aromatic compounds to larger molecules during the synthesis of certain dyes (9.3.5).

### 9.2.2.5  Oxidation of Polynuclear Aromatic Systems

An increase in the number of rings in polynuclear aromatic carbocyclics facilitates their conversion to the corresponding quinones by means of chromic acid in an acetic acid medium or by means of a solution of sodium dichromate in sulfuric acid. This is illustrated in Scheme 9.10 by the formation of anthraquinone from anthracene with additional examples mentioned in 9.3.5.

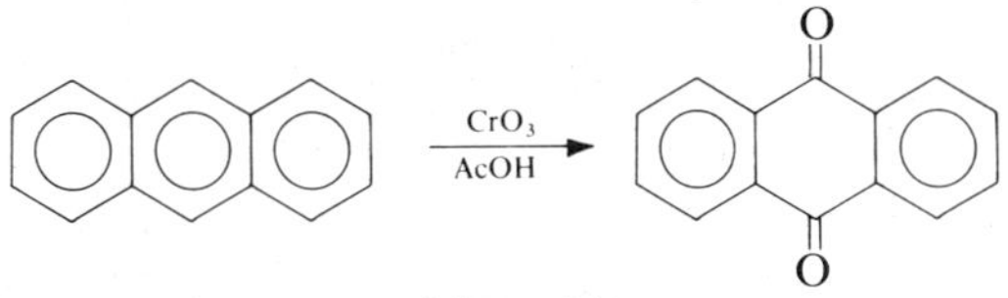

**Scheme 9.10**

### 9.2.2.6  Ammoxidation of Methyl-Substituted Aromatic Systems

The ammoxidation of methyl-substituted aromatic rings is mentioned in connection in with acrylonitrile (5.4.3.1) and in 8.3.1 in connection with the hydrogenation of the resulting benzonitriles to the corresponding amines on the way to nonaromatic iso-cyanates or to polymers of the nylon family. An illustration of the latter case is Mitsubishi's production of the semiaromatic Nylon MXD 6 derived from $m$-xylylene-diamine and adipic acid—a thermoplastic with exceptionally low oxygen and $CO_2$ permeability (Scheme 9.11).

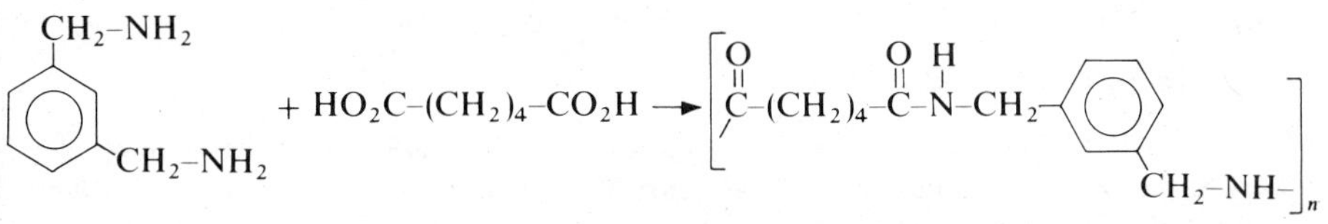

**Scheme 9.11**

### 9.2.2.7   Side-Chain Chlorination of Aromatic Systems

The side-chain chlorination of toluene (9.2.3) and similar systems, is also equivalent to an oxidation process as witnessed by the fact that water transforms the chlorinated products to the corresponding alcohols, carbonyl compounds, or carboxylic acids depending on the number of chlorine substituents available for hydrolysis.

### 9.2.2.8   Direct Hydroxylation of Aromatic Systems

The direct hydroxylation of aromatic C–H bonds to phenols is a highly desirable transformation that can be achieved by means of air or hydrogen peroxide in the presence of an EDTA chelate of iron. Other catalysts such as cuprous chloride and palladium or platinum chlorides can also be used especially in the case of aromatic carbocycles larger than benzene. A good yield in the oxidation of naphthalene to $\alpha$-naphthol in acetonitrile by means of oxygen in the presence of $Mo(CO)_6$ is reported by Metelitsa and collaborators in *Dokl. Akad. Nauk USSR* (Journal of the Soviet Academy of Science), **213**, 1079 (1973).

### 9.2.3   Chlorination Products

The introduction of halogen substituents other than chlorine is discussed elsewhere (9.3.1).

Benzene is chlorinated in the dark at about 40°C and in the presence of iron (that is converted *in situ* to ferric chloride) to give a whole spectrum of substitution products ranging from monochlorobenzene ($\sim$22MM lb, 42.5¢) to o- and p-dichlorobenzenes ($\sim$45MM and 80MM lb and 50¢ and 44¢, respectively) and to still smaller quantities of 1,2,4-trichlorobenzene (70¢), and 1,2,4,5-tetra-, penta-, and hexachlorobenzenes. The chlorination reaction can be carried out under oxychlorination conditions in order to recycle the HCl by-product.

Monochlorobenzene, or simply chlorobenzene (b.p. 132°C), is a rather inert and somewhat polar solvent used, for example, in the production and handling of TDI (interlude in Section 3.8) in the formulation of pesticides, and as a degreasing agent of metals during machining operations. These uses consume about one-third of its production. The major chemical uses of chlorobenzene are the production of o- and p-nitrochlorobenzene, ($\sim$75¢), which absorb 40% of the demand and the conversion to diphenyl ether or diphenyl oxide ($1.26 technical grade, $1.36 refined) and to o- and p-phenylphenols (Mobay, $2.64 and $1.85, respectively). o-Phenylphenol is used as an intermediate in the synthesis of dyes and as a germicide and fungicide.

In recent years, sulfonation of chlorobenzene by means of chlorosulfonic acid to yield primarily 4,4'-dichlorodiphenyl sulfone or di-p-chlorophenyl sulfone has become of interest. The chlorine-substituted sulfone and the analogous di-p-hydroxyphenyl sulfone (obtained by sulfonation of phenol), as well as other diphenols (e.g., BPA, p-hydroquinone or simply hydroquinone, and 2,6-dihydroxynaphthalene), are the building blocks of one of the families of *high-performance specialty thermoplastics* (HPSTs).

*High-Performance Specialty Thermoplastics*

Generally, the HPSTs are priced at $4 or higher and are produced at a relatively low level (a few to a dozen or so million pounds per year). Those derived from sulfones such as those mentioned above are referred to as "polysulfones," or more appropriately, as "poly(aryl ether sulfones)" (PES). The assembly of these materials is

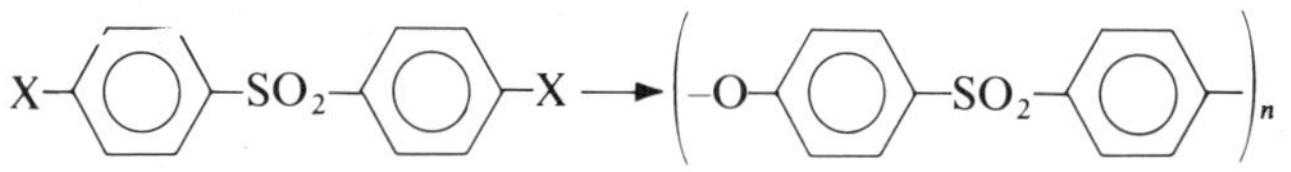

(where X = Cl and/or OH and the product is a PES, a member of ICI's Victrex family of HPSTs).

A similar PES is obtained by a reaction of 4,4'-substituted diphenyl sulfone with the anion of BPA to produce Union Carbide's Udel:

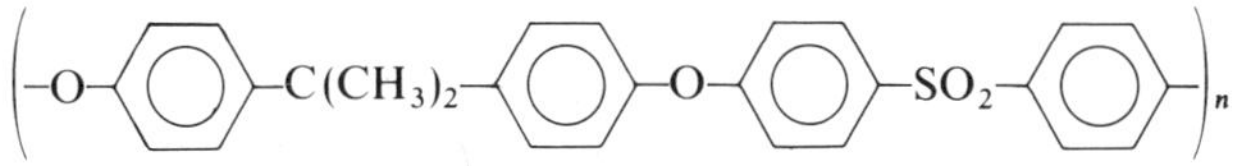

This material is priced at about $4.00 and recent has been produced at a level of 15MM lb.

The electron-withdrawing sulfone group (unlike the nitro group) exerts a relatively weak activating effect in nucleophilic substitution reactions of aromatic halogen compounds and in that sense resembles a ketone function. For this reason, and in order to relate the above-mentioned member of ICI's Victrex with other members derived from aromatic ketones, we introduce here the structures of poly(etherether ketone), (PEEK) and poly(ether ketone) (PEK):

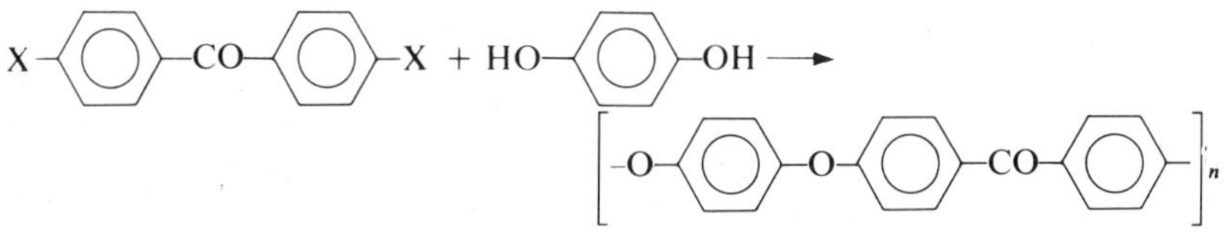

(where X represents chlorine substituents).

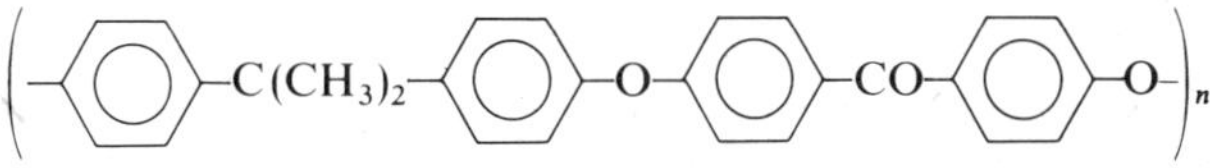

All the preceding polymerizations require the use of dipolar aprotic solvents and elevated temperatures in order to accelerate the otherwise reluctant nucleophilic displacements of halogen, and a glimpse at the reaction conditions is revealed by the following announcement (*CW* 12/2/87). The $14MM ICI plant near Blackpool, England has an annual capacity of 2.2MM lb of PEEK and PEK, and the condensation of dichlorobenzophenone and hydroquinone is carried out at 250–300°C in an unidentified, costly, high-boiling dipolar aprotic solvent in the presence of a metallic catalyst. The key to commercial success is said to depend on the recovery of the solvent and the removal of metallic impurities from the product to a concentration smaller than 50 ppm.

Variations of the preceding theme that permit a control of molecular weights and the linearity (i.e., crystallinity and processing characteristics) of the products, as well as some other thermomechanical properties, are the uses of (1) diphenols other than hydroquinone and BPA, namely, 4,4'-dihydroxydiphenyl or *p,p'*-diphenol, 2,6-dihydroxynaphthalene, and so on; (2) meta- rather than para-substituted halogen-containing sulfones or ketones; and (3) building blocks that contain both halogen and phenolic substituents on adjoining benzene rings separated by sulfone or ketone functions. The formation of poly(phenylene sulfide) is shown below and examples of other families of HPSTs are mentioned elsewhere (9.3.2).

The reactivity of 4,4'-dichlorodiphenyl sulfone in nucleophilic substitution reactions has been utilized for some time to prepare 4,4'-diaminodiphenyl sulfone [bis-(4-amino-phenyl)sulfone, or dapsone], a leprostatic medicinal. The latter can be also obtained by chlorosulfonation of acetanilide (technical grade $1.29), followed by hydrolysis.

Except for the formation of diphenyl ether and the phenylphenols, the conversion of monochlorobenzene to the remaining aforementioned chemicals follows common behavior patterns of benzenoid compounds in electrophilic substitution reactions. The exceptional case is an outgrowth of one of the traditional processes for the preparation of phenol: a vigorous alkaline "hydrolysis" of chlorobenzene to give sodium phenolate or phenoxide carried out at temperatures as high as 390°C, under pressures as high as 300 atm, and by means of caustic soda. As we understand the chemistry today, such extreme conditions produce, at least in part, a transient dehydrohalogenation intermediate known as benzyne that, while pictured here in a simplistic manner as a triple-bond-containing benzene ring, accounts for all the products of the reaction (see Scheme 9.12). Let us note in passing that benzyne, and substituted benzynes, can be generated by other means and can be trapped to give derivatives other than those shown in Scheme 9.12. However, this chemistry is beyond the scope of this book.

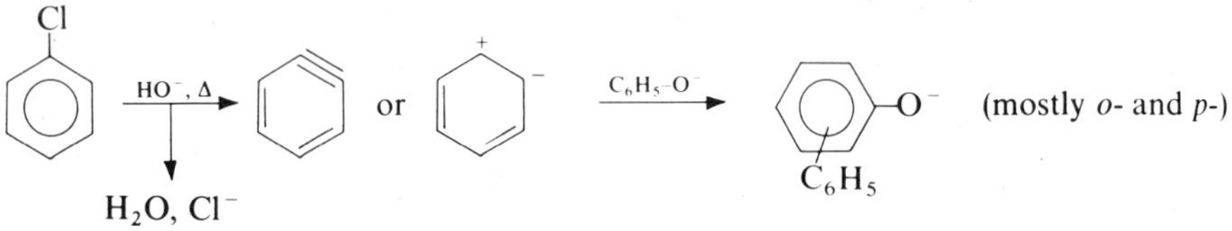

**Scheme 9.12**

Mixtures of diphenyl oxide and diphenyl are used as heat-transfer fluids distributed by Dow as Dowtherms ($14.92–$21.92/gal).

*p-Dichlorobenzene*, or "para," is a volatile solid known to many consumers as a substitute for naphthalene-based moth repellent and deodorant (urinals) and, until Dow abandoned the practice of benzene chlorination, also as "Paradow." Currently, one-third of the demand for para is still employed as a space deodorant and for control of moths, but, chemically speaking, more interesting is the 30% used in the production of *poly(phenylene sulfide)* (PPS; $3.30–$3.86) and another member of the HPSTs. PPS is assembled by means of a nucleophilic substitution of the chlorines by the sulfide ions of sodium sulfide (24¢). Again, as in the preceding cases the polysulfones and poly(ether ketones), the replacement of the reluctant chlorine substituents is facilitated by the use of dipolar aprotic solvents, such as DMF (3.5.3.1). In favor of these nucleophilic substitution reactions is the rather high nucleophilicity of sulfide and arylmercapto anions. PPS was introduced in 1973 by Phillips under its tradename "Ryton" at an initial production level of about 9MM lb. Since then, PPS has also been produced in the United States by Celanese (*MP* 9/86), Toyo Soda, and Phillips and Toray are collaborating on PPS project in Japan. A fiberglass-reinforced and mineral-filled Phillips product known as "Ryton R-7" is now selling for $1.45/lb (*MP* 11/86).

$o$-Dichlorobenzene is a rather inert, somewhat polar solvent (b.p. 180°C), also used for the handling TDI as well as for the removal of inks, paints, cleaning of engines, and so on. Its major chemical use is nitration to give 3,4-dichloronitrobenzene—the intermediate in the synthesis of herbicides derived from the corresponding 3,4-dichloroaniline and used in the cultivation of rice and other crops.

The chlorination of benzenoid compounds under conditions that generate chlorine radicals (atomic chlorine), namely, elevated temperatures and/or photochemical activation, either saturates the aromatic ring (see benzene hexachloride, 8.3.1) or leads to substitution of benzylic hydrogens if the latter are available. On a large scale, side-chain chlorination produces significant amounts of both side-chain and ring-substitution products. Some transformations of these materials are shown in Fig. 9.3, which appears later in this chapter (p. 551).

Benzyl chloride (70MM lb, 54¢) is an important building block of plasticizers such as $n$-butyl benzyl phthalate (59¢), a plasticizer used commonly in the production of vinyl flooring and other structural materials and benzyl quaternary ammonium salts such as $N$-benzyl-$N,N$-dimethylamine ($2.30, 0.35MM lb; see below), $N$-benzyl-$N,N,N$-tri-$n$-butylammonium chloride—a phase-transfer agent, and $N$-benzyl-$N,N$-dimethyl-$N$-alkyl-ammonium chloride (benzalkonium chloride)—a topical antiseptic and germicide (when the alkyl group contains 8–18 carbon atoms) commonly known as Winthrop's "Zephiran chloride." About 10% of the benzyl chloride demand is channeled into the production of quaternary ammonium compounds. Also, benzyl chloride is the precursor of benzyl alcohol (technical grade $1.26), which, in turn, is converted to esters such as benzyl acetate ($1.00), a valuable ingredient of gardenia and jasmine fragrances (see 9.2.1), benzyl benzoate ($1.65, 1.5MM lb, used as a miticide), benzyl cinnamate ($8.50), benzyl formate ($10.50), benzyl propionate ($3.35), benzyl salicylate ($2.90, a perfume fixative and sunscreen ingredient), and other fine chemicals used by the fragrance and flavor industry.

Some of the benzyl esters may be obtained more economically from benzyl chloride and the sodium salts of the appropriate carboxylic acids, especially when the ester is to be used as a plasticizer. As mentioned in 9.2.1, in recent years the United States was inundated by imports of both benzyl alcohol and benzyl acetate, not only from Mexico and Italy, but also from England, Spain, and the People's Republic of China. In 1986 a customs duty of 30% was imposed on the acetate except that imported from Mexico, which enjoyed a GSP status and exported duty-free about 0.5MM lb.

The demand for benzyl alcohol has increased in recent years because of its use in the synthesis of phenylalanine for the production of aspartame.

The reaction of benzyl chloride with sodium sulfide gives the expected dibenzyl sulfide and oxidation of the latter with dilute nitric acid or hydrogen peroxide stops at the level of dibenzyl sulfoxide [$(C_6H_5-CH_2)_2SO$], which turns out to be a corrosion inhibitor during "metal pickling" operations, that is, the cleaning of metallic surfaces by acidic solutions.

The conversion of benzyl chloride to benzyl cyanide is the route to $\beta$-phenylalanine ($1.50) and phenylacetic acid (pure, $4.50). The latter reacts with phenethyl alcohol (NF $2.65) to produce phenethyl phenylacetate ($5.50). In this context it is noteworthy that phenethyl alcohol can be prepared by a Friedel–Crafts alkylation of benzene with EO:

$$C_6H_6 + EO(+AlCl_3 \text{ at } 5°C) \rightarrow C_6H_5-CH_2-CH_2-OH$$

For use as the "oil of roses" fragrance, phenethyl alcohol may be prepared from the Grignard reagent, $C_6H_5-MgBr$ and EO, or by means of the Bouveault–Blanc reduction

(*ONR*-14) of an ester of phenylacetic acid:

$$C_6H_5-CH_2-CO_2Et + 3EtOH + 4Na \rightarrow C_6H_5-CH_2-CH_2-ONa + 3EtONa$$

A useful derivative of phenethyl alcohol is phenethyl acetate ($3.35).

Benzal chloride can be hydrolyzed to benzaldehyde, although the latter is produced by direct oxidation of toluene (9.2.2).

Benzotrichloride (80¢) and analogous trichloromethyl derivatives of aromatic compounds play a special role among industrial organic chemicals: they serve as precursors of the corresponding trifluoromethyl-substituted chemicals that benefit from the properties induced by the presence of the $-CF_3$ substituents (compare 3.2.3 and *SZM-I*, 205–212). These properties are:

1. Stabilization of the whole molecules toward oxidative degradation (thermal and biological).
2. Increased volatility.
3. Decreased solubility in both aqueous and nonaqueous media (as compared with analogous structures that contain the $-CH_3$ group).
4. Activation of halogen substituents to nucleophilic substitutions.

Some of these properties favor the presence of $-CF_3$ substituents in medicinals, pesticides, dyes, surfactants, and so on.

The conversion of trichloromethyl to trifluoromethyl substituents is achieved by anhydrous HF (69¢) according to the Simon procedure or by means of the Swarts reaction (*ONR*-88), which employs $SbF_3$ (b.p. 376°C), together with some antimony pentachloride (Sb metal is priced at $1.35).

Examples of three herbicides that contain a trifluoromethyl groups are

- Fluometuron, or [*N*-(3-trifluoromethylphenyl)-*N'*,*N'*-dimethylurea, or Ciba-Geigy's Cotoran], prepared by nitration of benzotrifluoride, reduction of the resulting *m*-nitrobenzotrifluoride, and the reaction of *m*-trifluoromethylaniline with methyl *N*,*N*-dimethylcarbamate (3.8.2):

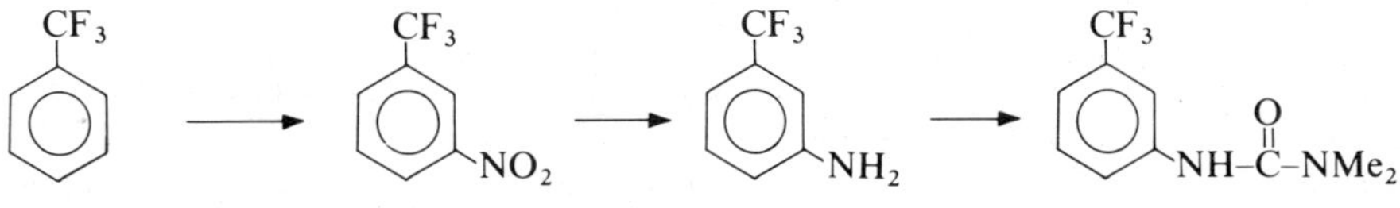

- Acifluorfen [or 5-[2-chloro-4(trifluoromethyl)-phenoxy]-2-nitrobenzoic acid, or Rohm & Hass's Blazer acquired recently by BASF; *CMR* 2/16/87, *CW* 2/16/76], assembled from the sodium salt of 2-chloro-4-trifluoromethylphenol and sodium 2-nitro-4-chlorobenzoate (9.3):

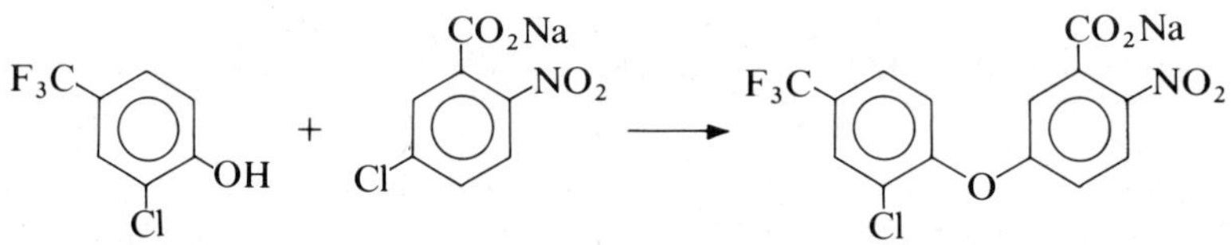

while 2-chloro-4-trifluoromethylphenol is obtained from *p*-cresol by controlled light catalyzed chlorination and the treatment of 2-chloro-4-trichloromethylphenol with HF.

- Trifluralin [or *N,N*-di-*n*-propyl-2,6-dinitro-4-trifluoromethylaniline, or Eli Lilly's Treflan], prepared by side chain chlorination of *p*-chlorotoluene, treatment of the corresponding trichloromethyl compound with HF, nitration and, finally, reaction of 2,6-dinitro-4-trifluoromethylchlorobenzene with di-*n*-propylamine:

The 1981 sales of Treflan herbicide for use in the cultivation of soybean and cotton totaled 41.8MM lb, valued at $375 million.

An example of the effect of the trifluoromethyl group on physiological properties is the trifluoromethyl derivative of aspirin, (acetyl 4-trifluoromethylsalicylic acid, or triflusal). This compound can be assembled from 3-trifluoromethylphenol by means of the Kolbe reaction in a manner analogous to the synthesis of salicylic acid from the unsubstituted phenol (9.3.4), followed by acetylation with acetic anhydride. Trifusal inhibits the aggregation of platelets and hence acts as an antithrombotic. 3-Trifluoromethylphenol is prepared from benzotrifluoride by nitration, reduction to the corresponding aniline, diazotization, and decomposition of the diazonium salt to the desired phenol (9.3.6).

Another of many trifluoromethyl-group-containing medicinals is trifluoperazine, a derivative of the phenothiazine ring system shown elsewhere (10.3.2) but assembled from 3-trifluoromethylaniline.

1,4-Bis-trichloromethylbenzene is obtained by side-chain chlorination of *p*-xylene and is used as an ingredient of flame-retardant polymers.

The light-catalyzed chlorination of diphenylmethane is the first step in the assembly of diphenhydramine, or *β*-dimethylaminoethyl benzhydryl ether, known to consumers as Parke-Davis's "Benadryl," the first antihistaminic to appear on the marketplace 20 years ago (see Scheme 9.92, 9.3.5.2). Since the Parke-Davis original patents have expired, diphenhydramine is now distributed as an OTC anti-allergy medication under a variety of other tradenames.

The multichlorination products of diphenyl and terphenyl known as "polychlorinated biphenyls" (PCBs) were used since the 1930s as transformer oils, hydraulic fluids, heat-transfer media, fire retardants, and so on at an annual rate as high as 72MM lb. Then, in 1966, thanks to the development of gas chromatography, which, at this particular period in time enabled their detection in ppm concentrations, it was discovered that they were accumulating in the environment because of resistance to biodegradation, and they became suspect carcinogens on the basis of some animal experimentation. The production and use of PCBs was banned by the Toxic Substances Control Act of 1976 (TSCA) and, on one hand, a wide-scale industry has developed to safely destroy the PCBs in existence, and, on the other hand, to replace existing transformer oils by safer products such as the polysiloxanes (3.3.2.6.B). It is estimated that over 10MM gal of such silicone fluids have been used for this purpose since 1974. Other PCB replacements are mineral oils and synthetic high-molecular-weight hydrocarbons (M. Reisch, *CEN* 6/29/87). The variety of methods of decomposing PCBs to harmless end products includes incineration of contaminated soils and a search of strains of microorganisms capable of metabolizing the material (*CE* 1/19/87, *CW* 4/1/87). The problem is also being pursued in Europe, and steam cracking at 1500°C is claimed by a French group to be effective (*CE* 2/16/87). An encouraging study by General Electric (*CMR* 6/5/87) indicates a gradual degradation of PCB spilled into the Hudson River and into Silver Lake, Pittsfield, Massachusetts by naturally occurring microorganisms (also see P. Zurer, "Researchers confirm bio-degradability of PCBs", *CEN* 11/21/88, p. 5). In the meantime, however, many millions of dollars are being spent for numerous cleanup projects designed to eliminate PCBs from the environment without converting them to dioxins (9.3.4) presumed to be harmful to humans.

### 9.2.4   Sulfonation Products

The introduction of the sulfonic acid (sulfo) group into an aromatic ring can be carried out by means of 95% sulfuric acid, ($\sim 3.6$¢–4.25¢, depending on location), fuming sulfuric acid ($SO_3$–$H_2SO_4$; priced $\sim 2$¢ higher than 95% sulfuric acid), chlorosulfonic acid ($ClSO_3H$, 18.5¢), and sulfur trioxide ($SO_3$—the anhydride of sulfuric acid). These sulfonating agents are listed in approximately ascending order of reactivity and are chosen, in part, to match the ease with which the benzenoid substrate undergoes electrophilic substitution reactions. Thus, the sulfonation of nitrobenzene, benzotrifluoride, benzenesulfonic acid, chlorobenzene, benzene, toluene, and the xylenes, for example, will require (in the order that these aromatic substrates are listed) less reactive sulfonating agents to bring about a sulfonation under otherwise equivalent reaction conditions.

The use of chlorosulfonic acid as a sulfonating agent is usually reserved for situations that call for the preparation of sulfonic acid esters or amides or for the direct formation of sulfones (see the uses of 4,4′-dichloro and 4,4′-dihydroxy derivatives of diphenyl sulfone described in 9.2.3). Thus, in the case of toluene, the reaction with chlorosulfonic acid leads to the formation of a mixture of o- and p-toluenesulfonyl chloride. The latter ($1.65) produces the plasticizer p-toluenesulfonamide ($3.55) on treatment with ammonia;

$$CH_3-\langle\bigcirc\rangle-SO_2-Cl + NH_3 \longrightarrow CH_3-\langle\bigcirc\rangle-SO_2-NH_2$$

Similarly, o- and p-toluenesulfonyl chloride can be treated with ethylamine or cyclohexyl-amine to produce the plasticizers o- or p-N-ethyltoluenesulfonamides and o- and p-N-cyclohexyltoluenesulfonamides, respectively.

p-Toluenesulfonyl chloride can be converted to the corresponding esters or tosylates, such as methyl tosylate ($p$-$CH_3$–$C_6H_4$–$SO_2$–$OCH_3$, also written as Ts–O–$CH_3$). The tosylates possess an excellent leaving group in the form of the toluenesulfonate anion, and for this reason they have played an important role in academic organic chemistry, but their use can be extended to transformations of industrial importance. Thus, for example, the tosylate of β-naphthol can be converted to ethyl 2-naphthyl ether, or nerolin, an ingredient of perfumes obtained in a fashion that avoids traces of halide impurities:

$$\langle naphthyl\rangle-O-Ts + EtO^- Na^+ \longrightarrow \langle naphthyl\rangle-O-Et + TsO^- Na^+$$

The conversion of $o$-toluenesulfonyl chloride to saccharin is discussed elsewhere (9.3.3).

In the presence of an excess of toluene and of some aluminum chloride, $p$-toluenesulfonyl chloride gives di-$p$-tolyl sulfone $[(p\text{-}CH_3\text{-}C_6H_5)SO_2]$. As indicated above, the reaction of chlorobenzene with chlorosulfonic acid can be carried all the way to the formation of di-$p$-chlorophenyl sulfone, the building block of certain HPSTs. In the case of aromatic substrates that are highly reactive toward electrophilic substitution reactions (e.g., phenols or substituted anilines), sulfone formation may occur even during sulfonation by means of sulfuric acid in the course of preparation of sulfonic acids unless the reaction conditions are carefully controlled. Additional examples of sulfone formation are shown in 9.3.3.

*Hydrotropes*

Sulfonic acids are rather strong acids, and their alkali metal salts constitute the highly hydrophilic group of (linear) alkylbenzenesulfonate (LAS) anionic surfactants. In the absence of the relatively large lipophilic alkyl groups, the alkali metal (or ammonium) salts of benzenesulfonic, toluenesulfonic, xylenesulfonic, and analogous compounds function as *hydrotropes*. Hydrotropes are water-soluble chemicals that increase the solubility in water of otherwise rather water insoluble organic materials, especially when the aqueous medium contains a relatively high concentration of alkaline salts such as phosphates, carbonates, and so on. Thus hydrotropes aid the dispersion of oily materials in aqueous solutions without the use of organic cosolvents. Commonly employed hydrotropes are, for example, the sodium or ammonium salts of (mixed) xylenesulfonic acid (40% solution, Witco 29.5¢ or 93% sodium salt powder, Ruetgers-Nease 70¢, and sodium cumenesulfonate, 90% powder, 70¢ Witco, or 93% powder 70% Ruetger-Nease). Hydrotropes are important ingredients of industrial and institutional cleaning formulations. A recent addition to the wide selection of hydrotropes is the disulfonate of dinonylnaphthalene, King Industries' Nacorr 3800 ($1.10); see Scheme 9.13.

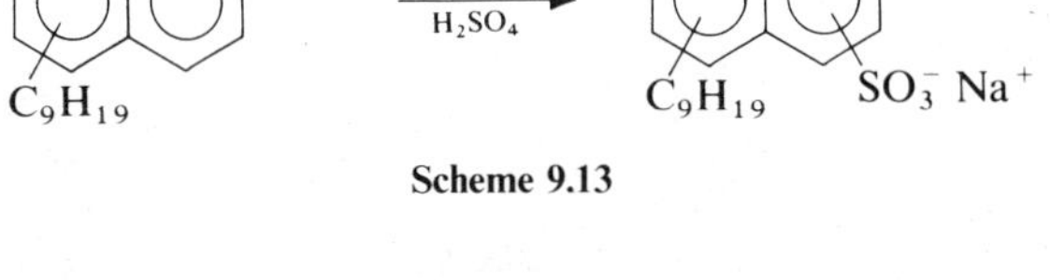

**Scheme 9.13**

Aromatic sulfonic acids per se are useful acidic catalysts in reactions such as esterification, hydrolysis of carboxylic acid derivatives, etherification of alcohols, and other acid-requiring transformations. Among the latter we must include the curing of PF, MF, UF, and of furan-based foundry resins (10.2.1). They are also employed as electrolytes in electroplating operations and may be used in place of sulfuric acid under circumstances where the latter may cause charring of the reaction mixture. These reasons explain the presence on the marketplace of compounds such as benzenesulfonic acid (90%, 42¢), toluenesulfonic acid (anhydrous 92% powder, 45¢ Cities Service), ethylbenzenesulfonic acid (64.5¢ Cities Service), chlorobenzenesulfonic acid (93%, $1.25, Jim Walter), phenolsulfonic acid (65%, 58¢, Jim Walter), and naphthalenesulfonic acids.

Also, we must keep in mind the availability of strongly acidic ion-exchange resins obtained by sulfonation of cross-linked polystyrene.

In the pricing of the above-mentioned nonpolymeric sulfonic acids we must consider the difficulties involved in the isolation of these highly water-soluble materials and separation from residual sulfuric acid. After the latter is removed by precipitation and filtration of calcium sulfate, addition of soda ash to the solution of the rather water soluble calcium salts of sulfonic acids precipitates calcium carbonate, and the free sulfonic acids are obtained (in solution) by means of ion-exchange resins.

Sulfonic acids, or more properly stated, the sodium salts obtained by means of soda ash (58%, 4.15¢–6.15¢, depending on physical state) are the traditional intermediates in the production of phenols (9.3.3, 9.3.4). Thus, fusion of the sodium benzenesulfonate with caustic soda (solid 76%, 25¢) produces sodium phenolate that must be acidified to obtain phenol (34¢) and a similar procedure applied to the disulfonation product of benzene, namely, sodium $m$-benzenedisulfonate, produces resorcinol (technical grade \$1.80, USP grade \$4.25–\$4.50, depending on physical state). In the case of phenol, this traditional energy-intensive route is definitely displaced by the Hock process (2.4.4.3, 5.4.6, 9.2.1), while resorcinol is now also prepared from the diisopropylation product of benzene (9.2.1). The sulfonation of naphthalene—an intermediate process en route to producing the naphthols, which are important building blocks of dyes—is interesting because while the initial sulfonation product obtained at 0°C is naphthalene–$\alpha$-sulfonic acid, the latter rearranges on heating in the presence of sulfuric acid to the isomeric naphthalene–$\beta$-sulfonic acid (Jacobsen rearrangement; $ONR$-48). However, either sulfonic acid can be converted to the corresponding $\alpha$- and $\beta$-naphthols (\$1.81 and \$1.20, respectively), as shown in Scheme 9.14.

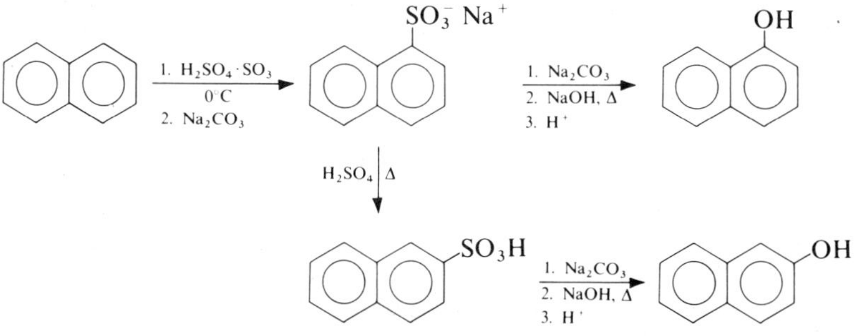

**Scheme 9.14**

Sulfonic acid groups are introduced in many molecules that possess color-producing moieties, (i.e., *chromophores*) to convert these structures into dyes, that is, molecules that have an affinity for the substrate subjected to dyeing operations. Thus, for example, Hoechst's sodium 2,3-dihydroxynaphthalene-6-sulfonate (\$9.75) is known as "Coupler 1" because it can be coupled with different diazonium salts (see 9.3.6) to produce a series of azo dyes. In this case we have an "acidic dye" by virtue of the presence of the sulfonic acid, or sulfo, group, and such a dye will be particularly suitable for the dyeing of nitrogen-containing substrates such as wool, silk, and nylons.

### 9.2.5  Nitration Products

Relatively few nitro derivatives of benzenoid structures are end products of the nitration process. Exceptions to this statement are found mostly among explosives, and, strangely, synthetic musklike perfume ingredients. The latter are known as "synthetic musks" because their scents imitate the natural products isolated from the glands of the male deer, the civet cat, and certain other animals. While the natural musks are macrocyclic rings containing about 15 carbon atoms, the synthetic materials deliver a similar, supposedly erotic odor by way of highly substitued nitro-group-containing compounds derived from benzene, *m*-xylene, pseudocumene, and *m*-cresol (see 9.3.6).

First, let us look at the high-volume nitroaromatics that are steppingstones to the corresponding reduction products, although this is no longer an exclusive route in the case of aniline (9.3.1) since the latter is now also produced from phenol (9.3.4).

Of the nearly 1B-lb demand for nitrobenzene, most of it is converted to aniline—the feedstock for the production of MDI and PMDI. The remainder is channeled into rubber chemicals, dyes, and other fine chemicals. Although the use of aniline in the production of hydroquinone has disappeared recently, fortunately for the nitrobenzene production capacity, the demand for MDI–PMDI continues to grow steadily.

Nitration, like sulfonation (9.2.4), can be carried out with nitric acid systems of increasing nitrating vigor. The latter can be controlled in part by varying the concentration of sulfuric acid, which is usually an ingredient of the nitration mixture. The role of sulfuric acid is to control the concentration of nitric acid and to catalyze the formation of the nitronium ion, $O_2\overset{+}{N}$. If the sulfonic acid group, or sulfo group, should substitute first a hydrogen of the aromatic ring, it becomes eventually displaced by the irreversible entry of the nitro group. Nitrobenzene can be obtained (*FKC*, 571–574) at 50–55°C with a mixture that contains about 32–39% nitric acid and 53–60% sulfuric acid. The concentration of the acids, as well as the temperature, can be raised for procedures involving less reactive benzenoid systems or dinitro products. The use of dilute nitric acid is not advisable for nitration unless the procedure in question involves very reactive benzeneoid substrates since dilute nitric acid tends to function as an oxidizing rather than as a nitrating agent. Nitration of benzene beyond the *m*-dinitro derivative is very difficult because of the deactivating effect of the two nitro substituents, and actually it is more practical to nitrate toluene to the military explosive 2,4,6-trinitrotoluene (TNT), oxidize this compound to the corresponding 2,4,6-trinitrobenzoic acid, and decarboxylate the latter to 1,3,5-trinitrobenzene. Clearly, the presence of the electron-releasing methyl substituent facilitates the triple nitration. A very powerful but dangerous nitrating agent is acetyl nitrate ($CH_3$–CO–O–$NO_2$), obtained *in situ* by mixing (in the cold!) concentrated nitric acid and acetic anhydride. It usually explodes at about 60°C and is used on a small scale for the preparation of fine chemicals in which the nitro group is being introduced in sterically hindered (ortho) positions.

Controlled nitration of toluene produces *o*- and *p*-nitrotoluene (40¢ and 60¢, respectively), and subsequent nitration of the mixture of the isomeric mononitrotoluenes gives a 80:20 mixture of 2,4- and 2,6-dinitrotoluene (30¢). Since most of these dinitrotoluenes are hydrogenated to the corresponding diaminotoluenes for conversion into TDI (3.8.1, but see also 3.8.4), and since the 2,6-diisocyanates of toluene react sluggishly during PUR formation, the price of pure 2,4-dinitrotoluene ($1.20) is considerably higher than the above-mentioned mixture of isomers. Pure 2,4-dinitrotoluene is obtained by nitration of 4-nitrotoluene (b.p. 238°C, m.p. 54°C), which can be separated from its *o*-isomer (b.p. 222°C, m.p. −10°C) by either fractionation and/or crystallization.

The nitration of naphthalene occurs preferentially in the 1- or alpha position, and the second nitro group enters in the alpha position of the adjoining benzenoid ring to yield 1,5- and 1,8-dinitronaphthalenes. The corresponding reduction products are mentioned in Section 9.3.6.

Chlorobenzene is nitrated to produce 2- and 4-nitrochlorobenzene, priced at about 74¢. The corresponding anilines, 2-chloroaniline ($1.55) and 4-chloroaniline ($1.70–$2.00, depending on the physical state), can be prepared by reduction of the nitro substituents. However, the reduction must avoid conditions that affect the chlorine group in view of the fact that halogen substituents are activated toward nucleophilic substitutions by nitro groups located at ortho and para positions. This behavior makes the o- and p-chlorine-containing nitro compounds important intermediates for the synthesis of a variety of derivatives of the nitroaromatics (9.3.6).

The nitration–oxidation of N,N-dimethylaniline ($1.03) produces N-methyl-N,2,4,6-tetranitroaniline (nitramine or tetryl, m.p. 130–132°C). *Caution*: material explodes at about 180°C.

Some additional functional products that either contain nitro substituents or require nitro group-containing intermediates for their synthesis are shown in Sections of 9.3.1–9.3.6.

## 9.3 PREPARATION AND USES OF REPRESENTATIVE BENZENOID CHEMICALS OF INDUSTRIAL IMPORTANCE

### 9.3.1 Benzenoid Halogen-Containing Chemicals

#### 9.3.1.1 Chlorine Compounds

Even though a chlorine substituent of a simple benzenoid structure resists nucleophilic substitution reactions, chlorobenzene was for a long time the building block for the industrial production of both phenol and aniline. The first-mentioned product is obtained by heating chlorobenzene at elevated temperatures ($\leqslant 450°C$) with aqueous alkali (under pressure $\leqslant 300$ bar) in the presence of various catalysts (e.g., copper phosphate). Aniline is obtained under similar forceful conditions (e.g., heating at 220°C with ammonia in the presence of cuprous catalysts). Phenol is also obtained from benzene by means of an oxidative chlorination (the Raschig phenol process; $ONR$-74) that uses hydrochloric acid, with oxygen at 230°C in the presence of cupric and ferric chlorides, but this route requires extensive recycling of the reactants because of low conversions. Actually the oldest source of phenol is the caustic fusion of sodium benzenesulfonate carried out in iron reactors because of the corrosive nature of molten caustic soda (m.p. 318°C). Since 1954 the Hock process applied to cumene is the predominant source of phenol (and of the coproduct acetone), and the reduction of nitrobenzene (9.2.5) as a source of aniline was always a strong competitor of chlorobenzene.

The decline of the role of monochlorobenzene as a precursor of phenol caused the production level of monochlorobenzene and its unit cost to rise and at current prices (48.5¢), chlorobenzene is no longer an economically attractive starting material for the production of phenol (46¢) as compared to the use of cumene (20¢), which also yields acetone (27¢). In recent years, aniline (62¢), also found a new precursor in phenol (9.3.4), a process that is strongly competitive with the traditional reduction of nitrobenzene (33¢).

The forceful alkaline hydrolysis of chlorobenzene is still employed for the production of the by-products isolated from the phenol process introduced by Dow 50 years ago, (viz., *o*- and *p*-phenylphenols, $2.64 (as of January 1989) and $1.85, respectively), as well as diphenyl ether or oxide (technical grade $1.24). *o*-Phenylphenol (OPP) is currently the subject of a controversy between Dow and the EPA (*CW* 11/11/87): Dow's petition that its "Dowicide" be dropped from a toxic emissions inventory under the "Right-to-Know" section of the Superfund law was denied by EPA, which claims that there is sufficient "evidence of potential carcinogenicity, developmental toxicity and environmental toxicity persistence." OPP has been used for several decades as an industrial antimicrobial in household, institutional, and industrial cleansers as a preservative of starch-based adhesives and a biocide of readily spoiled oranges during shipments. Both phenylphenols are intermediates for the preparation of certain dyes.

4,4-Biphenol, or 4,4′-dihydroxybiphenyl, is rising in importance as the building block of thermoplastic polyesters known as *polyarylates* (9.3.2.2), in which the diphenol, together with some other phenolic components, replaces EG in the reaction with PTA. Schenectady Chemicals was expected to become the first domestic source of 4,4′-biphenol during 1988 (*CMR* 9/28/87). Most likely, this material is obtained by means of one of the traditional phenol processes, such as the caustic fusion of the corresponding disulfonate salt of biphenyl. The desired biphenol can also be obtained by means of the hydrolytic decomposition of the bisdiazonium salt of benzidine (9.3.6), although the latter is a potent carcinogen.

The presence of a chlorine substituent deactivates somewhat the benzene ring toward electrophilic reagents, but when the reaction is achieved under conditions more forceful

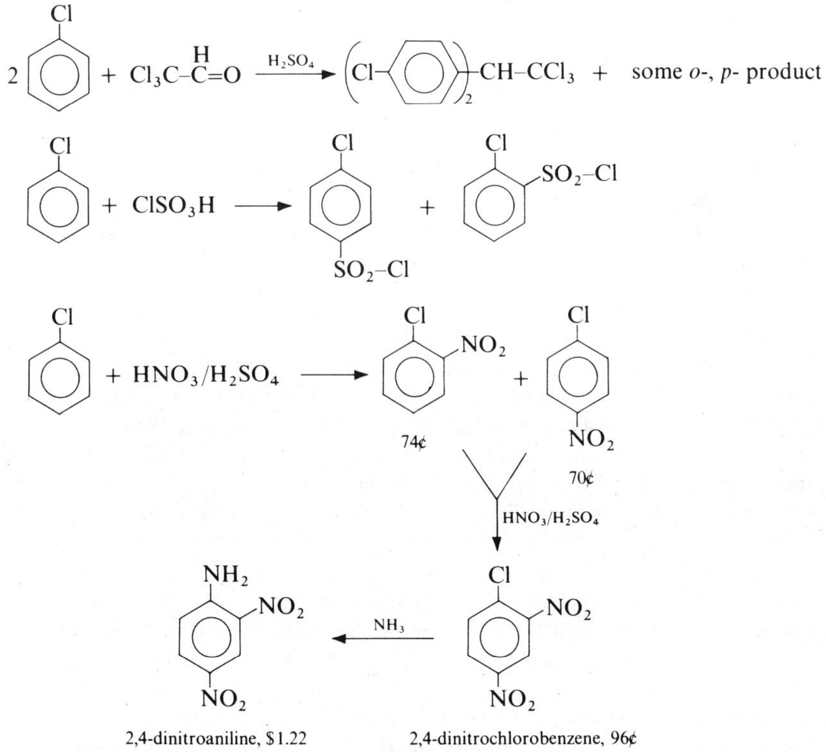

2,4-dinitroaniline, $1.22          2,4-dinitrochlorobenzene, 96¢

**Scheme 9.15**

than those required in the case of benzene, substitution occurs preferentially in the ortho and para positions. This behavior is illustrated by the regioselective substitutions during the formation of DDT, as well as during the formation of a series of chlorosulfonation and nitration products mentioned, in part, elsewhere (see Scheme 9.15).

On the other hand, multiple chlorine substituents promote nucleophilic substitution reactions. Phillips took advantage of this behavior in the case of p-dichlorobenzene to introduce PPS the engineering thermoplastic Ryton (PPS) (9.2.3). An extreme illustration of the effect of multiple chlorine substituents is the behavior of hexachlorobenzene, a product of exhaustive chlorination of benzene, in which the cumulative effect of six chlorine substituents sacrifices one of its kind to displacement by hydroxide or mercaptide ions under relatively mild reaction conditions (Scheme 9.16). The resulting pentachloro-phenol (55¢) has been used for quite some time as a common wood preservative that prevents damage caused by fungi, mildew, and other destructive pests. Since 1984 the EPA limits the use of this and other wood preservatives (such as creosote—a phenolic coal-tar fraction, 70¢/gal; salts of chromium, arsenic, copper, and tin) to persons state-licensed to deal with these chemicals.

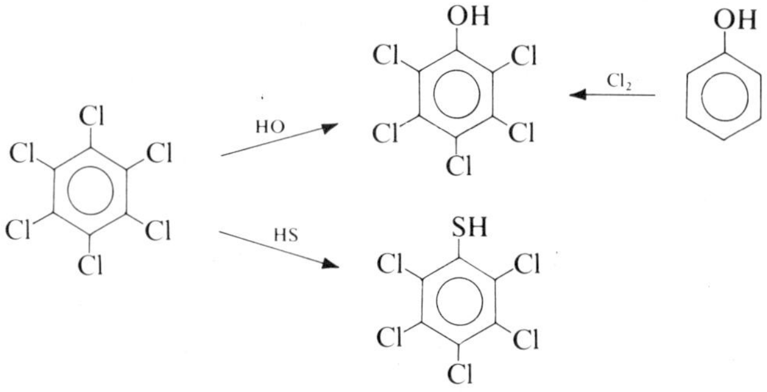

**Scheme 9.16**

*The Case of the Dioxins*

An intermediate situation with respect to the activation effect of chlorine substituents involves the conversion of 1,2,4,5-tetrachlorobenzene to 2,4,5-trichlorophenol required for the preparation of 2,4,5-trichlorophenoxyacetic acid (2,4,5-T, previously listed $1.10; (suppliers are no longer listed in OPD *Chemical Buyers Directory*, 1989 ed., except for Aldrich Chemical Co., which sells 250 g for $41.20 with the condition that "this quantity cannot be shipped by UPS or parcel post"), as well as its butyl and isooctyl esters. 2,4,5-Trichlorophenol is also the precursor of α-2,4,5-trichlorophenoxypropionic acid—known as "silvex," and of its methyl and ethyl esters (Scheme 9.17). These materials function as plant-growth stimulants and are sufficiently powerful to cause the exhaustion and death of plants and hence were used as herbicides and, especially, for the clearing of undesired woody growth from untilled land. During the Vietnam War the U.S. Armed Forces contracted seven different producers to supply 2,4,5-T for the clearing of jungle enemy hideouts, and the defoliant herbicide became known as "Agent Orange." The Armed Forces discontinued the use of Agent Orange in 1970 as the result of a finding by the National Cancer Institute that an impurity detected in Agent Orange, identified as 2,3,7,8-tetrachlorodibenzo-p-dioxin (TCDD, or simply dioxin) was responsible for birth defects

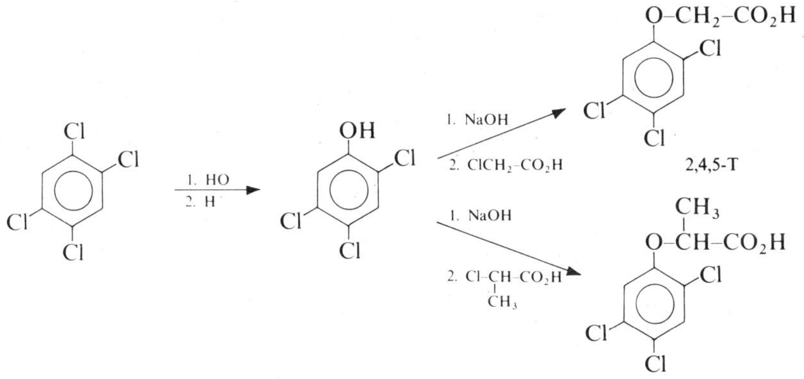

**Scheme 9.17**

in certain strains of laboratory mice. This finding, coupled with a claim in the Saigon press that the use of 2,4,5-T defoliants was responsible for damage to human health, led in 1984 to a class-action suit of Vietnam War veterans against the manufacturers of Agent Orange. An out-of-court settlement established a $180-million trust fund to benefit Vietnam veterans. In the meantime, neither the Center for Disease Control—a federal government—operated research organization that monitors all sorts of health problems, nor different sectors of the Armed Forces, nor Dow's study of 2192 employees of the Midland, MI plant who may have been exposed to TCDD during the manufacturing operations, were able to substantiate a correlation between exposure to TCDD and an incidence of cancer, deformities of children fathered by Vietnam veterans, and other health problems attributed to TCDD except a skin disease known as *chloracne*. A review of the Agent Orange controversy is offered by D. J. Hanson, "Science Failing to Back up Vietnam Concerns about Agent Orange," *CEN* 11/9/87, pp. 7–14.

Actually, the most extensive known industrial accident involving TCDD occurred on July 10, 1976 in Seveso, Italy, when a runaway reaction released somewhere between 0.5 and 5 kg of TCDD into the environment. Five years later, toxicologists and other medical experts were still unable to affirm that this massive exposure to TCDD had produced long-range ill effects to the human population (*CEN* 6/29/81; H. J. Sanders, "Herbicides," *CEN* 8/3/81). This conclusion was confirmed in another report published in the *Journal of the American Medical Association* (*JAMA*) (D. Hanson, "No Rise in Birth Defects after Italy's Dioxin Leak," *CEN* 3/28/88). (For previous evaluations of the Seveso accident, see *CW* 1/27/82; *CEN* 2/82; and J. Sambeth, *CE* 5/16/83.) A study of 1500 children exposed to the highest dose of TCDD during the Seveso accident (published by a team of American and Italian scientists in *JAMA*; reported in *CMR* 11/24/86) also was unable to reveal any serious effects on the children's health.

In April 1987 (*CMR* 4/27/87) the Federal Appeals Court in Manhattan, New York denied any liability of the seven manufacturers of Agent Orange or the U.S. Government for damages related to exposure of Vietnam veterans to TCDD but approved the distribution of moneys of the trust fund ($180 million plus ~ $50 million of accumulated interests) among all disabled persons or their survivors regardless of cause. On June 30, 1988 the Supreme Court disposed of the final challenges filed against the preceding settlement (*CEN* 7/6/88). The dispute concerning harmful effects of TCDD to humans is likely to continue, and it is fueled by a study of 24,253 veterans conducted by the Veterans

Administration that indicates a higher cancer death rate among army personnel who served in Vietnam (*CEN* 9/14/87). Thus, prudence mandates that derivatives of 2,4,5-trichlorophenol be avoided.

For a critical, quantitative evaluation of some of the TCDD toxicity data (in animals and humans), see John L. Wong, "Controlling Environmental Carcinogens: A Critical Look at Alleged Culprits," *CT* 1/87, pp. 46–53 and an adaptation of that paper published by the National Council for Environmental Balance in the form of a pamphlet "Cancer: What to Fear—What Not to Fear." Professor Wong also analyzes the related situations that deal with DDT, EDB, and methylene dichloride and the tendency of regulating agencies to prohibit the use of a given chemical in view of the lengthy and costly animal tests. A concise, illuminating introduction to toxicology and the difficult, unambiguous determination of carcinogenic effects of chemicals in humans is found in Chapter 4, "Toxicity and Dosage: The Varying Shades of Grey" of H. D. Crone's *Chemicals & Society*, Cambridge University Press, Cambridge, UK, 1986.

The formation of TCDD during the conversion of 1,2,4,5-tetrachlorobenzene to 2,4,5-trichlorophenol can be accounted for by two likely mechanisms. In the first, a nucleophilic attack of the 2,4,5-trichlorophenoxide ion on a molecule of unreacted tetrachlorobenzene initially gives the expected hexachlorodiphenyl ether, followed by another nucleophilic substitution of a chlorine substituent and a cyclization step. The second likely mechanism involves a bimolecular reaction of two 2,4,5-trichlorophenoxide ions, which leads directly, or in two stages, to TCDD (Scheme 9.18). It is reasonable to expect that the activation energy of a nucleophilic substitution reaction by the 2,4,5-trichlorophenoxide ion on tetrachlorobenzene, and especially on another trichlorophenoxide ion, is higher than that of the desired reaction in which the hydroxide ion reacts with tetrachlorobenzene. Thus, the formation of TCDD by-product is favored at temperatures above and beyond those required to accomplish the desired reaction. It is also reasonable to expect that the bimolecular reaction of two 2,4,5-trichlophenoxide ions is favored at higher concentrations of the initial tetrachlorobenzene than those required to accomplish the desired reaction. These considerations suggest that exceptionally high yields of TCDD were likely obtained by those manufacturers of Agent Orange who failed to control the temperature of the process and/or employed relatively high initial concentrations of the organic substrate.

Now that our society is conditioned to pay attention and fear TCDD, and since our analytical capabilities are increasing annually, traces of TCDD are being discovered "in

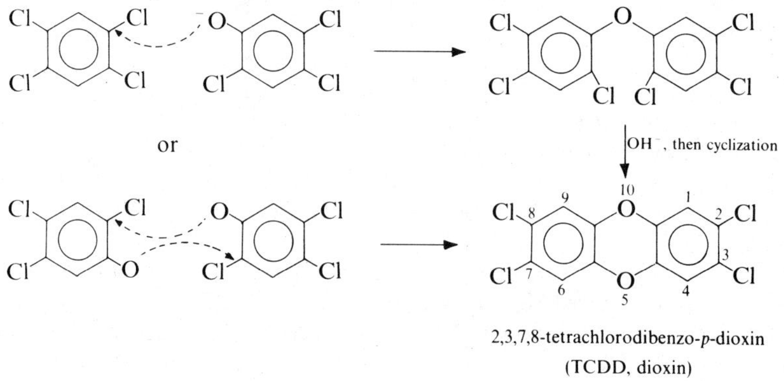

2,3,7,8-tetrachlorodibenzo-*p*-dioxin
(TCDD, dioxin)

**Scheme 9.18**

the pulp and sludge of paper mills, as well as in some finished paper products" (*CMR* 9/28/87) such as "tissue, toilet paper, sanitary pads, and paper towels. The concentrations of TCDD seem to "range from 2.1 ppt in disposable diapers to 13 ppt in writing paper." The latter figure is equivalent to a "baker's dozen" of dollars in the current federal budget, and, with further improvements in analytical capabilities, we are likely to detect a handful of any kind of molecules wherever we look. In the end, the TCDD problem, and analogous problems with harmful effects of chemicals, may well boil down to the wisdom of Aureolus Philippus Theophrastus Bombastus von Hohenheim, better known as Paracelsius (1493–1541), a Swiss-born alchemist and physician (J. Bronowski, *The Ascent of Man*, Little, Brown and Co., Boston, 1973, pp. 139–144) who some 400 years ago observed in so many words, "it is all a question of concentrations!"

One of the victims of the dioxin scare is hexachlorophene, a bacteriocide employed until a few years ago in personal deodorant preparations and obtained from 2,4,5-trichlorophenol and formaldehyde (3.4.2.5) (Scheme 9.19).

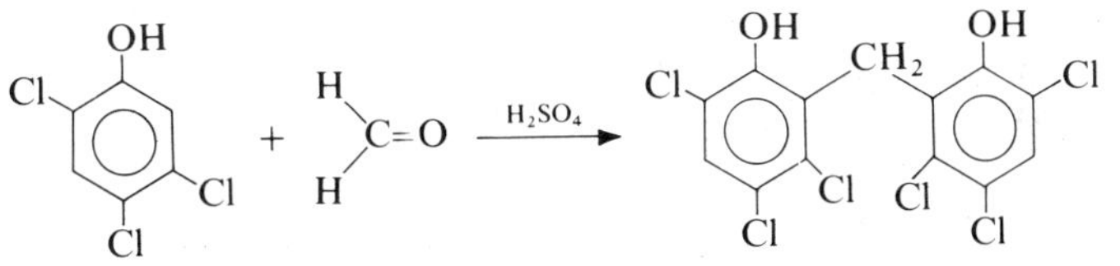

**Scheme 9.19**

A family of herbicides related in function to 2,4,5-T but free of suspicion of containing TCDD by-product is derived from 2,4-dichlorophenol and chloroacetic acid (Scheme 9.20). 2,4-Dichlorophenoxyacetic acid (2,4-D; $1.22), used in the form of its *n*-butyl ester ($1.19) or as the dimethylamine salt (solution $8.02/gal). 2,4-Dichlorophenoxyacetic acid and its propionic and butyric acid analogs are the most commonly used herbicides (> 60MM lb for weed control), and in 1988, because of lack of supporting evidence, the EPA dropped (*CMR* 3/21/88) the special review of their carcinogenic potential to the dismay of the National Coalition against the Misuse of Pesticides, which labeled the EPA decision as being "outrageous."

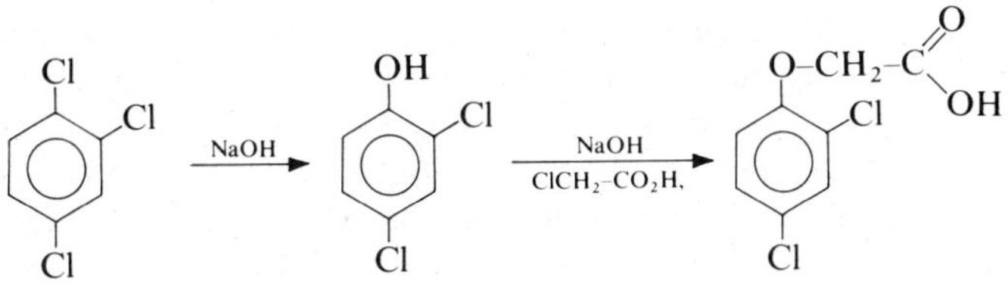

(also produced from phenol)                2,4-D

**Scheme 9.20**

Another selective weed killer of the phenoxy herbicide family is 2-methyl-4-chlorophenoxyacetic acid (MCPA; $1.55). It is assembled from 4-chloro-2-methylphenol obtained, in turn, by chlorination of *o*-cresol (99%, 75¢).

### 9.3.1.2   Bromine, Iodine, and Fluorine Compounds

Chlorine-containing aromatic compounds are preferred by industry as synthetic inter-mediates over the corresponding bromine derivatives because of the difference in the cost of these two halogens:

$Cl_2$      9.75¢/lb   or   1.485¢/mol
$Br_2$      43¢/lb   or   15.2¢/mol

Naturally, the relatively high cost of bromine can be readily absorbed in the production of fine chemicals.

The bromination of benzene in the presence of iron catalyst (that is converted to ferric bromide and functions as a Lewis acid) proceeds as does the analogous chlorination reaction:

$$C_6H_6 + Br_2 \rightarrow C_6H_5\text{–Br} \rightarrow p\text{-dibromobenzene, \$5.00}$$

Bromobenzene is converted more readily to the Grignard reagent than is chloro-benzene, and a 2.0 $M$ solution of phenylmagnesium bromide in THF is available commercially (Orsynex, \$3.59/lb of Grignard reagent).

On the other hand, bromine derivatives analogous to some of the chlorine compounds mentioned elsewhere in this book are employed as *flame-retardants.*

*Flame Retardants (FR)*

These materials are used as either additives or reactive components that become incorporated into the polymeric structure. Flame-retardants are usually used in con-junction with antimony oxide ($Sb_2O_3$; \$1.35), zinc borate (technical grade, 43% ZnO; 37% $B_2O_3$, 55¢), and ferric oxide—materials that exert a synergistic function and allow the use of decreased amounts of the relatively costly bromine-containing ingredients. Examples of typical flame retardants obtained from aromatic as well as nonaromatic precursors are shown as follows:

- Tetrabromophthalic anhydride, Ethyl's "Saytex RB-49", \$1.35, obtained by bromination of o-xylene, followed by oxidation of the corresponding 3,4,5,6-tetrabromo-o-xylene; it is incorporated into UPRs and alkyds. Also, it is converted to corresponding phthalate esters, or it is incorporated in hydroxyl group-termin-ated esters derived from diethylene and other glycols for use as "polyols" in the assembly of flexible PUR foams (3.8.2);

- 3,4,5,6-Tetrabromo-1,2-dimethylolbenzene, Ethyl Corporation's Saytex RB-79 Diol (\$1.35), derived from the preceding anhydride by esterification and hydrogenation of the resulting ester. This leads to polyols used in the production of rigid PURs.

- Ethylene-bis(tetrabromophthalimide), Ethyl's Saytex BT-93 (\$2.55), obtained from the preceding anhydride and ethylenediamine (\$1.30); used as a flame-retardant additive for (PS), polyolefins (PET), nylons, EPDM, PC, ionomers, and so on.

- Tetrabromobisphenol A, Ethyl's Saytex RB-100 (\$1.16), obtained by bromination of bisphenol A (at the ortho positions adjacent to phenolic functions); incorporated into epoxy, PC, and other polymers that can be derived from phenolic substituents by way of allylic ethers, acrylate esters, and so on.

- Decabromodiphenyl oxide (DBDPO), Ethyl's Saytex 102 (\$1.70), a fully brominated derivative of diphenyl oxide, flame-retardant additive for PS, polyolefins, PET, elastomers, and so on.

- Octabromodiphenyl oxide, Ethyl's Saytex 111 ($1.45), a partially bromination product of diphenyl oxide used in a similar fashion as the fully brominated analog.
- Pentabromoethylbenzene, 2,4,6-tribromostyrene dibromide, derived by addition and ring bromination of styrene; a flame-retardant additive for thermoset polyesters, textile fibers, adhesives, PURs, and so on.
- Dibromoethyldibromocyclohexane, obtained by addition of bromine to the cyclic butadiene dimer, vinylcyclohexene (6.8.1.4); used as a flame-retardant additive for SAN, PS, PURs, adhesives, coatings, textiles, and so on.
- Ethylene-bis(5,6-dibromonorbornane-2,3-carboximide), obtained by bromine addition to the Diels–Alder adduct of MA and cyclopentadiene (6.8.1.4); used as a flame-retardant additive for polyolefins, polyesters, nylons, and so on.
- Hexabromocyclododecane (HBCD), obtained by addition of bromine to the cyclic butadiene trimer (6.8.2); used as a flame-retardant additive for PS foams, SAN, adhesives, coatings, and so on.
- Vinyl bromide, obtained by high-temperature bromination of ethylene and a flame-retardant comonomer of vinyl polymers.
- 2,3-Dibromopropanol, obtained by addition of bromine to allyl alcohol; used in the form of ethoxylation or propoxylation derivatives as a flame-retardant component of polyesters, and so on.

Until a few years ago the bromine-containing analog of PCB (polybrominated biphenyl; PBB) was used as a fire-retardant additive. The major component of PBB was the hexabromo derivative (Michigan Chemicals' Firemaster BP-6), which is degraded by exposure to UV light and hence not as environmentally persistent as the corresponding chlorine compound. Unfortunately, an idiotic mistake perpetrated in Michigan during 1973 that consisted of mixing some PBB with cattle feed led to a massive slaughter of cows and aroused public alarm concerned with health hazards associated with the consumption of contaminated milk and beef [L. J. Carter, *Science* **192**, 240 (1976)]. Although no permanent deleterious effects on the affected human population have been proved thus far, PBB was withdrawn from the market.

Another famous fire-retardant that caused a societal perturbance was Tris-BP [tris(2,3-dibromopropyl)phosphate, or tris for short], obtained by addition of bromine to triallyl phosphate:

$$(CH_2=CH-CH_2-O)_3PO + 3Br_2 \rightarrow (Br-CH_2-CHBr-CH_2-O)_3PO$$

First, an overzealous but poorly advised Congress of the United States mandated the use of tris as a flameproofing material for children's nightwear with the noble intention of reducing or preventing deaths in the event of accidental fires. Then, by 1981 the EPA listed tris as a carcinogen—not at all a surprising finding in view of the fact that $\beta$-halogen-substituted thioethers and amines are known to decompose to give powerful alkylating agents of the mutagenic sulfur and nitrogen mustard family:

$$-Q-CH_2-\underset{\underset{R}{|}}{C}H-X- \rightarrow -Q^+\!\!\underset{\underset{R}{|}}{\overset{/CH_2}{\diagdown CH}} \qquad X^- \rightarrow \text{alkylation}$$

(where Q represents a nucleophilic sulfur or nitrogen group and X represents a relatively weakly bonded halogen) and R represents a residual moiety of the molecule.

Extensive, classical studies published following World War II by S. Winstein of UCLA demonstrated that the reactivity of these systems extended to structure in which Q also represents an acetate group:

$$
\underset{\underset{\text{O–CH}_2\text{–CH–X}}{|}}{\overset{\text{CH}_3\text{–C=O}}{|}} \longrightarrow \quad \underset{\underset{\text{O–CH}_2\text{–CH}}{|}}{\overset{\text{CH}_3\text{–}\overset{+}{\text{C}}\text{–O}}{|}} \diagdown \quad X^- \longrightarrow \text{alkylation}
$$

Clearly, this system is analogous to the phosphate ester structure of tris. An additional dangerous alkylating potential of tris is the possibility that dehydrohalogenation of the $\beta$-bromine gives rise to the notoriously reactive allylic bromide system:

$$ \text{Br–CH}_2\text{–CHBr–CH}_2\text{–O–} \rightarrow \text{Br–CH}_2\text{–CH=CH–O–} + \text{HBr} $$

The tris incident ended when Congress was forced to rescind the mandated use of this fireproofing additive, and in January 1983 President Reagan signed a bill that indemnified losses suffered by textile manufacturers who were left holding about $50 million worth of useless inventory (*CMR* 1/10/83).

To do at least partial justice to the topic of flame-retardants, we must point out that chlorine compounds analogous to some of the above-mentioned bromine-containing materials, as well as a series of phosphorus-containing compounds, are also used for this purpose. Some examples of these compounds are mentioned elsewhere in this book (see Index).

The importance of flame-retarding materials is bound to grow in the future as government regulations are tightened in order to decrease or prevent human losses suffered during disastrous fires in hotels, airplanes, theaters, nightclubs, and so on.

### 9.3.1.3 Iodine Compounds

The direct iodination of aromatic hydrocarbons is rather difficult because the by-product of iodination, namely, HI, is a strong reducing agent and tends to reverse the reaction. Iodination of benzene to iodobenzene can be achieved successfully when carried out in the presence of dilute nitric acid that destroys HI as it is formed. Thus, iodo substituents are usually introduced by way of diazonium intermediates (9.3.6). An exception to this statement are the aromatic systems that contain strongly electron-donating groups and phenols offer the most common examples of direct iodination (see below).

Iodobenzene is priced at about $14.55/lb in part because of its multistep preparation (by way of aniline) and also because of the relatively high cost of sodium iodide ($10.15), aggrevated by limited demand, and problems with storage because of decomposition on exposure to light. On the other hand, iodobenzene is a model of unique chemistry peculiar to iodoaromatics and not exhibited by chloro- or bromobenzene (Scheme 9.21).

An example of the facile iodination of phenolic systems is the conversion of phenolphthalein to the sodium salt of tetraiodophenolphthalein—a diagnostic aid because of its opacity to X rays (Scheme 9.22). Several other radiopaque chemicals are derived from phenols and amines.

Another class of iodinated aromatics is represented by antiamebic compounds. The two example shown here are derivatives of quionoline (10.3.1), but the iodination reaction

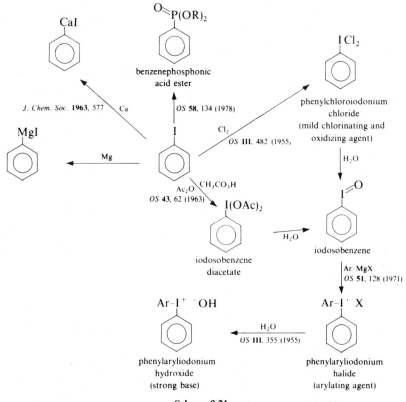

**Scheme 9.21**

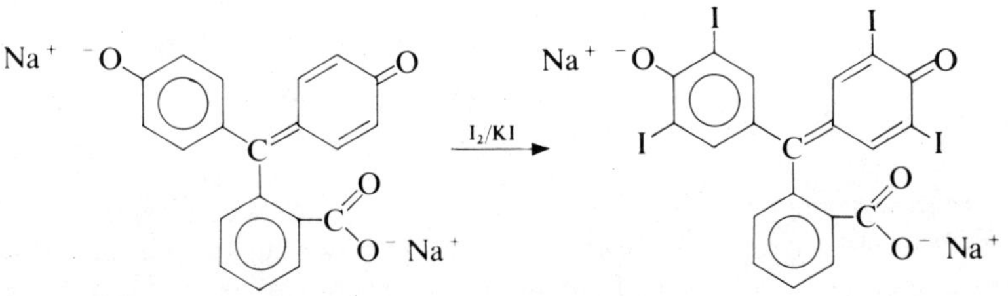

**Scheme 9.22**

occurs by virtue of their phenolic substituent:

Iodoquinol, or 5,7-diiodo-8-
    hydroxyquinoline

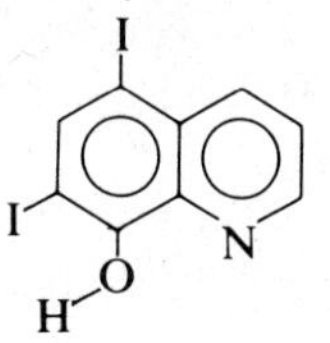

Iodochlorhydroxyquin, or 5-chloro-8-
hydroxy-7-iodoquinoline, USP $15.90

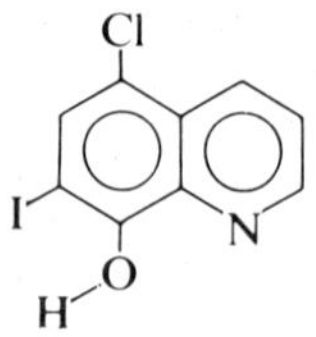

Iodine is an antiseptic substance (i.e., "tincture of iodine," a solution of iodine in ethanol found in most household medicine cabinets) and antiseptic products are obtained by iodination of suitable precursors such as either

- The readily substituted pyrrole ring (10.2.1) to give 2,3,4,5-tetraiodopyrrole

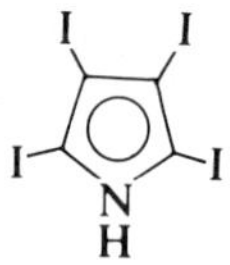

- Thymol, NF 3.75 (*SZM*-II, 124, 131), which forms a complex mixture of derivatives known as "thymol iodide" ($52.30) when an alkaline solution of thymol is treated with a solution of iodine in $KI$; shown in Scheme 9.23 is the structure of the carrier of iodine in the form of hypoiodide groups (positive iodine).

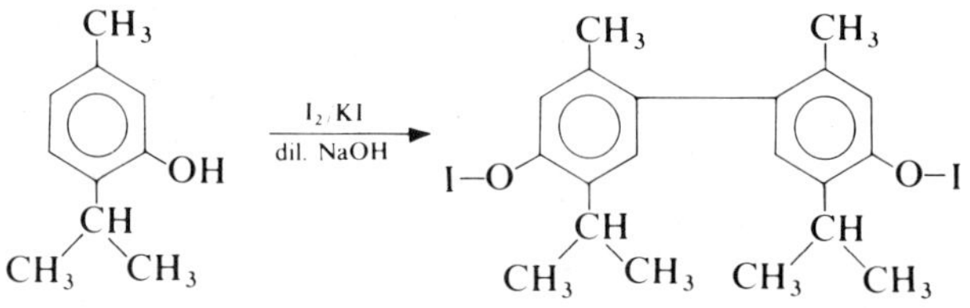

**Scheme 9.23**

- Poly($N$-vinyl-2-pyrrolidinone) (see 4.7.2), which forms a complex with iodine known as povidone–iodine that is used during surgery as a topical antiseptic; most likely, the formation of the complex is related to the ease with which iodine gives charge-transfer complexes:

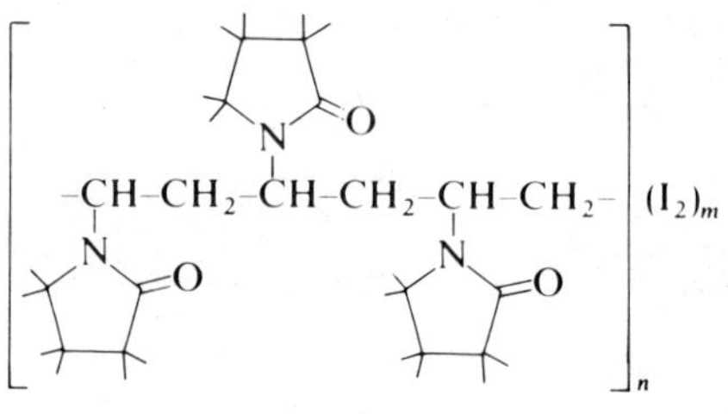

Finally, there are iodine-containing amino acids and related iodine-containing carboxylix acids, which are involved in the physiology of the thyroid gland:

Throxine, or 3,5,3′,5′-
tetraiodothyronine

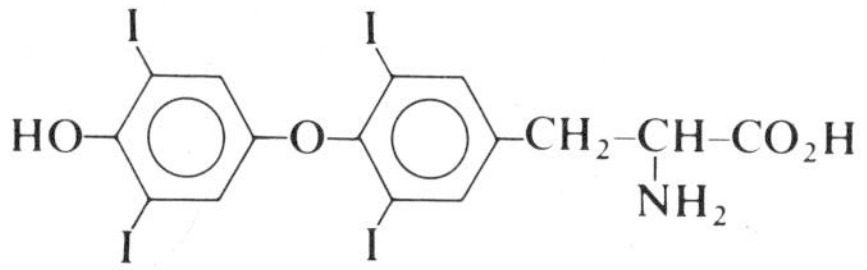

$$HO-\text{(ring)}-O-\text{(ring)}-CH_2-\underset{\underset{NH_2}{|}}{CH}-CO_2H$$

3,5,-Diiodotyrosine, a
thyroid inhibitor

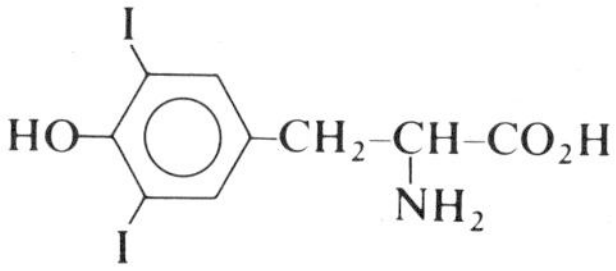

$$HO-\text{(ring)}-CH_2-\underset{\underset{NH_2}{|}}{CH}-CO_2H$$

Thyropropic acid, an
anticholesteremic

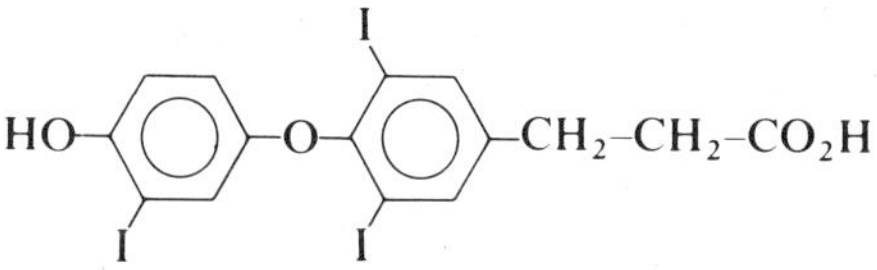

$$HO-\text{(ring)}-O-\text{(ring)}-CH_2-CH_2-CO_2H$$

3,5-Diiodosalicylic acid, an
animal growth promotor

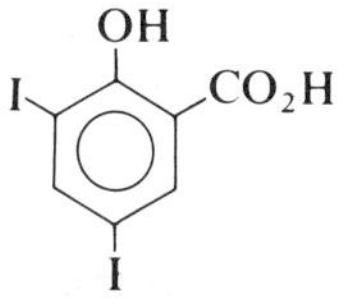

$$\text{(ring with OH, }CO_2H\text{, I, I)}$$

### 9.3.1.4   Fluorine Compounds

Elementary fluorine is a powerful oxidizing agent, and direct fluorination of most carbon compounds can result in an explosive formation of carbon tetrafluoride (see 3.2.3). While the difficulties associated with a direct introduction of fluorine substituents onto aromatic rings are being overcome [see *CEN* 1/27/86 for the use of the *N*-fluoro-*N*-methyltrifluoromethanesulfonamide reagent, $F_3C-SO_2-N(CH_3)F$, and the direct fluorination of benzenoid compounds by means of *N*-fluoroperfluorosulfonimides, $(R_f-SO_2)_2N-F$, where $R_f = C_nF_{n+1}$, reported by D. D. DesMarteau and coworkers (*CEN* 1/29/86, 11/30/87; *JACS* **109**, 7194 (1987))], the desired fluorine-containing aromatic compounds are being obtained by means of the traditional Schiemann reaction (*ONR*-81). This procedure is illustrated by the preparation of 3,3′- and 4,4′-difluorobenzophenones— interesting building blocks of some HPSTs (see 9.2.3). The process starts with the nitration of either benzophenone or diphenylmethane, reduction of the dinitro derivatives to the corresponding diamines, diazotization of the diamines (9.3.6) and, finally, the application of the Schiemann procedure, as shown in Scheme 9.24. The modern modification of the Schiemann procedure is based on the decomposition of a given diazonium salt in hydrofluoric acid (see 9.3.6).

Fluorine substituents of benzenoid structures exhibit a dual influence: on one hand, they are responsible for resonance stabilization of the whole molecule; on the other hand, they promote nucleophilic substitutions because of their excellent leaving-group character. The latter behavior is exacerbated by the presence of additional electron-withdrawing groups such as the ketone function in the above-mentioned difluorobenzophenones and, in particular, by the two nitro groups in 2,4-dinitrofluorobenzene (DNFB).

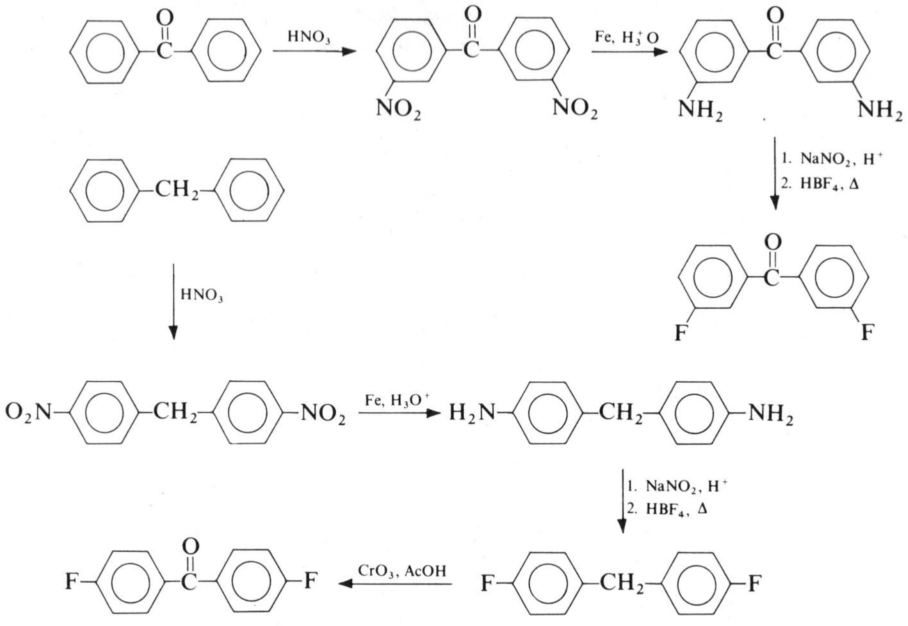

**Scheme 9.24**

The latter compound became the classical reagent for derivatization of amino acids and peptides, under suitably mild reaction conditions, which paved the way for the elucidation of the composition of proteins (starting with the work by 1958 Nobel laureate F. Sanger and coworkers on the sequence of amino acid building blocks of insulin—a protein containing a total of 51 amino acids of 16 different structures (see Scheme 9.25).

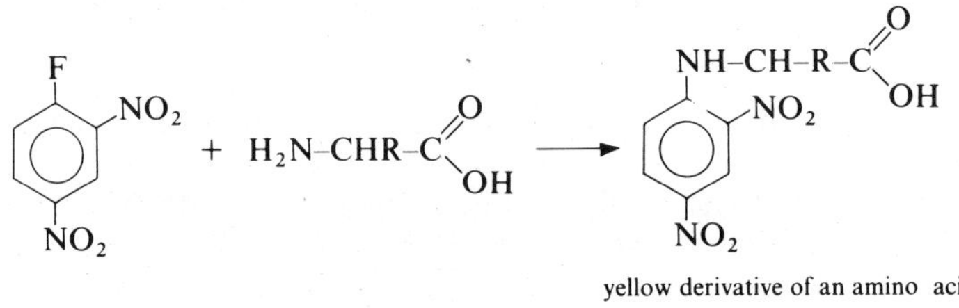

yellow derivative of an amino acid

**Scheme 9.25**

The special route to the introduction of trifluoromethyl groups, and some uses of the resulting materials, are mentioned elsewhere (9.3.1.1). In 1982 the costs of *p*-chlorobenzo-trifluoride and 3,4-dichlorobenzotrifluoride were $2.60 and $2.74, respectively (as reported by Hooker—now part of Occidental Petroleum), and the same company was a source of *p*-trifluoromethylbenzaldehyde and *p*-trifluorobenzoic acid. These compounds were obtained by starting with a large-scale chlorination of either toluene or *p*-xylene that yielded side-chain and ring-substitution products. On separation of the reaction products, the trichloromethyl groups of individual components were converted to trifluoromethyl substituents, the dichloromethyl groups were hydrolyzed to carboxaldehydes by means of sulfuric acid, and so on. For a recent update and a short review of aromatic fluorine compounds, see C. D. Hewitt and M. J. Silveste, *Aldrichimica Acta*, **21** (1), 3–10 (1988).

### 9.3.2  Benzenoid Carboxylic Acids and Derivatives

#### 9.3.2.1  Monocarboxylic Acids

The production of benzoic acid by oxidation of toluene and its major industrial uses were described in Section 9.2.2. Some additional transformations of benzoic acid are shown in Scheme 9.26.

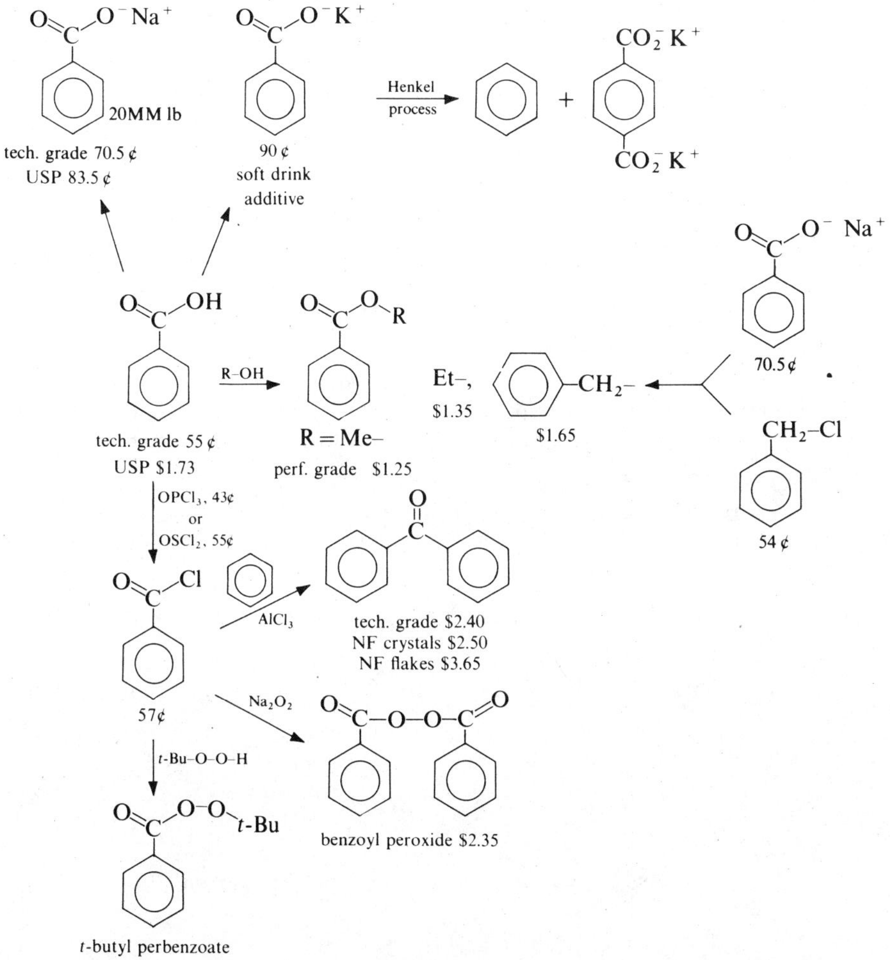

**Scheme 9.26**

Since nitration of benzoic acid produces *m*-nitrobenzoic acid, the production of the isomeric *p*-nitrobenzoic acid and its desirable reduction product *p*-aminobenzoic acid (PABA; USP $7.65) requires that the nitration of toluene precede the oxidation step that introduces the carboxylic acid group (as shown above). PABA is a component of the vitamin B complex, its ethyl ester is the local anesthetic benzocaine (USP $9.35), and it is an ingredient of sunburn lotions. Until recently, the list prices of PABA and USP benzocaine were $5.80 and $7.50, respectively, and a likely reason for their steep increases is the increased demand for sunburn lotions due to the rising awareness of skin cancer. Also, PABA is a building block of such fine chemicals as folic acid—another component of the vitamin B complex used in the treatment of certain kinds of anemias—assembled, as can be seen from the structure shown in Scheme 9.27, by combining glutamic acid, PABA, and the pteridine nucleus (10.3.3)

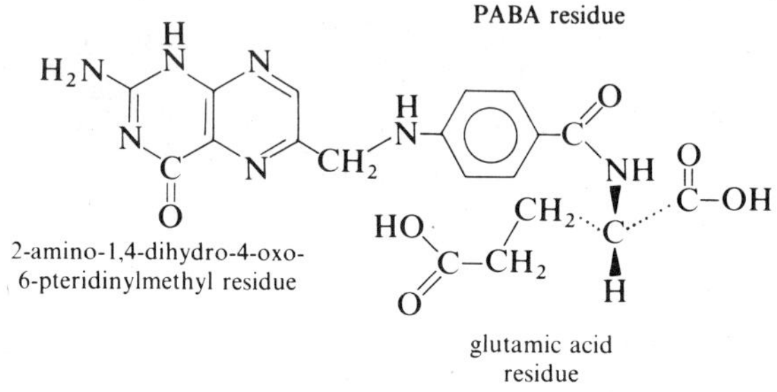

2-amino-1,4-dihydro-4-oxo-
6-pteridinylmethyl residue

glutamic acid
residue

**Scheme 9.27**

Derivatives of *N,N*-dimethylaminobenzoic acid, such as its octyl ester, are also ingredients of sunscreen formulations.

The isomeric toluic acids (9.2.2) are the starting materials for the preparation of such diverse materials as the mosquito repellent *N,N*-diethyl-*m*-toluamide (DEET; $2.75) and the building block of organic pigments such as 3-amino-4-methylbenzamide (Hoechst-Celanese $12.50) (Scheme 9.28).

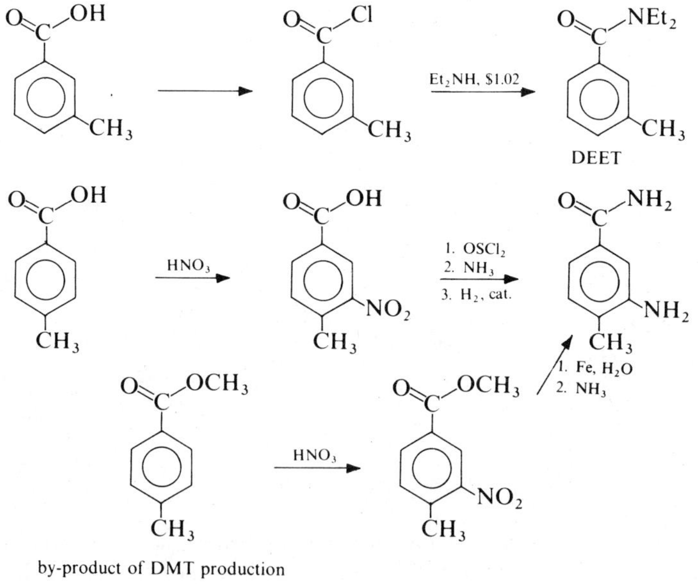

by-product of DMT production

**Scheme 9.28**

The EPA is currently investigating ($CW$ 9/30/87) incidents of poisonings of animals exposed to Hartz Mountain's Blockade—a flea and tick repellent that contains a mixture of DEET and fenvalerate, the ester of α-4-chlorophenylisovaleric acid and the cyano-hydrin of 3-phenoxybenzaldehyde (Scheme 9.29). Each individual component of the mixture is actually relatively harmless to mammals, although DEET is known to be an irritant to eyes and mucous membranes (but not to skin), while fenvalerate is known to be a skin irritant and cause eye damage. It appears that the formulators of Blockade overlooked the possibility that DEET can function as a solvent of fenvalerate, promote its absorption through the skin, and may even facilitate the gradual hydrolysis of the ester that would result in the liberation of HCN.

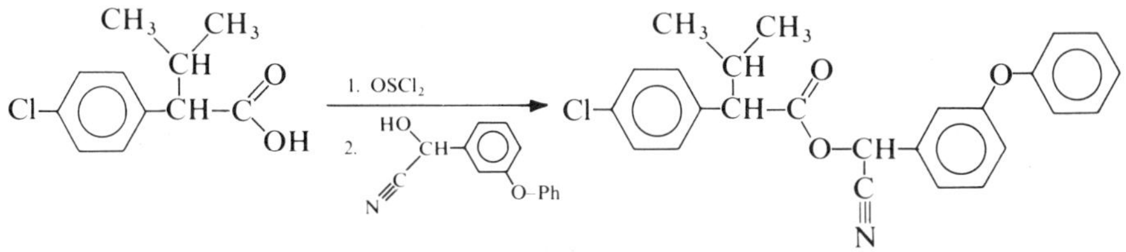

**Scheme 9.29**

Hydroxy-substituted benzoic acids cannot be obtained by oxidation of the cresols because of the susceptibility of phenols to oxidation (9.3.4). Fortunately, the Kolbe–Schmitt reaction ($ONR$-52) provides an entry to $o$- and $p$-hydroxybenzoic acids. The former, better known as "salicylic acid" (technical $1.00, USP crystals $1.33, powder $1.68), is the building block of the analgesic (see below) aspirin (or acetyl salicylic acid, USP crystals $2.20, 10 and 16% starch granulation $2.22 and $2.30, respectively). Also, various esters of salicylic acid such as methyl salicylate, oil of wintergreen (NF $1.45), and phenyl salicylate (or salol; $4.80), are in common use. Because of a difference in hydrolysis of salol as a function of pH, salol is used to coat medications intended to reach the alkaline environment of the intestine while passing through the acidic environment of the stomach. Small amounts of salicylic acid are used in the formulation of foundry resins, and this compound is also subjected to transformations typical of phenolic compounds ex-emplified by the formation of the antibacterial sulfasalazine, or salicylazosulfapyridine (see below). $p$-Hydroxybenzoic acid (PHBA; $1.95) is the minority product of the Kolbe–Schmitt reaction carried out under traditional conditions, but it can be obtained by thermal rearrangement of the potassium salt of salicylic acid above 150°C. It is the parent of two common esters that function as bacteriostatic and fungicidal food preservatives known as "parabens" and used to an extent of about 3MM lb. Thus, we have methylparaben (techn. $4.50, NF $4.60) and ($n$-)propylparaben (techn. $4.80, NF $4.90). Currently interest in PHBA is increasing because of its role as a building block of some high-performance thermoplastics of the polyarylate family (see 9.3.2.2). The formation and transformations of the above-mentioned hydroxybenzoic acids are shown in Scheme 9.30.

*Aspirin and Its OTC Painkiller Competitors*

About half of the salicylic acid production is acetylated below 90°C with an excess of acetic anhydride to produce a 90% yield of aspirin ($FKC$, 117–120). This analgesic has been available since 1898 and is also offered to the public as the sodium salt or as a

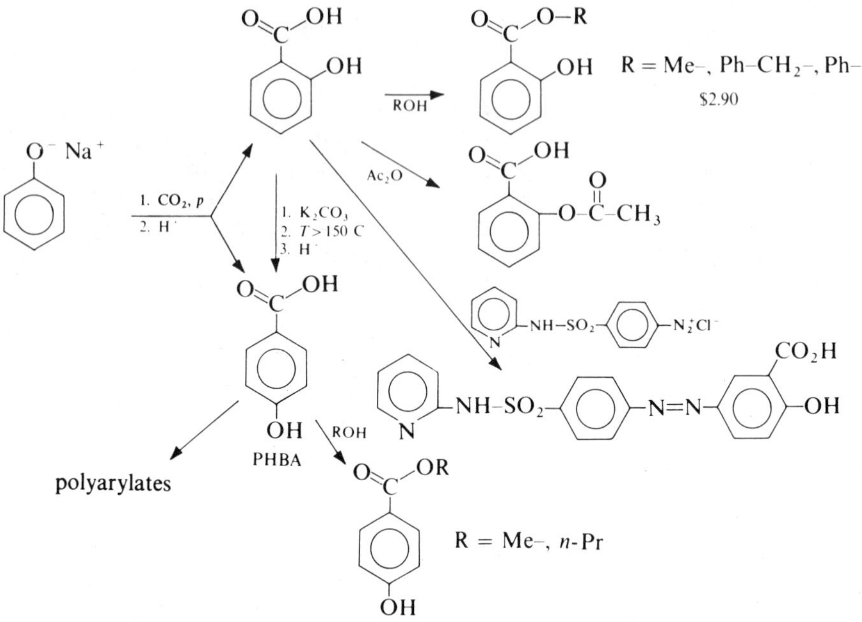

**Scheme 9.30**

mixture with sodium carbonate buffer. The demand of some 33MM lb in 1980 decreased in 1981 to 29.6MM lb because of the negative publicity associated with the incidence of the often fatal Reye's syndrome suffered by children with viral diseases. Since then, an aggressive marketing compaign by the three major competitors for the OTC analgesic retail market has been under way: aspirin, acetaminophen, and ibuprofen. A fourth, relatively minor competitor is acetophenetidine or phenacetin. The chemical structures of these four analgesics are shown as follows, with a cross-reference to their syntheses:

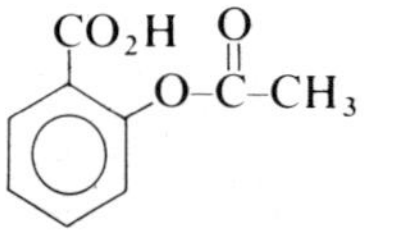

Aspirin
(acetyl salicyclic acid;
USP crystal powder, $2.20)

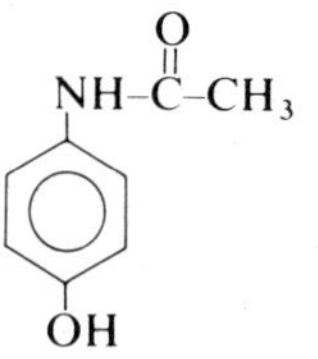

Acetaminophen
[$N$-acetyl-$p$-aminophenol (APAP);
$2.71] (see 9.3.5)

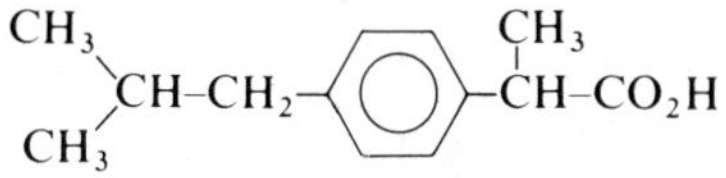

Ibuprofen or
[2-($p$-isobutylphenyl)-propionic acid]
(see 9.3.5)

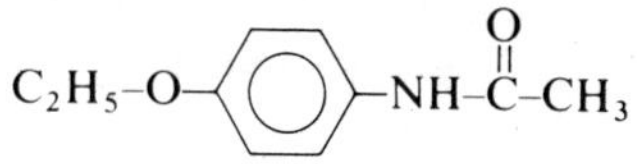

Phenacetin, (acetophenetedine;
USP $2.20)
(see 9.3.4)

Sales of materials such as OTC pain-killers are highly sensitive to impressions of the consumers. Thus, the recognition of a relationship between Reye's syndrome and aspirin caused the sales of aspirin to drop by about 25% between 1980 and 1981, while the sale of APAP grew at an annual rate of 5–7%, which is considerably higher than the annual increase of 1.5% in population. In June 1982 the U.S. Secretary of Health and Human Services mandated a warning label cautioning against the use of aspirin by children, and this adversely affected aspirin sales. Then, in 1983, cyanide-spiked acetaminophen distributed under the McNeil tradename "Tylenol" was responsible for the deaths of six unsuspecting persons in Chicago, and within a 1-year period the price of aspirin rose from $1.65 to $2.14 (October 1983) while that of APAP dropped from $3.71 to $3.30. A withdrawal of all suspect Tylenol from the consumer market, accompanied by a successful publicity campaign, caused a recovery in the sales of APAP, and ever since, television advertisements of the common pain-killers have become more aggressive in the promotion of the major chemical ingredients of the analgesics as well as different tradenames of identical chemicals. A by-product of the Chicago cyanide-spiked analgesic deaths, followed by similar tainted-product-related deaths in the New York area, was the development of tamperproof packaging.

By 1986 there appeared on the U.S. market the generic form of ibuprofen since the original patents issued during 1964–1968 to Boots Pure Drug of England had expired. Ethyl announced (*CMR* 11/3/86) an expansion of its ibuprofen capacity to more than 4.4MM lb. The appearance of this third major competitor for the $2-billion OTC nonsteroidal antiinflammatory drug (NSAID) market added new fuel to the fire of vigorous competition. We now hear and watch entertainment personalities radiating sincerity and trustworthiness, telling us how one particular brandname is used most by the "majority of hospitals" or how the "majority of doctors" (not "physicians") would choose this and only that particular product if they were stranded on a deserted island or climbing Mount Everest, how one manufacturer discovered a "better formula for aspirin," and other such figments of Madison Avenue imagination. Also, in recent years aspirin competitors coalesced to campaign against the manufacture of aspirin by advertising their "aspirin-free pain-killers," whereupon some aspirin distributors launched a counterattack (May 1987) that proclaims a life-saving effect of a one-per-day dose of aspirin tablets for potential victims of heart disease and stressing the wonder-drug qualities of the material. Actually, in October 1985 the FDA approved the use of aspirin as a cardiac medication. For reports of other beneficial effects of aspirin, see *CMR* 4/7/86 and 12/15/86. In January 1988 a study in the *New England Journal of Medicine* reports that one aspirin tablet per day prevents heart attacks in healthy men, and this widely publicized report (Associated Press, 2/1/88) will, no doubt, compensate for the negative effect of the Reye's syndrome connection (*CB* 4/88) and may even increase the demand in the United States by as much as 9MM lb/yr. In January 1988 the price of aspirin rose momentarily by 25¢/lb (*CMR* 2/8/88) to $2.45/lb USP crystal powder. Another study (*CMR* 4/4/88) that reports a 64% reduction of deaths due to heart attacks when the patient recieves a combination of aspirin and streptokinase—an enzyme that dissolves clots in the coronary artery—may provide an additional boost to the consumption of aspirin.

After many decades of uncertainties concerning the mechanism of aspirin's physiological function(s), it is now believed that this chemical inhibits the enzyme cyclooxygenase that is responsible for the conversion of polyunsaturated fatty acids to prostaglandins. The latter, when present in excessive amounts, cause fever and pain.

### 9.3.2.2  Dicarboxylic Acids

The major large-volume uses of the three dicarboxylic acids derived from the xylenes are outlined in 9.2.2. Here we add a short interlude about the importance of plasticizers that are derived to a large extent from phthalic anhydride.

### The Importance of Plasticizers

The U.S. capacity to produce phthalate plasticizers is about 1.7B lb, and the demand hovers around 1.4B lb. The most frequently used plasticizer is dioctyl phthalate (DOP), priced at 46¢ and derived from 2-EH. It is used extensively to plasticize vinyl polymers and, in particular, flexible PVC, or "artificial leather" products. The U.S. capacity for DOP production is 870MM lb. Next in importance are diisodecyl phthalate (DIDP) and diisonyl phthalate (DINP), both derived from oxo alcohols and priced competitively with DOP at 46¢–47¢. Priced somewhat more attractively at 42¢ is DIOP, diisooctyl phthalate that differs from DOP by being derived from $C_8$ oxo alcohols.

Apart from economic considerations, the choice of plasticizers depends on their effect on the temperature–viscosity and temperature–flexibility relationships of the plasticized polymer mass. These properties affect the ease of processing (extrusion, calendering, etc.) and the ultimate uses of consumer products such as the artificial leather goods used in the manufacture of upholstery, clothing, and the like. Also, the volatility of the plasticizer and its compatibility with the polymeric mass (that also contains pigments, stabilizers, and other additives) are significant since these parameters can vary with time and temperature. With these and other variables in mind, we can accept the existence of numerous plasticizers on the marketplace because, in addition to those mentioned above, we also have di-*n*-butyl phthalate (DBP; 54¢), diisobutyl phthalate (DIBP; 55¢), dicyclohexyl phthalate ($1.25), diundecyl phthalate (DUP; 61¢), benzyl octyl phthalate (BOP), and still others.

Dimethyl phthalate (DMP; b.p. 284 °C, 65¢) and diethyl phthalate (DEP; 69¢; odorless cosmetic grade 97.5¢) are used more as solvents than plasticizers, especially, DMP, which functions as an insect repellent and/or vehicle for other repellents.

In addition to the large family of phthalate plasticizers that constitute the majority of the 2B-lb plasticizer demand, there are ester plasticizers derived from the above-mentioned alcohols and other dicarboxylic acids. Thus, adipic, azelaic, and sebacic acids give rise to dioctyl esters priced at 66¢, 99¢, and $1.47, respectively, as well as tricresyl phosphate (TCP, $1.60) and still other special-purpose compounds. Some aromatic sulfonamide plasticizers are mentioned in 9.2.4.

Phthalic anhydride has some other interesting industrial uses not mentioned in 9.2.2. Thus, while the original discovery and synthesis of saccharin (water-soluble sodium salt FCC granules $2.75, powder $3.70) was based on the chlorosulfonation product of toluene (9.3.3), Sherwin-Williams manufactured saccharin starting with phthalimide (85¢). Actually, phthalimide is the building block of several other interesting materials;

- Phthalonitrile—the key intermediate of phthalocyanine pigments (10.3.3) available in blue and green shades and valued at about $9.25/lb and sufficiently stable for use in pigmentation of ceramic products.
- Methyl anthranilate—an ingredient of perfumes and flavors (grape) ($1.41), obtained by esterification of anthranilic acid ($1.71).

- Anthranilic acid—the intermediate in the preparation of two traditional dyes; indigo (Vat Blue 1) and thioindigo (Vat Red 41).
- The fungicide (N-trichloromethylthio)phthalimide, or folpet—the aromatic analog of captan (8.3.2.2).

The preceding transformations are outlined below, and included is the mechanism of the classical Hofmann reaction (ONR-45) that converts phthalimide to anthranilic acid by way of a rearrangement of an intermediate nitrene (Scheme 9.31).

Additional relatively simple reaction products of phthalic anhydride of industrial interest are

- Tetrachlorophthalic and tetrabromophthalic anhyrides prepared from about 7MM lb of phthalic anhydride and serving as intermediates for the production of some flame-retardants mentioned in 9.3.1. In the case of tetrachloro derivative, its conversion to the phthalocyanine Pigment Green 7 (10.3.3) is analogous to that of the chlorine-free phthalocyanines mentioned above except that it is carried out in one step by heating phthalic anhydride, urea, and cuprous chloride in the presence of suitable catalysts.
- Diallyl phthalate, 1.5MM lb, a component of UPRs especially formulated for use in insulating coatings.
- Phenolphthalein monosodium salt, 750MM lb, the active ingredient of laxative preparations; the product of an acid-catalyzed condensation of phthalic anhydride and phenol (see below).
- o-Benzoylbenzoic acid—a Friedel–Crafts acylation reaction (ONR-33; 9.2.1) of phthalic anhydride in an excess of benzene; cyclized by means of sulfuric acid to anthraquinone (see below), the building block of certain dyes shown in 9.3.5.
- 3-Propylidinephthalide ($45), the condensation product of phthalic anhydride and butyric acid formed in the presence of phosphorus pentachloride; it is a valuable flavoring agent used in meats, tobacco, maple flavor, and spices.

The three last-mentioned condensation reactions of phthalic anhydride are shown in Scheme 9.32.

The strong commercial interest in terephthalic acid motivated the development of the Henkel I process, which consists of an isomerization of phthalic to terephthalic acid. It is carried out in a continuous fashion by heating the dipotassium salts at temperatures as high as 400°C in the presence of zinc and cadmium catalysts. Another, somewhat analogous route to terephthalic acid is the disproportionation of the potassium salt of benzoic acid developed mainly by Henkel, Phillips, and Rhône-Poulenc (CEN 6/12/78). Both of these processes are represented in Scheme 9.33.

Rigid foams extend the use of polymers for structural purposes because of the economic advantage when a relatively costly dense material is converted into a structurally sound product of low density. Numerous blowing agents are available for this purpose, and one of them, N,N'-dinitroso-N,N'-dimethylterephthalamide (NTA), is derived from DMT (Scheme 9.34).

Ammoxidation of the xylenes converts the hydrocarbons into nitriles that can be fully hydrogenated to the corresponding aminomethylcyclohexanes or hydrogenated selectively to give the corresponding benzylamines. With phosgene, the latter are converted to

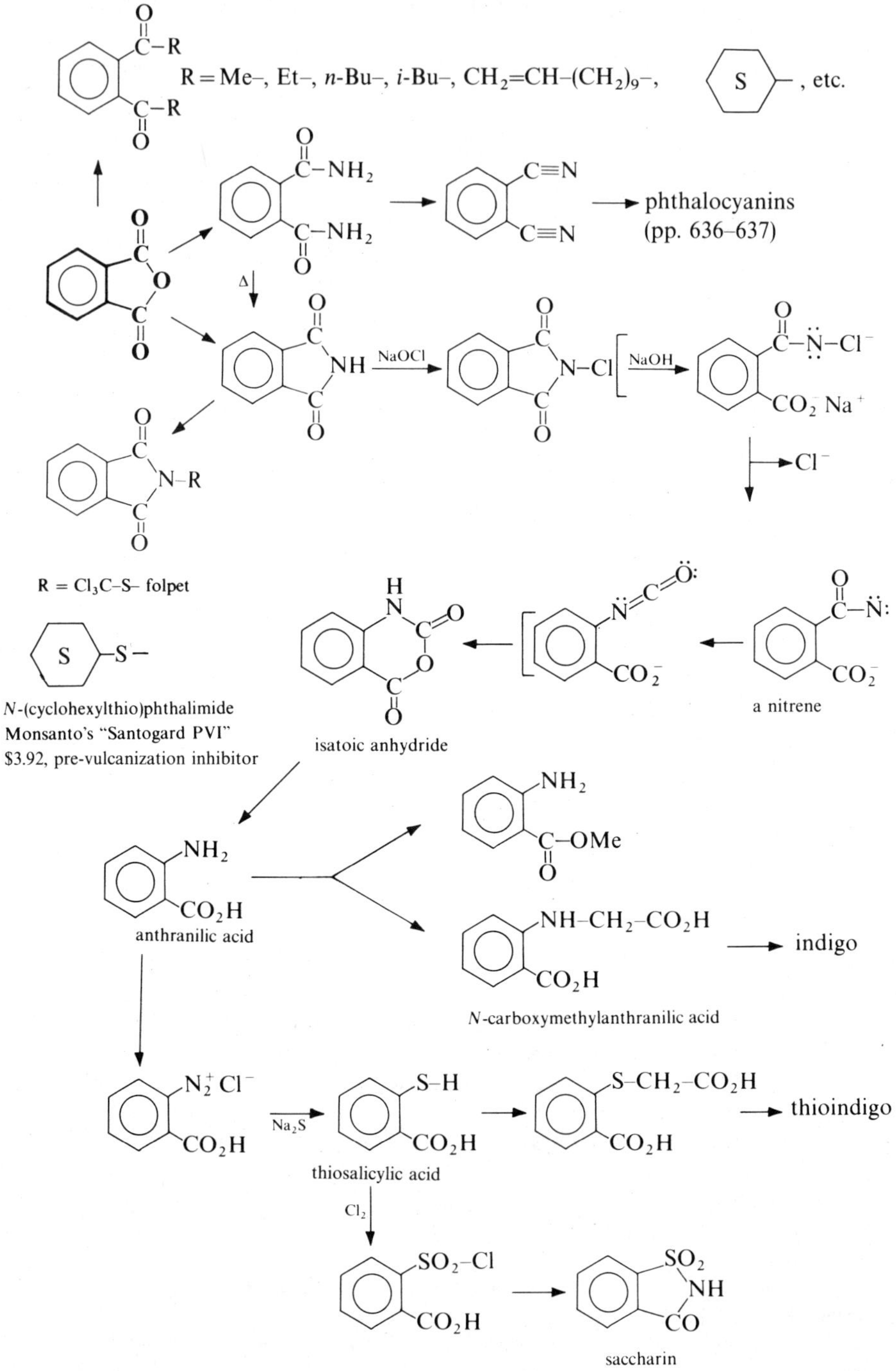
R = Me–, Et–, n-Bu–, i-Bu–, CH2=CH–(CH2)9–, etc.
phthalocyanins
(pp. 636–637)
NaOCl
NaOH
a nitrene
R = Cl3C–S– folpet
N-(cyclohexylthio)phthalimide
Monsanto's "Santogard PVI"
$3.92, pre-vulcanization inhibitor
isatoic anhydride
anthranilic acid
indigo
N-carboxymethylanthranilic acid
Na2S
thiosalicylic acid
thioindigo
Cl2
saccharin

Scheme 9.31

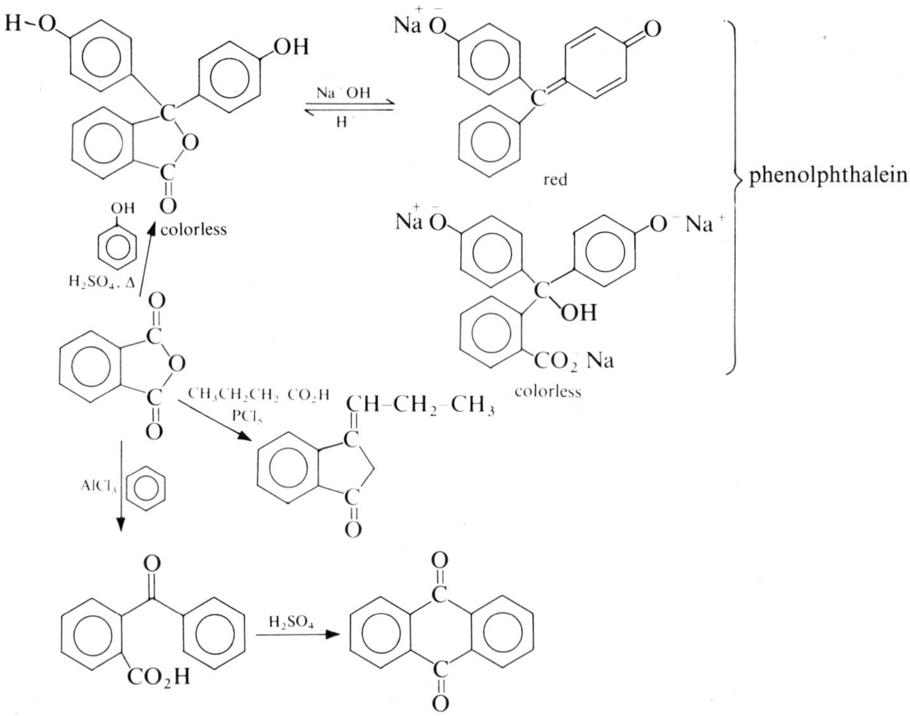

**Scheme 9.32**

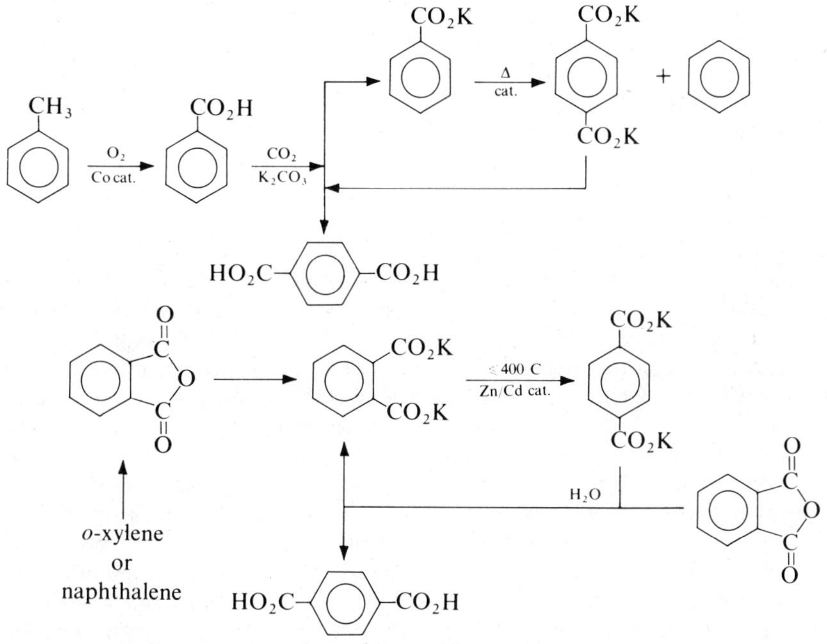

**Scheme 9.33**

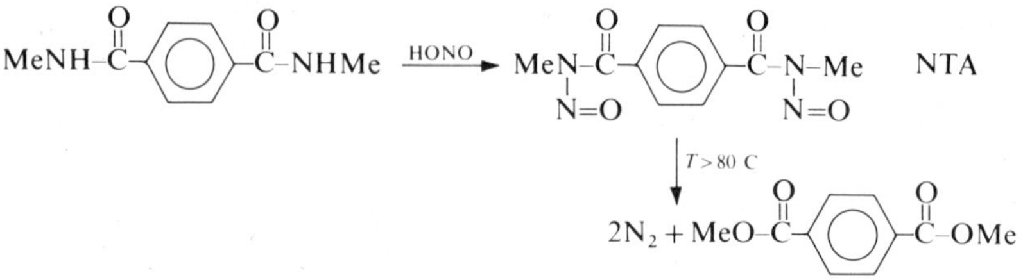

**Scheme 9.34**

benzylic isocyanates (interlude in Section 3.8) as illustrated by the transformation products that were offered until recently by Sherwin-Williams (Scheme 9.35).

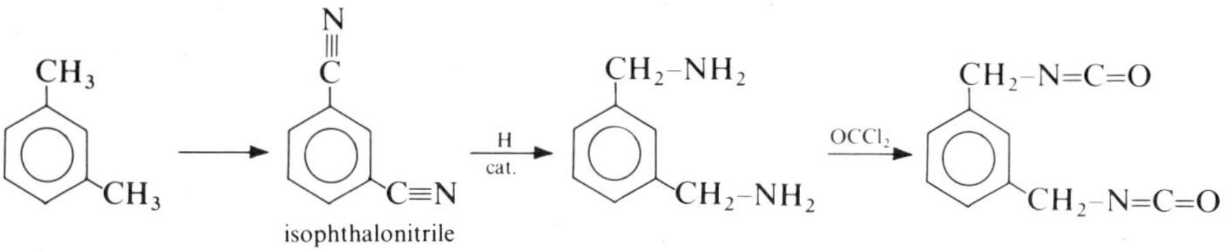

**Scheme 9.35**

*Thermoplastic Elastomers*

The importance of terephthalic acid, PTA, and/or DMT as the building block of PET is emphasized elsewhere and the engineering–plastic properties of PBT are mentioned in Section 6.7. At this point we may add that PBT shows thermal and mechanical properties superior to those of PET because the "hard" segments provided by the linear terephthalate moiety alternates with "soft" segments introduced by the $-O-CH_2-CH_2-CH_2-CH_2-O-$ residues of the 1,4-butanediol component. The latter structural moiety can be expanded to even more flexible segments by the use of poly(ethylene glycols), PEG and PTMEG. When the $(-CH_2-CH_2-CH_2-CH_2-O-)_m$ segments become long enough ($m > 12$), then one obtains a material that belongs to the family of thermoplastic elastomers (TPEs). The TPEs must not to be confused with "thermoplastic esters," a term that describes polymeric esters such as PET, PBT, and other esters that lack elastic properties, and 'thermoset polyesters,' which are cross-linked alkyd (ex-glyptal) resins and may include MA, styrene, α-methylstyrene, and dicyclopentadiene, in which case they become UPRs. Thus, the unique feature of TPEs is that they are elastomers that are not cross-linked by primary chemical bonds, but rather by intermolecular attractive forces that survive over a reasonable temperature range and are responsible for elastic memory. However, at a given temperature threshold, the intermolecular attractive forces are challenged by thermally induced molecular motions and the material becomes plastic. In the case of the above-mentioned material, it behaves like an elastomer between −50°C and about 150°C, but above 200°C it begins to show viscoelastic behavior; that is, it can be made to flow under pressure and be molded into desired objects that behave like typical elastomers on cooling. The concept of thermoplastic elastomers is applicable to other assemblies of structural moieties that are combined in a two-dimensional polymer, provided that (1) rigid, "hard" segments contribute elastic memory through relatively strong intermolecular attractive forces and

(2) lengthy, flexible, "soft" segments contribute an opportunity for an increase in entropy by virtue of chaotic arrangements within these soft molecular regions.

We must keep in mind that the spontaneous return of an elastomeric material to its original shape (from a distorted shape induced by the application of an external force—pull or squeeze) is driven by a gain in positive entropy that accompanies the return from a generally organized (or more chaotic) state of matter.

Actually, the first examples of TPEs were the PUR elastomers discussed elsewhere. The more commonly used TPEs are polymers in which a polystyrene segment provides the hard block while the two common dienes—1,3-butadiene or isoprene, or a copolymer of ethylene and propylene—represent the soft block. However, many other TPEs can be designed (C. P. Rader, *CT* 1/88, 54–59) by combining hard and soft segments or dispersing particles of traditional, cross-linked elastomers (e.g., SBR) in a matrix of thermoplastics.

An example of a TPE is Du Pont's Hytrel (see 6.7). Another interesting family of TPEs is derived from carboxylated polymers, that is, predominately nonpolar vinyl polymers in which carboxylic acids are incorporated by means of the acrylic acid comonomer. These potential ionomers are converted into salts of multivalent metals (zinc, calcium, magnesium, aluminum, etc.), and now strong intermolecular attractive forces are provided by ionic bonds.

### Aramids, LCPs, and PARs

It is not surprising that Du Pont—the pioneer of nylon 6/6 in 1938 and the pioneer of PET film (Mylar) and fiber, as well as PBT—would not engage in the development of polymers that combine aromatic dicarboxylic acids, such as terephthalic as well as isophthalic, with aromatic diamines. In 1965 this effort led to the discovery of the family of aramids, represented by Kevlar and Nomex.

The *a*romatic poly*amide*s, or *aramids*, were the first in a series of high-performance, specialty thermoplastics (9.2.3), derived from aromatic dicarboxylic acids introduced in this chapter. For the sake of an orderly presentation, the materials derived from tri- and tetracarboxylic acids are discussed in Section 9.3.2.3.

The "specialty" nature of all these materials resides in their relatively high unit cost, and a consequent relatively low-volume demand (a few dozen million pounds in the U.S. market) based on the need for high performance that justifies a high investment. One may say that an arbitrary unit-cost dividing line between specialty and other still highly useful but less costly materials is about \$2.00–\$3.00/lb, but that most HPSTs are priced higher. This arbitrary unit-cost criterium must be used with a grain of salt because of volatile pricing policies. For example, because of growing competition in the PPS area, Phillips announced (*CW* 10/8/86) a 31% reduction in the price of its Ryton R-7 grade of PPS (from \$2.21 to \$1.45) while a year later (*CMR* 9/7/87) it introduced an improved formulation of its Ryton R-404 priced at \$3.86. Phillips's production capacity of PPS has also been expanded to a 16MM-lb/yr level.

In this highly competitive market, the relatively high cost of specialty polymers is usually a function of costly building blocks from which they are assembled and special polymerization and processing technology required for their utilization. The focus on unit prices, however, can result in a blurred picture because of existing price ranges that depend on the presence of reinforcing materials, blends with other polymers (polymer alloys), and other factors that affect the useful properties of the materials.

Kevlar and Nomex are obtained from the acid chlorides of terephthalic and isophthalic acids, respectively, matched up with *p*- and *m*-phenylenediamines, respectively. It is clear that for obtaining polymers of sufficiently high molecular weight, the polymerization of these acid chlorides and diamines is more troublesome than, for example, vinyl polymerizations (catalyzed by either free radicals or by cationic or anionic intermediates), the stepwise condensation of carboxylic acids, or their simple esters to yield polymeric esters by elimination of either water or of an alcohol or phenol in a transesterification reaction. Interfacial polymerization is a suitable method for the production of aramids and, in this case, consists of a rapidly stirred two-phase system of alkaline water (that neutralizes hydrogen chloride as it is liberated) and a solution of the acid chloride in an organic solvent that is immiscible with water but one that dissolves well the polymeric product so that it may "grow" as the polymerization progresses. The amine component can be added to such a mixture and, depending on its solubility in water, may distribute itself between the two phases. Most importantly, however, the aromatic amine—a weak nucleophile to start with—must be liberated by the alkaline solution from being an unreactive hydrochloride salt. Interfacial polymerization can also be used for the preparation of polycarbonates (PCs), but this system is less complicated because the phenolic component is soluble in the aqueous phase as a reactive phenolate salt, and all it needs is some phase-transfer catalyst to migrate into the organic phase, where it encounters the carbonate-producing reactant.

Recently, the production of Kevlar fiber caused an international dispute between Du Pont and Holland's Akzo, who began manufacturing a similar fiber in 1986 under its tradename "Twaron." Actually, the patent suit revolved around the use of the solvent system rather than the chemical structure of the material. Originally, Du Pont utilized a mixture of two dipolar aprotic solvents (3.5.3), NMP, and hexamethylphosphoramide (HMPA) according to U.S. Patent 3,767,756. However, the mutagenic properties of HMPA caused Du Pont to switch to a NMP–$CaCl_2$ solvent system that had been patented by Akzo, and the courts decided the patent infringement case in favor of Akzo (*CW* 4/6/88, *CE* 4/25/88). Fortunately, the two manufacturers negotiated an amicable settlement (*CW* 5/18/88) and intend to cooperate in the promotion of this aramid fiber, which already enjoys a worldwide demand of over 33MM lb with an annual growth in excess of 10%.

The unit prices of the aramids Kevlar and Nomex are affected strongly by the unit costs of the respective diamine since, while the two acid chlorides are priced at about the same level (Du Pont, technical grade $1.95); *p*- and *m*-phenylenediamine cost $4.00 and $2.07, respectively. This price differential is accounted for by the number of steps required to obtain the diamine from benzene (9.3.6). It is noteworthy that the unit price of Kevlar, $4.00, reflects the cost of its diamine building block.

The linear, rather rigid structure of Kevlar and the less linear but still fairly rigid structure of Nomex explain the major differences in their properties. Kevlar fibers are strong enough for use in the manufacture of bulletproof vests, and small amounts of Kevlar pulp (an aggregate of short-length fibers) can be used to reinforce other polymers; replace asbestos fibers in the manufacture of brakes, gaskets, coatings, and so on; and induce thixotropic properties in dispersions of solutions of epoxy and other resins.

Thixotropic effects are caused by attractive forces that are strong enough to create a solid network of linear particles dispersed within a liquid medium. The linear particles thus reduce viscous flow of the mixture; however, the attractive forces are not strong enough to resists turbulence created by vigorous mixing of the mixture.

The more linear structure of Kevlar, as compared to that of Nomex, explains the four- to sixfold greater tensile strength of the former since it provides greater opportunities for intermolecular hydrogen-bonding between adjacent macromolecules and their aggregates. On the other hand, the less linear structure of Nomex is responsible for its six- to eightfold greater stretching capability than its more crystalline cousin. In an advantage-ous combination of properties of both polymers, strands of the two materials woven into cables result in a nonrusting, lightweight product that is stronger than steel (on the basis of relative strength per unit of mass), and such cables are used to anchor off-shore oil-drilling platforms. Nomex is molded into structurally strong objects of low density. Du Pont's sales of both aramids is estimated at about 30MM lb, valued at about $300 million. In 1987 Du Pont announced (*CW* 3/11/87) the development of Kevlar 149 fibers. They are 40% more stiff, absorb less than half the moisture than fibers made from the original material, and are suitable, for example, for the construction of high-performance components of helicopters. Apparently Du Pont was able to improve its fiber-processing technology to enhance intermolecular hydrogen bonding between macromolecules and their aggregates, and this improvement reduces hydrogen bonding between exposed amide groups and moisture.

Actually, Kevlar also enjoys the distinction of having been the first commercially available liquid crystal polymer (LCP), which is a material in which rigid, linear macromolecular aggregates exist not only in the liquid as well as the solid state. Such orderly orientation of rodlike macromolecules is induced either by flow of the molten material or when the solid is obtained in a controlled manner from a flowing solution that is being concentrated at the same time. These two methods of achieving a highly ordered aggregate of macromolecules are referred to as *thermotropic* and *lyotropic* processing.

In 1984 Dart and Kraft introduced the LCP copolyester Xydar, in which terephthalic acid is combined with 4,4'-biphenol and *p*-hydroxybenzoic acid to yield a highly rigid structure (Scheme 9.36). In late 1987 Amoco acquired Xydar from Dartco—a subdivision of Premark International (the manufacturer of Tupperware) that had bought out Dart and Kraft.

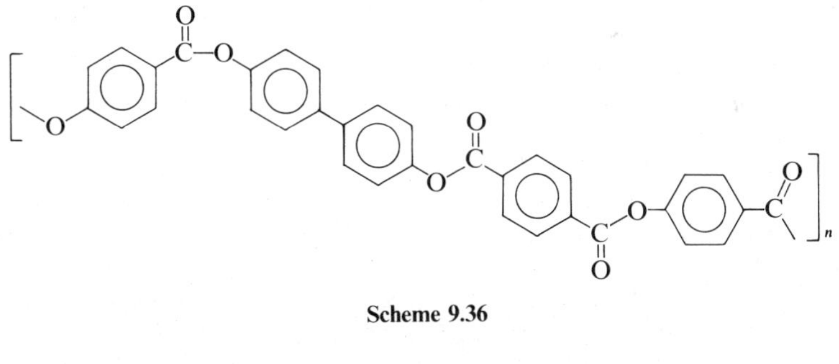

**Scheme 9.36**

A whole series of Vectra high-performance polyester LCPs was introduced in 1985 by Hoechst Celanese. They are assembled from *p*-hydroxybenzoic acid and 6-hydroxy-2-naphthoic acid.

Another recent addition to the variety of high-performance thermoplastics are the polyarylates (PARs)—amorphous phenolic esters derived from aromatic dicarboxylic acids and diphenols. The current selection of PARs is priced somewhat higher than $2.00/lb, and the present total production is the United States of about 35–40MM lb is in the hands of Amoco (Ardel), Celanese (Durel), and Du Pont (Arylon). The first representatives of this family of amorphous engineering resins are obtained by assembling mixtures of isophthalic and terephthalic acids with bisphenol A, and the polymerization methods involve either interfacial condensation of the respective acid chlorides or the transesterification of the respective diphenyl esters. The diphenyl esters of isophthalic and terephthalic acids are available from Thiokol at $5.75 and $5.00, respectively, and as mentioned above, Du Pont can supply the acid chlorides of terephthalic and isophthalic acids at $1.95. A representative structure of a polyarylate assembled from bisphenol A and isophthalic and terephthalic acids is shown in Scheme 9.37.

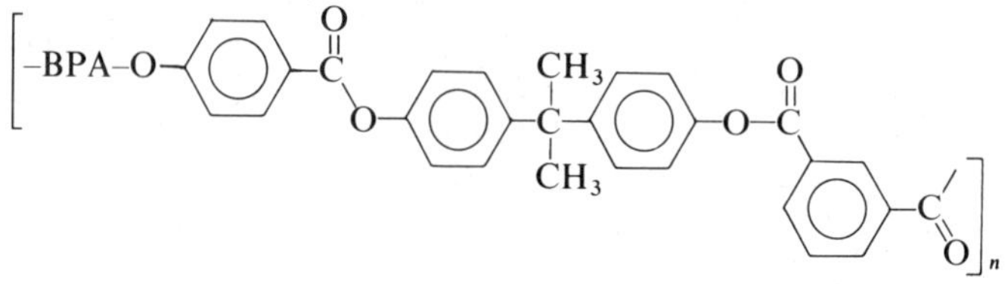

Scheme 9.37

Throughout this discussion we stress the current situation of PARs, LCPs, and other families of HPSTs because there are, obviously, many ways of varying building blocks and obtain new materials that are advantageous for given applications. A boost in this intraproduct competition among the HPSTs was provided by Amoco Chemicals' "cultural revolution" (K. Portnoy, *CW* 7/29/87, 22–24), which was marked by the 1986 acquisition of Union Carbide's operations in this area of polymers and the creation of Amoco Performance Products, Inc. (APPI), who thus far has produced some new aromatic "fine carboxylic acids" such as

- 1,1,3-Trimethyl-3-phenylindane-4′,5-dicarboxylic acid (PIDA) the oxidation product of the alkylation of two molecules of toluene with 2-methyl-1,4-pentadiene (see 5.4.1).
- 2,6-Naphthalenedicarboxylic acid (2,6-NDA) and the corresponding dimethyl ester, DM-2,6-NDA. This dicarboxylic acid is likely obtained by a selective diisopropylation of naphthalene (possibly facilitated by the use of a zeolite catalyst that promotes the formation of beta-substituted products in a process that recycles alpha-substituted naphthalenes), followed by oxidation. While such a process accounts for the relatively high unit cost of 2,6-NDA, it does provide an option to produce 2,6-dihydroxynaphthalene—an interesting variant of 4,4′-diphenol.
- t-Butyl-1,3-benzenedicarboxylic acid or 5-t-butylisophthalic acid (5-t-BIA) by isobutylation of m-xylene (with isobutylene or t-butyl alcohol), followed by oxidation.
- Di(p-carboxyphenyl) ether obtained by oxidation of di-p-tolyl ether most likely derived from p-chlorotoluene ($1.00) and p-hydroxybenzoic acid.

Unfortunately, according to an Amoco representative, unit prices cannot be estimated at this time because "there could be substantial pricing changes even by the time your book is published."

Naturally, once such building blocks are available commerically, they can be used to make products other than those indicated above. Thus, the rigid, unsymmetrical, and forcibly nonplanar structure of PIDA can be used to assemble thermoplastic and thermosetting (alkyd) polyester and polyamide resins, especially for use as coating materials or for modifying such products obtained from more traditional ingredients. The same is true of 2,6-NDA, which offers a symmetrical, rigid structural moiety in which the relatively large molecular area of the naphthalene moiety, as compared with that of terephthalic acid, provides a greater opportunity for intermolecular associations by a vertical stacking of aromatic rings from neighboring macromolecules that promotes the operation of London dispersion forces. The PET analog obtained from 2,6-NDA, or Amoco's PEN, is a component of magnetic tapes manufactured by an improved vapor deposition of magnetic oxides.

Besides serving as a building block of polymers, the naphthalene ring of 2,6-NDA offers the opportunity to prepare derivatives such as sulfonic acids at the unsubstituted alpha positions. Thus, the sulfonation of 2,6-NDA followed by a reaction with 2-amino-*p*-cresol is reported to give bis(benzoxazole)-substituted sulfonaphthalenes that function as optical brighteners by virtue of their fluorescence. These materials are incorporated in fibers or films in order that they may absorb a yellow tinge intrinsic to these materials; they then emit the absorbed light at a longer wavelength and thus brighten the appearance of the materials to the observer (Scheme 9.38).

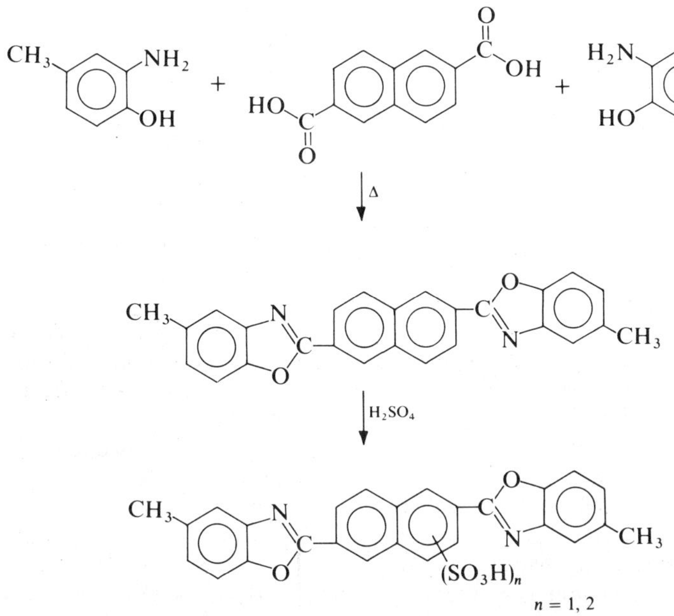

$n = 1, 2$

**Scheme 9.38**

Aromatic dicarboxylic acids in which the functional groups are isolated from each other can be converted to polymers of the benzimidazole family by reaction with 3,3′,4,4′-tetraaminobiphenyl (TAB) or 3,3′-diaminobenzidine (DAB; see 9.3.6) and analogous systems symbolized here by $[o\text{-}(H_2N)_2C_6H_3]_2Q$ (where Q represents either a bond that links the substituted biphenyl moiety, or such moieties as –O–, –CH$_2$–, and –CO–) (Scheme 9.39).

In 1983 Celanese began the production of its *poly(benzimidazole)* (PBI) in a 1MM-lb plant using diphenyl isophthalate and imported DAB. The resulting high-temperature- and flame-resistant fiber was found to be useful in the production of safety gloves and where Q = $-O-$, $-CH_2-$, $-\overset{\overset{\displaystyle O}{\|}}{C}-$, $-$, etc.

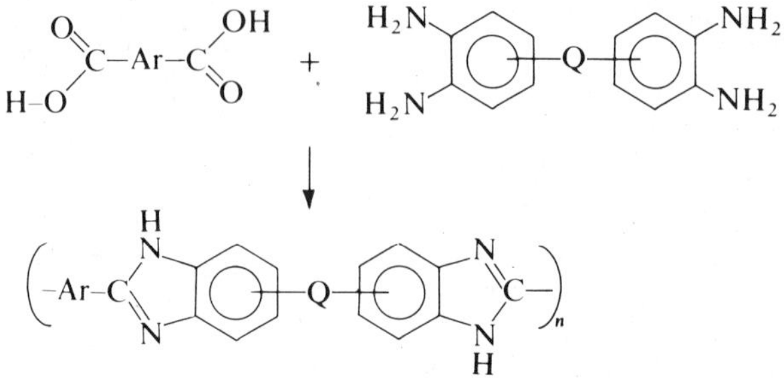

Ar = aromatic moiety of dicarboxylic acid

**Scheme 9.39**

other protective clothing. In 1987 Cleanese and Osmonics announced ($CW$ 4/7/87) a joint effort to develop PBI-based membranes for reverse osmosis and ultrafiltration applications.

### 9.3.2.3  Trifunctional and Higher Functionality Carboxylic Acids

The presence of more than two carboxylic acid groups broadens the horizon of chemical opportunities for these benzenoid building blocks.

In the case of 1,3,5-benzenetricarboxylic, trimesic acid, another member of Amoco's "fine acid" family derived—as the name suggests—from mesitylene, the trifunctionality can be utilized to modify linear polyesters, such as PET, by introducing either a small or a high degree of cross-linking that results in a thermosetting polymer. The former modification basically retains the thermoplastic, mostly two-dimensional molecular structure except for a significantly increased average molecular weight of the final product, and the resulting fibers and films exhibit higher thermomechanical resistance and decreased permeability to gases. Naturally, trimesic acid can be used to modify any polymer that contains functional groups such as amino and phenolic capable of interacting with the three independently functioning carboxylic acid constituents.

Trimesic acid esters of medium-sized alcohols ($C_7$–$C_8$) function as plasticizers.

The chemical possibilities of tricarboxylic and higher benzenoid carboxylic acids become even more interesting in the case of vicinal carboxylic acid groups in which both functional groups can react in unison. These analogs of phthalic acid readily form anhydrides that can serve as intermediates in the formation of polymeric amide–imides and polyimides (PIs). These possibilities are illustrated by the formation of polymers from the two prototypes of such chemical systems, namely, trimellitic anhydride (Amoco's TMA, 80¢) and pyromellitic dianhydride (PMDA, Allco, $6.95), obtained by oxidation of 1,2,4-trimethylbenzene or pseudocumene, and 1,2,4,5-tetramethylbenzene or durene, respectively (Scheme 9.40).

In the case of TMA, while the isolated carboxylic acid group forms amides by reaction with isolated diamines (such as the above-mentioned phenylenediamine building blocks of Du Pont's Kevlar and Nomex), the diacarboxylic anhydride yields an imide moiety. The result is a polyamide–imide such as Amoco's Torlon (Scheme 9.41).

Trimellitic anhydride can also react with vicinal aminophenols to produce the benzoxazole system (as illustrated above in the case of the optical brightener) and with

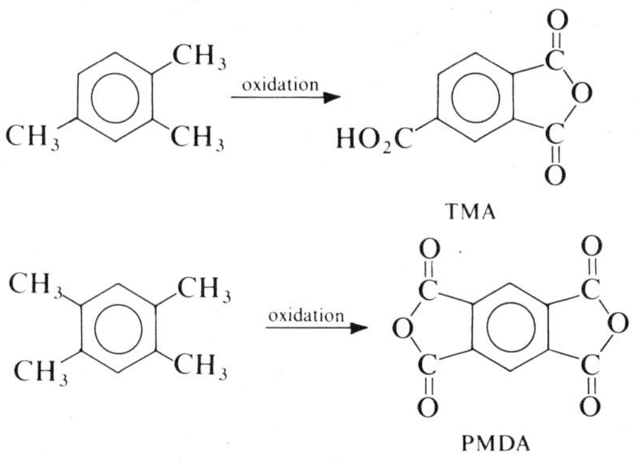

TMA

PMDA

**Scheme 9.40**

vicinal diamines it yields benzimidazoles (as shown in 9.3.2.2). With di-functional and higher functionality alcohols TMA gives rise to cross-linked, thermoset polyesters.

All three carboxylic acid groups of TMA participate in the formation of trimellitate plasticizers such as the Stayflex products manufactured by Reichold (acquired recently by Dai Nippon Ink and Chemicals of Tokyo) shown here with 1984 prices:

- Triisooctyl trimellitate (TIOTM), 74¢–80¢
- Tri-2-ethylhexyl or trioctyl trimellitate (TOTM), 74¢–80¢;
- n-Octyl/n-decyl trimellitate, NONDTM, 88¢–94¢.

The triallyl trimellitate cross-linking agent is Alcolacs's Sipomer TATM ($2.04).

Derivatives of TMA can be prepared in a selective manner by first converting it to an acid chloride–anhydride. For example:

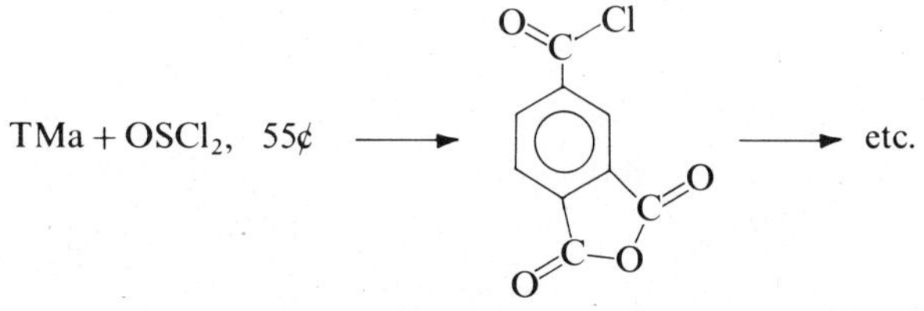

TMa + OSCl$_2$,  55¢  ⟶   ⟶ etc.

In the United States Du Pont has pioneered the production of the high-performance polyimide film Kapton assembled from PMDA and bis(4-aminophenyl) ether. The material is priced at about $50/lb, resists temperatures as high as 400°C, and is infusible, flame-resistant, and insoluble in practically all organic solvents. It is obvious that processing of materials of this kind is a major technological challenge, and while Du Pont has mastered the processing of Kapton, other polyimides have been developed that are more tractable. This is achieved by using either a combination of less symmetrical, less planar dianhydrides than PMDA and/or less symmetrical diamines of greater flexibility than bis(4-aminophenyl) ether as a result of greater rotational freedom at the central link.

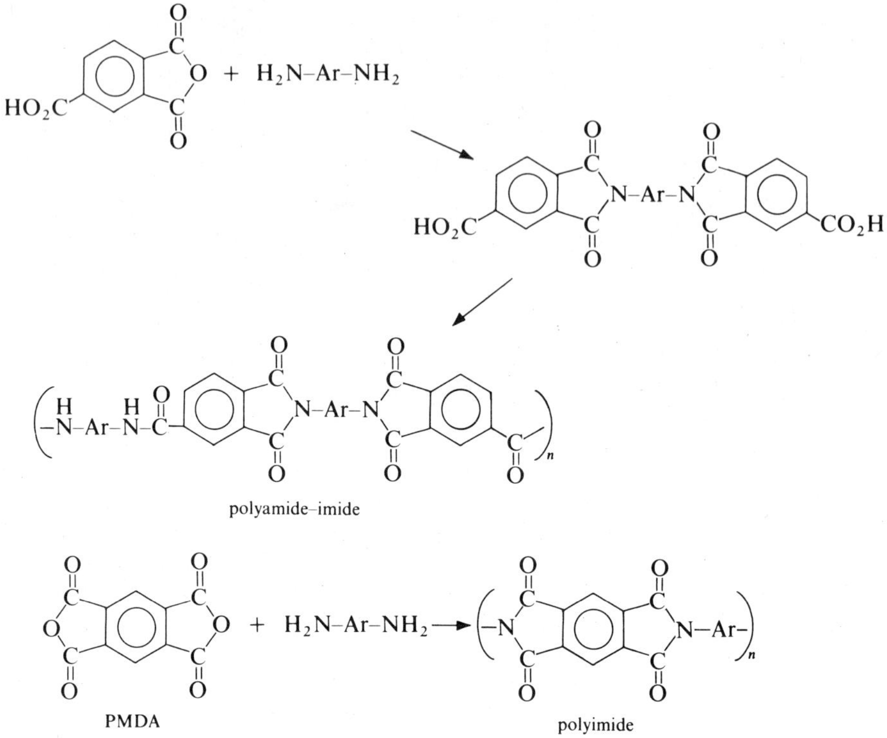

polyamide–imide

PMDA                    polyimide

where Ar represents an aromatic moiety in

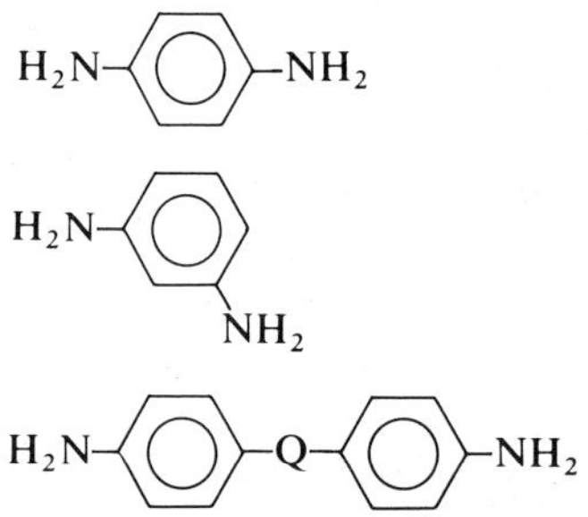

$Q = -, -CH_2-, -O-,$ etc.

**Scheme 9.41**

Two examples of building blocks that attain the first-mentioned objective are

- 3,3′,4,4′-Benzophenonetetracarboxylic dianhydride, Allco's BTDA, \$5.55, obtained by oxidation of the alkylation product of *o*-xylene and acetaldehyde (9.2.1):

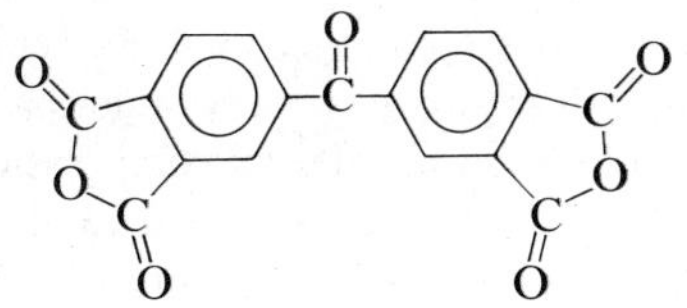

- 2,2-Bis(3′,4′-anhydrodicarboxyphenyl)-hexafluoropropane,  or  Allco's  6-FDA, $200.00, obtained by oxidation of the alkylation product of *o*-xylene with costly hexafluoroacetone:

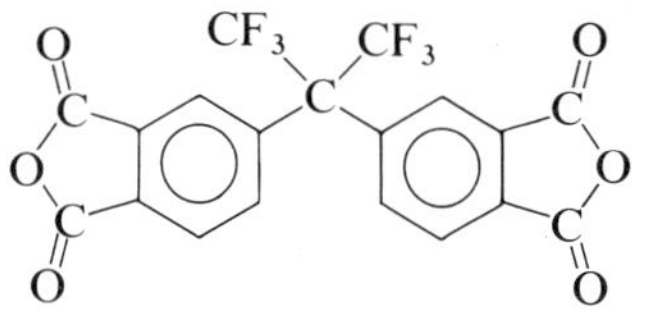

Examples of building blocks that attain the second objective are diamines in which two or more *m*- or *p*-aminophenyl groups are linked by groups other than an ether linkage, such as $-CH_2-$, $-C(CH_3)_2-$, $-CH_2-CH_2-$, and $-S-$. Also, one can employ a mixture of diamines such as is obtained [Ciba-Geigy's diaminophenylindane (DAPI)] obtained by nitration of either the condensation product of benzene and 2-methyl-1,4-pentadiene (see 5.4.1), or the dimerization product of isopropenylbenzene, followed by reduction. See Scheme 9.42.

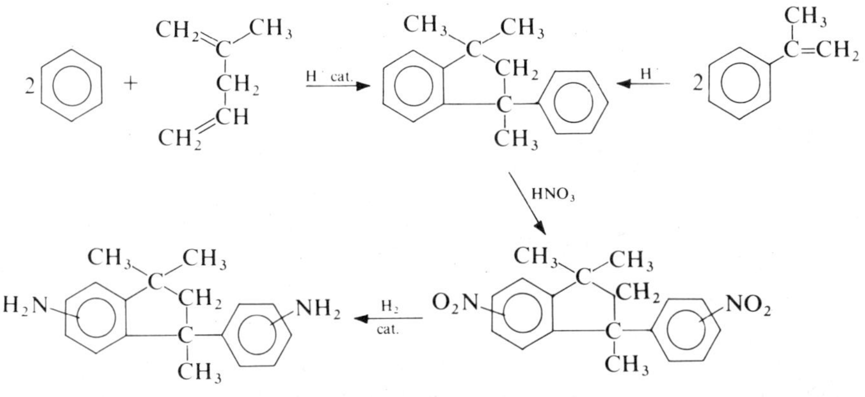

DAPI

**Scheme 9.42**

While still on the subject of polyimides, a somewhat different purpose is served by 5-norbornene-2,3-dicarboxylic acid anhydride—the Diels–Alder adduct of cyclopentadiene and MA (6.6.9, 6.8.3) or its methyl ester. This material is used as a comonomer to terminate chain growth during the formation of an oligoimide derived from BTDA and a suitable diamine. The resulting prepolymer (shown in Scheme 9.43) to be formed from the methyl ester of BTDA (or BTDE) and 4,4′-methylenedianiline is then thermally cross-linked by cracking the terminal norbornene ring at about 200 atm and above 325°C. Polymers of this type are members of the PMR family of thermoset polyimides developed at the NASA Lewis Research Center as part of a program designed to produce highly temperature resistant materials that, however, do not behave like "ground firebrick." In other words, that resist very high temperatures but can still be processed, that is, shaped, into desired objects.

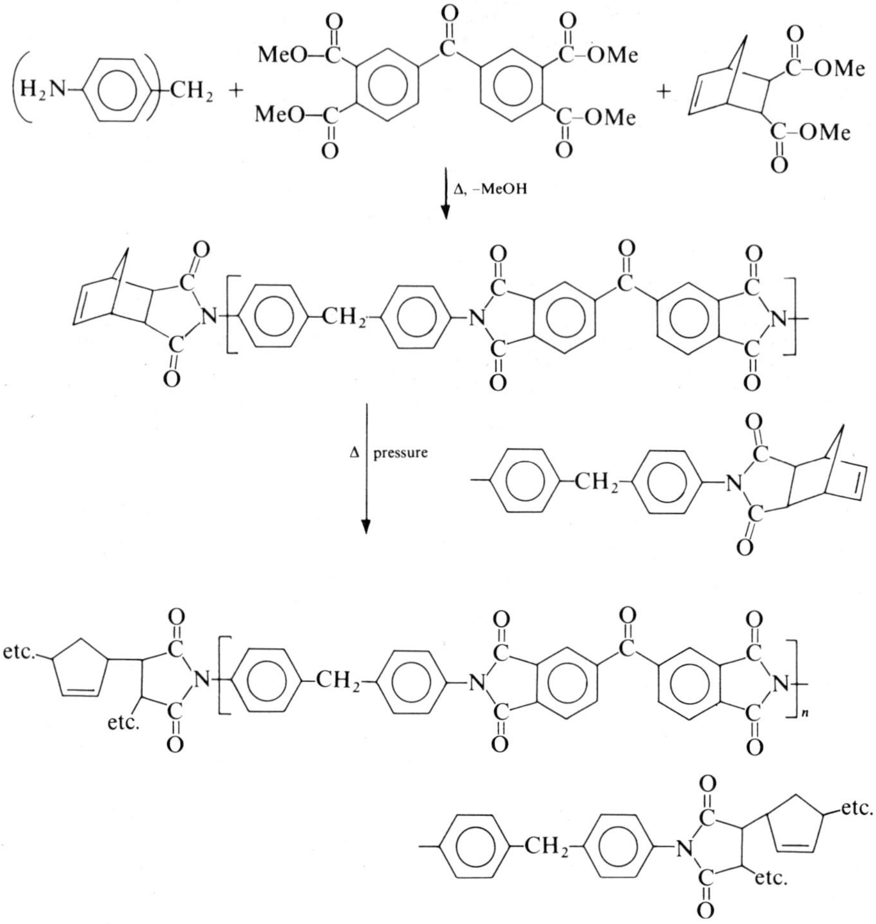

(where $n = 2$ and "etc." indicates bond formation in final, cross-linked polymer)

**Scheme 9.43**

Finally, it is possible to employ an isocyanate in place of an amine in order to prepare the desired imide from a dicarboxylic anhydride with a potentially advantageous liberation of $CO_2$ gas, as shown in Scheme 9.44.

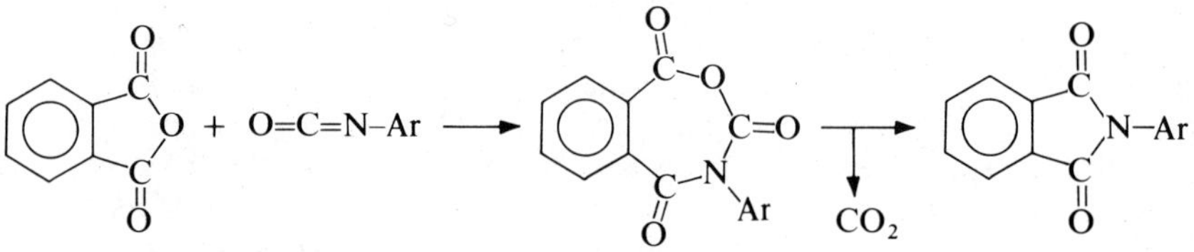

**Scheme 9.44**

Considering the multitude of factors that can be varied in the design of polyimide materials, it is not surprising that new versions of PIs have appeared in recent years.

The prospect of a larger and less costly supply of durene by-product of the MTG process may stimulate the production of materials other than the high-performance specialty polymers described above.

### 9.3.3  Benzenoid Sulfur-Containing Chemicals

The introduction of the sulfo ($-SO_3H$) and chlorosulfonyl ($-SO_2-Cl$) substituents, as well as the formation of aryl sulfones, is mentioned in 9.2.4 with emphasis on the large-volume production of sulfonate surfactants. Additional examples of sulfur-containing chemicals of industrial interest are offered at this point.

The customary scarcity of sucrose (table sugar) during wartime (due to endangered shipping of sugarcane-derived sucrose from tropical areas) led to the foundation of the Monsanto Chemical Company in St. Louis, Missouri during World War I. This two-man operation (the imported Swiss chemist Gaston Dubois and the capital provider, Mr. Queeny, who was married to a woman whose maiden name was Monsanto) grew into a company that we now know as "Monsanto"—one of the largest chemical businesses in the United States (1987 sales of some $7.64 billion) and, for that matter, in the world. The embryonic enterprise developed around the idea of supplying the synthetic sweetener saccharin discovered in 1879 by the American chemist Ira Remsen, who had the curious habit of eating salami sandwiches while performing chemical experiments—a practice that today's OSHA and EPA would definitely frown on. Remsen's serendipitous discovery occurred during preparation of a mixture of *o*- and *p*-isomers of chlorosulfonyl-toluene, which involved converting the somewhat oily product into the corresponding crystalline toluenesulfonamides (melting points of *o*- and *p*-isomers are 156 and 139°C, respectively) and separating them by crystallization. It is fortunate that Remsen did not know that the *o*-isomer would eventually carry a label "Cancer suspect agent." Each isomer was oxidized separately with potassium permanganate (currently priced at $1.13), and to Remsen's surprise, it was the derivative of *o*-toluenesulfonamide that radically changed the taste of the salami sandwich. The transformations that convert toluene to saccharin or *o*-benzoicsulfimide and other related derivatives are summarized in Scheme 9.45. Actually, soluble saccharin (i.e., its sodium salt; FCC $ granular $2.75, powder $3.70) is about 500 times as sweet as sucrose, which has greatly benefited diabetics (and diet-conscious consumers) over the past 60 years. An experiment with rats that gorged themselves with food containing high concentrations of saccharin and subsequently developed bladder tumors caused the FDA in 1977 to ban saccharin for human use—a repeat performance of the 1969 ban of cyclamate. Protests by an aroused public that included decade-long consumers of this sweetener convinced the U.S. Congress to rescind the FDA ban and to declare the saccharin issue as unresolved even though it carried the above-mentioned warning label. In a yearly ritual, each Congress since 1977 has extended the moratorium on the FDA ban of saccharin, presumably out of respect for their colleague—the author of the Delaney Amendment to the Federal Food, Drug, and Cosmetic Act of 1938—and without a hearing passes the bill to the Senate. At about the same time Monsanto abandoned the production of saccharin and acquired Searle, the proprietor of 1968 and 1970 U.S. patents for the sweetener aspartame. While aspartame has displaced the use of saccharin in most soft drinks, the superior stability of saccharin to heat is the reason for its continued use by the canned food industry as well as its price, which is lower than that of aspartame: $3.00–$4.00, depending on physical form, as

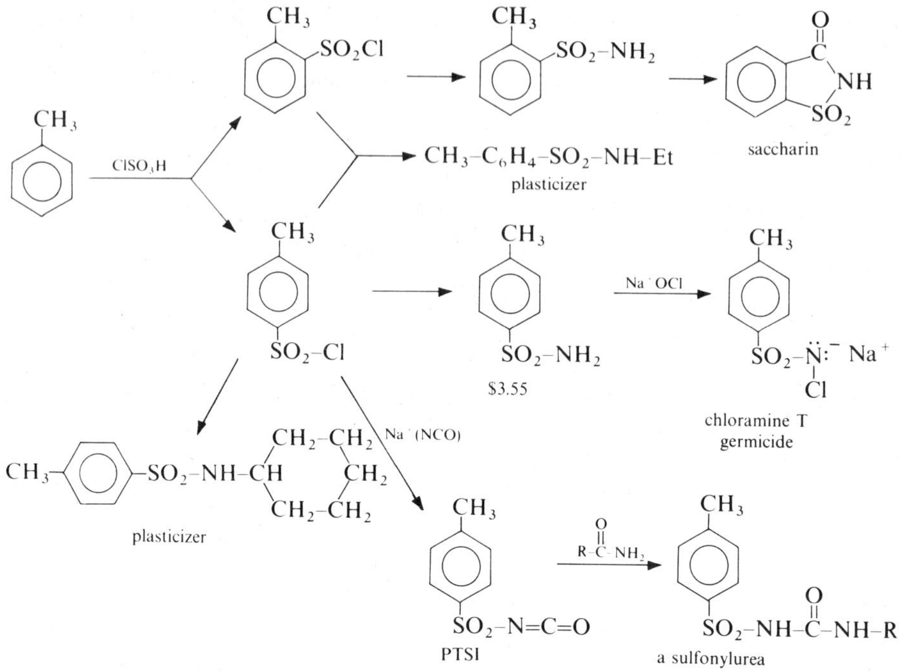

CH₃ — $SO_2Cl$  →  CH₃ — $SO_2$–NH₂  →  saccharin

CH₃ (toluene)  —ClSO₃H→

$CH_3$–$C_6H_4$–$SO_2$–NH–Et  (plasticizer)

CH₃ — $SO_2$–Cl  →  CH₃ — $SO_2$–NH₂  ($3.55)  —Na OCl→  CH₃ — $SO_2$–Ṅ:⁻ Na⁺ with Cl (chloramine T, germicide)

$CH_3$—⟨⟩—$SO_2$–NH–CH(CH₂–CH₂)(CH₂–CH₂)  plasticizer   Na (NCO)

CH₃ — $SO_2$–N=C=O  (PTSI)  —R–C(=O)–NH₂→  CH₃ — $SO_2$–NH–C(=O)–NH–R  (a sulfonylurea)

**Scheme 9.45**

compared to 10–15-fold higher unit cost of aspartame. The price of the better-tasting calcium salt of saccharin is about $5.50 (*CMR* 5/16/88). Saccharin is still produced in the United States by PMC Specialties, but much is imported about 1.6MM lb in 1987, mostly from Japan and South Korea. Clearly, the unwarranted warning label discourages some consumers who tend to believe what they see in print.

The fascinating topic of artificial sweeteners is beyond the scope of this book, but suffice to say that more powerful synthetic sweeteners than saccharin and aspartame are on the industrial horizon.

The by-product of the original saccharin process, namely, *p*-toluenesulfonyl chloride, or tosyl chloride, is converted to some amide plasticizers (see 9.2.4), it is used as a building block of antidiabetic diuretics of the sulfonyl urea family (see below) and as a potential intermediate in the production of *p*-toluenesulfonyl isocyanate (PTSI; Upjohn, $5.00). The latter can be also prepared from *p*-toluenesulfonyl chloride and sodium cyanate (Na⁺ ⁻(NCO), 85¢), except that this ambident nucleophile may give undesired by-products. In any case, PTSI is employed to remove traces of water from water-sensitive reactions (such as the preparation of PURs):

$$Ar-SO_2-N=C=O + H_2O \rightarrow Ar-SO_2-NH-CO_2H \rightarrow Ar-SO_2-NH_2 + CO_2$$

where Ar = *p*-tolyl.

The preparation of a new class of sulfonylurea herbicides by Du Pont (Glean and Oust) starts with *o*-chloro and *o*-carbomethoxy derivatives of benzenesulfonyl isocyanate, respectively (10.3.3).

The production of most traditional antibacterials of the sulfa family commences with the chlorosulfonation of aniline, or more accurately speaking, with the chlorosulfonation of acetanilide in which the amino function of aniline is temporarily protected by acetylation in order to avoid its undesirable reactions with chlorosulfonic acid. The resulting $p$-chlorosulfonylacetanilide is then allowed to react with ammonia or a variety of amino compounds (symbolized here by $Q-NH_2$, where Q most often represents a heterocyclic moiety mentioned in Chapter 10) to produce a series of sulfa drugs. The simplest sulfa drugs are sulfanilamide ($p$-$H_2N-C_6H_4-SO_2-NH_2$; NF \$3.33) and sulface-tamide ($p$-$H_2N-C_6H_4-SO_2-NH-CO-CH_3$; USP \$9.08) (Scheme 9.46). Literally thousands of structural modifications of sulfanilamide were studied in order to determine antibacterial selectivity, human toxicity or undesirable side effects, and solubility. The latter property can affect painful depositions of crystalline particles in the kidney; hence combinations of sulfa drugs are often prescribed in place of a larger dose of a single substance. The glamor associated with the use of sulfa drugs was diminished considerably by the introduction of antibiotics (starting with penicillin during World War II; *SZM-II*, 43–47), but, as is evident from the industrial market prices cited above, the use of the sulfa drugs in human and, particularly, in veterinary medicine survives to this day.

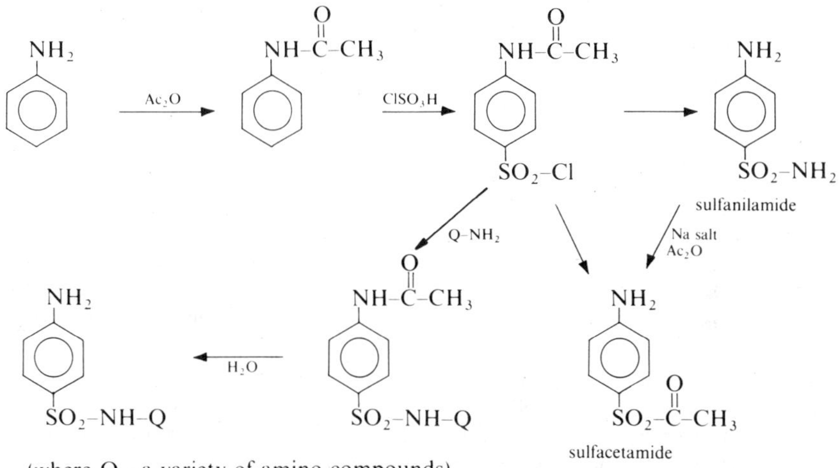

(where Q = a variety of amino compounds)

**Scheme 9.46**

Sulfanilic acid is a low-priced chemical because it is obtained directly from aniline (62¢) and sulfuric acid (100% virgin grade South East 3.4¢, East Coast 3.6¢, Midwest 4.01¢, West Coast 4.25¢) by heating the initial aniline monosulfate salt to about 180°C when the intermediate phenylsulfamic acid rearranges first to orthoanilic and then to sulfanilic acid (Scheme 9.47). Sulfanilic acid is a precursor of many dyes because its amino group can be diazotized and coupled with appropriate structural components to yield the desired azo chromophore (9.3.6), while the sulfo group induces water-solubility and provides a strongly acidic function that promotes an attachment of the dye to somewhat basic substrates.

The unsymmetrical chlorine-containing diphenyl sulfone, tetradifon (2,4,5,4'-tetra-chlorodiphenyl sulfone; Chemical Dynamics, ~ \$10), is an acaricide and ovicide used on

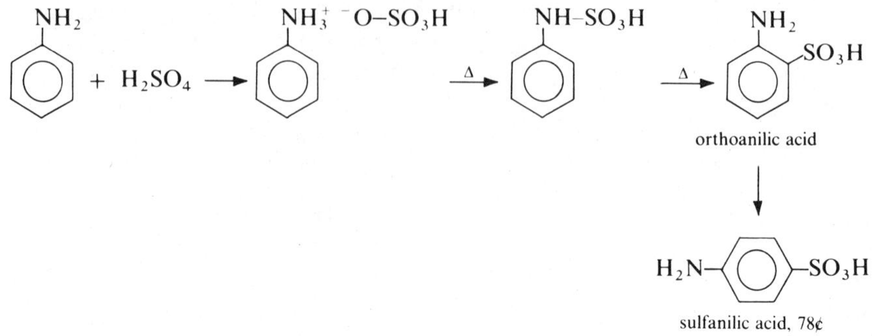

**Scheme 9.47**

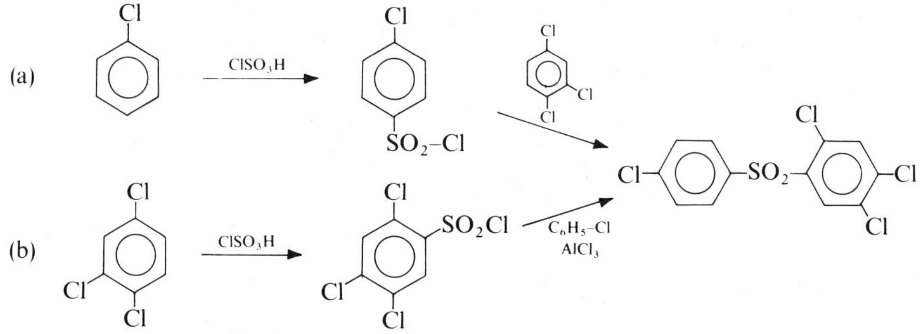

**Scheme 9.48**

deciduous fruit trees, citrus, and cotton, and there are two possible routes to its synthesis. The first (a in Scheme 9.48) starts with the chlorosulfonation of chlorobenzene and is followed by a Friedel–Crafts reaction of p-chlorobenzenesulfonyl chloride with 1,2,4-trichlorobenzene (70¢), while the second (b in Scheme 9.48) starts with a chlorosulfonation of 1,2,4-trichlorobenzene and is followed by the Friedel–Crafts reaction of the resulting 2,4,5-trichlorobenzenesulfonyl chloride with chlorobenzene (Scheme 9.48). Route b seems to be more practical than a because

1. The Friedel–Crafts reaction requires a higher energy of activation than chloro-sulfonation, and an excess of chlorobenzene serves a dual purpose as a reactant and as a solvent.
2. The formation of undesirable ortho-substituted derivatives of chlorobenzene is minimized in route b because the attacking reactant is a bulky sulfonyl chloride and its entry at the position ortho to the chlorine substituent is sterically inhibited.

*Sulfone-Containing Polymers*

As noted in 9.2.4, vigorous chlorosulfonation of chlorobenzene leads to the direct formation of di(p-chlorophenyl) sulfone or 4,4′-dichlorodiphenyl sulfone, and this material serves as a building block of some sulfone-containing HPSTs mentioned in 9.2.4. At this point, we may add to this topic that the use of DMSO (b.p. 189°C) as the dipolar aprotic solvent is a good choice because this excellent solvent is stable under alkaline conditions of the reaction carried out at about 135°C. Union Carbide's Udel PSO ($3.82) was introduced in 1965, and, obviously, it must be processed above its glass-transition temperature $T_g \sim 190°C$ and competes with some of the ICI's Victrex poly(aryl ether–sulfones) ($4.25–$6.50).

In all the sulfone-containing polymers, the diphenyl sulfone moiety provides chemical stability because of the delocalization of the $\pi$ electrons of the benzene rings into the $d$-orbitals of the sulfone group, and, in this particular structure, the nonbonded electrons of the $p,p'$ ether functions reinforce this resonance phenomenon. The same resonance phenomenon accounts for the rigidity and hence crystallinity of the diaryl sulfone moiety (the melting point of diphenyl sulfone itself is 128–129°C as compared to the $-40°C$ m.p. of the related diphenyl sulfide). The flexible ether and/or isopropylidene moieties provide sufficient rotational freedom so that the materials become plastic and can be processed, albeit at rather high temperatures; by the same token, however, this is evidence that the end products are temperature-resistant.

Another sulfone-containing polymer is the condensation product of $m,m'$-diamino diphenyl sulfone and the tetracarboxylic ketone BTDA (9.3.2.3). The $m,m'$ location of the amino groups in the diphenyl sulfone reveals that its precursors are the $m,m'$-dinitrodiphenyl sulfone and the corresponding diamine (Scheme 9.49). This sulfone-containing thermoplastic is best described as a poly(aryl imide–sulfone). (We ignore in this name the presence of the ketone function because it is not directly responsible for the assembly of the polymer.) In 1987 it was still under development by Celanese under the acronym "PISO" (where the "O" represents the oxygenated state of the sulfur moiety—a source of future confusion if the corresponding sulfoxide should appear on the industrial horizon. Generally, diaryl sulfoxides are not as stable as sulfones but do offer unique properties that could be of interest. In any case, the $T_g$ of PISO is reported to be 273°C and requires processing temperatures as high as 350°C. The higher $T_g$ of this poly(aryl imide–sulfone), as compared to that of Udel, can be attributed to the planarity of the aromatic imide moiety and the presence of the ketone function in place of the isopropylidine group. The $m,m'$ orientation of substituents in the diphenyl sulfone moiety (in place of the $p,p'$ orientation in Udel) prevents an even higher $T_g$ and thus greater processing difficulties.

Excessive processing difficulties may have contributed to the Union Carbide's withdrawal of the poly(aryl sulfone) derived from $p$-biphenylsulfonyl chloride subjected

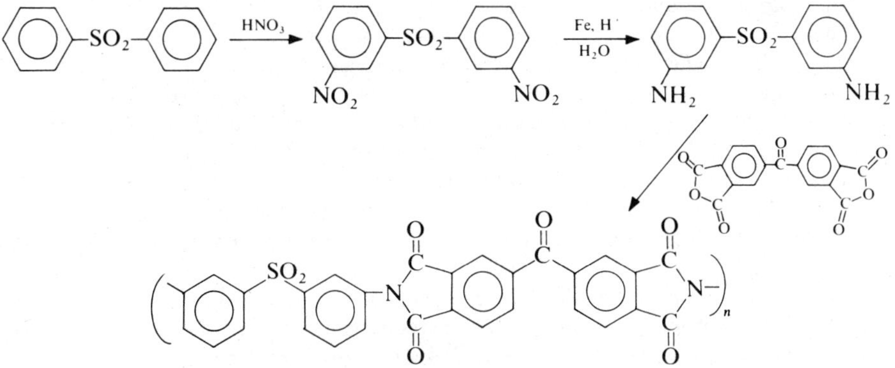

**Scheme 9.49**

to a Friedel–Crafts reaction (Scheme 9.50). The rigidity of the macromolecules in this poly(p-biphenyl sulfone), introduced as Union Carbide's Radel, may be decreased somewhat by promoting some ortho substitution or by including some flexible or less rodlike aromatic comonomers. Since UC plans to reintroduce this type of poly(aryl sulfone) at a projected unit price of about \$7/lb, it seems reasonable to assume that such macromolecules can be assembled by Friedel–Crafts reactions of arylsulfonyl chlorides, provided the presence of the chlorosulfonyl group does not interfere with electrophilic substitutions at another end of the monomer. This structural requirement is necessary in view of the deactivating effect of the chlorosulfonyl group in electrophilic reactions when the latter are attempted with a chlorosulfonyl-substituted benzene moiety. An example of a situation that avoids this problem is the synthesis of a poly(aryl sulfone) by the Friedel–Crafts reaction of the chlorosulfonyl derivative of diphenyl ether (Scheme 9.51). Here the site of the Friedel–Crafts reaction is isolated by the ether structure from the deactivating effect of the chlorosulfonyl group. Even though the final product can be described as an aromatic poly(ether–sulfone), the term "poly(aryl sulfone)" is more appropriate since the ether function is not directly involved in the assembly of the macromolecules. Amorphous poly(aryl sulfones) can be obtained by the judicious choice of a chlorosulfonyl derivative of an unsymmetrical aromatic hydrocarbon such as the condensation product of benzene and 2-methyl-1,4-pentadiene, which is utilized by Ciba-Geigy for the production of DAPI.

**Scheme 9.50**

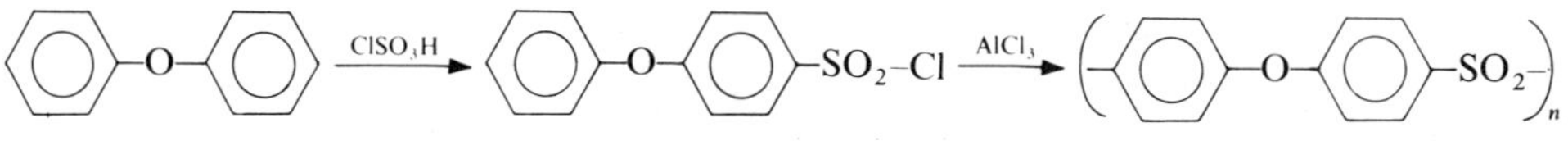

**Scheme 9.51**

The preceding examples of nucleophilic substitution reactions of p,p'-dichlorodiphenyl sulfone also suggest its conversion with ammonia to p,p'-diaminodiphenyl sulfone. However, this compound can be also obtained by rather vigorous chlorosulfonation of acetanilide (compare with the preparation of sulfa aritibacterials), followed by hydrolysis of the acetyl groups. In any case, p,p'-diaminodiphenyl sulfone (dypsone) and its glucose–sodium bisulfite adduct (glucosulfone sodium) are used to combat Hansen's disease, or leprosy. Another medicinal derived by chlorosulfonation, in this case of p-chloroacetanilide, is diazoxide (3-methyl-7-chloro-1,2,4-thiadiazine-1,1-dioxide). This antihypertensive is obtained by treating the sulfonyl chloride with ammonia and a cyclization reaction. The structures of these medicinals are shown in Scheme 9.52.

The incorporation of p-sulfostyrene monomer into vinyl polymers improves the mechanical and dyeing properties of the otherwise chemically inert products (e.g., PAN). This monomer can be prepared by means of a temporary protection of the vulnerable carbon–carbon double bond of styrene (Scheme 9.53).

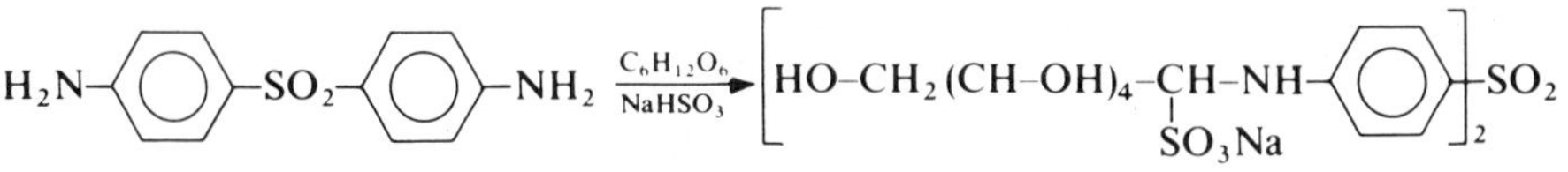

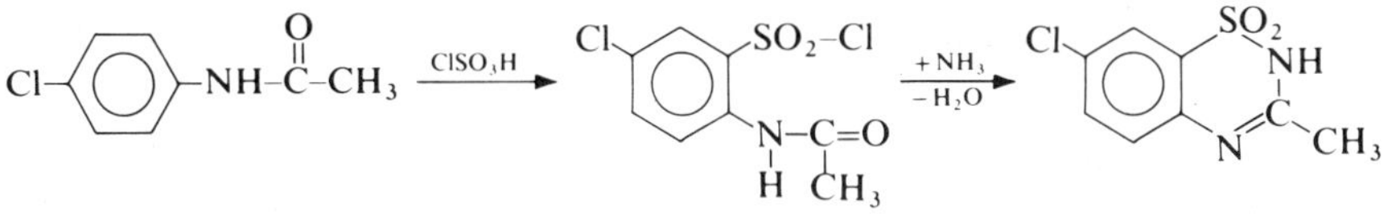

**Scheme 9.52**

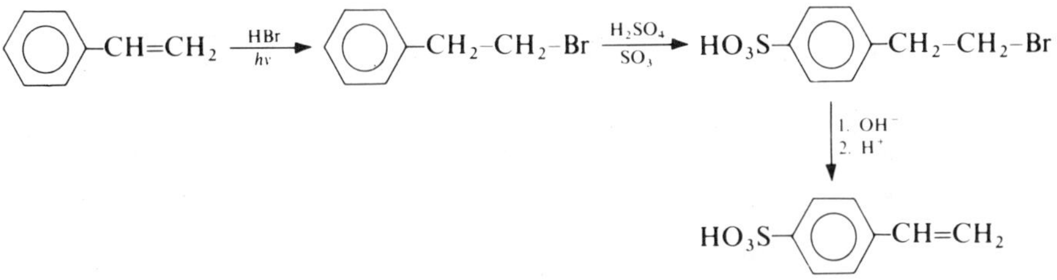

**Scheme 9.53**

An effective blowing agent of the sulfonylhydrazide family is prepared by starting with a double chlorosulfonation of diphenyl ether (Scheme 9.54).

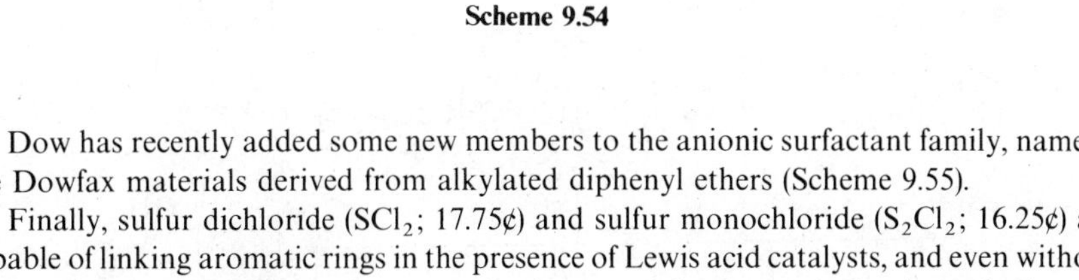

4,4′-oxy-bis(benzenesulfonylhydrazide)

decomposition products $+2N_2$

**Scheme 9.54**

Dow has recently added some new members to the anionic surfactant family, namely, the Dowfax materials derived from alkylated diphenyl ethers (Scheme 9.55).

Finally, sulfur dichloride ($SCl_2$; 17.75¢) and sulfur monochloride ($S_2Cl_2$; 16.25¢) are capable of linking aromatic rings in the presence of Lewis acid catalysts, and even without the need of the latter when the aromatic rings contain electron-releasing substituents. Thus, for example, phenol and $SCl_2$ react to produce thiodiphenol or $p,p'$-dihydroxydiphenyl sulfide, an effective larvicide and an intermediate in the preparation of some dyes and polymers. It is significant that when Crown-Zellerbach—a 2MM-lb/yr producer of this compound, priced at \$3.35—announced the intention of ceasing its production, this decision was abandoned on demand of the thiodiphenol customers (*CMR* 8/11/86). Other examples of electrophilic substitution reactions of the sulfur chlorides are presented in 10.3.2.

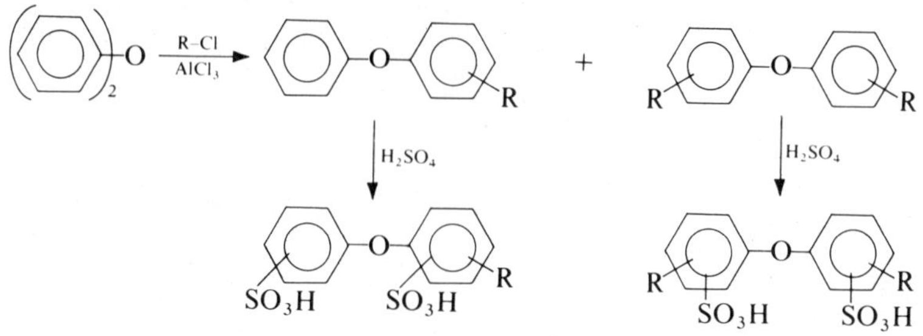

(where R– represents linear $(C_{10}-C_{16}$ branched $C_{12})$

**Scheme 9.55**

The preparation of PPS from *p*-dichlorobenzene and sodium sulfide bears a superficial resemblance to the preceding formation of a diaryl sulfide.

### 9.3.4 Phenols and Derivatives

#### 9.3.4.1 Monofunctional Phenols

Phenol is one of the large-volume, third-generation commodity chemicals (see Table 1.1). The demand for phenol in the United States exceeds 3B lb and is matched by a production capacity of about 3.3B lb. Only about 0.5B lb of the production utilizes the toluene–benzoic acid–phenol route rather than the cumene-based Hock process (Scheme 9.56).

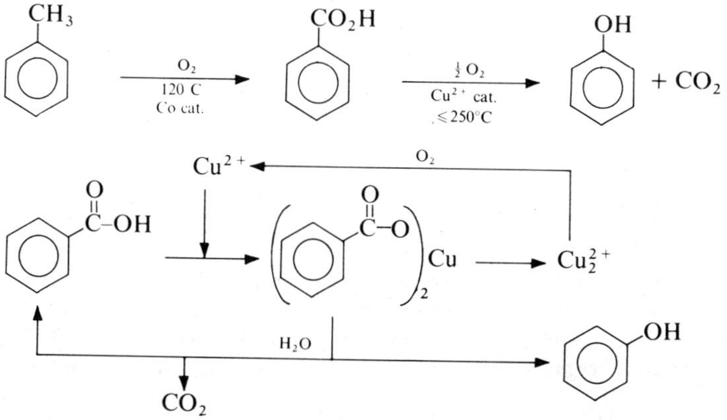

**Scheme 9.56**

The energy-intensive alkaline hydrolysis of chlorobenzene to phenol requires the use of copper catalysts and temperatures as high as 450°C. This procedure is obsolete except when modified to produce *o*-phenylphenol, diphenyl ether, and other useful by-products (9.2.3, Scheme 9.12). Some phenol is isolated during petroleum-refining operations.

Still in the developmental stage is the catalytic hydrodealkylation (HDA) of coal-tar cresylic acids announced by Hydrocarbon Research (AICHE Meeting, November 1981) and represented here in a simplest possible fashion as

$$CH_3-C_6H_4-OH + H_2 \rightarrow C_6H_5-OH + CH_4$$

Once the technology of this HDA process is worked out, it may become a source of simple phenolics from abundant lignins (*SZM-II*, 148–157).

The three greatest uses of phenol are

1. The production of phenol–formaldehyde (PF) resins, which accounts for somewhat over one-third of the total demand but it is dependent on the economic health of the construction industry.
2. The production of bisphenol A (BPA), which absorbs about 25% of the phenol demand and is growing steadily (the production of 823.4MM lb in 1986 rose to 1.00B lb in 1987) because of the key role of this intermediate in the production of PCs, epoxy resins, and specialty products such as several HPSTs and halogenated flame-retardants.
3. The production of caprolactam by way of the reduction of phenol to cyclohexanol (8.3.1).

Smaller quantities of phenol are consumed for the production of

• Alkylphenols that end up mostly as surfactants except for 2,6-dimethylxylenol—the building block of PPE—and the alkyl phenols used widely as disinfectants (see below).
• Aniline (since 1982).
• Salicylic and *p*-hydroxybenzoic acids.
• Phenolphthalein and numerous other fine chemicals, some of which are mentioned below.

The currently weak dollar is stimulating exports of domestic phenol and its major derivatives.

Some recent developments that illustrate the dynamics of phenol-based chemicals are

• The traditional PF materials, with sales of about 2.7B lb, are challenging the PUR-dominated $400-million foam-insulation market (*CW* 10/21/87) on the strength of less costly material costs.
• The traditional $C_8$–$C_{12}$-alkylated phenols that have been sulfonated, ethoxylated, and otherwise chemically modified to produce a variety of surfactants. These are now subjected to additional chemical transformations; for example, Texaco is announcing the availability of surface-active polyether amines derived from nonylphenol:

$$p\text{-}C_9H_{19}-C_6H_4-(O-CH_2CH_2)_n-(O-CH_2CHMe)_2-NH_2$$

(Surfonamine MNPA, where $n = 1$–10). These products are assembled by means of ethoxylation and propoxylation of nonylphenol and are also subjected to alcohol amination. The resulting surface-active materials function as corrosion inhibitors, fuel and lubricant additives, oil-field chemicals, reactive ingredients of plastics by way of attachment through the amino function, and so on.

- The introduction by Hoechst Celanese of the large-scale acylation of phenol to *o*- and *p*-hydroxyacetophenones. Both isomers are being exploited in an imaginative fashion (see below for details).

The currently primary source of phenol—the Hock process applied to cumene—yields some interesting by-products besides the coproduct acetone. Thus, the reduction of the intermediate hydroperoxide yields α-hydroxycumene, which, in turn, can be dehydrated to α-methylstyrene. The heterolytic dissociation of cumene hydroperoxide and a re-arrangement of the resulting cationic species also gives rise to valuable acetophenone (technical grade 76¢, perfume grade $2.15) in addition to methanol. Another by-product derived from α-hydroxycumene is the rather stable cumyl carbocation, which survives long enough to react with phenol to yield 4-hydroxydiphenyldimethylmethane, an abbreviated form of BPA. These by-products are significant in view of the large-scale process from which they are derived. The chemical changes are summarized in Scheme 9.57.

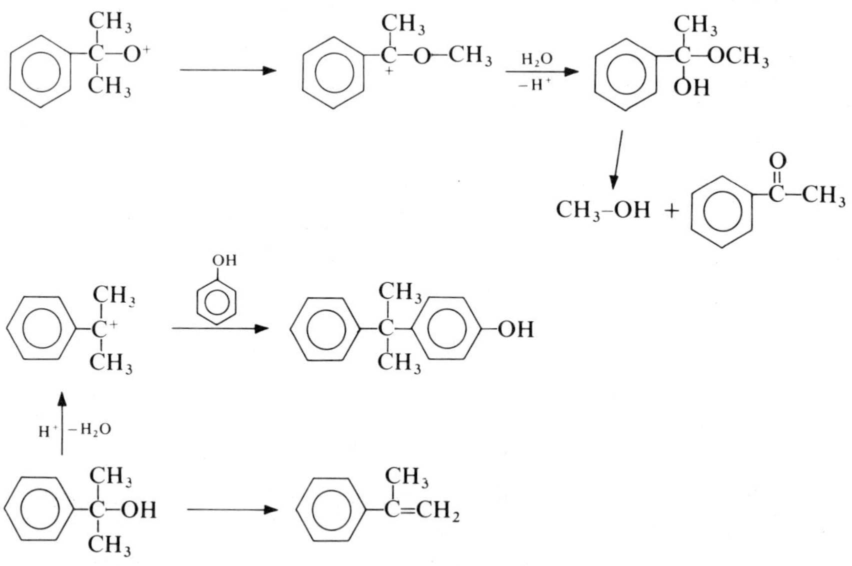

**Scheme 9.57**

The alkylation of phenols is a source of numerous products. A simple example is *p-t-*amylphenol (*t*-AMP), prepared from isoamylene or isoamyl alcohol and phenol. As Pennwalt's Pentaphen 67 ($1.08), it is used extensively as an antimicrobial in a variety of disinfecting formulations. *t*-AMP, *o*-phenylphenol (OPP), and *o*-benzyl-*p*-chlorophenol (OBCP) are particularly effective antimicrobials when used in the presence of anionic and nonionic surfactants (*CMR* 5/16/88). Other products of possibly $BF_3$ ($3.47)-catalyzed alkylation reactions of phenol and the dimer of isobutylene or the trimer and tetramer of propylene α-olefins or oxo alcohols include octylphenol (76¢), nonylphenol (53¢;

~200MM lb, and dodecylphenol (48¢). These and similar (primarily para-substituted) products are not only converted to anionic and nonionic surfactants as indicated above but are also used as lubricating-oil additives and become ingredients of PF resins in order to impart oil-solubility when these modified phenolic resins are used in coatings. Nonoxynol-9, a vaginal contraceptive that provides some protection against gonorrhea and chlamydial infections, is derived from nonylphenol and nine EO residues.

In the case of substituted phenols, the phenolic group dominates the orientation of entering alkyl groups as demonstrated by the following examples:

- *m*-Cresol (94¢) gives 2-*t*-butyl-5-methylphenol (54¢), and double alkylation produces 2,4-di-*t*-butyl-5-methylphenol, which, on methylation of the phenolic function, becomes the intermediate for the formation of 2,4-di-*t*-butyl-5-methoxy-benzaldehyde—a synthetic musk perfume ingredient known as "ambral"

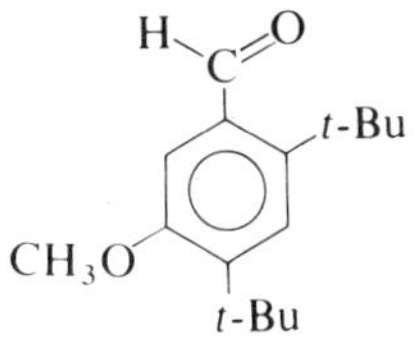

- *p*-Cresol ($1.20) gives 2,6-di-*t*-butyl-4-methylphenol, 2,6-di-*t*-butyl-*p*-cresol, or butylated hydroxytoluene (BHT; $1.28)—a common antioxidant used in plastics, rubber, functional liquids derived from petroleum, and feed and food products, with an annual demand of about 20MM lb.
- *p*-Methoxyphenol (hydroquinone monomethyl ether) gives (mainly) 2-*t*-butyl-4-methoxyphenol or butylated hydroxyanisole (BHA) (and some 3-*t*-butyl product, $8.80) used as a food antioxidant.

The original ortho-alkylation process of phenol is believed to occur by means of an aluminum alkoxide complex that brings together the phenolic substrate and the alcohol reactant and thus facilitates a concerted, cyclic electron flow with a corresponding shift in the bonding of some structural moieties. In the case of an ortho alkylation of phenol with methanol, there is first formed *o*-cresol as depicted in Scheme 9.58 and then 2,6-dimethylphenol or 2,6-xylenol. In place of a higher alcohol, one may employ the

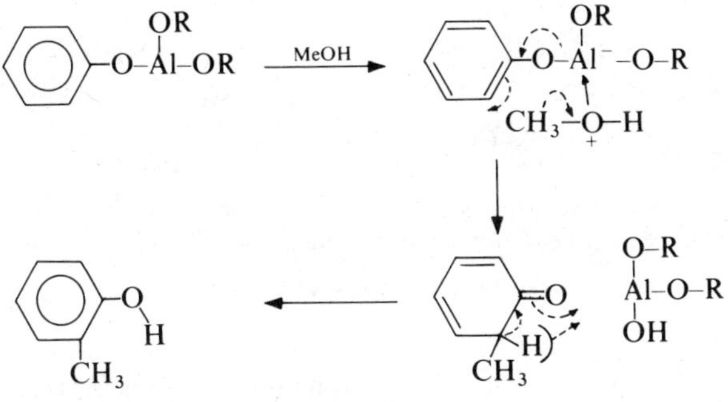

**Scheme 9.58**

corresponding olefin, and the aluminum phenoxide catalyst can be prepared by either heating the phenol with aluminum powder at about 180°C or adding aluminum alkyls. Catalysis by alumina can also be employed, but temperatures as high as 360°C may be required. Ortho alkylation of phenols supplies a variety of intermediates derived from higher alcohols or olefins:

|  |  |
|---|---|
| *o*-Isopropylphenol | Ethyl Corp., 96¢ |
| *o-sec*-Butylphenol | Ethyl Corp., 76¢ |
| *o-t*-Butylphenol | Ethyl Corp., 86¢ |
| *p-t*-Butylphenol | Ethyl Corp., 68¢ |
| 2,6-Di-*t*-butylphenol | Ethyl Corp., 96¢ |

Some examples of preparative uses of phenolic (and analogous aniline) ortho-alkylation products are shown in Scheme 9.59.

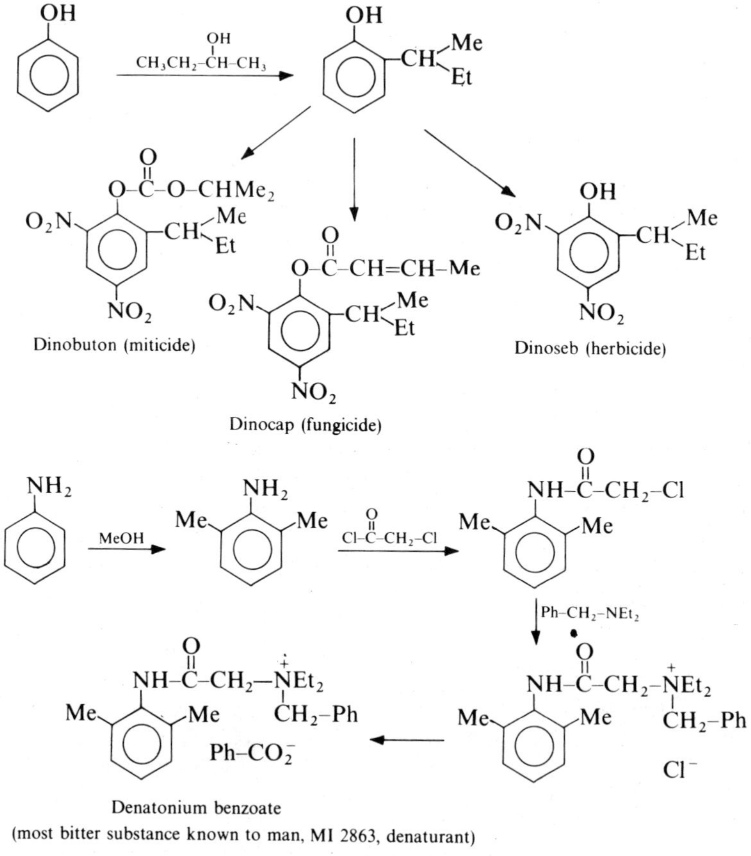

Dinobuton (miticide)

Dinocap (fungicide)

Dinoseb (herbicide)

Denatonium benzoate
(most bitter substance known to man, MI 2863, denaturant)

**Scheme 9.59**

One of the most important ortho-alkylation products of phenol is 2,6-xylenol because it is the building block of the HPST poly(phenylene ether) (PPE). This polymer was introduced originally by General Electric under the tradename "Noryl PPO", or poly(phenylene oxide), but now there are other manufacturers and the acronym "PPE" is generally accepted. The polymerization of the doubly ortho-substitued phenol occurs by means of the generally facile oxidation of phenols (see below for other examples). The ortho substituents block any reactions at positions that would otherwise result in a nonlinear assembly of the macromolecules (Scheme 9.60). The reaction is carried out by means of molecular oxygen and is catalyzed by cupric ions. The relatively high unit cost of PPE can be accounted for by the cost of 2,6-xylenol and the unorthodox polymerization technology; fortunately, however, PPE is capable of forming polymer blends or alloys with materials of lower cost. The first such blend (discovered accidentally) was PPE–PS— marketed as GE's Noryl and as Borg-Warner's Previx. These engineering polymer alloys cost about $1.80 and are still the most commonly employed materials of this kind except for high-impact PS (HIPS). Another common polymer blend is that of PPE–nylon. The total consumption of PPE blends in the United States reached about 225MM lb in 1987. Most of the PPE polymers are converted into automotive structural components.

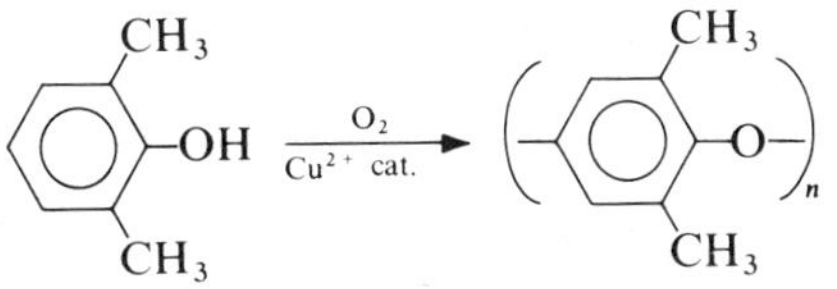

**Scheme 9.60**

The ease with which phenol undergoes electrophilic substitution reactions accounts for the formation of BPA when the products of the cumene Hock process are allowed to react in the presence of acid catalysts. However, the quality of the PCs obtained from BPA and either phosgene or diphenyl carbonate depends on maximum linearity of the macromolecules, and hence crude BPA is purified in order to isolate $p,p'$-BPA, while the fraction that contains some $o,p'$-isomer is utilized for the production of cross-linked epoxy resins and in the formulation of UPRs. About half of the BPA production is consumed by the PC market. The purification requirement explains the somewhat higher cost of the polycarbonate grade BPA (86¢) as compared with 82¢ for the nonpurified epoxy grade. An exceptionally pure PC is required for the manufacture of compact disks and similarly demanding electronic and optical devices. Until recently this material, derived from highly purified BPA and free of inorganic impurities, was imported from West Germany, but now Mobay is producing it domestically. The annual production of BPA during 1979–1981 averaged about 525MM lb, but the current 1B-lb production in the United States is expected to grow annually by about 9% in view of the large consumer demand for audio and optical devices such as compact laser–read disks. Another factor that contributes to the relatively high demand for PC are various polymer blends or alloys with ABS, PET, PBT, PPE, PAR, and other materials. Even though this does not represent a high-volume consumption of BPA, it is still the favorite difunctional phenol used in the assembly of HPSTs such as PARs, poly(ether sulfones), and PEI.

An interesting new aryldicyanate monomer, RDX 64826, derived from BPA was announced recently (R&D 10/87) by Interez. Most likely, RDX 64826 is obtained by the nucleophilic displacement of the ditosylate (9.3.3) of BPA with sodium cyanate (85¢). Apparently, the ambidental cyanate ion attacks the electron-poor carbon of the aromatic system with its "hard" anionic terminal. The new monomer cyclotrimerizes on heating to

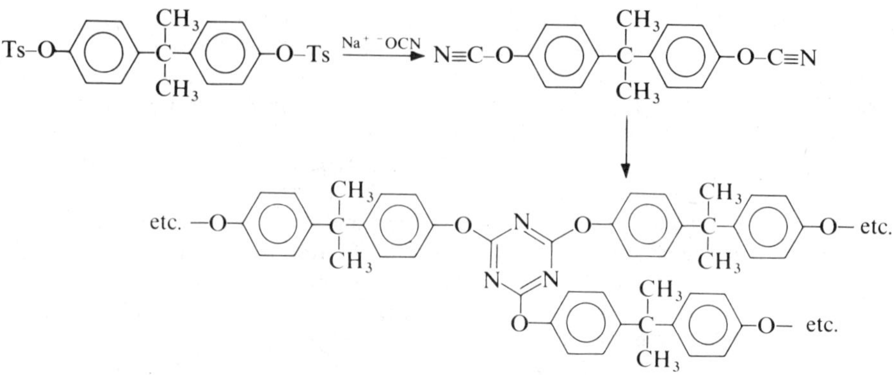

**Scheme 9.61**

produce an amorphous resin priced at $15–$20. The chemistry of this system is shown in Scheme 9.61.

The nonsubstituted positions ortho to the phenolic functions in BPA invite additional electrophilic reactions, and thus one obtains have the tetrahalogenated derivatives manufactured for use as flame retardants (9.3.1.2).

Another alkylation reaction of phenol is its reaction with cyclohexanol, which produces a mixture of o- and p-cyclohexylphenol and illustrates the use of polyphosphoric acid (PPA), a mild acidic catalyst.

The alkylation of phenol with octyl alcohol or capryl alcohol

$$[CH_3-CH(OH)-(CH_2)_5-CH_3; \$1.40]$$

derived from ricinoleic acid (*SZM-II*, 72), followed by nitration, and the formation of the ester of crotonic acid, yields the fungicide and miticide known as *dinocap*:

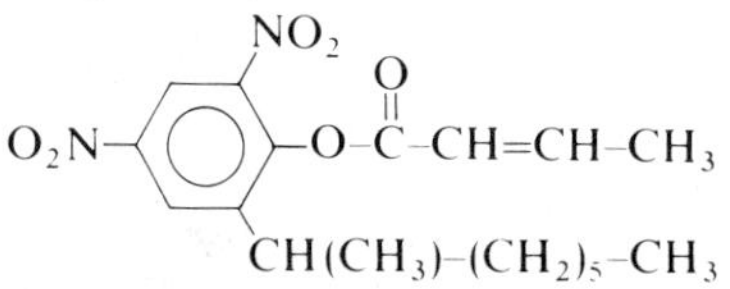

This material is best known as Rohm & Haas's "Karathane" and since the mid-1950s has been employed to protect about 20% of the domestic apple crop from mites and powdery mildew. On the basis of recent laboratory tests indicative of birth defects in rabbits, future use of dinocap is being scrutinized by the EPA (*CW* 11/84).

The chloromethylation of phenol occurs very readily, but the resulting hydroxy-activated benzyl chloride may react further to give polymeric products. In order to avoid such problems, phenyl acetate may be chloromethylated in place of phenol, with removal of the acetate group by hydrolysis after completion of the synthetic sequence, which involves the substituted benzyl chloride. A synthetic use of the chloromethylation reaction is illustrated by the formation of tyramine and p-hydroxyphenylacetic acid and some of its derivatives (Scheme 9.62).

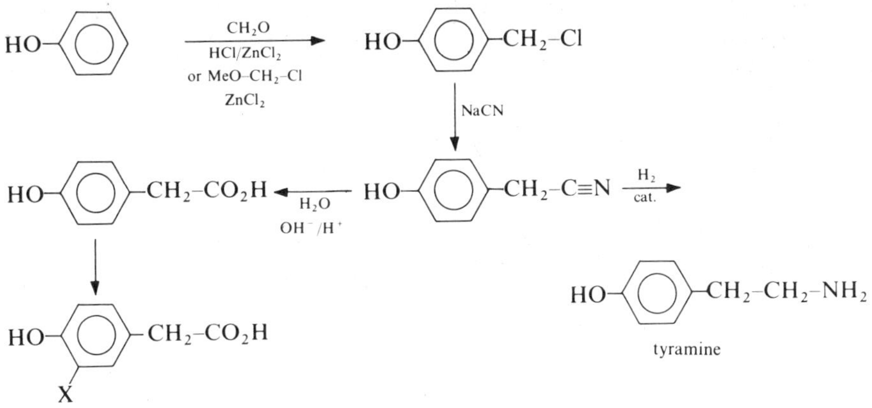

HO—⟨⟩  $\xrightarrow[\begin{array}{c}\text{HCl/ZnCl}_2\\ \text{or MeO–CH}_2\text{–Cl}\\ \text{ZnCl}_2\end{array}]{\text{CH}_2\text{O}}$  HO—⟨⟩—$CH_2$–Cl

↓ NaCN

HO—⟨⟩—$CH_2$–$CO_2H$  $\xleftarrow[\begin{array}{c}\text{H}_2\text{O}\\ \text{OH}^-/\text{H}^+\end{array}]{}$  HO—⟨⟩—$CH_2$–C≡N  $\xrightarrow[\text{cat.}]{\text{H}_2}$

↓

HO—⟨⟩—$CH_2$–$CO_2H$
⎮
X

HO—⟨⟩—$CH_2$–$CH_2$–$NH_2$

tyramine

X = Cl–, Br–, $O_2$N–, etc.

**Scheme 9.62**

The bridging of two phenolic rings under conditions essentially similar to those employed in chloromethylation occurs during the formation of BPA by means of acetone. However, when the more reactive carbonyl group of formaldehyde is utilized, it is advisable that the formation of polymers be avoided by blocking the reactive ortho and para positions of the phenol with substituents that do not interfere with the alkylation reaction. Thus, p-chlorophenol is converted to 2,2'-dihydroxy-5,5'-dichlorodiphenyl-methane (dichlorophene)—a germicide commonly employed in personal-care products such as soaps, and shampoos, as well as in agricultural fungicide formulations. Dichloro-phene replaces the analogous hexachlorophene derived from 2,4,5-trichlorophenol that is no longer manufactured in the United States because of the "dioxin scare" (9.3.1.1). Instead, it is imported from South Korea at a unit price of $36.50 (*CMR* 5/18/87). Also, 2-t-butylphenol is converted to 2,2'-methylene-bis(6-t-butylcresol) (Sherwin-Williams, $3.25), and 2,6-di-t-butylphenol forms bis(2,6-di-t-butyl-4-hydroxyphenyl)methane. Mild oxidation of the last-mentioned compound produces a thermodynamically stable galvin-oxyl radical that can be used to trap other, small-sized radicals (Scheme 9.63).

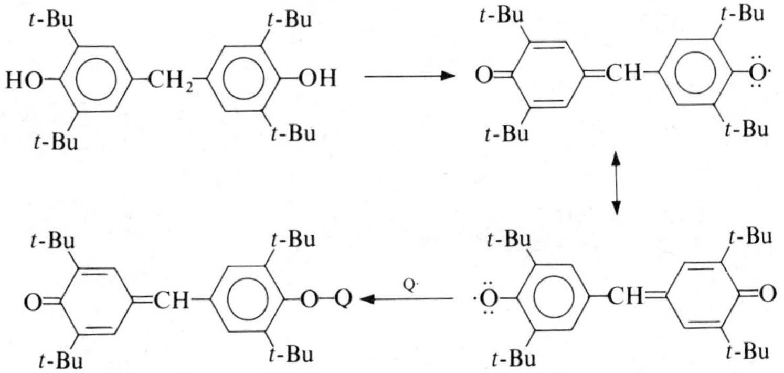

**Scheme 9.63**

The traditional Kolbe–Schmitt carboxylation of phenol to salicyclic acid and its *p*-isomer and some of their derivatives is discussed elsewhere in connection with a glimpse at "aspirin and its competitors." The carboxylation occurs by virtue of the delocalization of the anionic charge in the phenolate ion (Scheme 9.64).

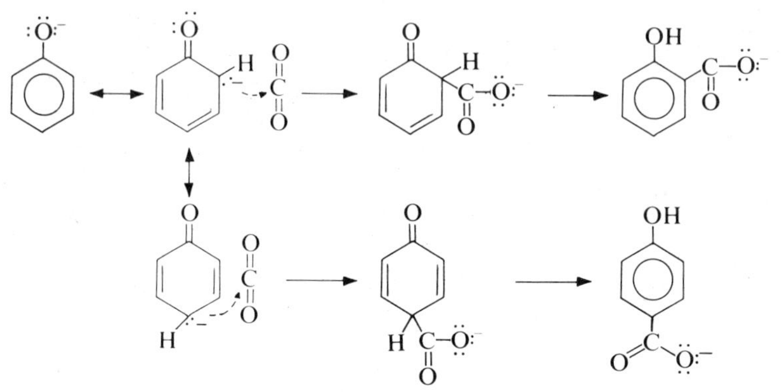

**Scheme 9.64**

Somewhat related to the Kolbe–Schmitt carboxylation of phenols is the Reimer–Tiemann reaction (*ONR*-75), which produces salicylaldehyde (Dow, $3.55) and *p*-hydroxybenzaldehyde (Dow, $3.25). The vicinal position of the functional groups in salicylaldehyde is conducive to cyclization of the corresponding cinnamic acid, obtained by means of the Claisen condensation, to the perfume ingredient coumarin (NF $7.10). However, the aldehyde function also reacts independently, as illustrated by means of the condensation reaction with nitroparaffins followed by their conversion to the aminoalkyl derivatives of phenol (Scheme 9.65). A useful application of *p*-hydroxybenzaldehyde is its conversion by means of the Strecker amino acid synthesis (*ONR*-87) to *p*-hydroxy-α-phenylglycine, one of the building blocks of synthetic penicillins and cephalosporins (*SZM-II*, 43–45).

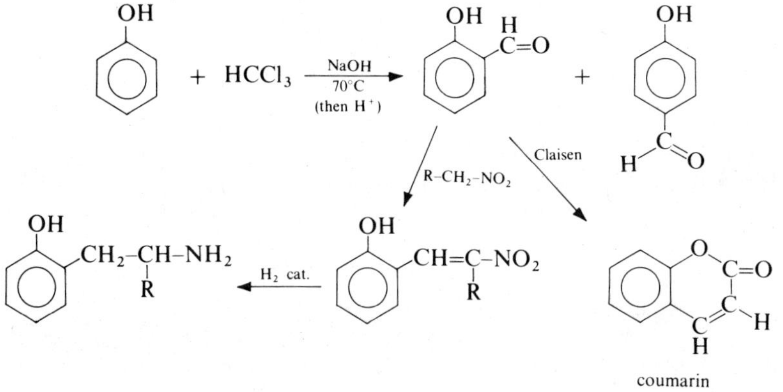

coumarin

**Scheme 9.65**

The ease of oxidation of phenolic systems is amply demonstrated by the formation of PPE, the oxidation of hydroquinones to the corresponding quinone systems (see 9.3.5), the oxidative coupling of two aromatic rings during the formation of thymol iodide (9.3.1.2), and the oxidative coupling of 2,6-di-$t$-butylphenol in the presence of free radicals (represented here by Q·) to yield the corresponding $t$-butyl-substituted biphenylquinone, which, in turn, is capable of capturing two additional radicals to give the corresponding derivative of 4,4′-dihydroxybiphenyl (Scheme 9.66).

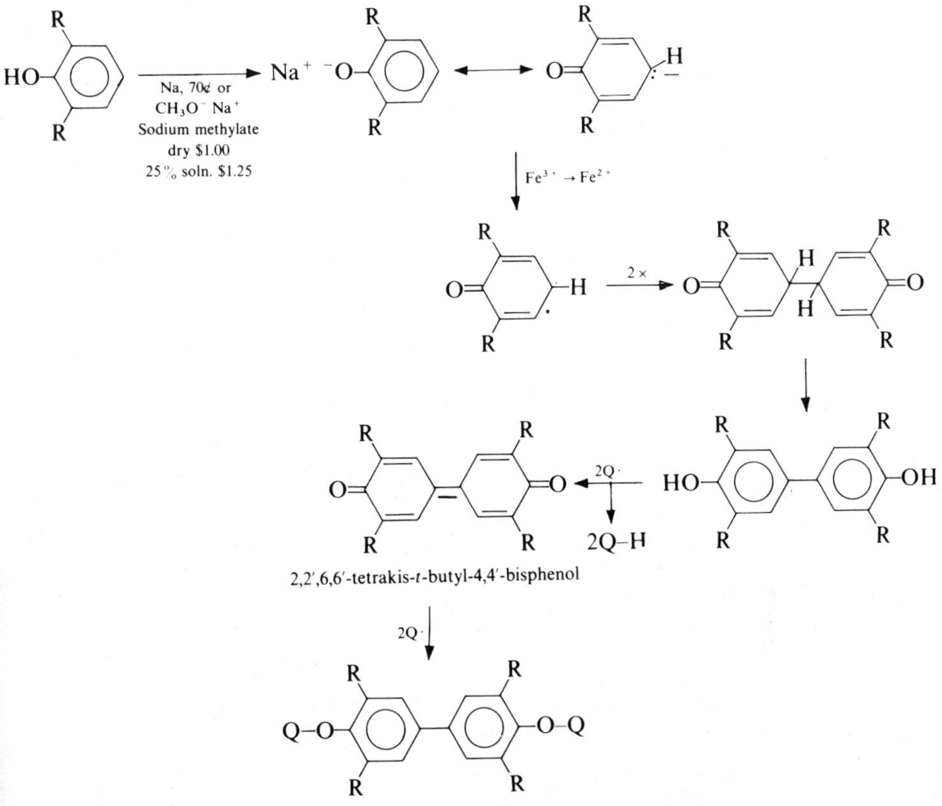

2,2′,6,6′-tetrakis-$t$-butyl-4,4′-bisphenol

where R $= t$-Bu $-$
 · Q $=$ radicals being scavanged

**Scheme 9.66**

*Antioxidants*

The function of antioxidants is to intercept free radicals formed during autooxidation of substrates, represented here by Q–H and which are sensitive to oxygen, and to prevent the autocatalytic destructive escalation of the oxidation process. The process is initiated as follows:

$$Q-H + :\ddot{O}-\dot{O}: \rightarrow (Q\cdot + H-O-O\cdot) \rightarrow Q-O-O-H$$

$$Q-O-O-H \rightarrow Q-O\cdot + H-O\cdot$$

Now, each reactive radical, Q–O· and H–O·, is likely to attack other molecules of the substrate Q–H unless it is captured by the antioxidant:

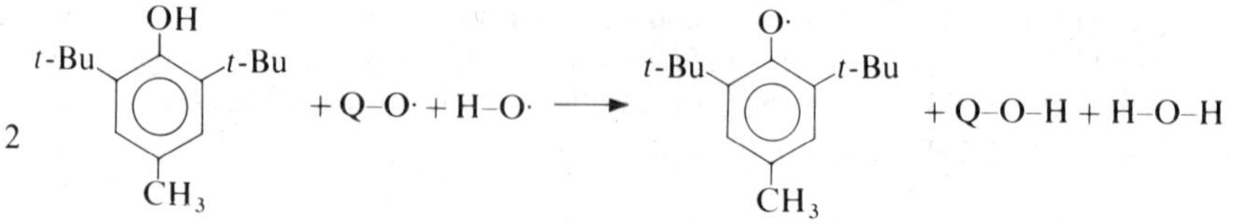

The newly formed radical derived from the antioxidant is relatively unreactive (because of the vicinal $t$-butyl substituents) and under normal circumstances does not initiate new chain reactions with the substrate Q–H. On the other hand, the newly formed radical may slowly undergo coupling and other reactions that give harmless, nonradical products such as these:

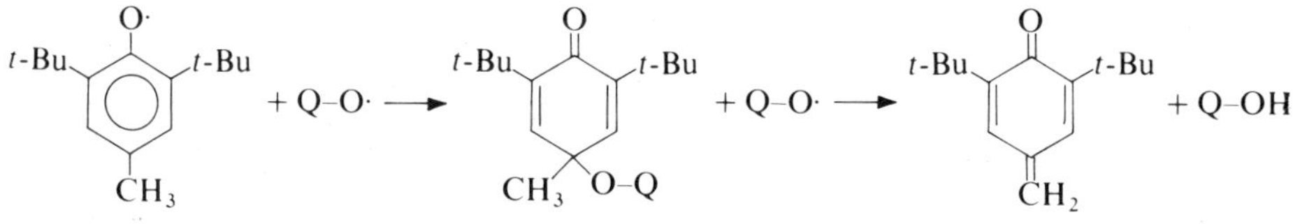

In addition to BHT (actually a derivative of $p$-cresol), the numerous family of industrial antioxidants include derivatives of

- Hydroquinones, such as butylated hydroxyanisole (BHA; food grade $8.80):

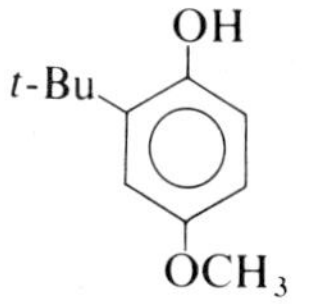

- 2,6-Di-$t$-butylphenol, such as stearyl 3-(3′,5′-di-$t$-butyl-4′-hydroxyphenyl)propionate, Borg-Warner's Ultranox 276:

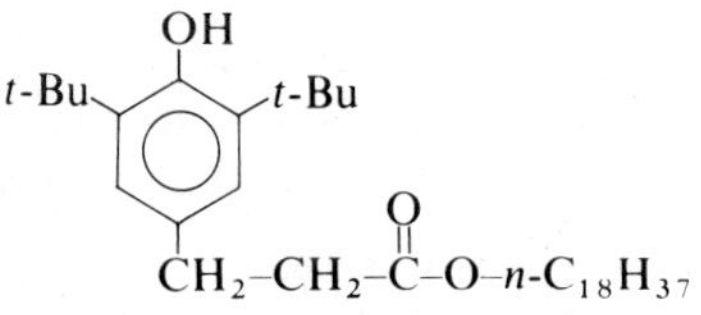

- $p$-Cresol, such as 2,2′-thiobis(4-methyl-6-$t$-butylphenol):

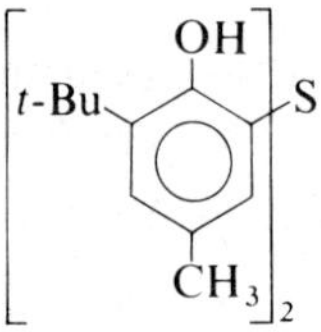

and many others. In addition to the *t*-butyl-blocked phenol family, the realm of antioxidants includes the families of hindered arylamines, phosphites, thioesters, quinolines, and ascorbic acid and its isomer erythorbic acid (*SZM-II*, 107). The sales of all antioxidants in the United States amount to about $550 million. Organic phosphites are known to have a synergistic effect on the antioxidant properties of hindered phenols, and mixtures of different antioxidants are often employed for the stabilization of polymers, rubbers, and other industrial products.

The novel phenol to aniline transformation (see 2.4.4.3) inaugurated in 1982 by Aristech (the former U.S. Steel Chemicals) at the Haverhill, Ohio plant was the first application of its kind of a direct conversion of a phenolic to an amino function. Most likely, the appropriate catalytic system induces a transient tautomerization of the phenol to a cyclohexadienone and addition of ammonia to the unstable keto system, followed by a dehydration step, as shown in Scheme 9.67. The proposed mechanism differs from a classical Bucherer reaction (*ONR*-16; 9.3.6) by not relying on an intervention of a sulfite intermediate. In any case, the marketing policy of the producer and the economic repercussions of the phenol to aniline conversion process are mentioned elsewhere. The conversion of resorcinol to *m*-aminophenol (see 9.3.4.2) seems to be an extension of this process.

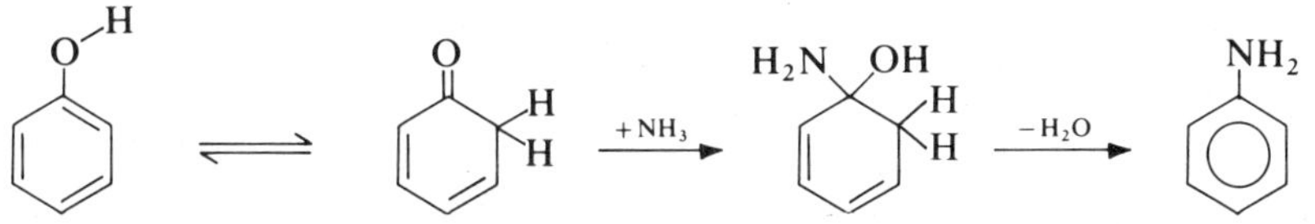

**Scheme 9.67**

An even more recent development is the phenol to APAP route designed by Hoechst Celanese. APAP is one of the serious competitors of aspirin (9.3.2.1), and the process starts with the preparation of *o*- and *p*-hydroxyacetophenones (2- and 4-HAP). Most likely, the hydroxyacetophenones are obtained by an acid-catalyzed Fries rearrangement (*ONR*-33) of phenyl acetate ($1.04) and the two isomers are readily separated on the basis of the greater volatility of the *o*-isomer because of its *intra*molecular hydrogen bonding, while the *p*-isomer is associated by *inter*molecular hydrogen bonds. The *p*-isomer is converted to the corresponding oxime by means of hydroxylamine sulfate (85¢), and, finally, the oxime is subjected to a Beckmann rearrangement (*ONR*-8) to yield the desired APAP. The production of the hydroxyacetophenones is carried out at a 300,000-lb/year level, while a multi-million-pound facility has been under construction for production planned to start in 1988. An admirable effort by chemists responsible for this development has demonstrated useful applications of both hydroxyacetophenones (Scheme 9.68). The Hoechst Celanese APAP process must be compared with the traditional route to APAP. The latter starts with phenol and can involve its nitrosation (with acidified sodium nitrite; USP 37.25¢), followed by reduction of *p*-nitrosophenol with iron filings and dilute hydrochloric acid to *p*-aminophenol and, finally, acetylation with acetic anhydride. The acetylation occurs preferentially at the more basic amino group to give the desired *N*-acetyl-*p*-aminophenol, APAP, acetaminophen, or *p*-hydroxyacetanilide ($2.71) (Scheme 9.69). Instead of the nitrosation step, a modified route to APAP may also start with phenyl acetate, followed by a carefully controlled mononitration, reduction of the *p*-nitrophenyl

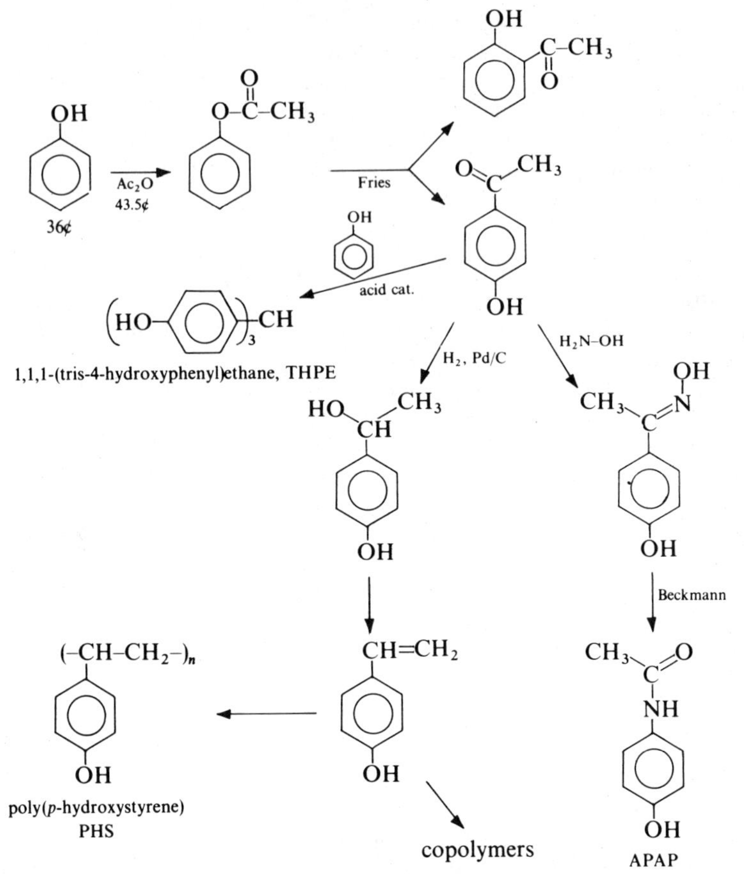

OH
36¢
Ac2O
43.5¢
Fries
acid cat.
1,1,1-(tris-4-hydroxyphenyl)ethane, THPE
H2, Pd/C
H2N–OH
Beckmann
poly(p-hydroxystyrene)
PHS
copolymers
APAP


**Scheme 9.68**

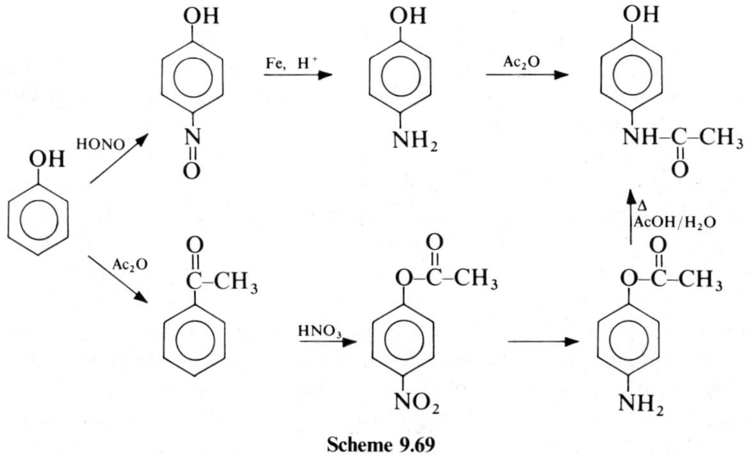

HONO
Fe, H+
Ac2O
HNO3
AcOH/H2O


**Scheme 9.69**

acetate to p-aminophenyl acetate, and a rearrangement of the latter to the desired acetanilide. Neither the nitrosation nor the nitration process promises a yield competitive with oxime formation and its rearrangement, and the separation of the isomeric acetophenones seems to be less troublesome than the isolation of the pure p-nitro product in the modified process.

An interesting and revealing footnote to the announcement of the new APAP process was the reaction of one of the traditional manufacturers of the material (*CMR* 9/1/86) who could not visualize the existence of a new route and thus confessed to be "dumbfounded" by the news!

Among the simple phenyl esters (other than the above-mentioned acetate) the more important chemicals are:

- Diphenyl carbonate, used in place of phosgene.
- Phenyl salicylate;
- Triphenyl phosphate ($1.64) used primarily as a plasticizer and flame retardant for cellulosic films.

Among the ethers of phenol we should first mention the methyl ether, or anisole, prepared by treatment of sodium phenoxide with either methyl chloride (25.5¢) or dimethyl sulfate (46¢). Anisole is almost as reactive as phenol itself when subjected to electrophilic substitution reactions, but unlike phenol, it is not subject to facile oxidation reactions. The preparation of methoxychlor ($2.05), a replacement for DDT; the chloromethylation of anisole; and the preparation of BHA (see above) illustrate the reactivity of anisole. The synthetic usefulness of p-methoxybenzyl chloride is illustrated by the preparation of the antihistaminic pyrilamine:

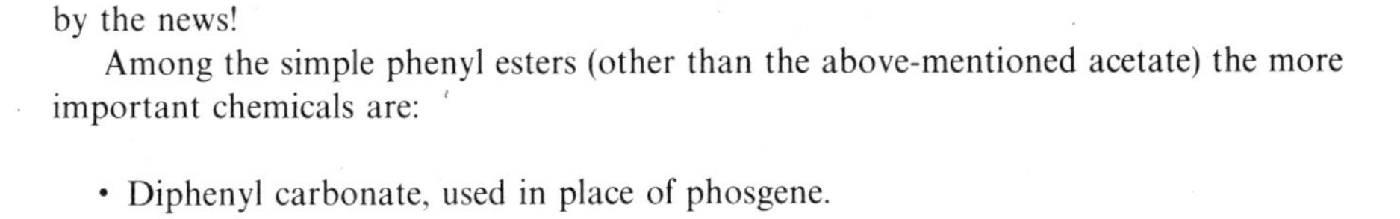

$$p\text{-MeO-C}_6\text{H}_4\text{-CH}_2\text{-Cl} + \text{Me}_2\text{NCH}_2\text{CH}_2\text{-NH-} \bigcirc_\text{N} \longrightarrow p\text{-MeO-C}_6\text{H}_4\text{-CH}_2\text{-N-}\bigcirc_\text{N}$$

and by the preparation of 4-methoxyphenethylamine, the cardiac depressant N-p-methoxybenzyl-N',N''-dimethylguanidine or meobentine, as well as 4-methoxyphenyl-acetic acid and 4-methoxyphenylacetone (Scheme 9.70).

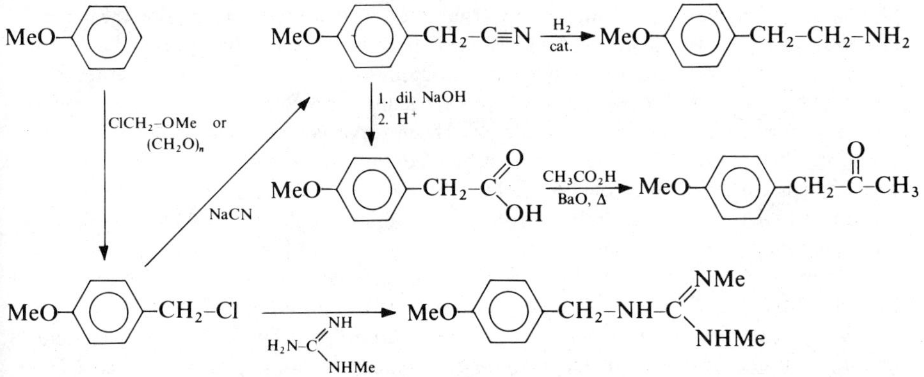

**Scheme 9.70**

Numerous derivatives of anisole are isolated from essential oils (*SZM-II*, 28). For example, anethole ( $p$-MeO–C$_6$H$_4$–CH=CH–CH$_3$, 200MM lb, natural USP \$4.75–\$5.25, synthetic \$4.60) is isolated from anise oil (\$6.13), which, in turn, is obtained from either anise seed (\$1.25–\$1.65, depending on origin) or fennel oil (sweet USP \$13.25), obtained, in turn, from fennel seed (\$0.56–\$1.05). Anethole is used by the food industry for its fragrance and flavor and is a common ingredient of toothpaste, soap, and similar products. The isomer of anethole in which the side chain is an allyl group is known as *methylchavicol* (\$18.00) and is responsible for the flavor of root beer. These and many similar compounds (*SZM-II*, 114–115) are also minority components of turpentine. Anethole can be cleaved oxidatively to $p$-anisaldehyde (\$5.00), and its reaction with sulfur gives a member of the trithione family, namely, anethole trithione—a choleretic (Scheme 9.71).

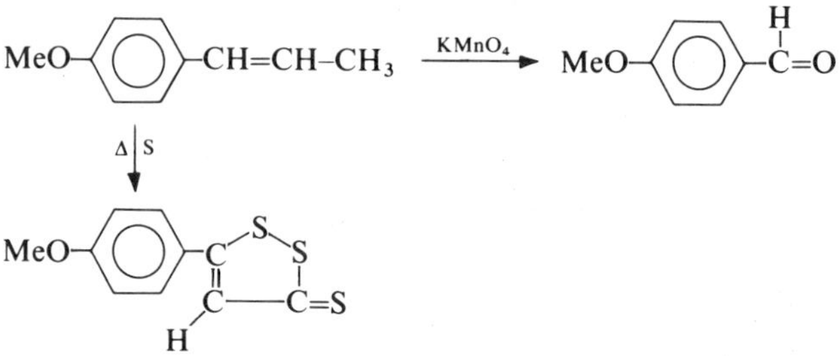

**Scheme 9.71**

The Perkin reaction (*ONR*-67) of $p$-anisaldehyde (or anisic aldehyde; \$5.00) results in the formation of $p$-methoxycinnamic acid, and the esterification of the latter with 2-ethoxyethanol yields a sunscreen agent known as *cinoxate*:

$$p\text{-MeO–C}_6\text{H}_4\text{–CH=CH–CO·O–CH}_2\text{CH}_2\text{–O–CH}_2\text{CH}_2\text{–OH}$$

Among the numerous chemicals of industrial interest derived from phenol by means of electrophilic substitution reactions other than those mentioned so far, there are the products of nitration.

Gentle nitration of phenol produces a mixture of $o$- and $p$-nitrophenol (\$1.00 and \$1.35, respectively). $p$-Nitrophenol is converted to esters of dimethyl or diethyl thiophosphates to give the insecticides methyl and ethyl parathion, respectively, priced at \$1.65 and \$1.75. These insecticides are quite toxic to humans, and, *as is true for all pesticides, they must be used by skilled and adequately protected operators*. Some cases of poisoning by these cholinesterase-inhibiting materials have been reported in developing countries in instances where illiterate persons used parathion containers for storage of drinking water. The chlorination of $o$-nitrophenol yields 2-nitro-4-chlorophenol (also obtained by nitration of $p$-chlorophenol; \$1.25), which gives 2-amino-4-chlorophenol (Mobay \$5.45) on reduction.

Somewhat more extensive nitration of phenol produces 2,4-dinitrophenol (Martin Marietta, \$1.71), used as a dyestuff intermediate, corrosion inhibitor, and antioxidant.

Finally, extensive nitration of phenol (often carried out in the presence of sulfuric acid to form sulfonic acid intermediates that are replaced by nitro groups) yields 2,4,6-trinitrophenol or picric acid, (paste technical grade \$5.00, pure \$6.00—dry basis prices).

Dry picric acid explodes on heating, on impact, or by the action of detonants such as heavy-metal azides (readily obtained from sodium azide—the gas-producing incredient of airbag devices). Picric acid is an antimicrobial and forms insoluble picrates with proteins, and, for both reasons, it is an ingredient of topical burn remedies. An example of such remedy is Abbott's Butesin picrate, the picrate of the local anesthetic *n*-butyl *p*-aminobenzoate. Silver picrate is used as a vaginal antimicrobial. The reduction of picric acid with hydrogen sulfide in the presence of ammonia yields 2-amino-4,6-dinitrophenol or picramic acid, a dyestuff intermediate. These nitration products of phenol and their derivatives are summarized in Scheme 9.72.

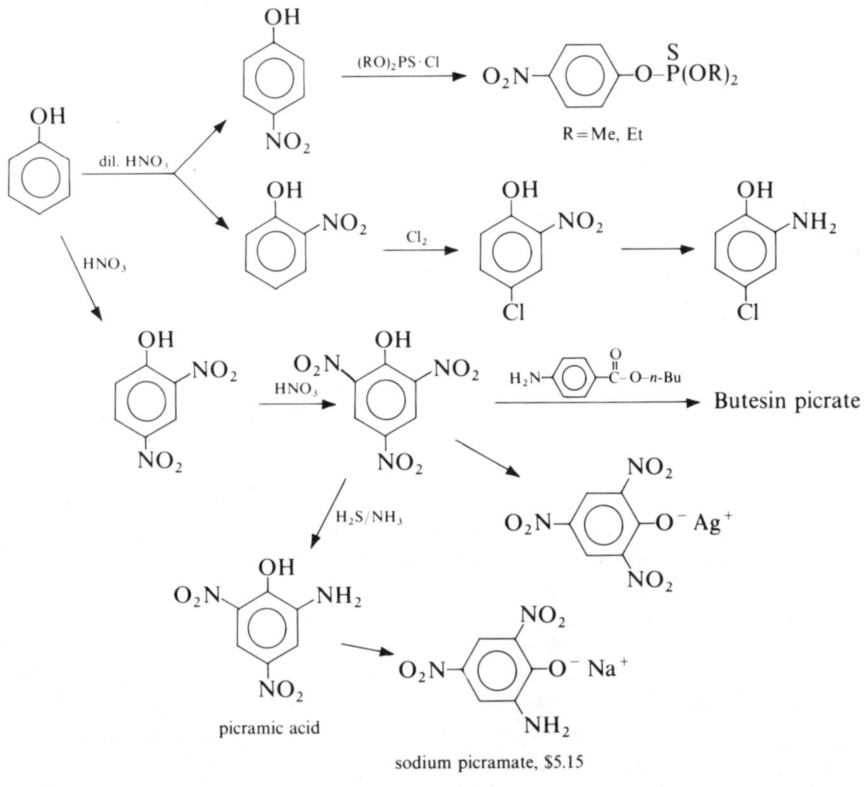

**Scheme 9.72**

The facile nitrosation of phenol (see APAP above) is another proof of the high reactivity of phenols toward electrophilic reactants. *p*-Nitrosophenol and the monooxime of *p*-benzoquinone constitute a pair of related tautomers (Martin Marrietta's Maroxol 10, sodium salt $3.30). The tautomers exhibit reactions expected of each component of the equilibrium mixture. Thus, with hydroxylamine sulfate (85¢), one obtains *p*-quinonedioxime (PQD; Martin Marietta, $4.10), while in the presence of some sulfuric acid and phenol one obtains the condensation product known as *indophenol*—a redox indicator that exhibits a distinct color change as a function of pH. The preceding chemical changes are shown in Scheme 9.73. Indophenol is one of many phenolic, nitrogen-containing, and

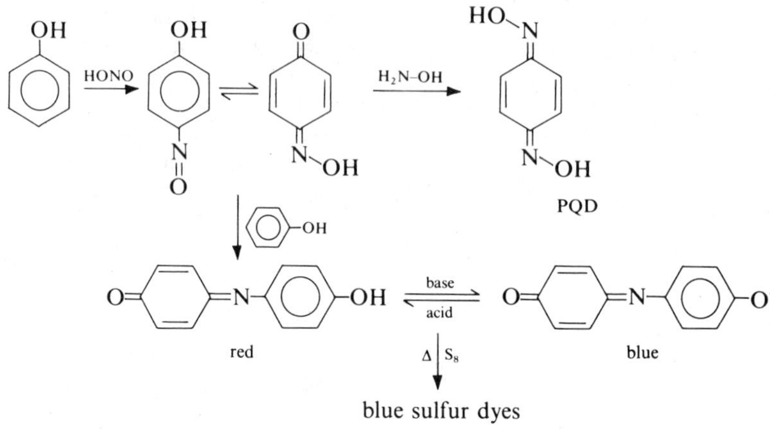

**Scheme 9.73**

otherwise substituted aromatic compounds that are converted to a member of the family of sulfur dyes on heating with sulfur and/or sodium polysulfide salts. These are complex dyes of ill-defined structure that have been in use since the last century. They are *vat dyes* in the sense that the reduced form is soluble in aqueous alkali, while air oxidation (of the textile impregnated with the dye solution) produces insoluble, that is, "fast" dye molecules. The color of the rather reproducible sulfur dyes depends on the substrate that is treated with sulfur and polysulfides, and, in the case of indophenol, such treatment gives rise to blue sulfur dyes.

The sulfonation of alkylated phenols is mentioned in connection with the discussion of surfactants. The sulfonation of phenol itself yields *o*- and *p*-phenolsulfonic acids or bis(*p*-hydroxyphenyl) sulfone as a function of the vigor of the reaction conditions. *p*-Phenolsulfonic acid (58¢), sodium salt (76¢), is converted to the zinc salt ($1.82), which functions as an astringent in personal care products such as styptic sticks.

The naphthols are prepared from naphthalene by sulfonation (9.2.4) and fusion of the sodium salt of the resulting sulfonic acids with caustic soda. The naphthols, as well as naphtholsulfonic acids, play an important role in the production of certain dyes, organic pigments, agrichemicals, antioxidants, pharmaceuticals, and other products. The price differential between the isomeric 1- or α-naphthol and the 2- or β-naphthol ($1.81 and $1.20, respectively) suggests that the sulfonation at 0°C required to limit the sulfonation to the alpha position is more difficult to control than the sulfonation at about 160°C employed in the preparation of the β-isomer (see 9.2.4 and 9.3.3). It is also possible to obtain 1-naphthol by a clever route that starts with tetrahydronaphthalene (8.3.1) and involves air oxidation to α-tetralone and catalytic dehydrogenation of the latter; see Scheme 9.74.

Propranolol is a cardiac depressant and beta-andrenergic blocker and can also be obtained, by treating 1-naphthol first with epichlorohydrin, followed by addition of isopropylamine to the resulting glycidyl ether to give the desired product (see 5.4.8).

Gentle sulfonation of 2-naphthol gives 2-hydroxy-6-naphthalenesulfonic acid, also known as "Schaeffer's beta acid" (the sodium salt is Schaeffer salt, $2.59), while excessive sulfonation yields a mixture of 2-hydroxy-6,8- and 2-hydroxy-3,6-naphthalenedisulfonic acids. The former, in the form of the disodium salt, is known as "G salt," while the latter—

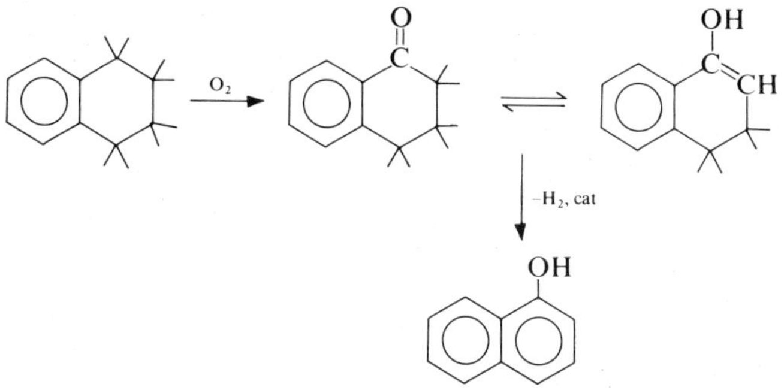

**Scheme 9.74**

a minority product of the mixture, is known as "R acid." Caustic soda fusion of the sodium salt of the R acid (R salt, $2.12) is the source of sodium 2,3-dihydroxynaphthalene-6-sulfonate (Hoechst-Celanese, $8.75).

The analogous sulfonation of 1-naphthol can lead to 1-hydroxy-2-naphthalenesulfonic acid (Schaeffer's alpha acid), 1-hydroxy-4-naphthalenesulfonic, and 1-hydroxy-8-naphthalenesulfonic acids. On heating, the last-mentioned compound forms the corresponding sultone (the sulfonic acid analog of a lactone), and sulfonation of the latter yields 1-hydroxy-4,8-naphthalene disulfonic acid.

The preceding examples of naphthol chemistry serve to illustrate the variety of dye intermediates and the traditional nomenclature that has evolved in that field. Additional examples are found in Section 9.3.6, which deals with nitro and amino derivatives of naphthalene.

The Kolbe carboxylation of phenols can be extended to the naphthol system: 1-naphthol produces the expected 1-hydroxy-2-naphthoic acid, but the isomeric 2-naphthol gives (instead of the expected 2-hydroxy-1-naphthoic acid) 3-hydroxy-2-naphthoic acid, also known as *β-oxynaphthoic acid* (or BON acid, 3.5MM lb, Hoechst $2.60). Both hydroxynaphthoic acids and their derivatives, such as the BON anilide—naphthol AS or azoic coupling component 7, are building blocks of azo dyes (by coupling with appropriate diazonium salts; see 9.3.6) and of organic pigments (by complex formation of azo structures) with transition metals. The parent naphthols are converted to the corresponding amines by means of the Bucherer reaction (9.3.6), and the naphthylamines, in turn, are also precursors of dyes. The carboxylation of naphthols is carried out at relatively high temperatures (130–250°C) and under pressure ($\sim 80$ psi).

The transformations of the above-mentioned naphthols and some of their derivatives are summarized in Scheme 9.75.

One of the most notorious derivatives of 1-naphthol is the insecticide 1-naphthyl-N-methylcarbamate, carbaryl, distributed by Union Carbide under its tradename "Sevin" and obtained by means of methyl isocyanate (MIC). The mishandling of MIC was the immediate cause of the Bhopal incident (3.5.3.6), but an independent, 3-year-long investigation by A. D. Little concluded that the massive release of MIC (caused by addition of water that produced an exothermic trimerization) was an act of sabotage by a disgruntled Indian employee (see W. Lepkowski, "Study Backs Carbide's Bhopal Sabotage Theory," *CEN* 5/16/88 and "Union Carbide Presses Bhopal Sabotage Theory," *CEN* 7/4/88, 8–11; *CMR* 5/16/88). The litigations in this extraordinary case are still in progress, and the outcome will undoubtedly have a profound and long-lasting effect on

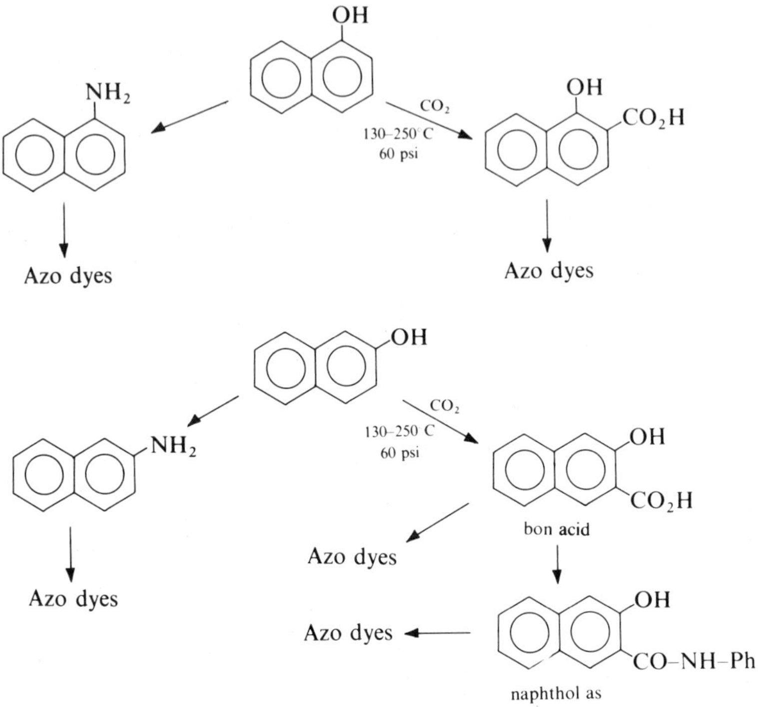

**Scheme 9.75**

the manner in which enterprises from industrially developed countries will invest in and manage the operations of chemical production facilities in developing countries.

The methyl and ethyl ethers of 2-naphthol are perfume ingredients known as "yara yara" ($2.81) and "new nerolin," respectively. The curious name of the last-mentioned chemical refers to the substitute of the naturally occurring *cis*-isomer of geraniol, nerol (*SZM-II*, 119, 120, 130)—a terpene constituent of neroli oil ($670 as of January 1989) imported from Tunisia.

Cresols, xylenols, and higher methylated phenols—usually referred to as *cresylics*, are discussed separately even though the above-mentioned ortho alkylation is shown to be a source of some methylated derivatives of phenol. The majority of the somewhat higher than 100MM-lb domestic consumption of these materials is supplied by coal- and petroleum-tar refiners, and a significant portion of the cresylics is consumed in the form of mixtures. The U.S. production capacity of synthetic methylated phenols is only 50MM lb and consumes about 5% of the demand for phenol.

The mixture of naturally derived *o-*, *m-*, and *p*-cresols can be separated into an *o*-cresol-rich fraction because this isomer has a lower boiling point (191°C) than its isomers (202°C). The 98 or 99% pure *o*-cresol is priced at 75¢, while the remaining mixture of *m-* and *p*-cresols sells for 82¢. The relatively low priced mixture of *m-* and *p*-cresols (as compared with the unit price of nearly pure compounds; see below) is converted with phosphorus oxychloride (O=PCl$_3$; 43¢), to tricresyl phosphate [TCP; (CH$_3$–C$_6$H$_4$–O)$_3$P=O, $1.60], a widely used plasticizer, hydraulic and other functional fluid component, flame-retardant, and antioxidant. Steric hindrance of *o*-methyl groups interferes with the formation of TCP. The coal-tar-derived cresylic acids are separated into two fractions according to whether the $m$-$p$-cresol content is above or below 25%, although both fractions are currently priced at 58¢. Pure (98%) $p$-cresol is priced at $1.20, and 95–98% pure $m$-cresol is priced at $1.65.

$o$-Cresol, the main product of ortho alkylation of phenol with methanol, is used in the production of formaldehyde-based resins that also contain some epoxy resin ingredients—epoxy cresol novolacs employed as encapsulants of computer chips. $o$-Cresol is also the starting material for the preparation of two herbicides: sodium 2-methyl-4-chlorophenoxyacetate (MCPA), a member of the phenoxyacetic acid herbicides, and 2-$sec$-butyl-4,6-dinitrophenol (dinoseb), which functions as an ovicide for dormant fruit trees and is consumed at a 7–11MM-lb level. In October 1986 the use of dinoseb was halted on the basis of an emergency ruling by the EPA because of health risks to unborn children of pregnant women workers and sterility risks to male field workers. Agricultural losses due to the dinoseb ban are estimated to cost American farmers as much as $90 million per year (*CMR* 10/13/86).

$p$-Cresol is a by-product of the ortho alkylation of phenol with methanol and, as shown above, is the precursor of the antioxidant BHT. $p$-Cresyl acetate and $p$-cresyl methyl ether ($6.00) are fine chemicals used by the perfume industry.

$m$-Cresol is the precursor of synthetic thymol (NF $3.75) and synthetic menthol ($9.00), even though both of these materials are also available from natural sources (*SZM-II*, 114) and sold for $3.75 and $8.00, respectively. Another product of interest because of its fragrance is the $t$-butyl-dinitro derivatives of $m$-cresol, known as "musk ambrette" ($6.00). The preparation of these materials is shown in Scheme 9.76.

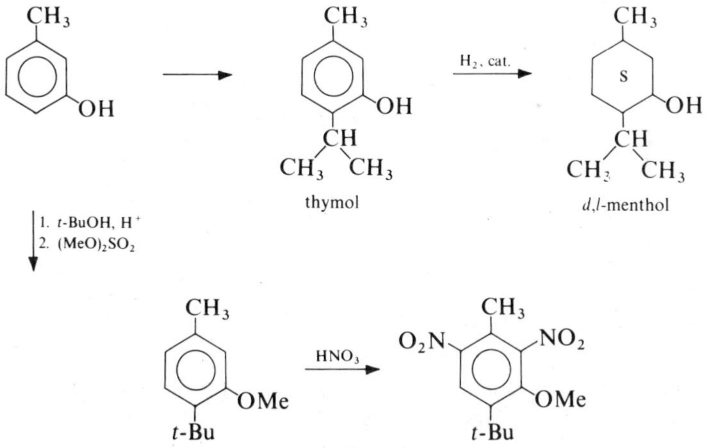

**Scheme 9.76**

The xylenols are the phenolic derivatives of the xylenes, and 2,6-xylenol—a product of the ortho methylation of phenol and the building block of the above-mentioned PPE polymers. The captive use of this product is estimated to exceed 200MM lb, and with the entry of PPE producers other than GE, the demand for 2,6-xylenol is bound to increase.

3,5-Xylenol, also known as $m$-*xylenol*, is an important building block for the synthesis of vitmain E or α-tocopherol and $dl$-α-tocopherol acetate (50% dry powder, $6.60; USP

$10.25). However, since this synthesis requires 2,3,5-trimethylphenol, 3,5-xylenol is subjected to ortho methylation with methanol in order to introduce the third methyl group. The problem of a source of 3,5-xylenol is solved by the cracking of isophorone. The chemistry leading to the final step in the assembly of α-tocopherol by combining the trimethylhydroquinone derived from trimethylphenol with isophytol (*SZM-II*, 127) is shown in Scheme 9.77.

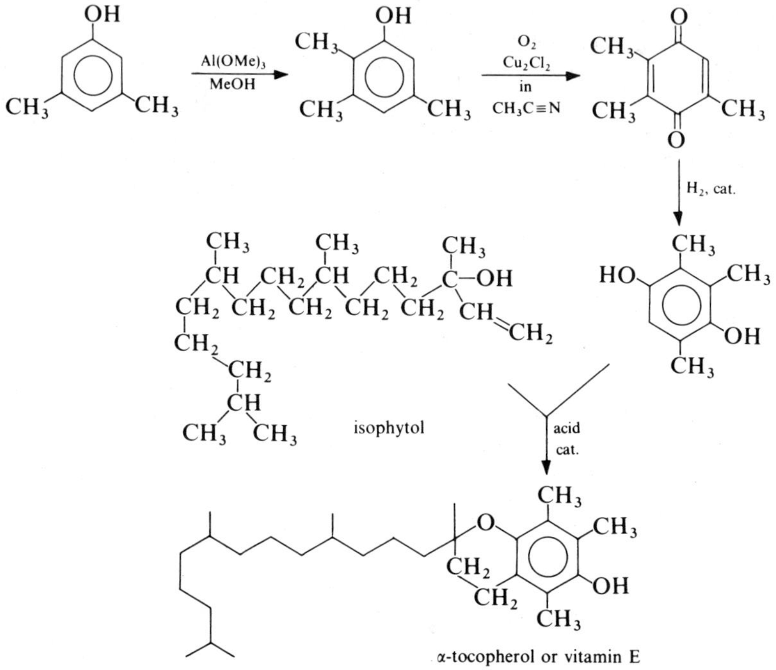

α-tocopherol or vitamin E
d,l-α-tocopheryl acetate, $10.25
d-α-tocopheryl acetate, 81% conc., $26.20

**Scheme 9.77**

### 9.3.4.2    Difunctional and Higher Functionality Phenolic Systems
There are three isomeric dihydroxybenzenes:

Catechol, or pyrocatechol
  technical grade $1.90, CP $3.60

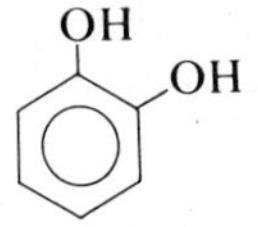

Resorcinol
  technical grade $1.95, USP crystals $4.26
                           USP powder $4.50

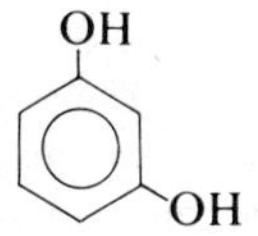

Hydroquinone
  technical grade $1.95, photo grade $2.54

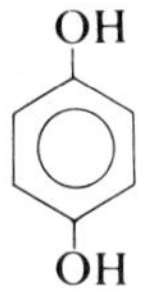

Catechol can be obtained by a cleavage of the naturally occurring monomethyl ether known as *guaiacol* (technical grade $2.70), isolated from guaiacwood oil ($3.30), but this route does not make much sense economically at this time. In any case, it is interesting that the cleavage of ethers is a function of both the acidic nature of the hydrogen halides and the nucleophilic character of the halide ions, and, since both parameters happen to coincide, the reactivity turns out to be H–I > **H–Br** > H–Cl. Guaiacol can be obtained by oxidation of lignin, but this technology is not sufficiently developed to be of industrial importance at this time.

*Direct Hydroxylation of Aromatic Rings*

Catechol is obtained on an industrial scale from coal tar and by direct hydroxylation of the aromatic ring of phenol. The latter route is of general interest and consists of a transition-metal catalyzed reaction ($Co^{2+}$, $Fe^{2+}$, etc.) with 70% hydrogen peroxide (45¢) at about 80°C. EDTA complexes of the metal catalysts improve the yields in the direct hydroxylation of phenol, and hydroquinone is a by-product of the reaction. It should be noted in passing that the Hock process applied to *o*-diisopropylbenzene does not lead to

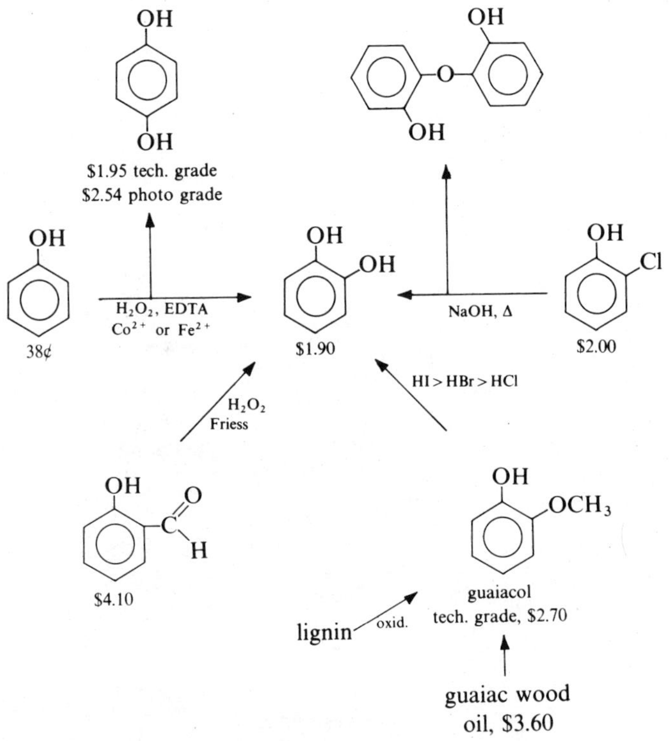

**Scheme 9.78**

the desired catechol, although it results in an interesting transformation by yielding 1,1,3-trimethylindane (compare with the products obtained from 2-methyl-1,4-pentadiene; 5.4.1). A forceful alkaline hydrolysis of $o$-chlorophenol ($2.00) was abandoned as a route to catechol for economic reasons— it may be revived if interest in a by-product of this process (viz., $o,o'$-dihydroxydiphenyl ether) should be awakened. Finally, catechol can be prepared from salicylaldehyde ($4.10) by means of the Fries peroxide rearrangement (see the conversion of cyclohexanone to caprolactone on 8.3.1), but again, the economics of this route is not competitive with the direct hydroxylation of phenol. The above-mentioned chemistry is summarized in Scheme 9.78.

Catechol can be monomethylated to guiacol and dimethylated to veratrol, converted to 4-$t$-butylcatechol (TBC; Dow, $2.32)—a vinyl polymerization inhibitor and anti-oxidant, subjected to the Reimer–Tiemann reaction (9.3.4.1, 9.3.5.1) to yield protocatechu-aldehyde, which, in turn, is an intermediate in the synthesis of $d,l$-epinephrine and dopa (see 5.2). The monoethyl ether of catechol can be used to prepare ethyl vanillin or bourbonal ($13.75), by means of the Reimer–Tiemann or another reaction that introduces a carboxaldehyde substituent (see 9.3.5.1).

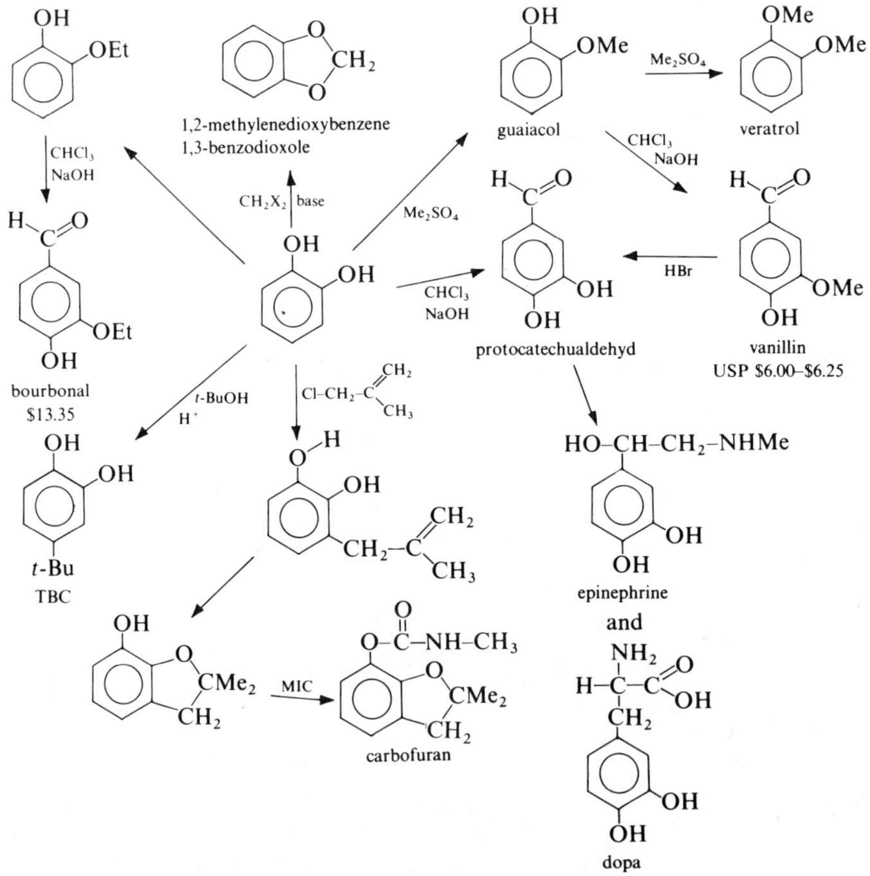

**Scheme 9.79**

With the participation of methallyl chloride, catechol is also the building block of carbofuran—a systemic insecticide of the carbamate family. These transformations of catechol are summarized in Scheme 9.79.

The dehydration of guaiacol and glycerine (99.5% natural, USP 72.5¢; 96% USP CP, 70.75¢; synthetic 99.5%, 78¢; 96%, 56¢) produces guaiacol glyceryl ether, an ingredient of cough medications. The reaction of guaiacol with an allylic alkylating reagent yields the p-allyl derivative known as *eugenol* (USP $2.73), and the analogous reaction with the dimethyl ether of catechol, or veratrol, produces methyl eugenol ($3.55). Both allylic derivatives are isomerized to the conjugated systems of isoeugenol and methyl isoeugenol ($6.60), respectively. All the $C_3$ derivatives of the catechol ethers are used as flavors and fragrances and are also isolated from essential oils. For example, eugenol is isolated from cloves ($1.80) or clove leaf oil ($\sim$ $1.00), and this particular compound is used as a dental analgesic. The oxidation of isoeugenol and methyl isoeugenol with potassium permanganate produces vanillin (USP $6.25; imported $6.00) and veratraldehyde (*Ald. F&F* $17.20), respectively. These and other derivatives of the preceding compounds are subjected to conventional synthetic transformations in order to arrive at downstream building blocks of pharmaceuticals and other fine chemicals as illustrated in Scheme 9.80.

Naturally, when the same process is applied to guaiacol, it supplies synthetic vanillin, and Rhône-Poulenc has recently announced (*CMR* 8/2/87, *CW* 8/5/87) plans to produce both of these flavor and frangrance chemicals.

Resorcinol has been traditionally prepared by means of an alkaline fusion of sodium benzenedisulfonate (9.2.4), but more recently the Hock process applied to m-diisopropylbenzene has begun to displace the former, energy-intensive route. In spite of its relatively high cost, about half of the demand for resorcinol is utilized as a component of PF resins in order to improve their adhesive properties to vinylpyridine-modifed SBR (10.3.1) in the manufacture of rubber tires. Resorcinol is also added to UF foams because of its high reactivity to formaldehyde. The most recent development concerned with the use of resorcinol is its conversion in Japan to m-aminophenol (*CE* 2/16/87) by a route analogous to the production of aniline from phenol (see 9.4.1 and 9.3.6).

Resorcinol is subjected to such conventional transformations as are the preparation of resorcinol monoacetate ($1.98), the Kolbe–Schmitt carboxylation to β-resorcylic acid, monochlorination to 4-chlororesorcinol (Rit-chem $9.10), and acylation with caproic acid or caproyl chloride to yield hexylresorcinol—the active ingredient Caprocol in MSD's Sucrets, a common medication for throat irritation. These transformations of resorcinol are summarized in Scheme 9.81.

Hydroquinone is currently produced by Eastman from p-diisopropylbenzene by means of the Hock process to replace the abandoned traditional route that started with aniline (Scheme 9.82). Eastman previously supplied over 70MM lb of manganese sulfate (out of the total demand for 100MM lb used mostly as a plant nutrient) from the aniline process, and the switch in the hydroquinone source caused a serious but only temporary perturbance in the manganous sulfate market.

Some of the alternative sources of hydroquinone are (1) electrochemical oxidation of benzene, (2) direct hydroxylation of phenol (see preceding discussion of catechol), (3) Reppe carbonylation of acetylene, and (4) oxidative degradation of p-hydroxybenzaldehyde (see Fries peroxide rearrangement mentioned above in the case of catechol).

The chemistry of these transformations is summarized in Scheme 9.83.

Hydroquinone is the classical developer of black-and-white photography since it discriminates between ordinary and photochemically excited silver ions ($Ag^{+*}$) in an exposed silver halide emulsion:

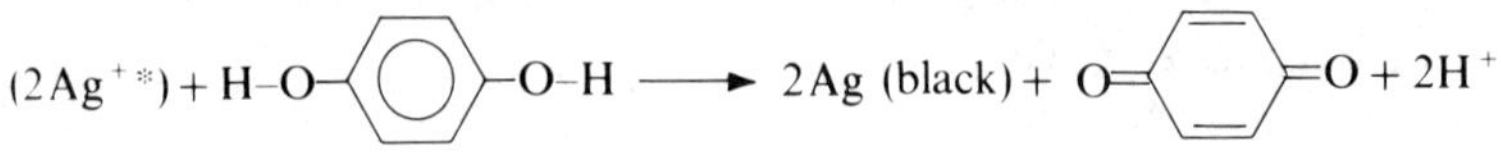

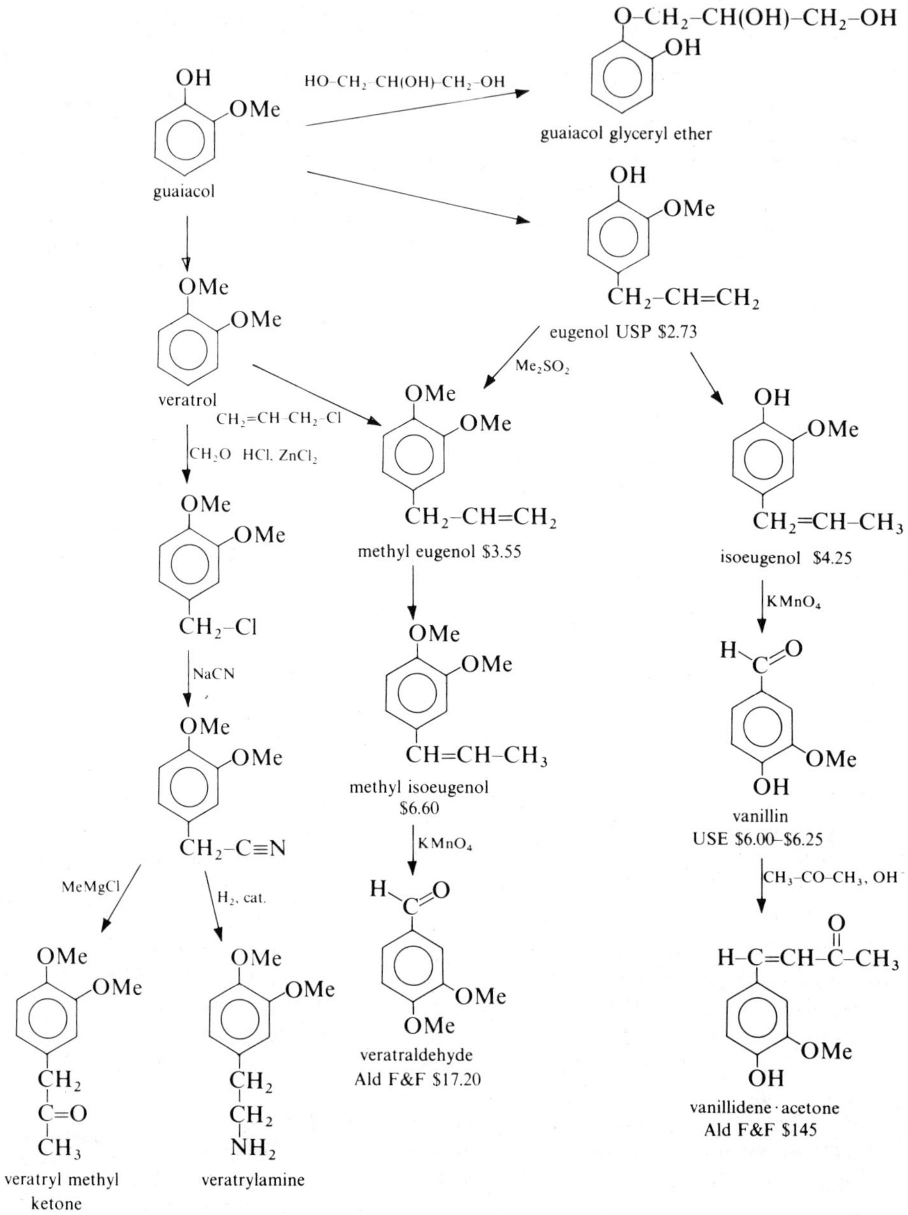

$$(2Ag^{+*}) + H{-}O{-}\langle\bigcirc\rangle{-}O{-}H \longrightarrow 2Ag\ (black) + O{=}\langle\bigcirc\rangle{=}O + 2H^{+}$$

**Scheme 9.80**

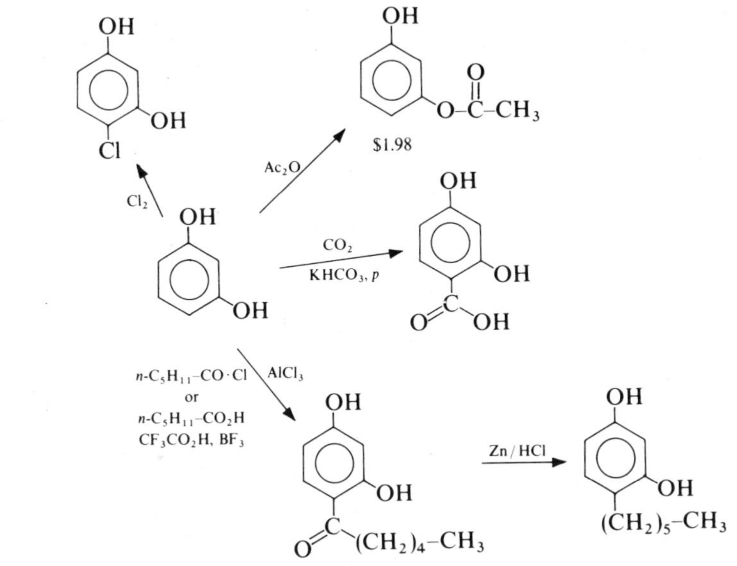

**Scheme 9.81**

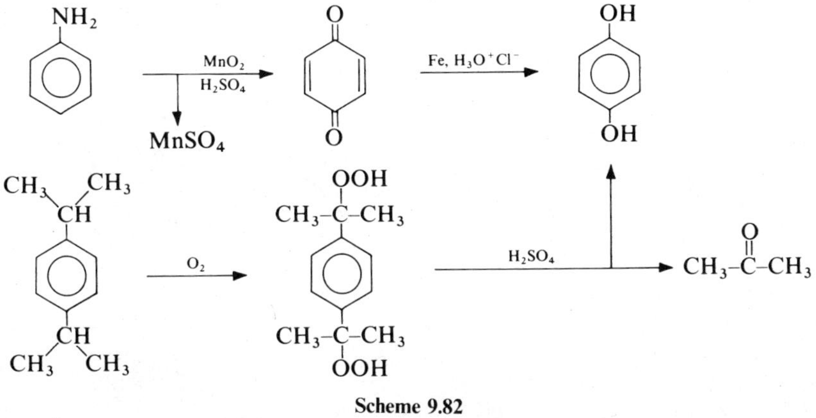

**Scheme 9.82**

The facile oxidation of hydroquinone systems accounts for the use of 2,5-di-$t$-amylhydro-quinone, Monsanto's Santovar A ($4.23) as an antioxidant. Eastman is currently announcing the availability of 2-phenylhydroquinone and potassium hydroquinone-2-sulfonate. The former compound could be derived from either 2,4-diisopropyldiphenyl subjected to a Hock process or $p$-benzoquinone subjected to a classical Gomberg–Bachmann reaction ($ONR$-36), while the latter hydroquinone may result from the homolytic addition of potassium bisulfite (or sodium bisulfite followed by cation exchange) to $p$-benzoquinone (9.84). Also currently, Rohner of Switzerland is advertising the availability of a family of 2-aryl-substituted hydroquinones—not only the phenyl derivative but also $p$-tolyl- and 2-chlorophenyl-substituted hydroquinone and, for that

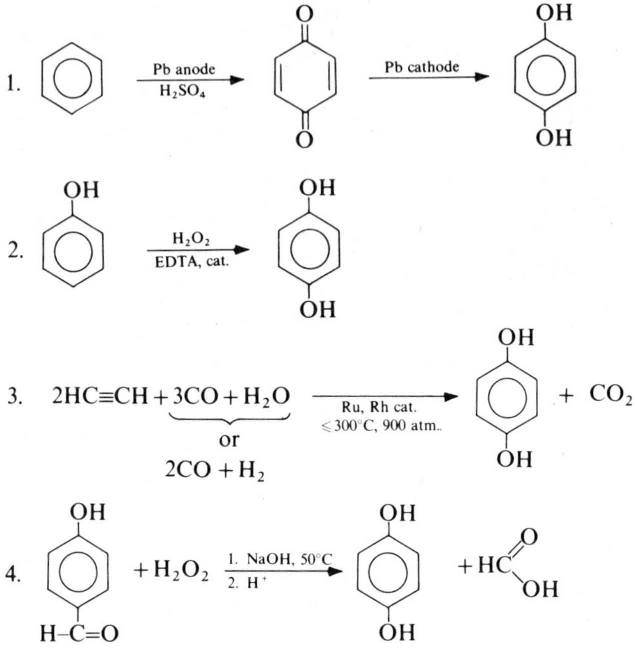

1.

2.

3. $2HC\equiv CH + 3CO + H_2O$ $\xrightarrow[\leqslant 300°C, 900 \text{ atm.}]{\text{Ru, Rh cat.}}$ + $CO_2$

or

$2CO + H_2$

4. + $H_2O_2$ $\xrightarrow[2.\ H^+]{1.\ \text{NaOH, 50°C}}$ + $HC\underset{OH}{\overset{O}{\big\|}}$

**Scheme 9.83**

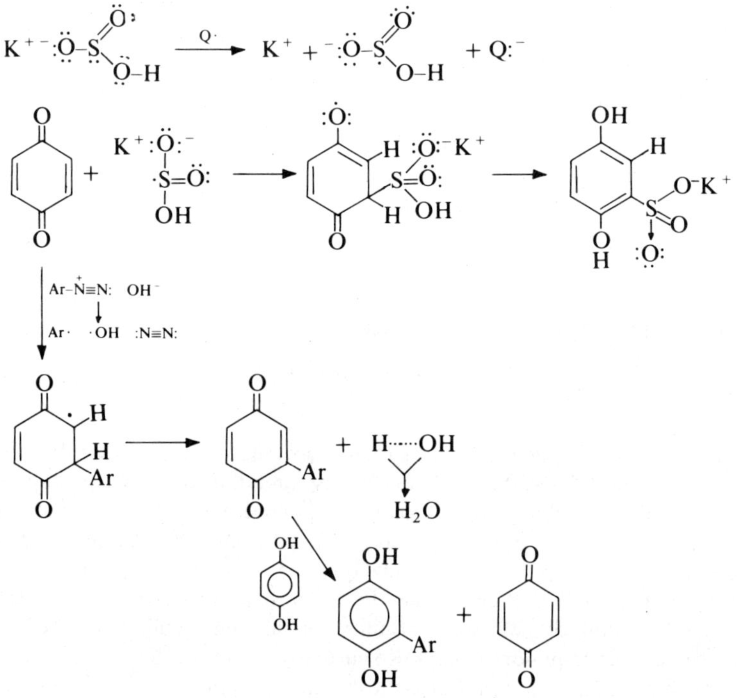

**Scheme 9.84**

matter, any other desired hydroquinone derivative of this type supplied by custom synthesis. This strongly suggests that the Gomberg–Bachmann reaction is being used as long as the diazotizable substituted aniline is available (9.3.6).

Hydroquinones are employed in concentrations as small as 0.1% as inhibitors of vinyl polymerizations, and a prospective larger volume use is the development of PARs, and LCPs.

Some simple derivatives of hydroquinone are the methyl ethers:

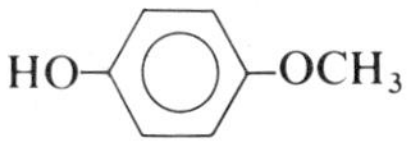

HO—⟨○⟩—OCH₃                    CH₃O—⟨○⟩—OCH₃

p-hydroxyanisole (HA) 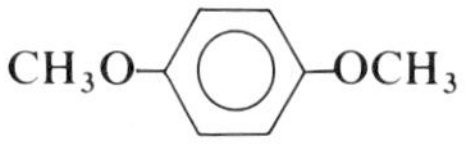          Hydroquinone dimethyl
                               ether ($2.50)

Of the three possible trifunctional phenols only the two symmetrical isomers seem to be of industrial importance at this time:

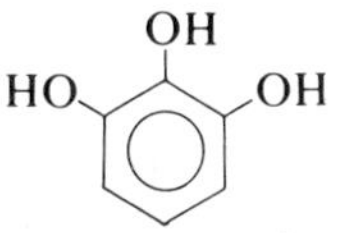

Pyrogallol (pyrogallic          Phloroglucinol
acid; $13.70)

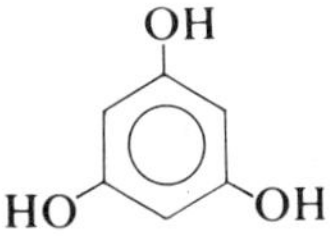

Until recently, pyrogallol was obtained commercially by the thermal decarboxylation of gallic acid ($10.50)—the symmetrical triphenolic carboxylic acid derived from tannins (*SZM-II*, 16, 157–160) and tara powder or gall nuts. It is potentially also available from lignins (*SZM-II*, 8, 83, 99, 148–157) (Scheme 9.85).

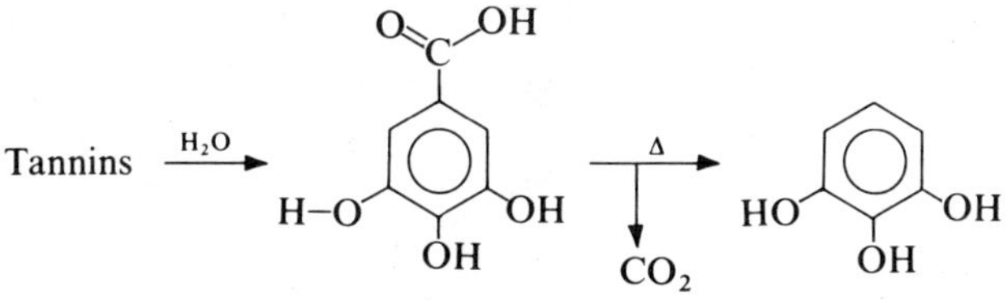

**Scheme 9.85**

The construction of the first synthetic pyrogallol production facility was announced in 1981 by BFC Chemicals, a subsidiary of FBC Ltd. of England, in connection with the expanded production of the propietary carbamate insecticide bendiocarb. The process involves the alpha chlorination of cyclohexanone to 2,2,6,6-tetrachlorocyclohexanone [in the presence of basic catalysts such as collidine (10.3.1) or triphenylphosphine oxide], and this is followed by a sequence of interesting reactions of the substituted cyclohexanone depicted in Scheme 9.86.

Pyrogallol is treated with 2,2-dimethoxypropane (i.e., the dimethylacetal or ketal of acetone), and the resulting ketal of pyrogallol is converted by MIC to Bendiocarb (Scheme 9.87).

Phloroglucinol can be obtained by the nitration of benzene to 1,3,5-trinitrobenzene, followed by reduction to 1,3,5-triaminobenzene and diazotization and hydrolytic decomposition of the triple diazonium salt (9.3.6).

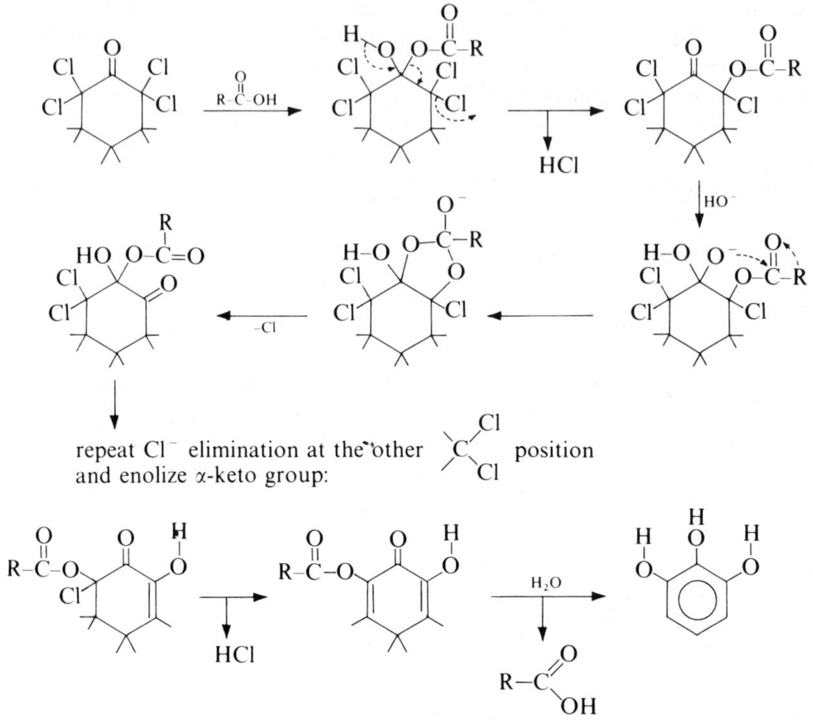

**Scheme 9.86**

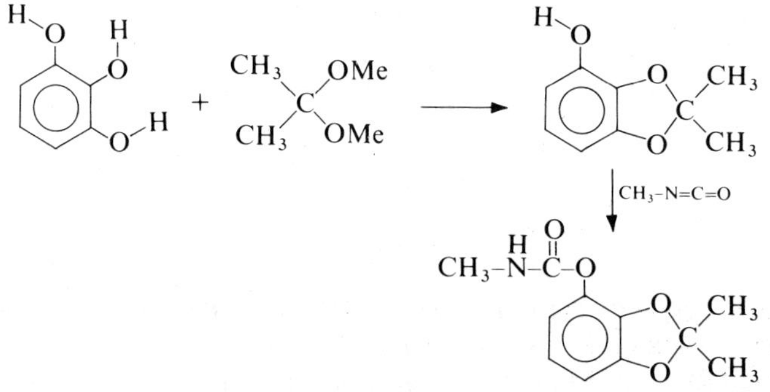

bendiocarb

**Scheme 9.87**

### 9.3.5   Benzenoid Aldehydes, Ketones, and Quinones

#### 9.3.5.1   Aldehydes

There are numerous synthetic procedures for the preparation of aromatic aldehydes, but from an industrial point of view the most practical are

1. Controlled oxidation of a methyl side chain (9.2.2).

2. Hydrolysis of a benzal chloride system (9.2.3).
3. The Reimer–Tiemann reaction of phenols (9.3.4).
4. The Friedel–Crafts-like Gattermann–Koch reaction (ONR-35) catalyzed by $AlCl_3$ and based on an *in situ* formation of formyl chloride [H–C(Cl=O)] from HCl and CO and the rather similar Gattermann aldehyde synthesis (ONR-35), which depends on the *in situ* formation of chloroformimide [HC(Cl)=NH] from HCl and HCN and the use of the $ZnCl_2$ catalyst.
5. The Rosenmund reduction (ONR-78) of an acid chloride in the presence of a Pd catalyst deposited on $BaSO_4$ and "poisoned" by quinoline or another additive that deactivates the Pd catalyst to prevent hydrogenation of the aldehyde to the corresponding benzyl alcohol.
6. The partial reduction of a nitrile to an imine (Ar–CH=NH), by means of stannous chloride ($6.06), a Stephen reduction (ONR-85), or lithium aluminum hydride ($LiAlH_4$), followed by hydrolysis to the desired aldehyde.
7. The reaction of a benzyl chloride with hexamethylenetetramine (HMTA) to yield a readily isolated quaternary ammonium salt of HMTA, a Delepine reaction (ONR-23). Hydrolysis of the salt under acidic conditions to yield the corresponding benzylamine of transient existence because formaldehyde gives rise to the methylol derivative, which, on dehydration and a tautomeric shift to the more conjugated system of a benzalimine of methylamine

$$Ar–CH_2–NH_2 \rightarrow Ar–CH_2–NH–CH_2–OH \rightarrow$$
$$Ar–CH_2–N=CH_2 \rightarrow Ar–CH=N–CH_3$$

is hydrolyzed to the desired aldehyde and methylamine (Sommelet reaction; ONR-84).
8. Direct formylation by transfer of the formyl group from $N,N$-dimethylformamide [H–C(=O)–NMe$_2$, DMF; 49¢) in the presence of phosphorus oxychloride (O=PCl$_3$; 43¢) to an aromatic ring that is rather susceptible to electrophilic substitution.

With the wide choice of these and other available synthetic routes to aromatic aldehydes, the selection of an optimum procedure depends primarily on the nature of substituents present in the aromatic starting material.

The parent compound, benzaldehyde (technical grade 73¢, NF 90¢) can be prepared by methods 1 and 2. As mentioned elsewhere, the presence of traces of chlorine-containing impurities is avoided if the product labeled "benzaldehyde FFC" (free from chlorine; e.g., *Ald. F&F* $1.85) is to be used as a flavor and fragrance enhancer in place of naturally derived "oil of bitter almonds." The latter contains a minimum of 95% of benzaldehyde and some HCN that, obviously, is removed before the essential oil is used for internal consumption.

Generally, substituted benzaldehydes are medium- to high-priced fine chemicals that serve as building blocks of higher-value-added end products. Some of the simpler benzenoid aldehydes and their likely methods of preparation are as follows:

- *o*-Chlorobenzaldehyde, $2.80, method 2
- *p*-Chlorobenzaldehyde, $3.84, method 2

- Salicylaldehyde, $4.20, phenol and method 3
- *p*-Hydroxybenzaldehyde, Dow $3.25, phenol and method 3
- Anisic aldehyde or *p*-anisaldehyde, $5.00, anisol and method 4, used, among other things, as an electroplating additive to obtain shiny surfaces on tin-plated objects
- Heliotropine or piperonal, $10.00, from 3,4-dioxymethylenecatechol and method 4, also isolated from *Ocotea cymbarum*, ($2.02), used in cherry and vanilla flavors and by the perfume industry
- Vanillin, 6MM lb, $6.00–$6.25, guaiacol and method 3, also isolated from seed pods or "vanilla beans" ($27–$37/lb) of the vinelike orchid *Vanilla planifolia* or obtained by oxidation of alkali lignin by means of nitrobenzene (*SZM-II*, 155)
- Ethyl vanilin or bourbonal, 1MM lb, $13.35, ethyl ether of catechol and method 3; duplicates the aroma of natural vanillin extract with 3–4 times greater intensity
- Methylvanillin or veratraldehyde, 3,4-dimethoxybenzaldehyde, Ald FF $17.20, veratrole and method 4, 6, or 7
- Methyl *p*-formylbenzoate (MFB), a by-product of the oxidation of *p*-xylene to DMT
- *N,N*-Dimethylaminobenzaldehyde, dye intermediate, also known as "Ehrlich's reagent"; *N,N*-dimethylaniline and method 8 [see *OS* **IV**, 331 (1963)].

Some interesting derivatives of benzaldehyde include

- Benzyl benzoate, $1.65, the product of the Tishchenko reaction (*ONR*-90); used as a perfume fixative, miticide, and flavoring agent of chewing gum.
- Mandelic acid, $3.63, a urinary antiseptic obtained by hydrolysis of the intermediate mandelonitrile (the latter, by the way, in the form of the glycoside is known as *amygdalin*. A few years ago this material was being isolated from apricot pits and employed as the alleged anticancer drug Laetrile).
- Phenylglycine, the product of the Strecker synthesis (*ONR*-87); Dane's salt [$C_6H_5$–C($CO_2K$)–NH–C(Me)=CH–$CO_2Et$], obtained by means of acetoacetic ester, is used in the production of semisynthetic penicillins (*SZM-II*, 45) because *Penicillium notatum* incorporates the $C_6H_5$–CH($NH_2$)–CO– moiety in the structure of the antibiotic; the analogous process that starts with *p*-anisaldehyde rather than benzaldehyde produces the antibiotic amoxicillin.
- Ephedrine, amphetamine, and phenylpropanolamine, products derived from the condensation of benzaldehyde with nitroethane (5.2).
- Cinnamic acid—the product of the Claisen condensation of benzaldehyde (4.6.4), the starting material for one of the commercial routes to *l*-phenylalanine that consists of a biocatalytic alpha amination and produces the desired stereoisomer (*CE* 11/25/86). Another synthetic route to *l*-phenylalanine employs the hydantoin synthesis; although there are several variations of the hydantoin synthesis, discussion of these routes is beyond the scope of this book.

The demand for *l*-phenylalanine arose with the popularity of the synthetic sweetener aspartame. The current production of *l*-phenylalanine is estimated at about 15MM lb, and its list price is about $38.50 (July 1987). The availability of *l*-aspartic acid [$HO_2C$–$CH_2$–CH($NH_2$)–$CO_2H$; $1.00–$1.35]—the partner building block required for the assembly of aspartame—is less problematic.

Some of the U.S. patents for aspartame are due to expire by 1992, and one can expect an intensified search for improved routes to this sweetener. However, there are also prospects for new and more powerful sweetener that will challange the commercial future of aspartame.

Cinnamic acid is converted to the methyl and ethyl esters ($4.65 and $18.65, respectively) and to the more complex (*SZM-II*, 120) linalyl cinnamate

$$Me_2C=CH-CH_2-CH_2-C(Me)(CH=CH_2)-O-CO-CH=CH-C_6H_5; \text{ } \$59.85$$

All these cinnamic acid esters are used by the perfumery industry.

Cinnamic aldehyde ($1.55) is the product of the aldol condensation (4.5.2) of benzaldehyde and acetaldehyde, and its reduction by means of sodium borohydride ($17.45 as a stabilized solution in water) gives cinnamyl alcohol ($4.50). α-Amylcinnamic aldehyde ($2.35) is the product of an aldol condensation of benzaldehyde and heptaldehyde: the latter is one of the thermal decomposition products of ricinoleic acid (*SZM-II*, 73). Basic Green 4, or Malachite Green Crystal ($6.90) is a dye obtained by the condensation of benzaldehyde and 2 mol of *N,N*-dimethylaniline, followed by oxidation of the resulting triarylmethane intermediate.

The preceding transformations of benzaldehyde are summarized in Scheme 9.88.

Some examples of products derived from substituted benzaldehydes are as follows:

- *o*-Chlorobenzaldehyde and malononitrile condense in the presence of a *t*-amine catalyst to yield *o*-chlorobenzalmalononitrile (CS), an irritant used as a riot control agent (Scheme 9.89).
- Vanillin is reduced by means of zinc amalgam and hydrochloric acid—the Clemmensen procedure (*ONR*-19)—to 4-methylguaiacol or creosol (Ald F&F $91), used by the perfume industry (Scheme 9.90).
- *p*-Nitrobenzaldehyde—the product of a controlled oxidation of *p*-nitrotoluene, is condensed with crotonaldehyde to give 5-(*p*-nitrophenyl)-2,4-pentadienal (NPPD), a common "tracking agent" apparently employed by the Soviet Union (*CW* 8/28/85) to monitor the movements of American diplomats stationed at the embassy in Moscow.
- *p*-Aminobenzaldehyde is a dye intermediate.

### 9.3.5.2 Aryl Ketones

The oxidation of substituted diarylmethane systems to diaryl ketones in the course of synthesis of multifunctional carboxylic acids is mentioned in 9.3.2.3 and is illustrated by the preparation of BTDA. Obviously, the simplest example of this method is the oxidation of diphenylmethane to benzophenone—the preparative method of choice for FFC (free from chlorine) benzophenone.

Benzenoid compounds are converted to alkyl aryl and diaryl ketones by means of the Friedel–Crafts acylation (*ONR*-33). In addition to the classical use of aluminum chloride, other catalytic systems include a large variety of Lewis and protic acids such as $ZnCl_2$, $SnCl_4$, HF, $H_2SO_4$, $BF_3$, $CF_3-CO_2H$, and polyphosphoric acid, selected according to the nature of substituents present on the benzenoid ring system (G. A. Olah, *Friedel–Crafts and Related Reactions*, Vols. 1–3, Interscience, New York, 1963–1965). A number of such acylation reactions and transformations of the products to useful derivatives are illustrated in Scheme 9.91.

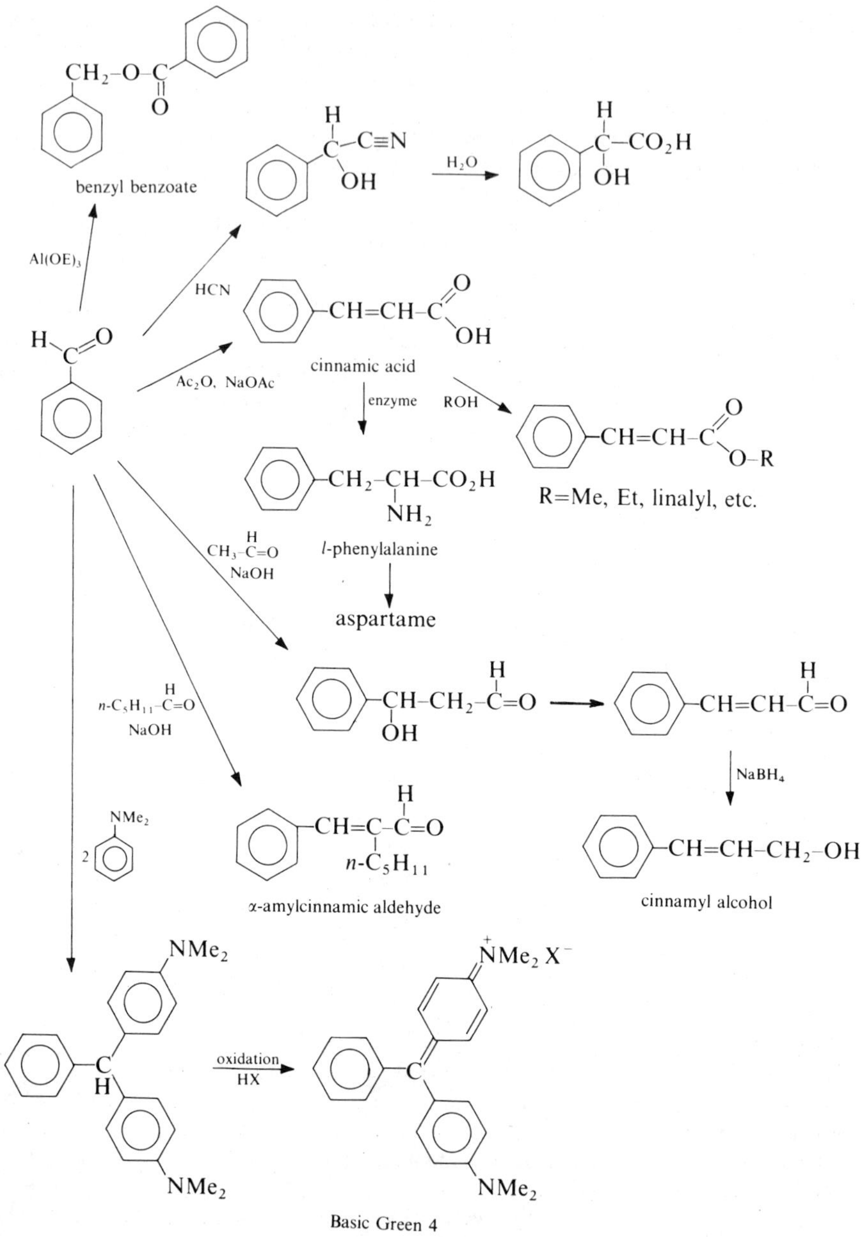
benzyl benzoate
Al(OE)3
HCN
H2O
Ac2O, NaOAc
cinnamic acid
enzyme
ROH
R=Me, Et, linalyl, etc.
l-phenylalanine
aspartame
CH3-C=O
NaOH
n-C5H11-C=O
NaOH
NaBH4
α-amylcinnamic aldehyde
cinnamyl alcohol
oxidation
HX
Basic Green 4

**Scheme 9.88**

Benzophenone (technical grade \$2.40, NF crystals \$2.50, NF flakes \$3.65) is a catalyst added to UV-cured polymers formulated to produce inks, adhesives, and coatings. The NF grade of benzophenone is used as an additive or intermediate in the preparation of medicinal substances and as a fixative in perfumes. Special grades of benzophenone

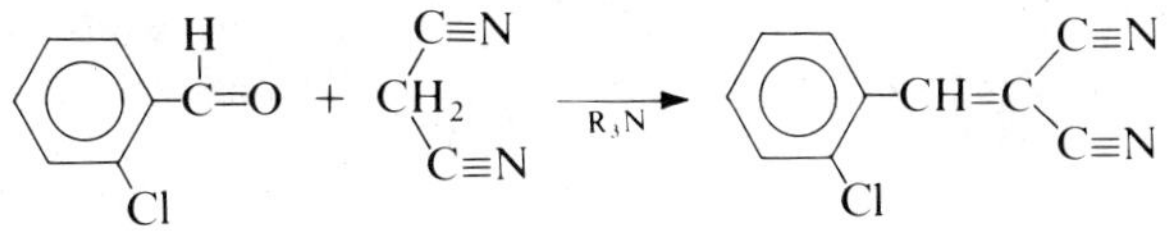
$R_3N$

CS

**Scheme 9.89**

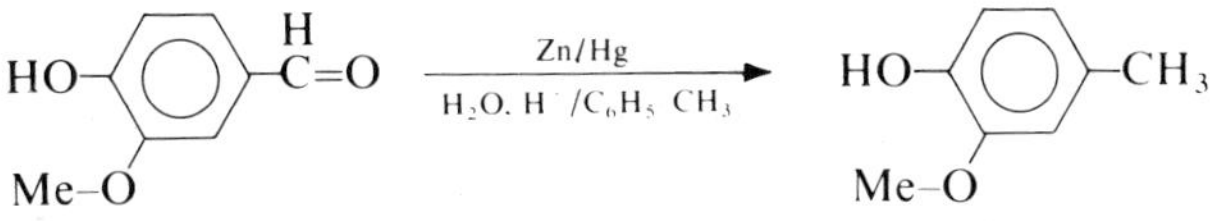
Zn/Hg
$H_2O$, $H^+$/$C_6H_5CH_3$

4-methylguaiacol

**Scheme 9.90**

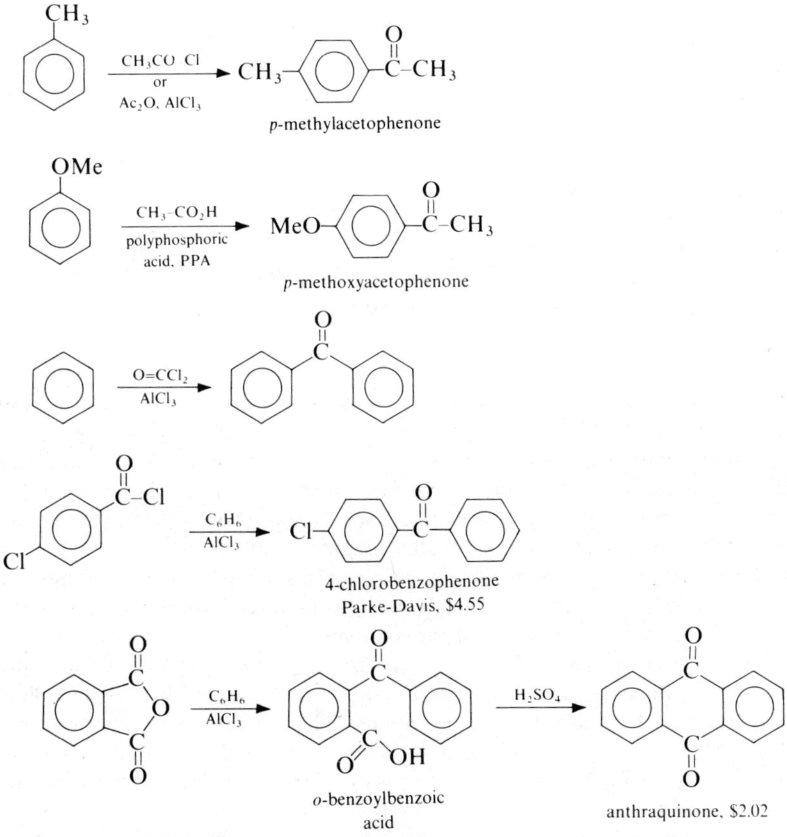
$CH_3CO$ Cl
or
$Ac_2O$, $AlCl_3$
*p*-methylacetophenone
$CH_3$–$CO_2H$
polyphosphoric
acid, PPA
*p*-methoxyacetophenone
O=$CCl_2$
$AlCl_3$
$C_6H_6$
$AlCl_3$
4-chlorobenzophenone
Parke-Davis, $4.55
$C_6H_6$
$AlCl_3$
$H_2SO_4$
*o*-benzoylbenzoic
acid
anthraquinone, $2.02

**Scheme 9.91**

prepared by the above-mentioned oxidation of diphenylmethane (9.2.2) are also available ($3.50 for UV-light-absorbent grade; Parke-Davis $3.65, aroma grade).

Benzophenone hydrazone (Parke-Davis $7.30) is used to generate diphenyldiazomethane:

$$Ph_2C{=}O + H_2N{-}NH_2 \rightarrow Ph_2C{=}NNH_2 \xrightarrow{HgO} Ph_2C{=}N_2 \; (+ Hg + H_2O)$$

$85\%, \$1.25$

Diphenyldiazomethane is a convenient reagent for the preparation of benzhydryl esters:

$$Ph_2C{=}N_2 + R{-}CO{\cdot}OH \rightarrow R{-}CO{-}O{-}CHPh_2 + N_2$$

A less costly route to benzhydrol—the intermediate in the preparation of Parke-Davis's Benadryl—the first antihistaminic available on the marketplace, is the reduction of benzophenone with zinc under alkaline conditions. The resulting benzhydrol is converted with HCl to benzhydryl chloride, and the latter is treated with dimethylaminoethanol (Scheme 9.92).

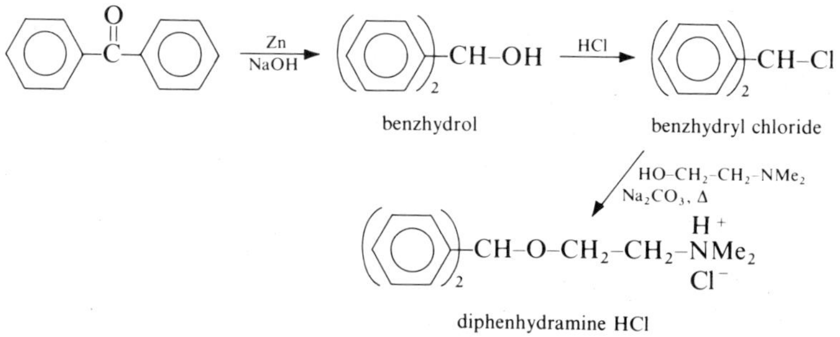

benzhydrol

benzhydryl chloride

diphenhydramine HCl

**Scheme 9.92**

Since the expiration of the 1946/47 Parke-Davis Benadryl patents, the antihistaminic known by its generic name "diphenhydramine hydrochloride" ($9.10) is distributed under a various tradenames, and as a mixture with 8-chlorotheophilline (10.3.3), it is also the anti-motion-sickness agent dimenhydrinate, best known as Searle's "Dramamine."

The above-mentioned 4-chlorobenzophenone (Parke-Davis $4.55) is used as a polymer additive in a manner similar to that of the parent ketone.

The benzoylation products of substituted phenols, such as those shown above, are excellent absorbents of UV light. Thus, 2-hydroxy-4-methoxybenzophenone, or oxybenzone, obtained by the benzoylation of the monomethyl ether of resorcinol, is used in cosmetic suncreen preparations that are becoming more popular as people become aware of the correlation between excessive sunbathing and skin cancer and the wrinkling of skin (*CW* 3/11/87). For reasons of compatability, 2-hydroxy-4-*n*-octoxybenzophenone, Cincinnati Milacron's Carstarb 700, is used as an UV-light-absorbing additive in polymers, and 2,3,4-trihydroxybenzophenone is used for the same reason in the manufacture of electronic products.

A totally different purpose is served by 2-hydroxy-4-nonylbenzophenone, which is converted to the oxime, and as Henkel's LIX 65 N ($3.30), used as an extractant of copper from mining residues as a result of selective complexing of cupric ions to produce kerosine-soluble chelates. This facilitates a recovery of about 1.6MM lb of copper per day.

4,4'-Bis(dimethylamino)benzophenone, or Michler's ketone, is formed from phosgene in the absence of a Lewis catalyst because of the great nucleophilic reactivity of $N,N$-dimethylaniline. It is an intermediate in the formation of dyes of the triarylmethane family such as, for example, Basic Green 4. Another example is the product of an acid-catalyzed reaction of Michler's ketone with another molecule of $N,N$-dimethylaniline that produces Gentian Violet, CI Basic Violet, which functions as an antimicrobial and anthelmintic (Scheme 9.93).

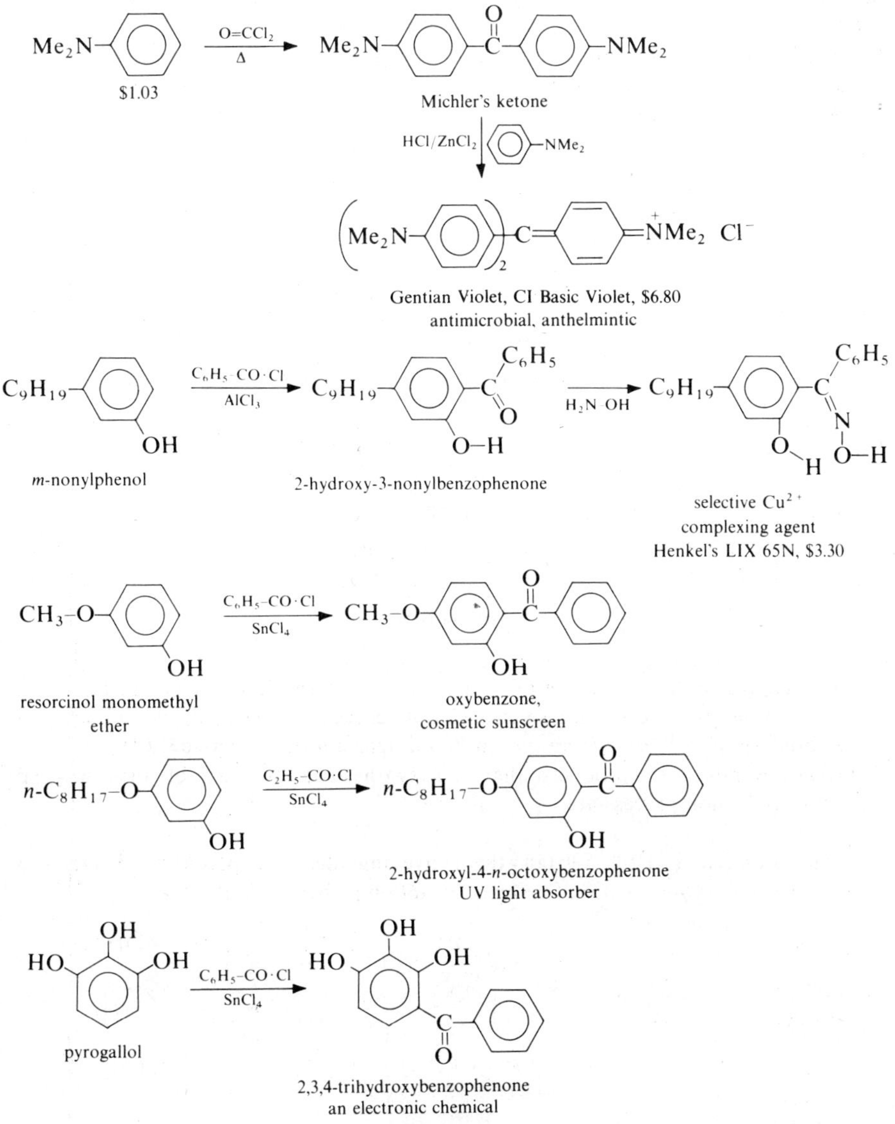

Scheme 9.93

For the preparation of musk ketone ($15.90), see 9.3.6.2.

### 9.3.5.3  Alkyl Aryl Ketones

Since attempted Friedel–Crafts alkylations of aromatic substrates usually lead to carbon skeleton rearrangements of the intermediate alkyl-derived carbocations, the acylation route is chosen instead in order to introduce an alkyl side chain of desired structure. The acylation, which employs an acid chloride or anhydride, initially yields the expected alkyl aryl ketone, and the latter must then be reduced to the alkyl substituent by means of either the Clemmensen reduction or the use of the Wolff–Kishner reaction (*ONR*-97). The latter depends on the base-catalyzed decomposition of an intermediate hydrazone (Scheme 9.94). The temperature requirement of the Wolff–Kishner reaction was found to be lowered by DMSO cosolvent, but a protic solvent must also be present to serve as a proton source in the rate-determining step of the reaction [H. H. Szmant, *Angew. Chem. Int. Ed.* **7**, 120–128 (1968)].

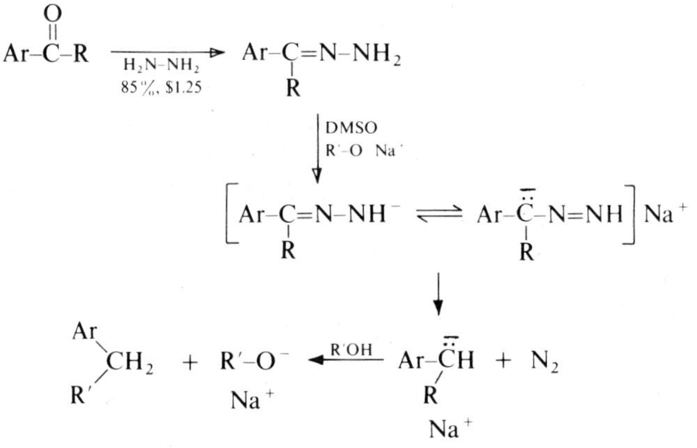

**Scheme 9.94**

Another reason for the use of alkyl aryl ketone intermediates in place of a direct alkylation is the greater ease with which acylation reactions can be controlled with regard to regioselectivity and for the prevention of multiple substitutions (see 9.2.1).

Some examples of products that are synthesized by way of the two-step acylation–reduction process are

- Hexylresorcinol (USP $30.00), the active ingredient of Beecham's Sucrets—a common OTC medication for throat irritation (Scheme 9.95).

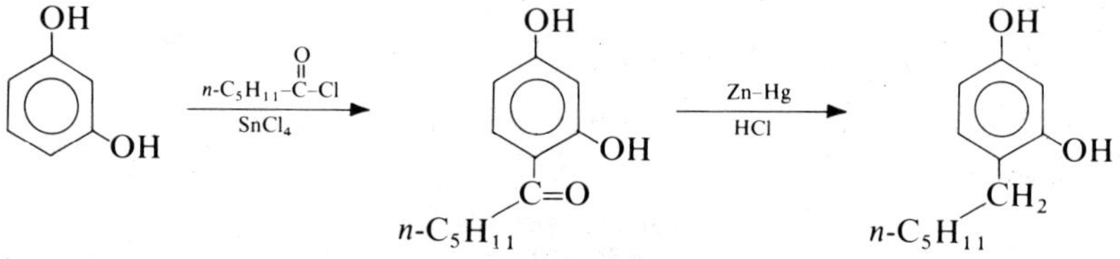

**Scheme 9.95**

• Piperonyl butoxide ($5.00), an insecticide synergist used in many domestic insect control formulations (Scheme 9.96).

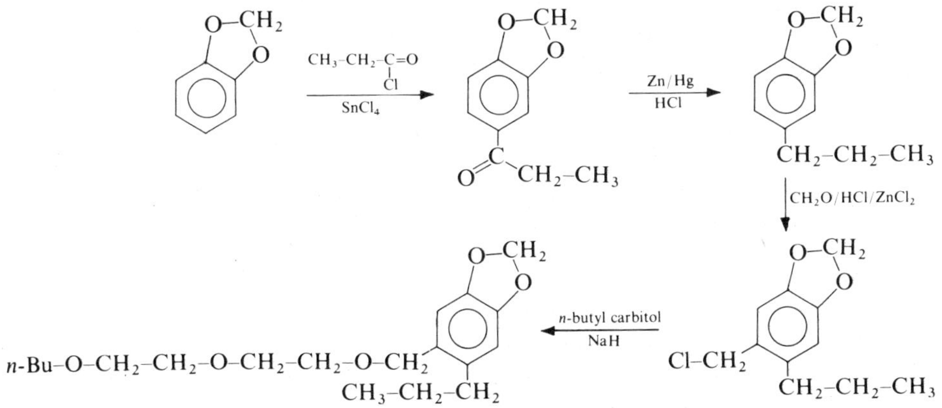

**Scheme 9.96**

Ibuprofen [2-(p-isobutylphenyl)propionic acid], the latest challanger of aspirin. Ibuprofen probably offers the best example of the two-step acylation–reduction procedure chosen to control the structure of the alkyl side chain of an aromatic end product. It can be prepared from benzene and isobutyroyl chloride (obtained from isobutyric acid (75¢) and thionyl chloride (55¢), and the resulting isopropyl phenyl ketone is reduced to isobutylbenzene by means of a Wolff–Kishner reaction. Another acylation with acetyl chloride yields p-isobutylacetophenone, and a Willgerodt–Kindler reaction (ONR-95) converts the ketone to the corresponding p-isobutylphenylacetic acid. To convert this arylacetic to an alpha-substituted propionic acid, one can esterify the acid and methylate the substituted arylacetate ester with methyl halide (after having generated a carbanion at the ester-activated α-methylene group) (Scheme 9.97).

This lengthy synthesis of a commercially promising medicinal stimulated a search for alternative synthetic routes to ibuprofen, and one of these utilizes a rather novel plumbilation reaction. This process involves the use of p-isobutyl phenyl lead triacetate and its reaction with the acetonide of methylmalonic acid in which there is a transfer of the aryl group from the lead-containing intermediate to the acetonide, and, finally, the arylation product decarboxylates on hydrolysis, as is expected of substituted malonic acids (Scheme 9.98). In view of Ethyl Corporation's extensive experience in sodium and lead chemistry, it is likely that Ethyl, currently the largest manufacturer of ibuprofen in the United States, is employing the last-mentioned process. Recently the unit price of ibuprofen dropped to about $11.80–$13.65/lb (CMR 3/7/88) because of overcapacity and imports from Japan, India, and Italy. Ibuprofen's share in the U.S. $2.3-billion OTC analgesic market is about 18% (vis-à-vis the 50% share for aspirin), with sales of about 5.5MM lb and an underutilized worldwide production capacity of about 17.5MM lb.

*Biocatalytic Membrane Technology*

Examination of the structure of ibuprofen reveals the existence of two stereoisomers, but so far the marketed product consists of the racemic mixture of

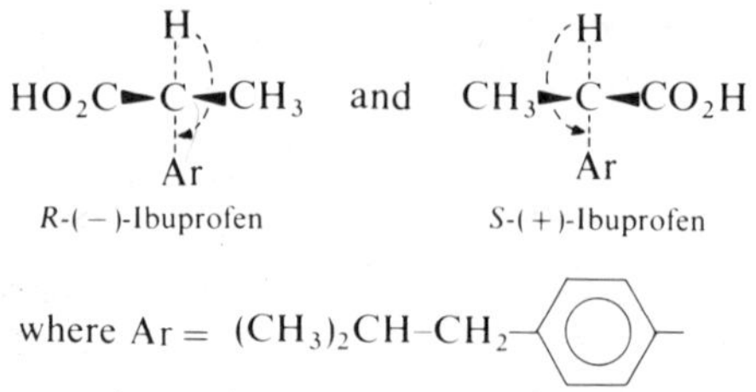

HO2C C CH3
and
CH3 C CO2H
H
Ar
R-(−)-Ibuprofen
S-(+)-Ibuprofen
where Ar = (CH3)2CH–CH2–


However, it turns out that the *S* enantiomer is pharmacologically about 100 times more active than its mirror image, and Sepracor Inc. of Marlborough, Massachusetts has developed a propietary biocatalytic membrane process capable of isolating the desired *S* enantiomer (*CW* 2/24/88).

The success of the biocatalytic membrane technology depends on hollow-fiber membranes in which selected enzymes are entrapped mechanically in the wall of fibers and that allow the entry, by diffusion, of an emulsion composed of an organic water-immiscible solvent and an aqueous solution. The membrane-containing system is assembled in a manner that separates the two components of the emulsion by selective diffusion as the mixture travels along the tubular reactor. In the case of the resolution of ibuprofen, it is used as an alkyl ester that is soluble in the chosen organic solvent, and the

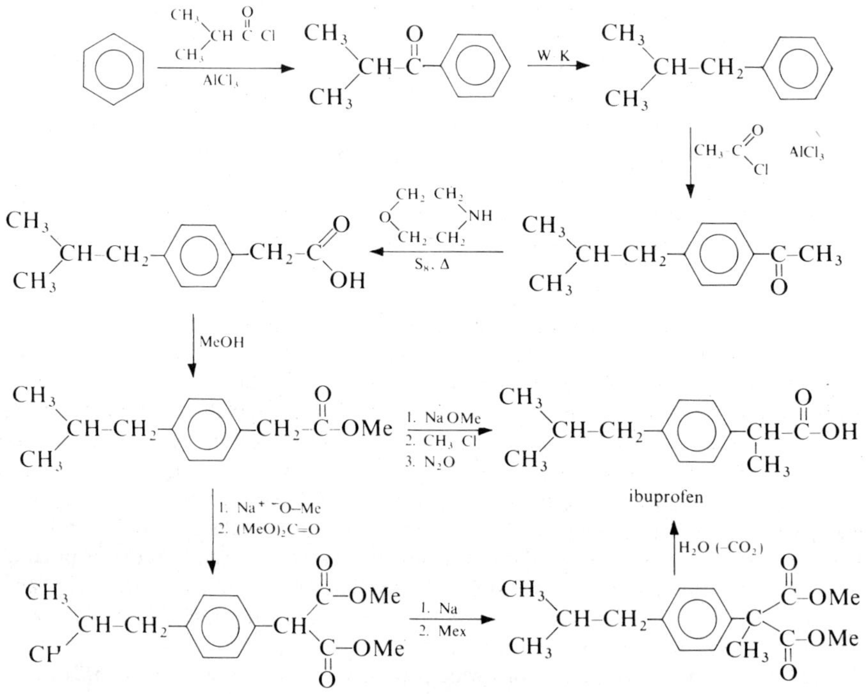

AlCl3
W K
MeOH
1. Na OMe
2. CH3 Cl
3. N2O
ibuprofen
1. Na+ −O–Me
2. (MeO)2C=O
H2O (–CO2)
1. Na
2. Mex
(where W–K represents the Wolff–Kishner reaction)


**Scheme 9.97**

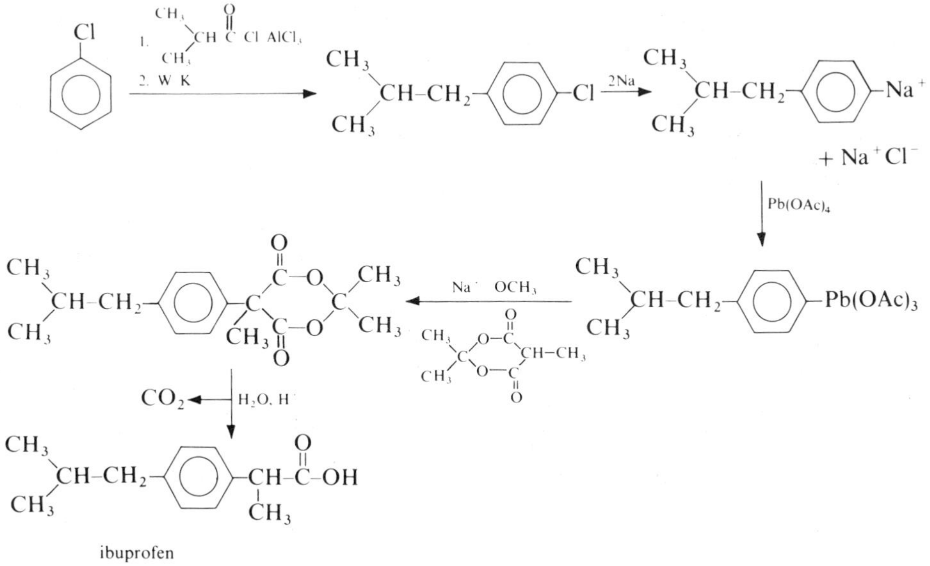

**Scheme 9.98**

ibuprofen

selected lipase or esterase enzyme hydrolyzes stereospecifically the $S(+)$-ibuprofen ester. The liberated $S(+)$ acid dissolves in the buffered aqueous component of the emulsion while the $R(-)$ ester remains in the organic solvent. In a separate operation, the $R(-)$ enantiomer can be recovered, racemized, and recycled.

Thus, we may expect to see in the near future the appearance on the marketplace of the highly active S-ibuprofen used in smaller doses that would decrease whatever undesired side effects are caused by the current use of the larger doses. The last-mentioned statement is intended not to imply that a prescribed use of ibuprofen is the cause for any special, undesired side effects but rather to remind the reader that *all* chemicals, natural or synthetic, tend to upset the delicate chemical balance of normal life processes and that the physiological burden of metabolism and detoxification of unnecessary intake of "strange" chemicals should be minimized whenever possible. Consequently, *all* chemicals, natural or synthetic, must be treated with a respect based on knowledge of the benefits and dangers that they produce.

This resolution of ibuprofen is only one of many successful separations possible by means of advances in membrane technology. For a concise introduction to this important topic, see E. K. Lee, "Membranes, Synthetic, Applications," in *Encyclopedia of Physical Science and Technology*, Vol. 8, Academic Press, San Diego, CA, 1987, pp. 20–55. Other, more extensive sources include

R. E. Kesting, *Synthetic Polymeric Membranes: A Structural Perspective*, 2nd ed., Wiley, New York, 1985.

J. H. Fendler, *Membrane Mimetic Chemistry: Characterization and Application of Micelles, Micro-emulsions, Monolayers, Bilayers, Vesicles, Host–Guest Systems, and Polyions*, Wiley, New York, 1982.

Next we cite some examples of alkyl aryl ketones prepared for reasons other than the introduction of an alkyl substituent:

1. Methyl 2-naphthyl ketone (2-acetonaphthone; $14.00) is obtained when the Friedel–Crafts reaction of naphthalene, acetyl chloride, and $AlCl_3$ and is carried out in nitrobenzene, while the use of $CS_2$ solvent yields a considerable amount of methyl 1-naphthyl ketone.

2. Celiprolol is a new-generation beta-blocker developed originally by Chemie Linz of Austria but acquired recently by Rorer (*CMR* 10/12/87), and used in the treatment of hypertension and angina pectoris. Its structure suggests that it is assembled from *o*-acetylphenol—the by-product of Hoechst-Celanese novel APAP process (Scheme 9.99).

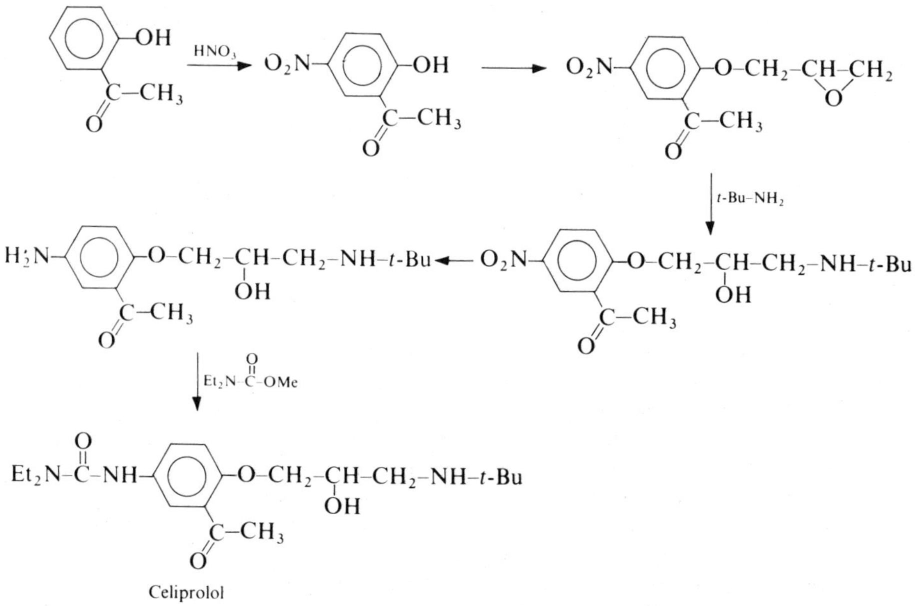

Celiprolol

**Scheme 9.99**

3. 2,2-Diethoxyacetophenone (DEAP) is an ingredient of rapidly cured, nonyellowing, UV-clear coatings; it is obtained by means of the Friedel–Crafts reaction of dichloroacetyl chloride and benzene, followed by the substitution of the chlorine substituents by sodium ethoxide:

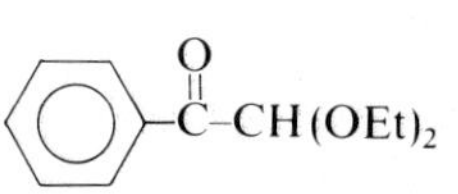

4. *p*-Hydroxypropiophenone is the starting material of one of several synthetic routes to diethylstilbesterol (DES) a synthetic estrogen.

One of the most direct approaches to the synthesis of DES is the reaction of the ketone with hydrazine to yield the corresponding azine after the initial phenolic function is stabilized by conversion to the methyl ether. Thermal decomposition of the azine is driven by formation of dinitrogen, and the removal of the methyl groups by means of HBr produces the desired *trans*-diethylstilbesterol (Scheme 9.100).

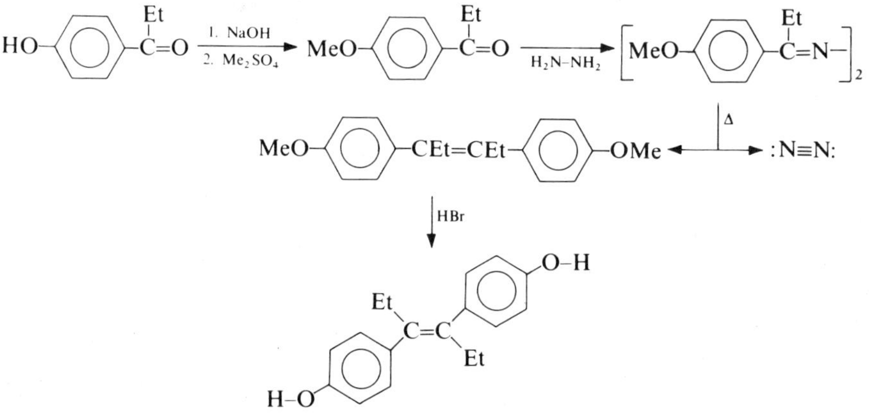

**Scheme 9.100**

Several decades ago DES was also frequently employed to prevent fetal human miscarriages, but this practice was abandoned because of incidents of vaginal cancer in daughters of the female patients. In 1979 the FDA banned the use of DES as a growth-stimulant for cattle. The comparison of the molecular anatomy of DES and $\beta$-estradiol reveals sufficient similarity to suspect that the mimicry of DES upsets the optimum concentration of the naturally occurring, highly powerful hormone.

### 9.3.5.4  Quinones

The preparation of p-benzoquinone, or simply benzoquinone (since the unstable, isomeric o-benzoquinone is not commonly encountered), is unavoidably mentioned in connection with difunctional phenols (9.3.4.2) and, specifically, hydroquinone.

Benzoquinone is subject to facile heterolytic 1,4-addition reactions that produce either substituted hydroquinones, or, if the latter are readily oxidized by the parent compound or by an external oxidizing agent, the corresponding substituted benzoquinones are obtained. The oxidizing power of benzoquinone increases with the presence of electron-withdrawing substituents (*CT* 9/87, 518).

The first mode of behavior is illustrated by the formation of substituted hydro-quinones such as 2-methoxy-, 2-acetoxy-, 2,3-dicyano-, and 2-chlorohydroquinones (Scheme 9.101).

The second mode of behavior is observed, for example, during the addition of HCl in the presence of the oxidizing agent chlorine (Scheme 9.102). In order to complete the conversion of benzoquinone to the tetrachloro-substituted benzoquinone, or chloranil, one employs potassium perchlorate rather than chlorine.

Benzoquinone is also subject to homolytic addition reactions as mentioned elsewhere (9.3.4.2) in connection with the preparation of the sulfonate of hydroquinone.

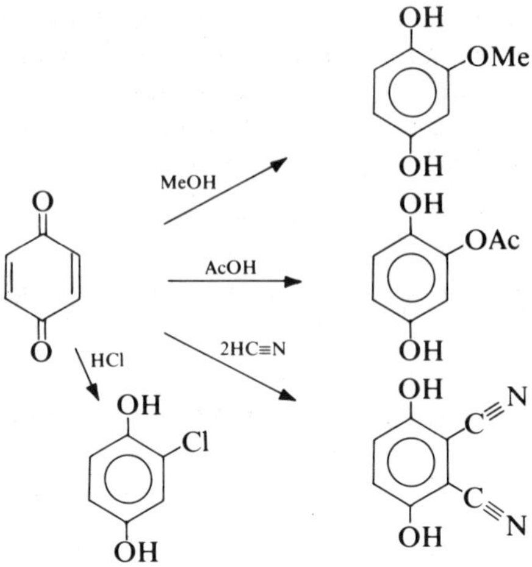

**Scheme 9.101**

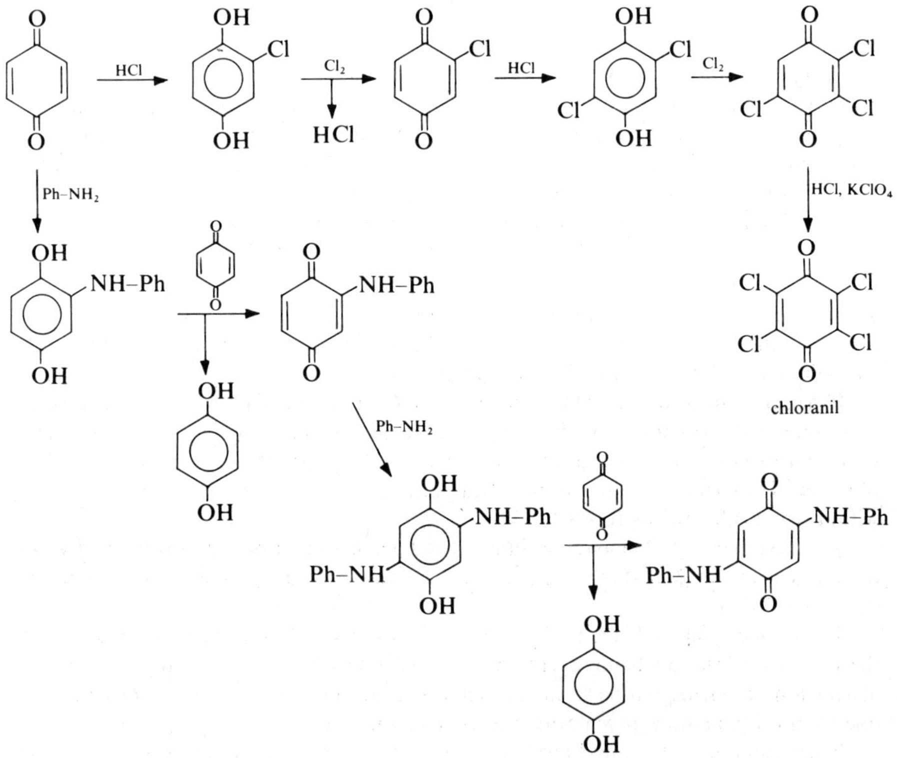

**Scheme 9.102**

Successive, acid-catalyzed reactions of benzoquinone with aminoguanidine [$H_2N-C(=NH)-NH-NH_2$] and thiosemicarbazide ($H_2N-NH-CS \cdot NH_2$) give rise to the antimicrobial ambazone:

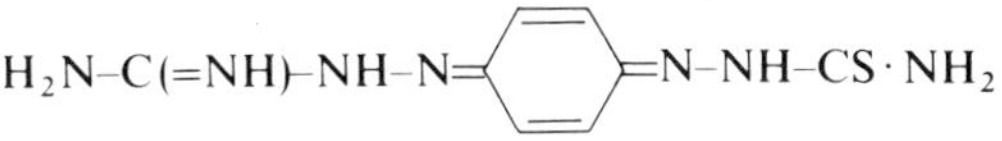

$$H_2N-C(=NH)-NH-N=\bigcirc=N-NH-CS \cdot NH_2$$

Another example of the ketone-like behavior of benzoquinone is its conversion to tetracyanoquinodimethane (TCNQ) by a condensation reaction with malononitrile (5.6.5) carried out in the presence of pyridine and some stannic chloride catalyst (Scheme 9.103). TCNQ, like TCNE (5.6.5), is a powerful electron acceptor in the formation of charge-transfer complexes with a great variety of electron donors (R. Foster, *Organic Charge-Transfer Complexes*, Academic Press, New York, 1969) and the charge-transfer complexes produced from TCNQ are of interest because of their conductivity of electricity.

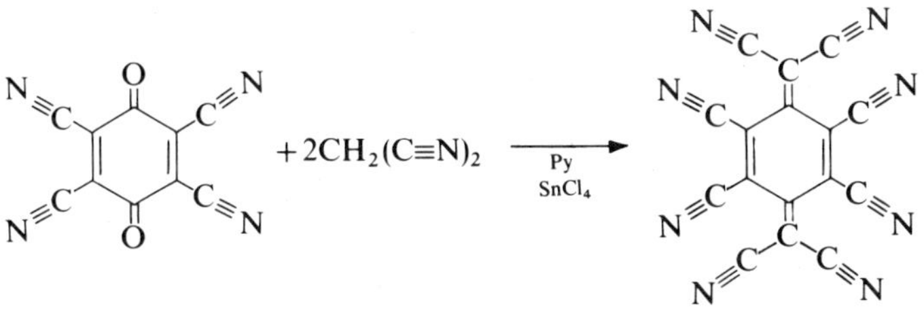

**Scheme 9.103**

Finally, benzoquinone functions as a dienophile in Diels–Alder reactions (6.8.1.4), and this behavior is illustrated in Scheme 9.104 in connection with 1,4-naphthoquinone.

Chloranil is a gentle oxidizing, or, better stated, dehydrogenating agent useful in certain synthetic operations. Thus, for example, very pure samples of *p*-terphenyl suitable for use in liquid scintillation measurements can be prepared from a *p*-biphenyl Grignard reagent and cyclohexanone with the last step involving chloranil (Scheme 9.104).

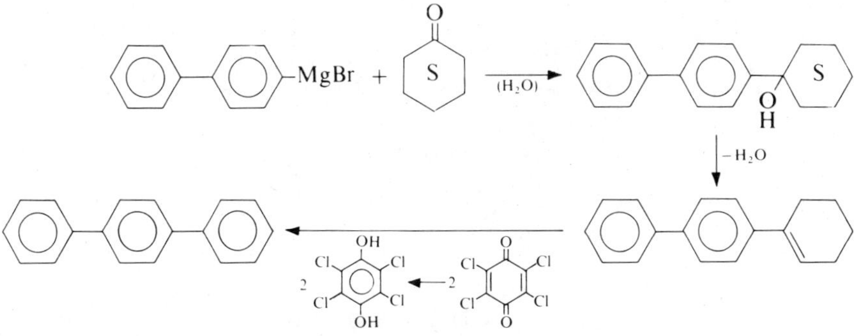

**Scheme 9.104**

The naphthoquinones offer a wider choice of isomeric possibilities:

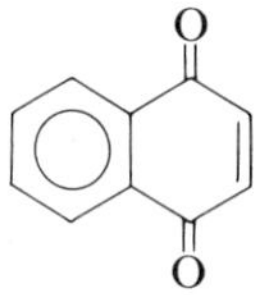

| 1,4-Naphthoquinone | 1,2-Naphthoquinone | 2,6-Naphthoquinone |
| α-Naphthoquinone | β-Naphthoquinone | amphi-Naphthoquinone |

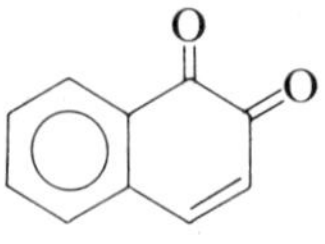

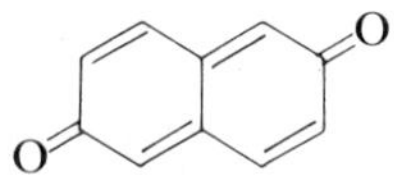

1,4-Naphthoquinone offers maximum stability because of optimum aromaticity and because of the a 1,4 arrangement of the two C=O moieties. 1,4-Naphthoquinone and its simple derivatives are obtained by means of mild oxidation of 1- and 4-hydroxy and/or amino-substituted naphthalenes. The susceptibility to oxidation of such substituted naphthalenes to quinones, in preference to side-chain oxidation, is illustrated by the preparation of 5-methyl-1,4-naphthoquinone from 1-methylnaphthalene. Generally, chromic acid in acetic acid is an excellent method for the conversion of condensed, polynuclear aromatic hydrocarbons to the respective quinones (Scheme 9.105).

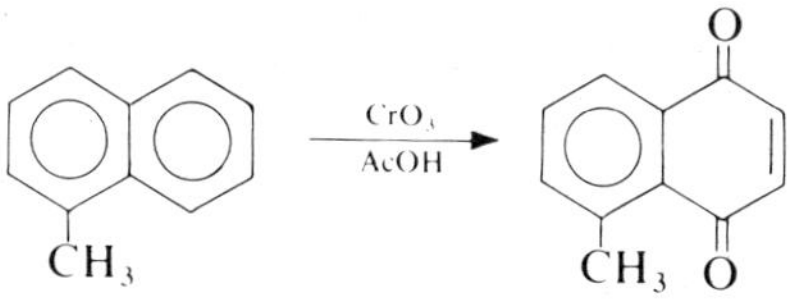

**Scheme 9.105**

The isomeric 2-methyl-1,4-naphthaquinone is the degradation product of the blood-clotting, prothrombogenic vitamin $K_1$, and it is prepared from 2-methylnaphthalene (Scheme 9.106). This simple degradation product of vitamin $K_1$, known as *menadione*, exhibits the physiological effect of the whole vitamin molecule, but, unfortunately, it also induces some undesirable side effects. To remedy this situation, menadione is converted to the oxime and the latter is carboxymethylated with sodium chloroacetate (4.6.3) to produce manadoxime, a useful antihemorrhagic agent (Scheme 9.107).

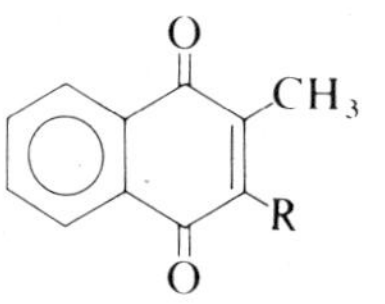

menadione  R = H

vitamin $K_1$  R = phytyl

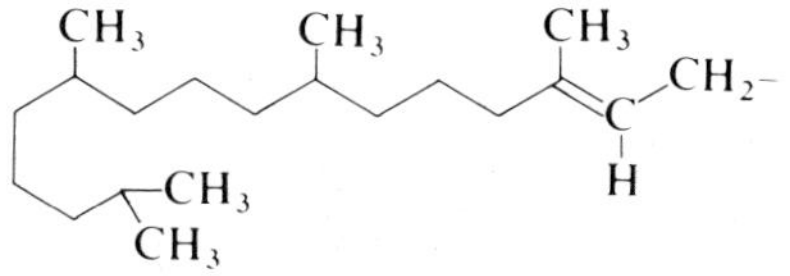

**Scheme 9.106**

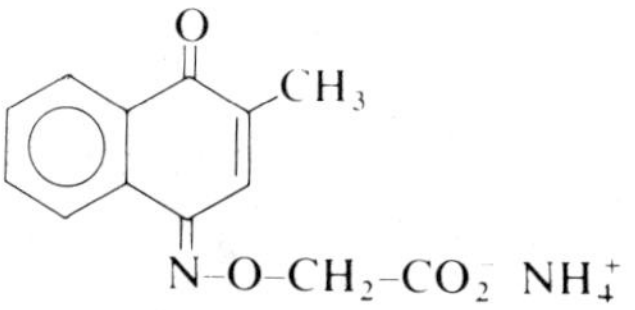

$$N-O-CH_2-CO_2^-\ NH_4^+$$

**Scheme 9.107**

The susceptibility to oxidation increases on progression from naphthalene to higher polynuclear aromatic hydrocarbons, and this is illustrated by the formation of 9,10-anthraquinone and 9,10-phenanthraquinone from the respective hydrocarbons by means of chromic acid–acetic acid or sodium dichromate in sulfuric acid (Scheme 9.108).

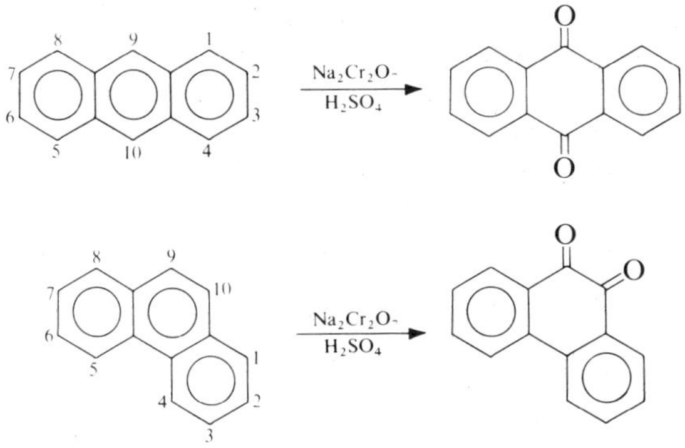

**Scheme 9.108**

In the case of substituted anthraquinones the ring closure of appropriate benzoyl-benzoic acids (9.3.5) is more competitive than oxidation of the respective precursors. For example, 2-chloroanthraquinone is obtained from phthalic anhydride and chlorobenzene, and similarly, 2-alkylanthraquinones are obtained from phthalic anhydride and the appropriate alkylbenzenes. The last-mentioned compounds are involved in an ingenious process to produce hydrogen peroxide in which the net chemical change is simply

$$H_2 + O_2 \rightarrow H\text{-}O\text{-}O\text{-}H$$

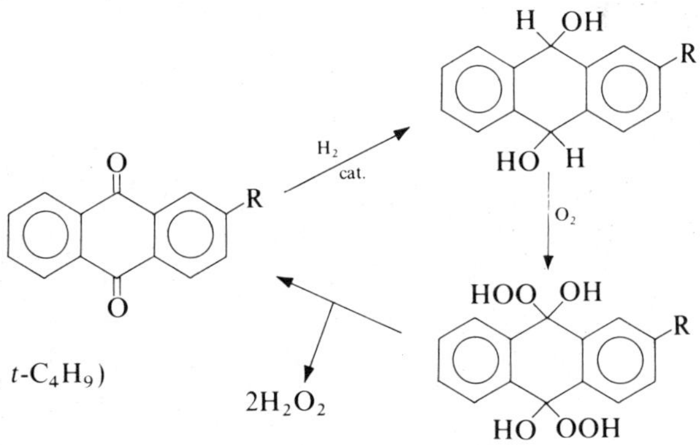

**Scheme 9.109**

This transformation is summarized in Scheme 9.109. Alkyl-substituted anthraquinones, rather than the parent compound, are used in the hydrogen peroxide process because of their more favorable solubility in a hydrocarbon solvent. This currently preferred hydrogen peroxide process (as compared to electrolysis of dilute sulfuric acid) is threatened by the recent Du Pont announcement (*CMR* 2/1/88) of a propietary process capable of combining directly the two elementary substances in the presence of an as yet undisclosed catalyst (*CW* 12/9/87). The economics of the oldest hydrogen peroxide process—the electrolytic formation of persulfuric acid followed by its hydrolysis—depends on the cost of electricity. Both the oldest and more recent anthraquinone processes can withstand for some time the challenge of the Du Pont process because of the rapidly growing demand for hydrogen peroxide ("Peroxide in New Growth Mode," *CMR* 6/29/87; J. Rivoire, "Peroxide Makers Seek Global Growth," *CW* 9/16/87, 38–39). In response to the growing demand for hydrogen peroxide is the installation of new, and the expansion of existing, production facilities by manufacturers such as FMC (*CEN* 6/8/87), Du Pont Canada (*CW* 6/10/87), and Degussa (*CEN* 8/31/87). The greatest impetus for the use of hydrogen peroxide comes from the pulp–paper industry, which attempts to depart from the use of chlorine-containing bleaching agents. However, whenever the federal government decides to impose serious restrictions on sulfur dioxide emissions from power plants and other sources of acid rain, the demand for hydrogen peroxide will be greatly expanded (*CE* 9/14/87). The same effect on the demand for hydrogen peroxide may result from its increased use in the treatment of wastewater. Of the current hydrogen peroxide demand of about 365MM lb in the United States, only 80MM lb are consumed in the production of chemicals (S. Stinson, *CEN* 6/1/87).

9,10-Anthraquinone, or simply anthraquinone ($2.00), can be obtained, in addition to the above-mentioned cyclization of benzoylbenzoic acid, by direct oxidation of anthracene with oxygen at about 400°C in the presence of appropriate catalysts or by means of chromic acid in acetic acid. In recent years it was discovered that tetrahydroanthraquinone (THAQ) accelerates the delignification of wood during the kraft pulping process (*SZM-II*, 151). The mechanistic implications of this interesting development cannot be discussed here, but suffice to say that it led to an attempt to prepare THAQ by a Diels–Alder reaction of naphthaquinone and butadiene. Unfortunately, the successful laboratory and pilot-plant experiences turned into a disaster during an attempted scaleup (2.3.2). Thus, for the moment, THAQ is still manufactured by controlled hydrogenation and partial dehydrogenation of anthraquinone. The preceding transformations involving anthraquinone are summarized in Scheme 9.110.

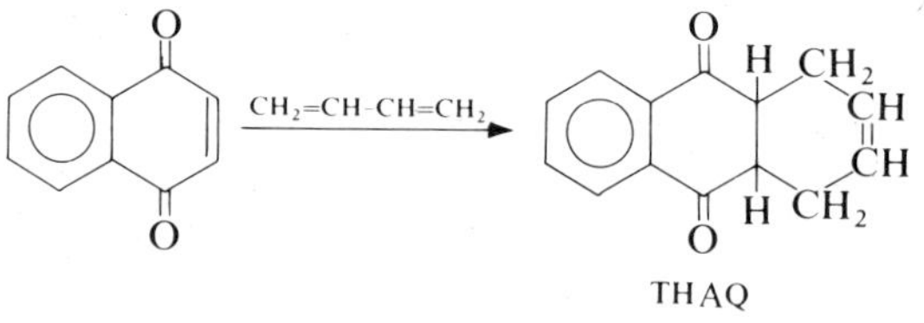

THAQ

**Scheme 9.110**

Anthraquinone is converted to sulfonic acids that serve as dye intermediates and produce water-soluble dyes that are easily attached ("fixed") to nitrogen-containing textile fibers and other, somewhat basic, substrates subjected to dyeing operations. Sulfonation by means of fuming sulfuric acid ($H_2SO_4$–$SO_3$) produces anthraquinone-2-sulfonic acid and the 2,6- and 2,7-anthraquinone disulfonic acids. Interestingly, anthraquinone is sulfonated at the 1 position in the presence of mercury salts. Alkaline, oxidative fusion of anthraquinone-2-sulfonic acid produces more than the expected phenol—it leads to 1,2-dihydroxyanthraquinone, better known as *alizarin*. Alizarin is an example of a

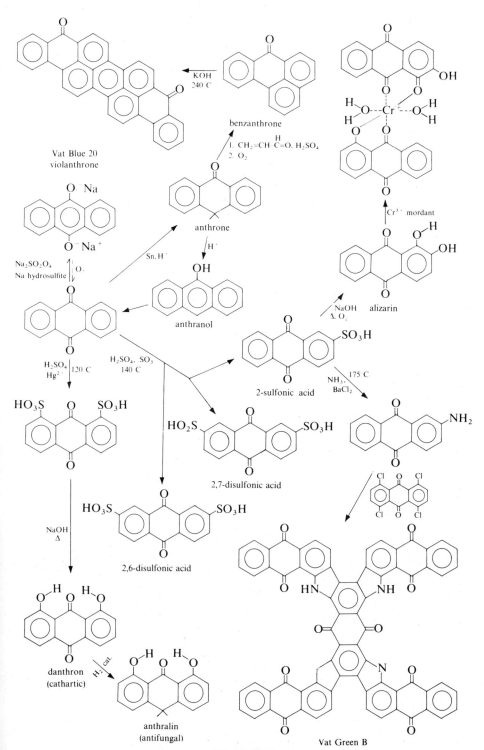

**Vat Blue 20**
violanthrone

benzanthrone

1. CH$_2$=CH–C=O, H$_2$SO$_4$
2. O$_2$

KOH
240 C

anthrone

Sn, H⁺

H⁺

OH

anthranol

Na$_2$SO$_2$O$_4$
Na hydrosulfite

O$_2$

Cr³⁺ mordant

NaOH
Δ, O$_2$

alizarin

H$_2$SO$_4$
Hg²⁺  120 C

H$_2$SO$_4$, SO$_3$
140 C

SO$_3$H

2-sulfonic acid

175 C
NH$_3$,
BaCl$_2$

NH$_2$

HO$_3$S      SO$_3$H

HO$_2$S           SO$_3$H

2,7-disulfonic acid

HO$_3$S           SO$_3$H

2,6-disulfonic acid

NaOH
Δ

danthron
(cathartic)

H$_2$/cat.

anthralin
(antifungal)

**Vat Green B**

**Scheme 9.111**

539

*mordant dye*, that is, it is fixed to the substrate such as wool and silk through the intervention of a metallic complex, and $Cr^{3+}$ is the preferred mordant. An isomer of alizarin, namely 1,8-dihydroxyanthraquinone, or danthron, is also a dye intermediate as well as a cathartic, and its partial hydrogenation to the deoxydihydro derivative yields anthralin, an antifungal claimed to be an effective treatment for psoriasis. The formation of anthraquinonesulfonic acids and the preceding transformations are summarized in Scheme 9.111.

The electron-withdrawing quinoid functions of anthraquinone activate the neighboring so-called peri substituents toward nucleophilic substitution reactions, and this behavior renders anthraquinone derivatives even more useful as dye intermediates. For example, the Friedel–Crafts acylation of *p*-chlorophenol with phthalic anhydride yields 1-chloro-4-hydroxyanthraquinone, which is hydrolyzed to 1,4-dihydroxyanthraquinone, and the latter is transformed to 1,4-bis(methylamino)anthraquinone by means of the Bucherer reaction (*ONR*-16) that is mentioned repeatedly in 9.3.6. The resulting Disperse Blue 3 dye is a member of the family of dyes that can diffuse into the interior of textile fibers because of their relatively small molecular size, but unfortunately they are not "fast," that is, strongly attached to the substrate because they are gradually lost upon repeated washings (Scheme 9.112).

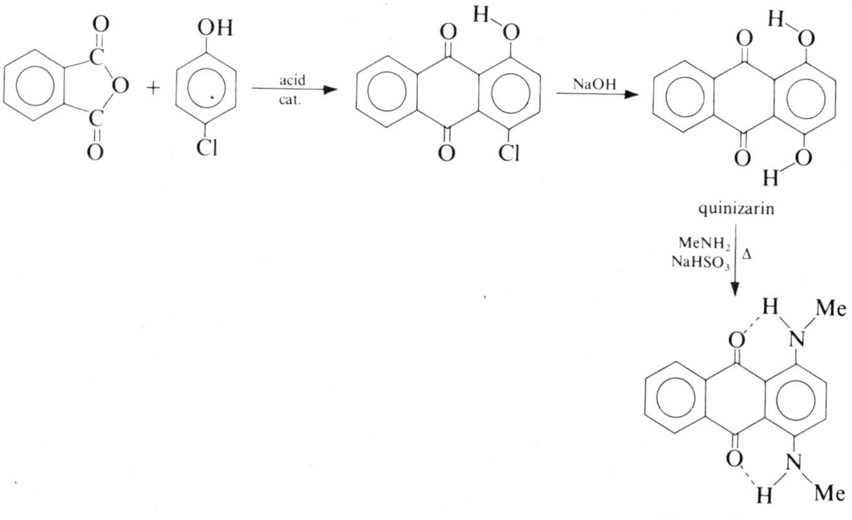

quinizarin

Disperse Blue 3

**Scheme 9.112**

Other examples of anthraquinone dyes are

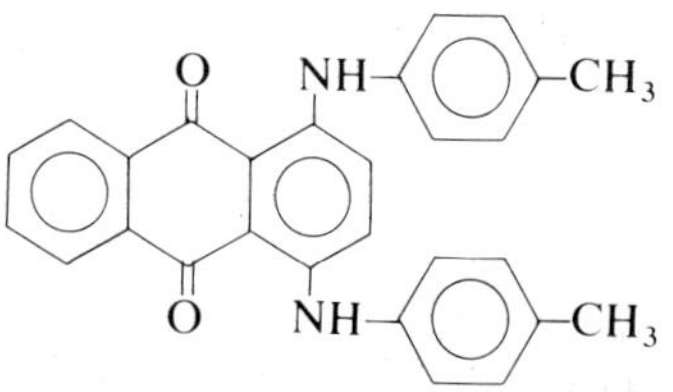

Solvent Green 3, Morton–Thiokol $13.00

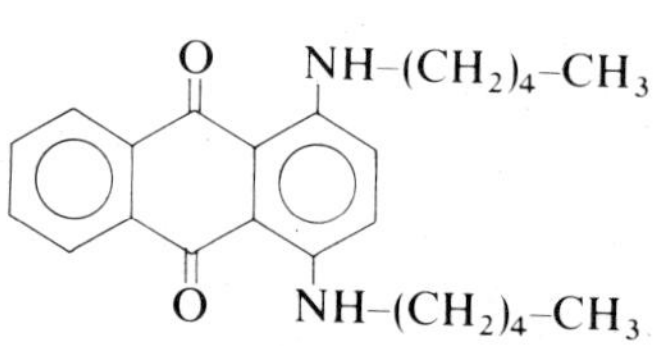

Solvent Blue 14

There are literally many thousands of industrially important dyes that cannot be mentioned in a book of this length. The reader is referred to specialized monographs such as E. R. Trotman, *Dyeing and Chemical Technology of Textile Fibers*, 6th ed., Wiley, New York, 1984, and especially to the *Colour Index* published jointly by the Society of Dyers and Colourists of England and the American Association of Textile Chemists and Colorists. The *Colour Index* (*CI*) lists most dyes of commercial importance, assigns them an internationally accepted code, and classifies them according to their chemical nature and/or method of application and their fundamental color. For example, "A Bl" represents an acidic blue dye, "B Br" a basic brown dye, "D R" a direct(ly applied) red dye, "Dis Y" a disperse yellow dye, and so on. A similar but significantly smaller list refers to coloring materials approved by the FDA for use in foods, drugs, and cosmetics, and another list refers to organic pigments. The weekly *CMR* price list cites a number of these materials and identifies them by means of the CI system.

Additional examples of dye molecules are found in 9.3.6 and in Chapter 10, but a glimpse at the chemistry of anthraquinone chemistry, as it relates to the formation of larger ring structures through fusion under oxidative and/or dehydrating conditions, is believed to be rather instructive and thus is offered in Scheme 9.113.

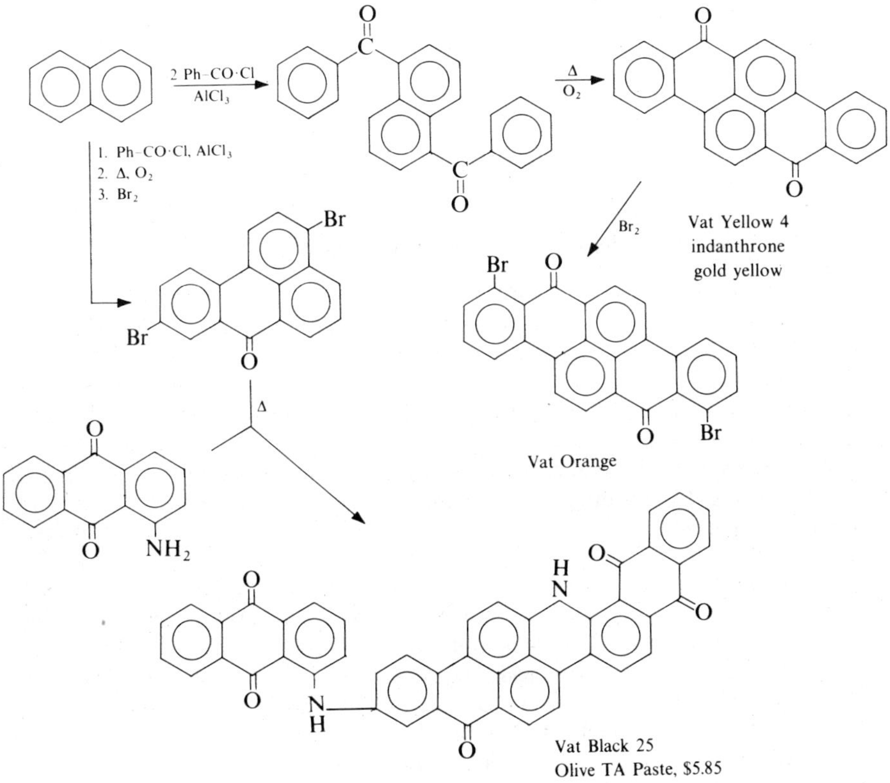

**Scheme 9.113**

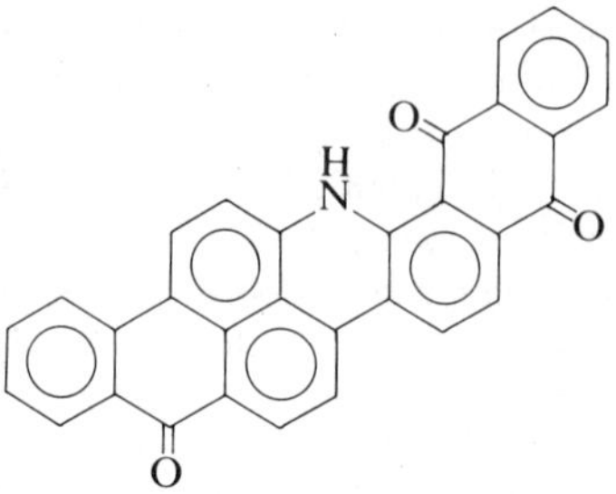

Vat Green 3
2MM lb, $2.00

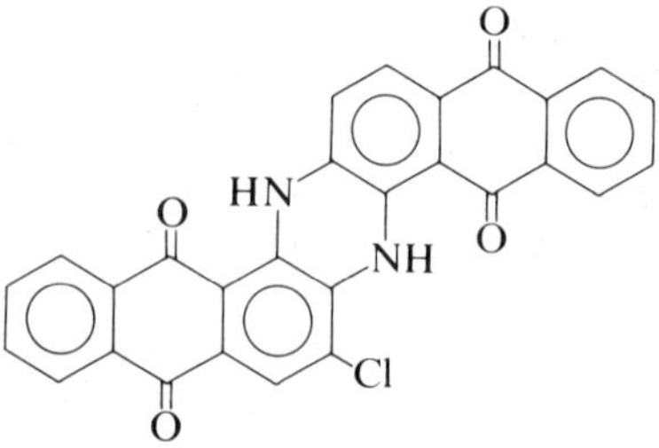

Vat Blue 6
1.2MM lb

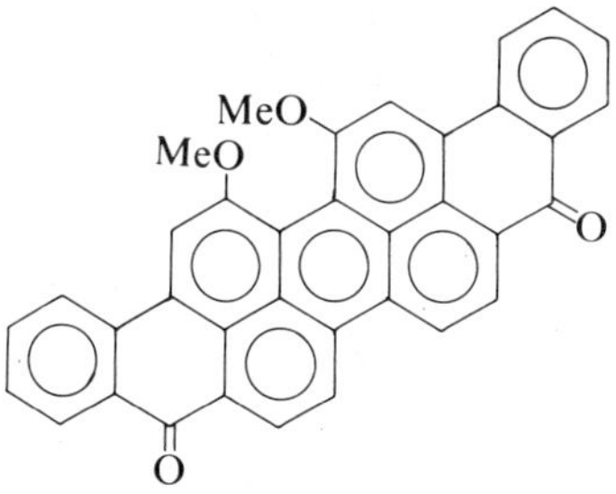

MeO
MeO

Caledon Jade Green
Vat Green 1, $5.50

**Scheme 9.113**    (*Continued*)

## 9.3.6    Benzenoid Nitro, Amino, and Related Structures

### 9.3.6.1    Introduction: Some Transformations of Aromatic Nitrogen Compounds

The nitration of benzene, toluene, chlorobenzene, and naphthalene is introduced in Section 9.2.5, and scattered references to other aromatic nitro compounds and their amino derivatives are unavoidably mentioned in other sections of the book. Here we expand the formation and utilization of benzenoid nitro compounds to additional cases of industrial interest.

An overview of the chemical possibilities associated with aromatic nitro compounds is presented in Fig. 9.1 and the essential methods to perform these transformations are elucidated as follows:

(a) Reduction of nitro compounds by means of hydrogen and Raney nickel catalyst (prepared by treatment of a Ni–Al alloy with caustic soda); by sodium hydrosulfite ($Na_2S_2O_4$; 71¢); or by means of iron filings–$H_2O$–$H^+$, tin, or zinc in the presence of dilute hydrochloric acid; $H_2S/NH_3$ is used to reduce *di*nitro to *mono*amino–nitro compounds. Actually, all nitrogen-containing substituents in which the nitrogen is attached to the aromatic ring are converted to anilines when exposed to vigorous reduction including the hydrazo compounds.

(b) Diazotization by means of sodium nitrite and hydrogen halides yields diazonium salts at 0–5°C (see c and d for exceptions).

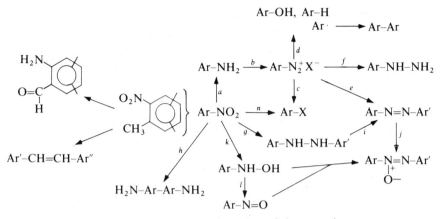

**Figure 9.1** Principal transformations of nitroaromatics.

(c) Decomposition of diazonium salts at moderate temperatures yields halogen-substituted derivatives, but iodo compounds are formed by spontaneous decomposition at low temperatures, and fluoro compounds require heating the corresponding fluoborate salts or the use of hydrofluoric acid.

(d) Decomposition of aqueous solutions of diazonium salts at moderate temperatures yields phenols, and addition of sulfide ions gives aromatic thiols, but in the presence of strong alkali the diazonium hydroxides decompose to aryl (and hydroxyl) radicals (Gomberg–Bachmann reaction; $ONR$-36).

(e) Coupling of diazonium salts with electron-rich aromatic systems (most often phenols and anilines) yields azo compounds.

(f) Reduction of diazonium salts by means of zinc and dilute acid yields arylhydrazines.

(g) Reduction of azo compounds with zinc and dilute alkali gives hydrazo compounds.

(h) Treatment of hydrazo compounds with strong mineral acids produces $p,p'$-diaminodiphenyl compounds, or benzidines.

(i) Mild oxidation (air, trace of alkali) of hydrazo compounds produces azo compounds.

(j) Oxidation of azo compounds with hydrogen peroxide or organic peracids produces azoxy compounds.

(k) Reduction of nitro compounds (see Fig. 9.1) with zinc and $NH_4^+Cl^-$ gives $N$-arylhydroxylamines.

(l) Low-temperature oxidation of $N$-arylhydroxylamines with $Na_2Cr_2O_7$–$H_2SO_4$ produces aromatic nitroso compounds (also formed directly by nitrosation of electron-rich aromatic systems such as phenols and anilines).

(m) Base-catalyzed condensation of $N$-arylhydroxylamines and aromatic nitroso compounds yields azoxy compounds.

(n) Nucleophiles (symbolized by Nu) replace halogen and nitro substituents, especially in the presence of activating electron-withdrawing groups (nitro, keto, sulfone, multiple chlorine substituents, etc.); the activating effect is stronger in ortho and para than in meta positions; the leaving-group character of $F- > Cl-$.

(o) Reduction product of nitro group when the latter oxidizes an ortho- or para-located methyl group to an aldehyde. Stilbene formation is possible by heating the reaction mixture in dilute alkali.

### 9.3.6.2 Nitro Compounds

As stated in 9.2, the two families of aromatic nitro compounds that are useful as such are the perfume ingredients known as nitro musks and explosives.

Synthetic musks are highly $t$-butyl- and nitro-substituted benzenoid compounds of the structures shown in Scheme 9.114.

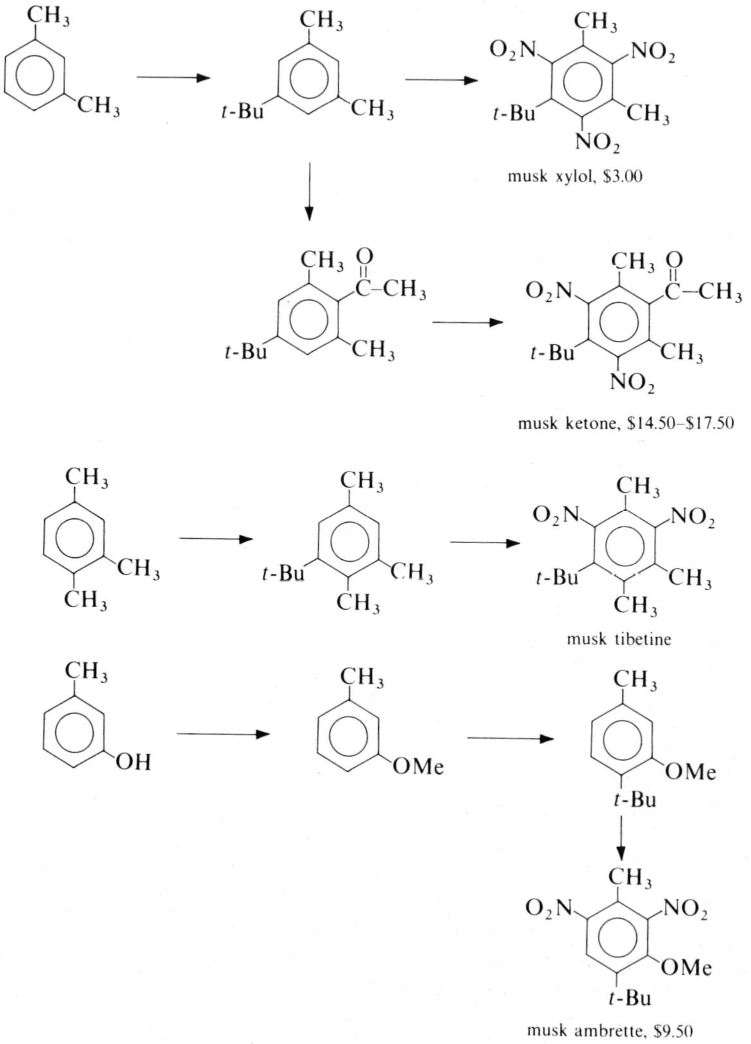

**Scheme 9.114**

Chemical explosives (as differentiated from the nuclear kind) are materials capable of a self-contained, rapid decomposition to gaseous products. The explosive decomposition almost instantly generates a strong pressure wave that ruptures the enclosure in which the materials are contained and propels fragments of the enclosure, and whatever other inert solid matter may have been added, into the surroundings. The same material may burn harmlessly (except for noxious gaseous products) when ignited in an open container unless it is sensitive to thermal activation. The explosive under discussion may be sensitive to mechanical or electric discharge activation that propagates the rapid decomposition throughout the bulk of the material. However, the relatively harmless explosive is usually "set off" by means of a detonator or a primary explosive [e.g., lead azide, $Pb(N_3)_2$—the detonator used in tips of "devastator bullets"; mercury fulminate, $Hg(ONC)_2$; diazodinitrophenol—the product of nitrosation of picramic acid (see below); lead styphnate—the lead salt of 2,4,6-trinitroresorcinol known as *styphnic acid*], and the explosion is then propagated throughout the mass of the secondary explosive.

Thus far we have mentioned many of the government-imposed regulations on the chemical industry; it is also appropriate to mention some regulations that have been overlooked. Millions of pounds of lead azide will soon be within reach of the general public as a result of the proliferation of sodium azide, the chemical used in automobile airbag devices. The canisters that activate the two airbags for the protection of the driver and front-seat passenger contain about 0.5 lb of sodium azide (see "Final Regulatory Impact Analysis, Passenger Car Front Seat Occupant Protection," National Highway Safety Transportation Agency (NHSTA), July 11, 1984). The water-soluble azide can be extracted and readily converted to water-insoluble lead, mercury, and other heavy-metal azide detonants (see description of a laboratory explosion in *CEN* 4/12/82). Apart of the uncontrolled access to airbag devices and mischievous manipulations of sodium azide obtained from "borrowed" or stolen vehicles (an estimated million cars are stolen each year according to an American Automobile Association article reprinted from *Protection Magazine*, 8/87), additional societal dangers are likely to result from the release of sodium azide during the scrap-yard operations. With regard to the latter point, the NHTSA report makes the optimistic assumption that "97 percent of the airbag systems in scrapped cars would be discharged by auto dismantlers prior to final disposal process" and that this would result in "about 230 pounds of sodium azide . . . released nationally each working day".

The toxicity of sodium azide and its hydrolysis product, hydrazoic acid, is far from being secret: the *1980 Registry of Toxic Effects of Chemical Substances*, Vol. II, U.S. Department of Health and Human Services, p. 621, reports for sodium azide a *lethal dose for humans* of 710 $\mu$g/kg of body weight, which is equivalent to 71 mg for a 220-lb person (by comparison, a standard aspirin tablet weighs 325 mg, and the $LD_{50}$ of sodium cyanide is only 15 mg/kg by oral application in rats). Hydrogen azide is reported in the *1980 Registry*, Vol. I, p. 903 to have a toxicity to humans at a concentration of 300 ppb by inhalation. Also, hydrogen azide is described by the *Merck Index*, p. 695 as being "extremely explosive" and having an $LD_{50}$ of 21.5 mg/kg in mice. It is reported (*CEN* 6/27/88) that the EPA has selected "21 extremely hazardous chemicals to review," and the list includes sodium azide.

Warnings concerning risks due to the proliferation of uncontrolled quantities of sodium azide were voiced in the media (*The Detroit News* 8/13/79; *CB* 7/84, p. 43; *CW* "An Azide Hazard May Deflate Air-bag Hopes," 12/83, 46–53; *CEN* 8/15/77, 10/23/78, 11/20/78, 10/27/80, 2/2/81, 1/26/87, 3/16/87, 3/30/87, 4/20/87, 7/13/87) and personally at a Public Hearing organized by the U.S. Department of Transportation and held in

Washington, DC on December 7, 1983. However, all these warning and dire predictions were in vain, and Ralph Nader was celebrating the announcement of millions of cars that are now being equipped with airbags by U.S. automobile manufacturers on the MacNeil–Lehrer News Hour of June 22, 1988. *The Wall Street Journal* of 1/20/88 reports that BDM International was awarded a 1-year contract for $10 million by a unit of Morton-Thiokol under which "BDM will automate the production of auto airbags."

Aromatic nitro compounds represent a relatively large family of chemical explosives because the oxidizing capability of nitro groups is coupled with the presence of oxidizable H and C constituents of the same molecule. The activation energy to bring about a self-contained redox reaction is sufficiently high, so that the most famous member of this family of explosives, namely, 2,4,6-trinitrotoluene (TNT; m.p. 80°C), can be handled in a molten state during the convenient filling operation of shells. TNT does not decompose explosively until about 275°C, when it forms a mixture of gaseous $H_2O$, CO, $NO_x$ ($N_2O$, NO, etc.) and possibly some carbon particles and $CO_2$.

1,3,5-Trinitrobenzene (TNB; m.p. 122.5°C) behaves in a similar way, except that it is difficult to prepare from benzene (see 9.2.5). The preparation of *N*-methyl-*N*,2,4,6-tetranitroaniline, nitramine or tetryl, is also described in 9.2.5. A simple exercise in balancing an equation that represents the decomposition of tetryl (to produce gaseous water, nitros oxide, and carbon monoxide) reveals that tetryl can produce 8 mol of gaseous products as compared to 6 mol from TNT or trinitrobenzene. This difference may explain why tetryl is a more powerful explosive than the other two materials.

Other nitro aromatic explosives are hexanitrostilbene (HNS), obtained from TNT by means of alkali and reoxidation of nitroso–nitro intermediates (see below) and the following nitro derivatives of the aniline family: 2,4,6-trinitroaniline, monoaminotrinitrobenzene (MATB); 1,3-diamino-2,4,6-trinitrobenzene (DATB); and 2,4,6-trinitro-1,3,5-triaminobenzene (TATB).

A recent study at Sandia National Laboratory shows (W. Worthy, CEN 8/10/87) that the sensitivity to shock among these explosives increases in the order TATB < DATB < MATB < TNT < TNB. These results are correlated with the stability of benzene rings that is a function of both the destabilizing effect of electron-withdrawing nitro groups (as long as they are coplanar with rings) and the stabilizing effect of electron-donating amino groups. The latter tends to compensate the former and induces an increase in the polarity of the conjugated nitroamino system, which, in turn, promotes shock-absorbing hydrogen bonding within the molecular aggregate.

The chemistry of the above-mentioned explosives is summarized in Scheme 9.115.

The preceding transformations illustrate

- The diazotization of picramic acid and the formation of an unstable diazooxy heterocycle.
- The need to protect anilino groups from oxidation reactions caused by nitric acid during nitration processes.
- The intramolecular redox behavior of the *p*-nitrotoluene system under basic conditions, which accounts for the formation of a *p*-nitrosobenzaldehyde.
- The reaction of the latter with the trinitro-activated methyl group of another TNT molecule to induce an aldol-type condensation.

Another example of the last-mentioned redox behavior is the formation of *p*-aminobenzaldehyde from *p*-nitrotoluene when the latter is heated in the alkaline solution

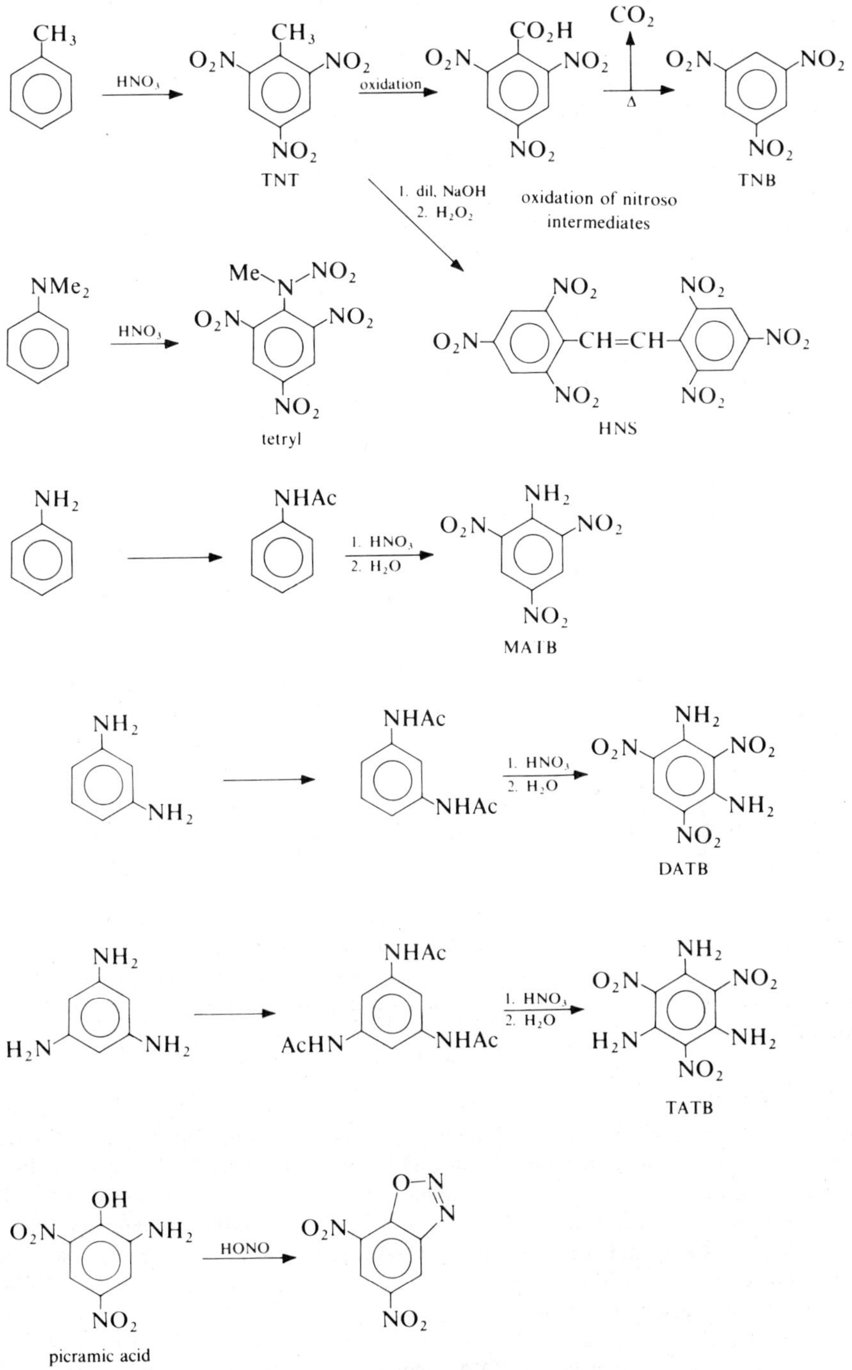
HNO3
oxidation
CO2
Δ
TNT
TNB
1. dil. NaOH
2. H2O2
oxidation of nitroso intermediates
HNO3
tetryl
CH=CH
HNS
1. HNO3
2. H2O
MATB
DATB
TATB
HONO
picramic acid

**Scheme 9.115**

of sodium polysulfide:

$$3(p\text{-}O_2N\text{-}C_6H_4\text{-}CH_3) + Na_2S_4 \rightarrow 3(p\text{-}H_2N\text{-}C_6H_4\text{-}CH=O) + Na_2SO_3 + 2S$$

The general mild oxidizing property of nitrobenzene and some of its derivatives is utilized in the course of the Skraup synthesis of quinolines ($ONR$-84; 10.3.1) and in the production of vanillin from lignin (9.3.5).

Other secondary explosives mentioned elsewhere in this book are: the trinitrate of glycerol, dinitrate of ethylene glycol, tetranitrate of pentaerythritol, cellulose nitrates, and RDX—the trinitro compound derived from hexamethylenetetramine (HMTA).

*Nitration at the Threshold of Higher-Value-Added Transformations*
Some higher-value-added transformations of several key aromatic compounds are presented here in chart form in order to illustrate:

- The economic gains that accompany such chemical changes;
- The preferred orientation during the different substitution reactions;
- The different synthetic maneuvers required in order to obtain the desired intermediates and end products. Figures 9.2–9.9 summarize some of the chemical transformations that start with benzene, toluene, chlorobenzene, the xylenes, phenol, the creosols, and some additional, miscellaneous starting materials that include naphthalene and some higher aromatic carboxylic compounds. For the sake of improved clarity, products of some benzidine rearrangements, indicated here by "bzdn," are shown separately in Fig. 9.8 even though the mechanisms of the benzidine, and the analogous semidine, rearrangements are briefly discussed below in connection with the reduction of nitrobenzene to hydrazobenzene (process $g$ in Fig. 9.1).

With reference to the chemical transformations, we should keep in mind that the rules of electrophilic substitutions in aromatic systems respond to preferred, and not exclusive, regioselectivity. Thus, on a large, industrial scale, minority products that "violate" the "rules" are also isolated. For example, large-scale nitration of toluene yields some *m*-nitro product in addition to the dominant *o*- and *p*-nitrotoluenes.

The preceding diagramatic summaries cannot include all significant chemical transformations that start with nitration, and thus some additional comments are offered as follows.

### 9.3.6.3    Anilines
As shown elsewhere (Table 1.1), the primary derivative of nitrobenzene is aniline, and, in addition to the traditional reduction route and the amination of chlorobenzene, the 1982 United States debut of the phenol to aniline conversion process (announced by USS Chemicals, now renamed "Aristech") had a profound effect on the aniline marketplace. Actually, this process had been operated by Mitsui in Japan since 1970 and was based on a Halcon development in the United States. In order to attract customers for its 200MM-lb production capacity, the manufacturer initially offered aniline at a price slightly less than 30¢ while the traditional suppliers of the same material were still listing their product at 46¢ [but were offering generous TVAs (temporary, voluntary allowances)] (2.4.1). A *CMR* report of 7/19/82 carried the headline "Aniline Trade Stunned by USS Pricing

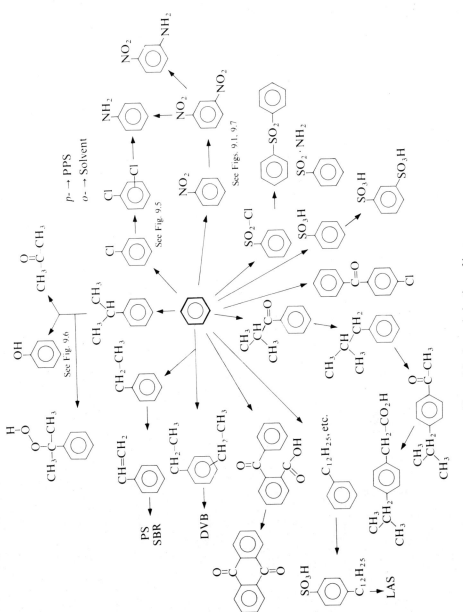

**Figure 9.2** Some derivatives of benzene.

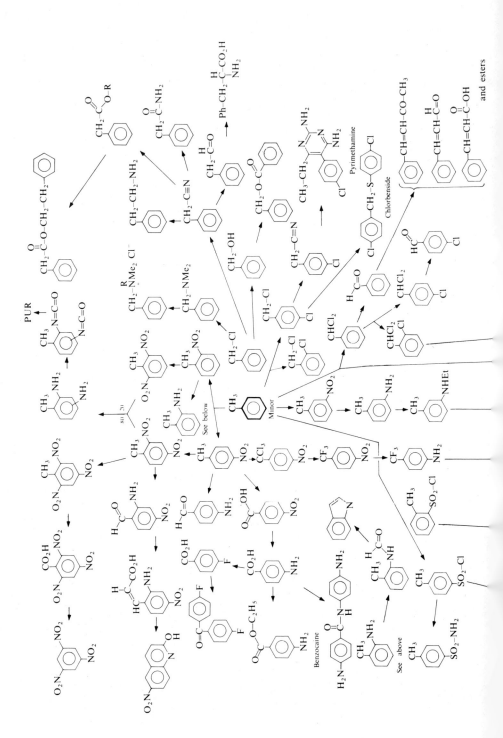

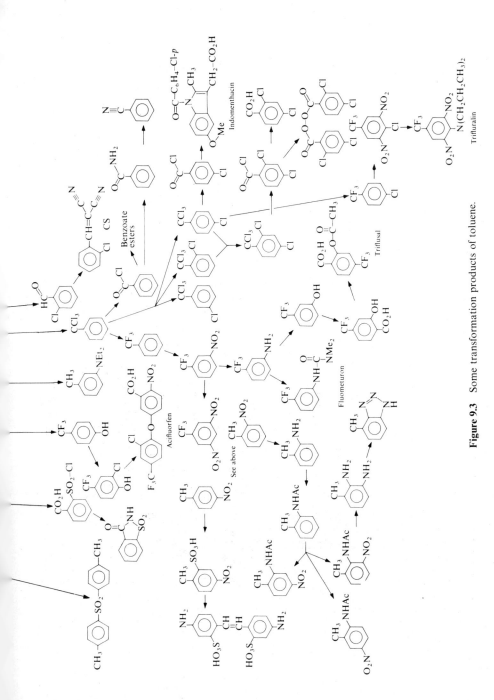

**Figure 9.3** Some transformation products of toluene.

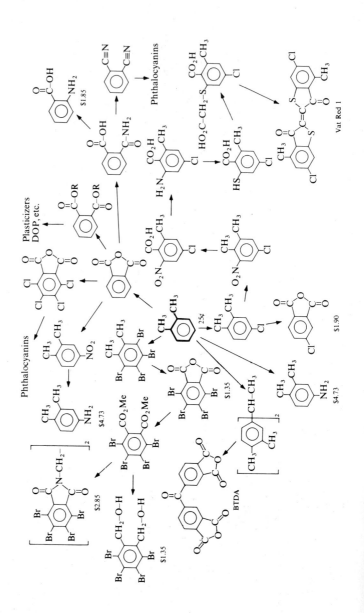

Vat Red 1

Phthalocyanins

Plasticizers
DOP, etc.

Phthalocyanins

BTDA

$1.85

$1.90

$4.73

$2.85

$1.35

$1.35

$4.73

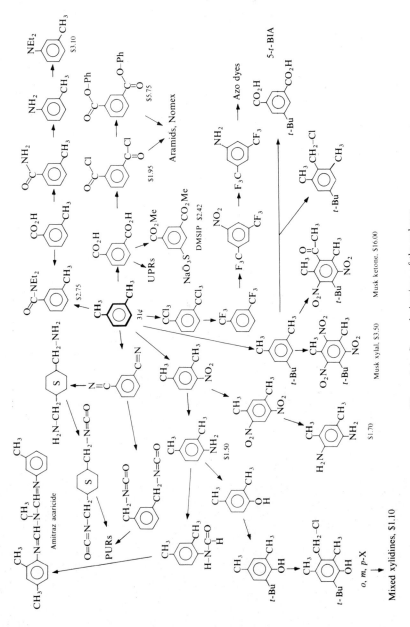

**Figure 9.4** Some derivatives of the xylenes.

553

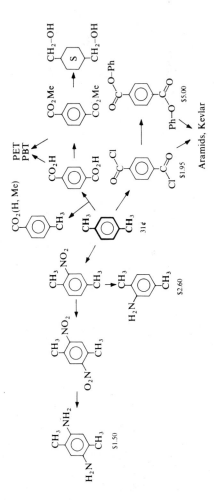

**Figure 9.4** *(Continued)*

554

Action" to reflect the dismay of traditional aniline suppliers. Gradually, the price of aniline rose (56¢ in July 1987; 58¢ in February 1988 and 62¢ in January 1989), driven primarily by the demand for MDI–PMDI required by the everincreasing PURs. Actually, about 70% of the aniline is consumed by the production of MDI–PMDI. While most of the intermediate aniline–formaldehyde condensation product, 4,4′-methylenedianiline (MDA; crude $1.75, purified $2.25), is produced for captive use for conversion to MDI and also for the introduction of urea linkages in the formulation of PUR's (interlude in Section 3.8), some 250MM lb of purified MDA reaches the marketplace for other uses. Thus, for example, reaction with epichlorohydrin yields tetraglycidyl methylenedianiline (TGMDA; 7MM lb)—the tetrafunctional building block of some epoxy resins.

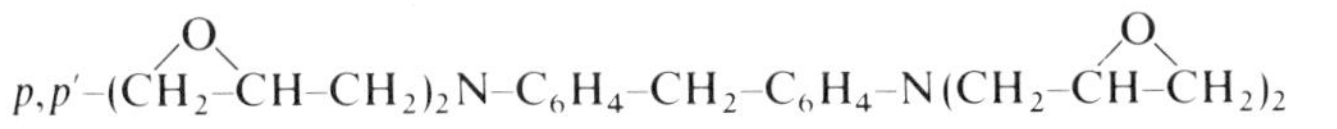

$$p,p'\text{-}(CH_2\text{-}CH\text{-}CH_2)_2N\text{-}C_6H_4\text{-}CH_2\text{-}C_6H_4\text{-}N(CH_2\text{-}CH\text{-}CH_2)_2$$

Also, MDA is an intermediate in the production of certain dyes and some high-performance specialty thermoplastics. Among the latter one can mention the polyimide–amide textile fiber derived from mellitic anhydride:

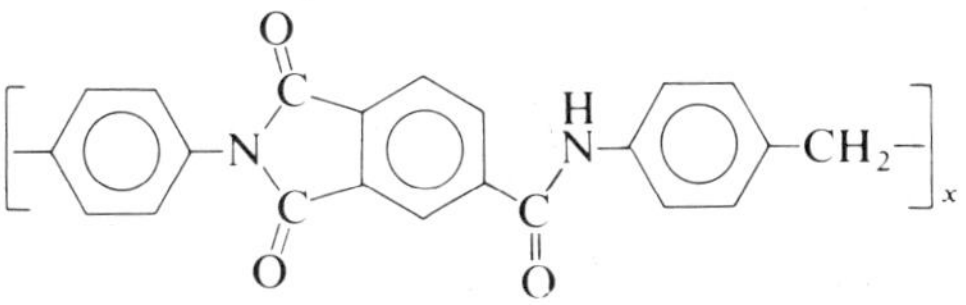

and Amoco's Torlon, in which MDA is concomitant with the analogous imide–amide unit derived from 4,4′-diaminodiphenyl ether (or oxydianiline):

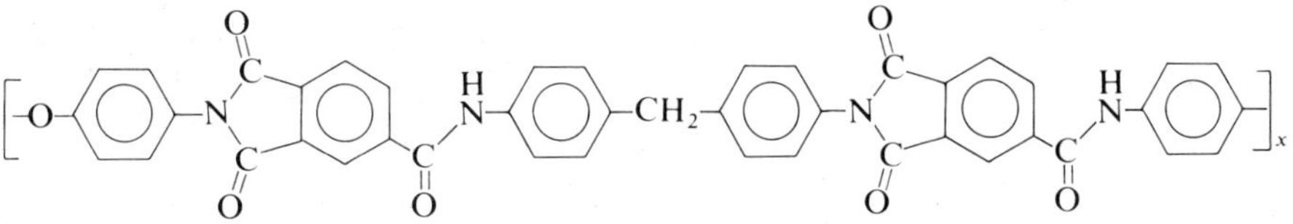

The phenol to aniline conversion process has now been extended by Sumitomo of Japan to the conversion of resorcinol to $m$-aminophenol (MAP), which is manufactured at a level of 2.2MM lb (*CMR* 11/13/85, *CW* 8/20/86). This process replaces the multistep route that starts with $m$-dinitrobenzene.

The phenol to aniline conversion process is reminiscent of the classical Bucherer reaction (*ONR*-16), which allows an interconversion of certain naphthols and naphthyl-amines in the presence of sulfites. The rationale for the ease with which this inter-conversion takes place is the occurrence of the keto–enol equilibrium in an aromatic system in which resonance stabilization is sufficiently large even if the aromatic system is partially decreased by the formation of the keto tautomer. This point can be illustrated in the case of $\beta$-naphthol—an example of a phenol in which the Bucherer reaction occurs with particular ease, and by means of examples that involve larger aromatic systems in which the loss of stability attributed to a formation of a keto tautomer is even smaller (Scheme 9.116).

The direct amination of benzene to aniline is claimed in the patent literature (U.S. Patents 3,919,155 and 4,031,106, 1975 and 1977, respectively, assigned to Du Pont), but

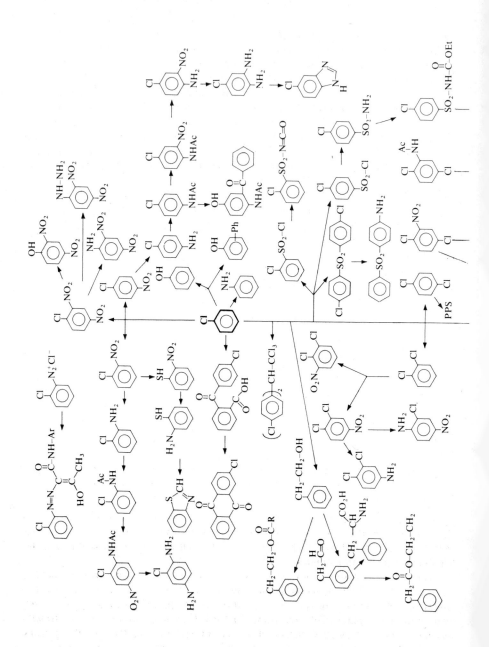

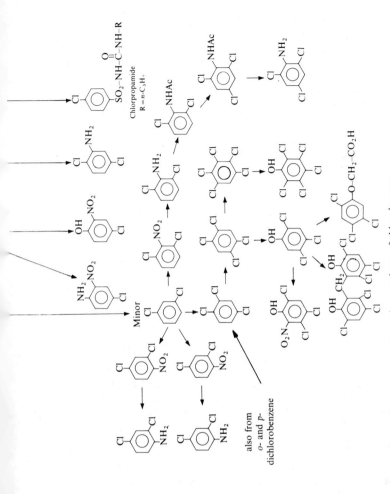

**Figure 9.5** Some transformation products of chlorobenzene.

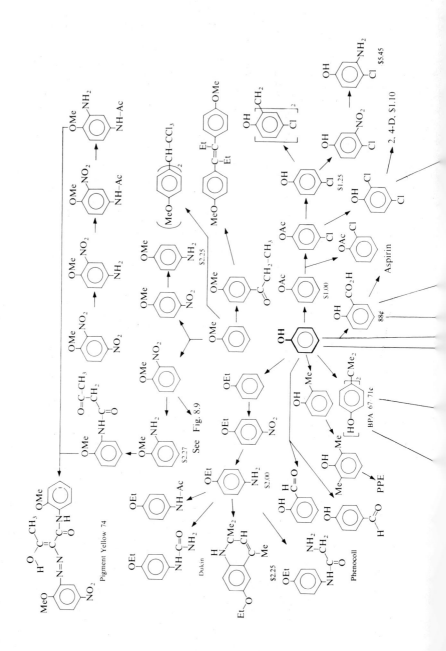

558

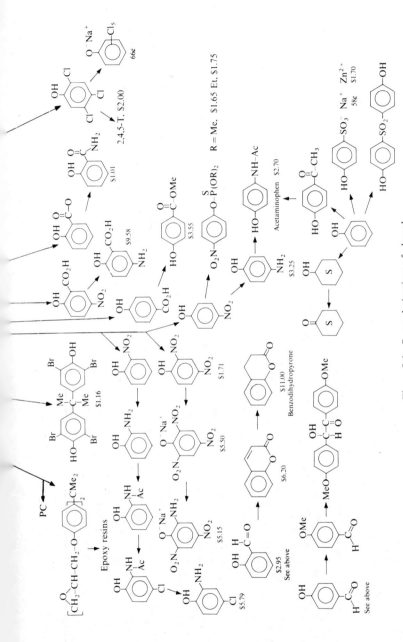

**Figure 9.6** Some derivatives of phenol.

559

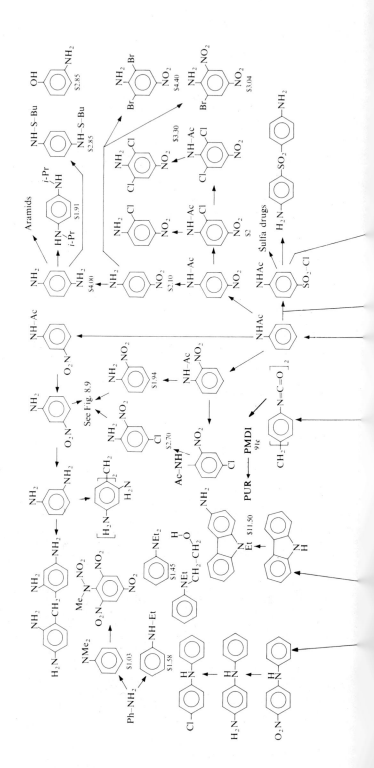

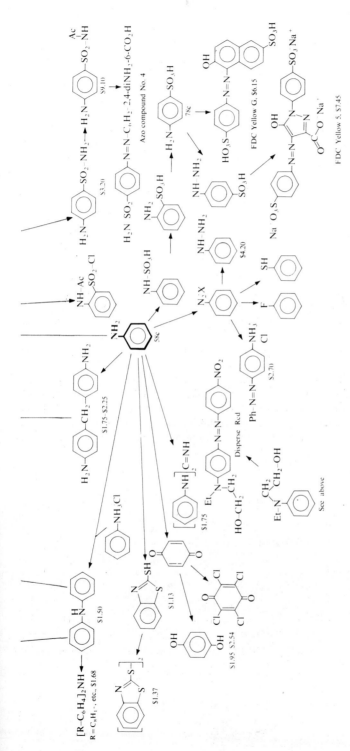

**Figure 9.7** Some derivatives of aniline.

561

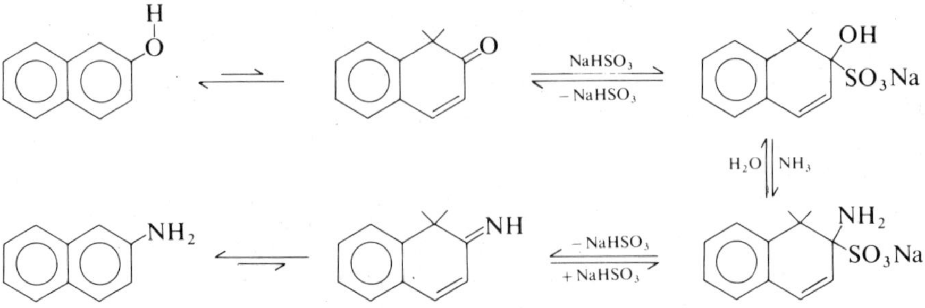

**Scheme 9.116**

evidence is lacking that such a process (350°C in the presence of NiO, Ni, and ZrO catalysts) is actually employed on an industrial scale.

*Rubber-Processing Chemicals*

About 20% of the current 880MM-lb demand for aniline is consumed by the production of the economically significant rubber-processing chemicals (see 1.1.1). There are approximately 200 rubber-processing chemicals utilized at a level of about 270MM lb and valued at about $390 million. Many are vulcanization control agents—both accelerators and retarders. Foremost among these are the derivatives of 2-mercaptobenzothiazole, obtained by a high-temperature reaction of aniline with carbon disulfide and elementary sulfur (Scheme 9.117). As indicated above, the *sym*-diphenylthiourea (or thiocarbanilide; vulcanization accelerator grade $2.20) is the intermediate on route to other accelerators such as diphenylguanidine ($2.52), 2-mercabenzothiazole (MBT; $1.25), and the MBT oxidation product—mercaptobenzothiazyl disulfide (MBTS; $1.33). On the other hand, the reaction of thiocarbanilide with cyclohexylamine (technical grade 95¢) (8.3.1) gives the vulcanization retarder *N*-cyclohexyl-2-benzothiazolesulfenamide, manufactured at a

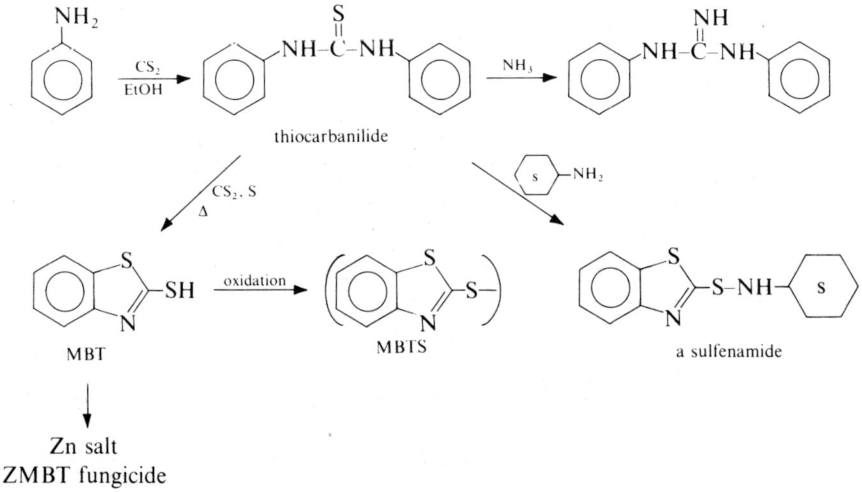

**Scheme 9.117**

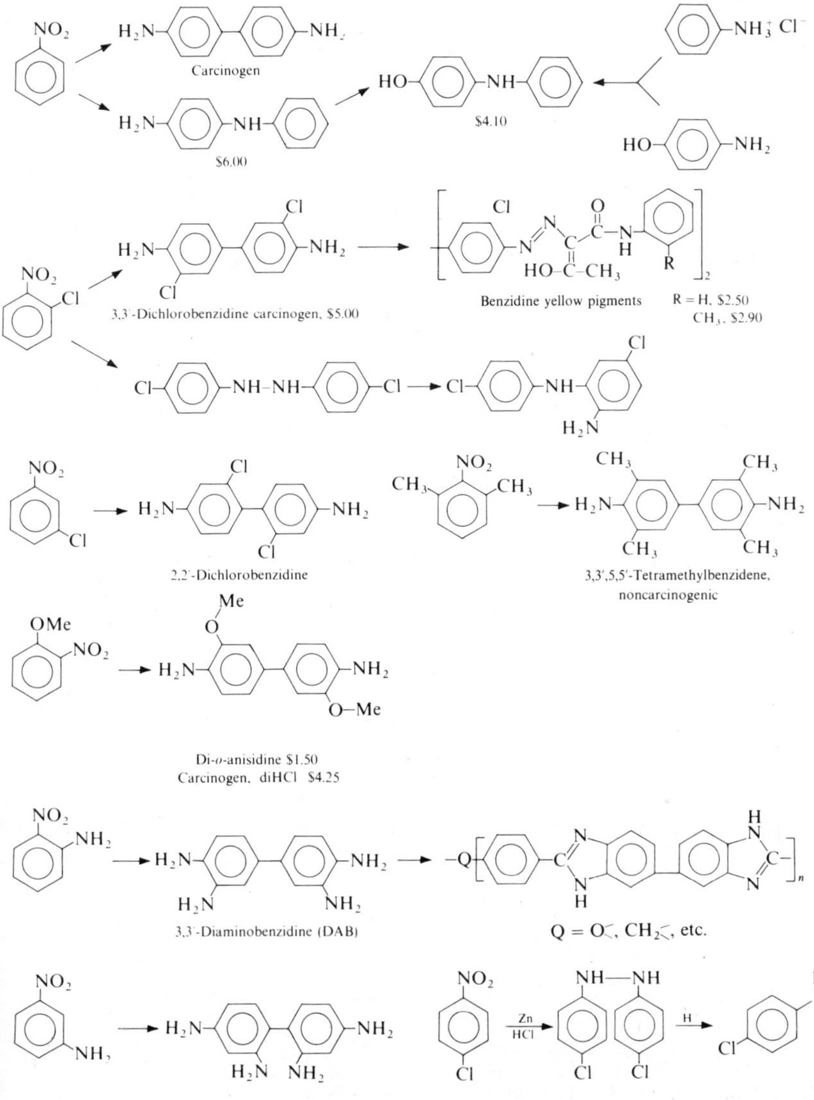

**Figure 9.8** Some benzidine and semidine transformation products.

level of about 3MM lb. The reason for the need for vulcanization accelerators of varying degrees of efficiency, in addition to the need for vulcanization retarders, is the fact that natural and the major synthetic rubber (SBR) are being cross-linked and processed to end products simultaneously, and thus these two activities must be carefully synchronized.

The zinc salt of MBT is also used as a fungicide and referred to as "ZMBT." Another important group of rubber-processing chemicals are the antioxidants and antiozonants. These additives prolong the useful life of rubber products since the partially unsaturated nature of the latter is subject to slow but persistent oxidative degradation. The simplest

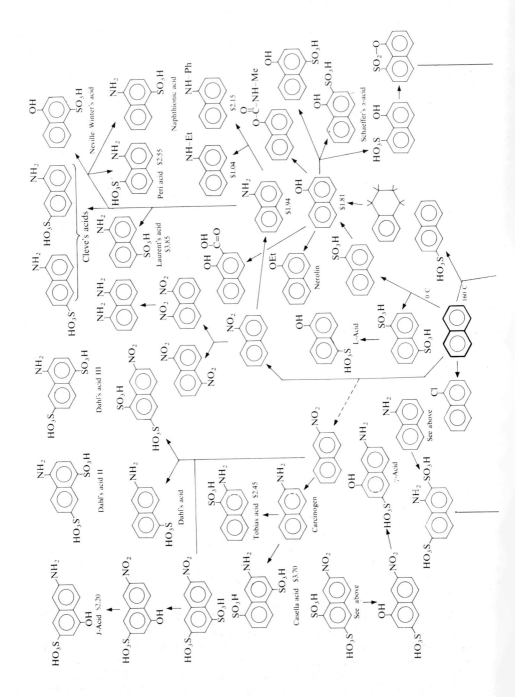

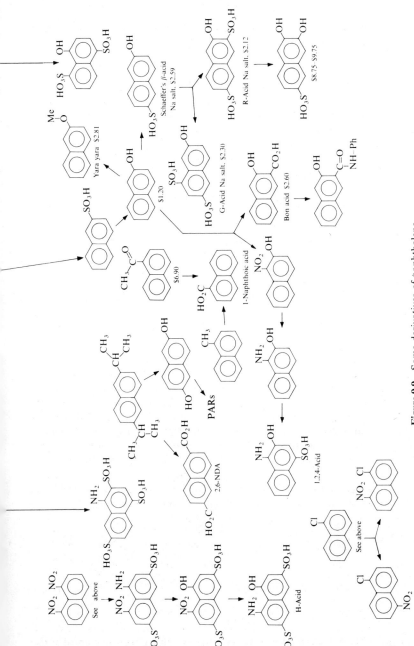

**Figure 9.9** Some derivatives of naphthalene.

antioxidant is diphenylamine (DPA; molten \$1.21, flake \$1.50, purified Uniroyal \$3.25), is obtained by heating aniline with aniline hydrochloride:

$$Ph–NH_2 + Ph–NH_3^+ \ Cl^- \rightarrow Ph_2NH + NH_4^+ \ Cl^-$$

and as a by-product of the traditional route to aniline by a high-temperature (200°C) amination of chlorobenzene carried out in the presence of cuprous chloride catalyst.

Other categories of rubber-processing chemicals include antiozonants of the phenyl-enediamine family, peptizers (which reduce the viscosity of the material as it is processed in a rubber mill), and are represented by thiophenols. The antidegradants and vulcaniza-tion control agents are the two largest categories of rubber-processing chemicals, used at levels of 120MM and 80MM lb, respectively.

Octylated diphenylamine (\$1.68), a derivative of the alkylation of aniline with 2-EH, is obtained in a similar fashion to that of DPA. Purified p-octyldiphenylamine and di(octylphenyl)amine are available from Uniroyal at about \$7 and \$2.05, respectively, and are employed as additives to petroleum products.

Another interesting derivative of aniline is p-hydroxydiphenylamine (\$4.22). This dye intermediate and vinyl polymerization inhibitor is obtained by B. F. Goodrich by means of a high-temperature reaction of aniline and hydroquinone:

$$C_6H_5–NH_2 + p\text{-}HO–C_6H_4–OH \rightarrow p\text{-}HO–C_6H_4–NH–C_6H_5$$

*Orthoalkylation*

The preceding mention of alkylated anilines provides an opportunity to mention the ortho-alkylation process as it applies to aniline. The traditional reaction conditions require temperatures of about 300°C and aluminum catalysts that are generated *in situ* and coordinates the aromatic substrate and the olefinic reactant in a way that facilitates the substitution in the ortho position(s) (Scheme 9.118). Some common ortho-alkylated derivatives of aniline and their applications in industrial organic chemistry are shown in Scheme 9.119. Recently (*CMR* 8/17/87, *CE* 9/14/87) Air Products announced a new acid-catalyzed ortho-alkylation process that introduces bulky alkyl groups to yield, for example, t-butyltoluenediamine (t-BTA; \$3.00) from a commercial mixture of TDA:

2,4-Diamino-5-*t*-butyltoluene (80%)          2,6-Diamino-*t*-butyltoluene (20%)

It is noteworthy that the differential reactivity of the two amino groups is magnified by the presence of the *t*-butyl substituent, and one can take advantage of that difference when the products are used as PUR extenders during a two-stage PUR process in RIM appli-cations (interlude in Section 3.8). Also, these amines serve as intermediates in the synthesis of antioxidants, as curing agents of epoxides, and so on.

The previously available 2,6-diethylaniline is the starting material for the production of the most widely used (80MM lb) herbicide, namely, alachlor, known best as

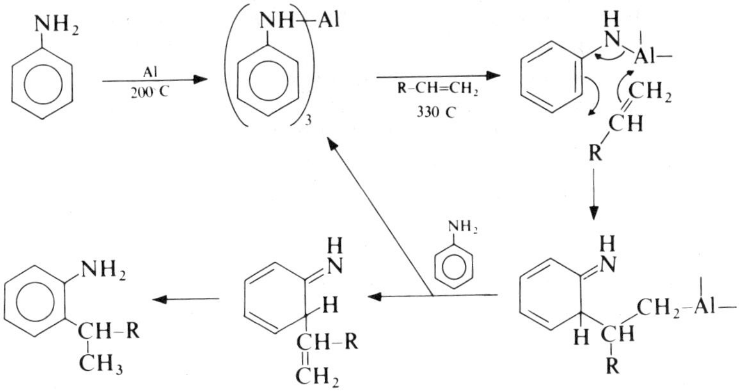
Al
200 C
R-CH=CH$_2$
330 C

**Scheme 9.118**

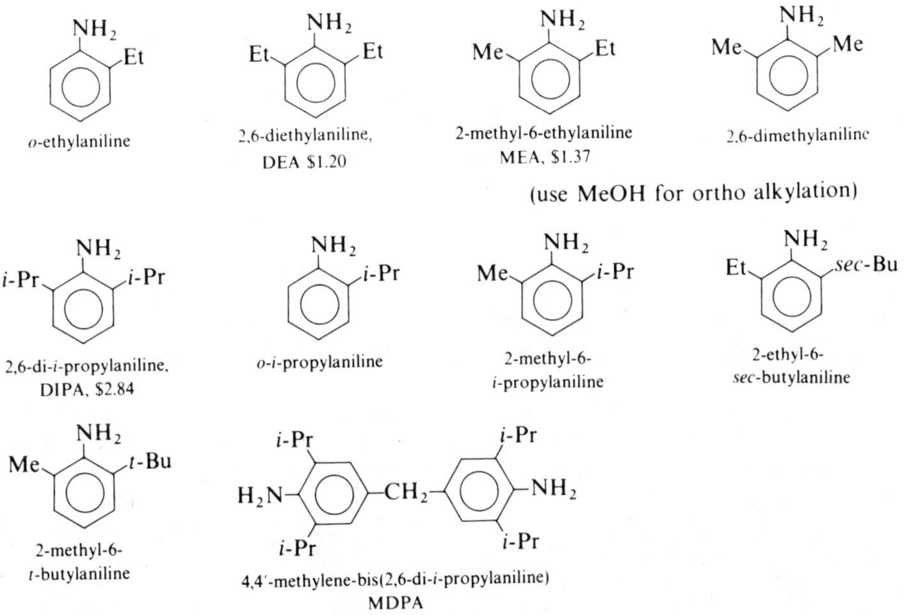
o-ethylaniline
2,6-diethylaniline,
DEA $1.20
2-methyl-6-ethylaniline
MEA, $1.37
2,6-dimethylaniline
(use MeOH for ortho alkylation)
2,6-di-i-propylaniline,
DIPA, $2.84
o-i-propylaniline
2-methyl-6-
i-propylaniline
2-ethyl-6-
sec-butylaniline
2-methyl-6-
t-butylaniline
4,4′-methylene-bis(2,6-di-i-propylaniline)
MDPA

**Scheme 9.119**

Monsanto's "Lasso" (4.6.3) and very useful in the cultivation of corn, soybean, and other crops. The aniline is treated with methyl chloromethyl ether and chloroacetyl chloride to give

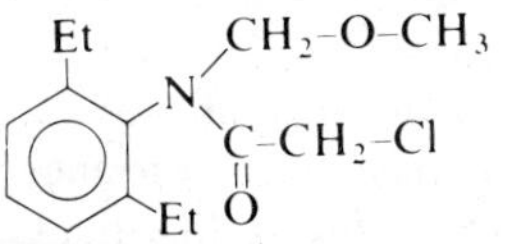

Even though the EPA has recently allowed the continued use of alachlor (*CMR* 12/21/87), the use of this herbicide is under vigorous attack by some environmentalists who have even resorted to demonstrations in front of Monsanto's Washington, DC offices (*CW* 5/18/88). The health hazards of alachlor were investigated rather thoroughly [M. Sun, *Science* **233**, 1143–1144 (9/12/86)]. It is estimated that the removal of alachlor from the market would result in a first-year loss of about $450 million to farmers because of increased costs of weed control and a reduction in crops.

N-Alkylation of aniline can be carried out by means of reductive amination or simple alkylating agents, as illustrated by the formation of propachlor—another herbicide (Dow's Bexton and Monsanto's Ramrod) and *N,N*-diethylaniline ($1.75) (Scheme 9.120).

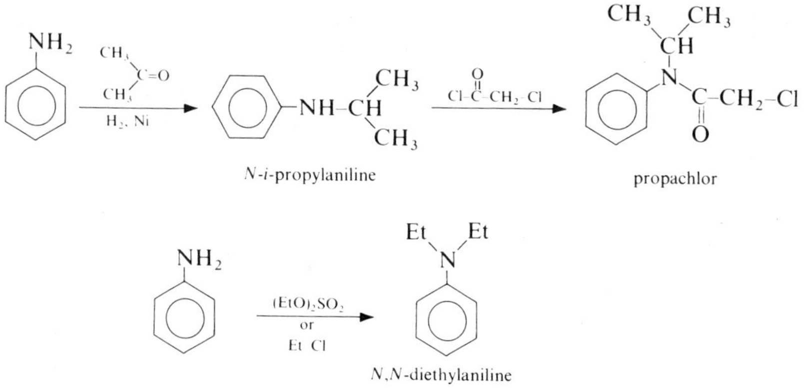

N-*i*-propylaniline

propachlor

N,N-diethylaniline

**Scheme 9.120**

The oxidation of aniline to dyes known as "aniline black" and used, among other things, in black shoe polish, has been practiced for over a century. Current interest in this complex material is revived because of its conductivity of electricity. New spectroscopic evidence (F. Wundl, University of California, Santa Barbara) confirms (*SZM-1*, 471–472) that the basic structure corresponds to a linear quinoid condensation product such as

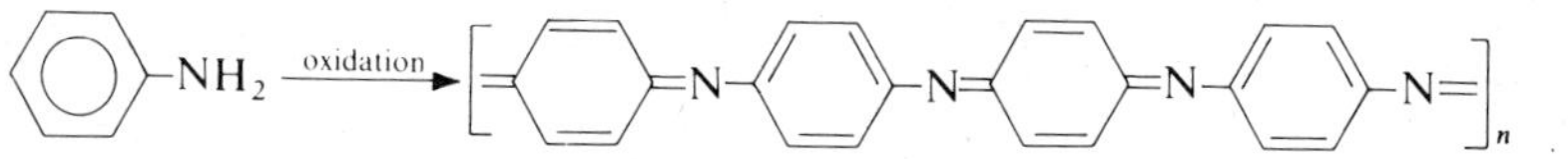

**Scheme 9.121**

A polyaniline of "special structure" and purity has been developed recently by Bridge-stone Corporation and commercialized in Japan jointly with Seiko Electronic Components (*CE* 9/28/87) in the form of a 1.6-mm-thick battery that weighs 1.7 g, is priced at $1.40, and lasts for a minimum of 5 years. In this battery polyaniline serves as an anode and a lithium–aluminum alloy functions as the cathode.

The reduction of nitrobenzene and its derivatives by means of zinc dust and aqueous alkali gives the corresponding hydrazo compounds, and probably the most significant reaction of this family of compounds is their rearrangement to *p,p'*-diaminodiphenyls or benzidines. There is experimental evidence that the benzidine rearrangement (*ONR*-9) requires the formation of a *di*protonated hydrazo intermediate and, apparently, such a diprotonated species folds under with the formation of a new, central carbon–carbon bond of biphenyl. On the other hand, a *mono*protonated hydrazo intermediate suffers a so-called semidine rearrangement (*ONR*-9) that gives rise to an amino derivative of diphenylamine. Naturally, both rearrangements require that the positions para (or ortho) to the original aryl–nitrogen bonds be unsubstituted. Examples of both types of rearrangements are shown in Fig. 9.8. The parent benzidine molecule is toxic, and its production as well as imports are forbidden in the United States. 3,3′-Diaminobenzidine (DAB) is of particular interest because it serves as a HPST building block for the assembly of poly(benzimidazoles). Thus, for example, the reaction of diphenyl isophthalate with DAB gives poly(benzimidazole) (PBI) a heat- and flame-resistant fiber developed by Celanese (Scheme 9.122).

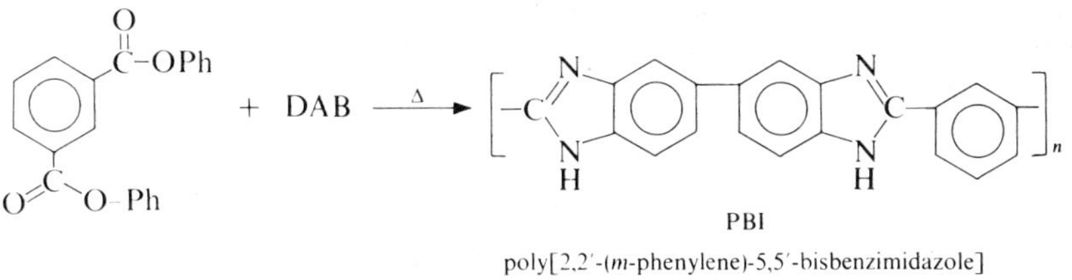

PBI

poly[2,2′-(*m*-phenylene)-5,5′-bisbenzimidazole]

**Scheme 9.122**

The conversion of anilines to diazonium salts is the bridge to a broad spectrum of useful derivatives (processes *c–f* in Fig. 9.1). The diazotization reaction occurs most readily with electron-rich anilines. The electron pair of the amino group functions as a nucleophilic reactant and replaces a water molecule of the protonated nitrous acid:

$$Ar–NH_2 + H_2O^+–N=O \rightarrow Ar–NH_2^+–N=O + H_2O$$

This rate-determining step is followed by a rapid proton shift and elimination of water to produce the resonance-stabilized aryldiazonium ion:

$$Ar–NH_2^+–N=O \rightarrow Ar–NH^+=N–O–H \rightarrow Ar–N^+\equiv N: + H_2O$$

Anilines that are electron-poor at the amino terminal can still be diazotized by generating aggressive nitrosonium ions, and once the aryldiazonium salts are formed, they react in the expected fashion.

The simplest reaction of the aryldiazonium ions is the replacement of the $N_2$ moiety by nucleophilic reagents (processes *c* and *d* in Fig. 9.1), with the reactivity decreasing in the order:

Mercaptide ions = iodide ions > bromide > hydroxide = chloride ions

and so on. The azide ion is also an excellent nucleophile and serves to introduce an azide substituent into an aromatic system:

$$p\text{-}CH_3–C_6H_4–N_2^+Cl^- + Na^+N_3^- \rightarrow p\text{-}CH_3–C_6H_4–N_3$$

p-Toluenediazonium chloride                    p-Tolyl azide

Another example of azide formation is the production of the photosensitive 4,4'-diazostilbene-2,2'-disodium disulfonate used for cross-linking of water-soluble gums and polymers:

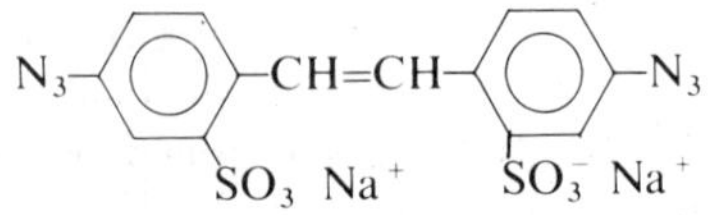

As mentioned elsewhere (9.3.1.2), the introduction of a fluorine substituent requires the use of either a forceful Schiemann procedure or the decomposition of a diazonium salt in liquid hydrogen fluoride (b.p. 19.5°C; extremely corrosive except in nickel equipment). Nevertheless, the desirable presence of a fluoro substituent in certain medicinal compounds and dyes explains the availability in commercial quantities of fluoroaromatics, and, indeed, a recently intensified manufacturing effort.

Since a few years ago, ISC Chemicals of Bristol, England has been supplying certain substituted fluorobenzenes:

- Alkyl fluorobenzenes derived from ring-substituted alkylanilines, priced at about £5–£7 (all prices in British pounds per kilogram).
- Nitro fluorobenzenes derived from the above, priced at about £6–£12.
- Amino derivatives of the above, priced at about £7–£15.

The use of a continuous process based on "hydrofluoric acid diazotization chemistry" is the basis (*CMR* 6/22/87, *CE* 6/22/87) of a new production facility established in the United States as well as in Italy (by a fine-chemical subsidiary of Eni Chem; *CE* 1/18/88).

3,3'- and 4,4'-Difluorobenzophenones are of interest as building blocks of certain HPSTs of the poly(ether–ketone) family since the ketone-activated fluorines are displaced by phenoxide ions in dipolar aprotic solvents. The fluorine-containing benzophenones can be synthesized from the appropriate dinitrobenzophenones by way of the diamine and the double diazonium salt.

Other organic fluorine-containing building blocks that are being announced by the American manufacturer include *p*-difluorobenzene, *p*-fluorophenol, the *p,p'*-difluoro derivatives of biphenyl and diphenyl ether, the *p*-chloro- and *o*-bromo derivatives of fluorobenzene, the 2,4,5-trifluoro derivative of acetophenone, the 2,5-difluoro derivative of 4-chloroacetophenone, and two fluorinated 4-chromanone derivatives—6-fluoro 4-chromanone and 6-fluoro-2-methyl-4-chromanone. All these intermediates for the synthesis of fine chemicals are derived from the corresponding anilines by means of the reaction of diazonium salts in hydrofluoric acid. The structures of these compounds are

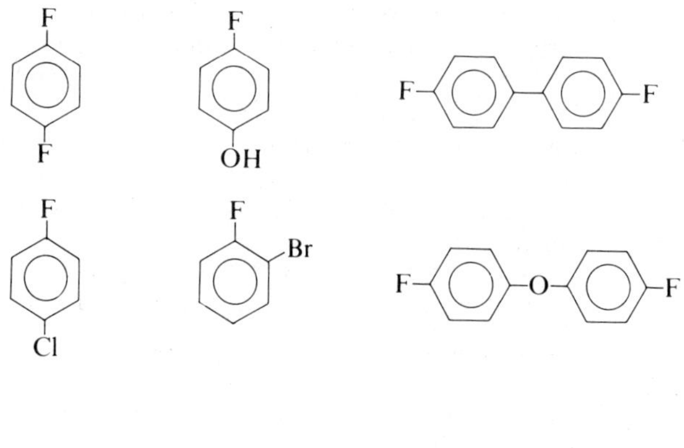

Diazonium salt derived from

Substrate subjected to coupling reaction

Product

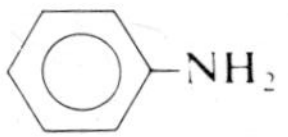

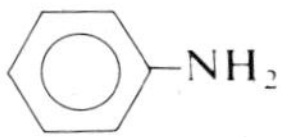

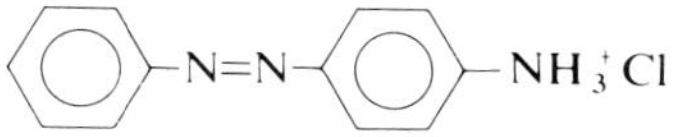

N=N—NH$_3^+$Cl$^-$

p-aminoazobenzene, HCl, $2.70

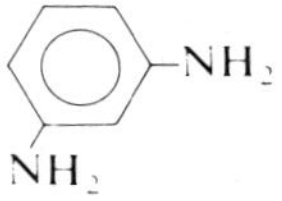

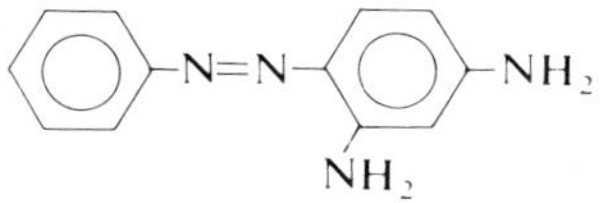

N=N—NH$_2$

NH$_2$

chrysoidine, Basic Orange 2, $5.50
silk, wool dye

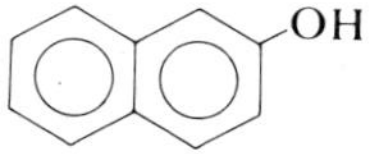

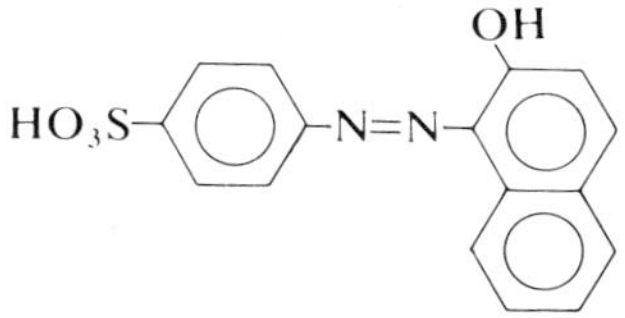

OH

HO$_3$S—N=N

Acid Orange 7, Orange II

AlCl$_3$

HO$_3$S—N=N—Al—O

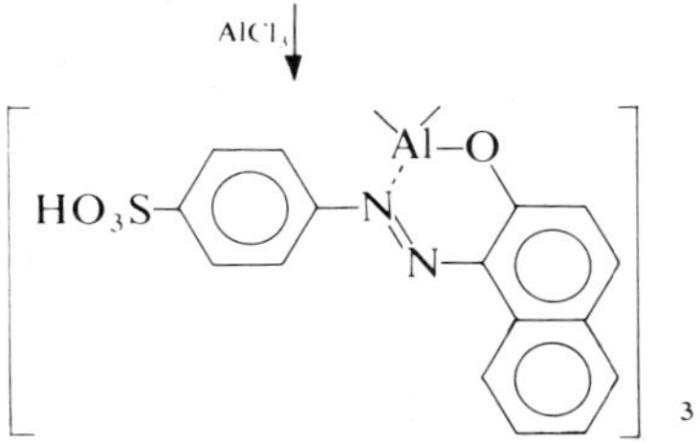

Pigment Orange 17
Persian Orange

**Scheme 9.123**

| Diazonium salt derived from | Substrate subjected to coupling | Product |
| --- | --- | --- |
| 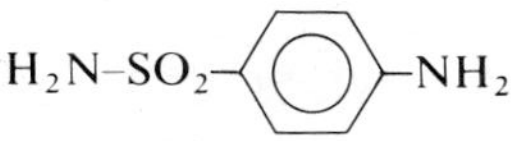 $H_2N-SO_2-$ (aryl) $-NH_2$ | 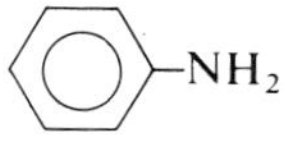 (aryl) $-NH_2$ | 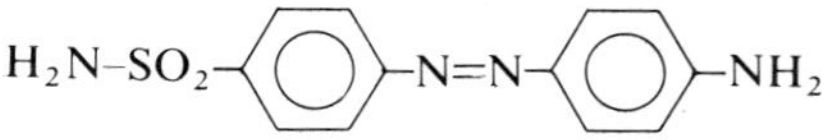 $H_2N-SO_2-$ (aryl) $-N=N-$ (aryl) $-NH_2$ <br> prontosil <br> sulfamidochrysoidine |
| 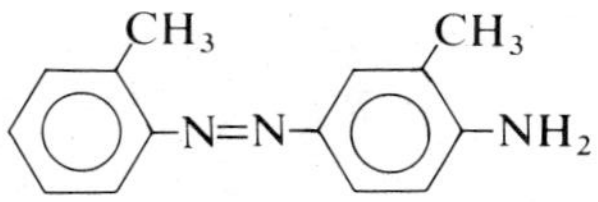 (aryl with $CH_3$) $-N=N-$ (aryl with $CH_3$) $-NH_2$ | 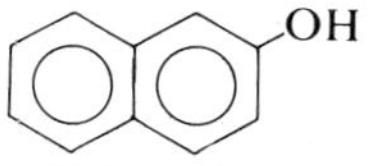 (naphthyl) $-OH$ | 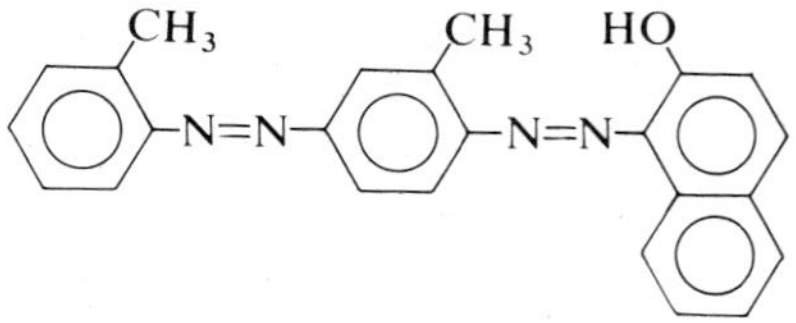 (naphthyl) $HO-N=N-$ (aryl with $CH_3$) $-N=N-$ (aryl with $CH_3$) <br> Scarlet Red, $9.55 |
| 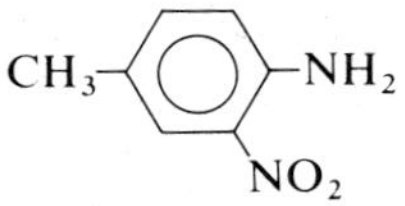 $CH_3-$ (aryl) $-NH_2$, $NO_2$ | 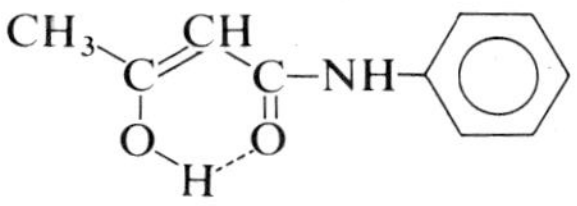 $CH_3-C=CH-C(=O)-NH-$ (phenyl) with $-O \cdots H-O$ | 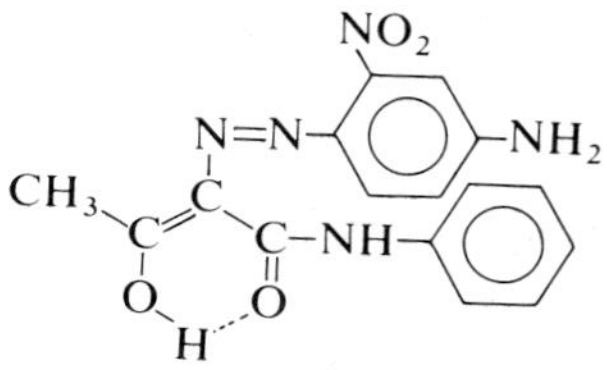 $NO_2$, (aryl) $-NH_2$, $-N=N-C-C(=O)-NH-$ (phenyl), $CH_3-C-O \cdots H-O$ <br> Hansa Yellow 6 or 1 <br> Pigment Yellow 1 |

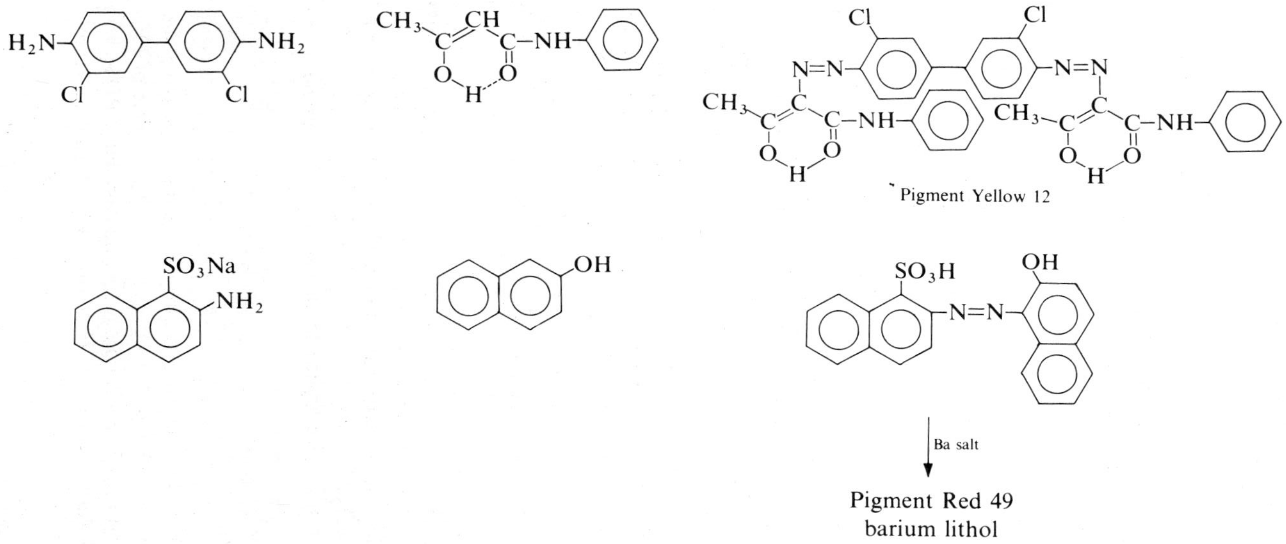

Pigment Yellow 12

Ba salt

Pigment Red 49
barium lithol

**Scheme 9.123** (*Continued*)

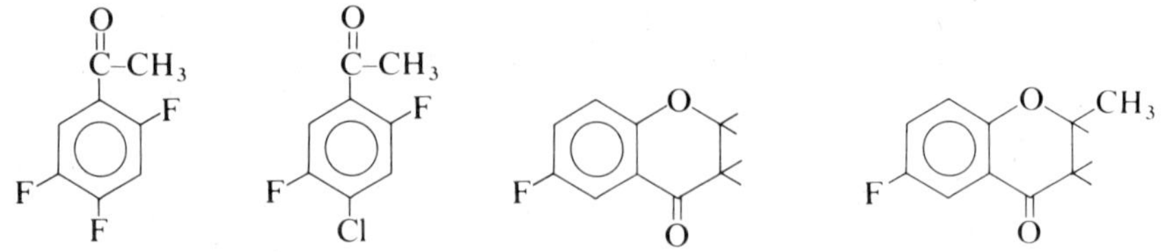

The above-mentioned chromanones are derived from *p*-fluorophenol by means of an acylation, followed by condensation of the resulting substituted acetophenone with formaldehyde or acetaldehyde:

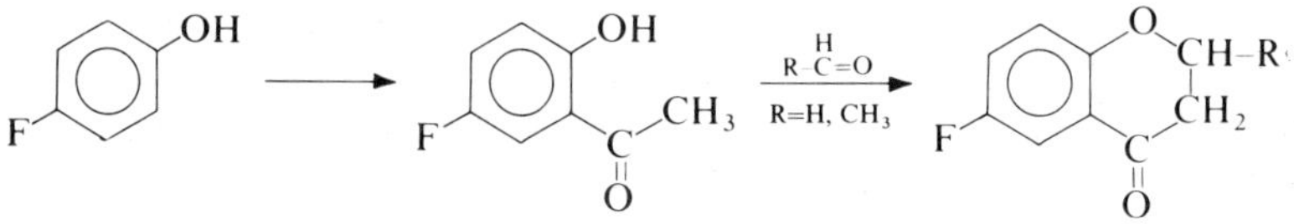

Direct fluorination of benzenoid compounds is mentioned in 9.3.1.4.

Another rather simple reaction of diazonium salts is their reduction to the corresponding arylhydrazines (process *f* in Fig. 9.1). Thus, the reduction of benzenediazonium chloride yields the fine chemical phenylhydrazine ($C_6H_5$–NH–$NH_2$, \$4.20), used in the synthesis of several dyes and medicinals that contain a N–N bond (see 9.2.2), and also employed as a stabilizer of explosives. The preparation of arylhydrazines from aromatic compounds that contain halogen substituents activated toward nucleophilic reactants (because of the presence of electron-withdrawing substituents such as nitro, sulfone, and keto), can be achieved more directly by means of hydrazine ($H_2N$–$NH_2 \cdot H_2O$, \$1.25). The hydrazino group of arylhydrazines provides another route to aromatic azides (see above) through the reaction with nitrous acid:

$$Ar-NH-NH_2 + HO-N=Q \rightarrow Ar-NH-NH-N=O \rightarrow Ar-N_3 + H_2O$$

The most versatile reaction of diazonium salts is the coupling reaction that occurs when the rather unstable diazonium ion is allowed to react with electron-rich aromatic substrates such as anilines, phenols, and analogous systems of the latter, namely, enols. The coupling reaction is the cornerstone of a great variety of azo dyes and is illustrated by the examples presented in Scheme 9.123.

Coupling of benzenediazonium ions with aniline to give *p*-aminoazobenzene,

$$C_6H_5-N_2^+ + C_6H_5-NH_2 \longrightarrow C_6H_5-N=N-\!\!\!\bigcirc\!\!\!-NH_2$$

and hydrogenation/hydrogenolysis of the latter to a readily separated mixture of aniline, b.p. 185°C, and *p*-phenylenediamine, PPD, b.p. 267°C, is the large-scale source of PPD required for the assembly of poly(*p*-phenylenediamine/terephthalamide), PPDT, best known as Kevlar (Du Pont) and Twaron (Akzo); (see H. Short, "A First Look at Aramid Processing," *CE* 4/89, pp. 37–41).

# 10

# HETEROCYCLIC BUILDING BLOCKS

## 10.1 INTRODUCTION

Countless heterocyclic structures contribute to the realm of medicinals, dyes, and photographic and other materials of industrial importance, including members of natural products that belong to families of alkaloids and nucleic acids described in multivolume monographs (*SZM-II*, 165–170).

Some of the treatises dealing with heterocyclic compounds include

A. R. Katritzky, Ed., *Advances in Heterocyclic Chemistry*, Vol. 41, Academic Press, San Diego, CA, 1987.

A. R. Katritzky, Ed., *Physical Methods in Heterocyclic Chemistry*, Vol. 3, Academic Press, New York, 1971.

S. W. Pelletier, *Alkaloids: Chemical and Biological Perspective*, Vols. 1–5, Wiley, New York, 1983–1987.

*Chemistry of Heterocyclic Compounds*, Wiley, New York (a series of monographs contributed by various editors, containing over 45 volumes).

C. B. Townsend and R. S. Tipson, *Nucleic Acid Chemistry: Improved and New Synthetic Procedures, Methods, and Techniques*, Wiley, New York, 1986.

A concise survey of heterocyclic chemistry that the reader may wish to consult is offered by A. Albert, *Heterocyclic Chemistry: An Introduction*, Athlone Press, London, 1959.

Only a limited number of this multitude of structures can be presented in a book of this length. The objective of the discussion that follows is to demonstrate (a) the synthetic relationship between the more commonly encountered heterocyclic ring systems and the simpler, usually acyclic, building blocks from which they can be assembled and (b) how the resulting heterocyclic structures, in turn, serve as building blocks of still more complex products of industrial interest.

Throughout the preceding chapters one encounters heterocyclic systems such as the epoxides or oxiranes, cyclic anhydrides (of maleic, succinic, phthalic acids, etc.), lactones and lactams (e.g., of the butyro and capro families), cyclic imines, imides, and hydrazides (of maleic acid, saccharin, *N*-methylpyrrolidone, etc.), and cyclic ethers and amines (e.g., dioxane, morpholine, triethylenediamine, hexamethylenetetramine). While such compounds also represent heterocyclic systems in the strictest sense, these structures are readily converted to acyclic compounds, and their discussion is minimized in this section of the book in which we concentrate on additional, especially, the generally resonance-stabilized aromatic heterocyclics. As is amply demonstrated in the pages that follow, electron delocalization is often the driving force that causes smaller building blocks to assemble into ring structures, and these, in turn, undergo almost spontaneous *in situ* air oxidation to produce resonance-stabilized aromatic heterocyclic compounds.

In the assembly of heterocyclic structures from smaller building blocks (irrespective of the actual reaction mechanisms that normally are much more complex than the simple schemes presented here for purely pragmatic reasons), it is useful to recognize certain common reaction patterns that involve either *elimination* and/or *addition* reactions.

The net result of elimination reactions is the loss of water, an alcohol, hydrogen sulfide, ammonia, or hydrogen halides from two (or more) acyclic building blocks or from a single acyclic building block that undergoes an intramolecular elimination reaction. In the application of this old-fashioned "lasso chemistry," it is important to consider (a) an active role played by all potential tautomers among the reacting species and (b) a spontaneous rearrangement of hydrogen substituents in order to generate the most resonance stabilized ring. The Hückel rule [planar, cyclic molecular systems that contain a continuous path for the delocalization of $(4n+2)$ electrons are highly resonance stabilized, where $n$ can have values of 0, 1, 2, 3, etc.) is a valid tool for the prediction of which ring is "most resonance stabilized." In the case of heterocyclic systems, we can ignore the case when $n=0$ and focus on the most commonly encountered systems that possess a total number of six or ten delocalized electrons.

Examples of elimination reactions that produce aromatic heterocyclic rings are shown in Scheme 10.1 *without* indicating details of tautomerization and hydrogen rearrangements.

Addition reactions that lead to the formation of heterocyclic systems occur most commonly between building blocks that contain either triple (nitrile or acetylenic) or double (carbonyl, thiocarbonyl, imine) bonds, and an active hydrogen of a H–O–, H–S, H–N=, and even H–$\overset{|}{\underset{|}{C}}$– functional group-containing building block. Addition reactions also occur with the participation of dipolar reactants such as

- Hydrazoic acid $\quad\quad$ $H-\overset{..}{\underset{..}{N}}-\overset{+}{N}\equiv N:\leftrightarrow H-\underset{.}{N}=\overset{+}{N}=\overset{..}{\underset{..}{N}}:$

- Diazomethane $\quad\quad$ $CH_2=\overset{+}{N}=\overset{..}{\underset{.}{N}}:^-\leftrightarrow :\bar{C}H_2-\overset{+}{N}\equiv N:$

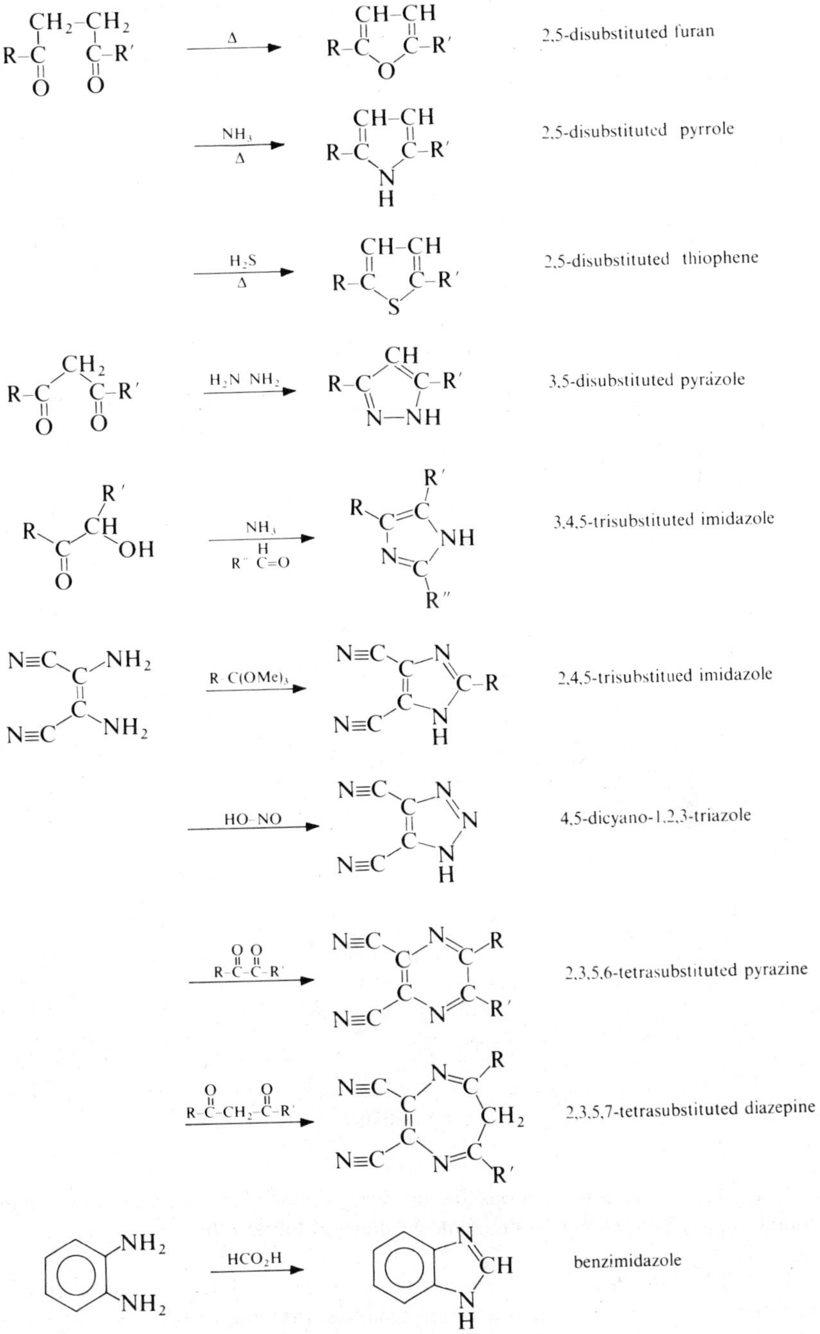

2,5-disubstituted furan

2,5-disubstituted pyrrole

2,5-disubstituted thiophene

3,5-disubstituted pyrazole

3,4,5-trisubstituted imidazole

2,4,5-trisubstitued imidazole

4,5-dicyano-1,2,3-triazole

2,3,5,6-tetrasubstituted pyrazine

2,3,5,7-tetrasubstituted diazepine

benzimidazole

**Scheme 10.1**

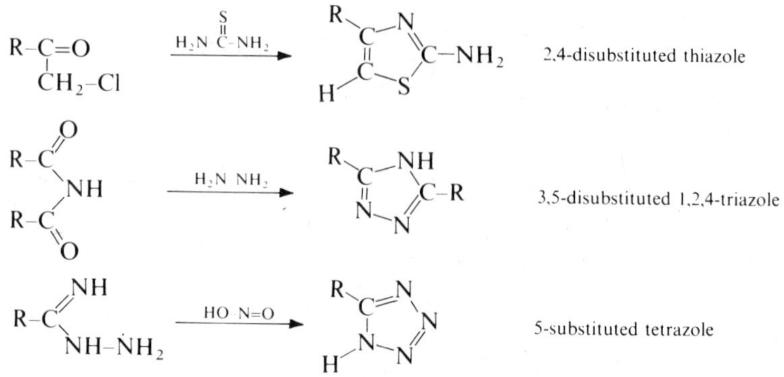

R–C=O / CH₂–Cl  →(H₂N–C(=S)–NH₂)→  2,4-disubstituted thiazole

R–C(=O)–NH / R–C(=O) (O)  →(H₂N NH₂)→  3,5-disubstituted 1,2,4-triazole

R–C(=NH)–NH–NH₂  →(HO N=O)→  5-substituted tetrazole

**Scheme 10.1** (*Continued*)

- Isocyanides or isonitriles such as *p*-toluenesulfonylmethyl isocyanide, tosylmethyl isocyanide, Nisso's TOSMIC, which, in effect, provides another polar reactant—the methylene isonitrile building block (Scheme 10.2).

$$:\bar{C}H_2{-}\ddot{N}::C: \leftrightarrow :\bar{C}H_2{-}\overset{+}{N}:::C:^-$$

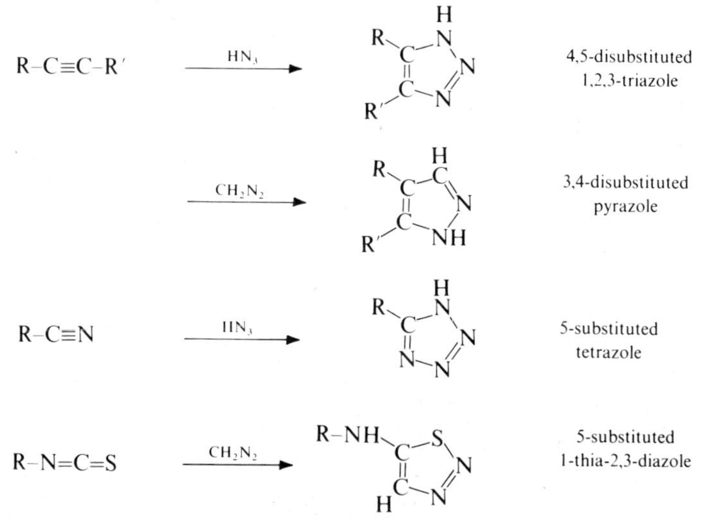

R–C≡C–R′  →(HN₃)→  4,5-disubstituted 1,2,3-triazole

→(CH₂N₂)→  3,4-disubstituted pyrazole

R–C≡N  →(HN₃)→  5-substituted tetrazole

R–N=C=S  →(CH₂N₂)→  5-substituted 1-thia-2,3-diazole

**Scheme 10.2**

Other examples of addition reactions that produce aromatic heterocyclic rings are shown as follows, again without any details of intermediate tautomerizations and hydrogen shifts (Scheme 10.3).

Finally, there are offered a few examples of the formation of heterocyclic structures by a combination of elimination and addition reactions (Scheme 10.4).

The heterocyclic families of compounds are presented in the order of increasing ring size and, within each ring size, in the order of the number of heteroatoms that constitute a

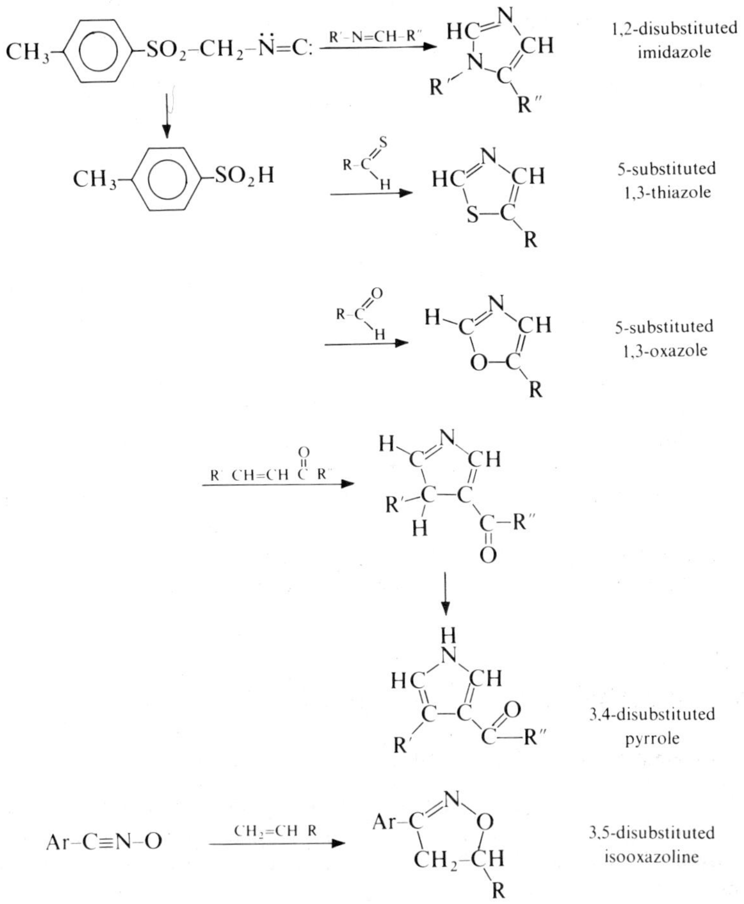
1,2-disubstituted imidazole
5-substituted 1,3-thiazole
5-substituted 1,3-oxazole
3,4-disubstituted pyrrole
3,5-disubstituted isooxazoline

**Scheme 10.3**

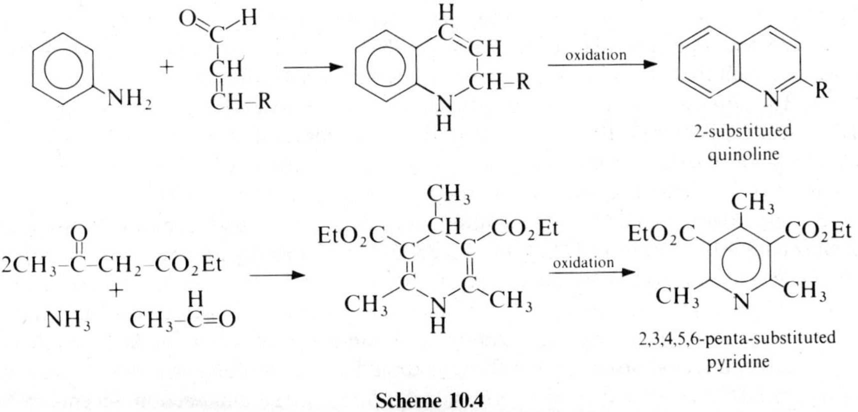
oxidation
2-substituted quinoline
oxidation
2,3,4,5,6-penta-substituted pyridine

**Scheme 10.4**

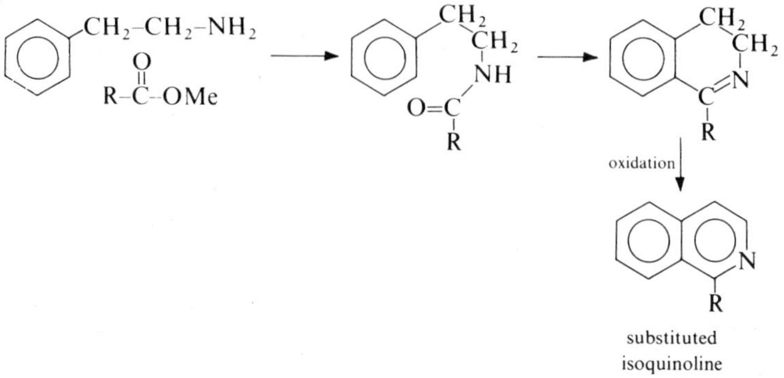

substituted
isoquinoline

**Scheme 10.4** *(Continued)*

given five-member, six-member, or higher-number ring structure. Within each category of ring size and a given number of heteroatoms, the oxygen-family-containing heterocycles are mentioned first in the order of increasing atomic number of the hetero atoms, and the nitrogen-family-containing ring compounds are mentioned next. In fused ring systems, the number of heteroatoms in the most-heteroatom-containing ring determines its place in the presentation rather than the total number of heteroatoms in the whole fused-ring system.

## 10.2 FIVE-MEMBERED HETEROCYCLIC SYSTEMS

### 10.2.1 One-Heteroatom-Containing Rings

Of the three heterocyclic systems that contain one heteroatom, thiophene resembles benzene in stability and even in some physical properties, but it is significantly more reactive toward electrophilic reagents. Furan and pyrrole barely pass the criterium of minimum resonance stabilization to qualify as being "aromatic," and their stability decreases in the presence of mineral acids.

Furan, tetrahydrofuran, (THF), and the acid-catalyzed polymerization of the latter to PTHF or PTMEG are mentioned elsewhere in connection with the chemistry of acetylene (4.7.4) and the use of poly(ether glycols) in the assembly of some PURs (interlude in Section 3.8). Also, furfuraldehyde (or furfural; 75¢), the underutilized product of the acid-catalyzed decomposition of pentoses (*SZM-II*, 98–103), is mentioned on several occasions (please consult Index). The decarbonylation of furfuraldehyde over soda lime is a source of furan. The latter is converted by catalytic hydrogenation to THF (96¢) and undergoes a Diels–Alder reaction with MA to yield the herbicide and defoliant endothall, and electrochemical oxidative methoxylation gives rise to a newcomer on the marketplace in the form of 2,5-dimethoxytetrahydrofuran (Hexagon, $8.65)—a hardener of the gelatin layer in photographic plates (by generating succinaldehyde *in situ* for the cross-linking of the protein). Also, traditional "hot tube reactions" of furan with ammonia and hydrogen sulfide (carried out over alumina at temperatures of ~400–450°C) produce pyrrole and thiophene, respectively, although the last-mentioned heterocycle can be obtained more economically by a cracking–sulfurization of *n*-butane or of linear butenes (Synthetic Chemicals of England produces thiophene at about $2.25, depending on the exchange rate of the dollar). The preceding transformations of furan are summarized in Scheme 10.5.

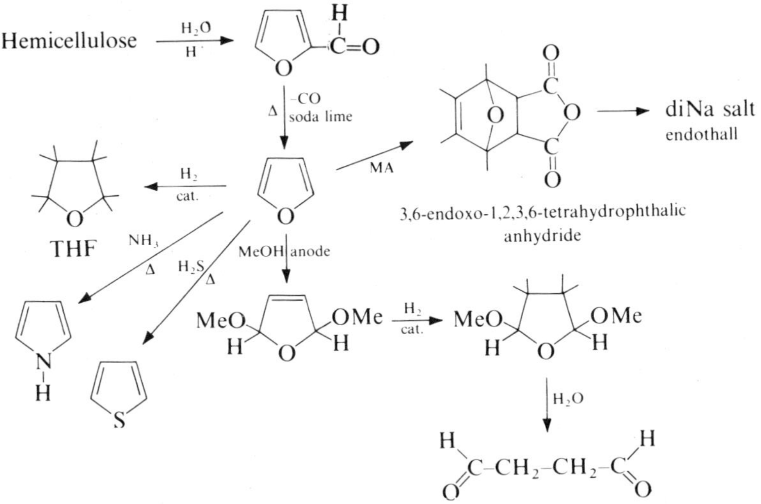

**Scheme 10.5**

Two other, but more complex, materials that belong to the furan family and are of considerable industrial importance are L-ascorbic acid (vitamin C, USP \$5.80) and the systemic insecticide carbofuran (9.3.4.2, Scheme 9.79). However, neither one of these compounds is prepared from furan.

Ascorbic acid is derived from sorbitol (anhydrous 68¢–70¢)—the hydrogenation product of glucose or dextrose (anhydrous 41¢–46.5¢)—by a combination of biological and chemical transformations described elsewhere (*SZM-II*, 107).

Another furan-related compound of current relevance is the antiviral agent ribavirin, which shows promise in the treatment of acquired immune deficiency syndrome (AIDS)-related condition (R. Dagani, *CEN* 1/26/87). The structure of this material also contains a triazole ring (10.2.3), but a synthesis of the whole molecule requires the use of 2,3,5-tri-*O*-benzoyl-1-*O*-acetyl-D-ribofuranose. The latter is currently manufactured on a relatively large scale (Pfanstiehl, \$98/lb in 100-kg quantities) because of an extensive ribavirin testing program. The discussion of the production of ribose by a degradation of six-carbon glucose or hydrolysis of nucleic acid of yeast (pure brewer's yeast, debittered, NF, *Saccharomyces*, \$1.10), is beyond the scope of this book. We limit ourselves here to illustrate the equilibrium between the isomeric structures or ribose and its conversion to the above-mentioned derivative required for the assembly of ribavirin (Scheme 10.6).

Pyrrole is an unstable material susceptible to air oxidation with the formation of intractable, resinous products. Protonation of the nitrogen electron pair destroys whatever aromatic resonance it possesses and converts pyrrole to a red polymer. While the pyrrole ring is the essential component of crucial natural products such as hemoglobin and chlorophyll, the industrial importance of simple pyrrole derivatives is limited to its complete hydrogenation product, pyrrolidine—a rather strong amine, and pyrrole-2-carboxaldehyde, obtained by means DMF (49¢) in the presence of phosphorus oxychloride (43¢). These transformations of pyrrole are summarized in Scheme 10.7.

The pyrrolidine building block is incorporated in some ultrafine chemicals such as the vasodilator buflomedil and the antihypertensive enalapril. The last-mentioned medicinal involves the use of the dipeptide *l*-alanyl-*l*-proline (Scheme 10.8).

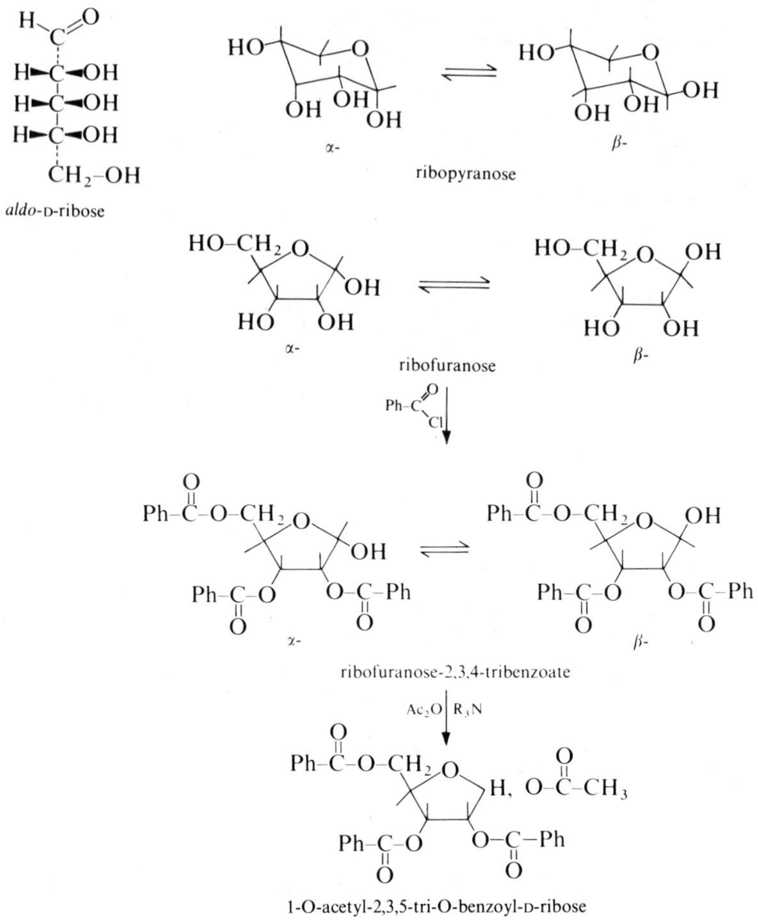
aldo-D-ribose
α-
β-
ribopyranose
α-
β-
ribofuranose
α-
β-
ribofuranose-2,3,4-tribenzoate
1-O-acetyl-2,3,5-tri-O-benzoyl-D-ribose

**Scheme 10.6**

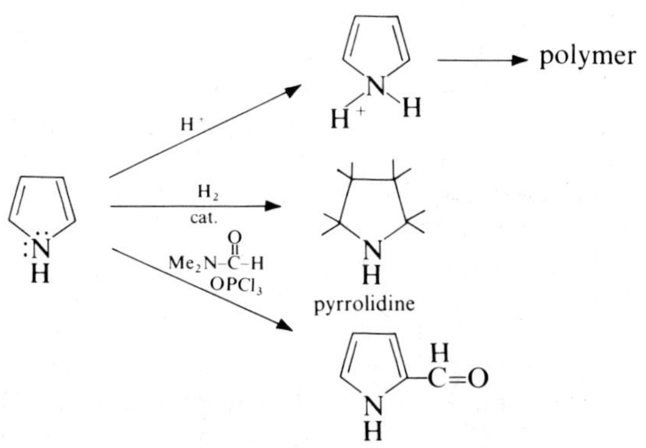
polymer
pyrrolidine

**Scheme 10.7**

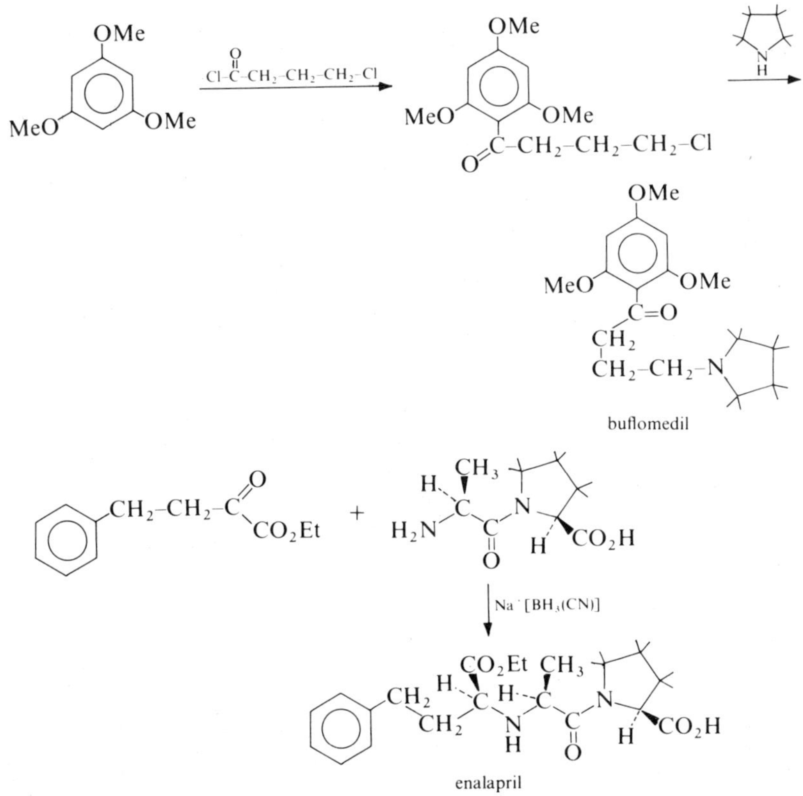

buflomedil

enalapril

**Scheme 10.8**

As shown elsewhere, pyrrolidone (BASF's Luviskol K-30, $3.64) and a series of *N*-substituted pyrrolidones, as well as PVP (BASF's Kollidon K-30, $6.24), are derived from butyrolactone, and the complex of PVP with iodine serves as a topical antiseptic applied before surgery. Aminoacetic acid or glycine (technical grade $1.88) condenses with succinic anhydride ($1.71) to yield *N*-carboxymethylpyrrolidine-2,5-dione, and the same reactants in the presence of formaldehyde give *N*-carboxymethyl-2-pyrrolidone-4-carboxylic acid, while ethyl acetoacetate, chloroacetaldehyde in the presence of ammonia condense to produce ethyl 2-methylpyrrole-3-carboxylate, and α-keto esters (R′–CH$_2$–CO–CO·OEt) and the imine of benzaldehyde, for example, yield 4,5-disubstituted 2,3-pyrrolidine-2,3-dione. These transformations are shown in Scheme 10.9.

Bismaleimides, are derivatives of MA and of readily available aromatic diamines such as MDA (crude $1.75, purified $2.25) and *m*-phenylenediamine ($2.07):

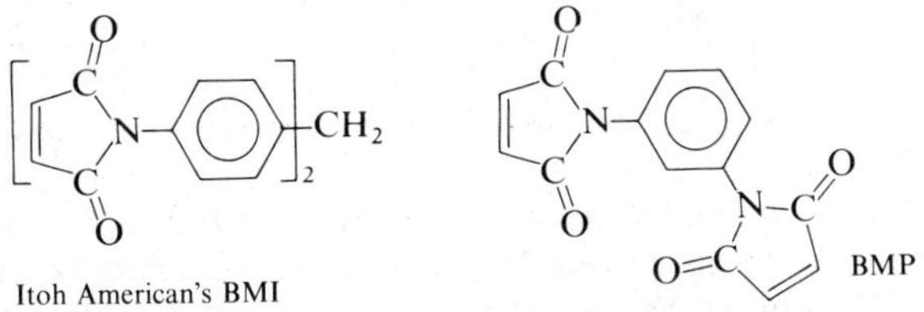

Itoh American's BMI

BMP

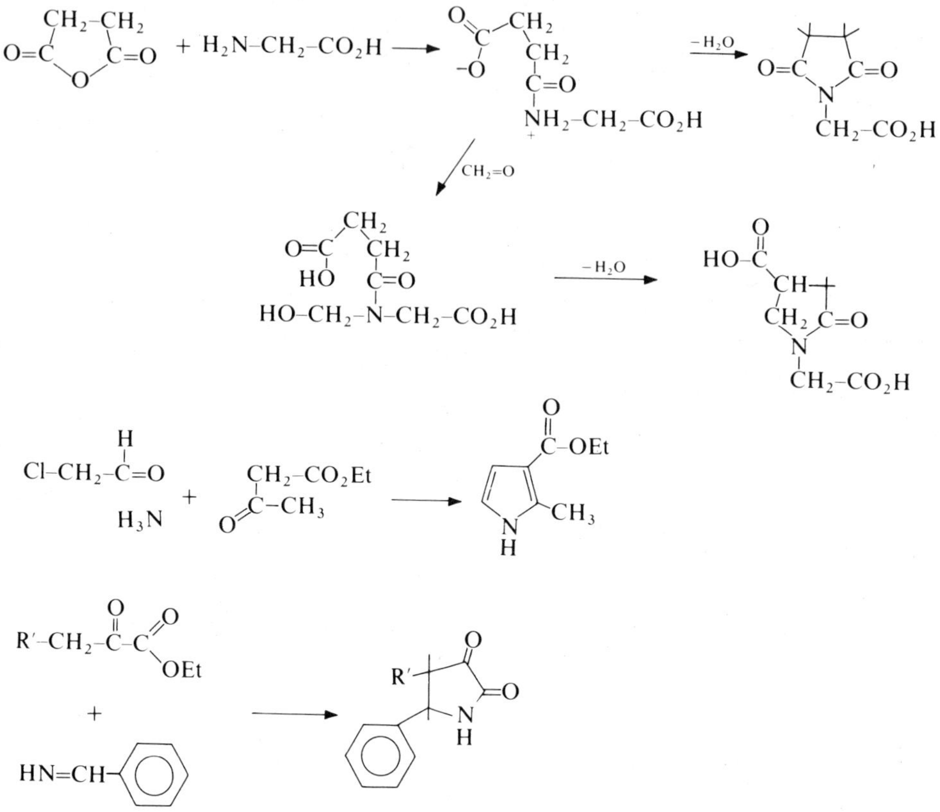

**Scheme 10.9**

These relatively new monomers are utilized for the preparation of high-performance specialty polymers. For example, Ciba-Geigy cures a bismaleimide prepolymer obtained from BMI and difunctional allylic monomers to form a high-performance structural adhesive identified as "XU 292" and priced at $25.

The stability of the pyrrole ring is increased in indole ($25.50) and carbazole because of the contribution of benzenoid aromaticity to the fused-ring systems:

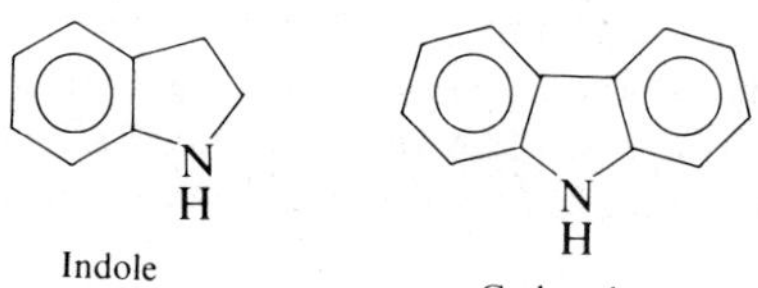

Indole                                      Carbazole

Indole is used as ingredient of perfumes since, in high dilution, it contributes to the scent of jasmine, lilac, and orange blossoms. It is synthesized by an alkaline fusion of N-o-tolylformamide (see Fig. 9.3). Commercially important derivatives of indole include the essential amino acid *l*-tryptophan ($28.25), the perfume ingredient skatole (3-methyl-indole, (Ald. F&F $225/lb in 1-kg quantity), the plant hormones (auxins) β-indoleacetic and β-indolebutyric acids (U.S. Biochem $155 and $310, respectively), and indoxyl (3-hydroxyindole)—the precursor of indigo (see below). Some of the synthetic routes to these materials are summarized in Scheme 10.10.

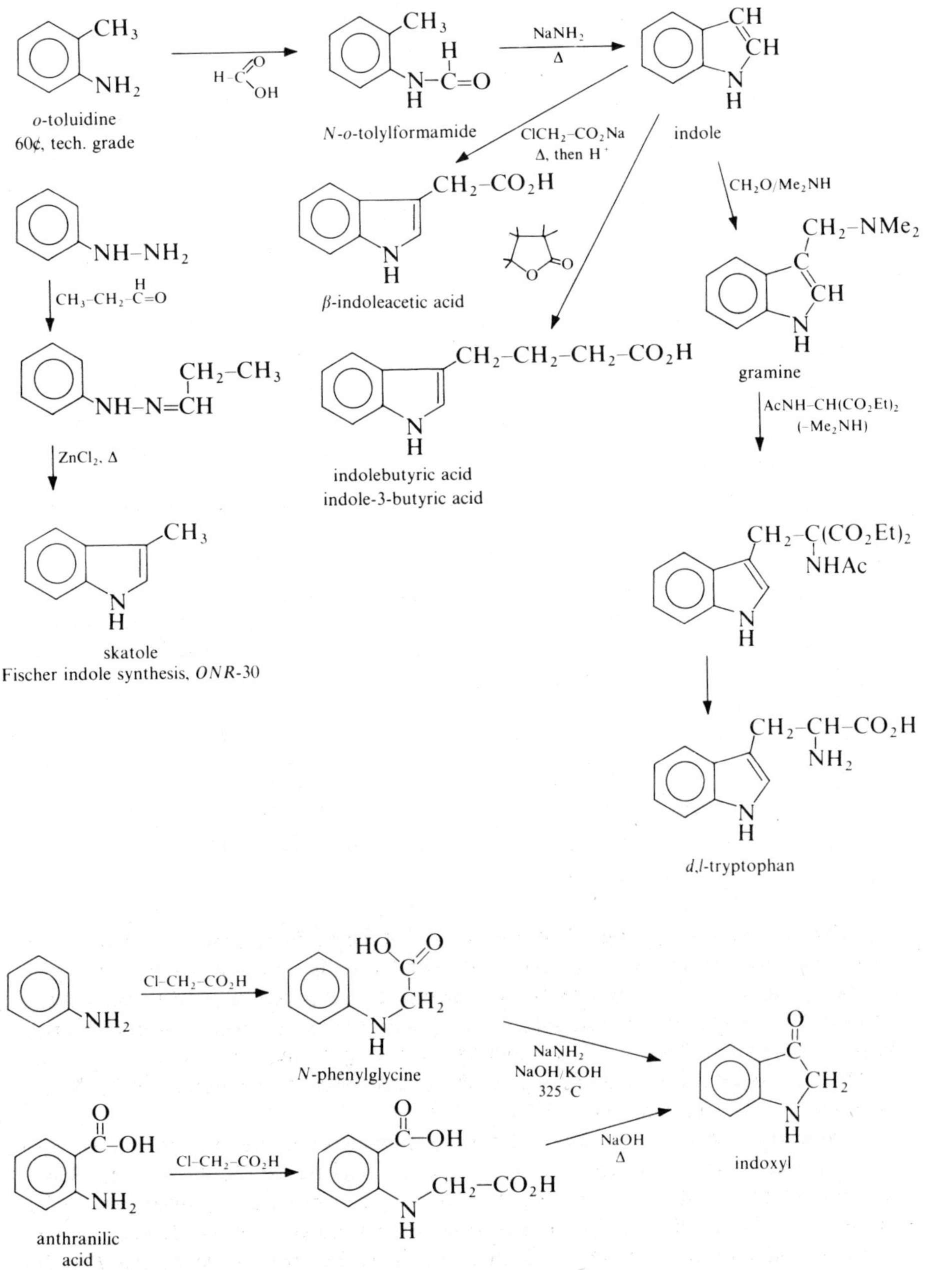

o-toluidine
60¢, tech. grade

N-o-tolylformamide

indole

β-indoleacetic acid

indolebutyric acid
indole-3-butyric acid

gramine

skatole
Fischer indole synthesis, *ONR*-30

d,l-tryptophan

N-phenylglycine

anthranilic
acid

indoxyl

**Scheme 10.10**

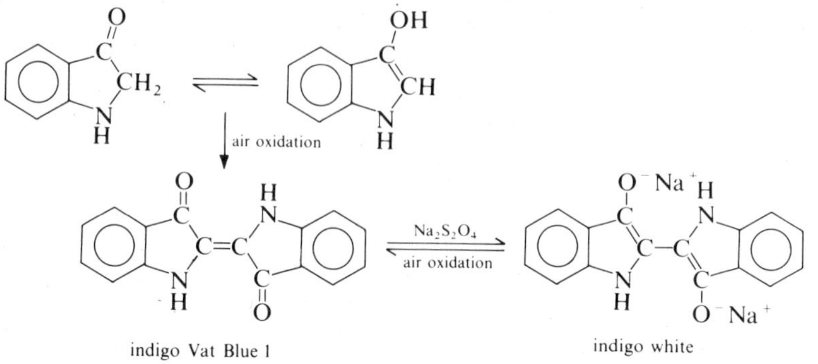

indigo Vat Blue 1  indigo white

**Scheme 10.11**

Indigo, Vat Blue 1 (imported at about \$1/lb) and its sulfur analog thioindigo—Vat Red 41—are, as suggested by their names, typical vat dyes; when in the reduced state they are soluble and colorless and thus are readily adsorbed by the substrate, but on oxidation, they develop a color and become insoluble and thus "fixed" to the substrate. Indigo, a dye known to ancient civilizations, was obtained from natural sources until a century ago, when it began to be manufactured competitively from either anthranilic acid (\$1.70) or from aniline (58¢). Both processes also involve the use of chloroacetic acid, 56¢ (see above) (Scheme 10.11). Obviously, the economics of the Hoechst process based on aniline is more favorable even though it employs sodium amide (prepared from anhydrous, liquefied ammonia, b.p. $-33°C$, 3.75¢) and metallic sodium (70¢). The 6,6'-dibromo derivative of indigo happens to be a synthetic replacement of the ancient Tyrian Purple. This dye was reserved for use by royalty because of difficulties in its isolation from the Mediterranean molusk *Murex*, but its modern equivalent is 5,5',7,7'-tetrabromoindigo, known as Vat Blue 5 or Bromindigo Blue 2BD:

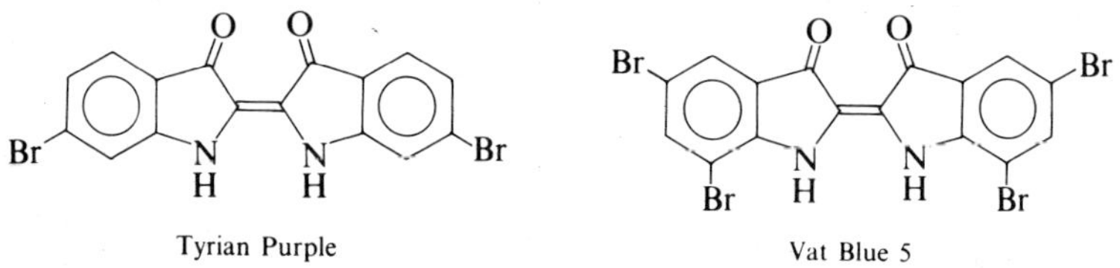

Tyrian Purple  Vat Blue 5

The analogous thioindigo, Vat Red 41, is obtained from 2-mercaptobenzoic or thiosalicylic acid and chloroacetic acid by way of thioindoxyl, as shown in Scheme 10.12.

Several substituted thioindigo dyes are prepared from the corresponding substituted anthranilic acids. Thus, for example, *p*-cresol can serve as a starting material for the production of Vat Orange 5—an important cotton dye—and *o*-xylene and 1-methylnaphthalene (see Figs. 9.4 and 9.9, respectively) can be the starting materials for Vat Red 1 and Vat Brown 5, respectively (Scheme 10.13).

Carbazole is obtained by air oxidation of diphenylamine (see Fig. 9.7) and is the intermediate in the production of various dyes (Scheme 10.14). Also, under basic conditions, the alkali salt of carbazole and acetylene produce *N*-vinylcarbazole—a monomer in the production of poly(vinylcarbazole) introduced in 1970 by IBM to serve as a photoconductor in its photocopying machines (J. M. Pearson and M. Stolka, *Poly(N-vinylcarbazole)*, Gordon & Breach, New York, 1982).

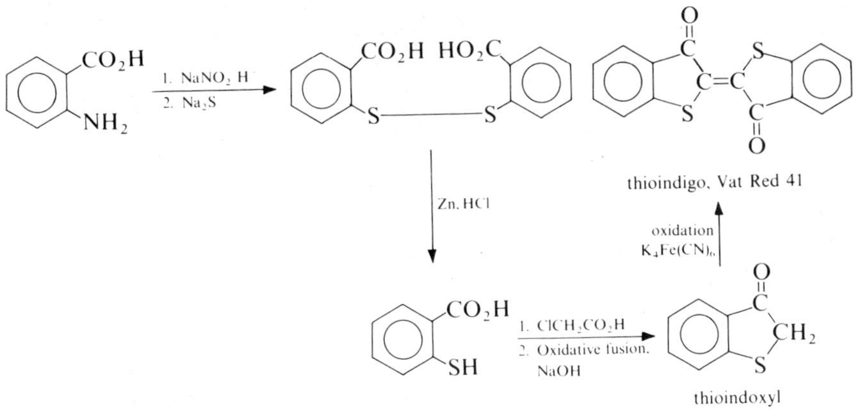

Scheme 10.12

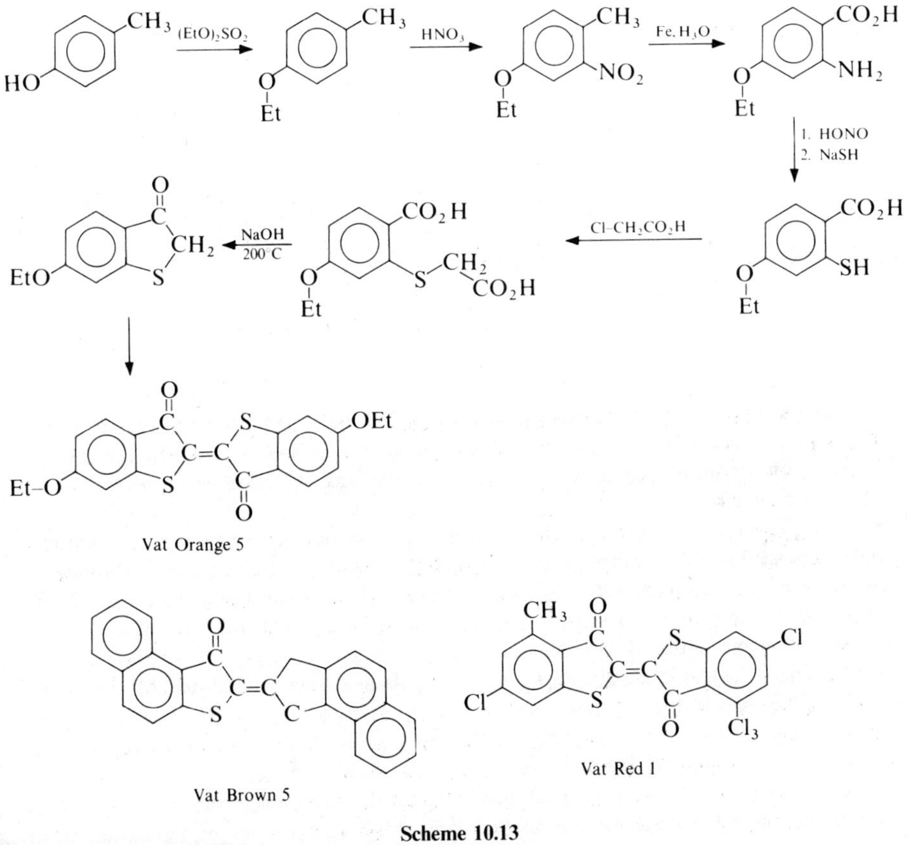

Vat Orange 5

Vat Brown 5

Vat Red 1

Scheme 10.13

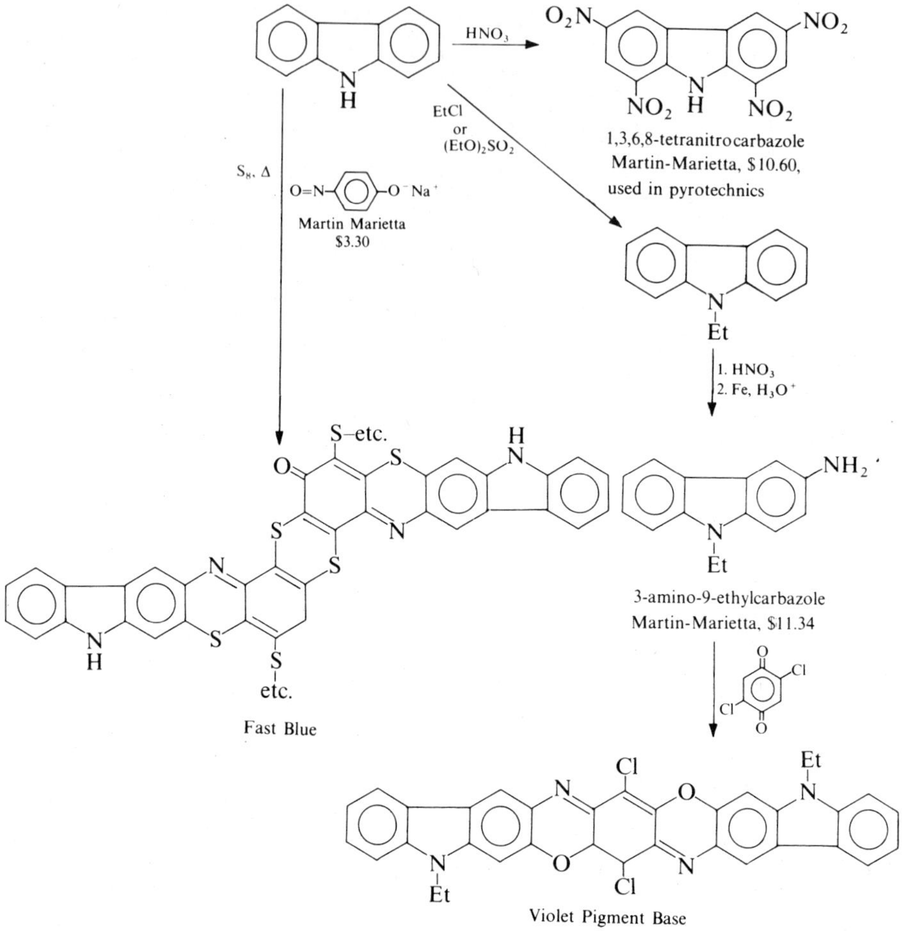

**Scheme 10.14**

Thiophene is incorporated in numerous medicinals because of its mimicry of benzene. The parent compound and its simple homologs, such as 2- and 3-methylthiophenes, are prepared on a commercial scale by means of a cracking–sulfurization of $n$-$C_4$ and $C_5$ alkanes or olefins.

The thiophene ring undergoes electrophilic substitution reactions most readily at the 2 and 2,5 positions. For example, chloromethylation produces 2-chloromethylthiophene, also known as 2-*thenyl chloride*—the building block of the antihistaminic methapyrilene. Also, one of the synthetic routes to the antiinflammatory medicinal suprofen (Suprol, McNeil Pharmaceuticals) involves a Friedel–Crafts reaction of thiophene with $p$-ethylbenzoyl chloride (catalyzed by a mild Lewis acid such as stannic chloride). The resulting ketone is then brominated with the Ziegler reagent (*ONR*-97), namely, *N*-bromosuccinimide (**NBS**), and the resulting product is treated with sodium cyanide to complete the assembly of the *alpha*-branched side chain (Scheme 10.15).

While the stabilization of the thiophene ring by electron delocalization is nearly that of benzene, the sulfur moiety is the Achilles' heel of the molecule: attempted oxidation to

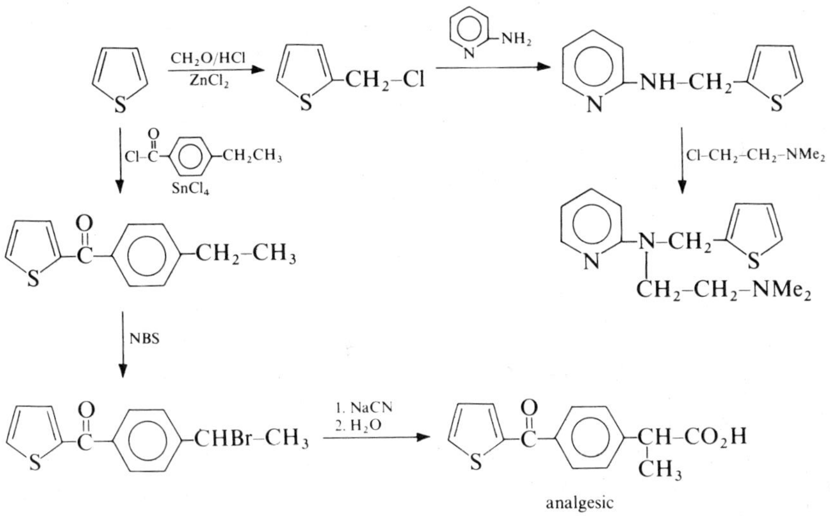

**Scheme 10.15**

sulfoxide or sulfone leads to complex decomposition products. We recall that the sulfone derivatives of dihydro- and tetrahydrothiophene (sulfolene and sulfolane, respectively) are obtained by the Diels–Alder reaction of 1,3-butadiene and sulfur dioxide (6.8.1.4). The vulnerability to oxidation of the sulfur moiety decreases as the thiophene ring is fused with stabilizing benzene rings. The simplest of these structures is 2-benzothiophene, benzo(*b*)thiophene, or thianaphthene, also available commercially from England. Thianaphthene can be prepared by a high-temperature ( ∼600°C) reaction of styrene and sulfur, and the next highest thiophene–benzene fused-ring system, namely, dibenzothiophene, is obtained in a similar fashion by a Friedel–Crafts-like reaction of biphenyl and sulfur:

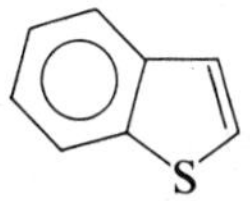

Thianaphthene 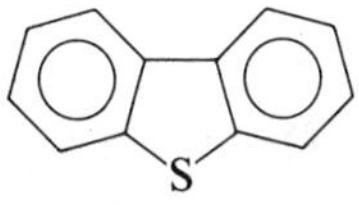          Dibenzothiophene

The latter compound is no longer vulnerable to oxidation and can be converted by means of chromic acid to dibenzothiophene *S,S*-dioxide.

Thioindigo, or Vat Red 41, and its analogs such as Vat Orange 5, Vat Red 1, and Vat Brown 5, are mentioned above in connection with indigo.

## 10.2.2   Two-Heteroatom-Containing Rings

1,3-Oxazolidines are readily accessible by means of a reaction of *N*-substituted ethanolamines with aldehydes and ketones, shown in Scheme 10.16. This reaction is carried out by heating the reactants in the presence of inert solvents (e.g., benzene or toluene) capable of removal of water by azeotropic means. The reversibility of the reaction suggests the use of 1,3-oxazolidines as a means of protecting the carbonyl compound during a sequence of synthetic operations.

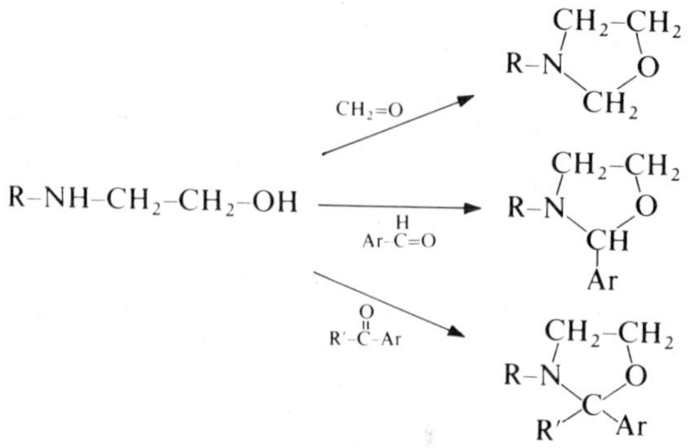

**Scheme 10.16**

The reaction of glycine with formic acid and acetic anhydride produces the corresponding azlactone—an oxazolinone. It contains an activated methylene group capable of aldol reaction, and the hydrogenation of the resulting condensation product constitutes a synthetic route to amino acids. An example of the Erlenmmeyer–Ploechl azlactone amino acid synthesis ($ONR$-28) is the preparation of leucine by starting with $N$-benzoylated glycine, or hippuric acid, and acetic anhydride, followed by an aldol condensation with isobutyraldehyde, hydrogenation, and hydrolysis (Scheme 10.17).

Hippuric acid is the detoxification product of benzoic acid—a discovery made by the German chemist Justus Liebig (1803–1873), who, together with the German chemist

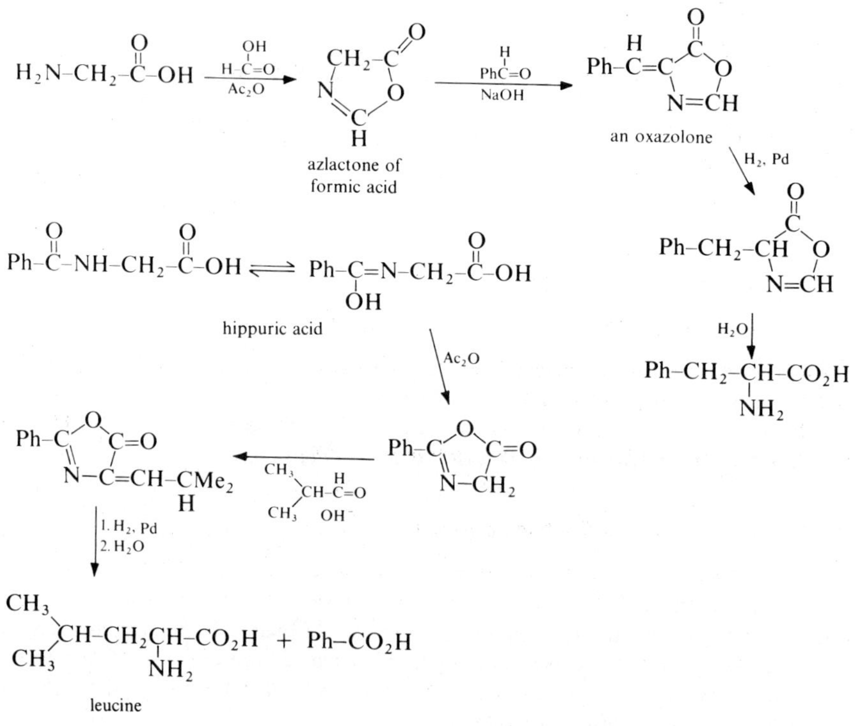

**Scheme 10.17**

Friedrich Wöhler (1800–1882), was one of the founders of the whole field of organic chemistry. About 150 years ago Liebig described how he administered substantial doses of sodium benzoate to a student, who, unprotected by the FDA, EPA, and OSHA, perspired profusely during the experiment but, in the end, delivered several liters of urine from which the professor isolated hippuric acid, thus proving that the detoxification mechanism in a human resembles that of a horse.

The reverse synthetic procedure, that is, one in which one starts with an α-amino acid and acetic anhydride and obtains an acetylated α-amino methyl ketone, constitutes the Dakin–West reaction (ONR-22). This reaction also involves an intermediate azlactone, and its special feature is the migration of the O-acetyl group donated by acetic anhydride to the nearby unsaturated carbon. A base such as pyridine catalyzes this rearrangement; see Scheme 10.18.

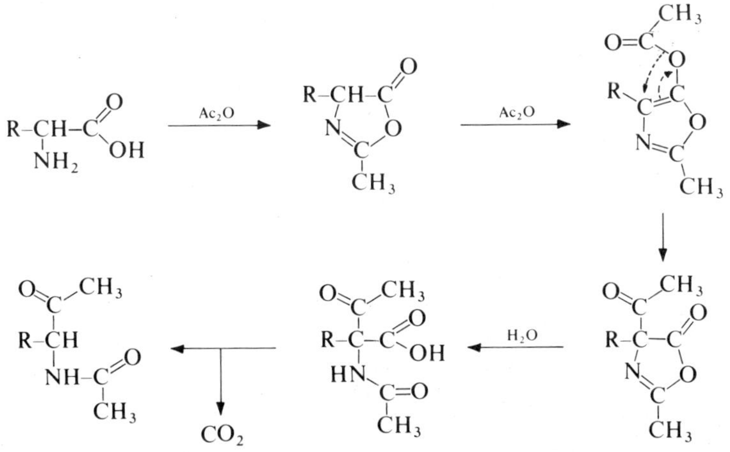

**Scheme 10.18**

Various 1,3-oxazoles are prepared from amides of aromatic acids and phenacyl chlorides and, in a highly purified state, serve as laser dyes. Typical structures of these costly, ultrafine products are shown in Scheme 10.19.

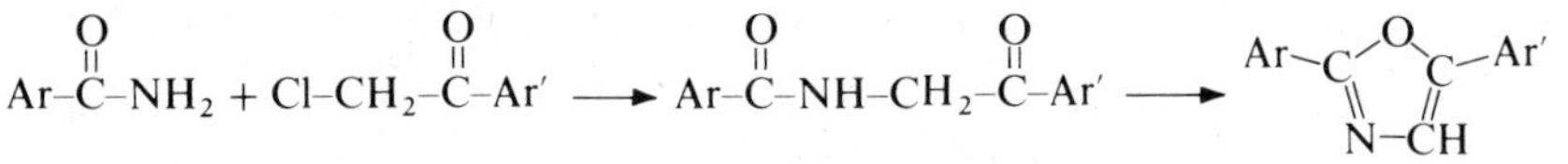

| Ar = | Ar' = | Name of substituted oxazole | Unit cost, $/lb (sigma) |
|---|---|---|---|
| $C_6H_5-$ | $C_6H_5-$ | 2,4-diphenyloxazole, PPO | 120 |
| $p$-$C_6H_5$-$C_6H_4-$ | $C_6H_5-$ | 2-(4'-biphenylyl)-5-phenyloxazole, PBO | 340 |
| $p$-$C_6H_5$-$C_6H_4-$ | $p$-$C_6H_5$-$C_6H_4-$ | 2,5-di(4'-biphenylyl)oxazole, BBO | 1670 |
| α-naphthyl- | $C_6H_5-$ | 2-(1'-naphthyl)-5-phenyloxazole, α-NPO | 2160 |

**Scheme 10.19**

Analogous laser dyes of the 1,3,4-oxadiazole family are described in 10.2.3.

An interesting recent discovery (*CEN* 6/13/88) is a family of new antivirals of a hydrophobic structure assembled from 1,3- and 1,2-oxazolines:

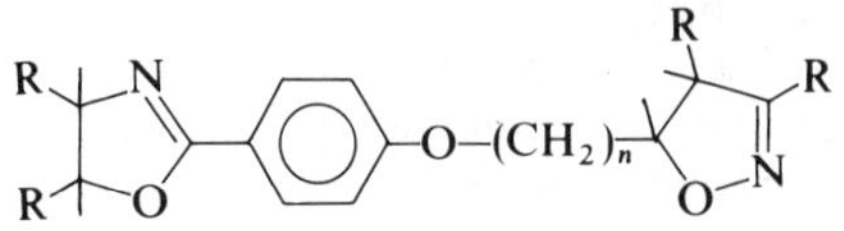

These materials inhibit the growth of viruses responsible for colds, polimyelitis, hepatitis A, and other ailments.

The reaction of acetaldehyde and cysteine produces 2-methyl*thiazolidine*-4-carboxylic acid (MTCA); as shown in Scheme 10.20. MTCA is claimed (*CEN* 8/2/82) to protect the liver against a variety of toxins including ethyl alcohol.

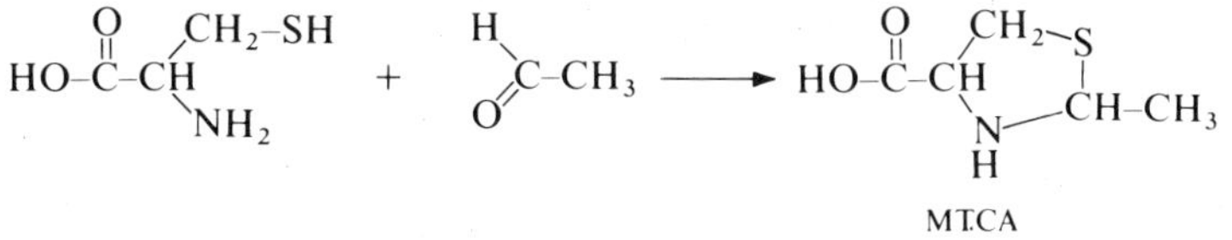

**Scheme 10.20**

2-Amino*thiazole* is obtained by the condensation of thiourea with chloroacetaldehyde, vinyl acetate, or—under more vigorous conditions—paraldehyde. It is the building block of the antibacterial sulfathiazole ($6.85). Another medicinal product derived from 2-aminothiazole by means of successive alkylations with the chloromethylation product of anisole and dimethylaminoethyl chloride is the antihistaminic zolamine. These transformations of 2-aminothiazole are summarized in Scheme 10.21.

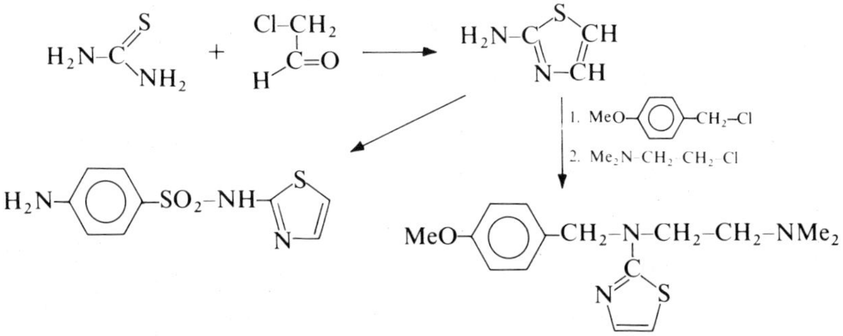

**Scheme 10.21**

The thiazole ring is considered to be an isostere of the pyridine ring. The term "isostere" is used to describe aromatic systems in which different moieties resemble each other on the basis of size and electronic features. In this case, the $-S-$ moiety (only one nonbonding electron pair of the two present on sulfur is involved in electron delocalization of the aromatic ring) is equivalent to the $-CH=CH-$ moiety. The thiophene–benzene isostere relationship accounts for the similarity of physical properties of benzene and thiophene derivatives suggested above. By the same token, zolamine is an isostere of another antihistaminic—pyrilamine, derived from 2-aminopyridine.

The condensation of chloroacetaldehyde with ammonium dithiocarbamate in place of thiourea gives 2-mercaptothiazole rather than 2-aminothiazole.

Among many other interesting thiazoles we find two ultrafine flavoring agents priced at $150 and $90/lb and used in chocolate and meat products. In addition to their chemical names—namely, ethyl 5-$\beta$-acetoacetonyl-4-methylthiazole and 4-methyl-5-$\beta$-hydroxy-ethylthiazole—these compounds are also identified by code numbers assigned by the Flavor and Extract Manufacturers Association, (FEMA), namely, 3204 and 3205, respectively. The simpler of the two substituted thiazoles, is actually an intermediate in the synthesis of vitamin $B_1$ or thiamine hydrochloride (see 10.2.2). It can be assembled from thioformamide and $\alpha$-bromoethyl hydroxyethyl ketone, as shown in Scheme 10.22.

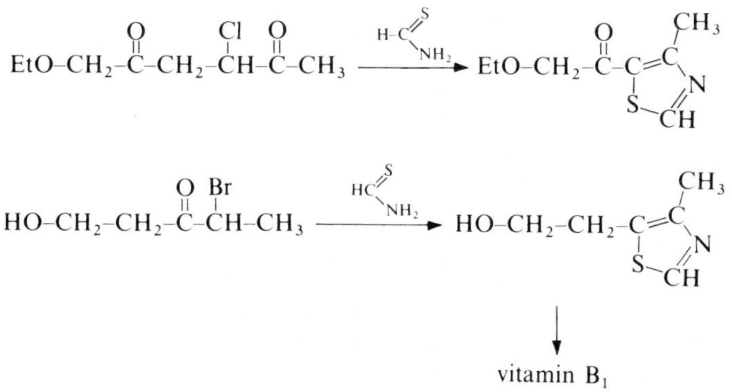

**Scheme 10.22**

The preparation of the 2-mercaptobenzothiazole system and its uses as rubber-processing chemicals is mentioned elsewhere (see Index).

An industrially important dyestuff intermediate of the pyrazole family (viz., 1-phenyl-3-methyl-5-aminopyrazole; Lonza $6.90) is obtained from the dimer of acetonitrile, diacetonitrile (Lonza $7.70) and aniline (Scheme 10.23).

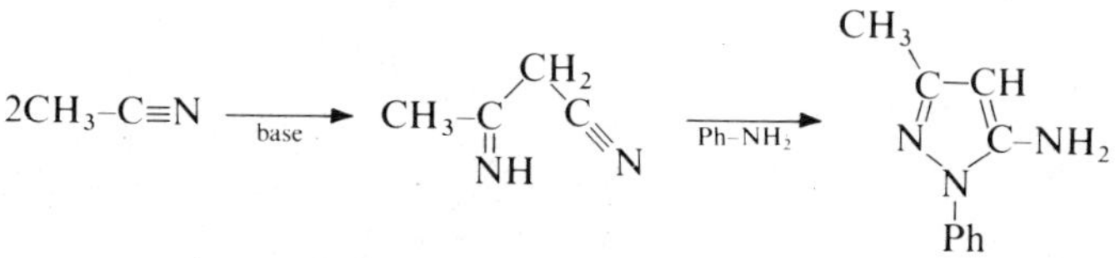

**Scheme 10.23**

The pyrazolone ring is readily assembled from acetoacetic esters (and similar $\alpha,\beta$-diketo, or $\beta$-keto ester systems) and hydrazine and its derivatives. The reaction product with hydrazine hydrate (85¢) gives rise to insecticides such as $O,O$-diethyl $O$-(3-methyl-5-pyrazolyl) phosphate (pyrazoxon) and the analogous thiophosphate (pyrazothion). The reaction product with phenylhydrazine ($4.20) gives 1-phenyl-3-methyl-5-pyrazolone (Hoechst-Celanese $3.00). It is the building block of two analgesics, namely, antipyrine and aminopyrine, but most of it is used in the manufacture of organic pigments. The pyrazolone starting material exists in three tautomeric forms. The above-mentioned transformations of the pyrazolones are illustrated in Scheme 10.24.

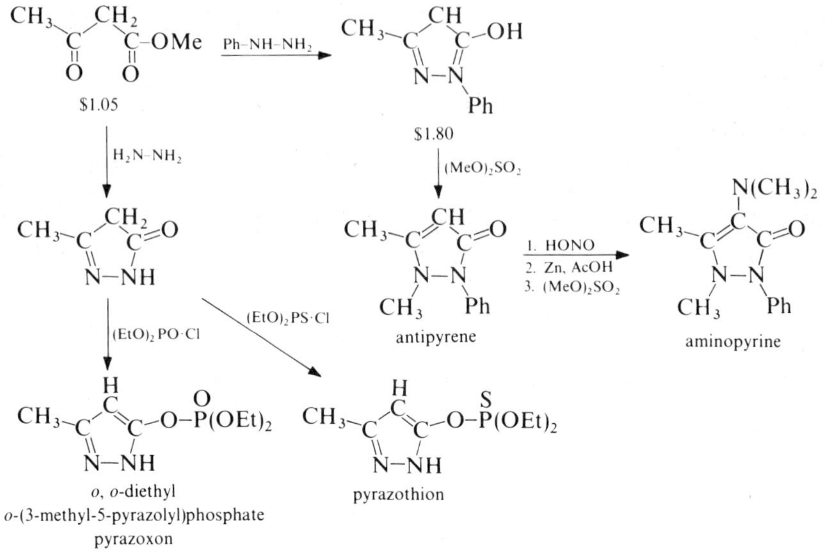

**Scheme 10.24**

The analogous oxaloacetic esters and phenylhydrazine produce 1-phenyl-3-carbethoxy-5-pyrazolone ($3.45). The latter, when coupled with the diazonium salt of sulfanilic acid, produces a water-soluble dye known as "Acid Yellow 23" (tartrazine; $6.18), approved for use in foods and cosmetics as FD&C No. 5 ($7.75). The formation of these compounds is shown in Scheme 10.25.

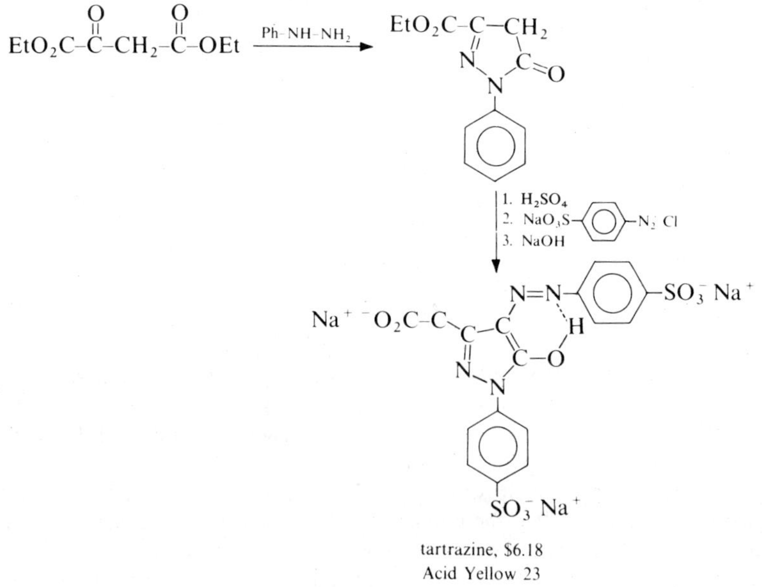

tartrazine, $6.18
Acid Yellow 23

**Scheme 10.25**

Substituted imidazoles are assembled readily by means of the Weidenhagen process, in which an appropriate 1,2-dicarbonyl system is condensed with an aldehyde in the presence of liquid ammonia. Thus, for example, glyoxal and acetaldehyde give 2-methylimidazole, which on alkylation with benzyl chloride yields 1-benzyl-2-methylimidazole (polyORGANIX's Curimid-NB; ~$12.80), a polymer catalyst for cross-linking of polyester–epoxy powder coatings. Similarly, pyruvic aldehyde, or pyruvaldehyde, obtained by a vapor-phase oxidation of 1,2-propanediol or glycerine, and propionaldehyde yield 2-ethyl-4-methylimidazole (polyORGANIX's Curimid-24; ~$15/lb)—an accelerator in the curing of epoxy resins. On cyanoethylation, the last-mentioned compound yields 1-(2-cyanoethyl)-2-ethyl-4-methylimidazole (poly-ORGANIX's Curimid-CN), which performs the same function as the preceding material with the advantage of an extended pot life.

Another example of· an imidazole of industrial interest, but utilized in a totally different manner, is butoconazolone nitrate—an ingredient of vaginal antifungal preparations. The preparations of the first-mentioned imidazoles and the structure of the last-mentioned compound are shown in Scheme 10.26.

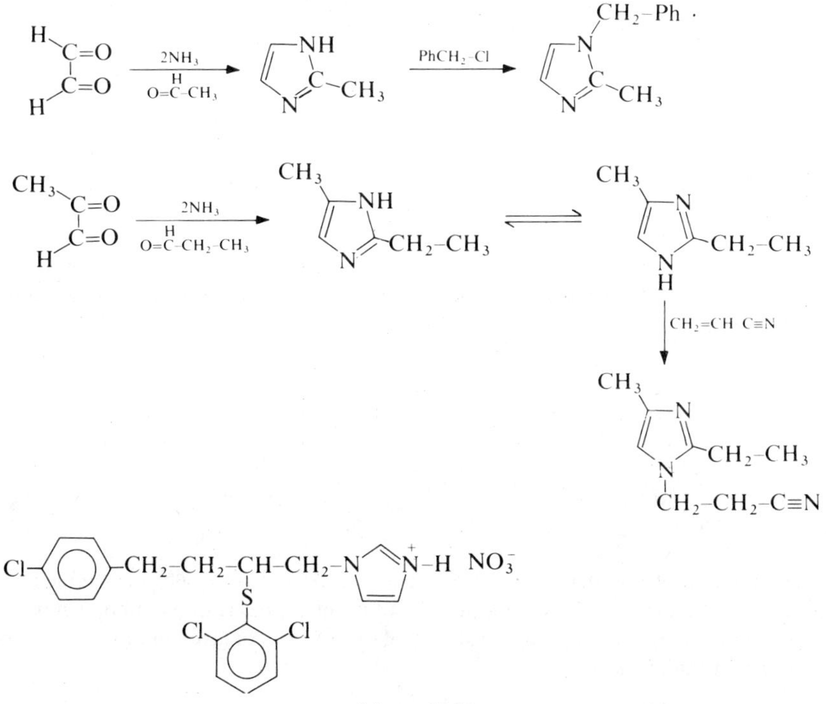

**Scheme 10.26**

One of the currently most prescribed medicinals in the United States is cimetidine, recognized by consumers as SK&F's Tagamet. It is used for treatment of gastric and duodenal ulcers and functions as an antagonist to histamine $H_2$ receptors that control the formation of pepsin—a proteolytic enzyme. The synthesis of cimetidine utilizes dicyandiamide, methylamine, 2-aminoethyl mercaptan assembled from the corresponding amino alcohols, and 4-hydroxymethylimidazole, assembled, most likely, from nitro-methane, acetaldehyde, formaldehyde, and formic acid (Scheme 10.27).

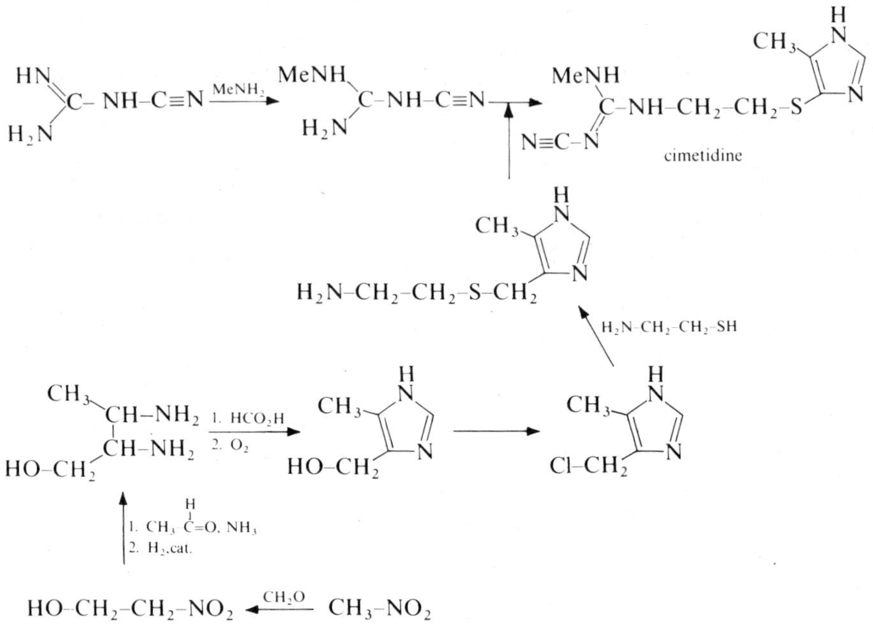

$$HN{=}C(NH_2){-}NH{-}C{\equiv}N \xrightarrow{MeNH_2} MeNH{-}C(NH_2){=}NH{-}C{\equiv}N \longrightarrow MeNH{-}C(N{\equiv}C{-}N){=}NH{-}CH_2{-}CH_2{-}S{-}[\text{4-methylimidazole}] \quad \text{cimetidine}$$

$$H_2N{-}CH_2{-}CH_2{-}S{-}CH_2{-}[\text{4-methylimidazole}]$$

$$\xleftarrow{H_2N{-}CH_2{-}CH_2{-}SH}$$

$$\underset{HO-CH_2}{\overset{CH_3}{CH{-}NH_2}}\,\underset{CH{-}NH_2}{} \xrightarrow[2.\ O_2]{1.\ HCO_2H} \underset{HO-CH_2}{[\text{CH}_3\text{-4-hydroxymethylimidazole}]} \longrightarrow \underset{Cl{-}CH_2}{[\text{CH}_3\text{-4-chloromethylimidazole}]}$$

$$\xleftarrow[2.\ H_2,\text{cat.}]{1.\ CH_3{-}CH{=}O,\ NH_3}$$

$$HO{-}CH_2{-}CH_2{-}NO_2 \xleftarrow{CH_2O} CH_3{-}NO_2$$

**Scheme 10.27**

A complex imidazole derivative is the orally active, broad-spectrum antifungal ketoconazole:

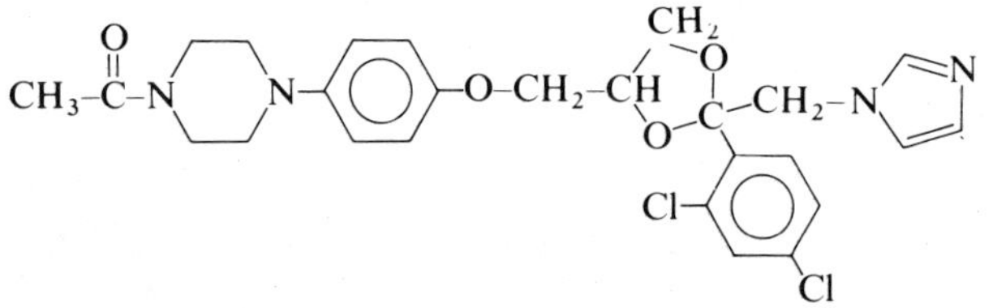

$$CH_3{-}\overset{O}{\overset{\|}{C}}{-}N\langle\text{piperazine}\rangle N{-}\langle\text{C}_6\text{H}_4\rangle{-}O{-}CH_2{-}CH{-}C(CH_2O)(O){-}CH_2{-}N\langle\text{imidazole}\rangle$$

An examination of this structure suggests that the desired molecule is assembled from chloroacetyl chloride, 2,4-dichlorophenol imidazole, glycerol, *p*-aminophenol, and *N*-acetodiethanolamine. The regioselectivity of the reaction of glycerol can be controlled by the use of the tosylated acetonide:

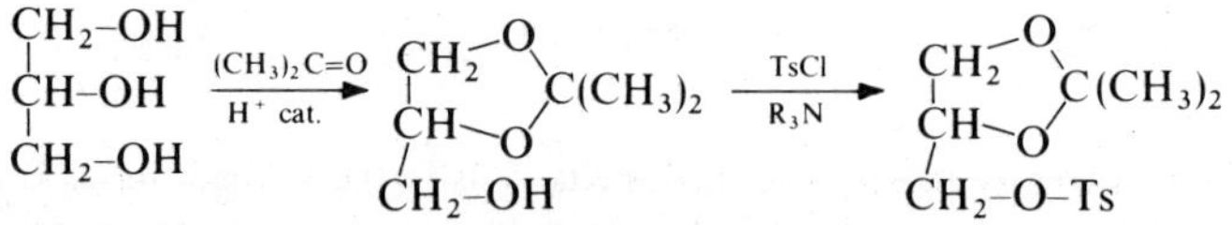

$$\underset{CH_2{-}OH}{\overset{CH_2{-}OH}{CH{-}OH}} \xrightarrow[H^+\ cat.]{(CH_3)_2C{=}O} \underset{CH_2{-}OH}{\overset{CH_2{-}O}{CH{-}O}}C(CH_3)_2 \xrightarrow[R_3N]{TsCl} \underset{CH_2{-}O{-}Ts}{\overset{CH_2{-}O}{CH{-}O}}C(CH_3)_2$$

(where Ts=*p*-CH$_3$-C$_6$H$_4$-SO$_2$-). A nucleophilic substitution reaction of the tosylate with the substituted phenolate and *trans*-ketalization of the acetonide with 2,4-dichlorophenol gives 2,4-dichloro-*β*-*N*-imidazolylacetobenzene, which leads to the desired end product.

An example of an industrially useful benzimidazole derivative is 2-phenyl-benzimidazole-5-sulfonic acid—a water-soluble sunscreening agent obtained by the condensation of sodium 3,4-diaminobenzenesulfonate and benzoic acid:

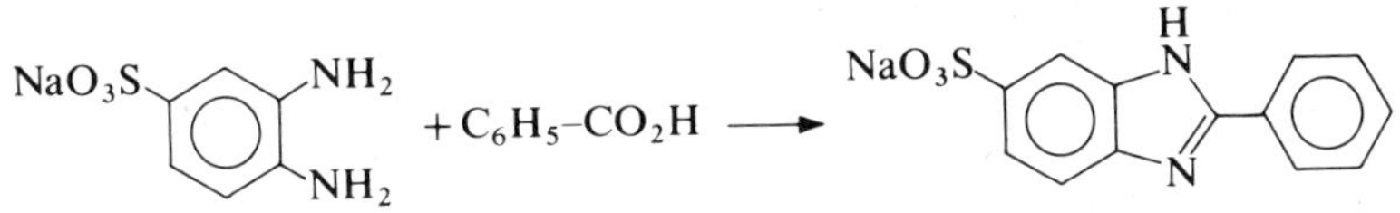

2-Substituted benzimidazoles are intermediates in the preparation of corrosion inhibitors for copper, brass, and bronze. Two examples are 2-(5'-amino-pentyl)benzimidazole and the ethoxylated derivative of 2-mercaptobenzimidazole, shown in Scheme 10.28. We recognize the use of the caprolactam building block in the assembly of the first-mentioned structure and the involvement of a sulfur-containing $C_1$ building block (e.g., COS, $CS_2$, a xanthate) in the assembly of the latter material.

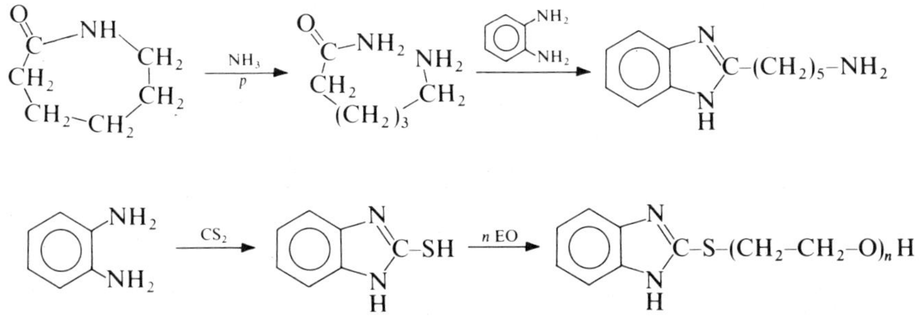

**Scheme 10.28**

A popular nasal decongestant is oxymetazoline hydrochloride ($136.50), retailed as a 0.5% solution under a variety of tradenames. It belongs to the imidazoline family of compounds, together with xylometazoline derived from the appropriate arylacetic acids and ethylenediamine. The arylacetic acids, in turn, are obtained by way of chloromethylation of either o-t-butyl-m-xylenol or butylated m-xylene, as illustrated in Scheme 10.29.

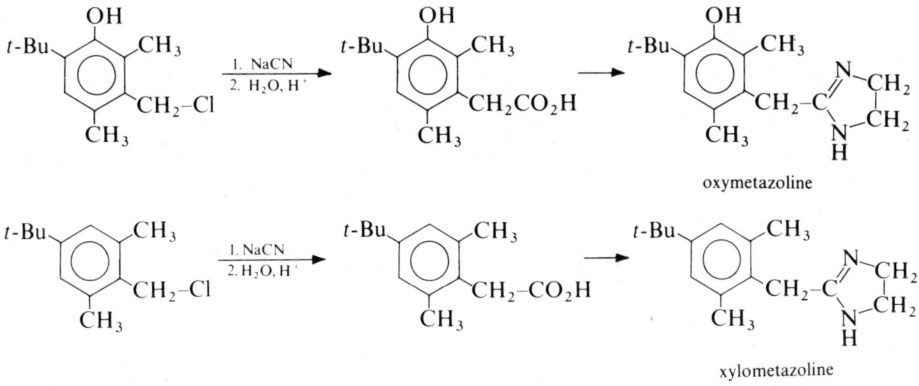

oxymetazoline

xylometazoline

**Scheme 10.29**

Many 2-alkyl-substituted imidazolines derived from fatty acids and ethylenediamine are either end products or intermediates of surface-active agents (*SZM-II*, 64), but they are also used as epoxy curing agents, additives in textile processing, flotation agents in the processing of ores, and so on.

The preparation of the 2-imidazolone system employs a urea building block as illustrated by the assembly of two prominent examples, namely, dimethylol dimethylhydantoin (DMDH)—a preservative used in cosmetic preparations, and the anticonvulsant diphenylhydantoin (sodium salt USP $5.00), commonly known as "dilantin" (J. Dreyfus, *A Remarkable Medicine Has Been Overlooked*, Simon & Schuster, New York, 1981). The former can be obtained from urea and α-chlororisobutyric acid or

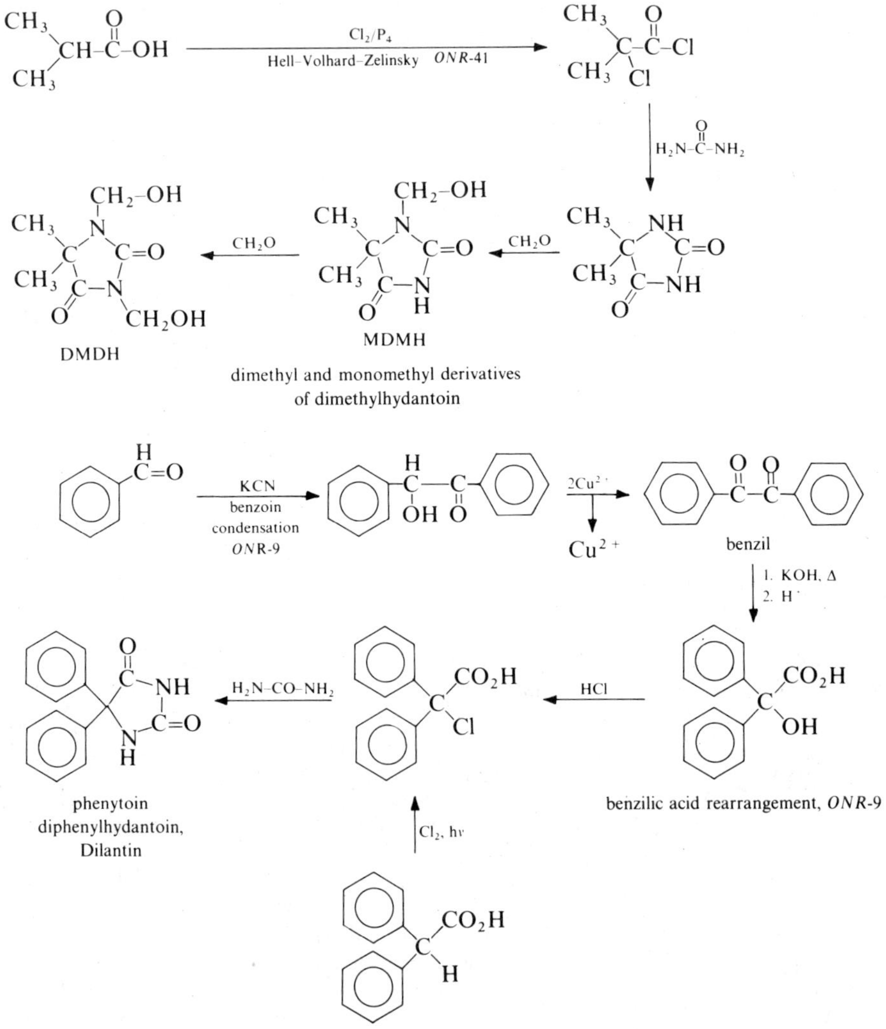

DMDH

MDMH

dimethyl and monomethyl derivatives
of dimethylhydantoin

phenytoin
diphenylhydantoin,
Dilantin

benzilic acid rearrangement, *ONR-9*

benzil

**Scheme 10.30**

the corresponding acid chloride, followed by the introduction of the methylol groups by means of formaldehyde, and the latter is obtained from urea and benzilic acid [$Ph_2C(OH)CO_2H$]. Benzilic acid is a classical fine chemical derived from benzaldehyde (technical grade 73¢) by means of the benzoin condensation (*ONR*-9), followed by a mild oxidation of benzoin to benzil, and another classical process, namely, the benzilic acid rearrangement (*ONR*-9). All these transformations are summarized in Scheme 10.30.

Ethylene urea is actually an imidazolidinone prepared from ethylenediamine and carbon dioxide and utilized as a finishing agent of textiles, in the treatment of leather, and as an additive during manufacture of resins. It is imported from West Germany by BASF ($4.20), while 1,3-dimethyl-2-imidazolidinone (imported by Kennedy & Klim and distributed at $6.85 if purchased in quantities > 5500 lb) is a rather stable, dipolar aprotic solvent. Closely related to the preceding imidazolidinones is 1,3-dimethylol-4,5-dihydroxyethyleneurea (DMDHEU), assembled from glyoxal, urea, and formaldehyde. This product was developed at the USDA Laboratory in New Orleans for treatment of cotton to render it fire-resistant and wrinkleproof. The structures of the preceding compounds are

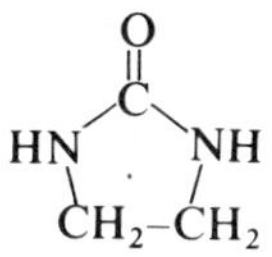

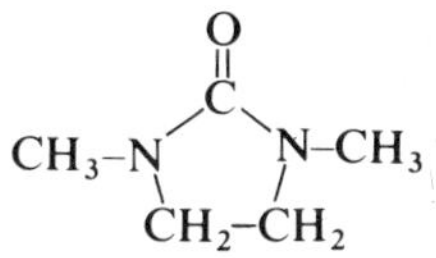

b.p. 106–108°C (17 mmHg)
m.p. 80°C

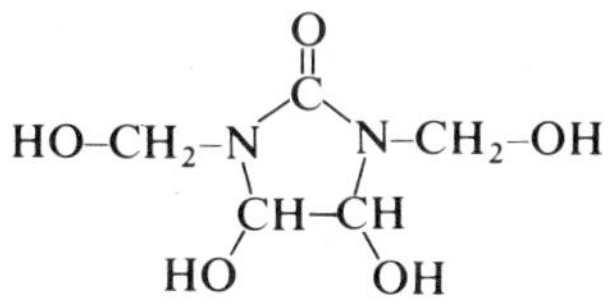

## 10.2.3  Three-Heteroatom-Containing Rings

Hydrazine is the common building block of the 1,3,4-oxadiazole and 1,3,4-thiadiazole rings since it contributes the two neighboring nitrogen moieties.

The 1,3,4-oxadiazole laser dyes are assembled in a similar fashion as shown in 10.2.2 in the case of the 1,3-oxazoles:

| Ar = | Ar′ = | | |
|------|-------|------|------|
| Ph– | Ph– | PPD | $32.65/5 g |
| Ph– | p-Ph–$C_6H_4$– | PBD | $103.95/100 g |
| p-t-Bu–$C_6H_4$– | p-Ph–$C_6H_4$– | Butyl-PBD | 96.50/100 g |
| p-Ph–$C_6H_4$– | p-Ph–$C_6H_4$– | BBD | 146.50/5 g |

The source of sulfur for the thiadiazoles can be either carbon disulfide or thiourea, as shown by the formation of 2,5-dimercapto-1,3,4-thiadiazole (Vanderbilt's Vanchem-DMTD, $3.08)—a corrosion inhibitor, or, for example, 2-amino-5-methyl-1,3,4-thiadiazole, respectively. The preparation of the last-mentioned thiadiazole also requires the participation of acetic acid and the product is the building block of the antibacterial sulfamethizole (p. 172). The syntheses of these thiazoles are summarized in Scheme 10.31. 2,5-Dimercapto-1,3,4-thiazole is readily converted into numerous derivatives by virtue of the reactivity of the two thiol functions.

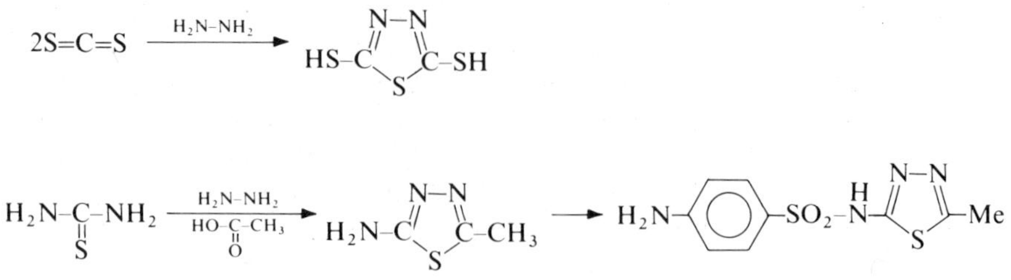

$$2S=C=S \quad \xrightarrow{H_2N-NH_2} \quad HS-C \underset{S}{\overset{N-N}{\diagdown}} C-SH$$

$$H_2N-\underset{\underset{S}{\parallel}}{C}-NH_2 \quad \xrightarrow[HO-\underset{\overset{\parallel}{O}}{C}-CH_3]{H_2N-NH_2} \quad H_2N-C \underset{S}{\overset{N-N}{\diagdown}} C-CH_3 \quad \longrightarrow \quad H_2N-\text{⟨O⟩}-SO_2-\underset{}{\overset{H}{N}}-\underset{S}{\overset{N-N}{\diagdown}}-Me$$

**Scheme 10.31**

The 1,2,3-triazole system is present in benzotriazole and its derivatives. These compounds are widely used as corrosion-inhibiting additives in antifreeze formulations, metal-cleaning and pickling solutions, metal polishes, and electroplating media. The benzotriazoles contain an acidic N–H bond, and aqueous solutions of their alkali metal salts protect copper and brass against corrosion. The benzene-ring-substituted benzo-triazoles are obtained from the corresponding o-phenylenediamines and nitrous acid as shown in the case of benzotriazole (Mobay, $5.90; photo grade $8.90), and tolyltriazol (Mobay, $2.40) (Scheme 10.32). The difference in price between the two benzenetriazoles reflects the cost of the respective o-phenylenediamines (compare Figs. 9.3 and 9.7).

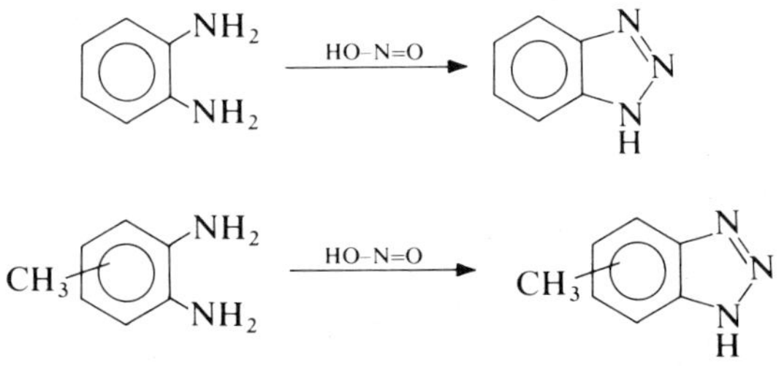

**Scheme 10.32**

The preparation of the analogous 2-substituted benzotriazoles requires the use of an o-phenylenediamine and a nitroso compound, and the latter are obtained most readily by means of the nitrosation of phenols. An example of such a compound is Ciba-Geigy's Tinuvin 327 ($11.72), 2-(3′,5′-di-t-butyl-2′-hydroxyphenyl)-5-chlorobenzotriazole, a stabilizer of light-sensitive polyolefins such as PP, PS, and PE (compare p. 275). The selective absorption of UV light and the degradation of the absorbed energy through intramolecular atomic and molecular motions (i.e., heat) is understandable in view of the structure of Tinuvin 327 (Scheme 10.33).

An example of a 1,2,4-triazole of current relevance is the antiviral ribavirin mentioned in connection with the ribose building block, (Scheme 10.6). The triazole portion of the molecule may be obtained from oxalamide, hydrazine, and formic acid, as shown in Scheme 10.34.

Finally, we can examine another 1,2,4-triazole system in which one the nitrogens is fused to pyridine ring, namely, trazodone hydrochloride—an antidepressant. A likely preparation of this ultrafine chemical starts with the condensation of 2-hydroxypyridine (10.3.1) with semicarbazide; see Scheme 10.35.

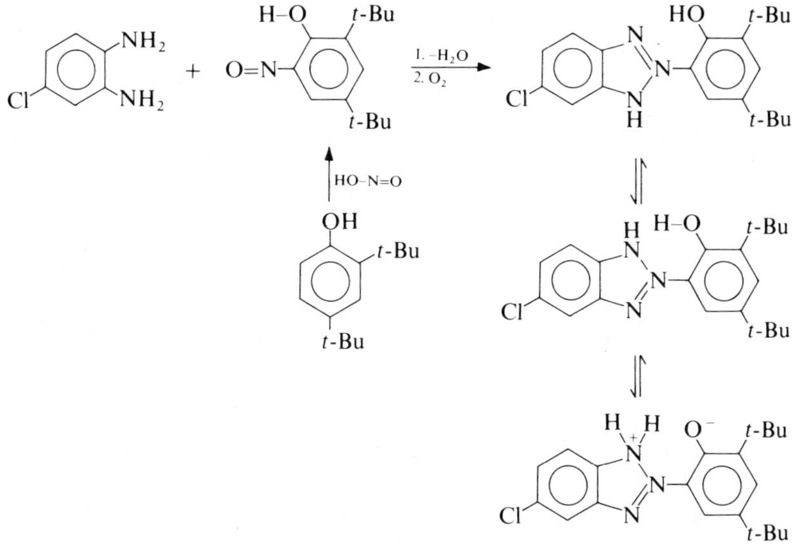

**Scheme 10.33**

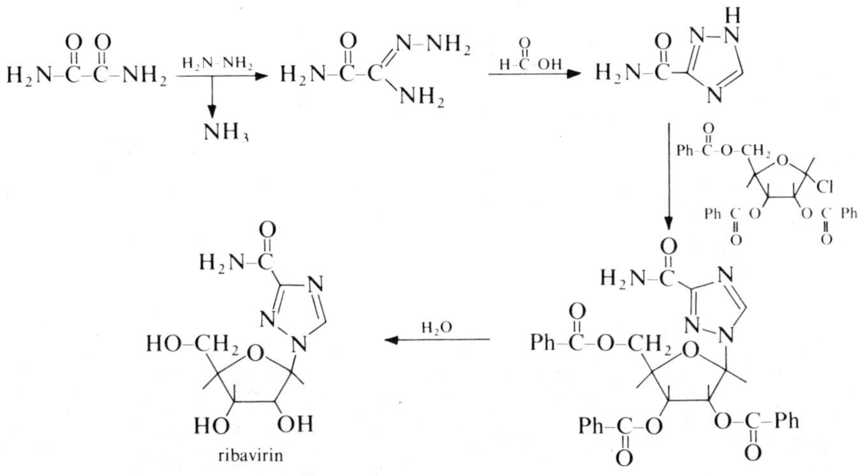

**Scheme 10.34**

Tetrazoles are usually obtained by means of hydrazoic acid, which contributes three of the four nitrogens, and the fourth nitrogen is introduced by the reactant that also carries the desired substituent. Thus, for example, the addition of hydrazoic acid to phenyliso-thiocyanate produces 1-phenyl-5-mercaptotetrazole (PMT), a chemical used in photography, and 5-phenyltetrazole is similarly derived from benzonitrile:

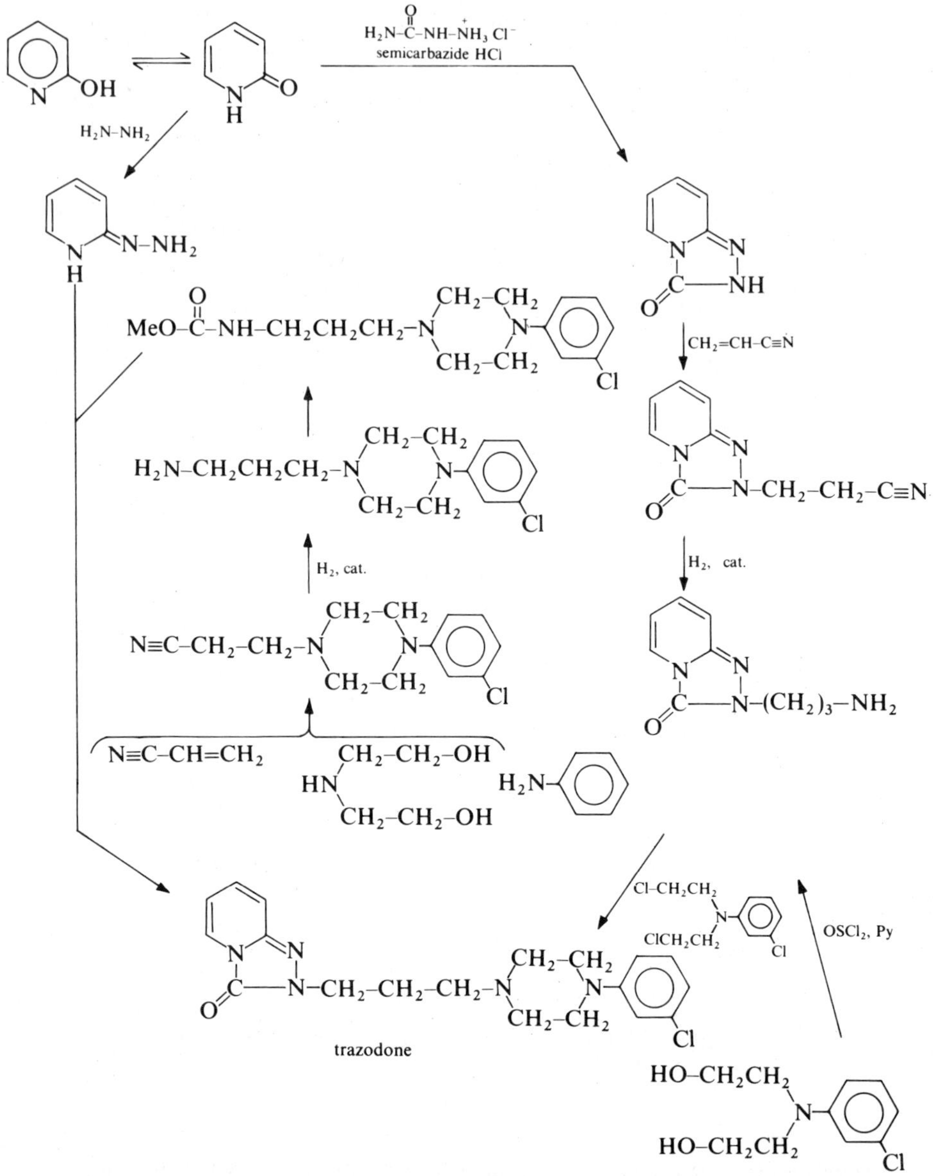

**Scheme 10.35**

The required phenylisothiocyanate is obtained by treatment of ammonium $N$-phenyldithiocarbamate with lead nitrate:

$$\text{Ph–NH–C(=S)–S}^- \text{NH}_4^+ + \text{Pb(NO}_3)_2 \rightarrow \text{Ar–N=C=S} + \text{PbS} + \text{NH}_4^+ \text{NO}_3^-$$

## 10.3  SIX-MEMBERED HETEROCYCLIC SYSTEMS

### 10.3.1  One-Heteroatom-Containing Rings

Pyridine occupies the dominant position among six-membered heterocyclics, but some oxygen-containing rings are also of industrial interest.

Dihydropyran, or more properly, 3,4-dihydro-$2H$-pyran, is a readily accessible third-generation derivative of furfuraldehyde (75¢) (10.2.1). Hydrogenation of furfuraldehyde to tetrahydrofurfuryl alcohol (90¢) and the acid-catalyzed rearrangement of the latter produce the desired fine chemical (Aldrich, $39.20/500 g) used primarily to protect or temporarily block alcohol functions during synthetic operations (Scheme 10.36). However, dihydropyran can also serve as a starting material for the synthesis of L-lysine hydrochloride ($1.83 feed grade; *SZM-II*, 161, 164), although this essential amino acid is obtained mostly by fermentation.

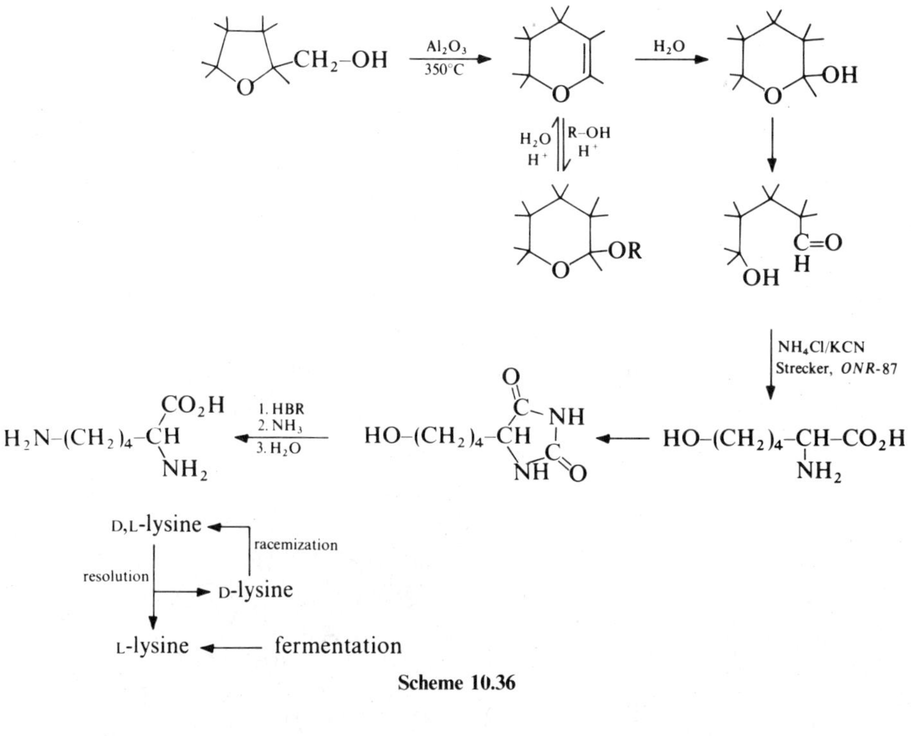

**Scheme 10.36**

The simplest benzopyran compounds of commercial interest are the fragrances coumarin, 1,2-benzopyrone—imported at a level of about 430,000 lb mostly from People's Republic of China and France and valued at $6.50–$6.75, and benzodihydro-pyrone valued, at $12.50 (Scheme 10.37).

Umbelliferone, or 7-hydroxycoumarin, is a common suncreen ingredient of cosmetic products, and a more complex coumarin derivative, namely 4-methyl-7-diethylaminocoumarin, is an optical bleach laundry additive because it emits a blue–white fluorescence at high dilution.

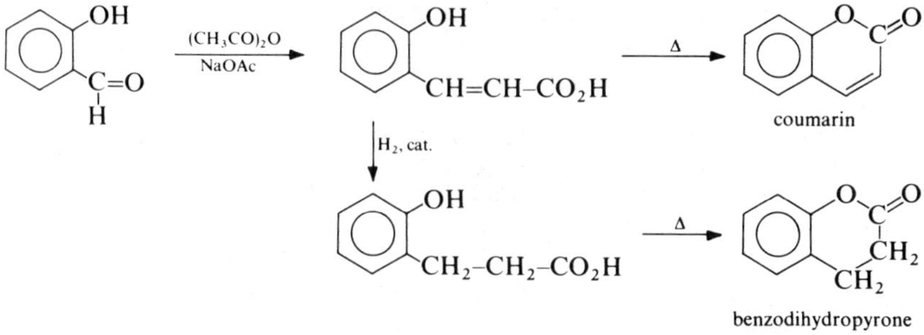

**Scheme 10.37**

The study of dicoumarin or dicoumarol—an ingredient of spoiled hay that was observed to cause internal hemorrhage in ruminants, led to the design of cyclocumarol—an anticoagulant useful in medicine and an intermediate in the preparation of coumadin. The latter, better known as "warfarin," is a rat poison (75¢ in 0.5% concentration). The preparation and structure of these coumarin derivatives is shown in Scheme 10.38.

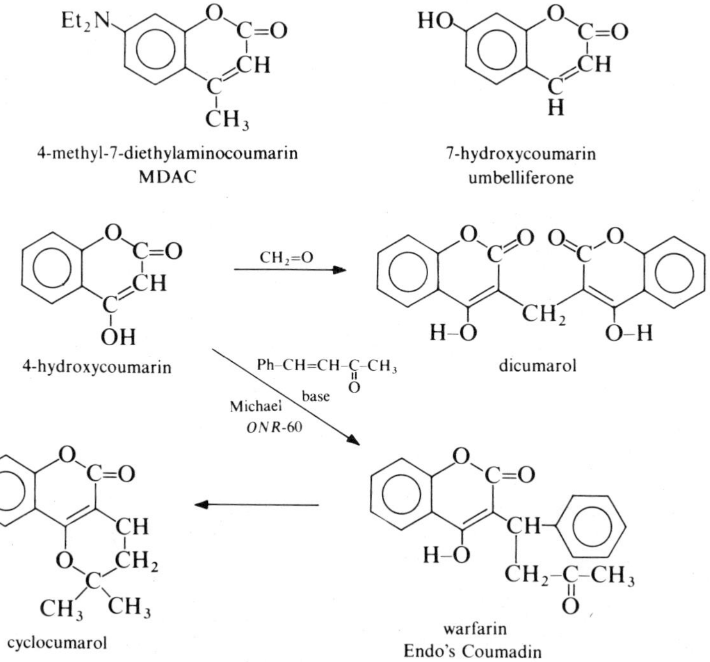

**Scheme 10.38**

Numerous coumarin derivatives serve as laser dyes that differ from each other in the lasing wavelengths (see *Kodak Laser Dyes*, Eastman Kodak Publication JJ-169, 12/87). Closely related to the benzo-2-pyrones are the members of the fluorescein family, in which the central pyran ring is fused to two benzene moieties. The parent structure fluorescein, also known as "Acid Yellow 73" and, in the form of the sodium salt, as "Uranine," is the product of the classical condensation reaction of phthalic anhydride and resorcinol:

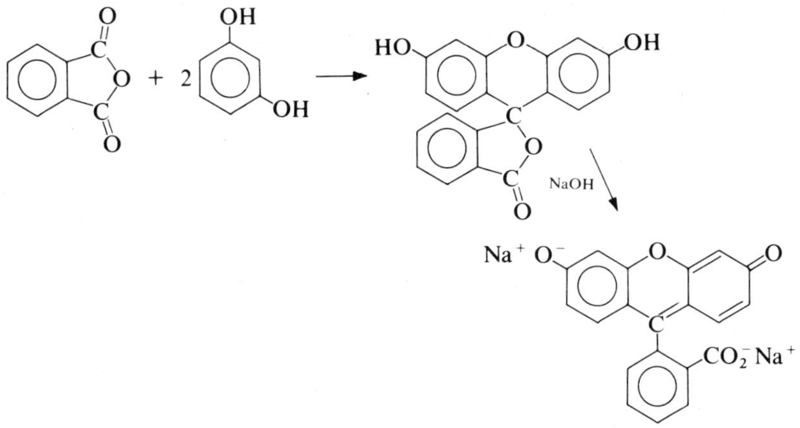

The laser-quality product is priced by Kodak at $19.80 for 500 mg.

The analogous condensation product of phthalic anhydride and *m*-diethylaminophenol is rhodamine B, Basic Violet 10 ($10.95), used as a luminescent pigment dispersed in plastics and a laser dye priced by Kodak at $19.60/g:

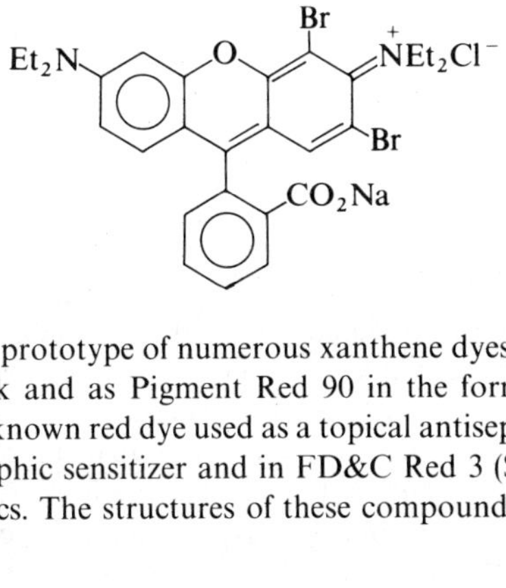

Fluorescein is also the prototype of numerous xanthene dyes such as eosin, Acid Red 87, used as dye in red ink and as Pigment Red 90 in the form of insoluble lead salt; mercurochrome, the well-known red dye used as a topical antiseptic; and erythrosin, Acid Red 51, used as photographic sensitizer and in FD&C Red 3 ($24), a coloring agent of foods, drugs, and cosmetics. The structures of these compounds are

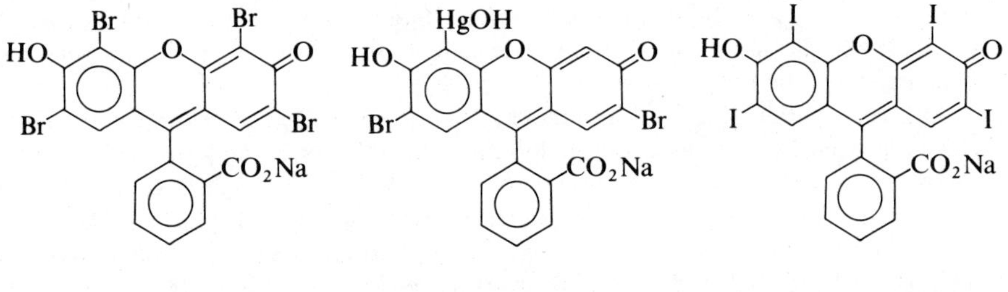

The thioxanthone ring system is readily assembled by means of a Friedel–Crafts reaction of phosgene and the appropriate diaryl sulfide, and the latter can be obtained by starting with an aryl mercaptan derived from a substituted aniline. The sequence of steps is illustrated in Scheme 10.39 with the preparation of 2-chlorothioxanthone—a UV-light-absorbing compound available commercially (at ~$45).

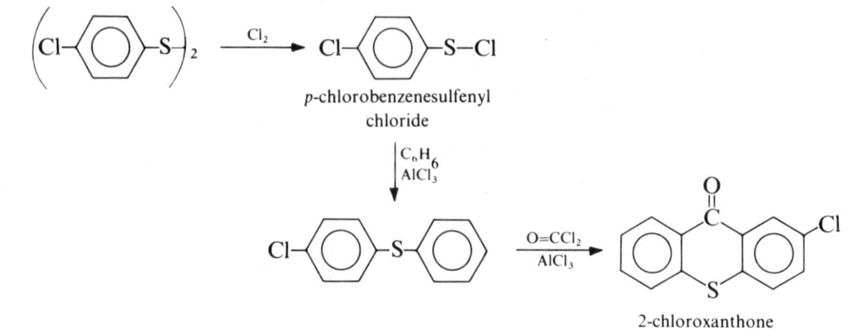

**Scheme 10.39**

The shrinkage of the U.S. steel production and the consequent decrease in coking operations on one hand and the considerable industrial interest in fine chemicals of the pyridine family on the other hand have stimulated the development of synthetic routes to a variety of substituted pyridines. Nevertheless, coal tar is still a source of the parent and methylated pyridines:

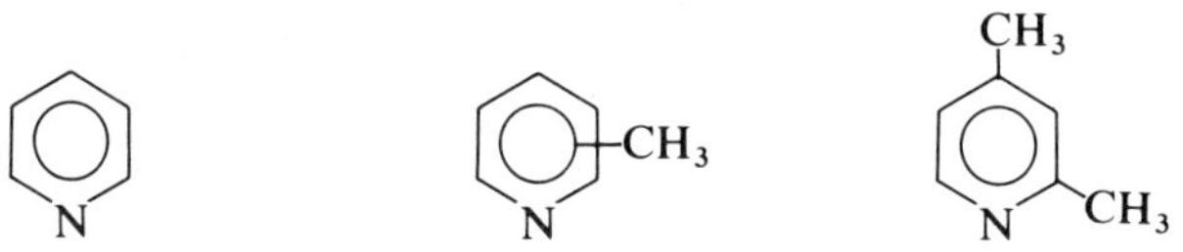

Pyridine (refined, $2.68)     Picolines (refined, $1.28)     2,4-Lutidine ($2.61)

Thermodynamically, the pyridine ring is a very stable aromatic ring system and withstands conventional oxidation conditions better than benzene. The electron-withdrawing nitrogen moiety deactivates the ring to electrophilic substitution reactions, but on application of more vigorous conditions than those required by benzene, the substitution occurs preferentially in the 2 and 4 positions. The behavior of the nitrogen moiety of the pyridine ring resembles somewhat the effect of a nitro group of benzene derivatives. For example, it activates the 2- and 4-methyl substituents toward base-catalyzed, aldol-type reactions and facilitates nucleophilic substitutions also in the 2 and 4 positions. The chemical properties of pyridine are illustrated by the preparation of some of its derivatives described below.

The participation of the nitrogen moiety of pyridine in electron delocalization of the whole ring decreases the basicity of pyridine as compared, for example, to the basicity of the saturated ring of piperidine (hexahydropyridine; $3.15). Nevertheless, pyridine is a traditional HCl-neutralizing ingredient in reactions that utilize acid chlorides. Alkylation with relatively high molecular weight alkyl chlorides, such as cetyl chloride ($n$-$C_{16}H_{33}$-Cl), derived from naturally occurring cetyl alcohol (NF 98.5¢) (*SZM-II*, 79), yields cetylpyridinium chloride. This and similar pyridinium compounds are employed as a germicidal cationic surfactant (active toward gram-negative bacteria), an algaecide, an antistatic agent in hair rinses, and so on.

The unpleasant odor and taste assures pyridine its role as a denaturant in certain ethanol formulations.

Apart from the potential production of pyridine from furfuryl alcohol, the vigorous reaction of ammonia with $C_2$ and $C_3$ building blocks such as acetaldehyde and acrylonitrile, respectively, provides a synthetic route to the six-membered ring system. The most significant product of such a reaction is 2-methyl-5-ethylpyridine because it is obtained in a relatively high ($\sim$70%) yield, and, furthermore, its oxidation product, namely, pyridine-2,5-dicarboxylic acid, is decarboxylated selectively at the 2 position, thus leaving behind the 3-carboxylic or nicotinic acid, or niacin (NF $3.40, feed grade $2.77). Niacin and its derivative nicotinamide (USP $3.18) are common food and feed additives that serve to prevent pellagra and are considered to be members of the numerous vitamin B family (as vitamin $B_3$). Historically speaking, nicotinic acid was isolated in the process of structure determinations of the tobacco alkaloid nicotine, and there exist several medicinal products that incorporate the nicotinic acid building block. One such example is the respiratory stimulant nikethamide, which is the $N,N$-diethylamide of nicotinic acid. The syntheses and structures of the preceding compounds are summarized in Scheme 10.40.

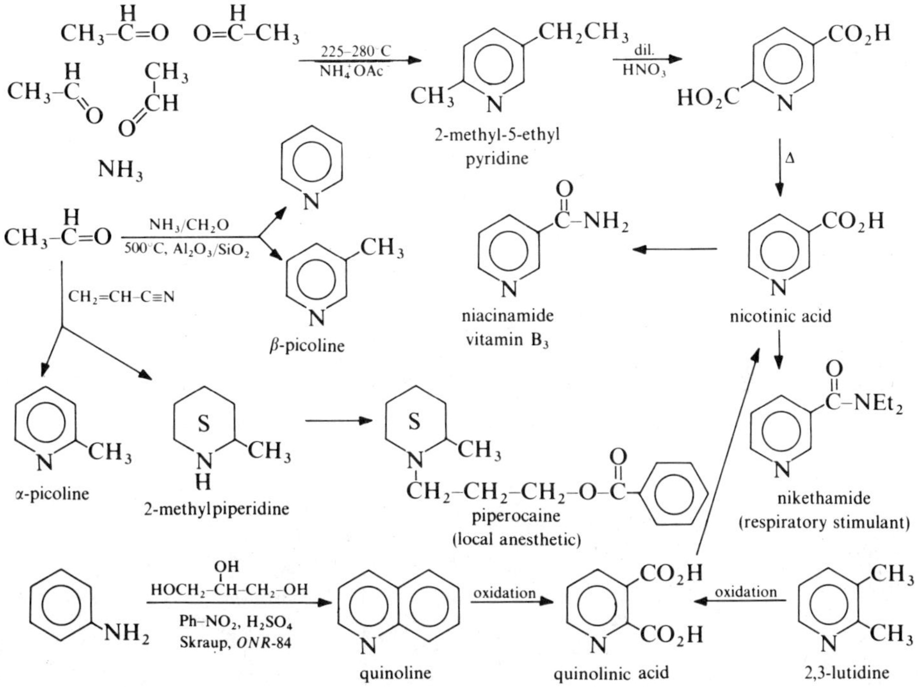

**Scheme 10.40**

The aldol-type condensation of 2- and 4-picolines with formaldehyde gives rise to the corresponding vinylpyridines. 2-Vinylpyridine carries a price tag of $3.47, and 4-vinylpyridine is converted to a somewhat cross-linked poly(4-vinylpyridine), PVP (Reilly's Reillex, $20.90), which functions per se as a basic catalyst and a polymeric matrix capable of coordinating metallic catalysts. The latter use of PVP prevents losses of high-priced metals and also modifies their catalytic behavior. Advantageous uses of Reillex have been demonstrated under the following circumstances:

- Carbonylation of methanol to acetic acid
- Rearrangement of methyl formate to acetic acid
- Air oxidation of cyclohexane to cyclohexanone and cyclohexanol
- Oxidation of tetrahydronaphthalene to 2- or $\beta$-1,2,3,4-tetrahydronaphthalene alcohol and ketone
- The production of 2-EH, initiated by an aldol condensation of two butyraldehyde molecules, and so on

PVP can be also used as a scavenger of trace quantities of undesirable proton concentrations, and as an acid-absorbing, ion-exchange resin.

The use of vinylpyridine comonomer in the production of modified SBR to improve the adhesive properties of the latter and as an additive to PAN to improve its dyeing properties, is mentioned elsewhere.

4-Picoline or $\gamma$-picoline is the source of the tuberculostatic drug known as *isoniazid* ($5.45). Isoniazid is the hydrazide of isonicotinic acid, and the synthetic route may start by subjecting $\gamma$-picoline to either oxidation or ammoxidation.

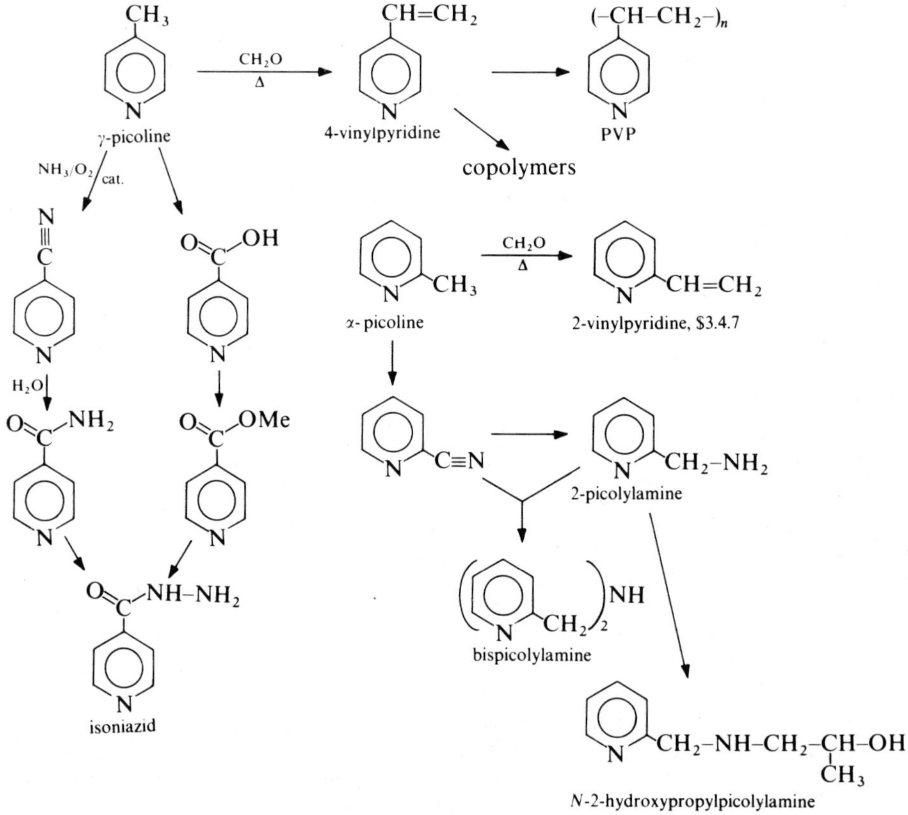

**Scheme 10.41**

The ammoxidation of α-picoline provides several pyridine building blocks such as 2-cyanopyridine (Nepera, \$9.95), 2-picolylamine (Nepera, \$28.50), N-hydroxypropylpicolylamine (Nepera, \$36.30), and bispicolylamine (Nepera, \$34.10). Although not specified as such, the two last-mentioned picolylamine compounds are most likely derivatives of α-picoline. The preceding transformations are summarized in Scheme 10.41.

α-Picoline is the starting material for the production of one of the most popular and effective present household pesticides, namely, chlorpyrifos, introduced in 1965. It is prepared by chlorination and chlorinolysis, followed by nucleophilic substitution of one of the 2-chlorosubstituents of the intermediate 2,3,5,6-tetrachloropyridine and conversion of resulting 2-hydroxypyridine to the O,O-diethylthiophosphate. The preparations of this compound and the related herbicide 3,5,6-trichloro-2-pyridinyloxyacetic acid, or triclopyr, are shown in Scheme 10.42.

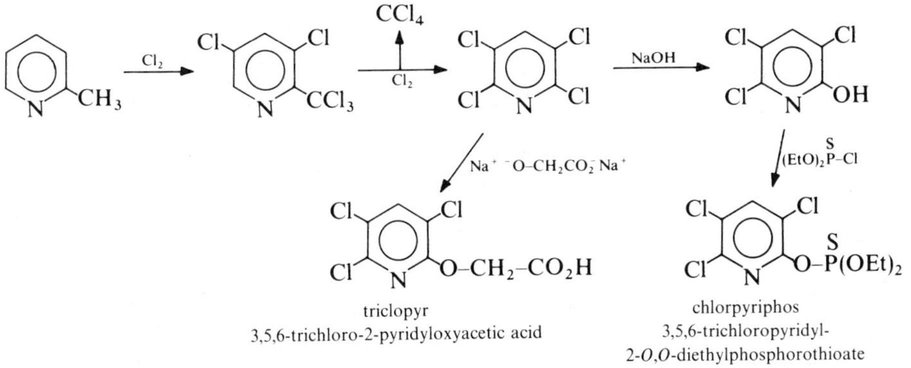

triclopyr
3,5,6-trichloro-2-pyridyloxyacetic acid

chlorpyriphos
3,5,6-trichloropyridyl-
2-O,O-diethylphosphorothioate

**Scheme 10.42**

The reaction of pyridine with a liquid ammonia solution of sodium amide yields the sodium salts of 2- and 4-aminopyridines, and the latter are intermediates in the preparation of various fine chemicals of industrial interest. Thus, 2-aminopyridine is the building block of the antibacterial sulfapyridine and the antihistaminics methapyriline hydrochloride and fumarate (\$27.64 and \$23.47, respectively), as well as pyrilamine. On the other hand, the methylation of 4-aminopyridine yields 4-dimethylaminopyridine (Nepera, \$29.60), which serves as an excellent catalyst in a variety of acylation and alkylation reactions [E. F. V. Scriven, *Chem. Soc. Rev.* **12**, 129–161 (1983)]. These and some other transformation products of pyridine are summarized in Scheme 10.43.

The reaction of pyridine with metallic sodium (also in a medium of liquid ammonia) causes the formation of pyridine radical anion, and in a manner characteristic of free radicals, the latter dimerizes to a dimer dianion that is readily oxidized to 4,4'-bipyridyl. The methylation of 4,4'-bipyridyl produces a powerful herbicide known as *paraquat* that was authorized in 1982 by the U.S. Drug Enforcement Administration for use in the eradication of illegal marijuana plantings. A somewhat similar herbicide diquat is derived from 2,2'-bipyridyl formed by a reaction of pyridine in the presence of nickel catalyst in an atmosphere of hydrogen, followed by alkylation of the bipyridyl with ethylene dibromide. The formation of the preceding pyridine derivatives is summarized in Scheme 10.44. Both bisquaternary pyridinium herbicides are rather quickly inactivated on contact with soil

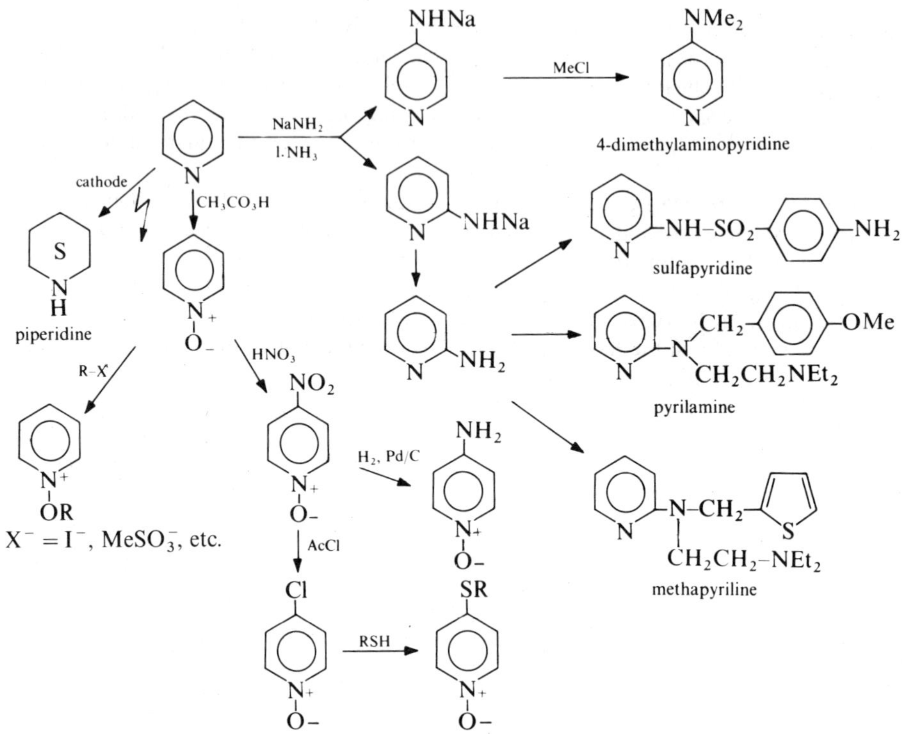

**Scheme 10.43**

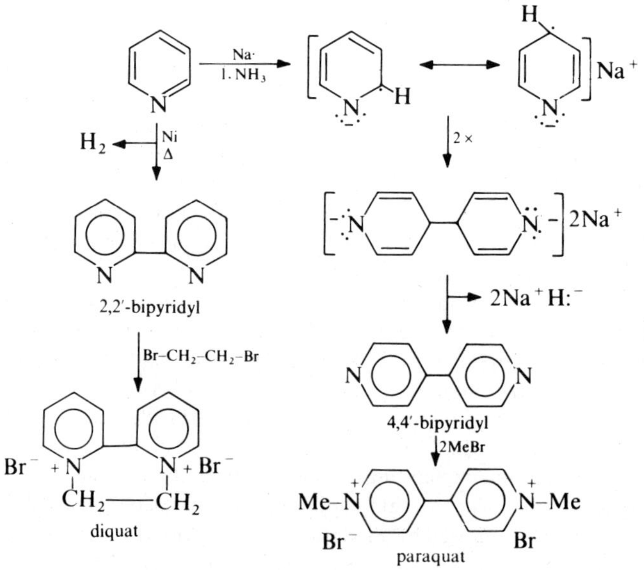

**Scheme 10.44**

and thus are promising energy-saving agents in clearance of weeds from land utilized solely agriculture. Also, it is noteworthy that paraquat stimulates the production of gum rosin by pine trees (*SZM-II*, 111–114).

Pyridoxine hydrochloride, (USP $19.55), usually referred to as *vitamin B_6*, can be synthesized from malononitrile and either an acetoacetic ester or a derivative of acetylacetone (Scheme 10.45).

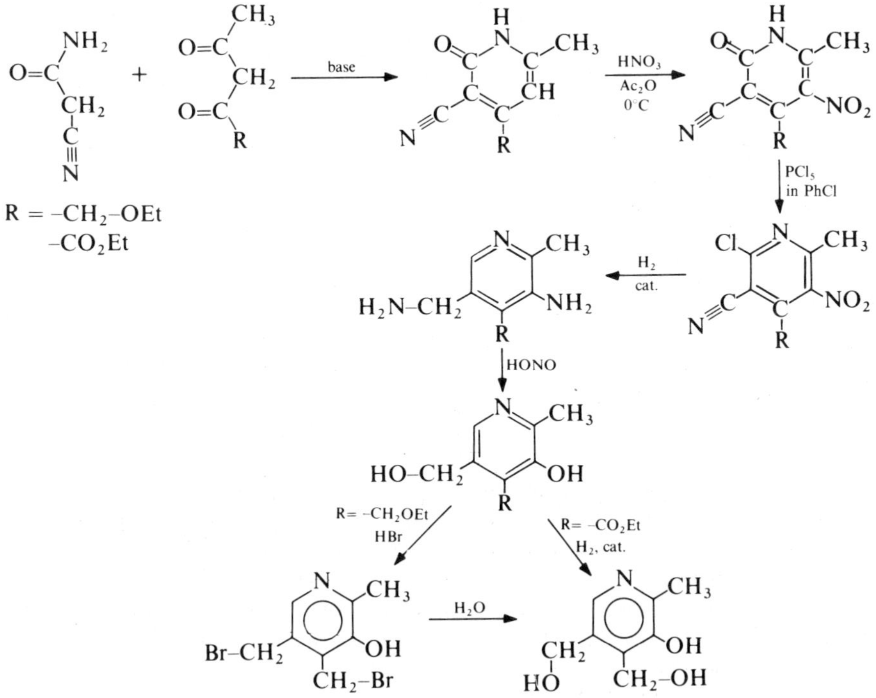

**Scheme 10.45**

Piperidine ($3.15) is a common building block of fine chemicals because it is a reactive secondary amine (as compared to diethylamine, e.g.) because its cyclic structure offers little steric hindrance at the nitrogen site. It is obtained by electrolytic reduction of pyridine ($2.59) or thermal decomposition of 1,5-pentanediamine hydrochloride. The latter may now be obtained from glutaric acid (crystalline $1.20, 50% aqueous solution priced at 37¢/lb of solution)—a recently available by-product of Du Pont's adipic acid process (*CMR* 5/26/86). Some typical derivatives of piperidine are

- The antispasmodic piperilate assembled from benzilic acid.
- The alpha-andregenergic blocker piperoxan that can be synthesized from catechol and the reaction product of piperidine and epichlorohydrin.
- Piperine, used to flavor brandy and synthesized from heliotropin or piperonal and crotonic acid.

The preparation of piperidine and of the above-mentioned derivatives is summarized in Scheme 10.46.

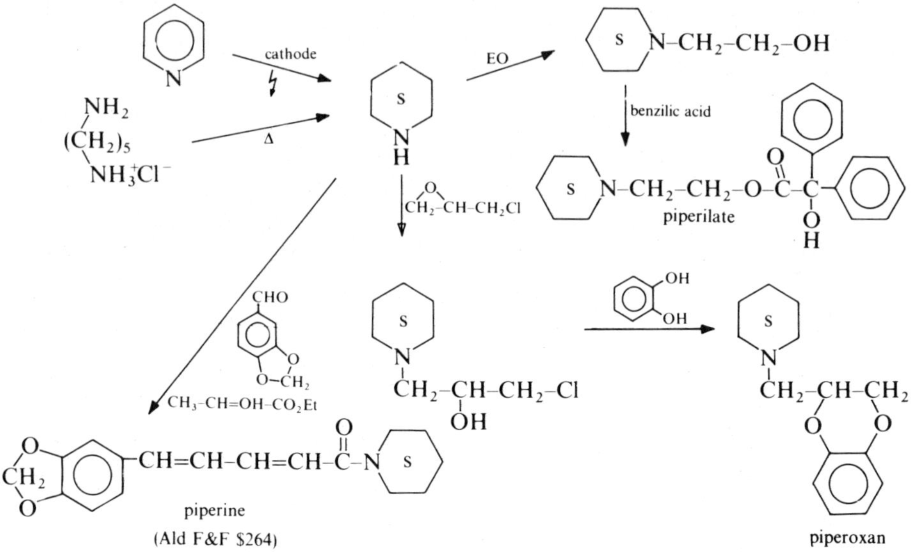

**Scheme 10.46**

As noted above, the preparation of α-picoline by hydrogenation of a mixture of acetaldehyde and acrylonitrile also produces some 2-methylpiperidine, and the latter can be converted to the above-mentioned local anesthetic piperocaine by means of γ-chloropropyl benzoate.

Most quinoline compounds are obtained by means of the Skraup synthesis (*ONR*-84), which consists of a one-step addition–cyclization–oxidation sequence of reactions involving an appropriately substituted aniline and glycerine. The glycerine can be considered to be an *in situ* source of acrolein;

$$HOCH_2CH(OH)CH_2-OH \rightarrow [HO-CH=CH-CH_2-OH \rightarrow O=CH-CH_2-CH_2-OH]$$

$$\rightarrow O=CH-CH=CH_2$$

The oxidation step may involve either the nitro compound related to the aniline reactant or a mild external oxidizing agent such as arsenic pentoxide. The application of the Skraup synthesis is illustrated by means of he preparation of

- The parent quinoline ($1.43).
- The antimalarials pamaquine and primaquine.
- The alkylbenzenesulfonyl amide of 8-aminoquinoline (Henkel's LIX 37), which forms a kerosene-soluble 2:1 complex with cupric ions and hence is employed in the extraction of copper from low-grade ores.
- The intestinal antiseptic iodochlorhydroxyquinoline (USP $15.90).
- 8-Hydroxyquinoline or oxyquinoline ($8.00), which functions as a fungistat in the form of its copper salt ($2.52 as a 10% emulsion) (Scheme 10.47).

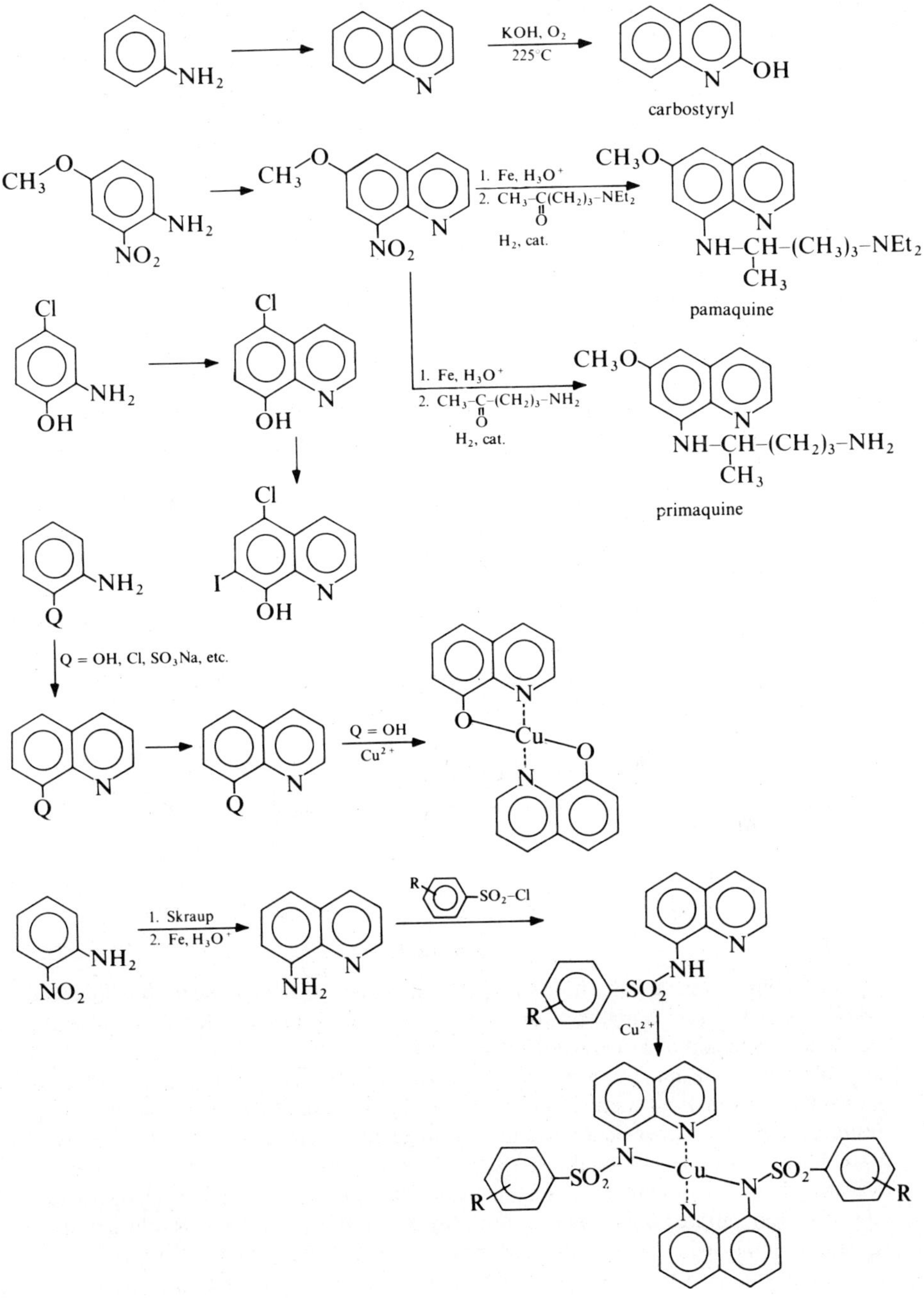
NH2
N
KOH, O2
225°C
OH
carbostyryl
CH3O
NO2
1. Fe, H3O+
2. CH3–C(CH2)3–NEt2
H2, cat.
NH–CH–(CH3)3–NEt2
CH3
pamaquine
Cl
1. Fe, H3O+
2. CH3–C–(CH2)3–NH2
NH–CH–(CH2)3–NH2
primaquine
I
Q
Q = OH, Cl, SO3Na, etc.
Q = OH
Cu2+
Cu
1. Skraup
2. Fe, H3O+
SO2–Cl
R
SO2
NH

Scheme 10.47

The reaction of 4,7-dichloroquinoline with 1-diethylamino-4-aminopentane yields chloroquine, an antimalarial of great initial promise. Unfortunately, certain strains of malaria-carrying mosquitoes acquired resistance to this medicinal, and thus the search for effective antimalarials continues. However, it is known that certain unrelated agents can reverse such acquired immunity, and, in the case of chloroquine, it was found [S. K. Martin et al. *Science* 899–900 (2/20/87)] that the cardiac drug verapamil—a derivative of veratrole

$$Ar-C (CN)(i-Pr) -CH_2-CH_2-CH_2-N(OMe)-CH_2-CH_2-Ar$$

[where $Ar = 3,4-(MeO)_2-C_6H_3-$] reverses the acquired immunity of mosquitoes to chloroquine in, at least some, of the chloroquine-resistant strains.

The relative stability of the fused benzene and pyridine rings in quinoline is demonstrated by the formation of pyridine-2,3-dicarboxylic acid, or quinolinic acid, on oxidation with potassium permanganate. The susceptibility of the nitrogen-containing moiety to nucleophilic substitution is shown above by the formation of 2-hydroxy-quinoline or carbostyryl on fusion of quinoline with KOH at about 225°C in the presence of oxygen.

In a manner similar to the addition of an aniline to acrolein during the Skraup reaction, a substituted aniline can also be condensed with other difunctional aliphatic building blocks. For example, diacetone alcohol (52¢) is condensed with p-phenetidine (see Fig. 9.4) to produce an antidegradent of rubber (Monsanto's Flectol Pastilles, $1.72), as well as an antioxidant used in feeds and dehydrated forage crops, and in rubber (Monsanto's Santoquin, $2.48). The presence of a tertiary carbon moiety in diacetone alcohol prevents the formation of a fully aromatic product in the resulting substituted 1,2-dihydroquinoline derivatives (Scheme 10.48).

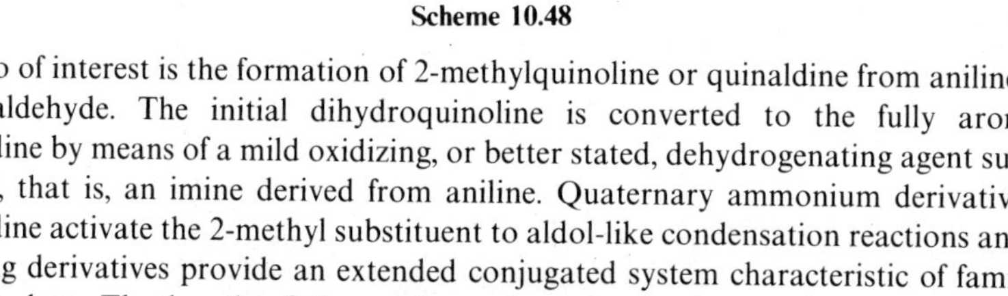

**Scheme 10.48**

Also of interest is the formation of 2-methylquinoline or quinaldine from aniline and crotonaldehyde. The initial dihydroquinoline is converted to the fully aromatic quinaldine by means of a mild oxidizing, or better stated, dehydrogenating agent such as an anil, that is, an imine derived from aniline. Quaternary ammonium derivatives of quinaldine activate the 2-methyl substituent to aldol-like condensation reactions and the resulting derivatives provide an extended conjugated system characteristic of family of cyanine dyes. The length of the conjugated system in the cyanines determines the wavelengths of radiation that is absorbed and a sufficiently extended conjugated structure absorbs even infrared light. The cyanine dyes are the prototypes of photographic sensitizers employed in black-and-white as well as in color photography. The formation of quinaldine and its conversion to a rather simple cyanine dye, namely, 1,1'-diethyl-2,2'-carbocyanine iodide, is shown in Scheme 10.49.

The condensation of a substituted aniline and of the enol ether of formylmalonic ester is involved in the preparation of norfloxacin—a relatively recent antibacterial (Scheme

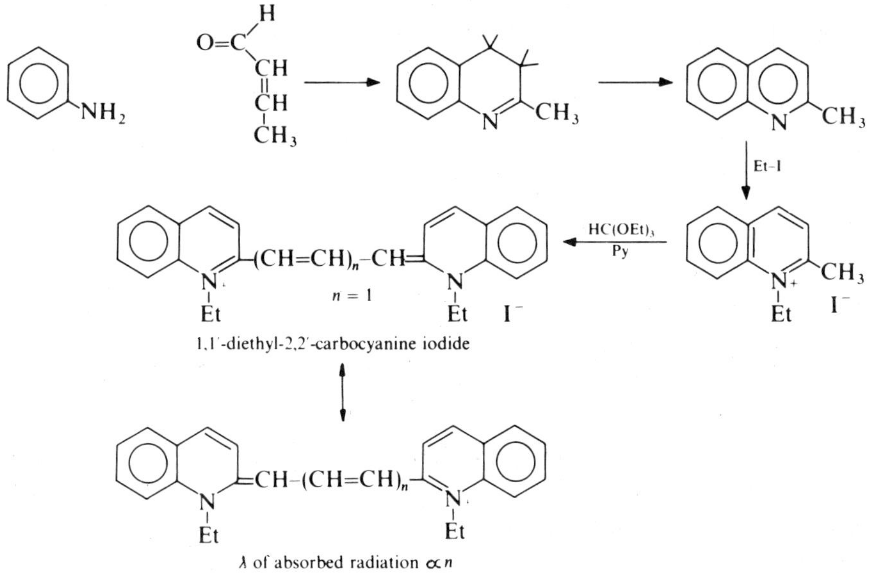

1,1'-diethyl-2,2'-carbocyanine iodide

$\lambda$ of absorbed radiation $\propto n$

**Scheme 10.49**

10.50). As suggested below, the preparation of the 3-chloro-4-fluoroaniline component of norfloxacin may involve the use of a phthalimide group to protect the intermediate aniline while the fluorine substituent is being introduced and, at the same time, to favor a chlorination in the ortho position relative to the fluorine substituent. A convenient

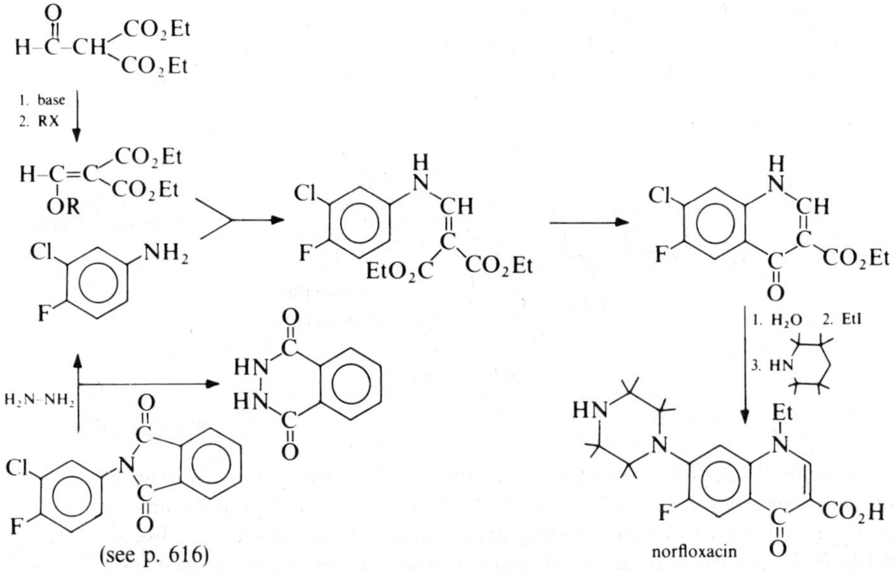

(see p. 616)

norfloxacin

**Scheme 10.50**

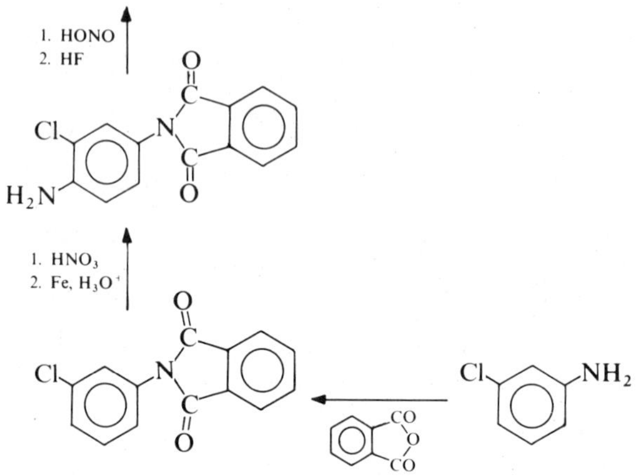

**Scheme 10.50**    (*Continued*)

method to remove the phthalyl group consists of the formation of the rather insoluble phthalhydrazide.

The acridine nucleus is the building block of a group of dyes that carry its name and that are also of medicinal value. For example, acriflavine is an orange material that fluoresces in dilute solution and previously was used as an antitrypanocidal to combat sleeping sickness. It is prepared from the methylene-bridged derivative of *m*-phenylenediamine (see Fig. 9.2), and the product actually consists of a mixture of 3,6-diaminoacridine and 3,6-diamino-10-*N*-methylacridinium chloride, as shown in Scheme 10.51.

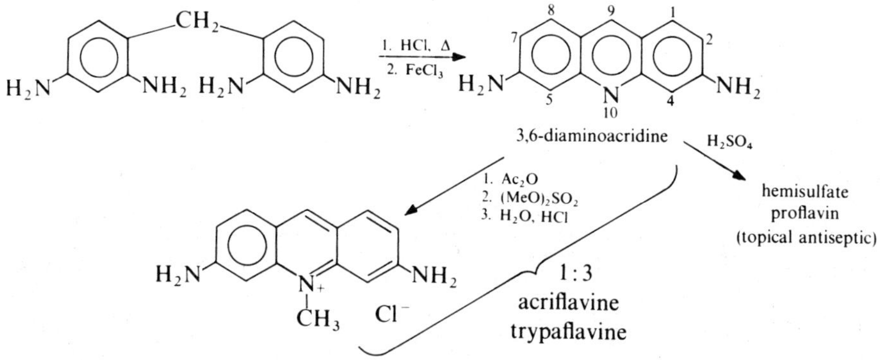

**Scheme 10.51**

Tacrine (1,2,3,4-tetrahydro-9-acridinamine, THA), also commonly referred to as *tetrahydroaminoacridine*, has been known for some time as a respiratory stimulant. It is currently subject to extensive testing in the treatment of Alzheimer's disease (*CMR* 8/10/87). It can be synthesized by starting with an intramolecular acylation of *N*-phenylanthranilic acid (Scheme 10.52).

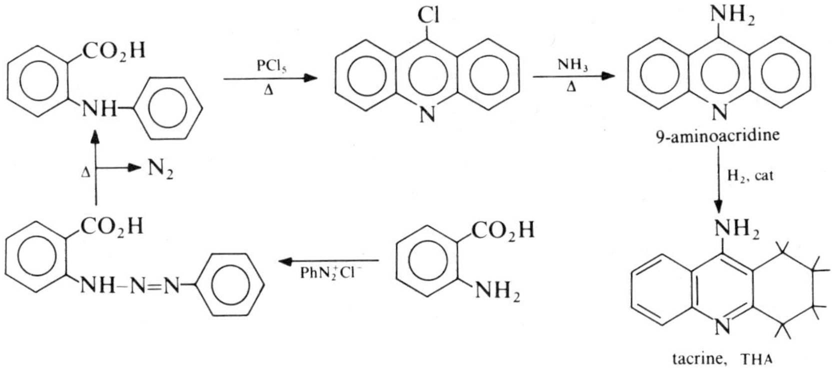

**Scheme 10.52**

## 10.3.2  Two-Heteroatom-Containing Rings

The most common six-membered oxygen-containing ring systems also contain a sulfur or a nitrogen moiety because, unlike oxygen, these hetero atoms are more capable of cyclic electron delocalization. Thus, the formation of a resonance-stabilized ring is the driving force in some Lewis acid-catalyzed cyclization reactions of sulfur such as the facile formation of phenoxathiin from diphenyl ether (Scheme 10.53).

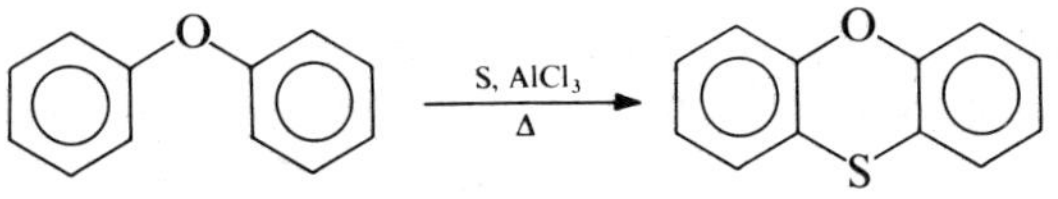

**Scheme 10.53**

The morpholine ring is readily obtained by a low-pressure, catalytic dehydration of diethanolamine (34¢), or diethylene glycolamine, and the latter route is apparently preferred by Air Products (*CMR* 11/26/84). In any case, morpholine (94¢) is a widely used solvent, corrosion inhibitor, a basic component of ionic surfactants, and so on. Its salicylate is used as an analgesic, and the *N*-methyl, ethyl, and propyl derivatives, as well as the dipolar aprotic solvent morpholine *N*-methyl oxide, are available commercially. The formation of these morpholine derivatives is summarized in Scheme 10.54.

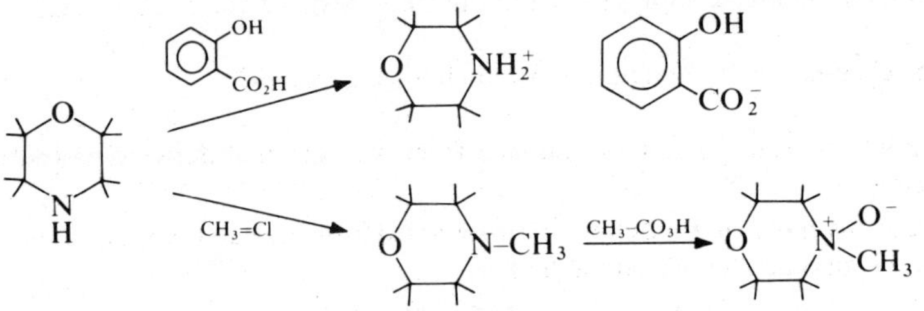

**Scheme 10.54**

An interesting six-membered ring system that contains two sulfur atoms is represented by the defoliant 2,3-dihydro-5,6-dimethyl-1,4-dithiin-1,1,4,4-tetroxide, Uniroyal's Harvade. It facilitates the mechanized cultivation of cotton, potatoes, soybeans, and rice, and the synthesis most likely involves the reaction of acetoin and ethane-1,2-dithiol (or the equivalent mixture of ethylene dibromide, 32¢, and sodium sulfhydrate, NaSH; 44–46% solution or 70–72% flakes, 25¢ (Scheme 10.55).

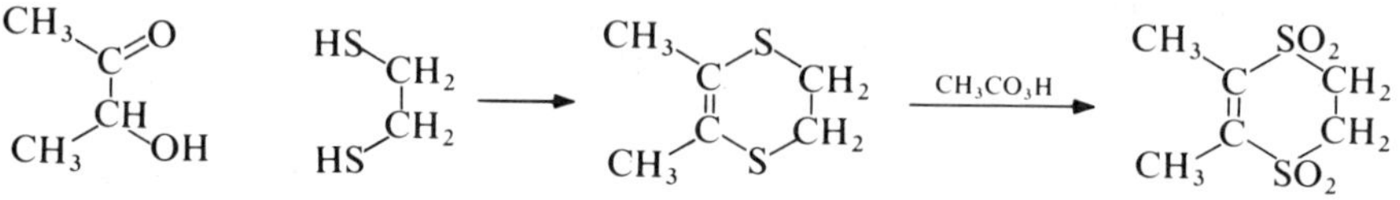

**Scheme 10.55**

The cyclization reaction of sulfur with diphenylamine and its derivatives provides a simple entry to phenothiazines. The parent compound, phenothiazine (PTZ; $2.57, industrial grade, purified ICI $2.87), is an anthelmintic and a building block of a variety of fine chemicals obtained either by substitution reactions of the secondary amine group or by starting with substituted diphenylamines (Scheme 10.56). About 70% of the total PTZ consumption is used as a polymerization inhibitor for acrylic monomers and as an additive in lubricants, rubber, gasoline, waxes, and other products. Also, PTZ is added to animal feed as an anthelmintic and to prevent manure from attracting flies. ICI is the only domestic producer of PTZ and is currently expanding its production by 20% (*CMR* 5/27/88).

The presence of two nitrogen atoms in six-membered rings gives rise to a vast number of important structures that, in theory, are derived from three different parent ring systems:

| | | |
|:---:|:---:|:---:|
| Pyridazine | Pyrimidine | Pyrazine |

Related compounds of these isomeric ring structures are discussed in the listed order.

Three examples of commercially significant pyridazines are the antihyperintensives 3-hydrazinopyridazine-6-carboxamide or hydracarbazine (also employed as a diuretic), 1-hydrazinophthalazine, or hydralazine, and the antimicrobial of the sulfa family sulfamethoxypyridazine. The vicinal arrangement of two nitrogen atoms calls for the use of hydrazine in the preparation of most pyridazines (Scheme 10.57, see p. 622).

Fine chemicals that contain the pyrimidine ring include

- Barbituric acid derivatives obtained from malonic acid derivatives and urea (see below).
- Derivatives of 2-aminopyrimidine (assembled from urea and propargyl alcohol, such as the antibacterial sulfadimethoxine.
- The antihistaminic thonzylamine hydrochloride (known to many consumers as the ingredient of Warner-Lambert's "Anahyst").

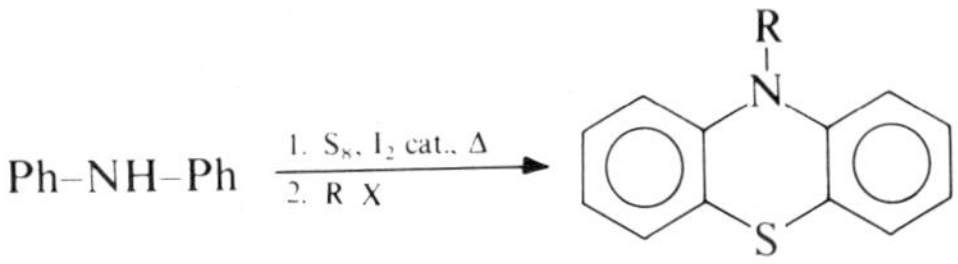

Ph–NH–Ph $\xrightarrow[\text{2. R X}]{\text{1. S}_x,\text{ I}_2\text{ cat., }\Delta}$

| R = | Name | Physiological function |
|---|---|---|
| –CH$_2$–CH$_2$–NEt$_2$ | Diethazine | Anticholinergic antiparkinsonian |
| –CH$_2$–CH$_2$–$\overset{H}{\underset{+}{N}}$Me$_2$ Cl$^-$ | Promethazine | Antihistaminic |
| –CH$_2$–$\underset{CH_3}{CH}$–$\overset{+}{N}$Me$_3$ MeOSO$_3^-$ | Thiazinamium methyl sulfate | Antihistaminic anticholinergic |
| –CH$_2$–$\underset{CH_3}{CH}$–$\overset{+}{N}$Me$_2$ Cl$^-$ $\underset{}{}$ CH$_2$–CH$_2$–OH | N-Hydroxyethyl-promethazine chloride | Antihistaminic |
| –$\underset{O}{\overset{\|}{C}}$–O–CH$_2$–CH$_2$–O–CH$_2$CH$_2$–NMe$_2$ | Dimethoxanate | Antitussive |

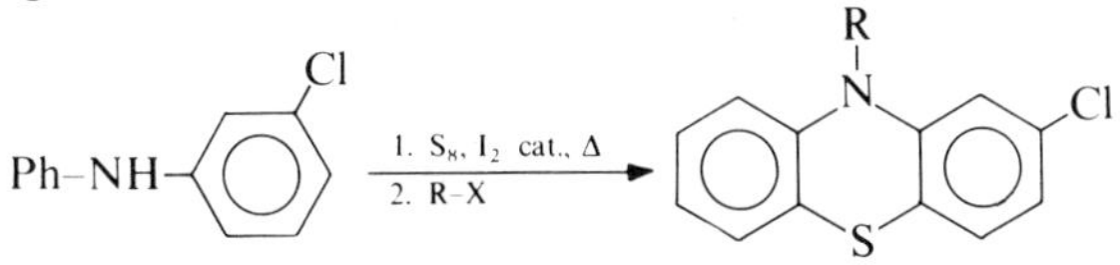

Ph–NH– $\xrightarrow[\text{2. R-X}]{\text{1. S}_x,\text{ I}_2\text{ cat., }\Delta}$

R = –CH$_2$–CH$_2$–CH$_2$–NMe$_2$ chlorpromazine (tranquilizer, antiemetic)

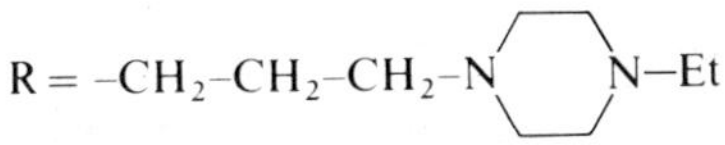

R = –CH$_2$–CH$_2$–CH$_2$–N⟨  ⟩N–Et  thiopropazate (antipsychotic)

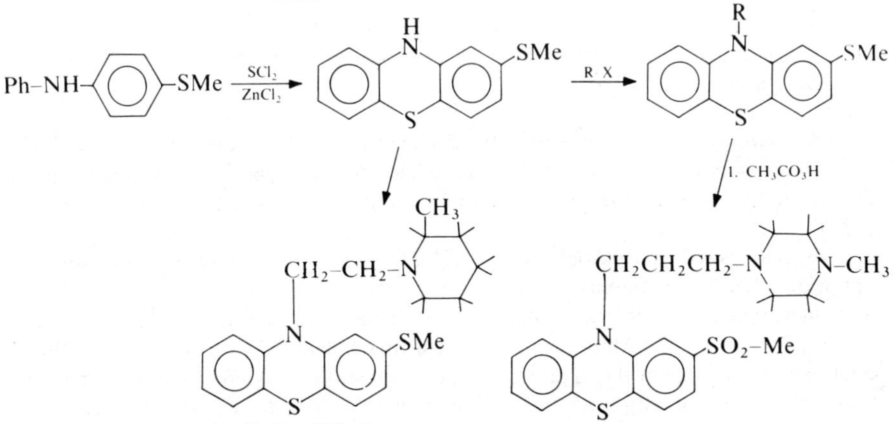

Ph–NH– –SMe $\xrightarrow[\text{ZnCl}_2]{\text{SCl}_2}$ $\xrightarrow{\text{R X}}$

$\xrightarrow{\text{1. CH}_3\text{CO}_3\text{H}}$

thioridazine, HCl salt
(antipsychotic)
0.03MM lb, $152.50
Sandoz's Mellaril

thioproperazine
(antiemetic,
neuroleptic)

**Scheme 10.56**

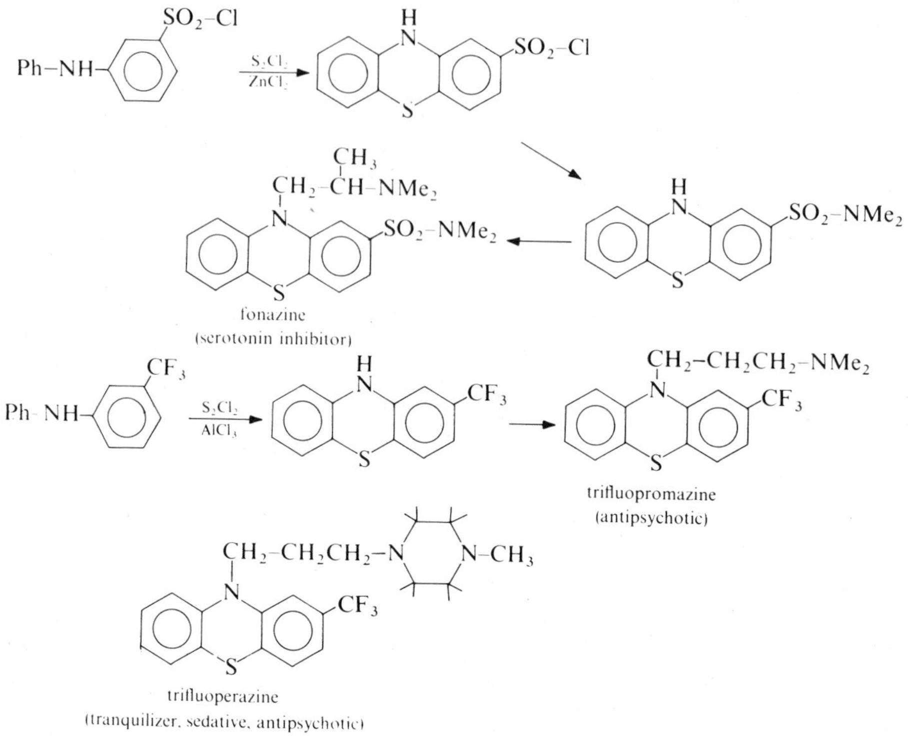

Scheme 10.56  *(Continued)*

- The bacteriocidal surfactant thonzonium bromide.
- Derivatives of malononitrile such as the coccidiostat amprolium that also contains the α-picoline residue.
- The thiophosphate-type insecticide best known to consumers as Geigy's "Diazinon" (Scheme 10.58, see p. 623).

The assembly of a relatively simple pyrimidine building block that employs a malonic ester (such as diethyl malonate (*Ald. F&F* $3.90 in 25-lb quantities) and guanidine is illustrated by the preparation of 2-amino-4,6-dihydroxypyrimidine (Dynamit Nobel, $12.90).

The synthesis of barbituric acid in 1864 by Adolph von Baeyer was of historical importance because it not only led to the discovery of various hypnotics and sedatives but also elucidated the tautomeric equilibria that dominate the chemistry of this heterocyclic system. The family of barbiturates is obtained by condensation of substituted malonic esters with either urea or thiourea. The use of a disubstituted malonic ester interferes with the development of a fully aromatic ring system, but the extended keto–enol tautomerism suffices to induce acidic properties in the condensation product. The syntheses of several barbiturates such as 5,5-diethylbarbituric acid, or barbital (NF $10.25, sodium salt NF $10.45); 5-ethyl-5-phenylbarbituric acid or phenobarbital (USP $8.85, sodium salt $12.25); 5-ethyl-5-(2′-pentyl)barbituric acid, or pentobarbital ($7.00, sodium salt $14.00);

and thiopental sodium, better known as Abbott's "Pentothal Sodium," are summarized in Scheme 10.59. It is noteworthy that thiopental sodium is used as a general anesthetic and that it is sometimes referred to as "truth serum."

A new family of potent weed killers that combine the presence of a pyrimidine or a 1,3,5-triazine ring with a sulfonylurea moiety was discovered recently by G. Levitt of Du Pont. The assembly of sulfometuron methyl, or Du Pont's Oust, involves the addition reaction of *o*-carbomethoxybenzenesulfonyl isocyanate and 2-amino-4,6-dimethyl-pyrimidine obtained from guanidine and acetylacetone. A companion sulfonylurea herbicide, namely, Du Pont's Chlorsulfuron (distributed as Glean), is assembled from *o*-chlorobenzenesulfonyl isocyanate (see Fig. 9.5) and a substituted 1,3,5-triazine (10.3.3). Glean is a very effective pre- and postemergence herbicide for the control of broad-leaf weeds that interfere with the cultivation of wheat and barley. An instructive analysis of investments in time and capital for the research that led to its development and its 1982 introduction on the marketplace was published in *CB* 2/85. The total project took 7 years at a total cost of $226.9 million. Glean sells for $220/lb, but the cost of application in the field is only about $2.50 per acre. The synthesis of the above-mentioned herbicides is summarized in Scheme 10.60. Recently (*CMR* 2/22/88) the USDA authorized Du Pont to conduct a field trial of tomato plants that were genetically engineered to tolerate the above-mentioned sulfonylurea herbicides. The successful results of such experiments pave the way for future high-tech agriculture.

The above-mentioned preparation of the amino-substituted pyrimidines bridges the family of barbituric acid derivatives and that of the fundamental building blocks of nucleic acids, the *nucleotides*. The four principal "bases" that constitute the genetic characteristics-determining nucleic acids are

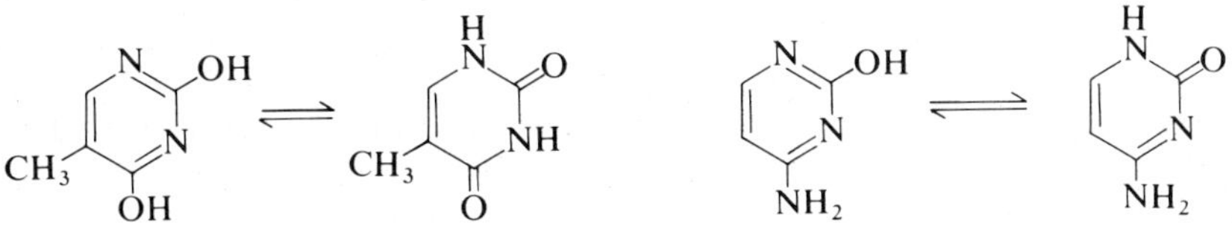

Thymine, T

Cytosine, C

Pyrimidines

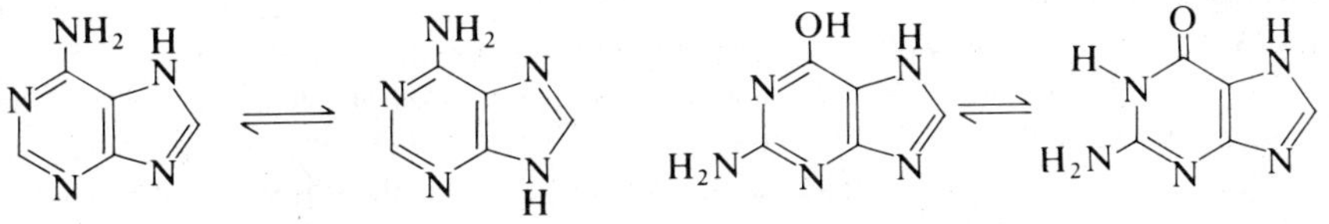

Adenine, A

Guanine, G

Purines

The nucleotides are composed of one of the preceding bases, a five-carbon monosaccharide either ribose or 2-deoxyribose (in RNA and DNA, respectively), and a phosphoric acid residue. The same combination of a base and the five-carbon monosaccharide, but lacking the phosphoric acid residue, is known as a *nucleoside* (Scheme 10.61, see p. 625).

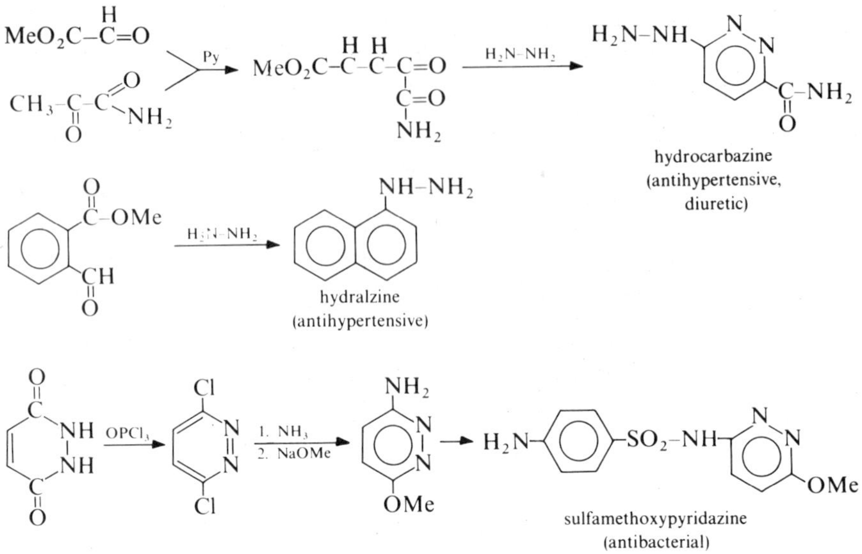

**Scheme 10.57**

The nucleotides are the monomers, which constitute the polymeric nucleic acids, and the sequence of three of such monomers (known as a *codon*) represents one "letter" in the "message" that spells out what capability (or lack thereof) the cell of a living organism possesses to synthesize the proteinaceous enzymes that control the chemistry of life processes. Actually, each codon represents the capability to synthesize one of the 20 amino acids that constitute naturally occurring proteins (which include the enzymes). A discussion of the double-helix arrangement of the polymeric nucleic acids that explains the pairing of bases in neighboring strands of the polymeric nucleic acids and the mechanism of cell reproduction (replication) is beyond the scope of this book. What is of importance to industrial chemistry, however, is the fact that humanity is at the threshold of determining, duplicating, and eventually modifying the sequence of nucleotides of the human genetic material (W. Lepkowski, *CEN* 2/15/88, 5; P. S. Zurer, *CEN* 3/14/88, 22–26)—a task that will consume several years of collective effort and will represent an investment of many billions of dollars (see "OTA Weighs in on a Gene-Mapping Project," *CW* 5/4/88). The enormity of this task can be appreciated if one considers that there are an estimated 3 billion pairs of bases in a given DNA (deoxyribonucleic acid), and that their sequential arrangement differs from one to another in the 23 chromosome pairs inherited by a newborn human from its parents. It is obvious that the identification of the chemical structure of each portion of the genetic material will require confirmation by means of a synthetic duplicate, and this will require large amounts of synthetic bases and auxiliary chemicals (R. M. Baum, "Mechanism of Sequence-Specific DNA Recognition Elucidated," *CEN* 1/4/88, 20–26). Fortunately, there exists already the automated capability to determine the sequence of individual bases by orderly, gradual cleavage of the polymeric nucleic acids and the capacity to reassemble oligonucleotides from their building blocks (D. Rotman, "Sequencing the Entire Human Genome," *Industrial Chemist* 12/87, 18–21). The same is true of the synthesis of the naturally occurring bases as well as their chemical modifications.

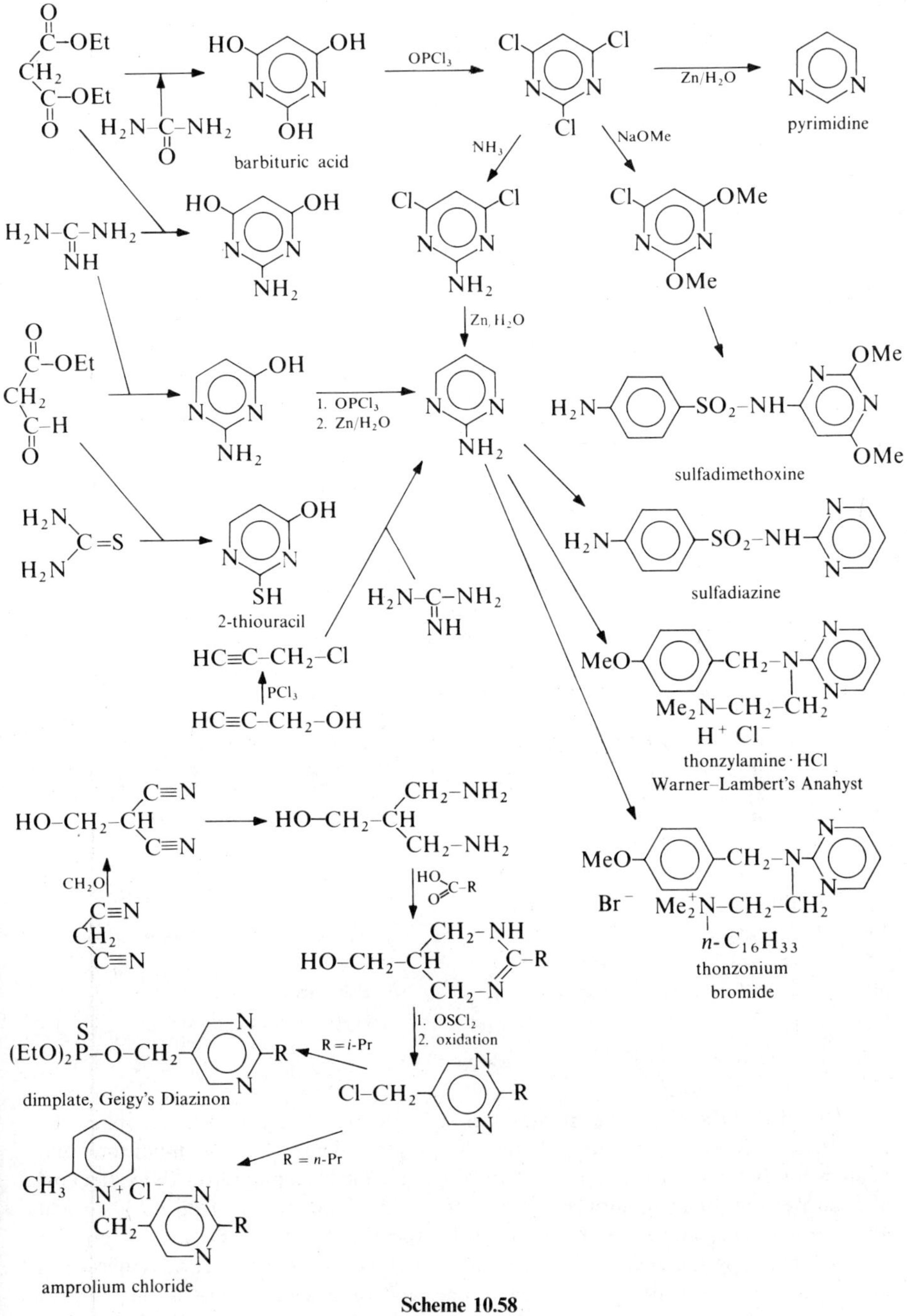
barbituric acid
OPCl3
Zn/H2O
pyrimidine
NH3
NaOMe
Zn, H2O
1. OPCl3
2. Zn/H2O
sulfadimethoxine
sulfadiazine
2-thiouracil
PCl3
thonzylamine · HCl
Warner–Lambert's Anahyst
CH2O
thonzonium
bromide
1. OSCl2
2. oxidation
R = i-Pr
dimplate, Geigy's Diazinon
R = n-Pr
amprolium chloride

**Scheme 10.58**

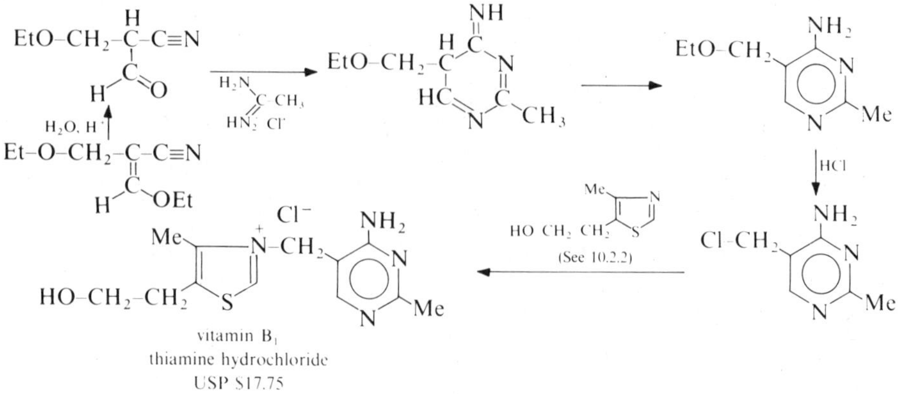

**Scheme 10.58**   (*Continued*)

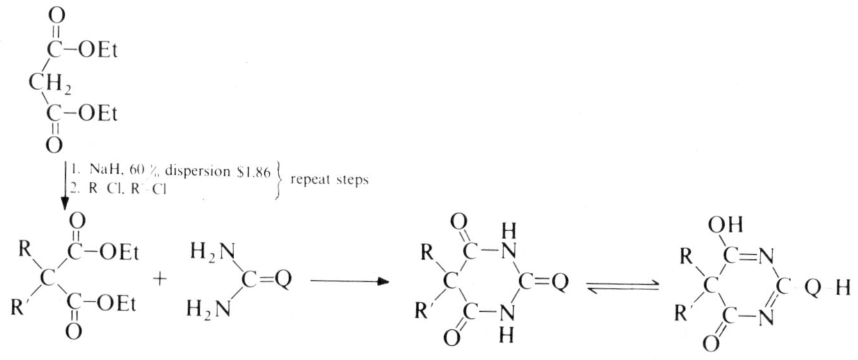

| R | R′ | Q: | Name | Major physiological effect |
|---|---|---|---|---|
| Et– | Et– | O | Barbital | Hypnotic |
| Et– | Ph– | O | Phenobarbital | Sedative, anticonvulsant |
| Et– | 2-pentyl | O | Pentobarbital | Anesthetic, anticonvulsant |
| Et– | iso-Am | O | Amobarbital | Hypnotic |
| allyl- | 2-pentyl | O | Secobarbital | Preanesthetic |
| Et– | iso-Am | S | Thiopental | Hypnotic, anesthetic |

**Scheme 10.59**

The current cost of the pyrimidines of interest is about $84 (supplied by polyORGANIX in 5-kg quantities), and the purines adenine (or 6-aminopurine) and guanine (or 2-amino-6-hydroxypurine) are also available from polyORGANIX at a cost of $136.50/lb in 10-kg quantities: sulfate $102.50/lb in quantities of up to 50 kg and guanine hydrochloride, $31.80/lb, in quantities above 50 kg, respectively.

Some of the pyrimidines and purines and their derivatives have practical applications other than the challenge of the genetic code, and of special interest is their use as antineoplastic agents. Thus, 5-fluorocytosine, or flucytosine, is an antifungal, and 5-fluorouracil as well as 5-fluoro-1-(tetrahydro-2′-furanyl)uracil, or tegafur, are

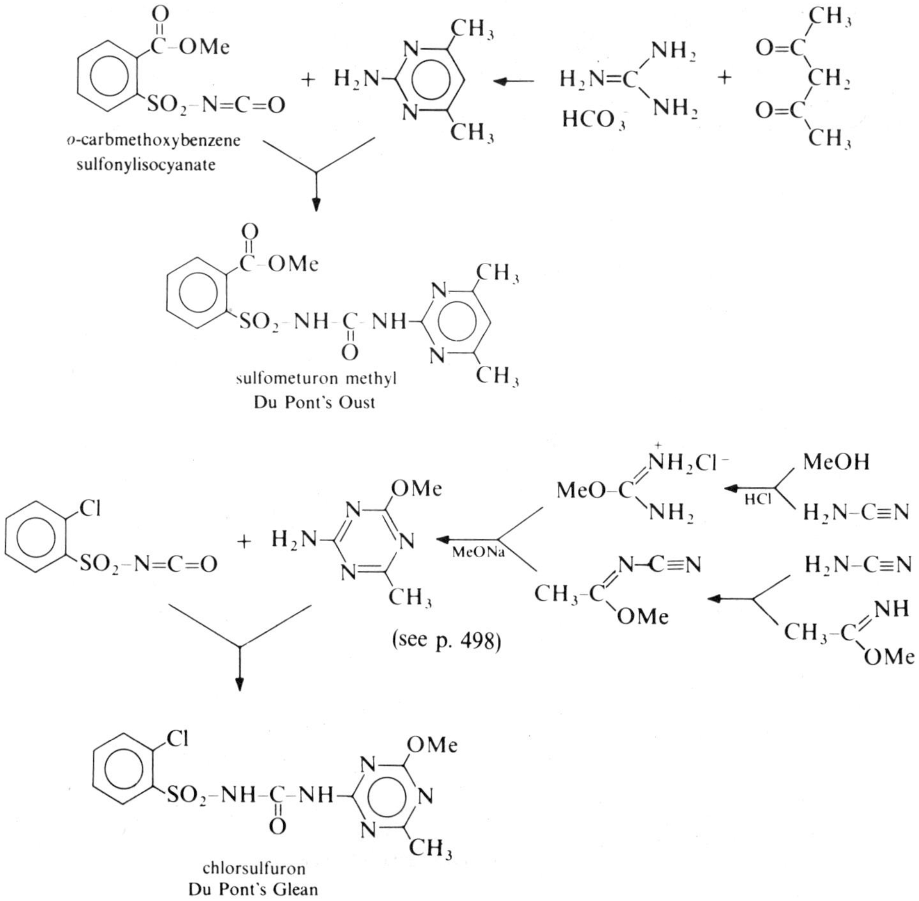

o-carbmethoxybenzene sulfonylisocyanate

sulfometuron methyl
Du Pont's Oust

(see p. 498)

chlorsulfuron
Du Pont's Glean

**Scheme 10.60**

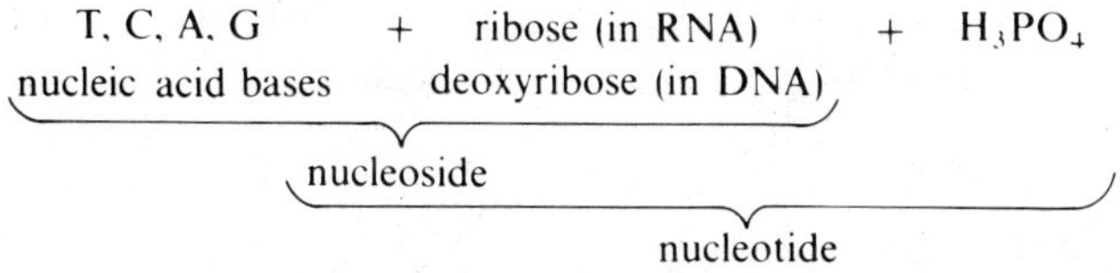

T, C, A, G    +    ribose (in RNA)    +    $H_3PO_4$
nucleic acid bases        deoxyribose (in DNA)

nucleoside

nucleotide

**Scheme 10.61**

antineoplastics. Currently, of great interest is the material known commonly as "AZT"—an acronym for 3'-azido-3'-deoxythymidine, or azidothymidine, claimed to retard the development of AIDS. Two related antiviral materials—2',3'-dideoxycytidine and 2',3'-dideoxyadenosine—are being evaluated as factors that may control the human immunodeficiency virus (HIV). Asahi Glass' 3'-deoxy-3'-fluoroadenosine is reported (*CW*

9/9/87) to be more effective than 5-fluorocil in arresting the propagation of leukemia cells in mice, and Asahi is investigating the fluoroadenosine as an intermediate for the synthesis of drugs used to combat cancer, AIDS, and malaria. Clinical trials are currently in progress (*CW* 10/28/87) designed to test the effectiveness of 2-chlorodeoxyadenosine (2-CdA) as an antilymphoidal cancer drug. Antivirals of interest are acyclovir—a relatively

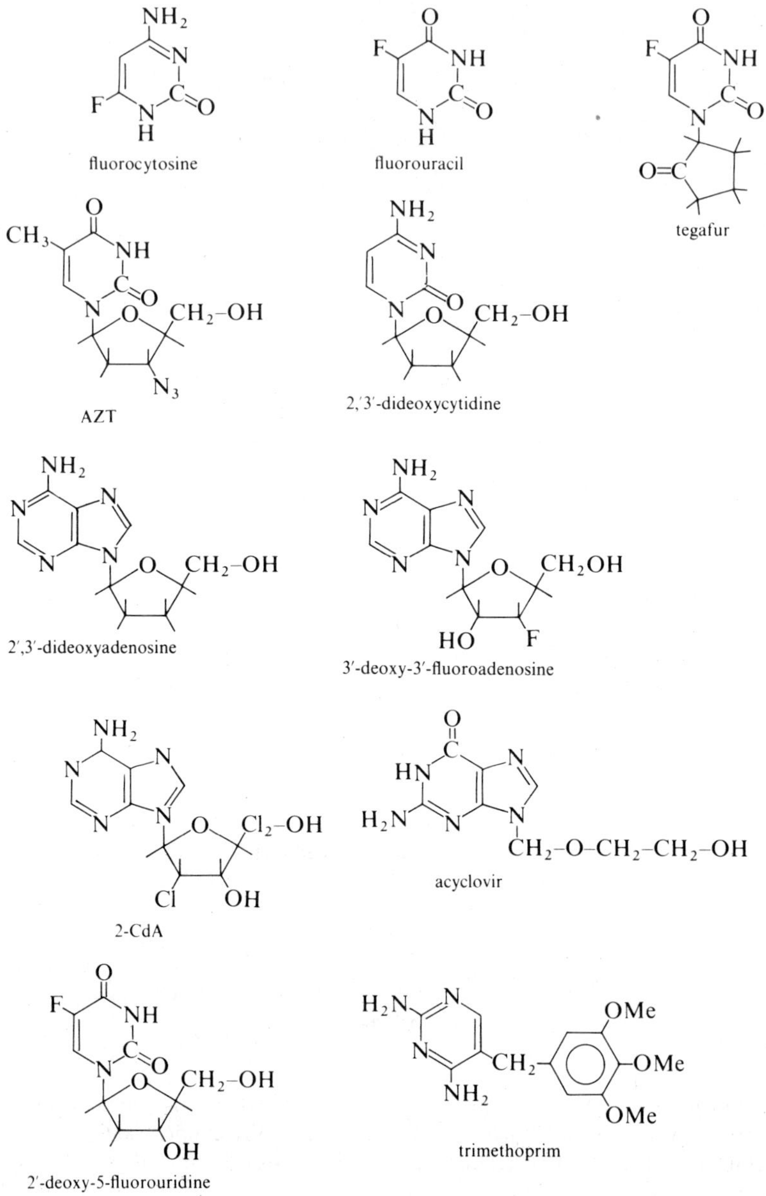

fluorocytosine

fluorouracil

tegafur

AZT

2,3'-dideoxycytidine

2',3'-dideoxyadenosine

3'-deoxy-3'-fluoroadenosine

2-CdA

acyclovir

2'-deoxy-5-fluorouridine

trimethoprim

**Scheme 10.62**

simple derivative of guanine (for control of herpes), the antibacterial trimethoprim related to cytosine, and the antineoplastic 2'-deoxy-5-fluorouridine. The structures of these ultrafine chemicals are shown in Scheme 10.62.

The mother compound pyrazine is rather readily accessible from o-phenylenediamine (see Fig. 9.1) and glyoxal by way of the oxidative degradation of the resulting condensation product, namely, quinoxaline, first to pyrazine-2,3-dicarboxylic acid and then to pyrazinoic acid, which is finally subjected to decarboxylation (Scheme 10.63). On the other hand, the flavor- and aroma-producing alkyl, alkoxyl, and other substituted pyrazines require tedious synthetic procedures and are, consequently, priced at $125–$7000/lb (*CMR* 5/23/88). It is estimated that about 100 pyrazines are used commercially, and the five most common ones cost between $250 and $500 per pound. Some examples of pyrazines in use follow. 2-Acetyl-3-ethylpyrazine and 2-mercaptoethylpyrazine or pyrazineethanethiol, are priced at $1680 and $4300/lb, respectively (*Ald. F&F*), while the simpler structures of 2,3,5,6-tetramethyl-, 2-ethyl-, 2-methyl-3-ethyl-, 2,5-dimethyl-, 2,3-dimethyl-, 2,3-diethyl-, and 2,3-diethyl-5-methylpyrazines are priced at $85, $250, $268, $108, $160, $268, and $309 per pound, respectively (*Ald. F&F*). The alkoxypyrazines tend to be more costly: 2-methoxy-3-methyl, 2-methoxy, and 2-isobutyl-3-methoxypyrazine are priced at $523, $1910, and $2130 per pound, respectively (*Ald. F&F*). The high unit prices of these ultrafine chemicals are compensated by the low concentrations (in ppm) required to provide the desired flavors and fragrances. Flavor- and aroma-producing thiazoles compete with pyrazines on the marketplace, and the most common ones are priced at $90–$150/lb (*CMR* 5/23/88).

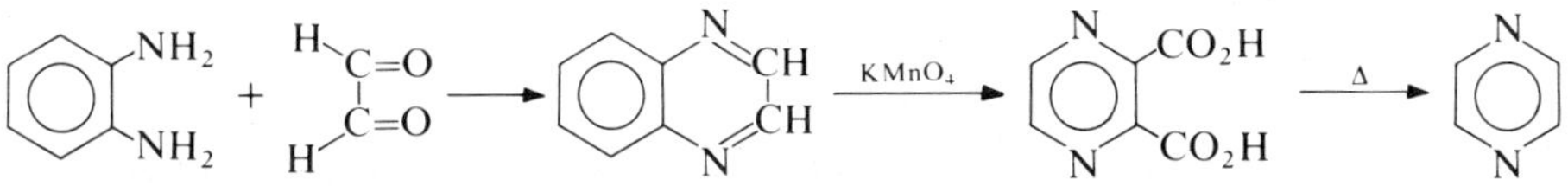

**Scheme 10.63**

As stated above, benzopyrazine, or quinoxaline, is prepared from o-phenylenediamine and glyoxal. Similarly, 2-aminoquinoxaline can be obtained from the same diamine and the amide of glyoxylic acid, which, in turn, can be generated from the amide of glycolic acid or by an oxidative cleavage of the diamide of maleic acid. 2-Aminoquinoxaline is the building block of the antimicrobial sulfaquinoxaline ($13.65), employed in veterinary medicine.

Dibenzopyrazine is commonly known as *phenazine*, and the mother compound is assembled from o-phenylenediamine and an o-benzoquinone generated *in situ* from the corresponding catechol. A phenazine of commercial interest is Janus Green B—a cotton and wool dye also used as an additive during the electrodeposition of copper. It is prepared by the coupling of the diazonium salt of 3-amino-7-diethylamino-5-phenylphenazonium chloride (or N,N-diethylphenosafranine) with N,N-dimethylaniline, while phenosafranine is obtained an oxidation of the mixture of the hydrochlorides of aniline and p-phenylenediamine with potassium dichromate (Scheme 10.64).

Piperazine ($1.80) is an intestinal anthelmintic used to eliminate pinworms or roundworks. It is used as piperazine citrate (36%, $2.25) or piperazine tartrate and is prepared by heating ethylenediamine monohydrochloride. On heating with ethylene dichloride the last-mentioned compound yields triethylenediamine—a strong base, a

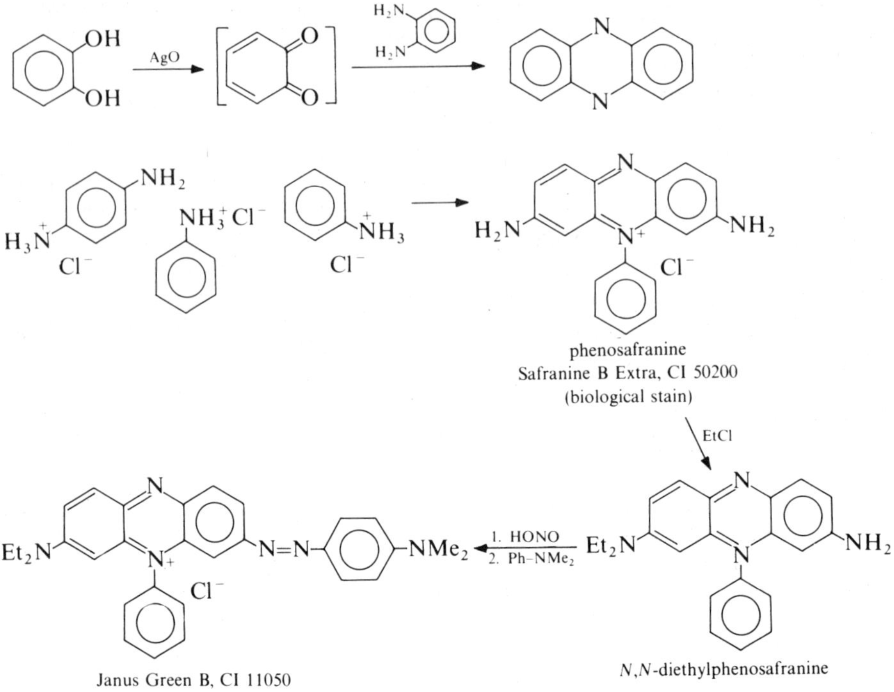

phenosafranine
Safranine B Extra, CI 50200
(biological stain)

Janus Green B, CI 11050

*N,N*-diethylphenosafranine

**Scheme 10.64**

nucleophile, and a PUR catalyst commonly known as "Dabco" [the acronym for 1,4-diazobicyclo(2.2.2)octane] (Scheme 10.65). An example of the use of the piperidine ring as a building block of a fine chemical is the tranquilizer, antidepressant, and hypotensive drug trazodone distributed by Mead Johnson as Desyrel (10.2.3).

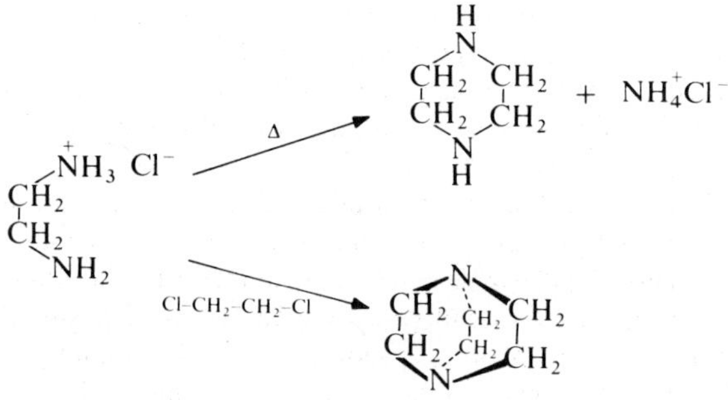

**Scheme 10.65**

### 10.3.3   Three- or More-Heteroatom-Containing Rings

The most common and industrially most important heterocyclic system of this category is the symmetrical triazine ring. It provides the optimum possibility for a circular electron delocalization and is readily assembled by trimerization of cyanogen chloride to cyanuric chloride ($1.25), or trimerization of cyanamide or urea to melamine, and trimerization of isocyanates to isocyanurates. The two first-mentioned derivatives of 1,3,5-*triazine* are crucial to the formation of additional triazines, and cyanuric chloride is of special significance because of the ease with which the chlorine substituents can be replaced by other structural moieties. This capability is exploited in the attachment of dyes to substrates subjected to the dyeing operation by utilizing the partially substituted cyanuric chloride as a chemical bridge. For example, ICI Americas provides a number of Procion dyes, shown below, that employ this strategy to create a strongly bonded, that is, "fast," dyed product (Scheme 10.66). An overview of the formation and typical uses of 1,3,5-triazines is presented in Fig. 10.1.

It is noteworthy that while cyanuric and isocyanuric acids are tautomers, and while there exist derivatives of both tautomers, the derivatives of the former are usually obtained from its acid chloride, cyanuric chloride.

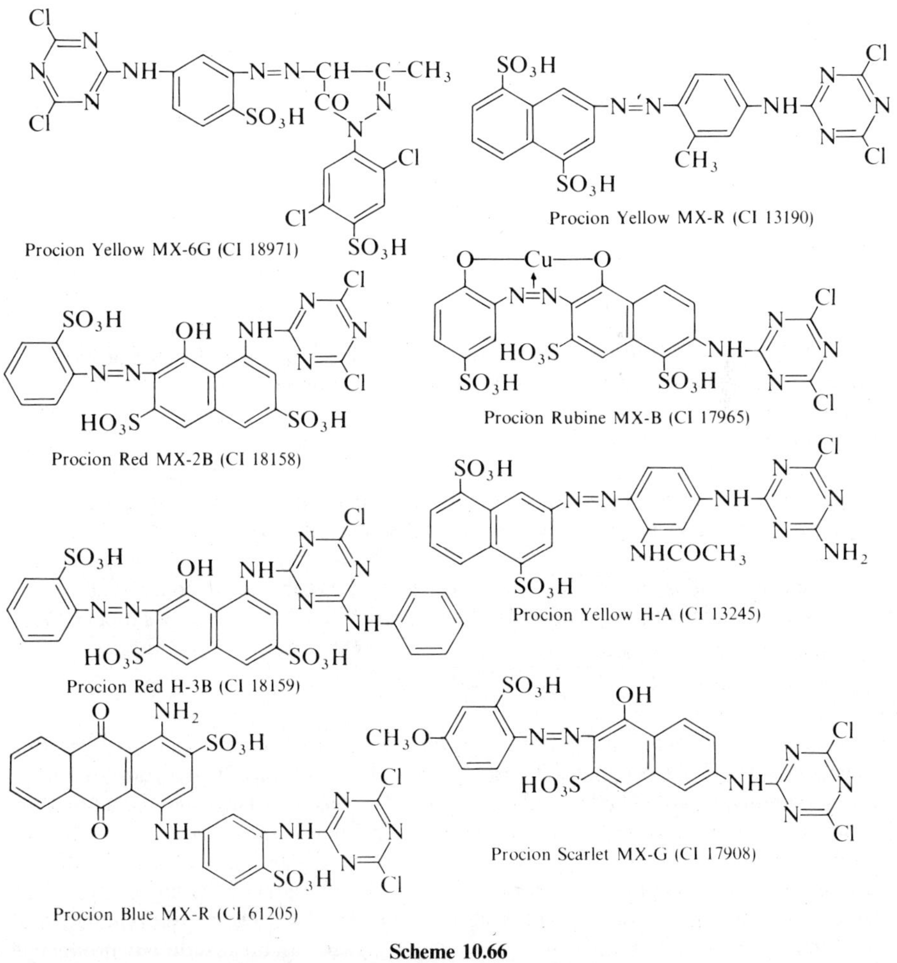

Procion Yellow MX-6G (CI 18971)

Procion Yellow MX-R (CI 13190)

Procion Red MX-2B (CI 18158)

Procion Rubine MX-B (CI 17965)

Procion Red H-3B (CI 18159)

Procion Yellow H-A (CI 13245)

Procion Blue MX-R (CI 61205)

Procion Scarlet MX-G (CI 17908)

**Scheme 10.66**

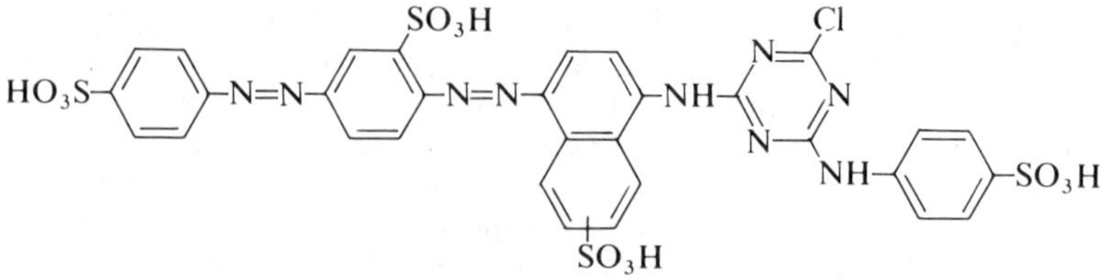

Procion Brown H-2G (CI 26440)

Procion Blue H-B (CI 61211)   SO₃H (*m–p*  mixture)

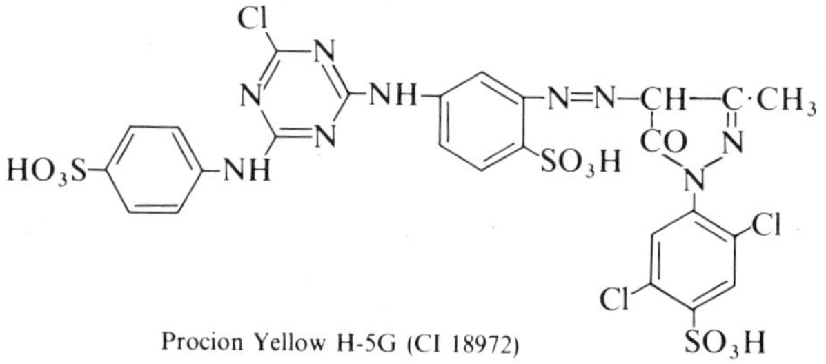

Procion Yellow H-5G (CI 18972)

**Scheme 10.66**   (*Continued*)

Melamine (110MM lb, 46¢) is, of course, a significant industrial building block of melamine–formaldehyde (MF) resins (55¢), used as molding compounds (46.5¢) 20% of which are employed for molding of a variety of consumer products, while 27% are used in the fabrication of laminates (27%). The rest are employed in coatings (22%), paper treatments (14%), adhesive formulations (6%), and textile treatments (3%).

The same four nitrogen-containing fused-ring system characteristic of the purines and barbiturates is found in three closely related alkaloids (*SZM-II*, 166):

- Caffeine, USP synthetic, or the by-product of decaffeinated coffee (and, more recently, also decaffeinated tea), $6.90 imported, $7.25 domestic, as of January 1989;
- Theophylline, USP $8.20
- Theobromine, $63.50

Caffeine is a stimulant of the central nervous system and a diuretic and it is present in some soft drinks sold commercially. Both theophilline and theobromine are diuretics, cardiac stimulants, vasodilators, and smooth-muscle relaxants. Theobromine is found in cacao beans (≤3%), cola nuts, and tea, and theophilline is most readily accessible by synthesis that starts with symmetrical dimethylurea and a cyanoacetic ester. Caffeine, theobromine, and theophilline are members of the xanthine alkaloid family because they have in common the 2,6-dioxopurine ring known as *xanthine*. The corresponding 2,6,8-trioxopurine, incidentally, is uric acid—excreted, together with urea, as the end product of

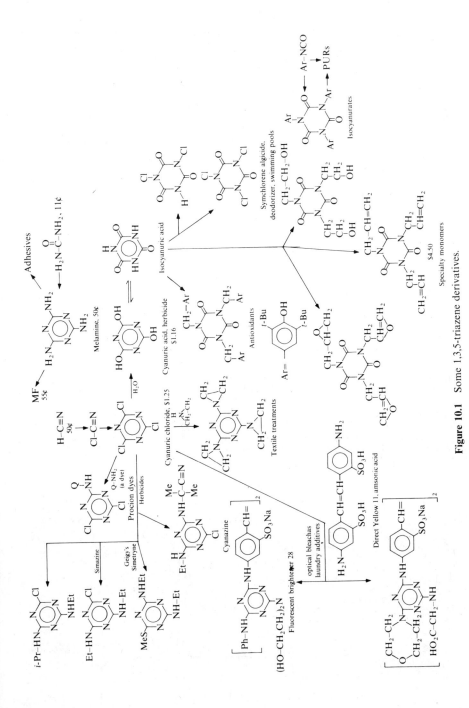

**Figure 10.1** Some 1,3,5-triazene derivatives.

631

nitrogen metabolism and the major metabolic end product of birds and many reptiles. The structures of the three last-mentioned alkaloids are arbitrarily shown below only in the tautomeric keto form, and the traditional numbering system is used throughout this discussion. To a large extent, the synthetic pathways are analogous to those used for the preparation of barbituric acid except that methylated ureas are condensed with convenient $C_3$ building blocks. The major challenge in the synthesis of the xanthine alkaloids is the addition of the imidazole moiety to the pyrimidine ring. The preparation of caffeine and theophylline can start with symmetrical dimethyl urea and employ cyanoacetic ester. The imidazole ring is then built up by means of a nitrosation of the unsubstituted carbon moiety (considered to be an activated methylene group or a carbon moiety ortho to a pseudophenolic hydroxyl), and reduction provides the two vicinal amino functions that are cyclized by means of formic acid (Scheme 10.67).

The preparation of theobromine is somewhat more devious because of the unsymmetrical structure of the product and because the unsubstituted nitrogen group can form undesired N-nitroso impurities. The synthesis can also begin with N-methylurea and a $C_3$ building block, but the isomeric products obtained on nitrosation, reduction, cyclization with formic acid, and methylation must be separated because the last-mentioned synthetic step can yield the desired 7- as well as some 9- and even some 1-methylated derivatives.

The interest in purines extends beyond the synthesis of xanthine alkaloids and the pyrimidine building blocks of nucleic acids. For example, the complex of diphenhydramine and 8-chlorotheophilline is the antimotion sickness remedy *dimenhydrinate*, known to many consumers as Searle's "Dramamine" (Scheme 10.68).

An example of a heterocyclic compound of the benzo-1,2,4-thiadiazide family is trichlormethiazide—a diuretic and antihyperintensive. It can most likely be synthesized from *m*-chloronitrobenzene (see Fig. 9.5), as shown in Scheme 10.69, p. 634.

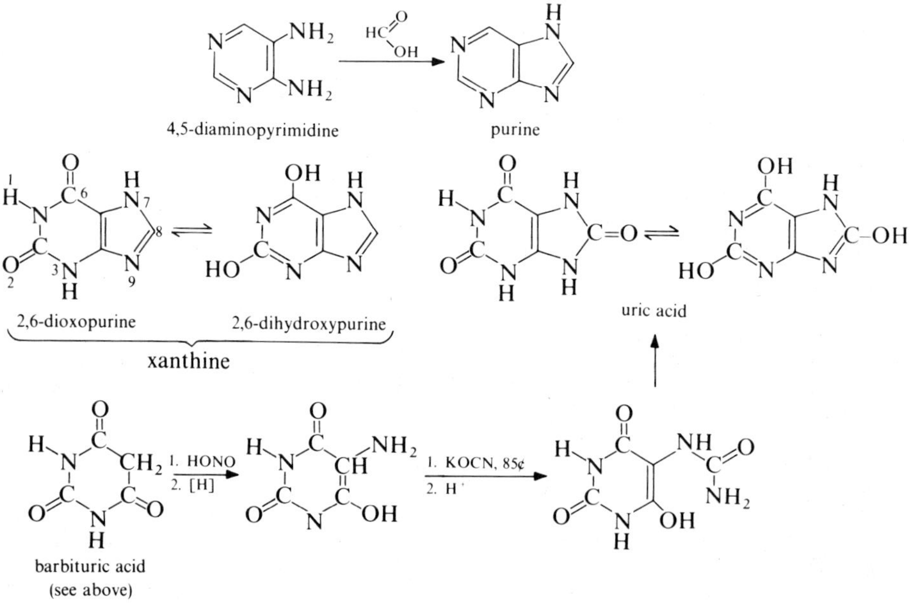

4,5-diaminopyrimidine          purine

2,6-dioxopurine          2,6-dihydroxypurine          uric acid

xanthine

barbituric acid
(see above)

**Scheme 10.67**

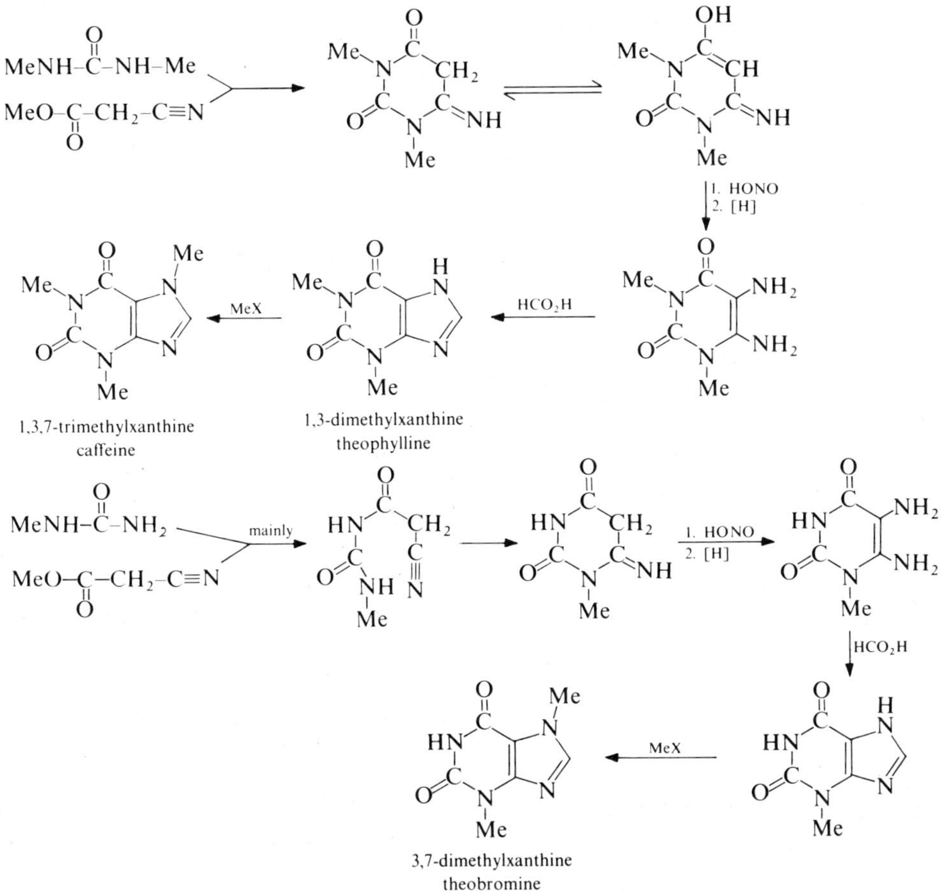

**Scheme 10.67**   (*Continued*)

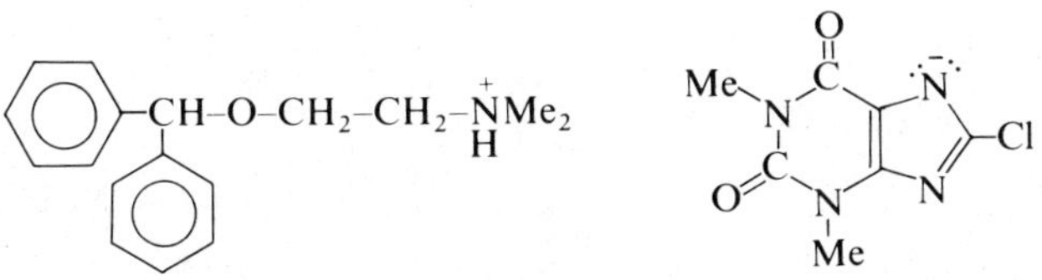

Searle's Dramamine

**Scheme 10.68**

Two examples of heterocyclic compounds that contain four nitrogen ring atoms in a fused pyrimidine and pyrazine system characteristic of the pteridine family are the diuretic triamterene or 6-phenyl-2,4,7-triaminopteridine and the more complex antineoplastic methotrexate (MTX) (Scheme 10.70).

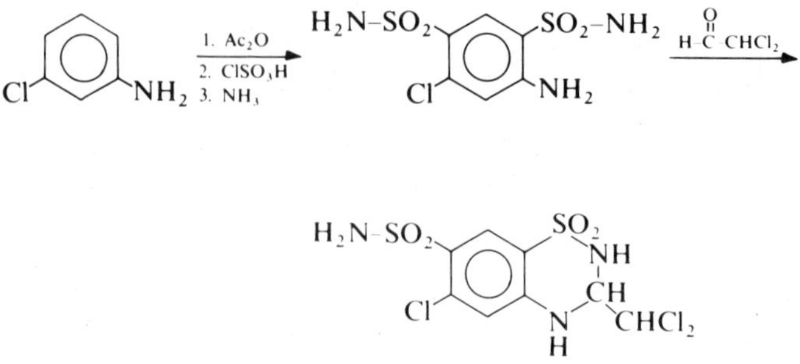
1. Ac2O
2. ClSO3H
3. NH3

Scheme 10.69

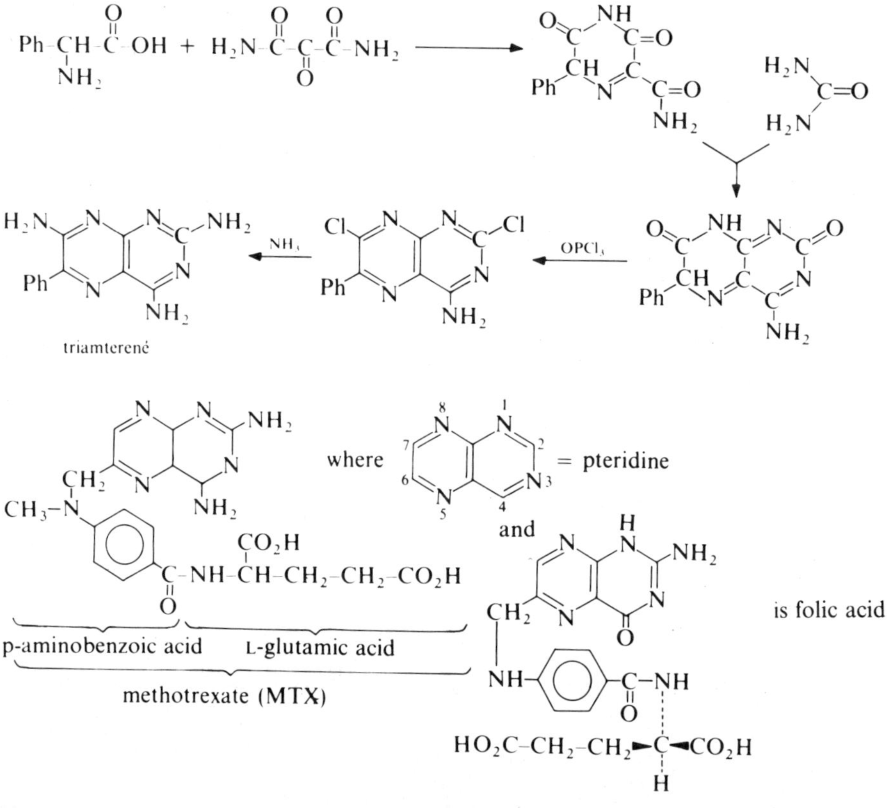
NH3
OPCl3
triamterene
where
= pteridine
and
is folic acid
p-aminobenzoic acid
L-glutamic acid
methotrexate (MTX)

Scheme 10.70

## 10.4 SEVEN-MEMBERED AND HIGHER-MEMBERED HETEROCYCLES

An industrially important seven-membered heterocycle that contains two nitrogen atoms is the tranquilizer diazepam—a member of the benzodiazepin family, and better known to many consumers as Hoffmann–La Roche's "Valium." This drug has been highly criticized

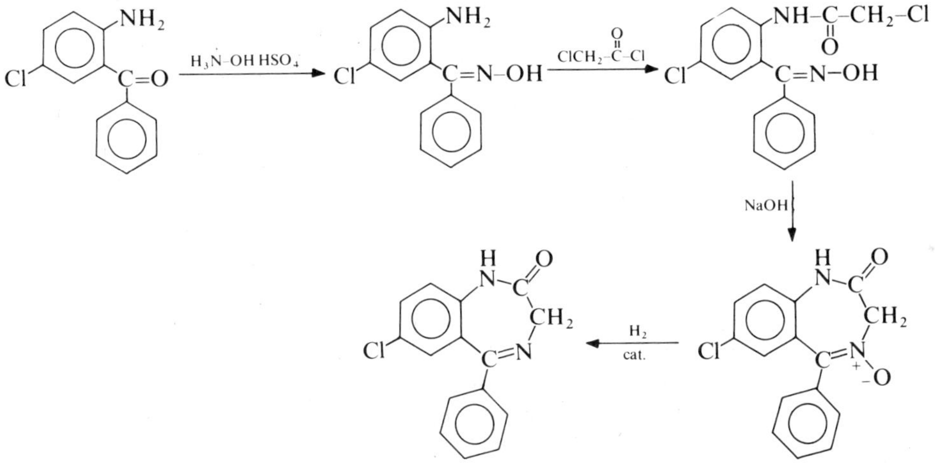

Scheme 10.71

because of rampant abuse by an edulging public (E. Bergmann, S. M. Wolfe, and J. Levin, *Stopping Valium*, Public Citizen's Health Research Group, Washington, DC, 1982). In the late 1970s, as many as 61 million prescriptions were issued annually in the United States by some lax physicians, and the annual level of prescriptions has now reached about 70 million. Currently, the retail value of generic diazepam is about $350/lb, and its synthesis starts with 2-amino-5-chlorobenzophenone (see Fig. 9.5), which is first converted to the oxime and subsequently treated with chloroacetyl chloride (Scheme 10.71). Closely related to diazepam are seven other tranquilizers. Their structures and tradenames that are most readily recognized by consumers are

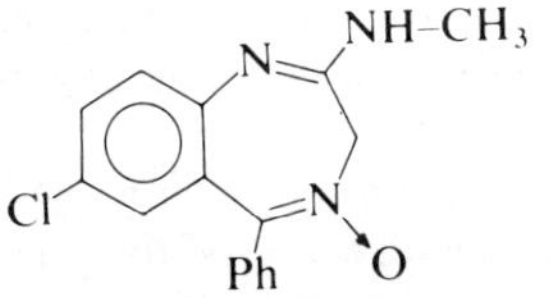

Chlordiazepoxide
Hoffmann–La Roche's Librium

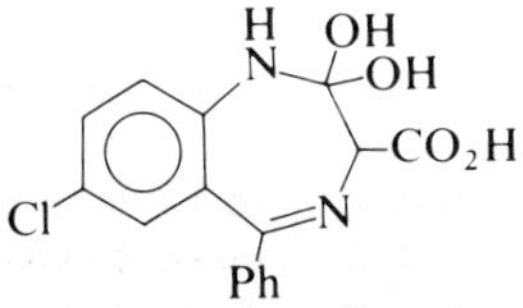

Chlorazepate diK salt
Abbott's Tranxene

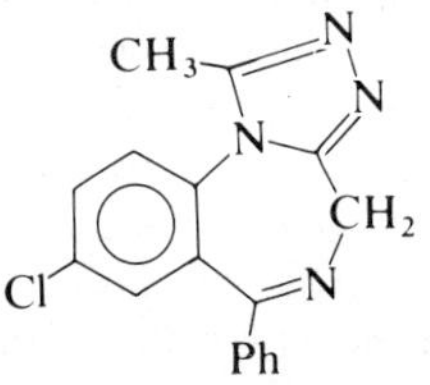

Alprazolam
Upjohn's Xanax

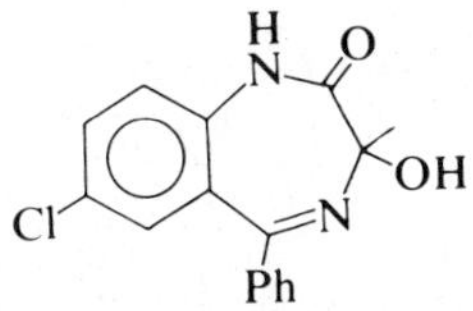

Oxazepam
Wyeth's Serax

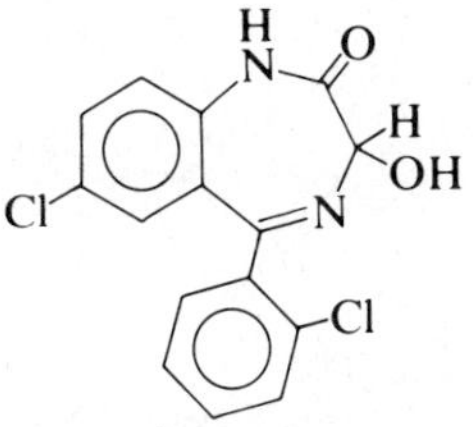

Lorazepam
Wyeth's Ativan

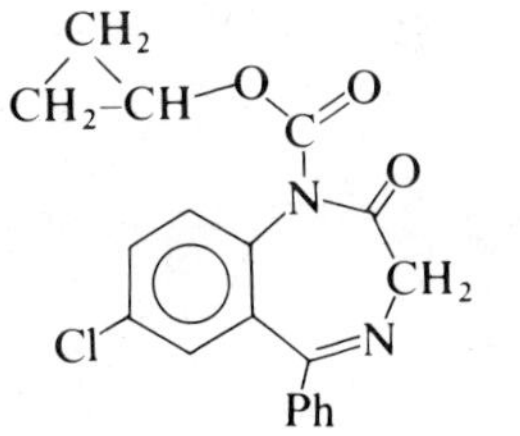

Prazepam
Parke-Davis's Centrax

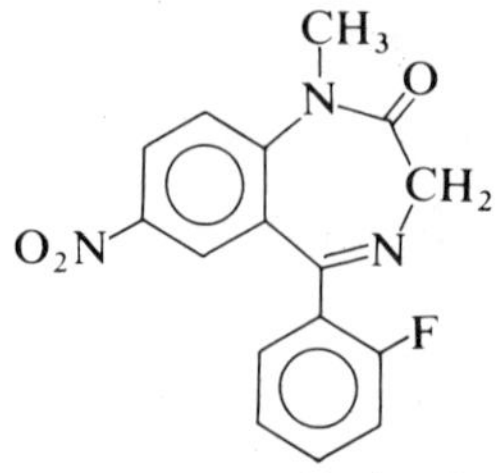

Flunitrazepam
Hoffmann–La Roche's Rohypnol

Included in this series of structures because of its benzodiazepam structure is the last-mentioned hypnotic, flunitrazepam.

The replacement of one nitrogen by sulfur in benzodiazepin yields benzothiazepin, and the members of this family of compounds are assembled from o-dinitrophenyl disulfides derived by substitution of the chlorine in o-nitrochlorobenzenes. In the case of the parent compound, the intermediate o-mercaptoaniline, or o-aminobenzenethiol (OABT), is cyclized by reaction with a $C_3$ building block such as acrylic or β-chloropropionic acid:

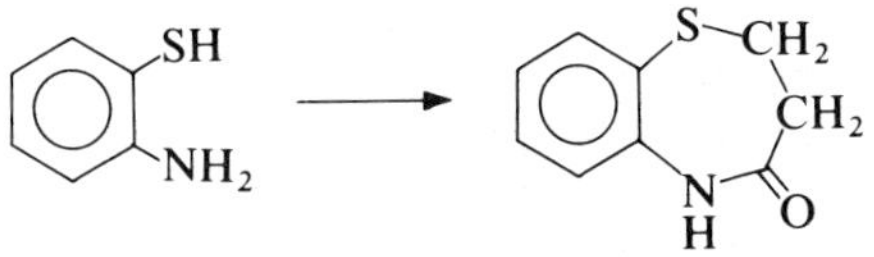

Two common members of the original family of crown ethers, dibenzo-18-crown and dibenzo-24-crown-8, are mentioned in connection with the cation-chelating capability of the glymes, that is, the dimethyl ethers of di-, tri-, and higher poly(ethylene glycols). Since the discovery by Pedersen of the important consequences caused by ·the trapping of cations of an ion pair in the cavity formed by a heterocycle of appropriate size, the field of cryptands, or cryptanols, has expanded enormously and now includes a variety of heteroatoms other than oxygen for selective trapping of certain cations. The structures of two of these ultrafine materials available in bulk quantities for large-scale applications are shown elsewhere (4.3.6.1, Scheme 4.11).

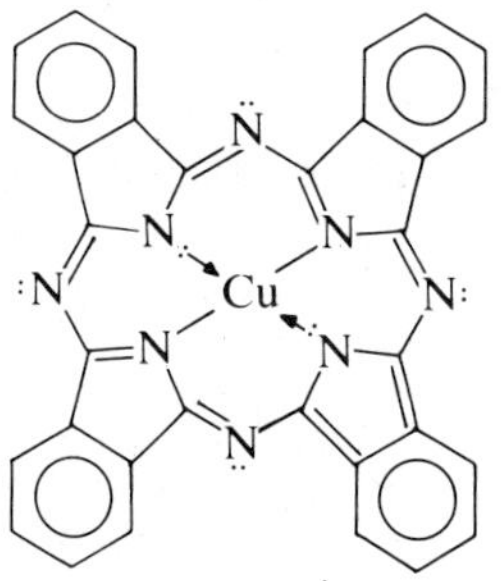

It is only fitting that this limited sampling of industrially significant heterocycles concludes with mention of the *phthalocyanines*. This synthetic modification of the ring system found in the life-giving pigments of blood and green plants, namely, heme and chlorophyl, is one of the most stable organic materials known—it can actually be used as a pigment incorporated in ceramics. Various blue, green, and red shades of the commercially available phthalocyanines are priced at \$8.50–\$12.50 and are assembled by heating cuprous chloride with phthalonitrile or phthalic anhydride in the presence of ammonia or urea (F. H. Moser and A. L. Thomas, *The Phthalocyanines*, CRC Press, Boca Raton, FL, 1983). The resulting 16-membered heterocycle can exist in various crystalline forms that are responsible for different shades of the pigment. Once formed, the phthalocyanine Pigment Blue can be chlorinated to Pigment Green 7, which contains 14 to 15 chlorine substituents per molecule (a similar chlorinated phthalocyanine is obtained from the reaction of tetrachlorophthalic anhydride and a source of ammonia). The estimated U.S. and worldwide consumptions of blue phthalocyanines are about 21 MM and 85MM lb, respectively.

# INDEX

A slash (/) is used to indicate a transformation or application utilizing two or more separate materials.
Examples:

"Cracking/sulfurization" means that the material is subjected simultaneously to two transformations which do not necessarily have to be carried out together.

"Diphenyl/diphenyl oxide" means that one uses a mixture of these two materials (in this case as an eutectic mixture for heat-transfer purposes).

"Diazotization/hydrolysis" means that in this particular case the diazotization process is followed by hydrolysis.

When interesting transformations are carried out in a series, the starting material, intermediate(s), and final product(s) are connected by means of "-to-".
Examples:

"3-Chlorodiphenylamine/sulfur -to- 3-chlorophenothiazine" means that a mixture of the two initial materials is converted to the final product.

"Starch -to- glucose -to- ethanol" means that one starts with starch, obtains the glucose intermediate, which, in turn, is converted to ethanol.

DTBP, 345
DTO, 386. *See also* Tall oil, distilled (DTO)
Dulcin, 558
Dumping, of imports, 48
Du Pont Company, 11, 21, 80, 207, 255, 330, 365, 414, 423, 429, 477, 478, 538, 611, 621. *See also* Kevlar (Du Pont); Nomex (Du Pont)
 and acquisition of Conoco, 37
 and Adiprene, PU rubber, 155
 and butadiene/HCN process, 180
 and CFC production, 69
 Delrins, 103
 diamines, linear $C_{12}$, 55
 dibasic esters (DBE), solvent, 55
 dicarboxylic acids, by-products, 55
 Dytek A, 55, 294, 382
 formaldehyde:
  exposure, 101
  plant, 102
 freon, 54
 glutaric acid, 55
 hydrogen peroxide process, 50
 Hypalon, 381
 Kalrez, 70, 75
 Lannate, 160
 Lucite, 86
 Lycra, 149, 155
 methanol production, rise-and-fall, 50, 80, 122
 MIC, 140
  from *N*-methylformamide, 163
 Mylar, 87
 Nafion, 76
 *New Horizons*, 55
 nylon 6/6 sources, 27
 PTFE, 70
 teflon, 69, 71
 Teflon PFA, 70, 76
 TEL production, 51
 XR resin, 76
Du Pont–Akzo dispute, 476
Du Pont Canada, vii, 538
Du Pont–Dixie Chem., 246
Duprene (Du Pont), 262
Durel (Celanese), 477
Durene, 96, 485
Dye intermediates, 508, 509, 523, 527, 538, 566, 571, 593
Dyes, 7, 269, 427, 450, 451, 453, 506, 507
 acidic, 450
 acridine, 616
 affinity for substrate, 450
 azo, 450, 487, 509, 510, 553, 572–574
 basic, 291

cuanine, 614, 625
disperse, 540
fast, 540
food & cosmetics, 57
laser, 591, 599, 605
mordant, 362, 522, 538–540
nomenclature, 509, 541
phenazine, 627
Procion (ICI Americas), 629, 631
red ink, 605
sulfur blue, 508
triarylmethane, 527
vat, 508, 586
xanthene, 605
Dynamit Nobel, 620

East Germany, 48, 166
Eastman Kodak, 386, 427, 434, 435, 515, 517, 604
Ecology, 139
Econometric indicators, 57
Economic impact, weeds in agriculture, 568
 incentives, dual products, 431
Economy:
 infrastructure, 17
 productive sectors, 16, 17
 of scale, 8, 438
 services, 17
EDB, 456. *See also* Ethylene dibromide (EDB)
EDET, 409
Edison, Thomas, vii
EDTA metal chelates, 442, 513
Edwards. J. G., 235
Efron, E., 53, 235
EG, 453. *See also* Ethylene glycol (EG)
2-EH, 470, 566, 608. *See also* 2-Ethylhexanol (2-EH); Octyl alcohol
EHD, 294, 382. *See also* Electrohydrodimerization (EHD)
Ehrlich, P., 235
Ehrlich's reagent, 522
Elastic memory, 154, 198, 204, 474
Elastomers, 156, 458
 modified, 399
Electric arc process, 255, 264
 conductors, 262
Electricity, cost, 431, 432
Electrochemistry, 610, 611, 612
 benzene -to- benzoquinone, 518
 benzoquinone -to- hydroquinone, 518
 oxidative methoxylation of furan, 580, 581
Electrodeposition, 627. *See also* Electroplating
Electrodes, battery, 256, 257